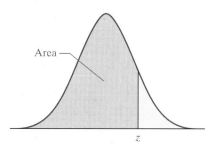

Area

z

| | | | | | Standard Normal Distribution | | | | | |
|---|---|---|---|---|---|---|---|---|---|---|
| z | .00 | .01 | .02 | .03 | .04 | .05 | .06 | .07 | .08 | .09 |
| 0.0 | 0.5000 | 0.5040 | 0.5080 | 0.5120 | 0.5160 | 0.5199 | 0.5239 | 0.5279 | 0.5319 | 0.5359 |
| 0.1 | 0.5398 | 0.5438 | 0.5478 | 0.5517 | 0.5557 | 0.5596 | 0.5636 | 0.5675 | 0.5714 | 0.5753 |
| 0.2 | 0.5793 | 0.5832 | 0.5871 | 0.5910 | 0.5948 | 0.5987 | 0.6026 | 0.6064 | 0.6103 | 0.6141 |
| 0.3 | 0.6179 | 0.6217 | 0.6255 | 0.6293 | 0.6331 | 0.6368 | 0.6406 | 0.6443 | 0.6480 | 0.6517 |
| 0.4 | 0.6554 | 0.6591 | 0.6628 | 0.6664 | 0.6700 | 0.6736 | 0.6772 | 0.6808 | 0.6844 | 0.6879 |
| 0.5 | 0.6915 | 0.6950 | 0.6985 | 0.7019 | 0.7054 | 0.7088 | 0.7123 | 0.7157 | 0.7190 | 0.7224 |
| 0.6 | 0.7257 | 0.7291 | 0.7324 | 0.7357 | 0.7389 | 0.7422 | 0.7454 | 0.7486 | 0.7517 | 0.7549 |
| 0.7 | 0.7580 | 0.7611 | 0.7642 | 0.7673 | 0.7704 | 0.7734 | 0.7764 | 0.7794 | 0.7823 | 0.7852 |
| 0.8 | 0.7881 | 0.7910 | 0.7939 | 0.7967 | 0.7995 | 0.8023 | 0.8051 | 0.8078 | 0.8106 | 0.8133 |
| 0.9 | 0.8159 | 0.8186 | 0.8212 | 0.8238 | 0.8264 | 0.8289 | 0.8315 | 0.8340 | 0.8365 | 0.8389 |
| 1.0 | 0.8413 | 0.8438 | 0.8461 | 0.8485 | 0.8508 | 0.8531 | 0.8554 | 0.8577 | 0.8599 | 0.8621 |
| 1.1 | 0.8643 | 0.8665 | 0.8686 | 0.8708 | 0.8729 | 0.8749 | 0.8770 | 0.8790 | 0.8810 | 0.8830 |
| 1.2 | 0.8849 | 0.8869 | 0.8888 | 0.8907 | 0.8925 | 0.8944 | 0.8962 | 0.8980 | 0.8997 | 0.9015 |
| 1.3 | 0.9032 | 0.9049 | 0.9066 | 0.9082 | 0.9099 | 0.9115 | 0.9131 | 0.9147 | 0.9162 | 0.9177 |
| 1.4 | 0.9192 | 0.9207 | 0.9222 | 0.9236 | 0.9251 | 0.9265 | 0.9279 | 0.9292 | 0.9306 | 0.9319 |
| 1.5 | 0.9332 | 0.9345 | 0.9357 | 0.9370 | 0.9382 | 0.9394 | 0.9406 | 0.9418 | 0.9429 | 0.9441 |
| 1.6 | 0.9452 | 0.9463 | 0.9474 | 0.9484 | 0.9495 | 0.9505 | 0.9515 | 0.9525 | 0.9535 | 0.9545 |
| 1.7 | 0.9554 | 0.9564 | 0.9573 | 0.9582 | 0.9591 | 0.9599 | 0.9608 | 0.9616 | 0.9625 | 0.9633 |
| 1.8 | 0.9641 | 0.9649 | 0.9656 | 0.9664 | 0.9671 | 0.9678 | 0.9686 | 0.9693 | 0.9699 | 0.9706 |
| 1.9 | 0.9713 | 0.9719 | 0.9726 | 0.9732 | 0.9738 | 0.9744 | 0.9750 | 0.9756 | 0.9761 | 0.9767 |
| 2.0 | 0.9772 | 0.9778 | 0.9783 | 0.9788 | 0.9793 | 0.9798 | 0.9803 | 0.9808 | 0.9812 | 0.9817 |
| 2.1 | 0.9821 | 0.9826 | 0.9830 | 0.9834 | 0.9838 | 0.9842 | 0.9846 | 0.9850 | 0.9854 | 0.9857 |
| 2.2 | 0.9861 | 0.9864 | 0.9868 | 0.9871 | 0.9875 | 0.9878 | 0.9881 | 0.9884 | 0.9887 | 0.9890 |
| 2.3 | 0.9893 | 0.9896 | 0.9898 | 0.9901 | 0.9904 | 0.9906 | 0.9909 | 0.9911 | 0.9913 | 0.9916 |
| 2.4 | 0.9918 | 0.9920 | 0.9922 | 0.9925 | 0.9927 | 0.9929 | 0.9931 | 0.9932 | 0.9934 | 0.9936 |
| 2.5 | 0.9938 | 0.9940 | 0.9941 | 0.9943 | 0.9945 | 0.9946 | 0.9948 | 0.9949 | 0.9951 | 0.9952 |
| 2.6 | 0.9953 | 0.9955 | 0.9956 | 0.9957 | 0.9959 | 0.9960 | 0.9961 | 0.9962 | 0.9963 | 0.9964 |
| 2.7 | 0.9965 | 0.9966 | 0.9967 | 0.9968 | 0.9969 | 0.9970 | 0.9971 | 0.9972 | 0.9973 | 0.9974 |
| 2.8 | 0.9974 | 0.9975 | 0.9976 | 0.9977 | 0.9977 | 0.9978 | 0.9979 | 0.9979 | 0.9980 | 0.9981 |
| 2.9 | 0.9981 | 0.9982 | 0.9982 | 0.9983 | 0.9984 | 0.9984 | 0.9985 | 0.9985 | 0.9986 | 0.9986 |
| 3.0 | 0.9987 | 0.9987 | 0.9987 | 0.9988 | 0.9988 | 0.9989 | 0.9989 | 0.9989 | 0.9990 | 0.9990 |
| 3.1 | 0.9990 | 0.9991 | 0.9991 | 0.9991 | 0.9992 | 0.9992 | 0.9992 | 0.9992 | 0.9993 | 0.9993 |
| 3.2 | 0.9993 | 0.9993 | 0.9994 | 0.9994 | 0.9994 | 0.9994 | 0.9994 | 0.9995 | 0.9995 | 0.9995 |
| 3.3 | 0.9995 | 0.9995 | 0.9995 | 0.9996 | 0.9996 | 0.9996 | 0.9996 | 0.9996 | 0.9996 | 0.9997 |
| 3.4 | 0.9997 | 0.9997 | 0.9997 | 0.9997 | 0.9997 | 0.9997 | 0.9997 | 0.9997 | 0.9997 | 0.9998 |

**TABLE II** (continued)

FUNDAMENTALS OF

# Statistics

## Michael Sullivan, III
Joliet Junior College

PEARSON
Prentice
Hall

Upper Saddle River, New Jersey 07458

**Library of Congress Cataloging-in-Publication Data**

Sullivan, Michael (date)
    Fundamentals of statistics / Michael Sullivan, III.
        p. cm.
    Includes index.
    ISBN 0-13-146449-3 (alk. paper)
      1. Statistics.   I. Title.
QA276.12.S846 2005
519.5--dc22                   2003068942

*Executive Acquisitions Editor*: Petra Recter
*Editor in Chief*: Sally Yagan
*Project Manager*: Jacquelyn Riotto
*Vice President/Director of Production and Manufacturing*:
   David W. Riccardi
*Executive Managing Editor*: Kathleen Schiaparelli
*Senior Managing Editor*: Linda Mihatov Behrens
*Production Editor*: Barbara Mack
*Assistant Managing Editor, Math Media Production*: John Matthews
*Media Production Editor*: Donna Crilly
*Assistant Manufacturing Manager/Buyer*: Michael Bell
*Manufacturing Manager*: Trudy Pisciotti
*Marketing Manager*: Krista M. Bettino
*Marketing Assistant*: Annett Uebel
*Development Editors*: Anne Scanlan-Rohrer and Frank Purcell
*Editor in Chief, Development*: Carol Trueheart
*Editorial Assistant/Print Supplements Editor*: Joanne Wendelken
*Art Director*: Jonathan Boylan
*Interior Designers*: Joseph Sengotta, Judy Matz-Coniglio
*Assistants to Art Director*: Geoffrey Cassar, Kristine Carney

*Cover Designer*: Jonathan Boylan
*Art Editor*: Thomas Benfatti
*Creative Director*: Carole Anson
*Director of Creative Services*: Paul Belfanti
*Director, Image Resource Center*: Melinda Reo
*Manager, Rights and Permissions*: Zina Arabia
*Interior Image Specialist*: Beth Boyd-Brenzel
*Image Permission Coordinator*: Debbie Hewitson
*Photo Researcher*: Melinda Alexander
*Art Studio*: Artworks
   *Managing Editor, AV Management and Production:* Patricia Burns
   *Production Manager*: Ronda Whitson
   *Manager, Production Technologies*: Matt Haas
   *Project Coordinator*: Jessica Einsig
   *Illustrators*: Jackie Ambrosius, Kathryn Anderson, Christopher
      Kayes, Mark Landis, Bill Neff, Audrey Simonetti, Stacy Smith
   *Quality Assurance*: Pamela Taylor, Kenneth Mooney,
      Timothy Nguyen
   *Composition*: Lithokraft

© 2005 Pearson Education, Inc.
Pearson Prentice Hall
Pearson Education, Inc.
Upper Saddle River, New Jersey 07458

Pearson Prentice Hall® is a trademark of Pearson Education, Inc.
Printed in the United States of America
10  9  8  7  6  5  4

ISBN 0-13-146449-3

Pearson Education LTD., London
Pearson Education Australia PTY, Limited, Sydney
Pearson Education Singapore, Pte. Ltd
Pearson Education North Asia Ltd, Hong Kong
Pearson Education Canada, Ltd., Toronto
Pearson Educación de Mexico, S.A. de C.V.
Pearson Education—Japan, Tokyo
Pearson Education Malaysia, Pte. Ltd

To My Wife
Yolanda

and

My Children
Michael, Kevin, and Marissa

# CONTENTS

## Chapter 1

## Data Collection    1

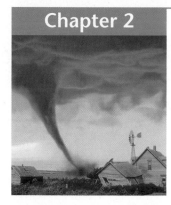

## Chapter 2

## Organizing and Summarizing Data    39

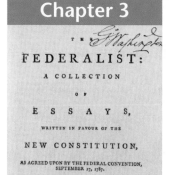

## Chapter 3

## Numerically Summarizing Data    83

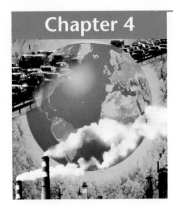
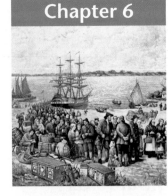
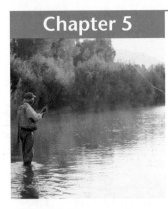

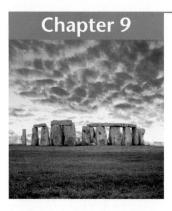
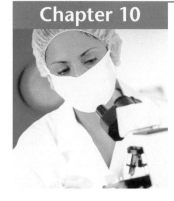
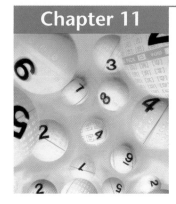

*Fundamentals of Statistics* is a brief version of *Statistics: Informed Decisions Using Data.* This shorter text is intended to meet the needs of those schools that offer a one-semester introductory statistics course that do not wish or do not have the time to cover some of the more advanced topics found in *Statistics: Informed Decisions Using Data.* Some of these more advanced topics that appear in *Statistics: Informed Decisions Using Data,* such as residual analysis, nonlinear regression, the Poisson distribution, and nonparametric statistics, have been removed from this briefer text. The coverage of topics has been adapted to make it more suitable for usage in a one-semester course, but this does not compromise the depth of coverage.

*Fundamentals of Statistics* retains the strong pedagogy, features, and writing style that have made *Statistics: Informed Decisions Using Data* so successful. In addition, it retains the popular problems and discussion that encourage problem solving and statistical thinking.

Having taught introductory statistics for the past 10 years at both 2-year and 4-year institutions has brought to my attention some basic needs and issues of the Introductory Statistics course. The diversity of both the students taking the course and the instructors teaching the course drives many of the challenges in presenting statistics at this level. Statistics is a powerful subject and one that I hold great passion for. It is the coupling of my passion for the subject with the desire for a text that would work for me, my students, and my school that led me to pursue writing this textbook.

Before getting started on this text, I spent many hours reflecting on the goals of the project. Overall, I wanted a textbook that instilled in students my enthusiasm and appreciation for the subject and supported the variety of approaches to teaching the subject. In particular, the following were my goals for the text:

- Students see and appreciate the usefulness of the subject
- Students can be more successful in the course
- Students are provided ample practice
- Instructors can present the material using multiple philosophies

## Keeping the Students' Interest

An immediate challenge for any instructor teaching this course is the motivation of the students. I always ask my students to pass in their class notes at the end of the semester. By studying these notes, I have been able to determine areas where my students lost interest and where they became excited about the subject. In addition, I was able to determine topics where the concept was lost and topics where the concept was completely understood. I also noticed that my students appreciated when I took a formal definition or theorem and expressed it in everyday language that they could understand. This view into how students read texts, attend lectures, and take notes was invaluable to me in writing this text.

Today's students demand that the material they study have relevance in their everyday lives. One of the joys of teaching a statistics course is that the question, "When will I ever use this stuff?" rarely comes up. However, students still say, "Okay, I can see the usefulness of this if I was a medical researcher, but that is not my goal in life!" Introductory statistics texts need to have interesting, real data sets that students can relate to. They need to see themselves analyzing the data and using the conclusions to make informed decisions that impact their lives. For example, what type of car should I buy? Where should I invest my retirement money? Where should I live? What major should I choose?

## Student Success

It is absolutely true that if students experience success early in a course, they will be inclined to work harder as the course progresses. However, if students struggle early, it will lead to disenchanted students who are more willing to "give up" when a difficult concept is introduced.

How can success be experienced while still maintaining the integrity of the course? After all, statistics is a discipline whose theory is deep and mathematically intensive. In addition, understanding the concepts of statistics can be elusive. This does not mean, however, that an introductory textbook should "spare students details" of statistical thought. It is possible, and indeed, desirable, to expose students to the intricacies of statistics. This can be done without delving deep into mathematical thought and losing most of your students.

The text has been written using a conversational tone that "speaks" to the student. This is accomplished through explanations that don't overly complicate the material. Because the text was written using the class notes obtained from my students, the presentation flows like a lecture. The pedagogy that is found in the text mirrors the presentation that I use in my lectures. When my lecture begins, I tie concepts learned earlier in the course to concepts that I am about to present. This is the "Putting It All Together" feature found at the beginning of each chapter. Before tackling a new section, I review topics that are going to be needed in order to succeed in the section. This is the "Preparing for This Section" feature. I then list the objectives that we are going to cover within the section. As I present the material, I provide students with statistically accurate definitions/theorems. I then restate the definitions/theorems using everyday language that doesn't compromise the definition or theorem. This is the "In Your Own Words" feature. I also mention some of the pitfalls in statistical analysis. These pitfalls are displayed in the "Caution" feature. After going over an example in class, I like to have my students attempt to solve a similar problem in class so that I am confident that they understand the concept before continuing. This is the "Now Work Problem xx" feature.

In addition, the text has been written to appeal to a variety of learning styles. There are many, many graphs and figures that allow students to visualize results. For students who learn best through discovery, there are Explorations that guide students through a series of questions that develop statistical concepts. For auditory learners there is a lecture series in which I present all the examples in the text on compact disk or video.

## Practice, Practice, Practice

The only way that students will learn statistics is by doing statistics. The exercise sets of a strong statistics text will help develop a student's confidence in, and understanding of, the material. This is accomplished in a twofold manner: (1) through graduated exercise sets, and (2) through problems that ask the student to think about concepts and encourage statistical thinking.

All the exercise sets begin with Concepts and Vocabulary, which are open-ended questions that ask students to define words and explain concepts using their own words. The next portion of the exercise sets is Skill Building. These are drill and practice type problems that develop computational skills that increase student understanding of formulas and concepts. These problems also serve as confidence building problems so that students experience early success. Finally, the Applying the Concepts problems are real data-based problems that ask a variety of questions that help develop solid statistical analysis. Not only do these problems ask the standard questions such as find the mean or compute a 95% confidence interval, they also ask for students to explain the results. In addition, there are questions that ask students to consider some additional questions. For example, what is the impact of an outlier on the arithmetic mean? Is the linear correlation coefficient resistant? What happens if we have outliers when constructing confidence intervals from small data sets? These "higher level of thinking" problems truly develop a student's understanding of statistical thinking.

## Flexibility

Just as there are many different learning styles, there are many different teaching styles. One of the challenges in writing a text is to create a product that "appeals to the masses." As I survey the halls of the Mathematics Department at Joliet Junior College, I see many high-quality instructors who present the same material in many different ways. Some like to incorporate as much technology as possible into their classroom, some prefer to minimize the technology. Some instructors prefer to use collaborative learning in order to present material, while others utilize lecture. For the Introductory Statistics course, some of the instructors are trained in the discipline of statistics, while others are trained in mathematics, but teach statistics. With these varied backgrounds, it is clear that a text needs to meet the needs of all these backgrounds and teaching philosophies.

Let's consider how technology is presented in the course. Every example in the text is presented using the "by hand" approach. The reason for this is twofold. First, and probably most importantly, by presenting a solution by hand, the student's ability to understand the concept is enhanced. How else can a student understand the concept of linear correlation, except by seeing the product of $z$-scores? Second, it allows for flexibility in philosophies. If your particular philosophy is to present statistics by utilizing "by hand" solutions, it is there for you. If you are more apt to use technology, then you can utilize the "Using Technology" feature. Following virtually every example is a gray "Using Technology" screen, which provides the output from a TI83+, Minitab, or Excel. In addition, problems that have 15 or more observations in the data set have a CD icon, which indicates to the instructor that the data set is available on the data CD.

## Using the Text Efficiently with Your Syllabus

To meet the varied needs of diverse syllabi, this book has been organized with flexibility of use in mind. When structuring your syllabus, notice the topics listed in the "Preparing for This Section" material at the beginning of the section, which will tip you off to dependencies within the course.

The two most common variations within an Introductory Statistics course are the treatment of Regression Analysis and the treatment of Probability.

- **Coverage of Correlation and Regression** The text was written with the descriptive portion of bivariate data (Chapter 4) presented after the descriptive portion of univariate data (Chapter 3). For instructors who prefer to postpone the discussion of bivariate data until later in the course, simply skip Chapter 4 and return to it prior to covering Sections 11.3 and 11.4. Within Chapter 4, an instructor may skip Section 4.3 without loss of continuity.

- **Coverage of Probability** The text allows for a course to present an extensive introduction to probability or light coverage of probability. For instructors wishing to present light coverage of probability, they may cover Section 5.1 and skip the remaining sections. A word of caution is in order, however. Instructors who will be covering the Chi-Square Test for Independence will want to cover Sections 5.1 through 5.3. In addition, any instructor who will be covering Binomial Probabilities will want to cover Independence in Section 5.3 and Combinations in Section 5.5.

## Chapter by Chapter Content

### Chapter 1  Data Collection

This chapter deals with the methods of obtaining data. There is a detailed presentation of the various sampling techniques along with circumstances under which each is used. In addition, there is an entire section dedicated to nonsampling errors and how to control them. The chapter ends with a detailed discussion of experimental design.

## Chapter 2 Organizing and Summarizing Data

This chapter addresses methods for summarizing qualitative data (Section 2.1) and quantitative data (Section 2.2). The chapter ends with a discussion of graphical misrepresentations of data. This section can be covered as a reading assignment.

## Chapter 3 Numerically Summarizing Data

Sections 3.1 and 3.2 present numerical measures of central tendency and dispersion. Section 3.3 is optional and can be skipped without loss of continuity. However, if it is skipped and Section 6.1 is covered, proceed slowly through the mean and standard deviation of a discrete random variable. Section 3.4 discusses measures of position including the $z$-score, percentiles, and outliers. Section 3.5 presents exploratory data analysis.

## Chapter 4 Describing the Relation between Two Variables

Section 4.1 introduces scatter diagrams and correlation. Section 4.2 presents the least-squares regression line. Section 4.3 presents the coefficient of determination. In addition, the material in this chapter can be postponed until before Section 11.3.

## Chapter 5 Probability

Section 5.1 introduces the basic concepts of probability and unusual events. Section 5.2 presents the Addition Rule. Section 5.3 presents the Multiplication Rule. Section 5.4 presents Conditional Probability and can be skipped without loss of continuity. Section 5.5 presents Counting Techniques and can be skipped without loss of continuity with the exception of Combinations.

## Chapter 6 The Binomial Probability Distribution

Section 6.1 introduces the concept of a random variable and discrete probability distributions along with expected value. Section 6.2 presents the binomial probability formula and an introduction to inference using binomial probabilities. If you intend to cover Section 6.2, then it is a good idea to also cover Section 5.3.

## Chapter 7 The Normal Probability Distribution

Sections 7.1 through 7.3 introduce the normal probability distribution. Section 7.4 presents normal probability plots as a means for assessing normality and is required in order to cover topics presented in Chapters 8–11. Section 7.4 does not require the use of technology because the output generated by technology is presented. This section is necessary in order to help students see that verifying normality is necessary before proceeding with inference for small samples. Section 7.5 introduces sampling distributions. Section 7.6 discusses the normal approximation to the binomial and is optional.

## Chapter 8 Confidence Intervals about a Single Parameter

Section 8.1 introduces the construction of a confidence interval when the population standard deviation is known, while Section 8.2 constructs confidence intervals when the population standard deviation is unknown. This approach is different from that in some other texts, but it is logical. It's as simple as "$\sigma$ known, use $z$; $\sigma$ unknown, use $t$." In both cases, small samples require that the population from which the sample was drawn must be normal–we check this requirement with a normal probability plot. Section 8.3 covers confidence intervals about a population proportion.

## Chapter 9 Hypothesis Testing

Section 9.1 provides an introduction to the language of hypothesis testing. Sections 9.2 and 9.3 test hypotheses regarding a population mean, again segmented by "$\sigma$ known, use $z$; $\sigma$ unknown, use $t$." Section 9.4 presents hypothesis testing about a population proportion. An interesting feature in this section is that it includes how to use the binomial probability distribution to compute exact $P$-values. This is especially important if the requirement for using the normal approximation to the binomial is not satisfied.

## Chapter 10 Inferences on Two Samples

Section 10.1 presents the analysis required for matched-pairs design. Section 10.2 presents the analysis for comparing two means from independent samples. Notice that the discussion regarding pooled estimates of $\sigma$ is absent. This is because the pooled estimate approach requires that the two populations have a common variance and this is an extremely difficult requirement to test. Because "pooling" versus "not pooling" often provide the same results, it is not necessary at this level to cloud the students' thought process any further. Section 10.3 discusses comparing two population proportions.

## Chapter 11 Additional Inferential Procedures

Section 11.1 presents the chi-square goodness of fit. Section 11.2 discusses chi-square tests for independence and homogeneity. Again, if Sections 5.3 and/or 5.4 were skipped, proceed slowly. Sections 11.3 and 11.4 are independent of Sections 11.1 and 11.2. Sections 11.3 and 11.4 require that Sections 4.1 and 4.2 were covered.

## Acknowledgments

Textbooks evolve into their final form through the efforts and contributions of many people. First and foremost, I would like to thank my family, whose dedication to this project was just as much as mine: my wife, Yolanda, whose words of encouragement and support were unabashed, and my children, Michael, Kevin, and Marissa, who would come and greet me every morning with smiles that only children can deliver. I owe each of them my sincerest gratitude.

I would also like to thank the entire Mathematics Department at Joliet Junior College, who provided support, ideas, and encouragement to help me complete this project. Countless ideas were bounced off of each of them and their responses are present throughout the text. Special thanks also go to the Joliet Junior College community, who also supported this project.

I sincerely appreciate the comments and suggestions of Erica Egizio, who had the courage to use the text in its nascent stages. Her comments and suggestions have proven to be invaluable. Erica is also to be commended for her "Note to the Instructor" annotations. Special thanks to Michael McCraith, who helped locate interesting data and also made the PowerPoint presentations friendly for both PCs and Macs. Thank you to Faye Dang and Priscilla Gathoni for checking all the answers. Thank you to Elena Catoiu, Kathleen Miranda, Gigi Williams, and Kevin Bodden for proofreading the text to help ensure its accuracy. Thanks to Lindsay Packer, who took on the daunting task of writing the solutions manuals; and to Beverly J. Dretzke, Kate McLaughlin, and Dorothy Wakefield, who wrote the technology guides; and David Lane, who wrote the applets workbook. I would also like to thank Max Linn, who wrote the Case Studies at the end of each chapter. A big thank you to the development editors: Frank Purcell of Twin Prime Editorial and Anne Scanlan-Rohrer. I would also like to thank my father, who has taught me the ropes. It is always nice to learn from the best.

There are many colleagues I would like to thank for their input. I owe each of them my sincerest thanks. I apologize for any omissions.

Fusan Akman, *Coastal Carolina Community College*

Grant Alexander, *Joliet Junior College*

Randall Allbritton, *Daytona Beach Community College*

Diana Joseph Asmus, *Greenville Technical College*

Margaret M. Balachowski, *Michigan Technological University*

Kari Beaty, *Midlands Technical College*

Charles Biles, *Humboldt College*

Kevin Bodden, *Lewis and Clark College*

Ken Bonee, *University of Tennessee*

Joanne Brunner, *Joliet Junior College*

Jim Butterbach, *Joliet Junior College*

Roxanne Byrne, *University of Colorado–Denver*

Joe Castillo, *Broward Community College*

Elena Catoiu, *Joliet Junior College*

Pinyuen Chen, *Syracuse University*

Faye Dang, *Joliet Junior College*

Julie DePree, *University of New Mexico—Valencia Campus*

Erica Egizio, *Joliet Junior College*

Richard Einsporn, *The University of Akron*

Jill Fanter, *Walters State Community College*

Franco Fedele, *University of West Florida*

William P. Fox, *Francis Marion University*

Joe Gallegos, *Salt Lake City Community College*

Frieda Ganter, *California State University–Fresno*

Cheryl B. Hawkins, *Greenville Technical College*

Marilyn M. Hixson, *Brevard Community College*

Kim Jones, *Virginia Commonwealth University*

Maryann E. Justinger, *Erie Community College–South Campus*

Iraj Kalantari, *Western Illinois University*

Donna Katula, *Joliet Junior College*

Mohammad Kazemi, *University of North Carolina–Charlotte*

Rita R. Kolb, *Community College of Baltimore County*

Melody Lee, *Joliet Junior College*

Mike Mays, *West Virginia University*

Jean McArthur, *Joliet Junior College*

Marilyn McCollum, *North Carolina State University*

Michael McCraith, *Joliet Junior College*

Jill McGowan, *Howard University*

Judy Meckley, *Joliet Junior College*

Jane Blake Millar, *Santa Rosa Junior College*

Glenn Miller, *Borough of Manhattan Community College*

Rachel L. Miller, *Portland State University*

Kathleen Miranda, *SUNY at Old Westbury*

Lindsay Packer, *College of Charleston*

Linda Padilla, *Joliet Junior College*

Ingrid Peterson, *University of Kansas*

Nancy Pevey, *Pellissippi State Technical Community College*

Philip Pina, *Florida Atlantic University*

David Ray, *University of Tennessee–Martin*

Mike Rosenthal, *Florida International University*

Robert Sackett, *Erie Community College–North Campus*

Karen Spike, *University of North Carolina–Wilmington*

John Squires, *Cleveland State Community College*

Timothy D. Stebbins, *Kalamazoo Valley Community College*

Patrick Stevens, *Joliet Junior College*

Sharon Stokero, *Michigan Technological University*

Cathleen M. Zucco Teveloff, *Trinity College*

Kirk Trigsted, *University of Idaho*

Robert Tuskey, *Joliet Junior College*

Farroll Tim Wright, *University of Missouri–Columbia*

James Zimmer, *Chattanooga State Technical Community College*

Recognition and thanks are due particularly to the following individuals for their valuable assistance in the preparation of this text: Petra Recter for her editorial skill and attention to detail; Sally Yagan for her support, interest, and friendship; Patrice Jones for his sage advice and marketing skills; Krista M. Bettino for her innovative marketing efforts; Barbara Mack for her organizational skills as a production supervisor; and the entire Prentice Hall sales staff for their confidence.

*Michael Sullivan, III*

As you begin your study of statistics, you may feel overwhelmed with the number of theorems, definitions, and formulas that confront you. Many students enter their first statistics course with a sense of anxiety. While these feelings are normal, I want to reassure you that statistics is an exciting discipline that can be learned, appreciated, and most importantly, applied, so that you can make informed decisions throughout your life.

My goal in writing this text was to provide students with a "how to" manual for studying and learning statistics. I have written the textbook using clear, uncomplicated language. In fact, the first draft of the text was written using my students' class notes. There are many learning aids included in the text to help make your study of the material easier and more rewarding. In addition, there are many supplements that accompany the text, including: a student solutions manual that presents completely worked out solutions to all the odd-numbered exercises; a CD lecture series that presents worked out solutions to all the examples in the text; and tutorial software (SullivanTutor) that will generate and grade as many problems as you like along with hints and worked out solutions. The questions are based on the objectives presented in the text. The goal is to minimize the number of frustrating experiences that inevitably come up in the learning process.

To help make your study of statistics more enjoyable and rewarding, I have included a list of time-tested "study tips" that promote successful learning. This list is a sequence of events that should occur prior to, during, and after each class meeting. While this list may seem overwhelming, personal experience indicates that it works.

## Before Class Begins

1. Read the section(s) to be covered in the upcoming class meeting.
2. Prepare a list of questions that you have based upon your reading. In many cases, your questions will be answered through the lecture. Those that are not answered completely can then be asked.

## During Class

1. Arrive early enough to prepare your mind and material for the lecture.
2. Stay alert. Do not doze off or daydream during class. It will be very difficult to understand the lecture when you "return to the class."
3. Take thorough notes. It is normal that certain topics will not be completely understood during the lecture. However, this does not mean that you throw your hands up in despair. Rather, continue to write your class notes. You can ask questions when appropriate. In my personal experience, I was amazed how often my confusion during class was alleviated after studying the in-class notes later when I had more time for reflection.

## After Class

1. Reread (and possibly rewrite) your class notes.
2. Reread the section.
3. Do your homework. Homework is not optional. There is an old Chinese proverb that says:

> I hear … and I forget
> I see … and I remember
> I do … and I understand

This proverb applies to anything where one wants to succeed. Would a pianist expect to be the best if she didn't practice? The only way you are going to learn statistics is by doing statistics. When you get problems wrong, ask for help.

**4.** If you have questions, visit your professor during office hours.

Just like your previous mathematics courses, statistics is a building process. This means the material is used throughout the course. If a topic is not understood, then this lack of understanding could come back to haunt you later in the class. This is the source of a lot of frustration in learning statistics or mathematics. You need to build a strong foundation before you continue to build the house.

## How to Use This Book Effectively and Efficiently

First, and foremost, this text was written to be read. You will find that the text has additional explanations and examples that will help you. As mentioned previously, be sure to read the text before attending class.

Many sections begin with "Preparing for This Section," a list of concepts that will be used in the section. Take the short amount of time required to refresh your memory. This will make the section easier to understand and will actually save you time and effort. Each section that has the "Preparing for This Section" feature will have a "Preparing for This Section" quiz on the companion Web site (www.prenhall.com/sullivanstats). The quiz asks questions related to the concepts that are listed. Objectives are provided at the beginning of each section. Read them. They will help you recognize the important ideas and skills developed in the section.

After a concept has been introduced and an example given, you will see **NW** *Now Work Problem xx.* Go to the exercises at the end of the section, work the problem cited, and check your answer in the back of the book. If you get it right, you can be confident in continuing on in the section. If you don't get it right, go back over the explanations and examples to see what you might have missed. Then rework the problem. Ask for help if you miss it again.

I have included an "In Your Own Words" feature that explains definitions, theorems, and concepts using everyday language. This is meant to help you understand the concepts presented in the text. There are also "Caution" statements. These are meant to make you aware of common errors that occur in statistics, so that you don't make these mistakes.

The chapter review contains a list of formulas and vocabulary introduced in the chapter. Be sure you understand how to use the formulas and that you know the definitions of the vocabulary. There is also a list of objectives along with review problems that correspond to the objective. If you can't do the problems listed for a particular objective, go back to the page indicated and review the material.

Please do not hesitate to contact me, through Prentice Hall, with any suggestions or comments that would improve the text. I look forward to hearing from you.

*Michael Sullivan, III*

From left to right, Michael IV, Michael III, Marissa, Yolanda, and Kevin

# A USER'S GUIDE TO THE TEXT

Utilizing the features found throughout this text will help you learn and study the content for this course. Many of these features were developed by the author's students for students.

## CHAPTER 2

### Organizing and Summarizing Data

**Outline**

 For additional study help, go to www.prenhall.com/sullivanstats

Materials include
- Projects:
  ■ Case Study: The Day the Sky Roared
  ■ Decisions: Tables or Graphs?
  ■ Consumer Reports Project
- Self-Graded Quizzes
- "Preparing for This Section" Quizzes
- PowerPoint Downloads
- Step-by-Step Technology Guide
- Graphing Calculator Help

**Putting It All Together**

In Chapter 1, we learned that statistics is a process. The process begins with asking a research question. In order to determine the answer to the question, information (data) must be collected. The information is obtained from a census, existing data sources, surveys or designed experiments. When data are collected from a survey or designed experiment, they must be organized into a manageable form. Data that are not organized are referred to as **raw data**.

Methods for organizing raw data include the creation of tables or graphs, which allow for a quick overview of the information collected. The organization of data is the third step in the statistical process. The procedures used in organizing data into tables and graphs depend upon whether the data are qualitative, discrete, or continuous.

## Putting It All Together

Each chapter opens with a discussion of topics that were covered and how they relate to topics that are about to be discussed. This allows students to see the "big picture" of how the topics relate to each other.

## Companion Web Site

Many additional free resources for this text including end of chapter case study, decisions and consumer reports can be found at: *www.prenhall.com/sullivanstats*

## Preparing for This Section

Most sections open with a referenced list (by section and page number) of key items to review in preparation for the section ahead. This provides a just-in-time review for the students.

**2.1** Organizing Qualitative Data

**Preparing for This Section** Before getting started, review the following:
✓ Qualitative data (Section 1.1, p. 4)

## Objectives

A numbered list of key objectives appears in the beginning of each section. As the topic corresponding to the objective is addressed, the number appears in the column.

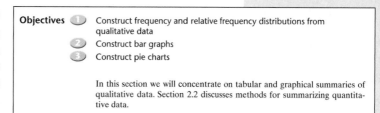

Objectives
1. Construct frequency and relative frequency distributions from qualitative data
2. Construct bar graphs
3. Construct pie charts

In this section we will concentrate on tabular and graphical summaries of qualitative data. Section 2.2 discusses methods for summarizing quantitative data.

PAGE 40

## In Your Own Words

When a definition or concept is presented, the "In Your Own Words" feature presents the definition or concept using everyday language while maintaining accuracy.

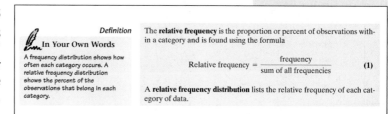

*Definition*

**In Your Own Words**

A frequency distribution shows how often each category occurs. A relative frequency distribution shows the percent of the observations that belong in each category.

The **relative frequency** is the proportion or percent of observations within a category and is found using the formula

$$\text{Relative frequency} = \frac{\text{frequency}}{\text{sum of all frequencies}} \quad \text{(1)}$$

A **relative frequency distribution** lists the relative frequency of each category of data.

PAGE 41

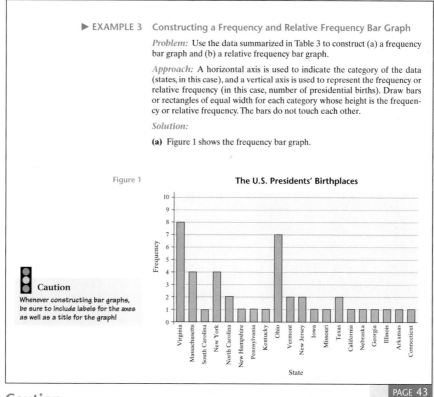

▶ EXAMPLE 3   Constructing a Frequency and Relative Frequency Bar Graph

*Problem:* Use the data summarized in Table 3 to construct (a) a frequency bar graph and (b) a relative frequency bar graph.

*Approach:* A horizontal axis is used to indicate the category of the data (states, in this case), and a vertical axis is used to represent the frequency or relative frequency (in this case, number of presidential births). Draw bars or rectangles of equal width for each category whose height is the frequency or relative frequency. The bars do not touch each other.

*Solution:*

(a) Figure 1 shows the frequency bar graph.

Figure 1

**The U.S. Presidents' Birthplaces**

**Caution**

Whenever constructing bar graphs, be sure to include labels for the axes as well as a title for the graph!

PAGE 43

## Caution

This alerts students to some of the pitfalls in statistical analysis.

## Historical Notes

For many of the concepts introduced, a short historical note regarding the individual responsible is presented. These historical notes are meant to "humanize" famous statisticians.

### Historical Note

Florence Nightingale was born in Italy on May 12, 1820. She was named after the city of her birth. Florence was educated by her father, who attended Cambridge University. Between 1849 and 1851, Florence studied nursing throughout Europe. In 1854, she was asked to oversee the introduction of female nurses into the military hospitals in Turkey. While there, she greatly improved the mortality rate of wounded soldiers. She collected data and invented graphs (the polar area diagram), tables, and charts to show that improving sanitary conditions would lead to decreased mortality rates. In 1869, Florence founded the Nightingale School Home for Nurses. Florence died on August 13, 1910.

*Solution:* We tally the data as shown in the second column of Table 13. The third column in the table shows the frequency of each class. From the frequency distribution, we conclude that a three-year rate of return between 15.0% and 19.9% occurs with the most frequency. The fourth column in the table shows the relative frequency of each class. So, 27.5% of the small-capitalization growth mutual funds had a three-year rate of return between 15.0% and 19.9%.

| TABLE 13 | | | |
|---|---|---|---|
| Class (Three-Year Rate of Return) | Tally | Frequency | Relative Frequency |
| 10.0–14.9 | ЖӀ ІІ | 7 | 7/40 = 0.175 |
| 15.0–19.9 | ЖӀ ЖӀ І | 11 | 11/40 = 0.275 |
| 20.0–24.9 | ЖӀ ІІІ | 8 | 8/40 = 0.2 |
| 25.0–29.9 | ЖӀ І | 6 | 0.15 |
| 30.0–34.9 | ІІІ | 3 | 0.075 |
| 35.0–39.9 | ІІІ | 3 | 0.075 |
| 40.0–44.9 | | 0 | 0 |
| 45.0–49.9 | ІІ | 2 | 0.05 |

Only 5% of the mutual funds had a three-year rate of return between 45.0% and 49.9%. We might conclude that mutual funds with a rate of return at this level outperform their peers and consider them worthy of our investment. This type of information would be difficult to obtain from the raw data. ◀◀

Notice that the choice of the lower class limit of the first class and the class width was rather arbitrary. While formulas and procedures do exist for creating frequency dist...
vide "better" summaries...
distribution is the...

## Examples

Examples are set up in "Problem," "Approach," and "Solution" structure. The "Problem" explains the situation and the question that is being asked. The "Approach" walks through the steps of how to analyze the information. Finally, the "Solution" applies the "Approach" and solves the "Problem."

▶ EXAMPLE 5    Constructing a Pie Chart

*Problem:* The data presented in Table 6 represent the educational attainment of residents of the United States 25 years or older, based upon data obtained from the 2000 United States Census. Construct a pie chart of the data.

| TABLE 6 | |
|---|---|
| Educational Attainment | 2000 |
| Less than 9th grade | 12,327,601 |
| 9th–12th grade, no diploma | 20,343,848 |
| High school diploma | 52,395,507 |
| Some college, no degree | 36,453,108 |
| Associate's degree | 11,487,194 |
| Bachelor's degree | 28,603,014 |
| Graduate/professional degree | 15,930,061 |
| **Totals** | **177,540,333** |

*Approach:* The pie chart will have seven parts, or sectors, corresponding to the seven categories of data. The area of each sector is proportional to the frequency of each category. For example,

$$\frac{12,327,601}{177,540,333} = 0.0694 = 6.94\%$$

of all U.S. residents 25 years or older have less than a 9th-grade education. The category "less than 9th grade" will make up 6.94% of the pie chart. Since a circle has 360°, the degree measure of the sector for the category "less than 9th-grade education" will be $(0.0694)360° \approx 25.0°$. Use a protractor to measure each angle.

*Solution:* We follow the approach presented for the remaining categories of data to obtain Table 7.

Figure 6
**Educational Attainment, 2000**

Bachelor's degree (16%)
Graduate/Professional degree (9%)
Less than 9th grade (7%)
Associate's degree (6%)
9th–12th grade (11%)
Some college, no degree (21%)
High school diploma (30%)

| TABLE 7 | | | |
|---|---|---|---|
| Education | Frequency | Relative Frequency | Degree Measure of Each Sector |
| Less than 9th grade | 12,327,601 | 0.0694 | 25.0 |
| 9th to 12th grade, no diploma | 20,343,848 | 0.1146 | 41.3 |
| High school diploma | 52,395,507 | 0.2951 | 106.2 |
| Some college, no degree | 36,453,108 | 0.2053 | 73.9 |
| Associate's degree | 11,487,194 | 0.0647 | 23.3 |
| Bachelor's degree | 28,603,014 | 0.1611 | 58.0 |
| Graduate/professional degree | 15,930,061 | 0.0897 | 32.3 |

To construct a pie chart by hand, we use a protractor to approximate the angles for each sector. See Figure 6. ◀◀

## Using Technology

Immediately following an example whenever appropriate, the output from a TI-83 Plus, MINITAB, or Excel is presented.

## Now Work Problem xxx

A NW icon **NW** along with a corresponding problem to be solved appears after a concept has been introduced. This directs the student to a problem in the exercises that tests the concept, ensuring that the concept has been mastered before moving on. The Now Work problems are identified in the exercises using orange numbers and a NW icon.

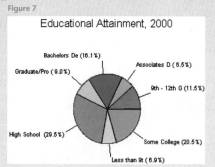

Using Technology: Certain statistical spreadsheets have the ability to draw pie charts. Figure 7 shows the pie chart of the data in Table 6, using Minitab.

Figure 7

Educational Attainment, 2000

Bachelors De (16.1%)
Associates D (6.5%)
Graduate/Pro (9.0%)
9th - 12th G (11.5%)
High School (29.5%)
Some College (20.5%)
Less than 9t (6.9%)

Pie charts can be created only if all the categories of the variable under consideration are represented. In other words, we could create a bar chart that lists the proportion of presidents born in the states of Virginia, Texas, and New York, but it would not make sense to construct a pie chart for this situation. Do you see why? Only 34.88% of the pie would be accounted for.

**NW** *Now Work Problem 19(e).*

PAGE 48

---

### 4.2 Assess Your Understanding

**Concepts and Vocabulary**

1. Explain the least-squares regression criterion.
2. What is a residual? What does it mean when a residual is positive?
3. Explain the phrase "outside the scope of the model." Why is it dangerous to make predictions "outside the scope of the model"?

4. If the linear correlation between two variables is negative, what can be said about the slope of the regression line?
5. What is meant by an unbiased estimator as it pertains to least-squares regression?
6. In your own words, explain the meaning of Legendre's quote given on page 166.

**Exercises**

• **Skill Building**

1. For the data set

| $x$ | 0 | 2 | 3 | 5 | 6 | 6 |
|---|---|---|---|---|---|---|
| $y$ | 5.8 | 5.7 | 5.2 | 2.8 | 1.9 | 2.2 |

   (a) Draw a scatter diagram. Comment on the type of relation that appears to exist between $x$ and $y$.
   (b) Given that $\bar{x} = 3.667$, $s_x = 2.42212$, $\bar{y} = 3.933$, $s_y = 1.8239152$, and $r = -0.9476938$, determine the least-squares regression line.
   (c) Graph the least-squares regression line on the scatter diagram drawn in part (a).

2. For the data set

| $x$ | 2 | 4 | 8 | 8 | 9 |
|---|---|---|---|---|---|
| $y$ | 1.4 | 1.8 | 2.1 | 2.3 | 2.6 |

   (a) Draw a scatter diagram. Comment on the type of relation that appears to exist between $x$ and $y$.
   (b) Given that $\bar{x} = 6.2$, $s_x = 3.03315$, $\bar{y} = 2.04$, $s_y = 0.461519$, and $r = 0.957241$, determine the least-squares regression line.
   (c) Graph the least-squares regression line on the scatter diagram drawn in part (a).

PAGE 171

## Exercises

The exercise sets are divided into three distinct categories: (1) Concepts and Vocabulary, (2) Skill Building, and (3) Applying the Concepts. The "Concepts and Vocabulary" problems are open-ended questions that ask students to define terms in their own words and also require students to think about statistical concepts. The "Skill Building" problems are drill and practice type problems that develop computational skills that increase student understanding of formulas and concepts.

• **Applying the Concepts**

*Problems 9–16 use the results from Problems 11–18 from Section 4.1.*

**9. Height versus Head Circumference** A pediatrician wants to determine the relation that exists between a child's height and head circumference. She randomly selects 11 children from her practice, measures their height and head circumference, and obtains the following data.

| Height (inches) | Head Circumference (inches) | Height (inches) | Head Circumference (inches) |
|---|---|---|---|
| 27.75 | 17.5 | 26.5 | 17.3 |
| 24.5 | 17.1 | 27 | 17.5 |
| 25.5 | 17.1 | 26.75 | 17.3 |
| 26 | 17.3 | 26.75 | 17.5 |
| 25 | 16.9 | 27.5 | 17.5 |
| 27.75 | 17.6 | | |

*Source:* Denise Slucki, Student at Joliet Junior College

(a) Find the least-squares regression line treating height as the predictor variable and head circumference as the response variable.
(b) Interpret the slope and intercept, if appropriate.
(c) Use the regression equation to predict the head circumference of a child who is 25 inches tall.
(d) Compute the residual based upon the observed head circumference of the 25-inch-tall child in the table. Is the head circumference of this 25-inch-tall child above average or below average?
(e) Draw the least-squares regression line on the scatter diagram of the data and label the residual.
(f) Notice that there are two children that are 26.75 inches tall. One of the children has a head circumference that is 17.3 inches, while the other has a head circumference that is 17.5 inches. How can this be?
(g) Would it be reasonable to use the least-squares regression line to predict the head circumference of a child who was 32 inches tall? Why?

**10. Round-cut Diamonds** A gemologist is interested in determining the relation that exists between the size of a diamond (carats) and the price. The following data represent the weight and price of D color, VS1 clarity, round-cut diamonds.

| Carats | Price | Carats | Price |
|---|---|---|---|
| 0.66 | $3282 | 0.80 | $4378 |
| 0.75 | $3950 | 0.90 | $5682 |
| 0.70 | $3543 | 0.91 | $6426 |
| 0.71 | $3788 | 1.18 | $9362 |
| 0.77 | $4108 | | |

*Source:* diamonds.com

(a) Find the least-squares regression line treating carats as the predictor variable and price as the response variable.
(b) Interpret the slope and intercept, if appropriate.
(c) Use the regression equation to predict the price of a color D, VS1 clarity round-cut diamond that weighs 0.91 carats.
(d) Compute the residual of a color D, VS1 clarity round-cut diamond that weighs 0.91 carats and costs $6426. Is this diamond priced above average or below average?
(e) Draw the least-squares regression line on the scatter diagram of the data and label the residual.
(f) Predict the price of a color D, VS1 clarity round-cut diamond that weights 0.85 carats.
(g) Suppose the gemologist had two 0.85 carat, color D, VS1 clarity round-cut diamonds, but one was priced at $4503 and the other at $5420. What might account for the difference in price of these two diamonds?
(h) Would it be reasonable to use the least-squares regression line to predict the price of a color D, VS1 clarity, round-cut diamond that weighs 0.25 carats? Why?

PAGE 172

---

(a) Construct a frequency distribution of the data.
(b) Construct a relative frequency distribution of the data.
(c) What percentage of the Saturdays had 10 or more customers waiting for a table at 6:00 P.M.?
(d) What percentage of the Saturdays had 5 or fewer customers waiting for a table at 6:00 P.M.?
(e) Construct a frequency histogram of the data.
(f) Construct a relative frequency histogram of the data.
(g) Describe the shape of the distribution.

**18. Highway Repair** The following data represent the number of potholes on 50 randomly selected 1-mile stretches of highway in the city of Chicago.

| | | | | | |
|---|---|---|---|---|---|
| 2 | 7 | 4 | 7 | 2 | 7 |
| 2 | 2 | 2 | 3 | 4 | 3 |
| 1 | 2 | 3 | 2 | 1 | 4 |
| 2 | 2 | 5 | 2 | 3 | 4 |
| 4 | 1 | 7 | 10 | 3 | 5 |
| 4 | 3 | 3 | 2 | 2 | |
| 1 | 6 | 5 | 7 | 9 | |
| 2 | 2 | 2 | 1 | 5 | |
| 3 | 5 | 1 | 3 | 5 | |

(a) Construct a frequency distribution of the data.
(b) Construct a relative frequency distribution of the data.
(c) What percentage of the 1-mile stretches of highway had seven or more potholes?
(d) What percentage of the 1-mile stretches of highway had two or fewer potholes?
(e) Construct a frequency histogram of the data.
(f) Construct a relative frequency histogram of the data.
(g) Describe the shape of the distribution.

**19. Tensile Strength** Tensile strength is the maximum stress at which one can be reasonably certain that failure will not occur. The following data represent the tensile strength (in thousands of pounds per square inch) of a composite material to be used in an aircraft.

(d) Construct a relative frequency histogram of the data.
(e) Describe the shape of the distribution.
(f) Repeat parts (a)–(e), using a class width of 15. Which frequency distribution seems to provide a better summary of the data?

**20. Miles on a Cavalier** A random sample of 36 three-year-old Chevy Cavaliers was obtained in the Miami, Florida area, and the number of miles on each car was recorded. The results are shown in the table below.

| | | | | | |
|---|---|---|---|---|---|
| 34,122 | 17,685 | 15,499 | 26,455 | 30,500 | 41,194 |
| 39,416 | 29,307 | 26,051 | 27,368 | 35,936 | 29,289 |
| 28,281 | 34,511 | 32,305 | 37,904 | 44,448 | 31,883 |
| 37,021 | 34,500 | 39,884 | 23,221 | 41,043 | 32,923 |
| 30,747 | 41,620 | 17,595 | 38,041 | 30,739 | 40,614 |
| 35,439 | 30,283 | 44,224 | 35,295 | 38,995 | 42,398 |

*Source:* cars.com

With the first class having a lower class limit of 15,000 and a class width of 4,000,

(a) Construct a frequency distribution.
(b) Construct a relative frequency distribution.
(c) Construct a frequency histogram of the data.
(d) Construct a relative frequency histogram of the data.
(e) Describe the shape of the distribution.
(f) Repeat parts (a)–(e), using a class width of 6000. Which class width seems to provide a better summary of the data?

**21. Serum HDL** Dr. Paul Oswiecmiski randomly selects 40 of his 20–29-year-old patients and obtains the following data regarding their serum HDL cholesterol:

| | | | | | | | |
|---|---|---|---|---|---|---|---|
| 70 | 56 | 48 | 48 | 53 | 52 | 66 | 48 |
| 36 | 49 | 28 | 35 | 58 | 62 | 45 | 60 |
| 38 | 73 | 45 | 51 | 56 | 51 | 46 | 39 |
| 56 | 32 | 44 | 60 | 51 | 44 | 63 | 50 |
| 46 | 69 | 53 | 70 | 33 | 54 | 55 | 52 |

PAGE 68

## Technology Step-by-Step
Drawing Histograms and Stem-and-Leaf Plots

**TI-83 Plus**

**Histograms**

*Step 1:* Enter the raw data in L1 by pressing STAT and selecting 1:Edit.

*Step 2:* Press 2ⁿᵈ Y = to access the StatPlot menu. Select 1:Plot1.

*Step 3:* Place the cursor on "On" and press ENTER.

*Step 4:* Place the cursor on the histogram icon (see the figure) and press ENTER.

*Step 5:* Press ZOOM and select 9:ZoomStat.

**MINITAB**

**Histograms**

*Step 1:* Enter the raw data in C1.

*Step 2:* Select the **Graph** menu and highlight **Histogram**.

*Step 3:* Select the data in C1 and press OK.

**Stem-and-Leaf Plot**

*Step 1:* With the raw data entered in C1, select the **Graph** menu and highlight **Stem-and-Leaf**...

*Step 2:* Select the data in C1 and press OK.

**Excel**

**Histograms**

*Step 1:* Enter the raw data in column A.

*Step 2:* Select **Tools** and **Data Analysis**...

*Step 3:* Select Histogram from the list.

*Step 4:* With the cursor in the Input Range cell, use the mouse to highlight the raw data. Select the Chart Output box and press OK.

*Step 5:* Double-click on one of the bars in the histogram. Select the Options tab from the menu that appears. Reduce the gap width to zero.

Excel does not draw stem-and-leaf plots.

**Technology Step-by-Step**
Brief step-by-step instructions for using the TI-83 Plus, MINITAB, and Excel follow each exercise set. This is meant as a supplement to the technology guides.

---

## CHAPTER 4 REVIEW

### Summary

In this chapter, we introduced techniques that allow us to describe the relation between two quantitative variables. The first step in identifying the type of relation that might exist is to draw a scatter diagram. The predictor variable is plotted on the horizontal axis, the corresponding response variable on the vertical axis. The scatter diagram can be used to discover whether the relation between the predictor and the response variable is linear. In addition, for linear relations, we can judge whether the linear relation shows positive or negative association.

A numerical measure for the strength of linear relation between two variables is the linear correlation coefficient. It is a number between −1 and 1, inclusive. Values of the correlation coefficient near −1 are indicative of a negative linear relation between the two variables; values of the correlation coefficient near +1 indicate a positive linear relation between the two variables. If the correlation coefficient is near 0, then there is little *linear* relation between the two variables.

Once a linear relation between the two variables has been discovered, we describe the relation by finding the least-squares regression line. This line best describes the linear relation between the predictor and the response variables. We can use the least-squares regression line to predict a value of the response variable for a given value of the predictor variable.

The coefficient of determination, $R^2$, measures the percent of variation in the response variable that is explained by the least-squares regression line. It is a measure between 0 and 1 inclusive. The closer $R^2$ is to 1, the more explanatory value the line has.

One item worth mentioning again is that a researcher should never claim causation between two variables in a study unless the data are experimental. Observational data allow us to say that two variables might be associated, but we cannot claim causation.

### Formulas

**Correlation Coefficient**

$$r = \frac{\sum \left( \frac{x_i - \bar{x}}{s_x} \right) \left( \frac{y_i - \bar{y}}{s_y} \right)}{n - 1}$$

**Equation of the Least-Squares Regression Line**
The equation of the least-squares regression line is given by

$$\hat{y} = b_1 x + b_0$$

where

$\hat{y}$ is the predicted value of the response variable,

$b_1 = r \cdot \dfrac{s_y}{s_x}$ is the slope of the least-squares regression

line, and

$b_0 = \bar{y} - b_1 \bar{x}$ is the intercept of the least-squares regression line.

**Coefficient of Determination, $R^2$**

$$R^2 = \frac{\text{variation explained by predictor variable}}{\text{total variation}}$$

### Vocabulary

| | |
|---|---|
| Bivariate data (p. 148) | Positively associate |
| Response variable (p. 149) | Negatively associat |
| Predictor variable (p. 149) | Linear correlation |
| Lurking variable (p. 149) | Residuals (p. 165) |
| Scatter diagram (p. 149) | Least-squares regre |

### Objectives

| Section | You should be able to ... | Review Exercises |
|---|---|---|
| 4.1 | 1 Draw scatter diagrams (p. 149) | 1–4, 9, 10, 20, 33–35(a) |
| | 2 Interpret scatter diagrams (p. 150) | 1–4(c) |
| | 3 Understand the properties of the linear correlation coefficient (p. 151) | 36 |
| | 4 Compute and interpret the linear correlation coefficient (p. 154) | 1–4(c), 9(d), 10(d) |
| 4.2 | 1 Find the least-squares regression line (p. 165) | 5–8(a) |
| | 2 Interpret the slope and y-intercept of the least-squares regression line (p. 167) | 5–8(c) |
| | 3 Predict the value of the response variable, based upon the least-squares regression line (p. 167) | 5–8(d) |
| | 4 Determine residuals, based upon the least-squares regression line (p. 168) | 5–8(e) |
| | 5 Compute the sum of squared residuals (p. 170) | 9, 10 (f) and (g) |
| 4.3 | 1 Compute and interpret the coefficient of determination (p. 176) | 11–14 |

### Review Exercises

**1. Engine Displacement versus Fuel Economy** The following data represent the size of a car's engine (in liters) versus its miles per gallon in the city for various 2001 domestic automobiles.

(a) Draw a scatter diagram treating engine displacement as the predictor variable and miles per gallon as the response variable.

(b) Compute the linear correlation coefficient between engine displacement and miles per gallon.

(c) Based upon the scatter diagram and the linear cor-

(a) Draw a scatter diagram treating temperature as the predictor variable and chirps per second as the response variable.

(b) Compute the linear correlation coefficient between temperature and chirps per second.

(c) Based upon the scatter diagram and the linear correlation coefficient, comment on the type of relation that appears to exist between the two variables.

**3. Apartments** The following data represent the square footage and rents for apartments in North Chicago and in the western suburbs of Chicago.

**Chapter Review**
The extensive review provides a list of important formulas and vocabulary and a list of objectives along with the corresponding Review Exercises that test the student's understanding of the objective.

PAGE 183

PAGE 182

**xxv**

# Student and Instructor Resources

## Student Resources

### Student Study-Pack
### (ISBN: 0-13-161580-7)

Everything a student needs to succeed in one place. Free packaged with the book, or available for purchase stand-alone. Study-Pack contains:

➤ *Student Solutions Manual*

Fully worked solutions to odd-numbered exercises

➤ *CD Lecture Videos*

This comprehensive set of videos covers each chapter and section and is presented by the text author, Michael Sullivan. They provide excellent support for students who require additional assistance, for distance learning and self-paced programs, or for students who missed class.

➤ *Pearson Tutor Center*

Tutors provide one-on-one tutoring for any problem with an answer at the back of the book. Students access the Tutor Center via toll-free phone, fax, or e-mail.

➤ *Technology Manuals on CD*

Contains tutorial instructions and worked out examples for:
- TI-83 Calculator
- Excel
- Minitab

➤ *Companion Website:*

*www.prenhall.com/sullivanstats*
Includes materials for each chapter and section – including a case study, decisions project, and Consumer Reports project for each chapter. Also, for each chapter there are tests, reading comprehension quizzes, STATLETS ™, PowerPoint slides, data files, and graphing calculator help.

## Instructor Resources

### Content Distribution Center

All instructor resources can be downloaded from a Website (URL and password can be obtained from your PH Sales Representative), or ordered individually:

➤ *Annotated Instructor's Edition*

Contains helpful teaching tips and answers to all the exercises throughout the text at appropriate moments (ISBN: 0-13-146450-7)

➤ *Instructor Solutions Manual*

Fully worked solutions to all textbook exercises. (ISBN: 0-13-146763-8)

➤ *TestGen*

Test-generating software—create tests from textbook section objectives. Questions are algorithmically generated, allowing for unlimited test versions. Instructors may also edit problems or create their own. (ISBN: 0-13-046503-8)

➤ *Test Item File*

A printed test bank derived from TestGen. (ISBN: 0-13-046493-7)

➤ *PowerPoint Lecture Slides*

Fully editable and printable slides that follow the textbook. Use during lecture or post to a Website in an online course. For download from the Companion Website and the Content Distribution Center.

➤ *Companion Website:*

*www.prenhall.com/sullivanstats*
Includes materials for each chapter and section – including a case study, decisions project, and Consumer Reports project for each chapter. Also, for each chapter there are tests, reading comprehension quizzes, STATLETS ™, PowerPoint slides, data files, and graphing calculator help. Contains pdf's of the following advanced topics:
- Confidence Intervals about a Population Standard Deviation
- Testing a Hypothesis about $\sigma$
- One-Way Analysis of Variance
- *F*- Distribution Critical Values Table

# LIST OF APPLICATIONS

# PHOTO AND ILLUSTRATION CREDITS

**Chapter 1**    Page 1, Fritz Polking/Bruce Coleman, Inc.; Page 5, PhotoEdit; Page 33, A. Barrington Brown/Photo Researchers, Inc.

**Chapter 2**    Page 39, Phil Degginger/PictureQuest; Page 55, PhotoEdit; Page 58, The Granger Collection, New York

**Chapter 3**    Page 83, The Pierpont Morgan Library, New York, NY/Art Resource, NY; Page 85, AP/Wide World Photos; Page 135, Alfred Eisenstaedt/TimePix

**Chapter 4**    Page 148, Felix Clouzot/Getty Images, Inc.-Hulton Archive Photos; Page 152, Photo Researchers, Inc.; Page 165, The Granger Collection, New York; Page 167, The Granger Collection, New York

**Chapter 5**    Page 185, Peter Beck/CORBIS; Page 192, Mary Evans Picture Library Ltd; Page 193, Photo Researchers, Inc.; Page 194, Photo Researchers, Inc.; Pages 196, 214, SuperStock, Inc.; Page 202, The Royal Society of London; Page 215, Sovfoto/Eastfoto; Page 228, AP/Wide World Photos; Page 231, AP/Wide World Photos

**Chapter 6**    Page 240, Courtesy of Schwenkfelder Library & Heritage Center, Pennsburg, PA; Page 249, Photo Researchers, Inc.; Page 254, The Granger Collection, New York; Page 255, AP/Wide World Photos

**Chapter 7**    Page 271, eStock Photography LLC; Page 274, CORBIS; Page 276, North Wind Picture Archives; Page 322, CORBIS

**Chapter 8**    Page 338, Peter Skinner/Photo Researchers, Inc.; Page 357, The Granger Collection, New York

**Chapter 9**    Page 382, Jan Stromme/Bruce Coleman, Inc.; Page 383, D. Graham/ H. Armstrong Roberts; Page 387, J. Neubauer/H. Armstrong Roberts; Page 392, Paul Bishop/University of California-Berkeley

**Chapter 10**   Page 435, Comstock Images

**Chapter 11**   Page 475, Reza Estakhrian/Getty Images, Inc.; Page 476, Richard B. Levine/ Frances M. Roberts

# Data Collection

## Outline

 **For additional study help, go to**
www.prenhall.com/sullivanstats

Materials include

- Projects:
  - Case Study: Chrysalises for Cash
  - Decisions: What Movie Should I Go To?
  - Consumer Reports Project
- Self-Graded Quizzes
- "Preparing for This Section" Quizzes
- STATLETs
- PowerPoint Downloads
- Step-by-Step Technology Guide
- Graphing Calculator Help

## Putting It All Together

Statistics is a discipline that plays a major role in many different areas. For example, it is used in sports to help a coach or manager make informed decisions about his/her competition. It is used to predict the outcome of elections and to help determine governmental policies. Statistics assists in determining the effectiveness of new medications. Statistics can be used to discover unlikely occurrences. In fact, statistics was used by the Democratic National Committee to determine that voters in Palm Beach County, Florida likely voted for Pat Buchanan incorrectly instead of their desired candidate, Al Gore. This statistical evidence led the Democrats to demand a recount of the ballots within that county and other counties in Florida.

Used appropriately, statistics can provide an understanding of the world around us. Used inappropriately, it can lend support to inaccurate beliefs. Understanding the methodologies of statistics will provide you with the ability to analyze and critique studies and experiments. With this ability, you will be an informed consumer of information, which will enable you to distinguish solid statistical analyses from the bogus presentation of numerical "facts."

## 1.1    Introduction to the Practice of Statistics

**Objectives**     ① Define statistics
② Understand the process of statistics
③ Distinguish between qualitative and quantitative variables
④ Distinguish between discrete and continuous variables

①    What is statistics? When asked this question, many people respond that statistics is numbers. This response is only partially correct.

*Definition*    **Statistics** is the science of collecting, organizing, summarizing and analyzing information in order to draw conclusions.

It is helpful to consider this definition in three parts. The first part of the definition states that statistics is the collection of information. The second part refers to the organization and summarization of information. Finally, the third part states that the information is analyzed in order to draw conclusions.

②    The definition implies that the methods of statistics follow a process.

### The Process of Statistics

1. *Identify the research objective.* A researcher must determine the question(s) he or she wants answered. The question(s) must be detailed so that it identifies a group that is to be studied and the questions that are to be answered. The group that is to be studied is called the **population**. An **individual** is a person or object that is a member of the population being studied. For example, a researcher may want to study the population of all 2002 model-year automobiles. The individuals in this study would be the cars.

### Caution

Many nonscientific studies are based on *convenience samples*, such as Internet surveys or phone-in polls. The results of any study performed using this type of sampling method are not reliable.

2. *Collect the information needed to answer the questions posed in (1).* Gaining access to an entire population is often difficult and expensive. In conducting research, we typically look at a subset of the population, called a **sample**. For example, the United States population of people 18 years or older is about 210 million. Many national studies consist of samples of size 1,100. The "collection of information" step is vital to the statistical process because if the information is not collected correctly, the conclusions drawn are meaningless. Do not overlook the importance of appropriate data collection processes.

3. *Organize and summarize the information.* This step in the process is referred to as *descriptive statistics*.

*Definition*    **Descriptive statistics** consists of organizing and summarizing the information collected.

Descriptive statistics describes the information collected through numerical measurements, charts, graphs and tables. The main purpose of descriptive statistics is to provide an overview of the information collected.

4. *Draw conclusions from the information.* This is the step in which the information collected from the sample is generalized to the population.

*Definition*    **Inferential statistics** uses methods that generalize results obtained from a sample to the population and measure their reliability.

For example, if a researcher is conducting a study based on the population of Americans aged 18 years or older, she might obtain a sample of 1,100 Americans aged 18 years or older. The results obtained from the sample would be generalized to the population. There is always uncertainty when using samples to draw conclusions regarding a population because we can't learn everything about a population by looking at a sample. Therefore, statisticians will report a level of confidence in their conclusions. This level of confidence is a way of representing the reliability of results. If the entire population is studied, then inferential statistics is not necessary because descriptive statistics would provide all the information that we need regarding the population.

The following example is a medical study to determine the effectiveness (medical researchers use the word "efficacy") of a drug. It will help illustrate the process of a statistical study.

▶ **EXAMPLE 1**   **Effectiveness of a Drug Treatment on the Common Cold**

According to researchers, little information exists on the cause and treatment of pain, pressure, and other discomfort attributed to the common cold.\* Researchers Steven J. Sperber, MD, and his associates wanted to determine the effectiveness of a new drug combination\*\* on symptoms of the common cold. The following procedures allowed the researchers to measure the effectiveness of these drugs:

1. *Identify the research objective.* Researchers wished to determine the effectiveness of a new drug combination on the symptoms associated with the common cold.

2. *Collect the information needed to answer the questions.* The researchers divided 430 subjects with cold symptoms into two groups. The first group (Group 1) had 216 subjects and the second group (Group 2) had 214 subjects. The Group 1 subjects were given two daily doses of the new drug combination. The Group 2 subjects were given two daily doses of a *placebo*. A **placebo** is an innocuous drug such as sugar water. Group 1 is called the **experimental group** and the drug is called the **treatment**. Group 2 is called the **control group**. Neither the doctor administering the drug nor the patient knew whether he or she was in the experimental or control group. This is referred to as a **double-blind** experiment. Prior to the first dose and after the second dose, the patients reported symptoms on a scale from 0 to 4 with 0 reflecting the absence of pain and 4 reflecting severe pain.

3. *Organize and summarize the information.* The results of the experiment indicated that both the experimental and control group had similar levels of pain prior to the administering of the drugs. However, after the second dose was administered, the reported pain of the experimental group decreased by 1.30 while the control group reported a decrease of 0.93.

4. *Draw conclusions from the data.* We wish to extend the results obtained from the sample of 430 subjects to all individuals that have a common cold. That is, the new drug combination is effective in relieving the symptoms attributable to the common cold. ◀◀

  *Now Work Problem 25.*

---

\*The discussion is based upon a study done by Steven J. Sperber, Ronald B. Turner, James V. Sorrentino, Robert R. O'Connor, James Rogers, and Jack M. Gwaltney that is published in the *Archives of Family Medicine*, 2000; 9:979–985.

\*\*The combination of pseudoephedrine hydrochloride with acetaminophen.

## Types of Data

Once a research objective is stated, a list of the type of information the researcher desires about the individual must be created. **Variables** are the characteristics of the individuals within the population. For example, this past spring my son and I planted a tomato plant in our backyard. We decided to collect some information about the tomatoes harvested from the plant. The individuals we studied were the tomatoes. The variable that interested us was the weight of the tomatoes. My son noted that the tomatoes had different weights even though they all came from the same plant. He discovered that variables vary. If variables did not vary, they would be constants and statistical inference would not be necessary. Think about it this way: If all the tomatoes had the same weight, then knowing the weight of one tomato would be sufficient in determining the weight of all tomatoes. However, the weights of tomatoes vary from one tomato to the next. One of the goals of research is to learn the causes of the variability so that we could learn to grow plants that yield the best tomatoes.

Variables can be classified into two groups.

*Definition*    **Qualitative or Categorical variables** allow for classification of individuals based on some attribute or characteristic.

**Quantitative variables** provide numerical measures of individuals. Arithmetic operations such as addition and subtraction can be performed on the values of a quantitative variable and provide meaningful results.

Many examples in this text will include a suggested **approach**, or a way to look at and organize a problem so that it can be solved. The approach will be a suggested method of "attack" toward solving the problem. This does not mean that the approach given is the only way to solve the problem because many problems have more than one approach leading to a correct solution. For example, if you turn the key on your car's ignition and it doesn't start, one approach would be to look under the hood and try to determine what is wrong. (Of course, this approach would work only if you know how to fix cars.) A second, equally valid approach would be to call an automobile mechanic to service the car.

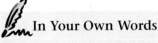
**In Your Own Words**

Typically, there is more than one correct approach to solving a problem.

▶ **EXAMPLE 2**    **Distinguishing between Qualitative and Quantitative Variables**

*Problem:* Determine whether the following variables are qualitative or quantitative.

**(a)** Gender
**(b)** Temperature
**(c)** Number of days during the past week a college student aged 21 years or older has had at least one drink
**(d)** Zip code

*Approach:* Quantitative variables are numerical measures such that arithmetic operations can be performed on the values of the variable, while qualitative variables describe an attribute or characteristic of the individual that allow researchers to categorize the individual.

*Solution:*

**(a)** Gender is a qualitative variable because it allows a researcher to categorize the individual as male or female. Notice that arithmetic operations cannot be performed on the attributes "male" and "female."
**(b)** Temperature is a quantitative variable because it is numeric and operations such as addition and subtraction provide meaningful results. For example, 70°F is 10°F warmer than 60°F.

**(c)** Number of days during the past week that a college student aged 21 years or older had at least one drink is a quantitative variable because it is numeric and operations such as addition and subtraction provide meaningful results.

**(d)** Zip code is a qualitative variable because it categorizes a location. Notice that the addition or subtraction of zip codes does not provide meaningful results. ◄◄

On the basis of the result of Example 2(d), we conclude that a variable may be qualitative while having values that are numeric. Just because a variable is numeric does not mean that the variable is quantitative.

**NW**  *Now Work Problem 3.*

**4**

We can further classify quantitative variables into two types.

*Definition*

A **discrete variable** is a quantitative variable that has either a finite number of possible values or a countable number of possible values. The term "countable" means that the values result from counting such as $0, 1, 2, 3$, and so on.

A **continuous variable** is a quantitative variable that has an infinite number of possible values that are not countable.

✎ **In Your Own Words**

Anytime that you count to get the value of a variable, it is discrete. If you measure to get the value of the variable, it is continuous. When deciding whether a variable is discrete or continuous, ask yourself if it is counted or measured.

Figure 1 illustrates the relationship among qualitative, quantitative, discrete, and continuous variables.

*Figure 1*

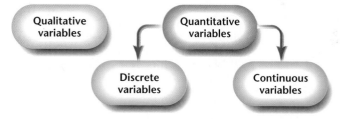

Recognizing the type of variable being studied is important because it dictates the type of analysis that can be performed.

An example should help to clarify the definitions.

► **EXAMPLE 3**

**Distinguishing between Discrete and Continuous Variables**

*Problem:* Determine whether the following quantitative variables are discrete or continuous.

**(a)** The number of heads obtained after flipping a coin five times.

**(b)** The number of cars that arrive at a McDonald's drive-through between 12:00 P.M. and 1:00 P.M.

**(c)** The distance a 2002 Pontiac Bonneville can travel in city driving conditions with a full tank of gas.

*Approach:* A variable is discrete if its value results from counting. A variable is continuous if its value is measured.

*Solution:*

**(a)** The number of heads obtained by flipping a coin five times would be a discrete variable because we would count the number of heads obtained. The possible values of the discrete variable are $\{0, 1, 2, 3, 4, 5\}$.

**(b)** The number of cars that arrive at a McDonald's drive-through between 12:00 P.M. and 1:00 P.M. is a discrete variable because its value would result from counting the cars. The possible values of the discrete variable are 0, 1, 2, 3, 4, and so on. Notice that there is no predetermined upper limit to the number of cars that may arrive.

**(c)** The distance traveled is a continuous variable because we measure the distance. ◄◄

Continuous variables are often rounded. For example, when the miles per gallon of gasoline for a certain make of car is given as 24 miles per gallon, it really means that the miles per gallon is greater than or equal to 23.5 and less than 24.5, or $23.5 \leq$ miles per gallon $< 24.5$.

**NW** *Now Work Problem 13.*

The list of observations a variable assumes is called **data**. While gender is a variable, the observations, male or female, are data. **Qualitative data** are observations corresponding to a qualitative variable. **Quantitative data** are observations corresponding to a quantitative variable. **Discrete data** are observations corresponding to a discrete variable and **continuous data** are observations corresponding to a continuous variable.

► **EXAMPLE 4** **Distinguishing between Variables and Data**

*Problem:* Table 1 represents a group of selected countries and information regarding these countries as of December 15, 2001. Identify the individuals, variables, and data in Table 1.

| TABLE 1 | | | |
|---|---|---|---|
| Country | Government Type | Life Expectancy (years) | Birth Rate per 1000 People |
| Australia | Democratic | 79.87 | 12.86 |
| Canada | Confederation | 79.56 | 11.21 |
| France | Republic | 78.9 | 12.1 |
| Morocco | Constitutional Monarchy | 69.43 | 24.16 |
| Poland | Republic | 73.42 | 10.2 |
| Sri Lanka | Republic | 72.09 | 16.58 |
| United States | Federal Republic | 77.26 | 14.2 |

*Source:* CIA World Factbook

*Approach:* An individual is an object or person that we wish to obtain data for. The variables are the characteristics of the individuals and the data are the specific values of the variables.

*Solution:* The individuals in the study are the countries: Australia, Canada, and so on (in red ink). The variables measured for each country are "government type," "life expectancy," and "birth rate" (in blue ink). The variable "government type" is qualitative because it categorizes the individual. The quantitative variables are "life expectancy" and "birth rate." Life expectancy is continuous because it is measured. Birth rate is discrete because we count the number of births. The observations are the data (in green ink). For example, the data corresponding to the variable "life expectancy" are 79.87, 79.56, 78.9, 69.43, 73.42, 72.09, and 77.26. The following data correspond to the individual Poland: a republic government with residents whose life expectancy is 73.42 years and a birth rate of 10.2 per 1,000 people.

"Republic" is qualitative data that results from observing the value of the qualitative variable "government type." The life expectancy of 73.42 years is quantitative data that results from observing the value of the quantitative variable "life expectancy."   ◀◀

**NW** *Now Work Problem 31.*

## 1.1 Assess Your Understanding

### Concepts and Vocabulary

1. Define statistics and list the requirements of the statistical process.
2. Explain the difference between a population and a sample.
3. Explain the difference between descriptive statistics and inferential statistics.
4. What does it mean when an experiment is double blind?
5. Define qualitative variable. Provide some examples. Define quantitative variable. Provide some examples.
6. Discuss the difference between discrete and continuous variables.

### Exercises

#### • Skill Building

*In Problems 1–10, classify the variable as qualitative or quantitative.*

1. Hair color
2. Salary
**NW** 3. Weight of cars
4. Religious affiliation
5. ACT score
6. Grams of fat in a cheeseburger
7. Number of customers served at Wendy's during lunch
8. Number of times "3" is observed after rolling a die 10 times
9. Types of surgical procedures offered at General Hospital
10. Amount of rainfall in Topeka, Kansas, during the summer of 2001

*In Problems 11–20, determine whether the quantitative variable is discrete or continuous.*

11. Home runs hit by Barry Bonds this season
12. Time spent studying for your first statistics exam
13. Strength of concrete in pounds per square inch
**NW**
14. Number of typos in a 500-page novel
15. Height of a tree
16. Number of people in a poll of 500 who believe Albert Einstein was the greatest scientist of the twentieth century
17. Number of flips of a coin needed until the first tail is observed
18. Speed of a car on the highway
19. Serum HDL cholesterol of a 32-year-old male
20. Annual income of a 45-year-old female executive

*In Problems 21–24, a research objective is presented. For each research objective, identify the population and sample in the study.*

21. The Gallup Organization contacts 1019 adult residents of the United States aged 18 years or older and asks whether the events of September 11, 2001 were a life-altering experience.
22. A quality control manager randomly selects 50 bottles of Coca-Cola that were filled on October 15 in order to assess the calibration of the filling machine.
23. A farmer wanted to learn about the weight of his soybean crop. He randomly sampled 100 plants and measured the weight of the soybeans on the plant.
24. Every year the United States Census Bureau releases the Current Population Report based on a survey of 50,000 households. The goal of this report is to learn demographic characteristics of all households within the United States such as income.

#### • Applying the Concepts

*For the studies given in Problems 25–30, (a) identify the research objective, (b) identify the sample, (c) list the descriptive statistics, and (d) state the conclusions made in the study.*

25. **Is Brain Volume Associated with Schizophrenia?** A **NW** study conducted by researchers was designed "to determine the genetic and nongenetic factors to structural brain abnormalities on schizophrenia." The researchers determined the brain volumes of 29 twins who were patients diagnosed with schizophrenia and compared

them to the brain volumes of 29 healthy twins. Based upon a high-resolution MRI, it was determined the whole-brain volumes were 2.2% smaller in the schizophrenic patients. The researchers concluded that an increased genetic risk to develop schizophrenia is related to reduced brain growth early in life.

*Source:* "Volumes of Brain Structures in Twins Discordant for Schizophrenia," William F.C. Baare, et al.; *Archives of General Psychiatry* 58 (2000): 33–40.

26. **Blood Pressure and Diet** A study was conducted by researchers to determine "if blood pressure level is associated with dietary micronutrients in adolescents at risk for hypertension." The researchers randomly selected adolescents aged 14 to 16 years that had high blood pressure. They measured the folic acid intake of the adolescents. The high-folate group had folic acid intake greater than the recommended daily allowance while the low-folate group had intake less than the recommended daily allowance. It was determined that the average diastolic blood pressure for the low-folate group was substantially higher than in the high-folate group. For boys it was 72 versus 67 mm Hg and for girls it was 76 versus 73 mm Hg. It was determined that for adolescents who are at risk for high blood pressure, the benefits of a diet rich in folic acid (such as fruit, vegetables, and low-fat dairy products) could contribute to the prevention of hypertension (high blood pressure) when instituted at an early age.

*Source:* "Dietary Nutrients and Blood Pressure in Urban Minority Adolescents at Risk for Hypertension," Bonita Falkner, et al.; *Archives of Pediatrics and Adolescent Medicine* 154 (2000): 918–922.

27. **The Mozart Effect** Researchers at the University of California, Irvine, wished to determine whether "music cognition and cognitions pertaining to abstract operations such as mathematical or spatial reasoning" were related. To test the research question, 36 college students listened to Mozart's sonata for two pianos in D major, K488, for 10 minutes and then were administered a spatial reasoning test using the Stanford-Binet intelligence scale. The same students were also administered the test after sitting in a room for 10 minutes in complete silence.

The mean score on the test following the Mozart piece was 119, while the mean test score following the silence was 110. The researchers concluded that subjects performed better on abstract/spatial reasoning tests after listening to Mozart.

*Source:* "Music and spatial performance," Frances H. Rauscher, et al.; *Nature* 365 (14 October 1993): 611.

28. **Propecia** Merck Pharmaceutical Company manufactures Propecia, a drug that claims to treat male pattern hair loss on the vertex (top of the head) and anterior midscalp area in men. For 12 months, doctors studied over 1800 men aged 18 to 41 with mild to moderate amounts of ongoing hair loss. All men, whether receiving Propecia or placebo (a pill containing no medication), were given a medicated shampoo. In general, men who took Propecia maintained or increased the number of visible scalp hairs and noticed improvement in their hair in the first year, with the effect maintained in the second year. Hair counts in men who did not take Propecia continued to decrease. Merck concluded that Propecia is effective in maintaining or increasing the amount of hair on the vertex and anterior midscalp area.

*Source:* www.merck.com

29. **Victims of Crime** Gallup News Service conducted a survey of 1012 adults aged 18 years old or older, August 29–September 5, 2000. The respondents were asked, "Has anyone in your household been the victim of a crime in the past 12 months?" Of the 1012 adults surveyed, 24% said they or someone in the household had experienced some type of crime during the preceding year. Gallup News Service concluded that 24% of all households had been victimized by crime during the past year.

30. **Global Warming** Gallup News Service conducted a survey of 1004 adults 18 years and older, April 3–9, 2000. The respondents were asked, "Do you have a great deal of concern regarding global warming (the greenhouse effect)?" Of the 1004 adults surveyed, 40% said they worried about global warming a great deal. Gallup News Service concluded that 40% of all Americans worry about global warming a great deal.

---

*In Problems 31–34, identify the individuals, variables, and data corresponding to the variables. Determine whether each variable is qualitative, continuous, or discrete.*

31. **Survey of Students in Sullivan's Business Calculus** (NW) **Course** Michael Sullivan surveyed five students in his business calculus course and obtained the following information:

| Student | Gender | Age | Number of Siblings |
|---|---|---|---|
| Neta Van Duyne | F | 19 | 1 |
| Dave Ebert | M | 19 | 1 |
| Kristin Bols | F | 19 | 2 |
| Michael Wirth | M | 19 | 1 |
| Jinita Desai | F | 20 | 1 |

32. **BMW Cars** The following information relates to the entire product line of BMW automobiles:

| Model | Body Style | Weight (pounds) | Number of Seats |
|---|---|---|---|
| M/Z3 Coupe | Coupe | 2945 | 2 |
| M/Z3 Roadster | Convertible | 2690 | 2 |
| 3 Series | Coupe | 2780 | 5 |
| 5 Series | Sedan | 3450 | 5 |
| 7 Series | Sedan | 4255 | 5 |
| Z8 | Convertible | 3600 | 2 |

*Source: Car and Driver magazine*

**33. Driver's License Laws** The following data represent driver's license laws for various states.

| State | Age for Driver's License | Blood Alcohol Concentration Limit | Mandatory Belt-use Law Seating Positions | Maximum Allowable Speed Limit, 1999 |
|---|---|---|---|---|
| Alabama | 16 | 0.08 | Front | 70 |
| Illinois | 18 | 0.08 | Front | 65 |
| Montana | 18 | 0.10 | All | 75 |
| New York | 17 | 0.10 | Front | 65 |
| Texas | 18 | 0.10 | Front | 70 |

*Source: Time Almanac,* 2000

**34. NCAA Division I Women Basketball Players** The following data represent the leading scorers in NCAA Division I Women's Basketball during the 1998/1999 season.

| Player | College | Number of Games | Points Scored |
|---|---|---|---|
| Tamika Whitmore | Memphis | 32 | 843 |
| Jackie Stiles | SW Missouri State | 32 | 823 |
| Kim Knuth | Toledo | 31 | 788 |
| Kristina Behnfeldt | Marshall | 26 | 621 |
| Jamie Cassidy | Maine | 31 | 738 |

*Source: Time Almanac,* 2000

**35.** Read a newspaper, magazine, or journal that contains a research study and identify (a) the research question the study addresses, (b) the population, (c) the sample, (d) any descriptive statistics, and (e) the inferences of the study.

## 1.2  Observational Studies; Simple Random Sampling

**Objectives**  **1**  Distinguish between an observational study and an experiment

**2**  Obtain a simple random sample

We are now familiar with some of the terminology used in describing data. Now, we need to determine how to obtain data. When we defined the word statistics, we said it is a science that involves the collection of data. Data can be obtained from four sources:

**1.** A census
**2.** Existing sources
**3.** Survey sampling
**4.** Designed experiments

We start by defining a census.

*Definition*   A **census** is a list of all individuals in a population along with certain characteristics of each individual.

The United States conducts a census every 10 years in order to learn the demographic make-up of the United States. Everyone whose usual residence is within the United States borders must fill out a questionnaire packet. The cost of obtaining the census in 2000 was approximately $6 billion. The census data provide information such as the number of members in a household, number of years at present address, household income, and more. Because of the cost of obtaining census data, most researchers obtain data through existing sources, survey samples, or designed experiments.

Have you ever heard the phrase, "There is no point in reinventing the wheel?" Well, there is no point in spending energy obtaining data that already exist either. If a researcher wishes to conduct a study, and a data set exists that can be used to answer the researcher's questions, then it would be silly to collect the data from scratch. For example, in the August 22, 2001 issue of the *Journal of the American Medical Association* ["Physical Activity, Obesity, Height, and the Risk of Pancreatic Cancer," Dominique S. Michaud, ScD, et al. Volume 286, No. 8], researchers did a study in which they attempted to identify factors that increase the likelihood of an individual getting pancreatic cancer. Rather than conducting their own survey, they used data from two existing surveys: the Health Professionals Follow-up Study and the Nurses' Health Study. By doing this, they saved time and money. The moral of the story: **Don't collect data that have already been collected!**

Survey sampling is used in research where there is no attempt to influence the value of the variable of interest. For example, we may want to identify the "normal" systolic blood pressure of American men aged 40–44. The researcher would obtain a sample of men aged 40–44 and determine their systolic blood pressure. No attempt is made to influence the systolic blood pressure of the men surveyed. Polling data is another example of data obtained from a survey sample because the respondent is simply asked his or her opinion. No attempt is made to influence this opinion. Data obtained from a survey sample lead to an *observational study*.

*Definition*   An **observational study** measures the characteristics of a population by studying individuals in a sample, but does not attempt to manipulate or influence the variable(s) of interest. Observational studies are sometimes referred to as *ex post facto* (after-the-fact) studies because the value of the variable of interest has already been established.

We distinguish an observational study with a *designed experiment*.

*Definition*   A **designed experiment** applies a treatment to individuals (referred to as **experimental units**) and attempts to isolate the effects of the treatment on a **response variable**.

### In Your Own Words

In observational studies, there is no control, while designed experiments try to maintain control.

Data obtained through experimentation will be thoroughly discussed in Section 1.5, but the main idea is that the researcher is able to control factors that influence the individuals. For example, suppose my son has two types of fertilizer and wants to determine which one is better. He might conduct an experiment in which he divides his garden into two plots. Plot 1 might receive the recommended amount of the first fertilizer and Plot 2 might receive the recommended amount of the second fertilizer. All other factors that affect plant growth (amount of sunlight, water, soil condition, and so on) would be the same for the two plots. The type of fertilizer is the treatment and the tomatoes are the experimental units. The weight of the tomatoes would be the response variable.

An example should clarify the difference between an observational study and a designed experiment.

▶ **EXAMPLE 1**   **Observational Study versus Designed Experiment**

In most types of research, the goal is to determine the relation, if any, that may exist between two or more variables. For example, a researcher may want to determine whether there is a connection between smoking and lung cancer.* This type of study would be performed using an observational study because it is *ex post facto* (after-the-fact) research. The individuals in the study are examined after they have been smoking for some period of time. The individuals are not controlled in terms of the number of cigarettes smoked per day, eating habits, and so on. A researcher simply interviews a sample of smokers and monitors their rate of cancer as it compares with a sample of nonsmokers. The nonsmokers serve as a *control group*; that is, the nonsmokers serve as the benchmark upon which the smokers' rate of cancer is judged. The hope is that nonsmokers and smokers are alike in all of their traits, such as diet and exercise, except for smoking. Then, if a significant difference between the two groups' rates of cancer exists, the researcher might claim smoking *causes* cancer.

In actuality, the researcher determined that smoking is *associated* with cancer. It may be that the population of smokers has a higher rate of cancer, but this higher incidence rate is not necessarily a direct result of the smoking. It is possible that smokers have some characteristic that differs from the nonsmoking group, other than smoking, that is the *cause* of cancer. The characteristics that may be related to cancer, but that have not been identified in the study, are referred to as **lurking variables**. For example, a lurking variable might be the amount of exercise. Maybe smokers generally exercise less than nonsmokers and the lack of exercise is the cause of cancer. In this observational study we might be able to state that smokers have a higher rate of cancer than nonsmokers and that smoking is *associated* with cancer, but we would not be able to definitively state that smoking causes cancer.

Obtaining this type of data through experimentation would require randomly dividing a sample of people into two groups. We would then require one of the groups to smoke a pack of cigarettes each day for the next 20 years (the treatment) to determine whether it has a higher incidence rate of lung cancer (the response variable) than the nonsmoking (control) group! By approaching the study in this way, we would be able to control many of the factors that were beyond our control in the observational study. For example, we could make sure each group had the same diet and exercise regiment. This would allow us to determine whether smoking is a cause of cancer. Of course, there are moral issues to consider in conducting this type of experiment. ◀◀

▮▮▮
**Caution**

*Beware of observational studies that claim causation!*

It is vital that we understand that observational studies do not allow a researcher to claim causation, only association.

> Observational studies are very useful tools for determining whether there is a relation between two variables, but it requires a designed experiment to isolate the cause of the relation.

Many observational studies are set up to learn characteristics of a population, such as polls. For example, the Gallup Organization routinely surveys Americans in an attempt to identify opinion.

Observational studies are performed for two reasons:

1. To learn characteristics of a population;
2. To determine whether there is an association between two or more variables where the values of the variables have already been determined. (Again, this is often referred to as *ex post facto* research.)

---

*The interested reader may wish to read Chapter 18, "Does Smoking Cause Cancer?" in David Salsburg's book *The Lady Tasting Tea*. W. H. Freeman and Co., 2001.

*Ex post facto* research is common in research where control of certain variables is impossible or unethical. For example, economists often perform research using observational studies because they do not have the ability to control many of the factors that affect our purchasing decisions. Some medical research requires that observational studies be conducted because of the risks thought to be associated with experimental study, such as the perceived link between smoking and lung cancer.

Experiments, on the other hand, are used whenever control of certain variables is desired. This type of research allows the researcher to identify certain cause and effect relationships among the variables in the study.

The bottom line to consider is control. If control is possible, an experiment should be performed. However, if control is not possible or necessary, then observational studies are appropriate.

**NW** *Now Work Problem 5.*

## Sampling

The goal in sampling is to obtain individuals that will participate in a study so that accurate information about the population can be obtained. For example, the Gallup Organization typically will poll a sample of about 1,000 adults aged 18 years or older from the population of adults aged 18 years or older. We want the sample to provide as much information as possible, but each additional piece of information has a price. So the question becomes "How can the researcher obtain accurate information about the population through the sample while minimizing the costs in terms of money, time, personnel, and so on?" There is a balance between information and cost. An appropriate sample design can maximize the amount of information obtained about the population for a given cost.

We will discuss four basic sampling techniques: simple random sampling, stratified sampling, systematic sampling, and cluster sampling. These sampling methods are based on planned randomness. Planned randomness means that the individuals selected for the study are chosen on the basis of some predetermined method, so that any biases that may be introduced (knowingly and unknowingly) by the surveyor during the selection process are eliminated. In other words, the surveyor does not have a choice as to who is in the study. We discuss simple random sampling now and discuss the remaining three types of sampling in the next section.

### ② Simple Random Sampling

The most basic sample survey design is **simple random sampling**, which is often abbreviated as **random sampling**.

*Definition*

A sample of size *n* from a population of size *N* is obtained through **simple random sampling** if every possible sample of size *n* has an equally likely chance of occurring. The sample is then called a **simple random sample**.

**In Your Own Words**

Simple random sampling is like selecting names from a hat.

The sample is always a subset of the population with $n < N$. Let's look at an example.

▶ EXAMPLE 2 Illustrating Simple Random Sampling

*Problem:* Sophia has four tickets to a concert. Six of her friends, Yolanda, Michael, Kevin, Marissa, Annie, and Katie, have all expressed an interest in going to the concert. Sophia decides to randomly select three of the six.

**(a)** List all possible samples of size $n = 3$ (without replacement) from the population of size $N = 6$.

**(b)** Comment on the likelihood of the sample containing Michael, Kevin, and Marissa.

*Approach:* We list all possible combinations of three people chosen from the six. Remember, in simple random sampling, each sample of size three is equally likely to occur.

*Solution:*

**(a)** The possible samples of size three are listed in Table 2.

| TABLE 2 | | | |
|---|---|---|---|
| Yolanda, Michael, Kevin | Yolanda, Michael, Marissa | Yolanda, Michael, Annie | Yolanda, Michael, Katie |
| Yolanda, Kevin, Marissa | Yolanda, Kevin, Annie | Yolanda, Kevin, Katie | Yolanda, Marissa, Annie |
| Yolanda, Marissa, Katie | Yolanda, Annie, Katie | Michael, Kevin, Marissa | Michael, Kevin, Annie |
| Michael, Kevin, Katie | Michael, Marissa, Annie | Michael, Marissa, Katie | Michael, Annie, Katie |
| Kevin, Marissa, Annie | Kevin, Marissa, Katie | Kevin, Annie, Katie | Marissa, Annie, Katie |

From Table 2, we see that there are 20 possible samples of size three from the population of size six. We use the term "sample" to mean the individuals in the sample.

**(b)** There is 1 sample that contains Michael, Kevin, and Marissa and 20 possible samples, so there is a 1 in 20 chance that the simple random sample would contain Michael, Kevin, and Marissa. In fact, all the samples of size three have a 1 in 20 chance of occurring.  ◄◄

**NW** *Now Work Problem 13.*

**Obtaining a Simple Random Sample**  The results of Example 2 leave one question unanswered: How do we actually select the individuals in a simple random sample? Simple random sampling is just like drawing names out of a hat. To obtain a simple random sample of size three from the population in Example 2, we could write the names of the six people in the population on different sheets of paper and then select three names from the hat. Simple random sampling is that easy!

Often, however, the size of the population is so large that performing simple random sampling in this fashion is not practical. Typically, random numbers are used by assigning each individual in the population a unique number between 1 and $N$, where $N$ is the size of the population. Then, $n$ random numbers from this list are selected. Because we must number the individuals in the population, we must have a list of all the individuals within the population, called a **frame**.

> **In Your Own Words**
>
> A frame lists all the individuals in a population. For example, a list of registered voters might be a frame.

► **EXAMPLE 3**  **Obtaining a Simple Random Sample**

*Problem:* The Village of Lemont wants to learn the opinion of its residents regarding the construction of a new shopping mall. The mayor wishes to obtain a simple random sample of size 10 from the 8791 residents of Lemont.

*Approach:* The mayor must first obtain a list of all residents in the village, the frame, and assign the residents numbers from 1–8791. Finally, the mayor will randomly select 10 residents to be in the sample. Once an individual is selected to be in the sample, he or she cannot be selected again. This is called **sampling without replacement**. We sample without replacement so that we don't select the same individual twice.

*Solution:* A table of random numbers is used to select the individuals to be in the sample. See Table 3.* We select a starting place in the table of random numbers. This can be done by closing our eyes and placing a finger on the table. That may sound haphazard, but it accomplishes our goal of being

---

*Each digit is in its own column. The digits are displayed sin groups of five for ease of reading. The digits in row 1 are 893922321274483 and so on. The first digit, 8, is in column 1; the second digit, 9, is in column 2; the ninth digit, 1, is in column 9.

completely random. Suppose we start in column 9, row 13. Since our data have four digits, we select four-digit numbers from the table.

Column 9

| | | | TABLE 3 | | | | | | |
|---|---|---|---|---|---|---|---|---|---|
| Row Number | | | | Number Column | | | | | |
| | 01–05 | 06–10 | 11–15 | 16–20 | 21–25 | 26–30 | 31–35 | 36–40 | 41–45 | 46–50 |
| 01 | 89392 | 23212 | 74**483** | **3**6590 | 25956 | 36544 | 68518 | 40805 | 09980 | 00467 |
| 02 | 61458 | 17639 | 96**252** | **9**5649 | 73727 | 33912 | 72896 | 66218 | 52341 | 97141 |
| 03 | 11452 | 74197 | 81**962** | **4**8433 | 90360 | 26480 | 73231 | 37740 | 26628 | 44690 |
| 04 | 27575 | 04429 | 31**308** | **0**2241 | 01698 | 19191 | 18948 | 78871 | 36030 | 23980 |
| 05 | 36829 | 59109 | 88**976** | **4**6845 | 28329 | 47460 | 88944 | 08264 | 00843 | 84592 |
| 06 | 81902 | 93458 | 42**161** | **2**6099 | 09419 | 89073 | 82849 | 09160 | 61845 | 40906 |
| 07 | 59761 | 55212 | 33360 | 68751 | 86737 | 79743 | 85262 | 31887 | 37879 | 17525 |
| 08 | 46827 | 25906 | 64708 | 20307 | 78423 | 15910 | 86548 | 08763 | 47050 | 18513 |
| 09 | 24040 | 66449 | 32353 | 83668 | 13874 | 86741 | 81312 | 54185 | 78824 | 00718 |
| 10 | 98144 | 96372 | 50277 | 15571 | 82261 | 66628 | 31457 | 00377 | 63423 | 55141 |
| 11 | 14228 | 17930 | 30118 | 00438 | 49666 | 65189 | 62869 | 31304 | 17117 | 71489 |
| 12 | 55366 | 51057 | 90065 | 14791 | 62426 | 02957 | 85518 | 28822 | 30588 | 32798 |
| 13 | 96101 | 3064**6** | **35**526 | 90389 | 73634 | 79304 | 96635 | 6626 | 94683 | 16696 |
| 14 | 38152 | 5547**4** | **30**153 | 26525 | 83647 | 31988 | 82182 | 98377 | 33802 | 80471 |
| 15 | 85007 | 1841**6** | **24**661 | 95581 | 45868 | 15662 | 28906 | 36392 | 07617 | 50248 |
| 16 | 85544 | 1589**0** | **80**011 | 18160 | 33468 | 84106 | 40603 | 01315 | 74664 | 20553 |
| 17 | 10446 | 2069**9** | **98**370 | 17684 | 16932 | 80449 | 92654 | 02084 | 19985 | 59321 |
| 18 | 67237 | 4550**9** | **17**638 | 65115 | 29757 | 80705 | 82686 | 48565 | 72612 | 61760 |
| 19 | 23026 | 8981**7** | **05**403 | 82209 | 30573 | 47501 | 00135 | 33955 | 50250 | 72592 |
| 20 | 67411 | 5854**2** | **18**678 | 46491 | 13219 | 84084 | 27783 | 34508 | 55158 | 78742 |

Row 13 — 13

We skip 9080 because it is larger than 8791

So, we start our sample by surveying individual 4635. We then proceed downward. The next individual in the sample would be 7430. If a number is repeated, we do not count it twice, since we are sampling without replacement. Because our population has individuals numbered from 1 to 8791, anytime we encounter 0 or a number larger than 8791, we skip it and continue. Continuing in this fashion, we have the following individuals in our sample:

4635, 7430, 1624, 0917, 1705, 4218, 4833, 2529, 3080, 1612

Each of the individuals selected in the sample is set in bold face type in Table 3 to help understand where the numbers came from. So the Village of Lemont would interview the residents corresponding to these numbers. ◀◀

**Caution**

Random-number generators are not truly random, because they are programs and programs do not act "randomly." The seed dictates the "random numbers" that are generated.

**Using Technology:** In practice, a random-number table is not used to obtain simple random samples. This is because many calculators and computers have random-number generators that use programs to obtain random numbers.

▶ EXAMPLE 4    Using a Graphing Calculator to Generate a Simple Random Sample

*Problem:* Use a graphing calculator to find a simple random sample of 10 residents for the Village of Lemont presented in Example 3.

*Approach:* The approach will be the same as that presented in Example 3. We number the residents 1–8791 and randomly select 10 residents to be in the sample, again without replacement.

*The Solution:* First, we must set the *seed*. The **seed** in a random-number generator provides an initial point for the calculator to start generating numbers. It is just like selecting the initial point in the table of random numbers from Example 3. The seed can be any nonzero number. See Figure 2(a). We are now ready to obtain the individuals in our simple random sample. Using the randInt( feature, we obtain the first individual in the random sample shown in Figure 2(b).

Figure 2

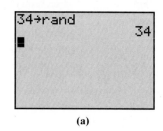  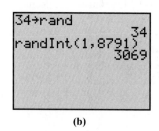

(a)                           (b)

The first person in our sample will be 3069. Continue pressing ENTER until we obtain our sample of size 10. The following people will be in our sample:

 Now Work Problem 15.

3069, 885, 5822, 8385, 3055, 7802, 6349, 1274, 2861, 8775    ◀◀

There is a very important consequence in comparing the individuals in the samples from Example 3 and Example 4. Notice that while both samples were constructed in a similar fashion, they resulted in different individuals in the sample! For this reason, each sample will likely result in different descriptive statistics. Any inference based on each sample may result in different conclusions regarding the population. Therein lies the nature of statistics. Inferences based upon samples will vary because the individuals in two different samples vary.

## 1.2 Assess Your Understanding

### Concepts and Vocabulary

1. Explain the difference between an observational study and an experiment. In your explanation, be sure to discuss the circumstances in which each is appropriate.
2. Explain why a frame is necessary to obtain a simple random sample.
3. Discuss why sampling is used in statistics.
4. What does it mean when sampling is done without replacement?

### Exercises

#### • Applying the Concepts

*In Problems 1–12, determine whether the study depicts an observational study or an experiment.*

1. A study to determine whether there is a relation between the rate of cancer and an individual's proximity to high-tension wires.
2. Rats with cancer are divided into two groups. One group receives 5 mg of an experimental drug that is thought to fight cancer, and the other receives 10 mg. After two years, the spread of the cancer is measured.
3. Seventh-grade students are randomly divided into two groups. One group is taught math using traditional techniques while the other is taught math using a reform

method. After one year, each group is given an achievement test to compare its proficiency with that of the other group.

4. A poll is conducted in which 500 people are asked whom they plan to vote for in the upcoming election.

5. A survey is conducted asking 400 people, "Do you prefer Coke or Pepsi?"
**NW**

6. While shopping, 200 people are asked to perform a taste test in which they drink from two unmarked cups. They are then asked which drink they prefer.

7. The article cited in Problem 25 from Section 1.1.

8. The article cited in Problem 26 from Section 1.1.

9. The article cited in Problem 27 from Section 1.1.

10. The article cited in Problem 28 from Section 1.1.

11. The article cited in Problem 29 from Section 1.1.

12. The article cited in Problem 30 from Section 1.1.

13. **Listing Possible Simple Random Samples** The United
**NW** States Senate finance committee has a subcommittee on long-term growth and debt reduction. As of 12/2001 the members of this subcommittee were Bob Graham, Frank Murkowski, Jon Kyl, Max Baucus, and Kent Conrad. Suppose two randomly selected committee members are asked to investigate the impact of a tax reduction bill on the federal deficit.

(a) List all possible simple random samples of size two.
(b) Comment on the likelihood that the two individuals selected are Bob Graham and Max Baucus.

14. **Listing Possible Simple Random Samples** During the presidency of Andrew Jackson, there were six cabinet positions: secretary of state, secretary of the treasury, secretary of war, attorney general, postmaster general, and secretary of the Navy. Suppose the president needs to send two members of the cabinet to represent the United States at a foreign dignitary's funeral.

(a) List all possible simple random samples of size two.
(b) Comment on the likelihood that the secretary of state and attorney general attend the funeral.

**NW** 15. **Obtaining a Simple Random Sample** The following table lists the 50 states:

| | | | | | | | | | |
|---|---|---|---|---|---|---|---|---|---|
| 1 | Alabama | 11 | Hawaii | 21 | Massachusetts | 31 | New Mexico | 41 | South Dakota |
| 2 | Alaska | 12 | Idaho | 22 | Michigan | 32 | New York | 42 | Tennessee |
| 3 | Arizona | 13 | Illinois | 23 | Minnesota | 33 | North Carolina | 43 | Texas |
| 4 | Arkansas | 14 | Indiana | 24 | Mississippi | 34 | North Dakota | 44 | Utah |
| 5 | California | 15 | Iowa | 25 | Missouri | 35 | Ohio | 45 | Vermont |
| 6 | Colorado | 16 | Kansas | 26 | Montana | 36 | Oklahoma | 46 | Virginia |
| 7 | Connecticut | 17 | Kentucky | 27 | Nebraska | 37 | Oregon | 47 | Washington |
| 8 | Delaware | 18 | Louisiana | 28 | Nevada | 38 | Pennsylvania | 48 | West Virginia |
| 9 | Florida | 19 | Maine | 29 | New Hampshire | 39 | Rhode Island | 49 | Wisconsin |
| 10 | Georgia | 20 | Maryland | 30 | New Jersey | 40 | South Carolina | 50 | Wyoming |

(a) Obtain a simple random sample of size 10 using Table I in Appendix A, a graphing calculator, or computer software.
(b) Obtain a second simple random sample of size 10 using Table I in Appendix A, a graphing calculator, or computer software.

16. **Obtaining a Simple Random Sample** The following table lists the 43 presidents of the United States.

| | | | | | | | | | |
|---|---|---|---|---|---|---|---|---|---|
| 1 | Washington | 10 | Tyler | 19 | Hayes | 28 | Wilson | 37 | Nixon |
| 2 | J. Adams | 11 | Polk | 20 | Garfield | 29 | Harding | 38 | Ford |
| 3 | Jefferson | 12 | Taylor | 21 | Arthur | 30 | Coolidge | 39 | Carter |
| 4 | Madison | 13 | Fillmore | 22 | Cleveland | 31 | Hoover | 40 | Reagan |
| 5 | Monroe | 14 | Pierce | 23 | B. Harrison | 32 | F.D. Roosevelt | 41 | George H. Bush |
| 6 | J.Q. Adams | 15 | Buchanan | 24 | Cleveland | 33 | Truman | 42 | Clinton |
| 7 | Jackson | 16 | Lincoln | 25 | McKinley | 34 | Eisenhower | 43 | George W. Bush |
| 8 | Van Buren | 17 | A. Johnson | 26 | T. Roosevelt | 35 | Kennedy | | |
| 9 | W.H. Harrison | 18 | Grant | 27 | Taft | 36 | L.B. Johnson | | |

(a) Obtain a simple random sample of size eight using Table I in Appendix A, a graphing calculator, or computer software.

(b) Obtain a second simple random sample of size eight using Table I in Appendix A, a graphing calculator, or computer software.

---

**17. Obtaining a Simple Random Sample** Suppose you are the president of the student government. You wish to conduct a survey in order to determine the student body's opinion regarding student services. The administration provides you with a list that contains the names and phone numbers of the 19,935 registered students.

(a) Discuss the procedure you would follow in order to obtain a simple random sample of 25 students.

(b) Obtain this sample.

**18. Obtaining a Simple Random Sample** Suppose the mayor of Justice, Illinois, asks you to poll the residents of the village. The mayor provides you with a list that contains the names and phone numbers of the 5832 residents of the village.

(a) Discuss the procedure you would follow in order to obtain a simple random sample of 20 residents.

(b) Obtain this sample.

---

## Technology Step-by-Step
### Obtaining a Simple Random Sample

***TI-83 Plus*** **Step 1:** Enter any nonzero number on the HOME screen.

**Step 2:** Press the STO ⟹ button.

**Step 3:** Press the MATH button.

**Step 4:** Highlight the PRB menu and select 1: rand.

**Step 5:** From the HOME screen press ENTER.

**Step 6:** Press the MATH button. Highlight PRB menu and select 5: randInt(.

**Step 7:** With randInt( on the HOME screen enter 1 , $N$ where $N$ is the population size. For example, if $N = 500$, enter the following:

randInt(1,500)

Press ENTER to obtain the first individual in the sample. Continue pressing ENTER until the desired sample size is obtained.

***MINITAB*** **Step 1:** Select the **Calc** menu and highlight **Set Base** ....

**Step 2:** Enter any seed number you desire. Note that it is not necessary to set the seed, because MINITAB uses the time of day in seconds to set the seed.

**Step 3:** Select the **Calc** menu and highlight **Random Data** and select **Integer** ...

**Step 4:** Fill in the window shown below with the appropriate values. To obtain a simple random sample for the situation in Example 3, we would fill in the following:

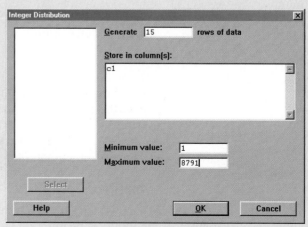

The reason we generate 15 rows of data (instead of 10) is in case any of the random numbers repeat. Select OK, and the random numbers will appear in column 1 (C1) in the spreadsheet.

*Excel*    ***Step 1:***  Be sure the Data Analysis Tool Pak is activated. This is done by selecting the **Tools** menu and highlighting **Add** − **Ins** ... Check the box for the Analysis ToolPak and select OK.

***Step 2:***  Select **Tools** and highlight **Data Analysis** ... Highlight **Random Number Generation** and select OK.

***Step 3:***  Fill in the window with the appropriate values. To obtain a simple random sample for the situation in Example 3, we would fill in the following:

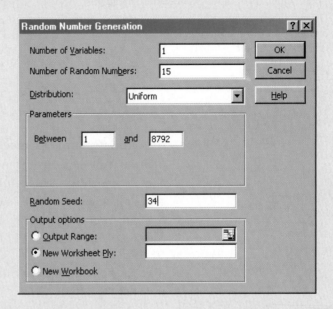

The reason we generate 15 rows of data (instead of 10) is in case any of the random numbers repeat. Notice also that the parameter is between 1 and 8792 so that any value less than or equal to that number is possible. In the unlikely event that 8792 appears, simply ignore it. Select OK, and the random numbers will appear in column 1 (A1) in the spreadsheet. Ignore any values to the right of the decimal place.

---

## 1.3   Other Types of Sampling

**Objectives**     Obtain a stratified sample

  Obtain a systematic sample

  Obtain a cluster sample

One of the goals of sampling is to obtain as much information as possible about the population at the least cost. Remember, we are using the word cost in a general sense. Cost includes monetary outlays, time, and other resources. With this goal in mind, we may find it advantageous to use sampling techniques other than simple random sampling.

### Stratified Sampling

Under certain circumstances, *stratified sampling* provides more information about the population for less cost than simple random sampling.

*Definition*

A **stratified sample** is obtained by separating the population into nonoverlapping groups called *strata* and then obtaining a simple random sample from each stratum. The individuals within each stratum should be homogeneous (or similar) in some way.

**In Your Own Words**

"Stratum" is singular while "strata" is plural. The word "strata" means division. So a stratified sample is a simple random sample of different divisions of the population.

For example, suppose Congress was considering a bill that abolishes estate taxes. In an effort to determine the opinion of her constituency, a senator asks a pollster to conduct a survey within her district. The pollster may divide the population of registered voters within the district into three strata: Republican, Democrat, and Independent. This is because the members within the three party affiliations may have the same opinion regarding estate taxes. The main criterion in performing a stratified sample is that each group (stratum) must have a common attribute that results in the individuals being similar within the stratum.

An advantage of stratified sampling over simple random sampling is that it may allow fewer individuals to be surveyed while obtaining the same or more information. This result occurs because individuals within each subgroup have similar characteristics, so opinions within the group do not vary much from one individual to the next.

▶ EXAMPLE 1    **Obtaining a Stratified Sample**

*Problem:* The president of DePaul University wants to conduct a survey in order to determine the community's opinion regarding campus safety. The president feels the DePaul community can be divided into three groups—resident students, nonresident (commuting) students, and staff (including faculty)—and so will obtain a stratified sample. Suppose there are 6,204 resident students, 13,304 nonresident students, and 2,401 staff, for a total of 21,909 individuals in the population. The president wants to obtain a sample of size 100, with the number of individuals selected from each stratum weighted by the population size. So resident students will make up $6,204/21,909 = 28\%$ of the sample, nonresident students account for 61% of the sample, and staff will constitute 11% of the sample. To obtain a sample of size 100, the president will obtain a stratified sample of $0.28(100) = 28$ resident students, $0.61(100) = 61$ nonresident students, and $0.11(100) = 11$ staff.

*Approach:* To obtain the stratified sample, conduct a simple random sample within each group. That is, obtain a simple random sample of 28 resident students (from the 6,204 resident students), a simple random sample of 61 nonresident students, and a simple random sample of 11 staff.

*Solution:* Using Minitab, with the seed set to 4032 and the values shown in Figure 3, we obtain the following sample of staff:

240, 630, 847, 190, 2096, 705, 2320, 323, 701, 471, 744

Figure 3

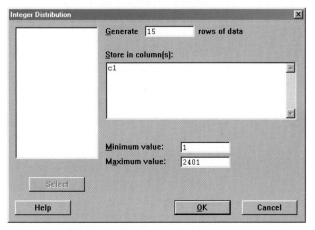

**Caution**

Do not use the same seed for all the groups in a stratified sample because we want the simple random samples within each stratum to be independent of each other.

Repeat this procedure for the resident and nonresident students.

An advantage of stratified sampling over simple random sampling is that it allows the researcher to determine characteristics within each stratum. This allows analysis to be performed on each subgroup to see if any significant differences between the groups exist. For example, we could analyze the data obtained in Example 1 to see if there is a difference in the opinions of students versus faculty.

 ## Systematic Sampling

In both simple random sampling and stratified sampling, it is necessary that a list of the individuals in the population being studied (the frame) exists. Therefore, these sampling techniques require some preliminary work before the sample is obtained. A sampling technique that does not require a frame is *systematic sampling*.

**Definition**

> A **systematic sample** is obtained by selecting every $k$th individual from the population. The first individual selected is a random number between 1 and $k$.

### In Your Own Words

Systematic sampling is like selecting every fifth person out of a line.

Because systematic sampling does not require a frame, it is a useful technique to employ when you can't obtain a list of the individuals in the population that you wish to study.

The idea behind obtaining a systematic sample is relatively simple: select a number $k$, randomly select a number between 1 and $k$ and survey that individual, and then survey every $k$th individual thereafter. For example, we might decide to survey every $k = $ 8th individual. We randomly select a number between 1 and 8 such as 5. This means we survey the 5th, $5 + 8 = $ 13th, $13 + 8 = $ 21st, $21 + 8 = $ 29th, and so on, individual until we reach the desired sample size.

▶ **EXAMPLE 2**    Obtaining a Systematic Sample Without a Frame

*Problem:* The manager of Jewel Food Stores wants to measure the satisfaction of the store's customers. Design a sampling technique that can be used to obtain a sample of 40 customers.

*Approach:* A frame of Jewel customers would be difficult, if not impossible, to obtain because it is unlikely that Jewel can maintain a list of all its customers. Therefore, it is reasonable to use systematic sampling by surveying every $k$th customer that leaves the store.

*Solution:* The manager decides to obtain a systematic sample by surveying every 7th customer. He randomly determines a number between 1 and 7—say, 5. He then surveys the 5th customer exiting the store and every 7th customer thereafter, until a sample of 40 customers exists. The survey will include customers 5, 12, 19, . . . , 278. *    ◀◀

But how do we select the value of $k$? If the size of the population is unknown, there is no mathematical way to determine $k$. It must be chosen simply by determining a value of $k$ that is not so large that we are unable to achieve our desired sample size, but is not so small that we do not obtain a sample that is representative of the population.

To clarify this point, let's revisit Example 2. Suppose we chose a value of $k$ that was too large—say, 30. This means that we will survey every 30th shopper, starting with the 5th. To obtain a sample of size 40 would require that 1175 shoppers visit Jewel on that day. If Jewel does

---

*Because we are surveying 40 customers, the first individual surveyed is the 5th, the second is the $5 + 7 = $ 12th, the third is the $5 + 2(7) = $ 19th, and so on, until we reach the 40th, which is the $5 + 39(7) = $ 278th shopper.

not have 1175 shoppers, the desired sample size will not be achieved. On the other hand, if $k$ is too small—say, 4—we would survey the 5th, 9th,...,161st shopper. It may be that the 161st shopper exits the store at 3 P.M., which means our survey did not include any of the evening shoppers. Certainly, this sample is not representative of *all* Jewel patrons! An estimate of the size of the population would certainly help determine an appropriate value for $k$.

To determine the value of $k$ when the size of the population, $N$, is known is relatively straightforward. Suppose we wish to survey a population whose size is known to be $N = 20{,}325$. We desire a sample of size $n = 100$. To guarantee that individuals are selected evenly from the beginning as well as the end of the population (such as early and late shoppers), we compute $N/n$ and round down to the nearest integer. For example, $20{,}325/100 = 203.25$, so $k = 203$. Then we randomly select a number between 1 and 203 and select every 203rd individual thereafter. So, if we randomly selected 90 as our starting point, we would survey the 90th, 293rd, 496th,..., 20,187th individuals.

We summarize the procedure as follows:

---

**Steps in Systematic Sampling, Population Size Known**

***Step 1:*** Determine the population size, $N$.

***Step 2:*** Determine the sample size desired, $n$.

***Step 3:*** Compute $N/n$ and round down to the nearest integer. This value is $k$.

***Step 4:*** Randomly select a number between 1 and $k$. Call this number $p$.

***Step 5:*** The sample will consist of the following individuals:

$$p, \ p + k, \ p + 2k, \ldots, p + (n - 1)k$$

---

Because systematic sampling does not require that the size of the population be known, it typically provides more information for a given cost than does simple random sampling. In addition, systematic sampling is easier to employ, so there is less possibility of interviewer error occurring, such as selecting the wrong individual to be surveyed.

 *Now Work Problem 17.*

### ③ Cluster Sampling

A fourth sampling method is called *cluster sampling*. The previous three sampling methods discussed have benefits under certain circumstances; so does cluster sampling.

*Definition*

A **cluster sample** is obtained by selecting all individuals within a randomly selected collection or group of individuals.

**✒ In Your Own Words**

Imagine a mall parking lot. Each subsection of the lot could be a cluster (Section F-4 for example).

Consider a quality control engineer who wants to verify that a certain machine is filling bottles with 16 ounces of liquid detergent. To obtain a sample of bottles from the machine, the engineer could use systematic sampling by sampling every $k$th bottle from the machine; however, it would be time consuming waiting next to the filling machine for the bottles to come off the line. Suppose that, as the bottles come off the line, they are placed into cartons of 12 bottles each. An alternative sampling method would be to randomly select a few cartons and measure the contents of all 12 bottles. This would be an example of cluster sampling. It is a good sampling method in this situation because it would speed up the data collection process.

▶ **EXAMPLE 3**   Obtaining a Cluster Sample

*Problem:* A sociologist wants to gather data regarding the household income within the City of Denver. Obtain a sample using cluster sampling.

*Approach:* The City of Denver can easily be set up so that each city block is a cluster. Once the city blocks have been identified, we obtain a simple random sample of the city blocks and survey all households on the blocks selected.

*Solution:* Suppose there are 10,493 city blocks in Denver. First, we must number the blocks from 1 to 10,493. Suppose the sociologist has enough time and money to survey 20 clusters (city blocks). Therefore, the sociologist should obtain a simple random sample of 20 numbers between 1 and 10,493 and survey all households from those clusters selected. Cluster sampling is a good choice in this example because it reduces the travel time to households that is likely to occur with both simple random sampling and stratified sampling. In addition, there is no need to obtain a detailed frame with cluster sampling. The only frame needed is one that provides information regarding city blocks.   ◀◀

Recall that in systematic sampling we had to determine an appropriate value for $k$, the number of individuals to skip between individuals selected to be in the sample. We have a similar problem in cluster sampling. The following are a few of the questions that arise:

- How do I cluster the population?
- How many clusters do I sample?
- How many individuals should be in each cluster?

First, it must be determined whether the individuals within the proposed cluster are homogeneous (similar individuals) or heterogeneous (dissimilar individuals). Consider the results of Example 3. City blocks tend to have similar households. Surveying one house on a city block is likely to result in similar responses from another house on the same block. This results in duplicate information. We conclude the following: If the clusters have homogeneous individuals, it is better to have more clusters with fewer individuals in each cluster.

What if the cluster is heterogeneous? Under this circumstance, the heterogeneity of the cluster likely resembles the heterogeneity of the population. In other words, each cluster is a scaled-down representation of the overall population. For example, a quality control manager might use shipping boxes that contain 100 light bulbs as a cluster, since the rate of defects within the cluster would closely mimic the rate of defects in the population. Thus, when each of the clusters is heterogeneous, fewer clusters with more individuals in each cluster is appropriate.

The four sampling techniques just presented are sampling techniques in which the individuals are selected randomly. Often, however, sampling methods are used in which the individuals are not randomly selected, such as *convenience sampling*.

## Convenience Sampling

Convenience sampling is probably the easiest sampling method.

*Definition*   A **convenience sample** is a sample in which the individuals are easily obtained.

There are many types of convenience samples, but probably the most popular are those in which the individuals in the sample are **self-selected** (i.e., the individuals themselves decide to participate in a survey). Examples of self-selected sampling include phone-in polling where a radio personality will ask his or her listeners to phone the station to submit their opinions. Another example is the use of the Internet to conduct surveys. For example, *Dateline* will present a story regarding a certain topic and ask its viewers to "tell us what you think" by completing a questionnaire online or phoning in an opinion. Both of these samples are poor designs because the individuals who decide to be in the sample generally have strong opinions about the topic. Additionally, a typical individual in the population will not bother phoning or logging on to their computer to complete a survey. Any inference made regarding the population from this type of sample should be made with extreme caution.

## Multistage Sampling

In practice, most large-scale surveys obtain samples using a combination of the techniques just presented.

As an example of multistage sampling, consider Nielsen Media Research. This company measures the number of households watching TV programs. Nielsen randomly selects households and monitors the programs these households are watching through a "People Meter." The meter is an electronic box placed on each TV within the household. The "People Meter" measures what program is being watched and who is watching it. Nielsen selects the households with the use of a two-stage sampling process.

*Stage 1:* Using U.S. Census data, Nielsen divides the country into geographic areas (strata). The strata are typically city blocks in urban areas and geographic regions in rural areas. About 6,000 strata are randomly selected.

*Stage 2:* Nielsen sends representatives to the selected strata and lists the households within the strata. The households are then randomly selected through a simple random sample.

As another example of multistage sampling, consider the sample conducted by the Census Bureau used for the Current Population Survey. This survey requires five stages of sampling:

*Stage 1:* Stratified sample
*Stage 2:* Cluster sample
*Stage 3:* Stratified sample
*Stage 4:* Cluster sample
*Stage 5:* Systematic sample

This survey is very important because it is used to obtain demographic estimates of the United States in noncensus years. A detailed presentation of the sampling method used by the Census Bureau can be found in *The Current Population Survey: Design and Methodology*, Technical Paper No. 40.

## Sample Size Considerations

Throughout the discussion of sampling, we did not mention how to determine the sample size. Determining the sample size is key in the overall statistical process. In other words, the researcher must ask the following question: "How many individuals must I survey in order to draw conclusions about the population within some predetermined margin of error?" The researcher must find the correct balance between the reliability of the results and the

Figure 4

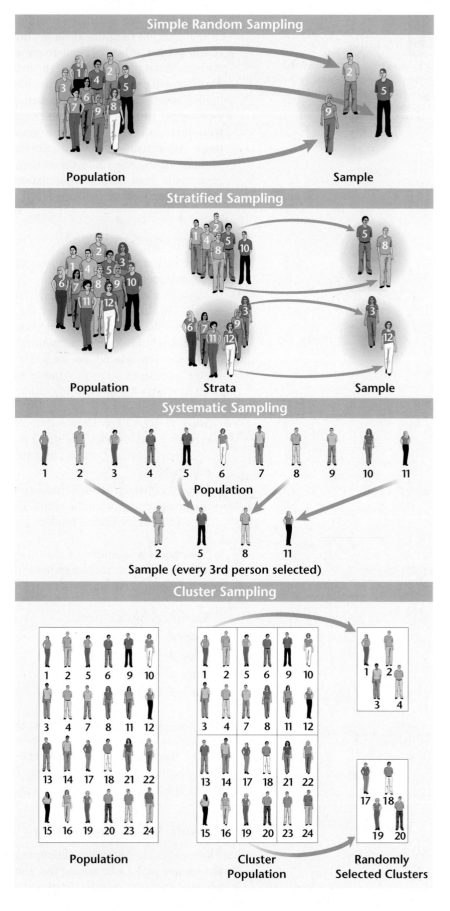

cost of obtaining those results. The bottom line is that time and money determine the level of confidence a researcher will place in the conclusions drawn from the sample data. The more time and money the researcher has available, the more accurate will be the results of the statistical inference.

Nonetheless, techniques do exist for determining the sample size required in order to estimate characteristics regarding the population within some margin of error. We will consider some of these techniques in Sections 8.1 and 8.3. (For a detailed discussion of sample size considerations, consult a text on sampling techniques such as *Elements of Sampling Theory and Methods* by Z. Govindarajulu, Prentice Hall, 1999.)

## Summary

Figure 4 provides a summary of the four sampling techniques presented.

## 1.3 Assess Your Understanding

### Concepts and Vocabulary

1. Describe a circumstance in which stratified sampling would be an appropriate sampling method.

2. Which sampling method does not require a frame?
3. Why are convenience samples ill advised?

### Exercises

#### • Skill Building

*In Problems 1–10, identify the type of sampling used.*

1. In order to estimate the percentage of defects in a recent manufacturing batch, a quality control manager at Intel selects every 8th chip that comes off the assembly line starting with the 3rd, until she obtains a sample of 140 chips.

2. In order to determine the average IQ of ninth-grade students, a school psychologist obtains a list of all high schools in the local public school system. She randomly selects five of these schools and administers an IQ test to all ninth-grade students at the selected schools.

3. In an effort to determine customer satisfaction, United Airlines randomly selects 50 flights during a certain week and surveys all passengers on the flights.

4. A member of Congress wishes to determine her constituency's opinion regarding estate taxes. She divides her constituency into three income classes: low-income households, middle-income households, and upper-income households. She then takes a random sample of households from each income class.

5. In an effort to identify whether an advertising campaign has been effective, a marketing firm conducts a nation-wide poll by randomly selecting individuals from a list of known users of the product.

6. A radio station asks its listeners to call in their opinion regarding the use of American forces in peacekeeping missions.

7. A farmer divides his orchard into 50 subsections, randomly selects 4 and samples all of the trees within the 4 subsections in order to approximate the yield of his orchard.

8. A school official divides the student population into five classes: freshman, sophomore, junior, senior, graduate student. The official takes a random sample from each class and asks the members' opinions regarding student services.

9. A survey regarding download time on a certain Web site is administered on the Internet by a market research firm to anyone who would like to take it.

10. A lobby has a list of the 100 senators of the United States. In order to determine the Senate's position regarding farm subsidies, they decide to talk with every seventh senator on the list starting with the third.

#### • Applying the Concepts

11. **Sample Design** The city of Naperville is considering the construction of a new commuter rail station. The city wishes to survey the residents of the city to obtain their opinion regarding the use of tax dollars for this purpose. Design a sampling method to obtain the individuals in the sample. Be sure to support your choice.

12. **Sample Design** A school board at a local community college is considering raising the student services fees.

The board wants to obtain the opinion of the student body before proceeding. Design a sampling method to obtain the individuals in the sample. Be sure to support your choice.

13. **Sample Design** Target wants to open a new store in the village of Lockport. Before construction, they want to obtain some demographic information regarding the area under consideration. Design a sampling method to

obtain the individuals in the sample. Be sure to support your choice.

14. **Sample Design** The county sheriff wishes to determine whether a certain highway has a high proportion of speeders traveling on it. Design a sampling method to obtain the individuals in the sample. Be sure to support your choice.

15. **Sample Design** A pharmaceutical company wants to conduct a survey of 30 individuals who have high cholesterol. The company has a list of 6,600 individuals who are known to have high cholesterol that they obtained from doctors throughout the country. Design a sampling method to obtain the individuals in the sample. Be sure to support your choice.

16. **Sample Design** A marketing executive for Coca Cola, Inc., wants to identify television shows that people in the Boston area who typically drink Coke are watching. The executive has a list of all households in the Boston area. Design a sampling method to obtain the individuals in the sample. Be sure to support your choice.

17. **Systematic Sample** The human resource department at a certain company wants to conduct a survey regarding worker morale. The department has an alphabetical list of all 4,502 employees at the company and wants to conduct a systematic sample.

(a) Determine $k$ if the sample size is 50.
(b) Determine the individuals who will be administered the survey. More than one answer is possible.

18. **Systematic Sample** In order to predict the outcome of a county election, a newspaper obtains a list of all 945,035 registered voters in the county and wants to conduct a systematic sample.

(a) Determine $k$ if the sample size is 130.
(b) Determine the individuals who will be administered the survey. More than one answer is possible.

19. Research the sampling methods used by a market research firm in your neighborhood. Report your findings to the class. The report should include the types of sampling methods used, number of stages, and sample size.

---

# 1.4 Sources of Errors in Sampling

**Objective** ① Understand the sources of error in sampling

①  Thus far, we have discussed *how* to obtain samples but have neglected to look at any of the pitfalls that inevitably arise in sampling. In this section we look at problems that can occur in sampling. Some of these problems can be remedied; some, however, have no solution. Collectively, these errors are called *nonsampling errors*.

*Definition*  **Nonsampling errors** are errors that result from the survey process. They are due to the nonresponse of individuals selected to be in the survey, to inaccurate responses, to poorly worded questions, to bias in the selection of individuals to be given the survey, and so on.

We contrast nonsampling errors with *sampling errors*.

*Definition*  **Sampling error** is the error that results from using sampling to estimate information regarding a population. This type of error occurs because a sample gives incomplete information about the population.

 **In Your Own Words**

We can think of sampling error as error that results from using a subset of the population to describe characteristics of the population. Nonsampling error is error that results from obtaining and recording information collected.

By incomplete information, we mean that the individuals in the sample cannot reveal all the information about the population because individuals are unique. Consider the following: Suppose the average age of 4 randomly selected students from a class of 30 is found to be 23.4 years. Further, assume that no students lied about their age, nobody misunderstood the question being asked, and the sampling was done appropriately. If the average age of all 30 students (the population) is 22.9 years old, then the sampling error is 23.4 − 22.9 = 0.5 year. Suppose now that the same survey is conducted, but this time one of the individuals in the survey lied about his age. Then the results of the survey will have nonsampling error.

Fortunately, we can control the amount of sampling error through an appropriately designed survey or experiment. Nonsampling error, on the other hand, can be more difficult to control.

When a sampling design is done poorly, the descriptive statistics computed from the data obtained in the sample may not be close to the values that would be obtained if the entire population were surveyed. For example, the *Literary Digest* predicted that Alfred M. Landon would defeat Franklin D. Roosevelt in the 1936 presidential election. The *Literary Digest* conducted a poll by mailing questionnaires based upon a list of its subscribers, telephone directories, and automobile owners. On the basis of the results, the *Literary Digest* predicted that Landon would win with 57% of the popular vote. However, Roosevelt won the election with about 62% of the popular vote. The incorrect prediction by the *Literary Digest* was the result of a poor sample design. In 1936, most of the subscribers to the magazine, households with telephones and automobile owners, were Republican, the party of Landon. Therefore, the choice of the frame that was used to conduct the survey led to an incorrect prediction. This is an example of non-sampling error.

We now list the various sources of nonsampling error. These sources include the frame, nonresponse, data entry error, and poorly worded questions.

## The Frame

Recall that the frame is the list of all individuals in the population under study. For example, in a study regarding voter preference in an upcoming election, the frame would be a list of all registered voters. Sometimes, obtaining the frame would seem to be a relatively easy task, such as obtaining the list of all registered voters. Even under this circumstance, however, the frame may be incomplete. People who recently registered to vote may not be on the published list of registered voters.

Often, it is difficult to gain access to a *complete* list of individuals in a population. For example, in general public opinion polls, random telephone surveys are frequently conducted, which implies that the frame is all households with telephones. This method of sampling will exclude any household that does not have a telephone as well as all homeless people. In such a situation, certain segments of the population are *underrepresented*.

In designing any sample, the hope is that the frame used is as complete as possible so that any results inferred regarding the population have as little error as possible.

## Nonresponse

Nonresponse means that an individual selected for the sample does not respond to the survey. Nonresponse can occur because individuals selected for the sample do not wish to respond or because the interviewer was unable to contact them.

This type of error can be controlled using callbacks. The type of callback employed typically will depend upon the type of survey initially used. For example, if nonresponse occurs because a mailed questionnaire was not returned, a callback might mean phoning the individual in order to conduct the survey. If nonresponse occurs because an individual was not at home, then a callback might mean returning to the home at other times in the day or other days of the week.

Another method that can be used to improve nonresponse is using rewards and incentives. Rewards may include cash payments for completing a questionnaire, made only upon receipt of the completed questionnaire. Incentives might also include a cover letter that states that the responses to the questionnaire will dictate future policy. For example, a village may send out questionnaires to households and state in a cover letter that the responses to the questionnaire will be used to decide pending issues within the village.

## Interviewers

A trained interviewer is essential in order to obtain accurate information from a survey. A good interviewer will have the skill necessary to elicit responses from individuals within a sample and be able to make the interviewee feel comfortable enough to give truthful responses. For example, a good interviewer should be able to obtain truthful answers to questions as sensitive as, say, "Have you ever cheated on your taxes?"

## Data Checks

Once data are collected, the results typically must be entered into a computer. Data entry inevitably results in input errors. It is imperative that data be checked for accuracy at every stage of the statistical analysis. In this text, we present some methodology that can be used to check for data entry errors.

## Questionnaire Design

Appropriate questionnaire design is critical in minimizing the amount of nonsampling error. We will concentrate on the main aspects in the design of a good questionnaire.

One of the first considerations in designing a question is determining whether the question should be open or closed.

An **open question** is one in which the respondent is free to choose his or her response. For example:

What is the most important problem facing America's youth today?

A **closed question** is one in which the respondent must choose from a list of predetermined responses.

What is the most important problem facing America's youth today?
(a) Drugs
(b) Violence
(c) Single parent homes
(d) Promiscuity
(e) Peer pressure

When designing an open question, be sure to phrase the question so that the responses are similar. (You don't want a wide variety of responses.) This allows for easy analysis of the responses. The benefit of closed questions is that they limit the number of respondent choices, and therefore, the results are much easier to analyze. However, this limits the choices and does not always allow the respondent to respond the way he or she might want to respond. If the desired answer is not provided as a choice, the respondent will be forced to choose a secondary answer.

Survey designers recommend conducting pretest surveys with open questions and then using the most popular answers as the choices on closed question surveys. Another issue to consider in the closed question design is the number of responses the respondent may choose from. It is recommended that the option "no opinion" be omitted, because this option does not allow for meaningful analysis. The bottom line is to try and limit the number of choices in a closed question format without forcing respondents to choose an option they otherwise would not.

## Wording of Questions

The wording of a survey question is vital in order to obtain data that are not misrepresentative. Questions must always be asked in balanced form. For example, the "yes/no" question

**Caution**

The wording of questions can significantly affect the responses and, therefore, the validity of a study.

**Do you oppose the reduction of estate taxes?**

should be written

**Do you favor or oppose the reduction of estate taxes?**

The second question is balanced. Do you see the difference? Consider the following report based on studies from Schuman and Presser (*Questions and Answers in Attitude Surveys*, 1981, p. 277), who asked the following two questions:

**(A)** Do you think the United States should forbid public speeches against democracy?

**(B)** Do you think the United States should allow public speeches against democracy?

For those respondents presented with question A, 21.4% gave "yes" responses, while for those given question B, 47.8% gave "no" responses. The conclusion you may arrive at is that most people are not necessarily willing to forbid something, but more people are willing not to allow something. These results imply that the wording of the question can alter the outcome of a survey.

Another consideration in wording a question is not to be vague. For example, the question, "How much do you study?" is too vague. Does the researcher mean how much do I study for all my classes or just for statistics? Does the researcher mean per day or per week? The question should be written, "How many hours do you study statistics each week?"

## The Order of the Questions, Words, and Responses

Many surveys will rearrange the order of the questions within a questionnaire so that responses are not affected by prior questions. Consider the following example from Schuman and Presser in which the following two questions were asked:

**(A)** Do you think the United States should let Communist newspaper reporters from other countries come in here and send back to their papers the news as they see it?

**(B)** Do you think a Communist country such as Russia should let American newspaper reporters come in and send back to America the news as they see it?

For surveys conducted in 1980 in which the questions appeared in the order (A, B), 54.7% of respondents answered yes to A and 63.7% answered yes to B. If the questions were ordered (B, A), then 74.6% answered yes to A and 81.9% answered yes to B. When Americans are asked if U.S. reporters should be allowed to report Russian news first, they are more likely to agree that Russians should be allowed to report American news. Questions should be rearranged as much as possible to help reduce the effects of this type.

Pollsters will also rearrange words within a question. For example, the Gallup Organization asked the following question of 1,017 adults aged 18 years or older:

**"Do you consider the first six months of the Bush administration to be a [rotated: success (or a) failure]?"**

Notice how the words "success" and "failure" were rotated. The purpose of this is to remove the effect that may occur by writing the word "success" first in the question.

Not only should the order of the questions and/or certain words within the question be rearranged, but, in closed questions, the possible responses should also be rearranged. The reason for this is that respondents are likely to choose early choices in a list rather than later choices.

## 1.4 Assess Your Understanding

### Concepts and Vocabulary

1. Why is it rare for frames to be completely accurate?
2. What are some solutions to nonresponse?
3. What is a closed question? What is an open question?
4. Discuss methods that can be used to improve response rate.
5. Discuss the benefits of having trained interviewers.
6. What are the advantages of having a pretest when constructing a questionnaire that has closed questions?
7. Discuss the pros and cons of telephone interviews that take place during the typical dinner hour.
8. Why is a high response rate desired? How would a low response rate affect survey results?
9. Discuss the advantages and disadvantages of open versus closed questions.

### Exercises

- **Skill Building**

*In Problems 1–8, the survey design is flawed. (a) Determine whether the sampling method or the survey itself is flawed. For flawed surveys, identify the cause of the error (wording of question, nonresponse, etc.). (b) Suggest a remedy to the problem.*

1. A college vice president wants to conduct a study regarding student achievement of undergraduate students. He selects the first 50 students who enter the building on a given day and administers his survey.

2. The Village of Oak Lawn wishes to conduct a study regarding the income level of households within the village. The village manager selects 10 homes in the southwest corner of the village and sends an interviewer to the homes to ascertain the household income.

3. An antigun advocate wants to estimate the percentage of people who favor stricter gun laws. He conducts a nationwide survey of 1,203 randomly selected adults 18 years old and older. The interviewer asks the respondents, "Do you favor harsher penalties for individuals who sell guns illegally?"

4. A magazine is conducting a study on the effects of infidelity in a marriage. The editors randomly select 400 women whose husbands were unfaithful and ask, "Do you believe a marriage can survive when the husband destroys the trust that must exist between husband and wife?"

5. A polling organization is going to conduct a study to estimate the percentage of households that speak a foreign language as the primary language. It mails a questionnaire to 1,023 randomly selected households throughout the United States and asks the head of household if a foreign language is the primary language spoken in the home. Of the 1,023 households selected, 12 responded.

6. Petland is considering opening a new store in Orland Park. Prior to opening the store, the company would like to know the percentage of households in Orland Park that own a pet. The market researcher obtains a list of households in Orland Park and randomly selects 100 of them. She mails a questionnaire that asks questions regarding pets in the house to the 100 households. Of the 100 questionnaires sent out, she receives 3 in return.

7. Suppose you are conducting a survey regarding illicit drug use among teenagers in the Baltimore School District. You obtain a cluster sample of 12 schools within the district and sample all sophomore students in the randomly selected schools. The survey is administered by the teachers.

8. Suppose you are conducting a survey regarding students' study habits. You obtain a list of full-time registered students and obtain a simple random sample of 90 students. One of the survey questions is "How many hours do you study?"

- **Applying the Concepts**

9. Consider the following two questions.

    A. *Suppose for a moment that a rape is committed in which the woman becomes pregnant. Do you think the criminal should or should not face additional charges if the woman becomes pregnant?*

    B. *Do you think abortions should be legal under any circumstances, legal under certain circumstances, or illegal in all circumstances?*

    Do you think the order in which the questions are asked will affect the survey results? If so, what can the pollster do to alleviate this response bias?

10. Consider the following question from a recent Gallup poll:

    *Thinking about how the abortion issue might affect your vote for major offices, would you vote only for a candidate who shares your views on abortion or consider a candidate's position on abortion as just one of many important factors? [rotated]*

    Why is it important to rotate the two choices presented in the question?

11. Write a survey question that contains strong wording and a survey question that contains tempered wording. Present the strongly worded question to 10 randomly

11. selected people and the tempered question to 10 different randomly selected people. How does the wording affect the response?

12. Write two questions that could have different responses, depending upon the order in which the questions are presented. Randomly select 20 people and present the questions in one order to 10 of the people and in the opposite order to the other 10 people. Did the results differ?

13. Plan your own survey. Start with your objective and take the survey all the way to the data collection step. Comment on the difficulties encountered.

14. Research a survey method used by a company or government branch. Present the methods used to your class. Your report should include sampling methods used, sample size, method of collection, and the frame used.

---

## 1.5 The Design of Experiments

**Objectives**  Define designed experiment

 Understand the steps in designing an experiment

 Understand the completely randomized design

 Understand the matched-pairs design

One of the major themes of this chapter has been data collection. Sections 1.2–1.4 discussed techniques for obtaining data through surveys. Obtaining data through an experiment is discussed in this section.

     When people hear the word "experiment," they typically think of a laboratory with a controlled environment. Statisticians have control when they perform experiments as well.

**Definition**

> A **designed experiment** is a controlled study in which one or more treatments are applied to *experimental units*. The experimenter then observes the effect of varying these *treatments* on a response variable. Control, manipulation, randomization, and replication are the key ingredients of a well-designed experiment.

The **experimental unit** (or **subject**) is a person, object, or some other well-defined item upon which a *treatment* is applied. The experimental unit is analogous to the individual in a survey. The **treatment** is a condition applied to the experimental unit. A **response variable** is a quantitative or qualitative variable that represents our variable of interest.

The goal in an experiment is to determine the effect the treatment has on the response variable. For example, a researcher may want to measure the effect of sleep deprivation on a person's fine-motor skills. The researcher might take a group of 100 individuals and randomly divide them into four groups. The first group might sleep eight hours a night for four nights, the second will sleep six hours per night for four nights, the third will sleep four hours per night for four nights, while the fourth group will sleep two hours per night for four nights. The experimental unit is the person, the treatment is the amount of sleep and the response variable might be the reaction time of the person to some stimulus.

Many designed experiments are **double blind**. This means that neither the experimental unit nor the experimenter knows what treatment is being administered to the experimental unit. For example, in clinical studies of the cholesterol-lowering drug Lipitor, researchers administered either 10 mg, 20 mg, 40 mg, 80 mg, or a **placebo**, an innocuous medication such as a sugar tablet, to patients with high cholesterol. Because the experiment was double blind, neither the patients nor the researchers knew which medication was being administered. It is important that double-blind methods be used in this case so that the patients and researchers do not behave in such a way as to affect the results. For example, the researcher might not give as much time to a patient receiving the placebo.

The process of conducting an experiment requires a series of steps.

**Steps in Conducting an Experiment**

*Step 1:* *Identify the problem to be solved.* The statement of the problem should be as explicit as possible. The statement should provide the experimenter with direction. In addition, the statement must identify the response variable and the population to be studied. Often, the statement is referred to as the **claim**.

*Step 2:* *Determine the factors that affect the response variable.* The factors that affect the response variable are referred to as **predictor variables**. These factors are usually identified by an expert in the field of study. In identifying the factors, we must ask, "What things affect the value of the response variable?" Once the factors (predictor variables) are identified, it must be determined which factors will be fixed at some predetermined level (the controls), which factors will be manipulated, and which factors will be uncontrolled.

*Step 3:* *Determine the number of experimental units.* As a general rule of thumb, choose as many experimental units as time and money will allow. Techniques do exist for determining sample size provided certain information is available.

*Step 4:* *Determine the level of the predictor variables.* There are three ways to deal with the predictor variables:

1. Control their levels so they remain fixed throughout the experiment. These are variables whose effect on the response variable is not of interest.

2. Manipulate or set them at predetermined levels. These are the variables whose effect on the response variable interests us. These variables constitute the treatment in the experiment.

3. Randomize so that the effects of variables whose level cannot be controlled is minimized. The idea is that randomization "averages out" the effects of uncontrolled predictor variables.

*Step 5:* *Collect and process the data.* This is the replication mentioned in the definition. In this step, we perform the experiment on each of the experimental units. The researcher then measures the value of the response variable and organizes the results. The idea is that any difference in the value of the response variable can be attributed to differences in the level of the treatment.

*Step 6:* *Test the claim.* This is the subject of inferential statistics. **Inferential statistics** is a process in which generalizations about a population are made on the basis of results obtained from a sample. In addition, a statement regarding our level of confidence in our generalization is provided. We study methods of inferential statistics in Chapters 8–11.

An example will help clarify the process of experimental design.

▶ **EXAMPLE 1**    **Designing an Experiment**

*Problem:* A farmer wishes to determine the optimal level of a new fertilizer on his soybean crop. Design an experiment that will assist him.

*Approach:* We perform the previously listed steps.

### Historical Note

Sir Ronald Fisher, often called the "Father of Modern Statistics," was born in England on February 17, 1890. He received a B.A. in astronomy from Cambridge University in 1912. In 1914, he took a position teaching mathematics and physics at a high school. He did this to help serve his country during World War I. (He was rejected by the army because of his poor eyesight.) In 1919, Fisher took a job as a statistician at Rothamsted Experimental Station, where he was involved in agricultural research. In 1933, Fisher became Galton Professor of Eugenics at Cambridge University, where he studied Rh blood groups. In 1943 he was appointed to the Balfour Chair of Genetics at Cambridge. He was knighted by Queen Elizabeth in 1952. Fisher retired in 1957 and died in Adelaide, Australia, on July 29, 1962. One of his famous quotations is "To call in the statistician after the experiment is done may be no more than asking him to perform a postmortem examination: he may be able to say what the experiment died of."

*Solution:*

**Step 1:**  The farmer wants to identify the optimal level of fertilizer in growing soybeans. We define "optimal" as the level that maximizes yield. Therefore, the response variable will be crop yield.

**Step 2:**  Some of the factors that affect crop yield are fertilizer, precipitation, sunlight, method of tilling the soil, type of soil, seed, and temperature.

**Step 3:**  In this experiment, we will divide one acre of land into three equal sized plots of land. On each plot, we will plant 500 soybean plants (experimental units).

**Step 4:**  We list the variables and their levels.

- **Fertilizer.**  We manipulate the level of this variable. We wish to measure the effect of varying the level of this variable on the response variable, yield. We will set the level of fertilizer (the treatment) as follows:

    Plot 1 receives no fertilizer.

    Plot 2 receives two teaspoons per gallon of water every two weeks.

    Plot 3 receives four teaspoons per gallon of water every two weeks.

- **Precipitation.**  Although we cannot control the amount of rainfall, we can control the amount of watering we do. This variable will be controlled so that each plot receives the same amount of water.

- **Sunlight.**  This is an uncontrollable variable, but it will be the same for each plot.

- **Method of tilling.**  We can control this variable. We agree to use the round-up ready method of tilling for each plot.

- **Type of soil.**  We can control certain aspects of the soil such as level of acidity. In addition, because each plot is within a one-acre area, it is reasonable to assume that the soil conditions of each plot are equivalent.

- **Plant.**  There may be variation from plant to plant. To account for this, we randomly assign the plants to a plot.

- **Temperature.**  This variable is not within our control, but will be the same for each plot.

**Step 5:**  Till the soil, plant the soybean plants, and fertilize according to the schedule prescribed. At the end of the growing season, determine the crop yield for each plot.

**Step 6:**  Determine whether any differences in yield exist between the three plots.  ◂◂

Figure 5 illustrates the experimental design presented in Example 1.

Figure 5

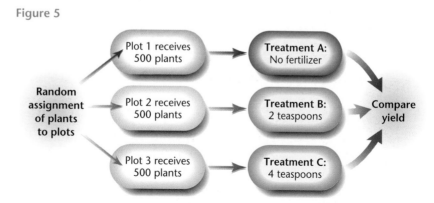

Example 1 presents an example of an experiment that is referred to as a **completely randomized design** because the experimental units (the plants) were randomly assigned the treatment. It is the most popular experimental design because of its simplicity, but it is not always the best. We discuss inferential procedures for the completely randomized design in which there are two treatments in Section 10.2. Inferential procedures for the completely randomized design in which there are three or more treatments require One-way Analysis of Variance, presented in Section C.3 on the compact disc that accompanies this text.

**NW** *Now Work Problem 1.*

### ④ Matched-Pairs Design

A **matched-pairs design** is an experimental design in which the experimental units are somehow related (that is, the same person before and after a treatment, twins, husband and wife, same geographical location, and so on). There are only two treatments in a matched-pairs design. In a matched-pairs design, one of the matched individuals will receive one treatment and the matched individual receives the second treatment. We then look at the difference in the results of each matched-pair. One common type of matched-pairs design is to measure a response variable on an experimental unit before a treatment is applied and then measure the response variable on the same experimental unit after the treatment is applied. In this way, the individual is matched against itself. These experiments are sometimes called before–after or pretest–posttest experiments.

▶ **EXAMPLE 2**  A Matched-Pairs Design

*Problem:* An employee of Joliet Junior College's health and fitness center wondered what effect exercise has on systolic blood pressure. Design an experiment that will determine this effect.

*Approach:* The design will be a matched-pairs design because we are measuring the response variable, systolic blood pressure (the blood pressure of an individual while the heart contracts), on the same experimental unit before and after a treatment.

*Solution:* This experiment will require that we measure the systolic blood pressure for each participant in the study before the exercise program and at the completion of the exercise program. We would then compare the difference in systolic blood pressures. Figure 6 shows a diagram illustrating the design.

Figure 6

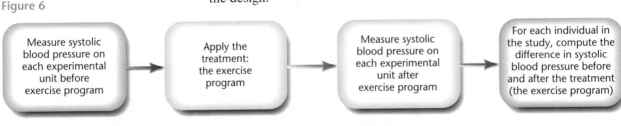

We discuss the statistical inference of matched-pairs design in Section 10.1.

**NW** *Now Work Problem 5.*

## 1.5 Assess Your Understanding

### Concepts and Vocabulary

1. Define the following:
   (a) experimental unit
   (b) treatment
   (c) response variable
   (d) predictor variable
   (e) double blind
   (f) placebo
   (g) confounding

2. Describe the difference between an observational study and an experiment.

### Exercises

#### • Applying the Concepts

1. **School Psychology** A school psychologist wants to test the effectiveness of a new method for teaching reading. She selects five hundred first grade students in District 203 and randomly divides them into two groups. Group 1 is taught by means of the new method, while Group 2 is taught via traditional methods. The same teacher is assigned to teach both groups. At the end of the year, an achievement test is administered and the results of the two groups compared.
   (a) What is the response variable in this experiment?
   (b) What is the treatment? How many levels does the treatment have?
   (c) Are any of the predictor variables controlled?
   (d) How does the researcher's design handle students from different socioeconomic levels?
   (e) What type of experimental design is this?
   (f) Identify the experimental units.
   (g) Draw a diagram similar to Figure 5 or Figure 6 to illustrate the design.

2. **Pharmacy** A pharmaceutical company has developed an experimental drug meant to relieve symptoms associated with the common cold. The company found 300 adult males in New York aged 25–29 years old and randomly divided them into two groups. Group 1 is given the experimental drug, while Group 2 is given a placebo. After one week of treatment, the percentage of each group that still has cold symptoms is compared.
   (a) What is the response variable in this experiment?
   (b) What is the treatment? How many levels does the treatment have?
   (c) Are any of the predictor variables controlled?
   (d) How does the researcher handle people from different parts of the country?
   (e) What type of experimental design is this?
   (f) Identify the experimental units.
   (g) Draw a diagram similar to Figure 5 or Figure 6 to illustrate the design.

3. **Diet** In a study to determine the benefits of tomatoes, a researcher randomly selects 600 adult males 30 years of age. She randomly divides the males into three groups of 200 each. Group 1 eats one serving of tomatoes per week, Group 2 eats three servings of tomatoes per week, and Group 3 receives five servings of tomatoes per week. The proportion of males in each group who contract prostate cancer is determined after 40 years.
   (a) What is the response variable in this experiment?
   (b) What is the treatment? How many levels does it have?
   (c) Are any of the predictor variables controlled? Which ones?
   (d) What type of experimental design is this?
   (e) Identify the experimental units.
   (f) Draw a diagram similar to Figure 5 or Figure 6 to illustrate the design.

4. **Social Work** A social worker wants to examine methodologies that can be used to improve truancy. She randomly samples 300 chronically truant students from District 103. The students are randomly divided into three groups. The students in Group 1 receive no intervention; the students in Group 2 are treated with positive reinforcement in which each day the student is not truant he or she receives a star that can be traded in for rewards; and the students in Group 3 are treated with negative reinforcement such that each truancy results in a one-hour detention. However, the hours of detention are cumulative, meaning that the first truancy results in one hour of detention, the second truancy results in two hours, and so on. After a full school year, the aggregate numbers of truancies are compared.
   (a) What is the response variable in this experiment?
   (b) What is the treatment? How many levels does it have?
   (c) Are any of the predictor variables controlled?
   (d) Can you think of any predictor variables that should be controlled?
   (e) What type of experimental design is this?
   (f) Identify the experimental units.
   (g) Draw a diagram similar to Figure 5 or Figure 6 to illustrate the design.

5. **Rats in Space** NASA conducted an experiment entitled "Regulation of Erythropoiesis during Spaceflight." Rats were placed in an Animal Enclosure Module 28 days prior to the flight. Two days prior to the flight, the weights of the rats were obtained. Upon return, the weights of the rats were again obtained. The difference

in before-flight and after-flight weights was recorded to determine the effect of spaceflight on weight.

(a) What is the response variable in this experiment?

(b) What is the treatment?

(c) What type of experimental design is this?

(d) Draw a diagram similar to Figure 5 or Figure 6 to illustrate the design.

6. For the experiment of Problem 27 in Section 1.1,

(a) What is the response variable in this experiment?

(b) What is the treatment?

(c) What type of experimental design is this?

(d) Draw a diagram similar to Figure 5 or Figure 6 to illustrate the design.

7. For the experiment of Problem 28 in Section 1.1,

(a) What is the response variable in this experiment?

(b) What is the treatment? How many levels does it have?

(c) What type of experimental design is this?

(d) Draw a diagram similar to Figure 5 or Figure 6 to illustrate the design.

8. **Designing an Experiment** Suppose you are interested in comparing Benjamin Moore's MoorLife Latex house paint with Sherwin Williams' LowTemp 35 Exterior Latex paint. Design an experiment that will answer the question, "Which paint is better for painting the wood siding on the exterior of a home?" In your design, be sure to identify the type of experimental design, population, experimental unit, response variable, treatment, factors, and their levels.

9. **Designing an Experiment** Suppose you are interested in determining the effectiveness of a new workout regimen. In particular, you want to know whether the new workout helps to increase lung capacity. Design an experiment that will answer the question, "Does this workout regimen increase lung capacity?" In your design, be sure to identify the type of experimental design, population, experimental unit, response variable, treatment, factors, and their levels.

10. **Completely Randomized Design** A pharmaceutical company wants to test the effectiveness of an experimental drug meant to reduce high cholesterol. The researcher at the pharmaceutical company has decided to test the effectiveness of the drug through a completely randomized design. She has obtained 20 volunteers with high cholesterol:

Ann, John, Michael, Kevin, Marissa, Christina, Eddie, Shannon, Julia, Randy, Sue, Tom, Wanda, Roger, Laurie, Rick, Kim, Joe, Colleen, and Bill.

Number the volunteers from 1–20. Use a random-number generator to randomly assign 10 of the volunteers to the experimental group. The remaining volunteers will go into the control group. List the individuals in each group.

11. **Effects of Alcohol** A researcher has recruited 20 volunteers to participate in a study. The researcher wishes to measure the effect of alcohol on an individual's reaction time. The 20 volunteers are randomly divided into two groups. Group 1 will serve as a control group in which participants drink four one-ounce cups of a liquid that looks, smells, and tastes like alcohol in 15-minute increments. Group 2 will serve as an experimental group in which participants drink four one-ounce cups of 80-proof alcohol in 15-minute increments. After drinking the last one-ounce cup, the participants sit for 20 minutes. After the 20-minute resting period, the reaction time to a stimulus is measured.

(a) What type of experimental design is this?

(b) Use Table I in Appendix A or a random-number generator to divide the 20 volunteers into Group 1 and Group 2 by assigning the volunteers a number between 1 and 20. Then randomly select 10 numbers between 1 and 20. The individuals corresponding to these numbers will go into Group 1.

12. Search a newspaper, a magazine, or some other periodical that describes an experiment. Identify the population, experimental unit, response variable, treatment, factors, and their levels.

13. Research the "placebo effect" and "Hawthorne effect." Write a paragraph that describes how each affects the outcome of an experiment.

## CHAPTER 1 REVIEW

### Summary

We defined statistics as a science in which data are collected, organized, summarized, and analyzed in order to infer characteristics regarding a population. Descriptive statistics consists of organizing and summarizing information, while inferential statistics consists of drawing conclusions about a population, based on results obtained from a sample. The population is a collection of individuals upon which the study is made, and the sample is a subset of the population.

Data are the observations of a variable. Data can be either qualitative or quantitative. Quantitative data are either discrete or continuous.

Data can be obtained from four sources: a census, existing sources, survey sampling, or a designed experiment. A census will list all the individuals in the population, along with certain characteristics. Due to the cost of obtaining a census, most researchers opt for obtaining a sample. In observational studies, the variable of interest has already been established. For this reason, they are often referred to as *ex post facto* studies. Designed experiments are used when control of the individuals in the study is desired in order to isolate the effect of a certain treatment on a response variable.

We introduced five sampling methods: simple random sampling, stratified sampling, systematic sampling, cluster sampling, and convenience sampling. All of the sampling methods, except for convenience sampling, are based on planned randomness, which allows for unbiased statistical inference to be made. Convenience sampling typically leads to an unrepresentative sample and biased results.

## Vocabulary

Be sure you can define the following . . .

| | | |
|---|---|---|
| Statistics (p. 2) | Data (p. 6) | Systematic sampling (p. 20) |
| Population (p. 2) | Qualitative data (p. 6) | Cluster sampling (p. 21) |
| Sample (p. 2) | Quantitative data (p. 6) | Convenience sampling (p. 22) |
| Descriptive statistics (p. 2) | Continuous data (p. 6) | Sampling error (p. 26) |
| Inferential statistics (p. 2) | Discrete data (p. 6) | Nonsampling error (p. 26) |
| Treatment (pp. 3, 31) | Census (p. 9) | Open question (p. 28) |
| Placebo (pp. 3, 31) | Observational study (p. 10) | Closed question (p. 28) |
| Double blind (pp. 3, 31) | Designed experiment (pp. 10, 31) | Experimental unit (p. 31) |
| Variable (p. 4) | Lurking variable (p. 11) | Response variable (p. 31) |
| Qualitative variable (p. 4) | Simple random sampling (p. 12) | Predictor variable (p. 32) |
| Quantitative variable (p. 4) | Frame (p. 13) | Completely randomized design (p. 34) |
| Continuous variable (p. 5) | Sampling without replacement (p. 13) | Matched-pairs design (p. 34) |
| Discrete variable (p. 5) | Stratified sampling (p. 19) | |

## Objectives

| Section | You should be able to . . . | Review Exercises |
|---|---|---|
| **1.1** | 1 Define statistics (p. 2) | 1 |
| | 2 Understand the process of statistics (p. 2) | 6 |
| | 3 Distinguish between qualitative and quantitative variables (p. 4) | 9–14 |
| | 4 Distinguish between discrete and continuous variables (p. 5) | 11–14 |
| **1.2** | 1 Distinguish between an observational study and an experiment (p. 10) | 15–18 |
| | 2 Obtain a simple random sample (p. 12) | 23 |
| **1.3** | 1 Obtain a stratified sample (p. 18) | 24 |
| | 2 Obtain a systematic sample (p. 20) | 25 |
| | 3 Obtain a cluster sample (p. 21) | 26 |
| **1.4** | 1 Understand the sources of error in sampling (p. 26) | 7 |
| **1.5** | 1 Define designed experiment (p. 31) | 5 |
| | 2 Understand the steps in designing an experiment (p. 32) | 8 |
| | 3 Understand the completely randomized design (p. 32) | 27, 28, 30 |
| | 4 Understand the matched-pairs design (p. 34) | 29, 31 |

## Review Exercises

*In Problems 1–5, provide a definition using your own words.*

1. Statistics

2. Population

3. Sample

4. Observational study

5. Designed experiment

6. What is meant by "the process of statistics"?

7. State some sources of error in sampling. Provide some methods for correcting these errors. Distinguish sampling and nonsampling error.

8. Describe the components in an appropriately designed experiment.

---

*In Problems 9–14, classify the variable as qualitative or quantitative. If the variable is quantitative, state whether it is discrete or continuous.*

9. State in which a person resides.

10. Marital status.

11. Amount of water consumed per day in ounces.

12. Total number of people in a household.

13. Frequency with which the word "statistics" appears in this text.

14. Number of people who visit the emergency room on Saturday.

*In Problems 15–18, determine whether the study depicts an observational study or a designed experiment.*

**15.** A poll of residents of the Village of Orland Park to obtain their opinion on a new bond issue.

**16.** A study in which 1,000 randomly selected arthritis patients are divided into two groups. One receives a placebo while the other receives an experimental drug. The patients from the two groups are then compared after six months.

**17.** A study in which the rates of cancer in people living near high-tension wires is compared with the general public's cancer rate to see if there are any differences between the two groups.

**18.** A study in which 500 households in the Houston metropolitan area are randomly selected. The head of household is asked to disclose the household income. The average household income in the Houston metropolitan area is compared with the average household income in the United States to determine whether there is any difference.

*In Problems 19–22, determine the type of sampling used.*

**19.** An interviewer in a mall is told to survey every fifth shopper, starting with the second.

**20.** A researcher randomly selects 5 of the 70 hospitals in a metropolitan area and then surveys all of the surgical doctors in each hospital.

**21.** A researcher segments the population of car owners into four groups: Ford, General Motors, Chrysler, and foreign. She obtains a random sample from each group and conducts a survey.

**22.** A list of students in elementary statistics is obtained in which the individuals are numbered 1 to 540. A professor randomly selects 30 of the students.

**23. Obtaining a Simple Random Sample** A credit card company wants to perform a study on the outstanding balances of its 1,403,032 cardholders. Obtain a simple random sample of 15 cardholders.

**24. Obtaining a Stratified Sample** A congresswoman wants to survey her constituency regarding public policy. She asks one of her staff members to obtain a sample of residents of the district. The frame she has available lists 9,012 Democrats, 8,302 Republicans, and 3,012 Independents. Obtain a stratified random sample of eight Democrats, seven Republicans, and three Independents.

**25. Obtaining a Systematic Sample** A quality control engineer wants to be sure that bolts coming off an assembly line are within prescribed tolerance levels. He wants to conduct a systematic sample by selecting every 9th bolt to come off the assembly line. The machine produces 30,000 bolts per day and the engineer wants a sample of 32 bolts. Which bolts will be sampled?

**26. Obtaining a Cluster Sample** A farmer has a 500-acre orchard in Florida. Each acre is subdivided into blocks of five. So altogether, there are 2,500 blocks of trees on the farm. After a frost, he wants to get an idea of the extent of the damage. Obtain a sample of 10 blocks of trees, using a cluster sample.

**27. Postpartum Depression** A total of 120 postpartum women meeting DSM-IV criteria for major depression were selected and randomly assigned 12 weeks of interpersonal psychotherapy or to a waiting-list condition-control group. The Hamilton Rating Scale for Depression scores of the women in each group were measured.
(a) What is the response variable in this experiment?
(b) What is the treatment? How many levels does it have?
(c) Create a list of predictor variables. Are any of them controlled?
(d) What type of experimental design is this?
(e) Identify the experimental units.
(f) Draw a diagram similar to Figure 5 or Figure 6 to illustrate the design.

**28. Acute Bipolar Mania** A total of 115 patients with a DSM-IV diagnosis of bipolar disorder was randomly divided into two groups. The 55 individuals in Group 1 received 5 to 20 mg/d of olanzapine and the 60 individuals in Group 2 received a placebo in this double-blind experiment. The effect of the drug was measured on the Young-Mania Rating Scale total score.
(a) What is the response variable in this experiment?
(b) What is the treatment? How many levels does it have?
(c) What type of experimental design is this?
(d) Identify the experimental units.
(e) What does "double blind" mean in the context of this experiment?
(f) Draw a diagram similar to Figure 5 or Figure 6 to illustrate the design.

**29. Skylab** The four members of Skylab had their lymphocyte count per cubic millimeter measured one day before liftoff and measured again upon their return to Earth.
(a) What is the response variable in this experiment?
(b) What is the treatment?
(c) What type of experimental design is this?
(d) Identify the experimental units.
(e) Draw a diagram similar to Figure 5 or Figure 6 to illustrate the design.

**30.** Describe what is meant by an experiment that is a completely randomized design. Contrast this experimental design with a randomized block design.

**31.** Describe what is meant by a matched-pairs design. Contrast this experimental design with a completely randomized design.

**Chapter 1 Projects located at www.prenhall.com/sullivanstats**

# Organizing and Summarizing Data

 **For additional study help, go to**
www.prenhall.com/sullivanstats

> **Materials include**
> - Projects:
>   - ■ Case Study: The Day the Sky Roared
>   - ■ Decisions: Tables or Graphs?
>   - ■ Consumer Reports Project
> - Self-Graded Quizzes
> - "Preparing for This Section" Quizzes
> - PowerPoint Downloads
> - Step-by-Step Technology Guide
> - Graphing Calculator Help

## Putting It All Together

In Chapter 1, we learned that statistics is a process. The process begins with asking a research question. In order to determine the answer to the question, information (data) must be collected. The information is obtained from a census, existing data sources, surveys or designed experiments. When data are collected from a survey or designed experiment, they must be organized into a manageable form. Data that are not organized are referred to as **raw data**.

Methods for organizing raw data include the creation of tables or graphs, which allow for a quick overview of the information collected. The organization of data is the third step in the statistical process. The procedures used in organizing data into tables and graphs depend upon whether the data are qualitative, discrete, or continuous.

## 2.1 Organizing Qualitative Data

**Preparing for This Section**    Before getting started, review the following:

     ✓ Qualitative data (Section 1.1, p. 4)

**Objectives**   **1**   Construct frequency and relative frequency distributions from qualitative data

     **2**   Construct bar graphs

     **3**   Construct pie charts

In this section we will concentrate on tabular and graphical summaries of qualitative data. Section 2.2 discusses methods for summarizing quantitative data.

**1**   **Organizing Qualitative Data**

Recall that qualitative data provide nonnumerical measures that categorize or classify an individual. When qualitative data are collected, we are often interested in determining the number of individuals that occur within each category.

**Definition**    A **frequency distribution** lists the number of occurrences for each category of data.

▶ **EXAMPLE 1**   **Organizing Qualitative Data into a Frequency Distribution**

*Problem:* Table 1 lists the presidents of the United States, along with their state of birth. Construct a frequency distribution of the state of birth.

### TABLE 1

Birthplace of U.S. President

| President | State of Birth | President | State of Birth | President | State of Birth |
|-----------|----------------|-----------|----------------|-----------|----------------|
| Washington | Virginia | Lincoln | Kentucky | Coolidge | Vermont |
| J. Adams | Massachusetts | A. Johnson | North Carolina | Hoover | Iowa |
| Jefferson | Virginia | Grant | Ohio | F.D. Roosevelt | New York |
| Madison | Virginia | Hayes | Ohio | Truman | Missouri |
| Monroe | Virginia | Garfield | Ohio | Eisenhower | Texas |
| J.Q. Adams | Massachusetts | Arthur | Vermont | Kennedy | Massachusetts |
| Jackson | South Carolina | Cleveland | New Jersey | L.B. Johnson | Texas |
| Van Buren | New York | B. Harrison | Ohio | Nixon | California |
| W.H. Harrison | Virginia | Cleveland | New Jersey | Ford | Nebraska |
| Tyler | Virginia | McKinley | Ohio | Carter | Georgia |
| Polk | North Carolina | T. Roosevelt | New York | Reagan | Illinois |
| Taylor | Virginia | Taft | Ohio | George H. Bush | Massachusetts |
| Fillmore | New York | Wilson | Virginia | Clinton | Arkansas |
| Pierce | New Hampshire | Harding | Ohio | George W. Bush | Connecticut |
| Buchanan | Pennsylvania | | | | |

*Approach:* To construct a frequency distribution, we create a list of the states (categories) and tally each occurrence. Finally, we add up the number of tallies to determine the frequency.

*Solution:* See Table 2. From the table, we can see that Virginia is the state in which most presidents were born—a total of eight.*

| TABLE 2 | | |
|---|---|---|
| **State** | **Tally** | **Frequency** |
| Virginia | ⅢⅢ Ⅲ | 8 |
| Massachusetts | ⅠⅠⅠⅠ | 4 |
| South Carolina | Ⅰ | 1 |
| New York | ⅠⅠⅠⅠ | 4 |
| North Carolina | ⅠⅠ | 2 |
| New Hampshire | Ⅰ | 1 |
| Pennsylvania | Ⅰ | 1 |
| Kentucky | Ⅰ | 1 |
| Ohio | ⅢⅢ ⅠⅠ | 7 |
| Vermont | ⅠⅠ | 2 |
| New Jersey | ⅠⅠ | 2 |
| Iowa | Ⅰ | 1 |
| Missouri | Ⅰ | 1 |
| Texas | ⅠⅠ | 2 |
| California | Ⅰ | 1 |
| Nebraska | Ⅰ | 1 |
| Georgia | Ⅰ | 1 |
| Illinois | Ⅰ | 1 |
| Arkansas | Ⅰ | 1 |
| Connecticut | Ⅰ | 1 |

◀◀

With frequency distributions, it is a good idea to add up the frequency column to make sure that it sums to the correct number of observations. In the case of the data in Example 1, the frequency column adds up to 43, as it should.

Often, rather than being concerned with the frequency with which categories of data occur, we would like to know the relative frequency of the categories.

*Definition*

**In Your Own Words**

A frequency distribution shows how often each category occurs. A relative frequency distribution shows the percent of the observations that belong in each category.

The **relative frequency** is the proportion or percent of observations within a category and is found using the formula

$$\text{Relative frequency} = \frac{\text{frequency}}{\text{sum of all frequencies}} \quad (1)$$

A **relative frequency distribution** lists the relative frequency of each category of data.

*History considers Cleveland to be two different presidents, even though he is the same individual, because he was not elected to two consecutive terms.

▶ EXAMPLE 2    **Constructing a Relative Frequency Distribution
of Qualitative Data**

*Problem:* Using the data in Table 2 on page 41, construct a relative frequency distribution.

*Approach:* Add all the frequencies, and then use Formula (1) to compute the relative frequency of each category of data.

*Solution:* We add the values in the frequency column from Table 2:

$$\text{Sum of all frequencies} = 8 + 4 + 1 + \ldots + 1 + 1 = 43$$

This result shouldn't be surprising, since there have been 43 presidents of the United States! We now compute the relative frequency of each category. For example, the relative frequency of the category "Virginia" is 8/43 ≈ 0.1860. After computing the relative frequency of the remaining categories, we obtain the relative frequency distribution shown in Table 3.

| TABLE 3 | | |
|---|---|---|
| **State** | **Frequency** | **Relative Frequency** |
| Virginia | 8 | 8/43 ≈ 0.1860 |
| Massachusetts | 4 | 4/43 ≈ 0.0930 |
| South Carolina | 1 | 1/43 ≈ 0.0233 |
| New York | 4 | 0.0930 |
| North Carolina | 2 | 0.0465 |
| New Hampshire | 1 | 0.0233 |
| Pennsylvania | 1 | 0.0233 |
| Kentucky | 1 | 0.0233 |
| Ohio | 7 | 0.1628 |
| Vermont | 2 | 0.0465 |
| New Jersey | 2 | 0.0465 |
| Iowa | 1 | 0.0233 |
| Missouri | 1 | 0.0233 |
| Texas | 2 | 0.0465 |
| California | 1 | 0.0233 |
| Nebraska | 1 | 0.0233 |
| Georgia | 1 | 0.0233 |
| Illinois | 1 | 0.0233 |
| Arkansas | 1 | 0.0233 |
| Connecticut | 1 | 0.0233 |

So 18.60% of all United States presidents were born in Virginia.  ◀◀

We would expect the relative frequency column in Table 3 to add to 1. However, the column adds up to 1.0004. The discrepancy is simply due to rounding error. Nonetheless, it is a good idea to add up the relative frequency column to verify your results.

**NW**  *Now Work Problems 19(a) and (b).*

Once raw data are organized in a table, we can create graphs. This allows us to "see" the data and therefore get a sense as to what the data are saying regarding the individuals in the study. In general, pictures of data result in a more powerful message than tables. Try the following exercise for yourself: Open a newspaper and look at a table and graph. Study each. Now put the paper away and close your eyes. What do you see in your mind's eye? Can you recall information obtained from the table or the graph? In general, people are more likely to recall information obtained from a graph than they are from a table.

### 2 Bar Graphs

One of the most common devices for graphically representing qualitative data is a bar graph.

*Definition*

A **bar graph** is constructed by labeling each category of data on a horizontal axis and the frequency or relative frequency of the category on the vertical axis. A rectangle of equal width is drawn for each category. The height of the rectangle is equal to the category's frequency or relative frequency.

▶ EXAMPLE 3    Constructing a Frequency and Relative Frequency Bar Graph

*Problem:* Use the data summarized in Table 3 to construct (a) a frequency bar graph and (b) a relative frequency bar graph.

*Approach:* A horizontal axis is used to indicate the category of the data (states, in this case), and a vertical axis is used to represent the frequency or relative frequency (in this case, number of presidential births). Draw bars or rectangles of equal width for each category whose height is the frequency or relative frequency. The bars do not touch each other.

*Solution:*

**(a)** Figure 1 shows the frequency bar graph.

Figure 1

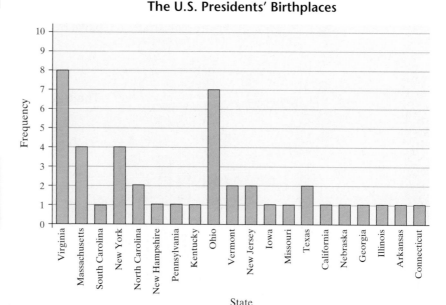

The U.S. Presidents' Birthplaces

🚦 **Caution**

Whenever constructing bar graphs, be sure to include labels for the axes as well as a title for the graph!

**(b)** Figure 2 shows the relative frequency bar graph.

Figure 2

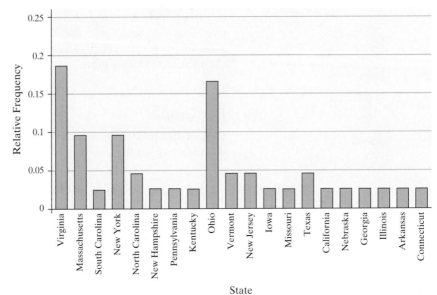

The U.S. Presidents' Birthplaces

### Caution

Watch out for graphs that start the scale at some value other than 0, have bars with unequal widths, or have bars with different colors, because they can misrepresent the data.

Notice that the frequency and relative frequency bar charts presented in Figures 1 and 2 look the same, except for the scale on the vertical axis.

Some statisticians prefer to create bar graphs with the categories arranged in decreasing order of frequency.

*Definition*   A **Pareto chart** is a bar graph whose bars are drawn in decreasing order of frequency or relative frequency.

Figure 3 illustrates a relative frequency Pareto chart for the data in Table 3.

Figure 3

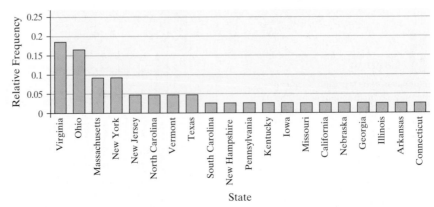

The U.S. Presidents' Birthplaces

**Using Technology:** Statistical spreadsheets such as Excel have the ability to draw bar charts and Pareto charts. Figure 4 shows the relative frequency bar chart of the data in Table 3, drawn with Excel.

Figure 4

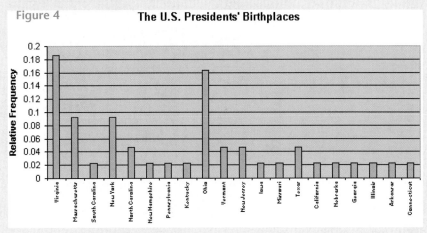

The U.S. Presidents' Birthplaces

 *Now Work Problems 19(c) and (d).*

## Side-by-Side Bar Graphs

Graphics provide insight when comparing two sets of data. For example, suppose we wanted to know if more people are finishing college today than in 1990. We could draw a **side-by-side bar graph** to compare the two data sets. Data sets should always be compared by using relative frequencies, because different sample or population sizes make comparisons using frequencies difficult.

**Caution**

When comparing two or more data sets, use relative frequencies!

▶ **EXAMPLE 4**   **Comparing Two Data Sets**

*Problem:* The data in Table 4 represent the educational attainment in 1990 and 2000 of adults 25 years and older who are residents of the United States.

**(a)** Draw a side-by-side relative frequency bar graph of the data.
**(b)** Are Americans becoming more educated?

| TABLE 4 | | |
|---|---|---|
| **Educational Attainment** | **1990** | **2000** |
| Less than 9th grade | 16,502,211 | 12,327,601 |
| 9th–12th grade, no diploma | 22,841,507 | 20,343,848 |
| High school diploma | 47,642,763 | 52,395,507 |
| Some college, no degree | 29,779,777 | 36,453,108 |
| Associate's degree | 9,791,925 | 11,487,194 |
| Bachelor's degree | 20,832,567 | 28,603,014 |
| Graduate/professional degree | 11,477,686 | 15,930,061 |
| **Totals** | **158,868,436** | **177,540,333** |

*Source:* United States Census Bureau

*Approach:* First, we determine the relative frequencies of each category for each year. To construct the side-by-side bar graphs, we draw two bars for each category of data. One of the bars will represent 1990 and the other will represent 2000.

*Solution:* Table 5 shows the relative frequency for each category.

| TABLE 5 | | |
| --- | --- | --- |
| **Educational Attainment** | **1990** | **2000** |
| Less than 9th grade | 0.1039 | 0.0694 |
| 9th–12th grade, no diploma | 0.1438 | 0.1146 |
| High school diploma | 0.2999 | 0.2951 |
| Some college, no degree | 0.1874 | 0.2053 |
| Associate's degree | 0.0616 | 0.0647 |
| Bachelor's degree | 0.1311 | 0.1611 |
| Graduate/professional degree | 0.0722 | 0.0897 |

**(a)** The side-by-side bar graph is shown in Figure 5.

Figure 5

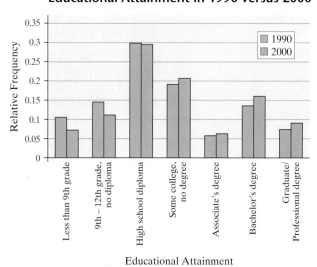

**(b)** From the graph, we can see that the proportion of Americans 25 years and older who are earning at least a bachelor's degree is increasing. This information is not clear from the frequency table, because the size of the population in the two years is different. So the increase in the number of Americans who earned a bachelor's degree is due partly to the increase in the size of the population. ◄◄

**NW** *Now Work Problem 15.*

### ③ Pie Charts

Pie charts are typically used to present the relative frequency of qualitative data.

*Definition*

A **pie chart** is a circle divided into sectors. Each sector represents a category of data. The area of each sector is proportional to the frequency of the category.

▶ **EXAMPLE 5**   **Constructing a Pie Chart**

*Problem:* The data presented in Table 6 represent the educational attainment of residents of the United States 25 years or older, based upon data obtained from the 2000 United States Census. Construct a pie chart of the data.

| TABLE 6 | |
| --- | --- |
| **Educational Attainment** | **2000** |
| Less than 9th grade | 12,327,601 |
| 9th–12th grade, no diploma | 20,343,848 |
| High school diploma | 52,395,507 |
| Some college, no degree | 36,453,108 |
| Associate's degree | 11,487,194 |
| Bachelor's degree | 28,603,014 |
| Graduate/professional degree | 15,930,061 |
| **Totals** | **177,540,333** |

*Approach:* The pie chart will have seven parts, or sectors, corresponding to the seven categories of data. The area of each sector is proportional to the frequency of each category. For example,

$$\frac{12,327,601}{177,540,333} = 0.0694 = 6.94\%$$

of all U.S. residents 25 years or older have less than a 9th-grade education. The category "less than 9th grade" will make up 6.94% of the pie chart. Since a circle has 360°, the degree measure of the sector for the category "less than 9th-grade education" will be $(0.0694)360° \approx 25.0°$. Use a protractor to measure each angle.

*Solution:* We follow the approach presented for the remaining categories of data to obtain Table 7.

| TABLE 7 | | | |
| --- | --- | --- | --- |
| **Education** | **Frequency** | **Relative Frequency** | **Degree Measure of Each Sector** |
| Less than 9th grade | 12,327,601 | 0.0694 | 25.0 |
| 9th to 12th grade, no diploma | 20,343,848 | 0.1146 | 41.3 |
| High school diploma | 52,395,507 | 0.2951 | 106.2 |
| Some college, no degree | 36,453,108 | 0.2053 | 73.9 |
| Associate's degree | 11,487,194 | 0.0647 | 23.3 |
| Bachelor's degree | 28,603,014 | 0.1611 | 58.0 |
| Graduate/professional degree | 15,930,061 | 0.0897 | 32.3 |

Figure 6

**Educational Attainment, 2000**

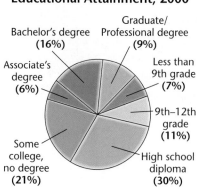

To construct a pie chart by hand, we use a protractor to approximate the angles for each sector. See Figure 6.   ◀◀

**Using Technology:** Certain statistical spreadsheets have the ability to draw pie charts. Figure 7 shows the pie chart of the data in Table 6, using Minitab.

Figure 7

### Educational Attainment, 2000

Bachelors De (16.1%)
Graduate/Pro ( 9.0%)
Associates D ( 6.5%)
9th - 12th G (11.5%)
High School (29.5%)
Some College (20.5%)
Less than 9t ( 6.9%)

Pie charts can be created only if all the categories of the variable under consideration are represented. In other words, we could create a bar chart that lists the proportion of presidents born in the states of Virginia, Texas, and New York, but it would not make sense to construct a pie chart for this situation. Do you see why? Only 34.88% of the pie would be accounted for.

**NW** *Now Work Problem 19(e).*

## 2.1  Assess Your Understanding

### Concepts and Vocabulary

1. What are raw data?
2. What is the relative frequency of a category of data?
3. What is a Pareto chart?

4. Why should relative frequencies be used in comparing two data sets?
5. Explain why Pareto charts might be preferred over bar graphs.

### Exercises

- **Basic Skills**

*In Problems 1–8, answer the questions on the basis of the graph.*

1. **Mutual Fund Owners**  The pie chart on the right depicts the percent of people within each age group who own mutual funds.

   (a) Which age group has the highest percentage of ownership?
   (b) Which age group has the lowest percentage of ownership?
   (c) What percentage of people 53 or older own mutual funds?

2. **Golf Driving Range**  The following pie chart depicts the bucket size golfers choose while at the driving range.

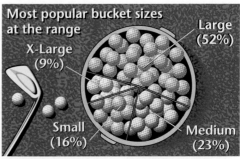

*Source: USA Today* Snapshot and the National Golf Foundation

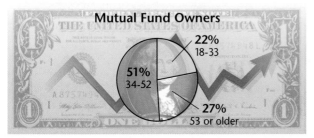

*Source: USA Today* Snapshot; Investment Company Institute, Summer 1999

(a) What is the most popular size? What percentage of golfers choose this size?

(b) What is the least popular size? What percentage of golfers choose this size?

(c) What percent of golfers choose a medium-sized bucket?

3. **Internet Users** The following bar graph represents the top eight countries in Internet users as of December 31, 2000, based on information provided by the Internet Industry Almanac.

### Top 8 Users of Internet

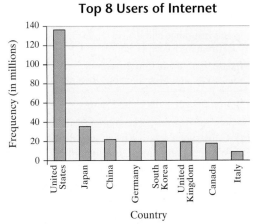

(a) Which country has the highest Internet usage?

(b) Approximately what is the Internet usage in Canada?

4. **Poverty** Every year the United States Census Bureau counts the number of people living in poverty. The Bureau uses money income thresholds as its definition of poverty, so that noncash benefits such as Medicaid and food stamps do not count toward poverty thresholds. For example, in 1999, the poverty threshold for two adults and one child was $13,410. The bar chart below represents the number of people living in poverty in the United States in 1999, by ethnicity:

### Number in Poverty

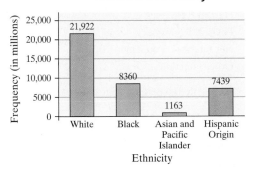

(a) How many whites were living in poverty in 1999?

(b) Of the impoverished, what percent are of Hispanic origin?

(c) Why might this graph be misleading?

5. **2000 Presidential Election** The following bar chart represents the number of voters who cast votes for the top six vote-getters in the 2000 presidential election. The number of votes received for each candidate appears above the bar.

### Votes in 2000 Presidential Election

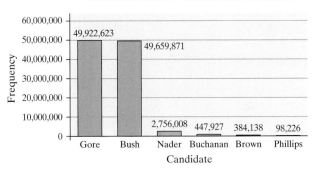

*Source:* cbsnews.com

(a) How many votes were cast for Pat Buchanan?

(b) What percent of the votes were cast for Al Gore? George W. Bush?

(c) In order to receive presidential election funds, a candidate must garner at least 5% of the vote. Will Ralph Nader receive presidential election funds?

6. **Birth Characteristics** The bar chart below represents the percentage of women who gave birth to a child in 1998, by educational attainment.

### Educational Attainment of Women Having a Child in 1998

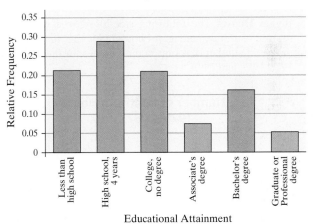

*Source:* U.S. Census Bureau, Current Population Reports

(a) What percentage of women who had a child in 1998 had four years of high school education?

(b) If 60,519,000 women gave birth to a child in 1998, how many had four years of high school education?

7. **Home Heating Fuel** The side-by-side bar graph on page 50 represents the percentage of households with the indicated source of home heating fuel for the years 1978, 1987, and 1997.

**Home Heating Fuel**

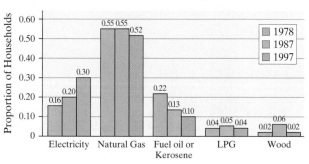

*Source:* Energy Information Administration, Residential Energy Consumption Survey, 1978, 1987, and 1997

(a) What percentage of households used electricity as their main source of home heating fuel in 1978?
(b) What was the most popular source of home heating fuel in 1987?
(c) What was the least popular source of home heating fuel in 1987?
(d) What might account for the rise in homes that use electricity as the main source of home heating fuel?
(e) What might account for the decline in homes that use fuel oil or kerosene as the main source of home heating fuel?

8. **Bigger Houses** The following side-by-side bar graph represents the percentage of households that have the indicated number of rooms in their homes for the years 1978, 1987, and 1997.

**Number of Rooms in House, 1978, 1987, 1997**

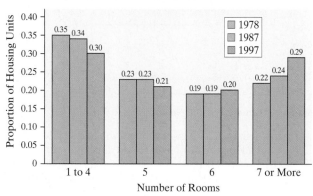

*Source:* Energy Information Administration, Residential Energy Consumption Survey, 1978, 1987, and 1997.

(a) What percentage of homes had five rooms in 1987?
(b) What percentage of homes had seven or more rooms in 1997?
(c) What was the most common home size in all three years of the survey?
(d) What might account for the fact that the percentage of homes with seven or more rooms has increased over the three years of the survey?

• **Applying the Concepts**

9. **Government Income** For fiscal year 2000 (October 1999–September 2000), the federal government collected $2,046.9 billion. The various sources of income are broken down in the following table.

| Source of Income | Amount (in billions of dollars) |
|---|---|
| Individual Income Tax and Tax Withholdings | 1,009.5 |
| Corporate Income Taxes | 234.7 |
| Social Insurance and Retirement Receipts | 691.5 |
| Excise, Estate and Gift Taxes, Customs, and Miscellaneous Receipts | 111.2 |

*Source:* Budget of the United States

(a) Construct a relative frequency distribution.
(b) What percentage of total income is attributable to individual income taxes?
(c) Construct a frequency bar graph.
(d) Construct a relative frequency bar graph.
(e) Construct a pie chart.
(f) In your opinion, which graph appears to place more emphasis on social insurance and retirement receipts as a source of income, the relative frequency bar graph or the pie chart? Why?

10. **Government Expenditures** For fiscal year 2000 (October 1999–September 2000), the federal government spent $1,998.8 billion. The breakdown of expenditures is given in the following table.

| Category | Expenditure (in billions of dollars) |
|---|---|
| National Defense and Foreign Affairs | 397.3 |
| Human Resources | 1,119.7 |
| Physical Resources | 121.0 |
| Net Interest on the Debt | 230.2 |
| Other Functions | 130.6 |

*Source:* Budget of the United States

(a) Construct a relative frequency distribution of the data shown.
(b) What percentage of total expenditures is attributable to net interest?
(c) Construct a frequency bar graph.
(d) Construct a relative frequency bar graph.
(e) Construct a pie chart.
(f) In your opinion, which graph appears to place more emphasis on net interest on the debt, the relative frequency bar graph or the pie chart? Why?

**11. College Survey** In a national survey conducted by the Centers for Disease Control in order to determine health-risk behaviors among college students, college students were asked, "How often do you wear a seat belt when riding in a car driven by someone else?" The frequencies are displayed below.

| Response | Frequency |
|---|---|
| Never | 125 |
| Rarely | 324 |
| Sometimes | 552 |
| Most of the time | 1257 |
| Always | 2518 |

(a) Construct a relative frequency distribution.
(b) What percentage of respondents answered "Always"?
(c) What percentage of respondents answered "Never" or "Rarely"?
(d) Construct a frequency bar graph.
(e) Construct a relative frequency bar graph.
(f) Construct a pie chart.

**12. College Survey** In a national survey conducted by the Centers for Disease Control in order to determine health-risk behaviors among college students, college students were asked, "How often do you wear a seat belt when driving a car?" The frequencies are displayed below.

| Response | Frequency |
|---|---|
| I do not drive a car | 249 |
| Never | 118 |
| Rarely | 249 |
| Sometimes | 345 |
| Most of the time | 716 |
| Always | 3093 |

(a) Construct a relative frequency distribution.
(b) What percentage of respondents answered "Always"?
(c) What percentage of respondents answered "Never" or "Rarely"?
(d) Construct a frequency bar graph.
(e) Construct a relative frequency bar graph.
(f) Construct a pie chart.
(g) Compare the relative frequencies of "Never," "Rarely," "Sometimes," "Most of the time," and "Always" with those in Problem 11. What might you conclude?

**13. Foreign-Born Population** The following data represent the region of birth of foreign-born residents of the United States in 2000.

| Region | Number (thousands) |
|---|---|
| Europe | 4772 |
| Asia | 8364 |
| Africa | 840 |
| Oceania | 180 |
| Latin America | 15,472 |
| North America | 836 |

*Source:* U.S. Bureau of the Census

(a) Construct a relative frequency distribution.
(b) What percentage of foreign-born residents were born in Africa?
(c) Construct a frequency bar graph.
(d) Construct a relative frequency bar graph.
(e) Construct a pie chart.

**14. Robbery** The following data represent the number of robberies in 1998.

| Type of Crime | Number of Offenses (in thousands) |
|---|---|
| Street or highway | 219 |
| Commercial house | 61 |
| Gas station | 10 |
| Convenience store | 26 |
| Residence | 54 |
| Bank | 9 |

*Source:* U.S. Federal Bureau of Investigation

(a) Construct a relative frequency distribution.
(b) What percentage of robberies were of a gas station?
(c) Construct a frequency bar graph.
(d) Construct a relative frequency bar graph.
(e) Construct a pie chart.

**15. Educational Attainment** On the basis of the 1999 Current Population Survey, there were 100.0 million males and 100.1 million females aged 25 years old or older in the United States. The educational attainment of males and females is listed below.

| Educational Attainment | Males (millions) | Females (millions) |
|---|---|---|
| Not a high school graduate | 16.6 | 16.6 |
| High school graduate | 31.8 | 34.8 |
| Some college, but no degree | 17.1 | 17.5 |
| Associate's degree | 7.0 | 8.0 |
| Bachelor's degree | 17.9 | 16.2 |
| Advanced degree | 9.6 | 7.0 |

(a) Construct a relative frequency distribution for males.
(b) Construct a relative frequency distribution for females.
(c) Construct a side-by-side relative frequency bar graph.
(d) Compare each gender's educational attainment. Make a conjecture about the reasons for the differences.

**16. Internet Access** The following data represent the number of people that had Internet access in the years 1997 and 2000, by level of education. Data are in thousands of U.S. residents.

(a) Construct a relative frequency distribution for each year.
(b) Draw a side-by-side relative frequency bar graph.
(c) Compare the access for the two years. Conjecture some reasons for the differences and similarities.

| Educational Attainment | 1997 | 2000 |
|---|---|---|
| No college | 7427 | 24,662 |
| Some college | 9742 | 31,462 |
| Graduated college | 10,367 | 34,379 |

*Source: United States Statistical Abstract, 2000*

**17. Space Launches** The following data represent the number of successful space launches in 1995 and 1998. A successful space launch is a launch in which the craft attains the Earth's orbit or escapes Earth.

| Country | Number of Launches, 1995 | Number of Launches, 1998 |
|---|---|---|
| Soviet Union/ Commonwealth of Independent States | 32 | 24 |
| United States | 27 | 34 |
| Japan | 2 | 2 |
| European Space Agency | 11 | 11 |
| China | 2 | 6 |
| Israel | 1 | 0 |

(a) Construct a relative frequency distribution for each year.
(b) Draw a side-by-side relative frequency bar graph.
(c) Compare space launches for the two years. Make a conjecture about the reasons for the differences.

**18. Transplants** The following data represent the number of transplants performed in 1990 and 1998.

| Transplant | 1990 | 1998 |
|---|---|---|
| Heart | 2108 | 2345 |
| Liver | 2690 | 4487 |
| Kidney | 9877 | 13,139 |
| Heart–lung | 52 | 47 |
| Lung | 203 | 862 |
| Pancreas/islet cell | 528 | 1221 |
| Intestine | 5 | 69 |

*Source: U.S. Department of Health and Human Services*

(a) Construct a relative frequency distribution for the years 1990 and 1998.

(b) What percentage of transplants in 1990 were heart transplants? What percentage of transplants in 1998 were heart transplants? What might account for the difference?
(c) Construct a side-by-side relative frequency bar graph.
(d) Compare the relative frequencies of each type of transplant for the two years. Make a conjecture about the reasons for the differences.

**19. 2000 Presidential Election** An exit poll was conducted in Palm Beach County, Florida, in which a random sample of 40 voters revealed whom they voted for in the presidential election. The results of the survey are listed below.

| | | | | | | |
|---|---|---|---|---|---|---|
| Bush | Bush | Gore | Bush | Gore | Gore | Gore |
| Gore | Gore | Gore | Gore | Bush | Gore | Gore |
| Bush | Bush | Gore | Nader | Gore | Gore | Gore |
| Bush | Bush | Gore | Gore | Bush | Gore | Bush |
| Gore | Gore | Bush | Gore | Gore | Bush | Gore |
| Bush | Gore | Gore | Gore | Bush | | |

(a) Construct a frequency distribution.
(b) Construct a relative frequency distribution.
(c) Construct a frequency bar graph.
(d) Construct a relative frequency bar graph.
(e) Construct a pie chart.
(f) On the basis of the data, make a conjecture about which candidate will win Palm Beach County. If George W. Bush wins Palm Beach County, what conclusions might be drawn, assuming that the sample was conducted appropriately? Would you be confident in making this prediction with a sample of size 40? If the sample consisted of 100 voters, would your confidence increase? Why?

**20. Hospital Admissions** The following data represent the diagnoses of a random sample of 20 patients admitted to a hospital.

| | | |
|---|---|---|
| Cancer | Motor Vehicle Accident | Congestive Heart Failure |
| Gun Shot Wound | Fall | Gun Shot Wound |
| Gun Shot Wound | Motor Vehicle Accident | Gun Shot Wound |
| Assault | Motor Vehicle Accident | Gun Shot Wound |
| Motor Vehicle Accident | Motor Vehicle Accident | Gun Shot Wound |
| Motor Vehicle Accident | Gun Shot Wound | Motor Vehicle Accident |
| Fall | Gun Shot Wound | |

*Source: Tamela Ohm, Student at Joliet Junior College*

(a) Construct a frequency distribution.
(b) Construct a relative frequency distribution.
(c) Which diagnosis has the most admissions?
(d) What percentage of diagnoses are motor vehicle accidents?
(e) Construct a frequency bar graph.
(f) Construct a relative frequency bar graph.
(g) Construct a pie chart.

**21. Which Position in Baseball Is the Highest Paying?** You are a prospective baseball agent and are in search of some clients. You would like to recruit the highest paid players as clients, so you perform a study in which you identify the 36 top-paid players as of February 2000 and their positions. The table below shows the results of your study.

| Player | Position | Player | Position | Player | Position |
|---|---|---|---|---|---|
| Kevin Brown | Pitcher | Derek Jeter | Shortstop | John Smoltz | Pitcher |
| Randy Johnson | Pitcher | Raul Mondesi | Right Field | Jim Thome | First Base |
| Albert Belle | Right Field | Gary Sheffield | Right Field | Todd Stottlemyre | Pitcher |
| Bernie Williams | Center Field | Tom Glavine | Pitcher | Robin Ventura | Third Base |
| Mike Piazza | Catcher | Shawn Green | Right Field | Chuck Finley | Pitcher |
| Larry Walker | First Base | Ken Griffey, Jr. | Center Field | Al Leiter | Pitcher |
| David Cone | Pitcher | Mark McGwire | First Base | Brian Jordan | Center Field |
| Pedro Martínez | Pitcher | Wilson Alvarez | Pitcher | Ray Lankford | Center Field |
| Mo Vaughn | First Base | Rafael Palmeiro | First Base | Juan González | First Base |
| Greg Maddux | Pitcher | Iván Rodriguez | Catcher | Kenny Rogers | Pitcher |
| Sammy Sosa | Right Field | Matt Williams | Third Base | Kenny Lofton | Center Field |
| Barry Bonds | Right Field | Andrés Gallarraga | First Base | Alex Rodríguez | Shortstop |

*Source:* SLAM! Baseball

(a) Construct a frequency distribution.
(b) Construct a relative frequency distribution.
(c) Which position appears to be the most lucrative? For which position would you recruit?
(d) Are there any positions for which you would avoid recruiting? Why?
(e) Draw a frequency bar graph.
(f) Draw a relative frequency bar graph.
(g) Draw a pie chart.

**22. Blood Type** A phlebotomist draws the blood of a random sample of 50 patients and determines their blood types as shown at right.

(a) Construct a frequency distribution.
(b) Construct a relative frequency distribution.
(c) According to the data, which blood type is most common?
(d) According to the data, which blood type is least common?
(e) Use the results of the sample to conjecture the percentage of the population that is blood type O. Is this statistic descriptive or inferential?
(f) Contact a local hospital and ask them the percentage of the population that is blood type O. Why might the results differ?
(g) Draw a frequency bar graph.
(h) Draw a relative frequency bar graph.
(i) Draw a pie chart.

| O | O | A | A | O |
|---|---|---|---|---|
| B | O | B | A | O |
| AB | B | A | B | AB |
| O | O | A | A | O |
| AB | O | A | B | A |
| O | A | A | O | A |
| O | A | O | AB | A |
| O | B | A | A | O |
| O | O | O | A | O |
| O | A | O | A | O |

## Technology Step-by-Step
Drawing Bar Graphs and Pie Charts

**TI-83 Plus**  The TI-83 Plus does not have the ability to draw bar graphs or pie charts.

**MINITAB**  **Bar Graphs**

*Step 1:* Enter the categories in C1 and the frequency or relative frequency in C2.

*Step 2:* Select the **Graph** menu and highlight **Chart**.

*Step 3:* Fill in the window with the appropriate values. In the Function cell, select Sum. In the "Y" cell, select the column containing the frequency or relative frequency. In the "X" cell, select the column containing the categories. The Display cell should be selected as "Bar." Click OK to obtain the bar graph.

**Pie Charts**

*Step 1:* With the categories entered in C1 and the frequencies entered in C2, select the **Graph** menu and highlight **Pie Chart** . . .

*Step 2:* Fill in the window as shown on the left to obtain the pie chart of Example 5. Click OK.

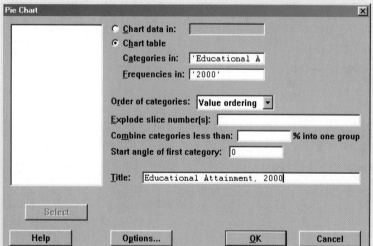

**Excel**  **Bar Graphs**

*Step 1:* Enter the categories in column A and the frequency or relative frequency in column B.

*Step 2:* Select the chart wizard icon. Click the "column" chart type. Select the chart type in the upper left-hand corner and hit "Next."

*Step 3:* Click inside the data range cell. Use the mouse to highlight the data to be graphed. Click "Next."

*Step 4:* Click the "Titles" tab to enter *x*-axis, *y*-axis, and chart titles. Click "Finish."

**Pie Charts**

*Step 1:* Enter the categories in column A and the frequencies in column B. Select the chart wizard icon and click the "pie" chart type. Select the pie chart in the upper left-hand corner.

*Step 2:* Click inside the data range cell. Use the mouse to highlight the data to be graphed. Click "Next."

*Step 3:* Click the "Titles" tab to the chart title. Click the "Data Labels" tab and select "Show label and percent." Click "Finish."

---

## 2.2  Organizing Quantitative Data

**Preparing for This Section**  Before getting started, review the following:

✓ Quantitative data (Section 1.1, p. 6)

✓ Continuous data (Section 1.1, p. 6)

✓ Discrete data (Section 1.1, p. 6)

**Objectives**   Summarize discrete data in tables

  Construct histograms of discrete data

3   Summarize continuous data in tables
4   Construct histograms of continuous data
5   Draw stem-and-leaf plots
6   Identify the shape of a distribution
7   Draw time series graphs

The first step in summarizing quantitative data is to determine whether the data are discrete or continuous. If the data are discrete, the categories of data will be the observations (as in qualitative data); however, if the data are continuous, the categories of data (called *classes*) must be created using intervals of numbers. We will first present the techniques required to summarize discrete quantitative data and then proceed to summarizing continuous quantitative data.

## Summarizing Discrete Data in Tables

The values of a discrete variable are used to create the categories of data.

▶ EXAMPLE 1   **Constructing Frequency and Relative Frequency Distributions from Discrete Data**

| TABLE 8 |
|---------|
| Number of Arrivals at Wendy's |

| 7 | 6 | 6 | 6 | 4 |
|---|---|---|---|---|
| 5 | 6 | 6 | 11 | 4 |
| 2 | 7 | 1 | 2 | 4 |
| 6 | 5 | 5 | 3 | 7 |
| 2 | 2 | 9 | 7 | 5 |
| 6 | 2 | 6 | 5 | 7 |
| 6 | 8 | 2 | 6 | 5 |
| 4 | 6 | 9 | 8 | 5 |

*Problem:* The manager of a Wendy's fast-food restaurant is interested in studying the typical number of customers who arrive during the lunch hour. The data in Table 8 represent the number of customers who arrive at Wendy's for 40 randomly selected 15-minute intervals of time during lunch. For example, during one 15-minute interval, seven customers arrived. Construct a frequency and relative frequency distribution.

*Approach:* The number of people arriving could be 0, 1, 2, 3, .... From Table 8, we see that there are 11 categories of data from this study: 1, 2, 3, ..., 11. We tally the number of observations for each category, add up each tally, and create the frequency and relative frequency distributions.

*Solution:* The frequency and relative frequency distributions are shown in Table 9.

| TABLE 9 | | | |
|---------|------|-----------|--------------------|
| Number of Customers | Tally | Frequency | Relative Frequency |
| 1 | I | 1 | $\frac{1}{40} = 0.025$ |
| 2 | ⊪ I | 6 | 0.15 |
| 3 | I | 1 | 0.025 |
| 4 | IIII | 4 | 0.1 |
| 5 | ⊪ II | 7 | 0.175 |
| 6 | ⊪ ⊪ I | 11 | 0.275 |
| 7 | ⊪ | 5 | 0.125 |
| 8 | II | 2 | 0.05 |
| 9 | II | 2 | 0.05 |
| 10 | | 0 | 0.0 |
| 11 | I | 1 | 0.025 |

On the basis of the relative frequencies, 27.5% of the 15-minute intervals had six customers arrive at Wendy's during the lunch hour.   ◀◀

**NW**   *Now Work Problems 17(a)–(d).*

## Histograms of Discrete Data

As with qualitative data, quantitative data may also be represented graphically. We begin our discussion with a graph called the *histogram*, which is similar to the bar graph drawn for qualitative data.

*Definition*

> A **histogram** is constructed by drawing rectangles for each class of data. The height of each rectangle is the frequency or relative frequency of the class. The width of each rectangle should be the same and the rectangles should touch each other.

▶ **EXAMPLE 2**    **Drawing a Histogram for Discrete Data**

*Problem:* Construct a frequency histogram and a relative frequency histogram using the data summarized in Table 9.

*Approach:* On the horizontal axis, we place the value of each category of data (number of customers). The vertical axis will be the frequency or relative frequency of each category. Rectangles of equal width are drawn, with the center of each rectangle located at the value of each category. For example, the first rectangle is centered at 1. For the frequency histogram, the height of the rectangle will be the frequency of the category. For the relative frequency histogram, the height of the rectangle will be the relative frequency of the category. Remember, the rectangles touch for histograms.

*Solution:* Figure 8(a) shows the frequency histogram. Figure 8(b) shows the relative frequency histogram.

**Caution**

The rectangles in histograms touch, while the rectangles in bar graphs do not touch.

Figure 8

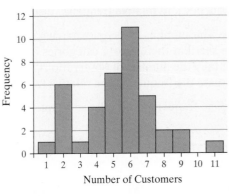

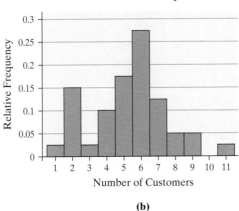

(a)                                                      (b)

**NW**  *Now Work Problems 17(e) and (f).*

## Summarizing Continuous Data in Tables

Raw continuous data do not have any predetermined categories that can be used to construct a frequency distribution; therefore, the categories must be created. Categories of data are created by using intervals of numbers called **classes**.

Table 10 is a typical frequency distribution created from continuous data. The data represent the number of United States residents between the ages of 25 and 74 that have earned a bachelor's degree. The data are based on the Current Population Survey conducted in 1998.

In the table, we notice that the data are categorized, or grouped, by intervals of numbers. Each interval represents a class. For example, the first class is 25–34-year-old residents of the United States who have a bachelor's degree. There are five classes in the table, each with a lower bound and an upper bound. The **lower class limit** of a class is the smallest value within the

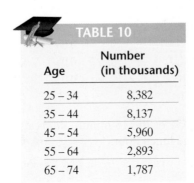

**TABLE 10**

| Age | Number (in thousands) |
|-----|------------------------|
| 25 – 34 | 8,382 |
| 35 – 44 | 8,137 |
| 45 – 54 | 5,960 |
| 55 – 64 | 2,893 |
| 65 – 74 | 1,787 |

| TABLE 11 | |
|---|---|
| Age | Number |
| 20 – 29 | 765 |
| 30 – 39 | 1,301 |
| 40 – 49 | 971 |
| 50 – 59 | 394 |
| 60 and older | 80 |

*Source:* United States Justice Department

class, while the **upper class limit** of a class is the largest value within the class. The lower class limit for the first class in Table 10 is 25, while the upper class limit is 34. The **class width** is the difference between consecutive lower class limits. The class width for the data in Table 10 is 35 − 25 = 10.

The classes in Table 10 do not overlap, so there is no confusion as to which class a data value belongs. Notice that the class widths are equal for all classes. One exception to this requirement is in open-ended tables. A table is **open ended** if the last class does not have an upper class limit. The data in Table 11 represent the number of persons under sentence of death as of December 31, 1999, in the United States. The last class in the table, "60 and older," is open ended.

▶ EXAMPLE 3

### Organizing Continuous Data into a Frequency and Relative Frequency Distribution

**In Your Own Words**

For qualitative and discrete data, the categories of data are formed by using the data. For continuous data, the categories are formed by using an interval of numbers, such as 30–39.

*Problem:* Suppose you are considering investing in a Roth IRA. You collect the data in Table 12, which represent the three-year rate of return (in percent) for a simple random sample of 40 small capitalization growth mutual funds. Construct a frequency and relative frequency distribution of the data.

| TABLE 12 | | | | |
|---|---|---|---|---|
| Three-Year Rate of Return of Mutual Funds | | | | |
| 27.4 | 16.7 | 10.8 | 24.1 | 35.9 |
| 12.7 | 28.5 | 22.2 | 18.4 | 17.4 |
| 22.6 | 29.6 | 11.6 | 45.9 | 16.6 |
| 32.1 | 47.7 | 10.9 | 18.4 | 23.3 |
| 18.2 | 32.0 | 25.5 | 23.7 | 38.1 |
| 23.7 | 14.7 | 12.8 | 31.1 | 21.9 |
| 18.4 | 21.3 | 27.0 | 19.6 | 15.8 |
| 14.7 | 37.0 | 19.2 | 18.5 | 29.1 |

*Source:* Morningstar.com

**Caution**

Watch out for tables with class widths that overlap, such as a first class of 20–30 and a second class of 30–40.

*Approach:* In order to construct a frequency distribution, we first create classes of equal width. There are 40 observations in Table 12, and they range from 10.8 to 47.7, so we decide to create the classes such that the lower class limit of the first class is 10.0 (a little smaller than the smallest data value) and the class width is 5. There is nothing magical about the choice of 5 as a class width. We could have selected a class width of 8 (or any other class width, as well). We simply choose a class width that we think will nicely summarize the data. If our choice doesn't accomplish this, we can always try another one. The lower class limit of the second class will be 10.0 + 5 = 15.0. Because the classes must not overlap, the upper class limit of the first class is 14.9. Continuing in this fashion, we obtain the following classes:

$$10.0 - 14.9$$
$$15.0 - 19.9$$
$$\vdots$$
$$45.0 - 49.9$$

This gives us eight classes. We tally the number of observations in each class, add up the tallies, and create the frequency distribution. The relative frequency distribution is created by dividing each class's frequency by 40, the number of observations.

### Historical Note

Florence Nightingale was born in Italy on May 12, 1820. She was named after the city of her birth. Florence was educated by her father, who attended Cambridge University. Between 1849 and 1851, Florence studied nursing throughout Europe. In 1854, she was asked to oversee the introduction of female nurses into the military hospitals in Turkey. While there, she greatly improved the mortality rate of wounded soldiers. She collected data and invented graphs (the polar area diagram), tables, and charts to show that improving sanitary conditions would lead to decreased mortality rates. In 1869, Florence founded the Nightingale School Home for Nurses. Florence died on August 13, 1910.

### In Your Own Words

Creating the classes for summarizing continuous data is an art form. There is no such thing as the correct frequency distribution. However, there can be incorrect frequency distributions. The larger the class width, the fewer classes a frequency distribution will have.

*Solution:* We tally the data as shown in the second column of Table 13. The third column in the table shows the frequency of each class. From the frequency distribution, we conclude that a three-year rate of return between 15.0% and 19.9% occurs with the most frequency. The fourth column in the table shows the relative frequency of each class. So, 27.5% of the small-capitalization growth mutual funds had a three-year rate of return between 15.0% and 19.9%.

| TABLE 13 | | | |
|---|---|---|---|
| Class (Three-Year Rate of Return) | Tally | Frequency | Relative Frequency |
| 10.0–14.9 | 卌 II | 7 | 7/40 = 0.175 |
| 15.0–19.9 | 卌 卌 I | 11 | 11/40 = 0.275 |
| 20.0–24.9 | 卌 III | 8 | 8/40 = 0.2 |
| 25.0–29.9 | 卌 I | 6 | 0.15 |
| 30.0–34.9 | III | 3 | 0.075 |
| 35.0–39.9 | III | 3 | 0.075 |
| 40.0–44.9 | | 0 | 0 |
| 45.0–49.9 | II | 2 | 0.05 |

Only 5% of the mutual funds had a three-year rate of return between 45.0% and 49.9%. We might conclude that mutual funds with a rate of return at this level outperform their peers and consider them worthy of our investment. This type of information would be difficult to obtain from the raw data. ◀◀

Notice that the choice of the lower class limit of the first class and the class width was rather arbitrary. While formulas and procedures do exist for creating frequency distributions from raw data, they do not necessarily provide "better" summaries. It is incorrect to say that one particular frequency distribution is the correct one. Constructing frequency distributions is somewhat of an art form in which the distribution that seems to provide the best overall summary of the data should be used.

Consider the frequency distribution in Table 14, which also summarizes the three-year rate of return data discussed in Example 3. Here, the lower class limit of the first class is 10.0 and the class width is 8.0. Do you think Table 13 or Table 14 provides a better summary of the distribution of three-year rates of return? In forming your opinion, consider the following: Too few classes will cause a "bunching" effect. Too many classes will spread the data out, thereby not revealing any pattern.

The goal in constructing a frequency distribution is to reveal interesting features of the data. With that said, when constructing frequency distributions, we typically want the number of classes to be between 5 and 20. When the data set is small, we want fewer classes; when the data set is large, we want more classes. Why do you think this is reasonable?

| TABLE 14 | | |
|---|---|---|
| Class | Tally | Frequency |
| 10.0–17.9 | 卌 卌 I | 11 |
| 18.0–25.9 | 卌 卌 卌 I | 16 |
| 26.0–33.9 | 卌 III | 8 |
| 34.0–41.9 | III | 3 |
| 42.0–49.9 | II | 2 |

**NW** *Now Work Problems 19(a) and (b).*

**Histograms of Continuous Data**

We're now ready to draw histograms of continuous data.

▶ **EXAMPLE 4** **Drawing a Histogram for Continuous Data**

*Problem:* Construct a frequency and relative frequency histogram of the three-year rate of return data discussed in Example 3.

*Approach:* To draw the frequency histogram, we will use the frequency distribution in Table 13. We label the lower class limits of each class on the horizontal axis. Then, for each class, we draw a rectangle whose width is the class width and whose height is the frequency. To construct the relative frequency histogram, we let the height of the rectangle be the relative frequency instead of the frequency.

*Solution:* Figure 9(a)* represents the frequency histogram, and Figure 9(b) represents the relative frequency histogram.

**Figure 9**
Frequency and Relative Frequency Histograms

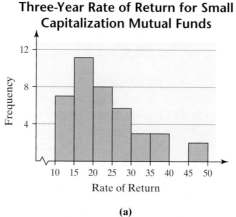

(a)

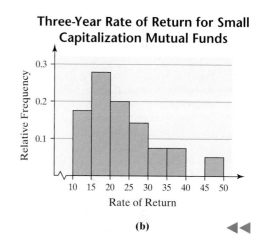

(b)

**NW** *Now Work Problems 19(c) and (d).*

◀◀

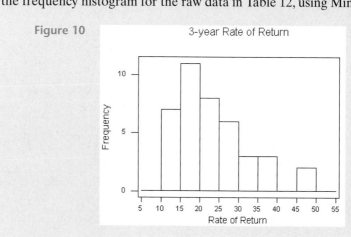

**Using Technology:** Statistical software and certain graphing calculators have the ability to draw histograms from raw data. Figure 10 shows the frequency histogram for the raw data in Table 12, using Minitab.

**Figure 10**

*The ∧∨ symbol on the horizontal axis indicates that part of the graph has been "cut out." This is done to avoid a lot of white space in the graph.

**⑤ Stem-and-Leaf Plots**

A **stem-and-leaf plot** is another way to represent quantitative data graphically. As we shall see, stem-and-leaf plots have some advantages over histograms. Use the following steps to construct a stem-and-leaf plot:

---

**Construction of a Stem-and-Leaf Plot**

*Step 1:* The **stem** of the graph will consist of the digits to the left of the rightmost digit. The **leaf** of the graph will be the rightmost digit. Sometimes it is necessary to modify the method of choosing the stem if a different class width is desired.

*Step 2:* Write the stems in a vertical column in increasing order. Draw a vertical line to the right of the stems.

*Step 3:* Write each leaf corresponding to the stems to the right of the vertical line. The leaves must be written in ascending order.

---

▶ **EXAMPLE 5** **Constructing a Stem-and-Leaf Plot**

*Problem:* The data in Table 15 represent the percentage of persons in poverty, by state, in 1997. Draw a stem-and-leaf plot of the data.

*Approach:*

*Step 1:* We will treat the integer portion of the number as the stem and the decimal portion as the leaf. For example, the stem of Alabama will be 14 and the leaf will be 8. The stem of 14 will include all data from 14.0 to 14.9.

| TABLE 15 | | | | | |
|---|---|---|---|---|---|
| *Percent of Persons Living in Poverty* | | | | | |
| **State** | **Percent** | **State** | **Percent** | **State** | **Percent** |
| Alabama | 14.8 | Kentucky | 16.4 | North Dakota | 12.3 |
| Alaska | 8.5 | Louisiana | 18.4 | Ohio | 11.8 |
| Arizona | 18.8 | Maine | 10.7 | Oklahoma | 15.2 |
| Arkansas | 18.4 | Maryland | 9.3 | Oregon | 11.7 |
| California | 16.8 | Massachusetts | 11.2 | Pennsylvania | 11.4 |
| Colorado | 9.4 | Michigan | 10.7 | Rhode Island | 11.9 |
| Connecticut | 10.1 | Minnesota | 9.7 | South Carolina | 13.1 |
| Delaware | 9.1 | Mississippi | 18.6 | South Dakota | 14.1 |
| D.C. | 23.0 | Missouri | 10.6 | Tennessee | 15.1 |
| Florida | 14.3 | Montana | 16.3 | Texas | 16.7 |
| Georgia | 14.7 | Nebraska | 10.0 | Utah | 8.3 |
| Hawaii | 13.0 | Nevada | 9.6 | Vermont | 10.9 |
| Idaho | 13.3 | New Hampshire | 7.7 | Virginia | 12.5 |
| Illinois | 11.6 | New Jersey | 9.2 | Washington | 10.5 |
| Indiana | 8.2 | New Mexico | 23.4 | West Virginia | 17.5 |
| Iowa | 9.6 | New York | 16.6 | Wisconsin | 8.5 |
| Kansas | 10.4 | North Carolina | 11.8 | Wyoming | 12.7 |

*Source: Poverty in the United States,* 1997 Current Population Reports

**Figure 11**

```
 7 | 7
 8 | 2 3 5 5
 9 | 1 2 3 4 6 6 7
10 | 0 1 4 5 6 7 7 9
11 | 2 4 6 7 8 8 9
12 | 3 5 7
13 | 0 1 3
14 | 1 3 7 8
15 | 1 2
16 | 3 4 6 7 8
17 | 5
18 | 4 4 6 8
19 |
20 |
21 |
22 |
23 | 0 4
```

**Step 2:** Write the stems vertically in ascending order, and then draw a vertical line to the right of the stems.
**Step 3:** Write the leaves corresponding to the stem in ascending order.

*Solution:*

**Step 1:** The stem from Alabama is 14 while the corresponding leaf is 8. The stem from Alaska is 8 and its leaf is 5, and so on.
**Step 2:** We write the stems vertically in Figure 11, along with a vertical line to the right of the stem.
**Step 3:** We write the leaves corresponding to each stem in ascending order. Figure 11 shows the stem-and-leaf diagram. ◄◄

If you look at the stem-and-leaf plot carefully, you'll notice that it is essentially a histogram turned on its side. The stem serves as the class. For example, the stem 15 contains all data from 15.0–15.9. The leaf serves as the frequency (height of the rectangle). Therefore, it is important to space the leaves equally when drawing a stem-and-leaf plot.

One of the advantages of the stem-and-leaf plot over frequency distributions and histograms is that the raw data can be retrieved from the stem-and-leaf plot.

**Once a frequency distribution or histogram of continuous data is created, the raw data are lost (unless reported with the frequency distribution); however, the raw data can be retrieved from the stem-and-leaf plot.**

The steps listed for creating stem-and-leaf plots sometimes must be modified in order to meet the needs of the data. Consider the next example.

**In Your Own Words**

The choice of the stem in the construction of a stem-and-leaf diagram is also an art form. It acts just like the class width. For example, the stem of 7 in Figure 11 represents the class 7.0–7.9. The stem of 8 represents the class 8.0–8.9. Notice that the class width is 1.0. The number of leaves is the frequency of each category.

► **EXAMPLE 6**   Constructing a Stem-and-Leaf Plot

*Problem:* Construct a stem-and-leaf plot of the three-year rate of return data in Table 12 on page 57.

*Approach:*

**Step 1:** If we use the approach from Example 5, the stems will be 10, 11, 12, ..., 47. However, this will not provide a very good visual display of the data (it will be too spread out). Using these stems would be like choosing a class width of 1. To address this problem, the data can be rounded. For example, we might round the data in Table 12 to the nearest integer. The stem can be the tens digits and the leaf can be the ones digits.
**Step 2:** We create a vertical column of the tens digits.
**Step 3:** The leaves will be the ones digits.

*Solution:*

**Step 1:** We round the data to the nearest integer as shown in Table 16.
**Step 2:** We use the tens digit for the stem as shown:

```
1 |
2 |
3 |
4 |
```

| TABLE 16 | | | | |
|---|---|---|---|---|
| 27 | 17 | 11 | 24 | 36 |
| 13 | 29 | 22 | 18 | 17 |
| 23 | 30 | 12 | 46 | 17 |
| 32 | 48 | 11 | 18 | 23 |
| 18 | 32 | 26 | 24 | 38 |
| 24 | 15 | 13 | 31 | 22 |
| 18 | 21 | 27 | 20 | 16 |
| 15 | 37 | 19 | 19 | 29 |

*Source:* Morningstar.com

***Step 3:*** Fill in the leaves in ascending order as shown in Figure 12. ◀◀

**Figure 12**

```
1 | 1 1 2 3 3 5 5 6 7 7 7 8 8 8 8 9 9
2 | 0 1 2 2 3 3 4 4 4 6 7 7 9 9
3 | 0 1 2 2 6 7 8
4 | 6 8
```

**Figure 13**

```
1 | 1 1 2 3 3
1 | 5 5 6 7 7 7 8 8 8 8 9 9
2 | 0 1 2 2 3 3 4 4 4
2 | 6 7 7 9 9
3 | 0 1 2 2
3 | 6 7 8
4 |
4 | 6 8
```

Looking at the stem-and-leaf plot in Figure 12, we notice that the data are bunched. To resolve this problem, we can use **split stems**. For example, rather than using one stem for mutual funds earning 10–19 percent, we could use two stems, one for the 10–14 interval and the second for the 15–19 interval. We do this in Figure 13.

The stem-and-leaf plot shown in Figure 13 reveals the distribution of the data better. As with the construction of class intervals in the creation of frequency histograms, judgment plays a major role. There is no such thing as a correct stem-and-leaf plot. However, a quick comparison of Figures 12 and 13 shows that some are better than others.

One final note regarding stem-and-leaf plots: They are best used when the data set is small.

**In Your Own Words**

Using split stems is like adding more classes to a frequency distribution.

**Using Technology:** Certain statistical software packages, such as Minitab, will draw stem-and-leaf plots. Figure 14 illustrates a stem-and-leaf plot for the data in Table 12.

Minitab truncated the data values (i.e., 10.8 is represented as 10) instead of rounding. The stem is found in the second column. The value (8) in the left column indicates that there are eight observations in the class containing the middle value (called the *median*). The values in the left column above the (8) represent the number of observations less than or equal to the upper class limit of the class. For example, 18 observations have a three-year rate of return of 19% or less. The values in the left column below the (8) indicate the number of observations greater than or equal to the lower class limit of the class. For example, 5 observations have a three-year rate of return of 35% or higher.

**Figure 14**

```
 7    1  0 0 1 2 2 4 4
18    1  5 6 6 7 8 8 8 8 9 9
(8)   2  1 1 2 2 3 3 3 4
14    2  5 7 7 8 9 9
 8    3  1 2 2
 5    3  5 7 8
 2    4
 2    4  5 7
```

**NW** *Now Work Problem 25.*

## 6 Distribution Shapes

One of the ways that a variable is described is through the shape of its distribution. Distribution shapes are typically classified as symmetric, skewed left, or skewed right. Figure 15 displays various histograms and the shape of the distribution.

Figure 15

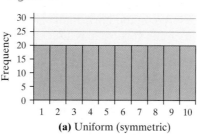

**(a)** Uniform (symmetric)

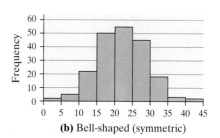

**(b)** Bell-shaped (symmetric)

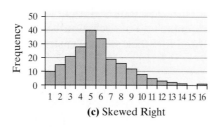

**(c)** Skewed Right

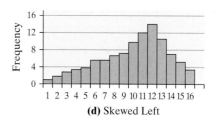

**(d)** Skewed Left

Figures 15(a) and (b) display symmetric distributions. These distributions are symmetric because, if we split the histogram down the middle, the right and left sides of the histograms are mirror images of each other. Figure 15(a) is a **uniform distribution**, because the frequency of each value of the variable is evenly spread out across the values of the variable. Figure 15(b) displays a **bell-shaped distribution**, because the highest frequency occurs in the middle and frequencies tail off to the left and right of the middle. Figure 15(c) illustrates a distribution that is **skewed right**. Notice that the tail to the right of the peak is longer than the tail to the left of the peak. Finally, Figure 15(d) illustrates a distribution that is **skewed left**, because the tail to the left of the peak is longer than the tail to the right of the peak.

▶ **EXAMPLE 7**  **Identifying the Shape of a Distribution**

*Problem:* Figure 16 displays the histogram obtained for the three-year rates of return for small capitalization stocks. Describe the shape of the distribution.

*Approach:* We will compare the shape of the distribution displayed in Figure 16 with those in Figure 15.

*Solution:* Since the tail to the right of the peak is longer than the tail to the left of the peak, the distribution is skewed right.

Figure 16

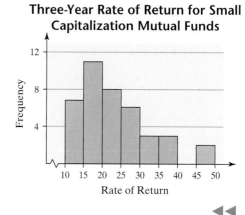

**Three-Year Rate of Return for Small Capitalization Mutual Funds**

◀◀

 *Now Work Problem 19(e).*

**⑦  Time Series Plots**

If the value of a variable is measured at different points in time, the data are referred to as **time series data**. The closing price of Cisco Systems stock each month for the past two years is an example of time series data.

*Definition*  A **time series plot** is obtained by plotting the time in which a variable is measured on the horizontal axis and the corresponding value of the variable on the vertical axis. Lines are then drawn connecting the points.

Time series plots are very useful in identifying trends in the data.

▶ **EXAMPLE 8**  **A Time Series Plot**

*Problem:* The data in Table 17 represent the closing price of Cisco Systems stock at the end of each month from January 2000 through November 2001, adjusted for stock splits. Construct a time series plot of the data.

**TABLE 17**

| Date | Closing Price | Date | Closing Price |
|------|---------------|------|---------------|
| 1/00 | 54.75 | 1/01 | 37.4375 |
| 2/00 | 66.0938 | 2/01 | 23.6875 |
| 3/00 | 77.3125 | 3/01 | 15.8125 |
| 4/00 | 69.3281 | 4/01 | 16.98 |
| 5/00 | 56.9375 | 5/01 | 19.26 |
| 6/00 | 63.5625 | 6/01 | 18.20 |
| 7/00 | 65.4375 | 7/01 | 19.22 |
| 8/00 | 68.625 | 8/01 | 16.33 |
| 9/00 | 55.25 | 9/01 | 12.18 |
| 10/00 | 53.875 | 10/01 | 16.92 |
| 11/00 | 47.875 | 11/01 | 20.44 |
| 12/00 | 38.25 | | |

*Source:* NASDAQ

*Approach:*

**Step 1:** Plot points for each month, with the date on the horizontal axis and the closing price on the vertical axis.
**Step 2:** Connect the points with straight lines.

*Solution:* Figure 17 shows the graph of the time series plot. The trend does not bode well for investors of Cisco Systems stock.

Figure 17

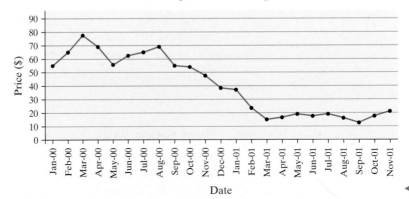

**Cisco Systems Closing Price**

**Using Technology:** Statistical spreadsheets, such as Excel or Minitab, and certain graphing calculators, such as the TI-83, have the ability to create time series graphs.

*Now Work Problem 29.*

## 2.2 Assess Your Understanding

### Concepts and Vocabulary

1. Discuss circumstances under which it is preferable to use relative frequency distributions, instead of frequency distributions, as a tabular summary.
2. Why shouldn't classes overlap when one summarizes continuous data?
3. The histogram at the right represents the total rainfall for each time it rained in Chicago during the month of August since 1871. The histogram was taken from the *Chicago Tribune* on August 14, 2001. What is wrong with the histogram?
4. State the advantages and disadvantages of histograms versus stem-and-leaf plots.
5. What are time series data?

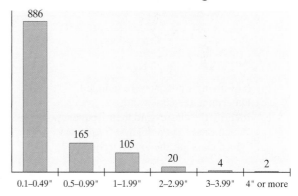

**Total August Rain Events Since 1871 in Chicago**

### Exercises

#### • Basic Skills

1. **Rolling the Dice** An experiment was conducted in which two fair dice were thrown 100 times. The sum of the dice was then recorded. The following frequency histogram gives the results.

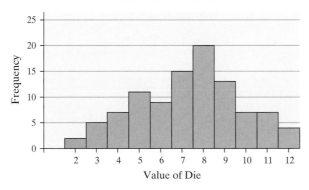

(a) What was the most frequent outcome of the experiment?
(b) What was the least frequent?
(c) How many times did we observe a seven?
(d) Determine the percentage of time a seven was observed.
(e) Describe the shape of the distribution.

2. **Car Sales** A car salesman records the number of cars he sold each week for the past year. The following frequency histogram shows the results.

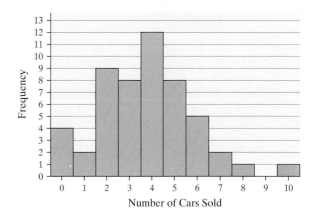

(a) What is the most frequent number of cars sold in a week?
(b) How many weeks were two cars sold?
(c) Determine the percentage of time two cars were sold.
(d) Describe the shape of the distribution.

3. **IQ Scores** The following frequency histogram represents the IQ scores of a random sample of seventh grade students. IQs are measured to the nearest whole number. The frequency of each class is labeled above each rectangle.

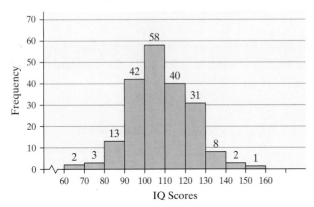

**IQs of 7th Grade Students**

(a) How many students were sampled?
(b) Determine the class width.
(c) Identify the classes and their frequency.
(d) Which class has the highest frequency?
(e) Which class has the lowest frequency?

(f) What is the relative frequency of an IQ between 120 and 129?
(g) Describe the shape of the distribution.

4. **Number of Licensed Drivers** The following frequency histogram represents the number of licensed drivers in the United States in 1997, by age.

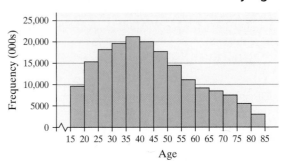

**Number of Licensed Drivers by Age**

(a) Determine the class width.
(b) Identify the classes.
(c) Which class has the highest frequency?
(d) Which class has the lowest frequency?
(e) Describe the shape of the distribution. Why do you think the distribution has this shape?

---

• **Applying the Concepts**

5. **Predicting School Enrollment** A local school district wants to know the number of children under the age of five in order to predict future enrollment. Toward that end, 50 households within the district are sampled, and the head of household is asked to disclose the number of children under the age of five living in the household. The results of the survey are presented in the table below.

| Number of Children Under Five | Number of Households |
|---|---|
| 0 | 16 |
| 1 | 18 |
| 2 | 12 |
| 3 | 3 |
| 4 | 1 |

(a) Construct a relative frequency distribution of the data.
(b) What percentage of households have two children under the age of five?
(c) What percentage of households have one or two children under the age of five?

6. **Free Throws** A basketball player habitually makes 70% of her free throws. In an experiment, a researcher asks this basketball player to record the number of free throws she shoots until she misses. The experiment is repeated 50 times. The table below lists the distribution of the number of free throws attempted until a miss is recorded.

| Number of Free Throws Until a Miss | Frequency |
|---|---|
| 1 | 16 |
| 2 | 11 |
| 3 | 9 |
| 4 | 7 |
| 5 | 2 |
| 6 | 3 |
| 7 | 0 |
| 8 | 1 |
| 9 | 0 |
| 10 | 1 |

(a) Construct a relative frequency distribution of the data.
(b) What percentage of the time did she miss on her 4th free throw?
(c) What percentage of the time did she make 9 in a row and then miss the 10th free throw?
(d) What percentage of the time did she make at least 5 in a row?

*In Problems 7 and 8, determine the original set of data. The stem represents the tens digit and the leaf represents the ones digit.*

**7.**

```
1 | 0  1  4
2 | 1  4  4  7  9
3 | 3  5  5  5  7  7  8
4 | 0  0  1  2  6  6  8  9  9
5 | 3  3  5  8
6 | 1  2
```

**8.**

```
4 | 0  4  7
5 | 2  2  3  9  9
6 | 3  4  5  8  8  9
7 | 0  1  1  3  6  6
8 | 2  3  8
```

*In Problems 9–12, find (a) the number of classes (b) the class limits of each class, and (c) the class width.*

**9. Health Insurance** The following data represent the number of people aged 25–64 covered by health insurance in 1998.

| Age | Number (in millions) |
| --- | --- |
| 25–34 | 29.3 |
| 35–44 | 37.0 |
| 45–54 | 30.4 |
| 55–64 | 19.5 |

***Source:*** *Current Population Reports*

**10. Earthquakes** The following data represent the number of earthquakes worldwide whose magnitude was less than 8.0 in 1998.

| Magnitude | Number |
| --- | --- |
| 0–0.9 | 2389 |
| 1.0–1.9 | 752 |
| 2.0–2.9 | 3851 |
| 3.0–3.9 | 5639 |
| 4.0–4.9 | 6943 |
| 5.0–5.9 | 832 |
| 6.0–6.9 | 113 |
| 7.0–7.9 | 10 |

***Source:*** National Earthquake Information Center

**11. Meteorology** The following data represent the high-temperature distribution in August in Chicago since 1872.

| Temperature | Days |
| --- | --- |
| 50–59° | 1 |
| 60–69° | 308 |
| 70–79° | 1519 |
| 80–89° | 1626 |
| 90–99° | 503 |
| 100–109° | 11 |

***Source:*** NOAA

**12. Multiple Births** The following data represent the number of live multiple delivery births (three or more babies) in 1999 for women 15–44 years old.

| Age | Number of Multiple Births |
| --- | --- |
| 15–19 | 66 |
| 20–24 | 411 |
| 25–29 | 1653 |
| 30–34 | 2926 |
| 35–39 | 1813 |
| 40–44 | 956 |

***Source:*** *National Vital Statistics Report* 49, (1): April 17, 2001

*In Problems 13–16 construct (a) a relative frequency distribution, (b) a frequency histogram, and (c) a relative frequency histogram for the data in the problem listed. Then answer the questions that follow.*

**13.** (Problem 9). Of the people covered by health insurance, what percentage are 25–34 years old? Of the people covered by health insurance, what percentage are 44 years or younger?

**14.** (Problem 10). What percentage of earthquakes registered 4.0–4.9? What percentage of earthquakes registered 4.9 or less?

**15.** (Problem 11). What percentage of the days had a temperature of 70–79°? What percentage of the days was it 79° or less?

**16.** (Problem 12). What percentage of multiple births were to women 40–44 years old? What percentage of multiple births were to women 24 years old or younger?

**17. Waiting** The following data represent the number of customers waiting for a table at 6:00 P.M. on Saturday for 40 consecutive Saturdays at Bobak's Restaurant:

| | | | |
| --- | --- | --- | --- |
| 11 | 5 | 11 | 3 |
| 4 | 5 | 13 | 9 |
| 13 | 10 | 9 | 6 |
| 10 | 8 | 7 | 3 |
| 7 | 9 | 10 | 4 |
| 6 | 8 | 6 | 7 |
| 6 | 4 | 14 | 11 |
| 8 | 10 | 9 | 5 |
| 8 | 8 | 7 | 8 |
| 8 | 6 | 11 | 8 |

(a) Construct a frequency distribution of the data.
(b) Construct a relative frequency distribution of the data.
(c) What percentage of the Saturdays had 10 or more customers waiting for a table at 6:00 P.M.?
(d) What percentage of the Saturdays had 5 or fewer customers waiting for a table at 6:00 P.M.?
(e) Construct a frequency histogram of the data.
(f) Construct a relative frequency histogram of the data.
(g) Describe the shape of the distribution.

**18.**  **Highway Repair** The following data represent the number of potholes on 50 randomly selected 1-mile stretches of highway in the city of Chicago.

| 2 | 7 | 4 | 7 | 2 | 7 |
|---|---|---|---|---|---|
| 2 | 2 | 2 | 3 | 4 | 3 |
| 1 | 2 | 3 | 2 | 1 | 4 |
| 2 | 2 | 5 | 2 | 3 | 4 |
| 4 | 1 | 7 | 10 | 3 | 5 |
| 4 | 3 | 3 | 2 | 2 | |
| 1 | 6 | 5 | 7 | 9 | |
| 2 | 2 | 2 | 1 | 5 | |
| 3 | 5 | 1 | 3 | 5 | |

(a) Construct a frequency distribution of the data.
(b) Construct a relative frequency distribution of the data.
(c) What percentage of the 1-mile stretches of highway had seven or more potholes?
(d) What percentage of the 1-mile stretches of highway had two or fewer potholes?
(e) Construct a frequency histogram of the data.
(f) Construct a relative frequency histogram of the data.
(g) Describe the shape of the distribution.

**19.** **Tensile Strength** Tensile strength is the maximum stress **NW** at which one can be reasonably certain that failure will not occur. The following data represent the tensile strength (in thousands of pounds per square inch) of a composite material to be used in an aircraft.

| 203.41 | 185.97 | 184.41 | 160.44 | 174.63 |
|--------|--------|--------|--------|--------|
| 209.58 | 190.67 | 200.73 | 180.95 | 185.34 |
| 213.35 | 207.88 | 206.51 | 201.95 | 205.59 |
| 218.56 | 210.80 | 209.84 | 204.60 | 212.00 |
| 242.76 | 231.46 | 212.15 | 219.51 | 225.25 |

*Source:* Vangel, Mark G., "New Methods for One-sided Tolerance Limits for a One-way Balanced Random-Effects ANOVA Model," *Technometrics* 34 (May, 1992) no. 2, 176-185

With the first class having a lower class limit of 160 and a class width of 10,
(a) Construct a frequency distribution.
(b) Construct a relative frequency distribution.
(c) Construct a frequency histogram of the data.

(d) Construct a relative frequency histogram of the data.
(e) Describe the shape of the distribution.
(f) Repeat parts (a)–(e), using a class width of 15. Which frequency distribution seems to provide a better summary of the data?

**20.** **Miles on a Cavalier** A random sample of 36 three-year-old Chevy Cavaliers was obtained in the Miami, Florida area, and the number of miles on each car was recorded. The results are shown in the table below.

| 34,122 | 17,685 | 15,499 | 26,455 | 30,500 | 41,194 |
|--------|--------|--------|--------|--------|--------|
| 39,416 | 29,307 | 26,051 | 27,368 | 35,936 | 29,289 |
| 28,281 | 34,511 | 32,305 | 37,904 | 44,448 | 31,883 |
| 37,021 | 34,500 | 39,884 | 23,221 | 41,043 | 32,923 |
| 30,747 | 41,620 | 17,595 | 38,041 | 30,739 | 40,614 |
| 35,439 | 30,283 | 44,224 | 35,295 | 38,995 | 42,398 |

*Source:* cars.com

With the first class having a lower class limit of 15,000 and a class width of 4,000,
(a) Construct a frequency distribution.
(b) Construct a relative frequency distribution.
(c) Construct a frequency histogram of the data.
(d) Construct a relative frequency histogram of the data.
(e) Describe the shape of the distribution.
(f) Repeat parts (a)–(e), using a class width of 6000. Which class width seems to provide a better summary of the data?

**21.** **Serum HDL** Dr. Paul Oswiecmiski randomly selects 40 of his 20–29-year-old patients and obtains the following data regarding their serum HDL cholesterol:

| 70 | 56 | 48 | 48 | 53 | 52 | 66 | 48 |
|----|----|----|----|----|----|----|----|
| 36 | 49 | 28 | 35 | 58 | 62 | 45 | 60 |
| 38 | 73 | 45 | 51 | 56 | 51 | 46 | 39 |
| 56 | 32 | 44 | 60 | 51 | 44 | 63 | 50 |
| 46 | 69 | 53 | 70 | 33 | 54 | 55 | 52 |

With the first class having a lower class limit of 20 and a class width of 10,
(a) Construct a frequency distribution.
(b) Construct a relative frequency distribution.
(c) Construct a frequency histogram of the data.
(d) Construct a relative frequency histogram of the data.
(e) Describe the shape of the distribution.
(f) Repeat parts (a)–(e), using a class width of 5. Which frequency distribution seems to provide a better summary of the data?

**22.** **Volume of Philip Morris Stock** The daily volume of a stock is the number of shares of the stock traded in a day. The following data represent a random sample of volume (in millions of shares) of Philip Morris stock traded for 35 days in 2001:

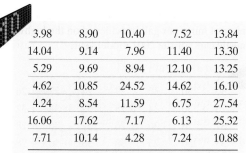

| 3.98 | 8.90 | 10.40 | 7.52 | 13.84 |
| 14.04 | 9.14 | 7.96 | 11.40 | 13.30 |
| 5.29 | 9.69 | 8.94 | 12.10 | 13.25 |
| 4.62 | 10.85 | 24.52 | 14.62 | 16.10 |
| 4.24 | 8.54 | 11.59 | 6.75 | 27.54 |
| 16.06 | 17.62 | 7.17 | 6.13 | 25.32 |
| 7.71 | 10.14 | 4.28 | 7.24 | 10.88 |

*Source:* http://finance.yahoo.com

With the first class having a lower class limit of 3 and a class width of 4,

(a) Construct a frequency distribution.
(b) Construct a relative frequency distribution.
(c) Construct a frequency histogram of the data.
(d) Construct a relative frequency histogram of the data.
(e) Describe the shape of the distribution.
(f) Repeat parts (a)–(e), using a class width of 6. Which frequency distribution seems to provide a better summary of the data?

**23. M&Ms** The following data represent the weights (in grams) of a simple random sample of 50 M&M plain candies.

| 0.87 | 0.88 | 0.91 | 0.92 | 0.86 |
| 0.91 | 0.90 | 0.94 | 0.82 | 0.89 |
| 0.89 | 0.88 | 0.91 | 0.86 | 0.84 |
| 0.83 | 0.88 | 0.88 | 0.86 | 0.85 |
| 0.91 | 0.94 | 0.88 | 0.87 | 0.90 |
| 0.86 | 0.93 | 0.91 | 0.84 | 0.83 |
| 0.87 | 0.79 | 0.87 | 0.88 | 0.90 |
| 0.93 | 0.90 | 0.82 | 0.88 | 0.86 |
| 0.89 | 0.86 | 0.76 | 0.85 | 0.84 |
| 0.79 | 0.93 | 0.84 | 0.93 | 0.87 |

With the first class having a lower class limit of 0.76 and a class width of 0.02,

(a) Construct a frequency distribution.
(b) Construct a relative frequency distribution.
(c) Construct a frequency histogram of the data.
(d) Construct a relative frequency histogram of the data.
(e) Describe the shape of the distribution.
(f) Repeat parts (a)–(e), using a class width of 0.04. Which frequency distribution seems to provide a better summary of the data?

**24. Old Faithful** The following data represent the length of eruption (in seconds) for a random sample of eruptions of "Old Faithful," a geyser in California.

| 108 | 113 | 102 | 106 | 107 |
| 102 | 100 | 100 | 109 | 105 |
| 103 | 108 | 116 | 105 | 90 |
| 110 | 99 | 101 | 103 | 111 |
| 110 | 103 | 120 | 110 | 101 |
| 102 | 104 | 103 | 101 | 108 |
| 105 | 101 | 92 | 104 | 94 |
| 110 | 106 | 103 | 90 | 110 |
| 104 | 109 | 99 | 95 | |

*Source:* Ladonna Hansen, park curator

With the first class having a lower class limit of 90 and a class width of 5,

(a) Construct a frequency distribution.
(b) Construct a relative frequency distribution.
(c) Construct a frequency histogram of the data.
(d) Construct a relative frequency histogram of the data.
(e) Describe the shape of the distribution.
(f) Repeat parts (a)–(e), using a class width of 10. Which frequency distribution seems to provide a better summary of the data?

*In Problems 25–28, construct stem-and-leaf diagrams.*

**25. Age at Inauguration** The following data represent the age of United States presidents on inauguration day.

| 57 | 61 | 57 | 57 | 58 |
| 54 | 68 | 51 | 49 | 64 |
| 65 | 52 | 56 | 46 | 54 |
| 47 | 55 | 55 | 54 | 42 |
| 55 | 51 | 54 | 51 | 60 |
| 55 | 56 | 61 | 52 | 69 |
| 57 | 50 | 49 | 51 | 62 |
| 61 | 48 | 50 | 56 | 43 |
| 64 | 46 | 54 | | |

**26. Divorce Rate** The following data represent the percentage of divorces (as a percentage of marriages in 1996) for selected countries. (*Note:* The U.S. figure is 49%. The highest is Belarus, at 68%, and the lowest is Macedonia, at 5%.)

| 68 | 65 | 64 | 63 | 63 |
| 56 | 55 | 53 | 52 | 49 |
| 43 | 43 | 41 | 41 | 40 |
| 38 | 35 | 34 | 28 | 26 |
| 24 | 21 | 19 | 18 | 18 |
| 15 | 15 | 13 | 13 | 12 |
| 61 | 46 | 39 | 26 | 18 |
| 56 | 45 | 39 | 25 | 17 |
| 7 | 6 | 5 | 12 | 12 |

*Source: Time Almanac,* 2000

Figure 18

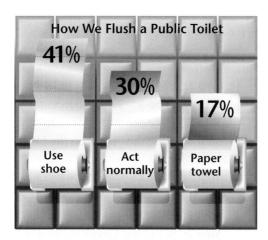

**How We Flush a Public Toilet**

41%

30%

17%

Use shoe

Act normally

Paper towel

*Approach:* We need to compare the vertical scales of each bar to see if they accurately depict the percentages given.

*Solution:* First, it is unclear whether the bars include the roll of toilet paper or not. In either case, the roll corresponding to "use shoe" should be 2.4(=41/17) times longer than the roll corresponding to "paper towel." If we include the roll of toilet paper, then the bar corresponding to "use shoe" is less than double the length of "paper towel." If we do not include the roll of toilet paper, then the bar corresponding to "use shoe" is almost exactly double the length of the bar corresponding to "paper towel." The vertical scaling is incorrect. ◀◀

▶ EXAMPLE 2    Misrepresentation of Data by Manipulating the Vertical Scale

*Problem:* The bar graph shown in Figure 19 depicts the average SAT math scores of college-bound seniors for the years 1989–1999, based on data from the College Board. Determine why this graph might be considered misrepresentative.

Figure 19

**Average SAT Math Score Over Time**

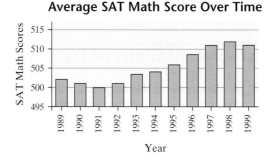

*Approach:* We need to look at the graph for any characteristics that may mislead a reader, such as manipulation of the vertical scale.

*Solution:* The graph in the figure may lead a reader to believe that SAT math scores have increased substantially between 1989 and 1999. While SAT math scores have increased between 1989 and 1999, they have not doubled or tripled, as may be inferred from the graph (since the bar for 1997 is three times as high as the bar for 1991). We notice in the figure that the vertical axis begins its labeling at 495 instead of 0. This type of scaling is common when the smallest observed data value is a rather large number. It is not necessarily done purposely to confuse or mislead the reader. Often, the main purpose in graphs is to discover a trend rather than the actual differences in the data. The trend is clearer in Figure 19 than in Figure 20, where the vertical axis begins at 0.

Often, instead of beginning the axis of a graph at 0 as in Figure 20, the graph is begun at a value slightly less than the smallest value in the data set. However, special care must be taken to make the reader aware of the vertical-axis scaling. Figure 21 shows the proper construction of the graph of the SAT scores, with the graph beginning at 495. The symbol ⌇ is used to signify that the graph has a "gap" in it.

Figure 20

**Average SAT Math Score Over Time**

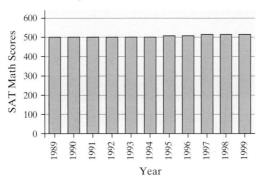

Figure 21

**Average SAT Math Score Over Time**

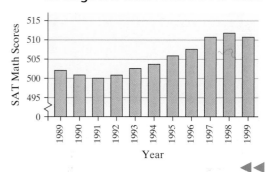

 *Now Work Problem 5.*

In addition to vertical-axis scaling, figures can be misleading through pictures. Consider the next example.

▶ **EXAMPLE 3**   **Misleading Graphs**

Figure 22

February 8, 1996          May 3, 1999

*Problem:* The Dow Jones Industrial Average (DJIA) is a collection of 30 stocks from the stock market that are thought to be representative of the United States economy. It includes companies such as Intel, Wal-Mart, and General Motors. Had you invested $10,000 in the DJIA on February 8, 1996, it would have been worth $20,000 May 3, 1999. To illustrate this investment, a brokerage firm might create the graphic shown in Figure 22. Describe how this graph is misleading.

*Approach:* Again, we look for characteristics of the graph that seem to manipulate the facts, such as an incorrect depiction of the size of the graphics.

*Solution:* The graphic on the right of the figure has been doubled in length, width, and height, causing an eight-fold increase in the size, thereby misleading the reader into thinking that the size of the investment increased by eight times instead of by two times.    ◀◀

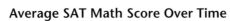 *Now Work Problem 13.*

There are many ways to create graphs that mislead. Two popular texts written about ways that graphs mislead or deceive are *How to Lie with Statistics* (W. W. Norton & Company, Inc., 1982), by Darrell Huff, and *The Visual Display of Quantitative Information* (Graphics Press, 2001), by Edward Tufte.

We conclude this section with some guidelines for constructing good graphics.

## Characteristics of Good Graphics

- Label the graphic clearly and provide explanations if needed.
- Avoid distortion. Don't lie about the data.
- Avoid three dimensions. Three-dimensional charts may look nice, but they distract the reader and often result in misinterpretation of the graphic.
- Do not use more than one design in the same graphic. Sometimes graphs use a different design in a portion of the graphic in order to draw attention to this area. Don't use this technique. Let the numbers speak for themselves.

## 2.3 Assess Your Understanding

### Exercises

**1.** The following graphic was taken from a *USA Today* "Snapshot." Explain how it is misleading.

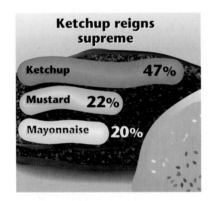

**2.** The following graphic was taken from the *Chicago Tribune* on Sunday, August 12, 2001. What is wrong with the pie chart on the right?

**Consumers Who Expect Within Six Months That...**

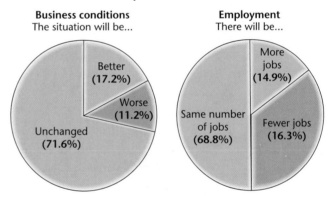

**3.** The following graphic was taken from http://www.unilever.com. Explain how it is misleading.

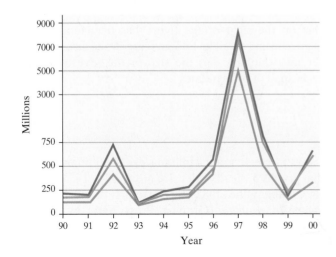

**4.** **Health Insurance** The following relative frequency histogram represents the percentage of people aged 25–64 years old covered by health insurance:

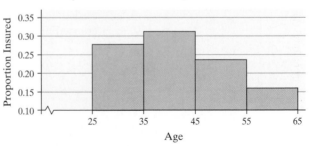

(a) Describe how this graph is misleading. What might a reader conclude from the graph?
(b) Redraw the histogram in two ways so that it is not misleading.

**5. Persons with Work Disability** The following frequency histogram represents the number of persons with work disability in 1998, based upon data from the U.S. Census Bureau.

**Number of People with Work Disability**

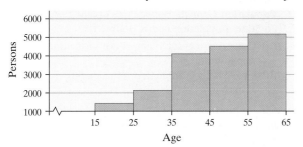

(a) Describe how this graph is misleading. What might a reader conclude from the graph?
(b) Redraw the histogram in two ways so that it is not misleading.

**6. Movie Production** The following time series plot shows the average production costs for making a movie in the years 1985–1997, based on data obtained from the Motion Picture Association of America:

**Average Cost for Making a Movie**

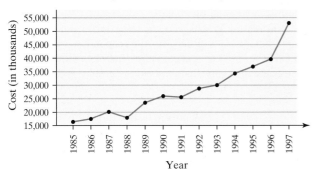

(a) Describe how this graph is misleading.
(b) What impression is the graph trying to convey?

**7. Health Care** The following time series plot displays health-care expenditures per person in the United States.

**Health-Care Expenditures per Person**

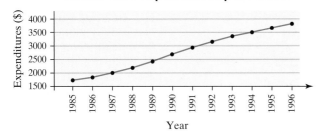

(a) Describe how this graph is misleading.
(b) What impression is the graph trying to convey?

**8. National Debt** The following graphic is from the *USA Today* "Snapshot."

How many times larger should the graphic for 1999 be than the 1900 graphic?

**9. Cost of Kids** The following graphic is from a *USA Today* "Snapshot" based on data from the Agriculture Department. It represents the percentage of income a middle-income family will spend on their children.

(a) How is the graphic misleading?
(b) What could be done to improve the graphic?

**10. Health-Care Expenditures** The following data represent health-care expenditures as a percentage of gross domestic product (GDP) from 1989–1996. Gross domestic product is the total value of all goods and services created during the course of the year.

| Year | Health Care as a Percent of GDP |
|------|-------------------------------|
| 1989 | 11.4 |
| 1990 | 12.2 |
| 1991 | 13.0 |
| 1992 | 13.4 |
| 1993 | 13.6 |
| 1994 | 13.6 |
| 1995 | 13.6 |
| 1996 | 13.6 |

*Source:* U.S. Health Care Financing Administration

(a) Construct a time series plot that a politician would create to support the position that health-care expenditures, as a percentage of GDP, are increasing and must be slowed.
(b) Construct a time series plot that the health-care industry would create to refute the opinion of the politician.
(c) Construct a time series plot that is not misleading.

**11. Landfill Utilization** The following data represent the number of pounds of solid waste generated per person per day that went into landfills in the United States from 1990–1997.

| Year | Waste Generated |
|------|-----------------|
| 1990 | 3.1 |
| 1991 | 2.9 |
| 1992 | 2.9 |
| 1993 | 2.9 |
| 1994 | 2.8 |
| 1995 | 2.5 |
| 1996 | 2.4 |
| 1997 | 2.4 |

*Source:* U.S. Environmental Protection Agency

(a) Construct a time series plot that supports the opinion that the amount of solid waste per person that goes into landfills is declining substantially.

(b) Construct a time series plot that supports the opinion that the amount of solid waste per person that goes into landfills is declining, but not substantially.

(c) Construct a time series plot that is not misleading.

**12. Population Growth** Between 1940 and 1998, the United States population nearly doubled.

(a) Construct a graphic that is not misleading to depict this situation.

(b) Construct a graphic that is misleading to depict this situation.

**13. Chlorofluorocarbons** The amount of chlorofluorocarbons (CFCs) emissions in 1990 was 193 metric tons, or approximately three times as much as the amount in 1997.

(a) Construct a graphic that is not misleading to depict this situation.

(b) Construct a graphic that is misleading to depict this situation.

## CHAPTER 2 REVIEW

### Summary

Raw data are first organized into tables. Data are organized by creating classes into which they fall. Qualitative data and discrete data have values that provide clear-cut categories of data. However, with continuous data, the categories, called classes, must be created. Typically, the first table created is a frequency distribution, which lists the frequency with which each class of data occurs. Another type of distribution is the relative frequency distribution.

Once data are organized into a table, graphs are created. For data that are qualitative, we can create bar charts and pie charts. For data that are quantitative, we can create histograms or stem-and-leaf plots.

In creating graphs, care must be taken not to draw a graph that misleads or deceives the reader. If a graph's vertical axis does not begin at zero, the symbol ⚡ should be used to indicate the "gap" that exists in the graph.

### Vocabulary

Raw data (p. 39)
Frequency distribution (p. 40)
Relative frequency distribution (p. 41)
Bar graph (p. 43)
Pareto chart (p. 44)
Side-by-side bar graph (p. 45)

Pie chart (p. 46)
Histogram (p. 56)
Class (p. 56)
Lower and upper class limit (pp. 56–57)
Class width (p. 57)
Stem-and-leaf plot (p. 60)

Uniform distribution (p. 63)
Bell-shaped distribution (p. 63)
Skewed right (p. 63)
Skewed left (p. 63)
Time series plot (p. 64)

### Objectives

| Section | You should be able to ... | Review Exercises |
|---------|---------------------------|------------------|
| **2.1** | 1 Construct frequency and relative frequency distributions from qualitative data. (p. 40) | 3(a), 4(a), 7(a),(b), 8(a),(b) |
| | 2 Construct bar graphs (p. 43) | 3(c) and (d), 4(c) and (d), 7(c), 8(c) |
| | 3 Construct pie charts (p. 46) | 3(e), 4(e), 7(d), 8(d) |
| **2.2** | 1 Summarize discrete data in tables (p. 55) | 9(a) and (b), 10(a) and (b) |
| | 2 Construct histograms of discrete data (p. 56) | 9(c) and (d), 10(c) and (d) |
| | 3 Summarize continuous data in tables (p. 56) | 11(a) and (b); 12(a) and (b), 13(a) and (b); 14(a) and (b) |
| | 4 Construct histograms of continuous data (p. 59) | 5(b) and (c); 6(b) and (c); 5(b), 6(b), 11(c) and (d); 12(c) and (d); 13(c) and (d); 14(c) and (d) |
| | 5 Draw stem-and-leaf plots (p. 60) | 15 and 16 |
| | 6 Identify the shape of a distribution (p. 62) | 5(b), 6(b), 11(c), 12(c), 13(c), 14(c); 15, 16 |
| | 7 Draw time series graphs (p. 63) | 17, 18 |
| **2.3** | 1 Understand the origins of misleading or deceptive graphs (p. 71) | 19, 20, 21 |

## Review Exercises

**1. Energy Consumption** The following bar chart represents the energy consumption of the United States (in quadrillion Btus) in 1999.

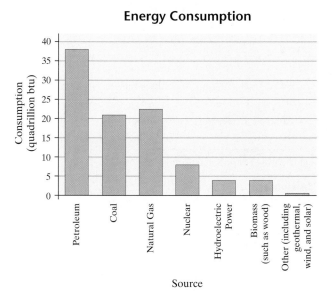

**Energy Consumption**

(a) Approximately how much energy did the United States consume through natural gas?
(b) Approximately how much energy did the United States consume through biomass?
(c) Approximate the total energy consumption of the United States in 1999.

**2. Taxi Time** The following histogram represents the taxi time (in minutes) on November 21, 2001, for Southwest Airlines at Nashville International Airport.

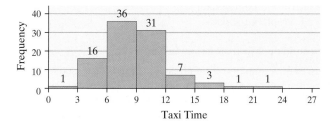

**Taxi Time Nashville Airport Southwest Airlines**

(a) How many Southwest flights departed from Nashville on November 21, 2001?
(b) Determine the class width.
(c) Identify the classes and their frequency.
(d) Which class has the highest frequency?
(e) What is the relative frequency of taxi time between 3 and 5.9 minutes?
(f) Describe the shape of the distribution.

**3. Weapons Used in Homicide** The following frequency distribution represents the cause of death in homicides for the year 1998.

| Cause of Death | Frequency |
|---|---|
| Gun | 9222 |
| Cutting/stabbing | 1890 |
| Blunt object | 753 |
| Personal weapons (hands, fists, etc.) | 952 |
| Strangulation | 313 |
| Other | 1080 |

*Source:* U.S. Federal Bureau of Investigation

(a) Construct a relative frequency distribution.
(b) What percentage of homicides were committed using a blunt object?
(c) Construct a frequency bar graph.
(d) Construct a relative frequency bar graph.
(e) Construct a pie chart.

**4. U.S. Greenhouse Gas Emissions** The following frequency distribution represents the total emissions in millions of metric tons of carbon equivalent in 1997 in the United States.

| Gas | Emissions |
|---|---|
| Carbon dioxide | 1487.9 |
| Methane | 179.6 |
| Nitrous oxide | 109.0 |
| Hydrofluorocarbons, perfluorocarbons, and sulfur hexafluoride | 37.1 |

*Source: Time Almanac,* 2000

(a) Construct a relative frequency distribution.
(b) What percent of emissions is due to carbon dioxide?
(c) Construct a frequency bar graph.
(d) Construct a relative frequency bar graph.
(e) Construct a pie chart.

**5. Vehicle Fatalities** The following frequency distribution represents the number of drivers in fatal crashes in 1999, by age, for males aged 20 to 84 years.

| Age | Number of Drivers | Age | Number of Drivers |
|---|---|---|---|
| 20–24 | 8067 | 55–59 | 2332 |
| 25–29 | 6195 | 60–64 | 1606 |
| 30–34 | 6274 | 65–69 | 1341 |
| 35–39 | 5130 | 70–74 | 1273 |
| 40–44 | 4631 | 75–79 | 1179 |
| 45–49 | 3636 | 80–84 | 902 |
| 50–54 | 3021 | | |

*Source:* National Highway Traffic Safety Administration

(a) Construct a relative frequency distribution.
(b) Construct a frequency histogram. Describe the shape of the distribution.
(c) Construct a relative frequency histogram.

(d) What percentage of males aged 20 to 84 involved in fatal crashes were 20–24?
(e) What percentage of males aged 20 to 84 involved in fatal crashes were less than 29 years old?

6. **Vehicle Fatalities** The following frequency distribution represents the number of drivers in fatal crashes in 1999, by age, for females aged 20 to 84 years old.

| Age | Number of Drivers | Age | Number of Drivers |
|---|---|---|---|
| 20–24 | 3190 | 55–59 | 1236 |
| 25–29 | 2446 | 60–64 | 956 |
| 30–34 | 2366 | 65–69 | 1015 |
| 35–39 | 2396 | 70–74 | 1050 |
| 40–44 | 2204 | 75–79 | 978 |
| 45–49 | 1813 | 80–84 | 779 |
| 50–54 | 1480 | | |

*Source:* National Highway Traffic Safety Administration

(a) Construct a relative frequency distribution.
(b) Construct a frequency histogram. Describe the shape of the distribution.
(c) Construct a relative frequency histogram.
(d) What percentage of females aged 20 to 84 involved in fatal crashes were 20–24?
(e) What percentage of females aged 20 to 84 involved in fatal crashes were less than 29 years old?

7. **Political Affiliation** A survey of 100 randomly selected registered voters in the city of Naperville were asked their political affiliation—Democrat (D), Republican (R), or Independent (I). The results of the survey are as follows:

| | | | | | | | | | |
|---|---|---|---|---|---|---|---|---|---|
| D | R | D | R | D | R | D | D | R | D |
| R | D | D | D | R | R | D | D | D | D |
| R | R | I | I | D | R | D | R | R | R |
| I | D | D | R | I | I | R | D | R | R |
| D | I | R | D | D | D | D | I | I | R |
| R | I | R | R | I | D | D | D | D | R |
| D | I | I | D | D | R | R | R | R | D |
| D | R | R | R | D | D | I | I | D | D |
| D | D | I | D | R | I | D | D | D | D |
| R | R | R | R | R | D | R | D | R | D |

(a) Construct a frequency distribution of the data.
(b) Construct a relative frequency distribution of the data.
(c) Construct a relative frequency bar graph of the data.
(d) Construct a pie chart of the data.
(e) What appears to be the most common political affiliation in Naperville?

8. **Educational Attainment** The Metra Train Company was interested in knowing the educational background of its customers. The company contracted a marketing firm to conduct a survey with a random sample of 50 commuters at the train station. In the survey, commuters were asked to disclose their educational attainment. The following results were obtained:

| | | | | |
|---|---|---|---|---|
| No high school degree | Some college | Advanced degree | High school graduate | Advanced degree |
| High school graduate | High school graduate | High school graduate | High school graduate | No high school degree |
| Some college | High school graduate | Bachelor's degree | Associate's degree | High school graduate |
| No high school degree | Bachelor's degree | Some college | High school graduate | No high school degree |
| Associate's degree | High school graduate | High school graduate | No high school degree | Some college |
| Bachelor's degree | Bachelor's degree | Some college | High school graduate | Some college |
| Bachelor's degree | Advanced degree | No high school degree | Advanced degree | No high school degree |
| High school graduate | Bachelor's degree | No high school degree | High school graduate | No high school degree |
| Associate's degree | Bachelor's degree | High school graduate | Bachelor's degree | Some college |
| Some college | Associate's degree | High school graduate | Some college | High school graduate |

(a) Construct a frequency distribution of the data.
(b) Construct a relative frequency distribution of the data.
(c) Construct a relative frequency bar graph of the data.
(d) Construct a pie chart of the data.
(e) What is the most common educational level of a commuter?

---

**9. Family Size** A random sample of 60 couples married for seven years were asked the number of children they have. The results of the survey are as follows:

| 0 | 0 | 3 | 1 | 2 | 3 |
|---|---|---|---|---|---|
| 3 | 4 | 3 | 3 | 0 | 3 |
| 1 | 2 | 1 | 3 | 0 | 3 |
| 4 | 2 | 3 | 2 | 2 | 4 |
| 2 | 1 | 3 | 4 | 1 | 3 |
| 0 | 3 | 3 | 3 | 2 | 1 |
| 2 | 0 | 3 | 1 | 2 | 3 |
| 4 | 3 | 3 | 5 | 2 | 0 |
| 4 | 2 | 2 | 2 | 3 | 3 |
| 2 | 4 | 2 | 2 | 2 | 2 |

(a) Construct a frequency distribution of the data.
(b) Construct a relative frequency distribution of the data.
(c) Construct a frequency histogram of the data. Describe the shape of the distribution.
(d) Construct a relative frequency histogram of the data.
(e) What percentage of couples married seven years have two children?
(f) What percentage of couples married seven years have at least two children?

**10. Waiting in Line** The following data represent the number of cars that arrive at McDonald's drive-through between 11:50 A.M. and 12:00 noon each Wednesday for the past 50 weeks:

| 1 | 7 | 3 | 8 | 2 | 3 | 8 | 2 | 6 | 3 |
|---|---|---|---|---|---|---|---|---|---|
| 6 | 5 | 6 | 4 | 3 | 4 | 3 | 8 | 1 | 2 |
| 5 | 3 | 6 | 3 | 3 | 4 | 3 | 2 | 1 | 2 |
| 4 | 4 | 9 | 3 | 5 | 2 | 3 | 5 | 5 | 5 |
| 2 | 5 | 6 | 1 | 7 | 1 | 5 | 3 | 8 | 4 |

(a) Construct a frequency distribution of the data.
(b) Construct a relative frequency distribution of the data.
(c) Construct a frequency histogram of the data. Describe the shape of the distribution.
(d) Construct a relative frequency histogram of the data.
(e) What percentage of the time did exactly three cars arrive between 11:50 A.M. and 12:00 noon?
(f) What percentage of the time did three or more cars arrive between 11:50 A.M. and 12:00 noon?

---

**11. Crime Rates by State** The following data represent the crime rate (crimes per 100,000 population) for each state in 1998.

| State | Crime Rate | State | Crime Rate | State | Crime Rate |
|---|---|---|---|---|---|
| Alabama | 4597 | Kentucky | 4859 | North Dakota | 2681 |
| Alaska | 4777 | Louisiana | 2889 | Ohio | 4328 |
| Arizona | 6575 | Maine | 6098 | Oklahoma | 5004 |
| Arkansas | 4283 | Maryland | 3041 | Oregon | 5647 |
| California | 4343 | Massachusetts | 5366 | Pennsylvania | 3273 |
| Colorado | 4488 | Michigan | 3436 | Rhode Island | 3518 |
| Connecticut | 3787 | Minnesota | 4683 | South Carolina | 5777 |
| Delaware | 5363 | Mississippi | 4047 | South Dakota | 2624 |
| D.C. | 8836 | Missouri | 4384 | Tennessee | 5034 |
| Florida | 6886 | Montana | 4826 | Texas | 5112 |
| Georgia | 5463 | Nebraska | 4071 | Utah | 5506 |
| Hawaii | 5333 | Nevada | 4405 | Vermont | 3139 |
| Idaho | 3715 | New Hampshire | 5281 | Virginia | 3660 |
| Illinois | 4873 | New Jersey | 2420 | Washington | 5867 |
| Indiana | 4169 | New Mexico | 3654 | West Virginia | 2547 |
| Iowa | 3501 | New York | 6719 | Wisconsin | 3543 |
| Kansas | 4859 | North Carolina | 3589 | Wyoming | 3808 |

*Source:* U.S. Census Bureau

In (a)–(d), start the first class at a lower class limit of 2000, and maintain a class width of 500:
(a) Construct a frequency distribution.
(b) Construct a relative frequency distribution.
(c) Construct a frequency histogram. Describe the shape of the distribution.
(d) Construct a relative frequency histogram.
(e) Repeat (a)–(d), using a class width of 1000. In your opinion, which class width provides the better summary of the data? Why?

12. **Disposable Income by State** The disposable personal income of individuals is their income after paying taxes. The following data represent the average disposable personal income (in dollars) for each state in the United States in 1999.

In (a)–(d), start the first class at a lower class limit of 18,000, and maintain a class width of 1000:
(a) Construct a frequency distribution.
(b) Construct a relative frequency distribution.
(c) Construct a frequency histogram. Describe the shape of the distribution.
(d) Construct a relative frequency histogram.
(e) Repeat (a)–(d), using a class width of 2000. In your opinion, which class width provides the better summary of the data? Why?

| State | Average Income | State | Average Income | State | Average Income |
|-------|----------------|-------|----------------|-------|----------------|
| Alabama | 20,068 | Kentucky | 19,930 | North Dakota | 20,842 |
| Alaska | 24,978 | Louisiana | 20,016 | Ohio | 23,018 |
| Arizona | 21,855 | Maine | 21,530 | Oklahoma | 19,800 |
| Arkansas | 19,412 | Maryland | 26,686 | Oregon | 22,964 |
| California | 25,100 | Massachusetts | 29,589 | Pennsylvania | 24,498 |
| Colorado | 26,801 | Michigan | 23,684 | Rhode Island | 25,686 |
| Connecticut | 31,797 | Minnesota | 26,003 | South Carolina | 20,491 |
| Delaware | 25,714 | Mississippi | 18,241 | South Dakota | 22,443 |
| D.C. | 31,457 | Missouri | 22,469 | Tennessee | 22,626 |
| Florida | 24,201 | Montana | 19,590 | Texas | 23,223 |
| Georgia | 23,225 | Nebraska | 23,805 | Utah | 20,013 |
| Hawaii | 24,305 | Nevada | 26,205 | Vermont | 22,308 |
| Idaho | 20,419 | New Hampshire | 26,732 | Virginia | 25,010 |
| Illinois | 26,519 | New Jersey | 30,251 | Washington | 26,203 |
| Indiana | 22,223 | New Mexico | 19,396 | West Virginia | 18,337 |
| Iowa | 22,252 | New York | 28,072 | Wisconsin | 22,213 |
| Kansas | 22,880 | North Carolina | 22,424 | Wyoming | 22,244 |

*Source:* U.S. Bureau of Economic Analysis

13. **Towing Capacity** The following data represent the towing capacity (in pounds) for 2001 sport utility vehicles.

| SUV | Towing Capacity | SUV | Towing Capacity | SUV | Towing Capacity |
|-----|-----------------|-----|-----------------|-----|-----------------|
| Acura MDX | 4500 | GMC Yukon | 8700 | Land Rover Range Rover | 7700 |
| BMW X5 | 6000 | GMC Yukon XL | 12000 | Lincoln Navigator | 8800 |
| Buick Rendezvous | 3500 | Honda Passport | 4500 | Mitsubishi Montero | 5000 |
| Chevrolet Blazer | 5600 | Hummer | 8300 | Nissan Pathfinder | 5000 |
| Chevrolet Suburban | 12000 | Infiniti QX4 | 5000 | Pontiac Aztek | 3500 |
| Chevrolet Tahoe | 8700 | Isuzu Axiom | 4500 | Suzuki XL-7 | 3000 |
| Dodge Durango | 7650 | Isuzu Rodeo | 4500 | Toyota 4Runner | 5000 |
| Ford Escape | 3500 | Jeep Cherokee | 5000 | Toyota Highlander | 3500 |
| Ford Expedition | 8100 | Jeep Grand Cherokee | 6500 | Toyota Land Cruiser | 6500 |
| Ford Excursion | 10000 | Jeep Liberty | 5000 | | |
| GMC Jimmy | 5900 | Land Rover Discovery | 7700 | | |

*Source:* Manufacturers

Start the first class at a lower class limit of 3000, and maintain a class width of 1000:
(a) Construct a frequency distribution.
(b) Construct a relative frequency distribution.
(c) Construct a frequency histogram. Describe the shape of the distribution.
(d) Construct a relative frequency histogram.

14. **Home Sales** The data to the right represent the closing price (in U.S. dollars) of homes sold in Joliet, Illinois, in December, 1999.

Start the first class at a lower class limit of 85,000, and maintain a class width of 10,000
(a) Construct a frequency distribution.
(b) Construct a relative frequency distribution.
(c) Construct a frequency histogram. Describe the shape of the distribution.
(d) Construct a relative frequency histogram.

| | | | | |
|---|---|---|---|---|
| 138,820 | 149,143 | 99,000 | 115,000 | 157,216 |
| 169,541 | 140,794 | 136,924 | 124,757 | 149,380 |
| 135,512 | 153,146 | 136,833 | 128,429 | 136,529 |
| 147,500 | 120,936 | 95,491 | 115,744 | 119,900 |
| 89,900 | 102,696 | 149,634 | 123,103 | 126,630 |
| 140,269 | 183,000 | 133,646 | 121,225 | 121,524 |
| 146,439 | 182,000 | 110,128 | 109,520 | 104,640 |
| 124,760 | 134,305 | 111,220 | 121,795 | 170,072 |
| 136,550 | 115,595 | 155,507 | 152,600 | 130,000 |
| 152,537 | 163,165 | | | |

*Source:* Transamerica Intellitech

15. **Air-traffic Control Errors** An air-traffic control error is said to occur when planes come too close to one another. The following data represent the number of air-traffic control errors in various regions around the United States for fiscal year 2000. Construct a stem-and-leaf diagram and comment on the shape of the distribution.

| Center | Errors, 2000 | Center | Errors, 2000 | Center | Errors, 2000 |
|---|---|---|---|---|---|
| Washington | 102 | Oakland | 17 | Kansas City, MO | 28 |
| New York ARTCC | 71 | Chicago TRACON | 14 | Albuquerque | 25 |
| Indianapolis | 54 | Oakland Bay | 8 | Boston | 21 |
| Memphis | 38 | Washington Dulles | 7 | Houston | 18 |
| Los Angeles | 33 | Cleveland | 74 | Minneapolis | 15 |
| Jacksonville, FL | 30 | Chicago ARTCC | 70 | Salt Lake City | 12 |
| New York TRACON | 27 | Atlanta | 40 | Miami | 8 |
| Miami | 21 | Dallas-Ft. Worth | 34 | | |
| Northern California | 18 | Denver | 33 | | |

*Source:* Associated Press

16. **Eat Your Vegetables!** The following data represent the number of servings of vegetables per day that a random sample of forty 20–39-year-old females consumes:

| | | | | | | | |
|---|---|---|---|---|---|---|---|
| 1.7 | 2.7 | 0.3 | 3.5 | 0.7 | 1.4 | 5.1 | 3.9 |
| 0.2 | 2.1 | 4.1 | 5.8 | 3.8 | 0.4 | 6.1 | 0.7 |
| 2.4 | 11.1 | 3.5 | 6.7 | 2.3 | 4.9 | 5.9 | 0.4 |
| 3.3 | 0.8 | 7.6 | 10.2 | 5.8 | 2.6 | 0.6 | 3.2 |
| 0.5 | 2.4 | 4.9 | 2.3 | 8.3 | 6.0 | 5.3 | 3.5 |

The data are based on a survey conducted by the United States Department of Agriculture. Construct a stem-and-leaf diagram of the data and comment on the shape of the distribution.

17. **Federal Minimum Wage Rates** The data to the right represent the value of the minimum wage for the years 1980–1997.

| Year | Minimum Wage | Year | Minimum Wage |
|---|---|---|---|
| 1980 | 3.10 | 1989 | 3.35 |
| 1981 | 3.35 | 1990 | 3.80 |
| 1982 | 3.35 | 1991 | 4.25 |
| 1983 | 3.35 | 1992 | 4.25 |
| 1984 | 3.35 | 1993 | 4.25 |
| 1985 | 3.35 | 1994 | 4.25 |
| 1986 | 3.35 | 1995 | 4.25 |
| 1987 | 3.35 | 1996 | 4.75 |
| 1988 | 3.35 | 1997 | 5.15 |

*Source:* U.S. Employment Standards Administration

(a) Construct a time series plot of the data.
(b) Comment on the apparent trend.

**18. Federal Minimum Wage** The following data represent the federal minimum wage for 1980–1997 in constant 1996 dollars. Constant dollars are dollars adjusted for inflation.

| Year | Minimum Wage | Year | Minimum Wage |
|------|--------------|------|--------------|
| 1980 | 5.90 | 1989 | 4.24 |
| 1981 | 5.78 | 1990 | 4.56 |
| 1982 | 5.45 | 1991 | 4.90 |
| 1983 | 5.28 | 1992 | 4.75 |
| 1984 | 5.06 | 1993 | 4.61 |
| 1985 | 4.88 | 1994 | 4.50 |
| 1986 | 4.80 | 1995 | 4.38 |
| 1987 | 4.63 | 1996 | 4.75 |
| 1988 | 4.44 | 1997 | 5.03 |

*Source:* U.S. Employment Standards Administration

(a) Construct a time series plot of the data.
(b) Comment on the apparent trend.
(c) Compare the time series plot in Problem 17 with that in this problem. Which graph is misleading? Why?

**19. Misleading Graphs** The following graph was found in a magazine advertisement for skin cream:

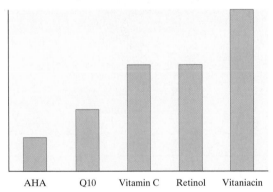

**Skin Health
(Moisture Retention)**

AHA    Q10    Vitamin C    Retinol    Vitaniacin

How is this graph misleading?

**20. Misleading Graphs** The following graphic is from a *USA Today* "Snapshot:"

Most popular ways to save for college

Savings account 37%
Mutual funds 31%
Bonds 21%
CDs 17%

Do you think the graph is misleading? Why? If you think it is misleading, what might be done to improve the graph?

**21. Misleading Graphs** In 1998, the average earnings of a high school graduate were $22,895. At $40,478, the average earnings of a recipient of a bachelor's degree were about 75% higher.

(a) Construct a graph that a college recruiter might create in order to convince high school students that they should attend college.
(b) Construct a graph that does not mislead.

**Chapter 2 Projects located at www.prenhall.com/sullivanstats**

# Numerically Summarizing Data

## Outline

**For additional study help, go to**
www.prenhall.com/sullivanstats

### Materials include

- Projects
  - Case Study: Who Was "A Mourner"?
  - Decisions: What Car Should I Buy?
  - Consumer Reports Project
- Self-Graded Quizzes
- "Preparing for This Section" Quizzes
- STATLETs
- PowerPoint Downloads
- Step-by-Step Technology Guide
- Graphing Calculator Help

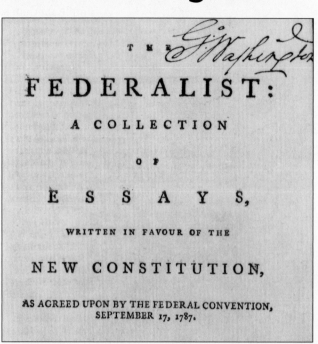

## Putting It All Together

The distribution of data refers to three characteristics of the data: its shape, its center, and its spread. In the last chapter, we discussed methods for organizing raw data into tables and graphs. These graphs (such as the histogram) allow us to identify the shape of the distribution. Recall that we describe the shape of a distribution as symmetric (in particular, bell-shaped or uniform), skewed right, or skewed left.

The center and spread are numerical summaries of the data. The center of a data set is commonly called the average. There are many ways to describe the "average" value of a distribution. In addition, there are many ways to measure the spread of a distribution. The most appropriate measure of center and spread to use depends upon the shape of the distribution.

Once these three characteristics of the distribution are known, we can analyze the data for interesting features, such as unusual data values, called *outliers*.

## 3.1    Measures of Central Tendency

**Preparing for This Section**    Before getting started, review the following:

   ✓ Quantitative data (Section 1.1, p. 6)

   ✓ Qualitative data (Section 1.1, p. 6)

   ✓ Population versus sample (Section 1.1, p. 2)

   ✓ Simple random sampling (Section 1.2, pp. 12–15)

**Objectives**  Determine the arithmetic mean of a variable from raw data

 Determine the median of a variable from raw data

 Determine the mode of a variable from raw data

 Use the mean and the median to help identify the shape of a distribution

We will first consider measures of central tendency. By this, we mean to numerically describe the average or typical data value. We hear the word "average" in the news all the time:

- The average miles per gallon of gasoline of the 2004 Chevrolet Camaro in city driving is 19 miles.
- According to the United States Census Bureau, the average commute time to work in 2000 was 24.3 minutes.
- According to the United States Census Bureau, the average household income in 2000 was $41,349.
- The average American woman is 5'4" tall and weighs 142 pounds.

**Caution**

Whenever you hear the word "average" used, be aware that there are many types of averages out there. One average could be used to support one position, while another average could be used to support a different position.

In this chapter, we discuss three averages: the *mean*, the *median*, and the *mode*. While other measures of central tendency exist, these three are the most widely used. When the word "average" is used in the media (newspapers, reporters, and so on) it usually refers to the mean. But beware! An average can be the mean, median or mode, and as we shall see, these three measures of central tendency can give very different results!

Before we discuss measures of central tendency, we must consider whether we are computing a measure of central tendency that describes a population or a sample.

*Definitions*

A **parameter** is a descriptive measure of a population.

A **statistic** is a descriptive measure of a sample.

**In Your Own Words**

To help you remember the difference between a parameter and a statistic, think of the following:
p = parameter = population;
s = statistic = sample

For example, if we determined the average test score for *all* the students in a statistics class, the average would be a parameter. If we computed the average based upon a random sample of five of the students, the average would be a statistic.

To obtain the value of a parameter we need a census. Recall that a census is a list of all the individuals in the population along with the values of the variables that we wish to study. Samples are used to compute statistics, which serve as estimates of the parameter.

### ① The Arithmetic Mean

When used in everyday language, the word "average" often stands for the arithmetic mean. To compute the arithmetic mean of a set of data, the data must be quantitative.

*Definitions*

The **arithmetic mean** of a variable is computed by determining the sum of all the values of the variable in the data set, divided by the number of observations. The **population arithmetic mean**, $\mu$ (read "mew"), is computed using all the individuals in a population. The population mean is a parameter. The **sample arithmetic mean**, $\bar{x}$ (read "x-bar"), is computed using sample data. The sample mean is a statistic.

While other types of means exist (see Problems 36 and 37), the arithmetic mean is generally referred to as the **mean**. We will follow this practice for the remainder of the text.

In statistics, Greek letters are used to represent parameters while Roman letters are used to represent statistics. Statisticians use mathematical expressions instead of words to describe the method for computing means.

*Definitions*

If $x_1, x_2, \ldots, x_N$ are the $N$ observations of a variable from a population, then the population arithmetic mean, $\mu$, is

$$\mu = \frac{x_1 + x_2 + \cdots + x_N}{N} = \frac{\sum x_i}{N} \tag{1}$$

If $x_1, x_2, \ldots, x_n$ are $n$ observations of a variable from a sample, then the sample arithmetic mean, $\bar{x}$, is

$$\bar{x} = \frac{x_1 + x_2 + \cdots + x_n}{n} = \frac{\sum x_i}{n} \tag{2}$$

**In Your Own Words**

To find the mean of a set of data, simply add up all the observations and divide by the number of observations.

Note that $N$ represents the size of the population while $n$ represents the size of the sample. The symbol $\Sigma$ (the Greek letter capital sigma) means the terms are to be added up. The subscript $i$ is used to make the various values distinct and does not serve as any mathematical operation. For example, $x_1$ is the first data value, $x_2$ is the second, and so on.

Let's look at an example to help distinguish the population mean and sample mean.

▶ **EXAMPLE 1**    Computing a Population Mean and Sample Mean

*Problem:* The data in Table 1 represent the number of home runs hit by all teams in the American League in 2001.

**(a)** Compute the population mean.
**(b)** Find a simple random sample of size $n = 5$.
**(c)** Compute the sample mean of the sample obtained in part (b).

| TABLE 1 | | | |
|---|---|---|---|
| **Team** | **Home Runs** | **Team** | **Home Runs** |
| 1. Anaheim Angels | 158 | 8. Minnesota Twins | 164 |
| 2. Baltimore Orioles | 136 | 9. New York Yankees | 203 |
| 3. Boston Red Sox | 198 | 10. Oakland Athletics | 199 |
| 4. Chicago White Sox | 214 | 11. Seattle Mariners | 169 |
| 5. Cleveland Indians | 212 | 12. Tampa Bay Devil Rays | 121 |
| 6. Detroit Tigers | 139 | 13. Texas Rangers | 246 |
| 7. Kansas City Royals | 152 | 14. Toronto Blue Jays | 195 |

*Source:* Major League Baseball

*Approach:*

**(a)** To compute the population mean, we add up all the data values and then divide by the number of individuals in the population.

**(b)** Recall from Section 1.2 that we can use either Table I in Appendix A, a calculator with a random number generator, or computer software to obtain simple random samples. We will use a TI-83 Plus graphing calculator.

**(c)** The sample mean is found by adding the data values that correspond to the individuals selected in the sample and then dividing by $n = 5$, the sample size.

*Solution:*

**(a)** We compute the population mean by adding the number of home runs hit by all 14 American League teams:

$$\Sigma x_i = 158 + 136 + 198 + \ldots + 195 = 2{,}506$$

We now take this result and divide by 14, the number of teams in the league:

$$\mu = \frac{\Sigma x_i}{N} = \frac{2{,}506}{14} = 179$$

**Figure 1**

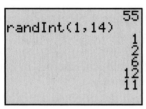

**(b)** To find a simple random sample of size $n = 5$ from a population whose size is $N = 14$, we will use the TI-83 plus random number generator with a seed of 55. (Recall that this is the starting point that the calculator uses to generate the list of "random" numbers.) Figure 1 shows the teams in the sample.

The Anaheim Angels (158 home runs), Baltimore Orioles (136 home runs), Detroit Tigers (139 home runs), Tampa Bay Devil Rays (121 home runs), and Seattle Mariners (169 home runs) are in our sample.

**(c)** We compute the sample mean of this sample by first adding the number of home runs hit by each of the teams in the sample:

$$\Sigma x_i = 158 + 136 + 139 + 121 + 169 = 723$$

We now take this result and divide by 5, the number of observations in the sample:

$$\bar{x} = \frac{\Sigma x_i}{n} = \frac{723}{5} = 144.6* \qquad \blacktriangleleft\blacktriangleleft$$

Notice from Example 1 that the mean does not necessarily have to equal the value of one of the actual observations. In addition, a sample mean will not necessarily equal the population mean.

**EXPLORATION** Using the data from Example 1, find two additional simple random samples of size $n = 5$ and compute the sample mean corresponding to these samples. Conclude that different samples lead to different sample means. $\qquad \blacktriangleleft\blacktriangleleft$

**NW** *Now Work Problem 11.*

It is helpful to think of the mean of a data set as the "center of gravity". In other words, the mean is the value such that a histogram of the data would be perfectly balanced with equal weight on each side of the mean. Figure 2 shows the histogram of the data in Table 1 with the mean labeled. The histogram balances at $\mu = 179$.

*When necessary, round to one more decimal place than that in the raw data.

Figure 2

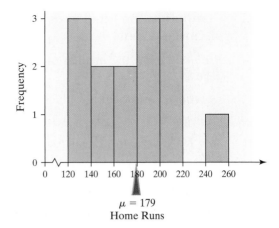

$\mu = 179$
Home Runs

 **EXPLORATION** Find a yardstick, a fulcrum, and three objects of equal weight (maybe obtain 1-kilogram weights from the physics department). Place the fulcrum at 18 inches so that the yardstick balances like a teeter-totter. Now place one weight on top of the yardstick at 12 inches, another at 15 inches and the third at 27 inches. See Figure 3.

Figure 3

Does the yardstick balance? Now compute the arithmetic mean of the location of the three weights. Compare this result with the location of the fulcrum. Conclude that the arithmetic mean is the center of gravity of the data set. ◄◄

## ② The Median

A second measure of central tendency is the median. To compute the median of a set of data, the data must be quantitative.

*Definition*

 **In Your Own Words**

To help remember the idea behind the median, think of the median of a highway; it divides the highway in half.

The **median** of a variable is the value that lies in the middle of the data when arranged in ascending order. That is, half the data are below the median and half the data are above the median. We use $M$ to represent the median.

To compute the median of a set of data, we use the following steps:

**Steps in Computing the Median of a Data Set**

*Step 1:* Arrange the data in ascending order.
*Step 2:* Determine the number of observations, $n$.
*Step 3:* Determine the observation in the middle of the data set.

- If the number of observations is odd, then the median is the data value that is exactly in the middle of the data set. That is, the median is the observation that lies in the $\frac{n+1}{2}$ position.

- If the number of observations is even, then the median is the arithmetic mean of the two middle observations in the data set. That is, the median is the arithmetic mean of the data values that lie in the $\frac{n}{2}$ and $\frac{n}{2} + 1$ positions.

► EXAMPLE 2 **Computing the Median of a Data Set with an Even Number of Observations**

*Problem:* Find the median of the population data listed in Table 1 on page 85.

*Approach:* We will follow the steps listed above.

*Solution:*

**Step 1:** Arrange the data in ascending order:

121, 136, 139, 152, 158, 164, 169, 195, 198, 199, 203, 212, 214, 246

**Step 2:** There are $n = 14$ observations.

**Step 3:** Because there are 14 observations (an even number), we compute the median, $M$, by determining the mean of the $\frac{n}{2} = \frac{14}{2} = 7$th observation and $\frac{n}{2} + 1 = \frac{14}{2} + 1 = 8$th observation. So the median is the mean of 169 and 195:

$$M = \frac{169 + 195}{2} = 182$$

121, 136, 139, 152, 158, 164, 169, 195, 198, 199, 203, 212, 214, 246

$$\uparrow$$
$$\boxed{M = 182}$$

Notice that there are seven observations to the left and seven observations to the right of the median. We conclude that 50% of American League teams hit fewer than 182 home runs and 50% hit more than 182 home runs. ◄◄

► EXAMPLE 3 **Computing the Median of a Data Set with an Odd Number of Observations**

*Problem:* The data in Table 2 represent the length (in seconds) of a random sample of songs released in the 70's.

| TABLE 2 | | | |
|---|---|---|---|
| Song Name | Length | Song Name | Length |
| "Sister Golden Hair" | 201 | "Stayin' Alive" | 222 |
| "Black Water" | 257 | "We Are Family" | 217 |
| "Free Bird" | 284 | "Heart of Glass" | 206 |
| "The Hustle" | 208 | "My Sharona" | 240 |
| "Southern Nights" | 179 | | |

*Approach:* We will follow the steps listed on page 87.

*Solution:*

**Step 1:** Arrange the data in ascending order:

179, 201, 206, 208, 217, 222, 240, 257, 284

**Step 2:** There are $n = 9$ observations.

**Step 3:** Since there are an odd number of observations, the median will be the observation exactly in the middle of the data set. The median, $M$, is

217 seconds (the $\dfrac{n+1}{2} = \dfrac{9+1}{2} = 5$th data value). We list the data in ascending order, with the median in blue.

$$179, 201, 206, 208, 217, 222, 240, 257, 284$$

Notice there are four observations to the left and four observations to the right of the median. So 50% of the songs of the 70's have lengths less than 217 seconds and 50% have lengths more than 217 seconds. ◀◀

 *Now compute the median of the data in Problem 3.*

**Using Technology:** Statistical spreadsheets and most calculators will compute the mean and median. Figure 4 shows the results of Example 1(a) and (2), using Excel.

Figure 4

| American League Home Runs | |
|---|---:|
| Mean | 179 |
| Standard error | 9.58019936 |
| Median | 182 |
| Mode | #N/A |

**3   The Mode**

A third measure of central tendency is the mode. The mode can be computed for either quantitative or qualitative data.

*Definition*

The **mode** of a variable is the most frequent observation of the variable that occurs in the data set.

To compute the mode, tally the number of observations that occur for each data value. The data value that occurs most often is the mode. A set of data can have no mode, one mode or more than one mode. If there is no observation that occurs with the most frequency, we say the data has **no mode**.

▶ **EXAMPLE 4   Finding the Mode of Quantitative Data**

*Problem:* The following data represent the number of children of the members of the Joliet Junior College mathematics department:

$$0, 0, 0, 0, 0, 0, 1, 1, 2, 2, 2, 3, 3, 3, 4$$

Find the mode number of children.

*Approach:* We tally the number of times we observe each data value. The data value with the highest frequency is the mode.

*Solution:* The mode is 0 because it occurs most frequently (six times). ◀◀

▶ **EXAMPLE 5   Finding the Mode of Quantitative Data**

*Problem:* Find the mode of the data listed in Table 1 on page 85.

*Approach:* Tally the number of times we observe each data value. The data value with the highest frequency is the mode. Although not necessary, it is helpful to find the mode of quantitative data by arranging the data in ascending order.

*Solution:* We arrange the data in ascending order:

$$121, 136, 139, 152, 158, 164, 169, 195, 198, 199, 203, 212, 214, 246$$

Since each data value occurs an equal number of times (once), there is no mode.

Notice that the output in Figure 4 on page 89 provided by Excel for this data set includes the mode. Why do you think the output for the mode is #N/A?

 *Now compute the mode of the data in Problem 3.*

A data set can have more than one mode. For example, the Boston Red Sox and New York Yankees played only 161 of the 162 scheduled games in 2001. Suppose Boston hit 1 home run and New York hit 9 home runs (the Yankees were feeling especially strong that day!) in the makeup game. Then Boston would have 199 home runs and New York would have 212 home runs, and the set of data in Table 1 would have two modes: 199 and 212. In this case, we say the data are **bimodal**. If a data set has three or more data values that occur with the highest frequency, the data set is **multimodal**. Typically, the mode is not reported for multimodal data because it is not representative of a central tendency or typical value. Data that has more than one mode might be an indication of some underlying relationship. For example, a histogram of heights in which the data set contained both males and females might show two modes. The two modes occur because of the variable gender. See Figure 5.

**Figure 5**
A bimodal distribution

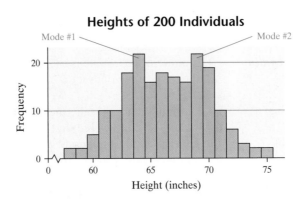

▶ **EXAMPLE 6**    Determining the Mode of Qualitative Data

*Problem:* Find the mode of the data listed in Table 1 in Section 2.1 on page 40, which lists the states in which the 43 presidents of the United States were born.

*Approach:* We look at the table to determine the state that has the highest frequency.

*Solution:* The state of Virginia occurs with the highest frequency, eight. The mode is therefore Virginia.    ◀◀

**④    The Shape of the Distribution and the Mean and the Median**

| TABLE 3 | |
| --- | --- |
| Mean | 179 |
| Median | 182 |

Often, the mean and the median provide different values. Table 3 shows the mean and median number of home runs hit by American League teams in 2001.

Notice that the median and the mean are close in value. Suppose the number of home runs hit by the Texas Rangers was 500 instead of 246. The median would not change, but the mean would increase from 179 to 197.1. In other words, the arithmetic mean is sensitive to extreme (very large or small) values in the data set, while the median is not. We say that the median is **resistant** to extreme values, but the arithmetic mean is not resistant. Therefore, when data sets have unusually large or small values relative to the entire set of data or when the distribution of the data is skewed, the median is the preferred measure of central tendency over the arithmetic mean because it is more representative of the typical observation.

In fact, the arithmetic mean and the median can be useful in determining the shape of a distribution. It can be shown that if a distribution is perfectly symmetric, then the median will equal the arithmetic mean (and the mode). So symmetric distributions will have a median and an arithmetic mean that are close in value. If the arithmetic mean is substantially larger than the median, the distribution will be skewed right. Do you know why? In distributions that are skewed right, there are a few data values substantially larger than the others. These larger data values cause the mean to be inflated while having little, if any, effect on the median. Similarly, distributions that are skewed left will have a mean that is substantially smaller than the median. We summarize these ideas in Table 4 and Figure 6.

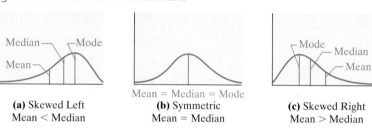

Figure 6 Mean/median versus skewness

**(a)** Skewed Left
Mean < Median

**(b)** Symmetric
Mean = Median

**(c)** Skewed Right
Mean > Median

**Caution**

Because the mean is not resistant, it should not be reported as a measure of central tendency when the distribution of data is highly skewed.

| TABLE 4 |
| --- |

**Relation Between the Mean, Median, and Distribution Shape**

| Distribution Shape | Mean versus Median |
| --- | --- |
| Skewed Left | Mean < Median |
| Symmetric | Mean = Median |
| Skewed Right | Mean > Median |

▶ **EXAMPLE 7**  **Using the Mean and the Median to Identify Distribution Shape**

*Problem:* The data in Table 5 represent the annual salary for players on the Los Angeles Lakers for the 2000–2001 season. The data are in thousands of dollars. For example, Shaquille O'Neal earned $19,290,000 during the 2000–2001 season.

| TABLE 5 | | | |
| --- | --- | --- | --- |
| Player | Salary (thousands of dollars) | Player | Salary (thousands of dollars) |
| Shaquille O'Neal | $19,290 | Chuck Person | $1,200 |
| Kobe Bryant | $10,130 | Tyronn Lue | $870 |
| Horace Grant | $6,500 | Devean George | $850 |
| Robert Horry | $4,800 | Mark Madsen | $710 |
| Rick Fox | $3,400 | Isaiah Ryder | $550 |
| Derek Fisher | $3,380 | Mike Penberthy | $320 |
| Brian Shaw | $2,250 | Stanislav Medvedenko | $320 |
| Ron Harper | $2,200 | Sam Jacobson | $270 |
| Greg Foster | $1,760 | | |

*Source:* bskball.com

**(a)** Find the mean and the median.

**(b)** Describe the shape of the distribution.

**(c)** Which measure of central tendency better describes the average salary of a player on the Los Angeles Lakers?

*Approach:*

**(a)** We will use statistical software to determine the mean and the median.

(b) We can identify the shape of the distribution using the frequency histogram and the relation between the mean and the median. (Refer to Figure 15 in Section 2.2 for identifying distribution shapes from histograms.)

(c) If the data are roughly symmetric, then the mean is a good measure of central tendency; if the data are skewed, then the median is a good measure of central tendency.

*Solution:*

(a) Using Minitab, we find $\mu = 3459$ and $M = 1760$. See Figure 7.

Figure 7

**Descriptive statistics**

| Variable | N | Mean | Median | TrMean | StDev | SE Mean |
|---|---|---|---|---|---|---|
| SE Means | | | | | | |
| Salaries | 17 | 3459 | 1760 | 2616 | 4852 | 1177 |

| Variable | Minimum | Maximum | Q1 | Q3 |
|---|---|---|---|---|
| Salaries | 270 | 19290 | 630 | 4100 |

The mean is much larger than the median; we conclude that the data are skewed right.

(b) Figure 8 shows a frequency histogram of the data. The shape of the histogram is skewed right.

Figure 8

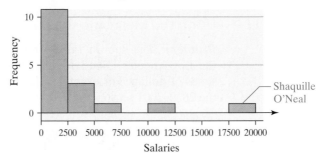

**Histogram of Salaries**

(c) Notice that only four players on the Lakers earned more than the mean salary of $3,459,000, while half the players earned less than the median of $1,760,000 and half earned more than the median. Therefore, the median better describes the typical salary on the Lakers. ◄◄

► EXAMPLE 8    Comparing the Mean and the Median

*Problem:* The data in Table 6 represent the birth weights (in pounds) of 50 randomly sampled babies.

| TABLE 6 | | | | | | | | |
|---|---|---|---|---|---|---|---|---|
| 5.8 | 7.4 | 9.2 | 7.0 | 8.5 | 7.6 | 7.3 | 7.1 | 7.6 |
| 7.9 | 7.8 | 7.9 | 7.7 | 9.0 | 7.1 | 6.9 | 7.0 | 6.7 |
| 8.7 | 7.2 | 6.1 | 7.2 | 7.1 | 7.2 | 6.9 | 7.0 | |
| 7.9 | 5.9 | 7.0 | 7.8 | 7.2 | 7.5 | 6.4 | 7.4 | |
| 7.3 | 6.4 | 7.4 | 8.2 | 9.1 | 7.3 | 7.8 | 8.2 | |
| 9.4 | 6.8 | 7.0 | 8.1 | 8.0 | 7.5 | 8.7 | 7.2 | |

(a) Find the mean and the median.

(b) Describe the shape of the distribution.

**(c)** Which measure of central tendency better describes the average birth weight?

*Approach:*

**(a)** Use a TI-83 Plus to compute the mean and the median.

**(b)** The histogram, along with the mean and the median, are used to identify the shape of the distribution.

**(c)** If the data are roughly symmetric, then the mean is a good measure of central tendency; if the data are skewed, then the median is a good measure of central tendency.

*Solution:*

**(a)** Using a TI-83 Plus, we find $\bar{x} = 7.49$ and $M = 7.35$. See Figure 9.

Figure 9

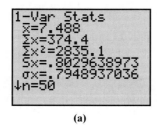

(a)

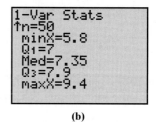

(b)

**(b)** See Figure 10 for the frequency histogram with the arithmetic mean and the median labeled.

Figure 10
Birth weights of 50 randomly selected babies

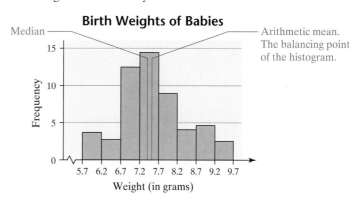

The distribution is bell shaped. We have further evidence of the shape because the arithmetic mean and the median are close to each other.

**(c)** Because the mean and the median are close in value, we use the mean as the measure of central tendency.  ◀◀

**NW**  *Now Work Problem 19.*

A question you may be asking yourself is "Why would I ever compute the mean?" After all, the mean and median are close in value for symmetric data and the median is the better measure of central tendency for skewed data. The reason we compute the mean is that much of the statistical inference that we perform is based upon the mean. We will have more to say about this in Chapter 7.

We conclude this section with the following chart, which addresses the circumstances under which each measure of central tendency should be used:

| Measure of Central Tendency | Computation | Interpretation | When to Use |
|---|---|---|---|
| Mean | Population Mean: $\mu = \dfrac{\sum x_i}{N}$  Sample Mean: $\bar{x} = \dfrac{\sum x_i}{n}$ | Center of gravity | When data are quantitative and the frequency distribution is roughly symmetric |
| Median | Arrange data in ascending order and divide the data set in half | Divides the bottom 50% of the data from the top 50% of the data | When the data are quantitative and the frequency distribution is skewed left or skewed right |
| Mode | | Most frequent observation | When most frequent observation is desired measure of central tendency or the data are qualitative |

## 3.1 Assess Your Understanding

### Concepts and Vocabulary

1. What does it mean if a statistic is resistant? Why is the median resistant, but the mean is not?
2. Describe how the mean and the median can be used to determine the shape of a distribution.
3. In the 2000 Census conducted by the United States Census Bureau there were two average household incomes reported: $41,349 and $55,263. One of these averages is the mean and the other is the median. Which is the mean? Support your answer.

4. The United States Department of Housing and Urban Development (HUD) uses the median to report the average price of a home in the United States. Why do you think they use the median?
5. A histogram of a set of data indicates that the distribution of the data is skewed right. Which measure of central tendency will be larger, the mean or the median? Why?

### Exercises

#### • Skill Building

1. **Crash Test Results** The Insurance Institute for Highway Safety crashed the 2001 Honda Civic four times at five miles per hour. The cost of repair for each of the four crashes is listed on the following line:

   $420, $462, $409, $236

   Compute the mean, median, and mode cost of repair.

2. **Cell Phone Usage** The following data represent the monthly cell phone bill for my wife's phone for six randomly selected months.

   $35.34, $42.09, $39.43, $38.93, $43.39, $49.26

   Compute the mean, median, and mode phone bill.

3. **Concrete Mix** A certain type of concrete mix is designed to withstand 3000 pounds per square inch (psi) of pressure. The strength of concrete is measured by pouring the mix into casting cylinders 6 inches in diameter and 12 inches tall. The cylinder is allowed to "set up" for 28 days. The cylinders are then stacked upon one another until the cylinders are crushed. The following data represent the strength of nine randomly selected casts.

   3960, 4080, 3200, 3100, 2940, 3830, 4090, 4040, 3780

   Compute the mean, median, and mode strength of the concrete (in pounds per square inch).

4. **Flight Time** The following data represent the flight time (in minutes) of a random sample of seven flights from Las Vegas, NV, to Newark, NJ, on Continental Airlines.

   282, 270, 260, 266, 257, 260, 267

   Compute the mean, median, and mode flight time.

#### • Applying the Concepts

5. **Corn Production** The following data represent the number of corn plants in randomly sampled rows (a 17-foot-by-5-inch strip) for various plot types. Compute the mean, median, and mode for each plot type. Which plot type appears to yield the most crops?

| Plot Type | Number of Plants |
|---|---|
| Sludge Plot | 25 27 33 30 28 27 |
| Spring Disk | 32 30 33 35 34 34 |
| No Till | 30 26 29 32 25 29 |
| Spring Chisele | 30 32 26 28 31 29 |
| Great Lakes Bt | 28 32 27 30 29 27 |

*Source:* Andrew Dieter and Brad Schmidgall, Joliet Junior College

**6. Soybean Yield** The following data represent the number of pods on a random sample of soybean plants for various plot types. Compute the mean, median, and mode for each plot type. Which plot type appears to yield the most pods per plant?

| Plot Type | Pods |
|-----------|------|
| Liberty | 32 31 36 35 41 34 39 37 38 |
| Fall Plowed | 29 31 33 22  7 19 30 36 30 |
| No Till | 34 30 31 27 40 33 37 42 39 |
| Chisele Plowed | 34 37 24 23 32 33 27 34 30 |
| Round-up Ready | 49 27 34 46 32 35 20 29 35 |

*Source:* Andrew Dieter and Brad Schmidgall, Joliet Junior College

**7. Day of Birth** A researcher wanted to know whether more births occur on Mondays than on Saturdays. She collects the following data that represent the number of live births on Mondays and Saturdays for the first 6 weeks in 1997, from the National Center for Health Statistics. Compute the mean, median, and mode number of births for each day. Does there appear to be a difference in the number of births on Monday versus Saturday? What might account for any difference?

| Births on Monday | | Births on Saturday | |
|------|------|------|------|
| 10,456 | 10,527 | 8,617 | 8,634 |
| 10,267 | 10,596 | 8,368 | 8,488 |
| 10,444 | 10,778 | 8,340 | 8,411 |

*Source:* National Center for Health Statistics

**8. Month of Birth** A researcher wanted to know whether more births occur in September than in March. She randomly selects seven weekdays in September, 1997 and seven weekdays in March, 1997, and records the number of births on these days. Compute the mean, median, and mode number of births for each month. Does there appear to be a difference in the number of births in March versus the number in September? What might account for any difference?

| Births in March | | Births in September | |
|------|------|------|------|
| 11,726 | 11,571 | 11,307 | 11,453 |
| 11,944 | 11,523 | 12,462 | 12,132 |
| 11,570 | 10,283 | 12,891 | 12,457 |
| 10,903 |  | 12,372 |  |

*Source:* National Center for Health Statistics

**9. Rats in Space** In the experiment "Regulation of Erythropoiesis During Spaceflight," led by NASA scientist Dr. Robert D. Lange, rats were sent to space on Spacelab Life Sciences I to measure the effect of space travel on weight. The following data represent the weights (in grams) of rats in an Animal Enclosure Module (AEM) on the flight return day for both a control (ground) group and experimental (flight) group. Compute the mean, median, and mode weights for both the control and experimental groups. Does there appear to be a difference in the weights? What might account for any difference?

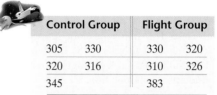

| Control Group | | Flight Group | |
|------|------|------|------|
| 305 | 330 | 330 | 320 |
| 320 | 316 | 310 | 326 |
| 345 |  | 383 |  |

*Source:* NASA Life Sciences Data Archive

**10. Reaction Time** In an experiment conducted on-line at the University of Mississippi, study participants are asked to react to a stimulus. In one experiment, the participant must press a key upon seeing a blue screen. The time (in seconds) to press the key is measured. The same person is then asked to press a key upon seeing a red screen, again with the time to react measured. The results for six study participants are listed below. Compute the mean, median, and mode reaction time for both blue and red. Does there appear to be a difference in the reaction time? What might account for any difference? How might this information be used?

| Participant Number | Reaction Time to Blue | Reaction Time to Red |
|------|------|------|
| 1 | 0.582 | 0.408 |
| 2 | 0.481 | 0.407 |
| 3 | 0.841 | 0.542 |
| 4 | 0.267 | 0.402 |
| 5 | 0.685 | 0.456 |
| 6 | 0.45 | 0.533 |

*Source:* PsychExperiments at the University of Mississippi (http://www.olemiss.edu/psychexps/)

**11. Pulse Rates** The following data represent the pulse rates (beats per minute) of the nine students enrolled in a section of Sullivan's Introductory Statistics course. Treat the nine students as a population.

| Student | Pulse |
|---------|-------|
| Perpectual Bempah | 76 |
| Megan Brooks | 60 |
| Jeff Honeycutt | 60 |
| Clarice Jefferson | 81 |
| Crystal Kurtenbach | 72 |
| Janette Lantka | 80 |
| Kevin McCarthy | 80 |
| Tammy Ohm | 68 |
| Kathy Wojdyla | 73 |

(a) Compute the population mean pulse.
(b) Determine two simple random samples of size three and compute the sample mean pulse of each sample.
(c) Which samples result in a sample mean that overestimates the population mean? Which samples result in a sample mean that underestimates the population mean? Do any samples lead to a sample mean that equals the population mean?

12. **Travel Time** The following data represent the travel time (in minutes) to school for the nine students enrolled in Sullivan's College Algebra course. Treat the nine students as a population.

| Student | Travel Time | Student | Travel Time |
|---------|-------------|---------|-------------|
| Amanda  | 39          | Scot    | 45          |
| Amber   | 21          | Erica   | 11          |
| Tim     | 9           | Tiffany | 12          |
| Mike    | 32          | Glenn   | 39          |
| Nicole  | 30          |         |             |

(a) Compute the population mean travel time.
(b) Determine three simple random samples of size four and compute the sample mean travel time of each sample.
(c) Which samples result in a sample mean that overestimates the population mean? Which samples result in a sample mean that underestimates the population mean? Do any samples lead to a sample mean that equals the population mean?

13. **Waiting for a Table** The following data represent the number of customers waiting for a table at 6:00 P.M. on Saturday for 40 consecutive Saturdays at Bobak's Restaurant:

| 11 | 6  | 5  | 8  | 11 | 6  | 3  | 7  |
|----|----|----|----|----|----|----|----|
| 4  | 6  | 5  | 4  | 13 | 14 | 9  | 11 |
| 13 | 8  | 10 | 10 | 9  | 9  | 6  | 5  |
| 10 | 8  | 8  | 8  | 7  | 7  | 3  | 8  |
| 7  | 8  | 9  | 6  | 10 | 11 | 4  | 8  |

(a) Compute the mean and the median number of customers waiting.
(b) Identify the shape of the distribution based on the histogram drawn in Problem 17 in Section 2.2, the arithmetic mean, and the median.

14. **Highway Repair** The following data represent the number of potholes on 50 randomly selected 1-mile stretches of highway in the city of Chicago:

| 2 | 7 | 4 | 7  | 2 | 7 |
|---|---|---|----|---|---|
| 2 | 2 | 2 | 3  | 4 | 3 |
| 1 | 2 | 3 | 2  | 1 | 4 |
| 2 | 2 | 5 | 2  | 3 | 4 |
| 4 | 1 | 7 | 10 | 3 | 5 |
| 4 | 3 | 3 | 2  | 2 |   |
| 1 | 6 | 5 | 7  | 9 |   |
| 2 | 2 | 2 | 1  | 5 |   |
| 3 | 5 | 1 | 3  | 5 |   |

(a) Compute the mean and the median number of potholes.
(b) Identify the shape of the distribution based on the histogram drawn in Problem 18 in Section 2.2, the arithmetic mean, and the median.

15. **Tensile Strength** Tensile strength is the maximum stress at which one can be reasonably certain failure will not occur. The following data represent tensile strength (in thousands of pounds per square inch) of a composite material to be used in an aircraft.

| 203.41 | 185.97 | 184.41 | 160.44 | 174.63 |
|--------|--------|--------|--------|--------|
| 209.58 | 190.67 | 200.73 | 180.95 | 185.34 |
| 213.35 | 207.88 | 206.51 | 201.95 | 205.59 |
| 218.56 | 210.80 | 209.84 | 204.60 | 212.00 |
| 242.76 | 231.46 | 212.15 | 219.51 | 225.25 |

*Source:* Vangel, Mark G., "New Methods for One-sided Tolerance Limits for a One-way Balanced Random-Effects ANOVA Model," *Technometrics*; May, 1992, Vol. 34, Issue 2, pp. 176–185.

(a) Compute the mean and median tensile strength.
(b) Identify the shape of the distribution based on the histogram drawn in Problem 19 in Section 2.2 and the relationship between the arithmetic mean and the median.

16. **Miles on a Cavalier** A random sample of 36 three-year-old Chevy Cavaliers was obtained in the Miami, FL area and the number of miles on each car was recorded. The data are shown the following table.

| 34,122 | 34,511 | 17,595 | 30,500 | 31,883 |
|--------|--------|--------|--------|--------|
| 39,416 | 34,500 | 44,224 | 35,936 | 32,923 |
| 28,281 | 41,620 | 26,455 | 44,448 | 40,614 |
| 37,021 | 30,283 | 27,368 | 41,043 | 42,398 |
| 30,747 | 15,499 | 37,904 | 30,739 |        |
| 35,439 | 26,051 | 23,221 | 38,995 |        |
| 17,685 | 32,305 | 38,041 | 41,194 |        |
| 29,307 | 39,884 | 35,295 | 29,289 |        |

*Source:* cars.com

(a) Compute the mean and median miles.
(b) Identify the shape of the distribution based on the histogram drawn in Problem 20 in Section 2.2 and the relationship between the arithmetic mean and, the median.

17. **Serum HDL** Dr. Paul Oswiecmiski randomly selects forty of his 20–29-year-old patients and obtains the following data regarding their serum HDL cholesterol:

| 70 | 56 | 48 | 48 | 53 | 52 | 66 | 48 |
|----|----|----|----|----|----|----|----|
| 36 | 49 | 28 | 35 | 58 | 62 | 45 | 60 |
| 38 | 73 | 45 | 51 | 56 | 51 | 46 | 39 |
| 56 | 32 | 44 | 60 | 51 | 44 | 63 | 50 |
| 46 | 69 | 53 | 70 | 33 | 54 | 55 | 52 |

(a) Compute the mean and the median serum HDL.
(b) Identify the shape of the distribution based on the histogram drawn in Problem 21 in Section 2.2 and the relationship between the arithmetic mean and the median.

**18. Volume of Philip Morris Stock** The volume of a stock is the number of shares traded on a given day. The following data, in millions, so that 3.98 represents 3,980,000 shares traded, represent the volume of Philip Morris stock traded for a random sample of 35 trading days in 2000:

| | | | | |
|---|---|---|---|---|
| 3.98 | 8.90 | 10.40 | 7.52 | 13.84 |
| 14.04 | 9.14 | 7.96 | 11.40 | 13.30 |
| 5.29 | 9.69 | 8.94 | 12.10 | 13.25 |
| 4.62 | 10.85 | 24.52 | 14.62 | 16.10 |
| 4.24 | 8.54 | 11.59 | 6.75 | 27.54 |
| 16.06 | 17.62 | 7.17 | 6.13 | 25.32 |
| 7.71 | 10.14 | 4.28 | 7.24 | 10.88 |

*Source:* http://finance.yahoo.com

(a) Compute the mean and the median number of shares traded.
(b) Identify the shape of the distribution based on the histogram drawn in Problem 22 in Section 2.2 and the relationship between the arithmetic mean and the median.

**19. M&Ms** The following data represent the weights (in grams) of a simple random sample of 50 M&M plain candies.

| | | | | | | |
|---|---|---|---|---|---|---|
| 0.87 | 0.88 | 0.82 | 0.90 | 0.90 | 0.84 | 0.84 |
| 0.91 | 0.94 | 0.86 | 0.86 | 0.86 | 0.88 | 0.87 |
| 0.89 | 0.91 | 0.86 | 0.87 | 0.93 | 0.88 | |
| 0.83 | 0.94 | 0.87 | 0.93 | 0.91 | 0.85 | |
| 0.91 | 0.91 | 0.86 | 0.89 | 0.87 | 0.93 | |
| 0.88 | 0.88 | 0.89 | 0.79 | 0.82 | 0.83 | |
| 0.90 | 0.88 | 0.84 | 0.93 | 0.76 | 0.90 | |
| 0.88 | 0.92 | 0.85 | 0.79 | 0.84 | 0.86 | |

*Source:* Michael Sullivan

(a) Compute the mean and the median weight.
(b) Identify the shape of the distribution based on the histogram drawn in Problem 23 in Section 2.2 and the relationship between the arithmetic mean and the median.

**20. Old Faithful** The following data represent the length of eruption (in seconds) for a random sample of eruptions of the Old Faithful Geyser in California.

| | | | | | | |
|---|---|---|---|---|---|---|
| 108 | 108 | 99 | 105 | 103 | 103 | 94 |
| 102 | 99 | 106 | 90 | 104 | 110 | 110 |
| 103 | 109 | 109 | 111 | 101 | 101 | |
| 110 | 102 | 105 | 110 | 106 | 104 | |
| 104 | 100 | 103 | 102 | 120 | 90 | |
| 113 | 116 | 95 | 105 | 103 | 101 | |
| 100 | 101 | 107 | 110 | 92 | 108 | |

*Source:* Ladonna Hansen, Park Curator

(a) Compute the mean and the median length of eruption.
(b) Identify the shape of the distribution based on the histogram drawn in Problem 24 in Section 2.2 and the relationship between the arithmetic mean and the median.

**21. Foreign-born Population** The following data represent the region of birth of foreign-born residents of the United States in 2000. Determine the mode region of birth.

| Region | Number (thousands) |
|---|---|
| Europe | 4,772 |
| Asia | 8,364 |
| Africa | 840 |
| Oceania | 180 |
| Latin America | 15,472 |
| North America | 836 |

*Source:* United States Census Bureau

**22. Robbery** The following data represent the number of offenses for various robberies in 1998. Determine the mode offense.

| Type of Crime | Number of Offenses (in thousands) |
|---|---|
| Street or Highway | 219 |
| Commercial house | 61 |
| Gas Station | 10 |
| Convenience Store | 26 |
| Residence | 54 |
| Bank | 9 |

*Source:* U.S. Federal Bureau of Investigation

**23. 2000 Presidential Election** An exit poll was conducted in Palm Beach County, Florida, in which a random sample of 40 voters revealed whom they voted for in the presidential election. The results of the survey are shown on page 98.

| | | | | |
|---|---|---|---|---|
| Bush | Gore | Bush | Gore | Bush |
| Gore | Gore | Gore | Bush | Bush |
| Gore | Bush | Bush | Gore | Gore |
| Gore | Gore | Gore | Bush | Gore |
| Gore | Gore | Gore | Gore | Bush |
| Gore | Gore | Gore | Gore | Gore |
| Gore | Nader | Gore | Bush | Bush |
| Gore | Bush | Bush | Gore | Bush |

Determine the mode.

24. **Hospital Admissions** The data to the right represent the diagnosis of a random sample of 20 patients admitted to a hospital:

| Cancer | Motor Vehicle Accident | Congestive Heart Failure |
|---|---|---|
| Gun Shot Wound | Fall | Gun Shot Wound |
| Gun Shot Wound | Motor Vehicle Accident | Gun Shot Wound |
| Assault | Motor Vehicle Accident | Gun Shot Wound |
| Motor Vehicle Accident | Motor Vehicle Accident | Gun Shot Wound |
| Motor Vehicle Accident | Gun Shot Wound | Motor Vehicle Accident |
| Fall | Gun Shot Wound | |

*Source:* Tamela Ohm, Student at Joliet Junior College

Determine the mode diagnosis.

25. **Which Position in Baseball Is the Highest Paying?** You are a prospective baseball agent and are in search of some clients. You would like to recruit the highest-paid players as clients so you perform a study in which you identify the 36 most highly paid players and their positions.

| Player | Position | Player | Position | Player | Position |
|---|---|---|---|---|---|
| Kevin Brown | Pitcher | Derek Jeter | Shortstop | John Smoltz | Pitcher |
| Randy Johnson | Pitcher | Raul Mondesi | Right Field | Jim Thome | 1st Base |
| Albert Belle | Right Field | Gary Sheffield | Right Field | Todd Stottlemyre | Pitcher |
| Bernie Williams | Center Field | Tom Glavine | Pitcher | Robin Ventura | 3rd Base |
| Mike Piazza | Catcher | Shawn Green | Right Field | Chuck Finley | Pitcher |
| Larry Walker | 1st Base | Ken Griffey, Jr. | Center Field | Al Leiter | Pitcher |
| David Cone | Pitcher | Mark McGwire | 1st Base | Brian Jordan | Center Field |
| Pedro Martinez | Pitcher | Wilson Alvarez | Pitcher | Ray Lankford | Center Field |
| Mo Vaughn | 1st Base | Rafael Palmeiro | 1st Base | Juan Gonzalez | 1st Base |
| Greg Maddux | Pitcher | Ivan Rodriguez | Catcher | Kenny Rogers | Pitcher |
| Sammy Sosa | Right Field | Matt Williams | 3rd Base | Kenny Lofton | Center Field |
| Barry Bonds | Right Field | Andres Gallarraga | 1st Base | Alex Rodriguez | Shortstop |

*Source:* SLAM! Baseball

Determine the mode position.

26. **Blood Type** A phlebotomist draws the blood of a random sample of 50 patients and determines their blood types as follows:

| | | | | | | |
|---|---|---|---|---|---|---|
| O | O | A | A | O | O | O |
| B | O | B | A | O | O | O |
| AB | B | A | B | AB | O | |
| O | O | A | A | O | A | |
| AB | O | A | B | A | O | |
| O | A | A | O | A | O | |
| O | A | O | AB | A | A | |
| O | B | A | A | O | A | |

Determine the mode blood type.

27. Refer to the tensile strength data from Problem 15. Suppose that the data value 185.97 was accidentally entered as 815.97. Recompute the mean and median with the incorrect data. How did this change affect the mean and median?

28. Refer to the mileage data from Problem 16. Suppose the data value 17,685 was accidentally recorded as 117,685. Recompute the mean and median for the incorrect data set. How did this change affect the mean and median?

**29.** For each of the three histograms shown, determine whether the mean is greater than, less than, or approximately equal to the median, and justify your answer.

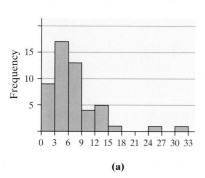

**(a)**

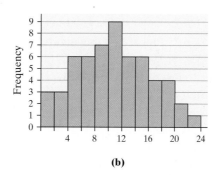

**(b)**

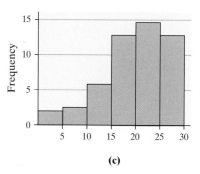

**(c)**

**30.** Match the histograms shown to the summary statistics:

|       | Mean | Median |
|-------|------|--------|
| (I)   | 42   | 42     |
| (II)  | 31   | 36     |
| (III) | 31   | 26     |
| (IV)  | 31   | 32     |

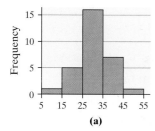

**(a)**

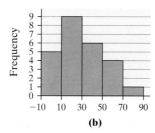

**(b)**

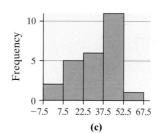

**(c)**

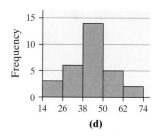

**(d)**

**31.** You are negotiating a contract for the Players' Association of the NBA. Which measure of central tendency will you use to support your claim that the average player's salary needs to be increased? Why? As the chief negotiator for the owners, which measure would you use to refute the claim made by the Player's Association?

**32.** In October of 2000, the mean amount of money lost per visitor to the Harrah's River Boat was $110. Do you think the median was more than, less than, or equal to this amount? Why?

**33.** Suppose that the mean of a set of six data values is 34. What is the sum of the six data values?

**34.** Use the five test scores of 65, 70, 71, 75, and 95 to answer the following questions:
   (a) Find the sample mean.
   (b) Find the median.
   (c) Which measure of central tendency best describes the typical test score?
   (d) Suppose the professor decides to curve the exam by adding 4 points to each person's score. Compute the sample mean based on the adjusted scores.
   (e) Compare the unadjusted test score mean with the curved test score mean. What effect did adding 4 to each score have on the mean?

**35.** For each of the following situations, determine which measure of central tendency is most appropriate:
   (a) Average price of a home sold in Pittsburgh, Pennsylvania, in 2002
   (b) Most popular major for students enrolled in a statistics course
   (c) Average test score when the scores are distributed symmetrically
   (d) Average test score when the scores are skewed right
   (e) Average income of a player in the National Football League

**36. Trimmed Mean** Another measure of central tendency is the trimmed mean. It is computed by determining the arithmetic mean of a data set after deleting the smallest and largest observed value. Compute the trimmed mean for the data in Problem 19. Is the trimmed mean resistant? Explain.

**37. Midrange** The midrange is also a measure of central tendency. It is computed by adding the smallest and largest observed value of a data set and dividing the result by 2; that is,

$$\text{Midrange} = \frac{\text{Largest Data Value} + \text{Smallest Data Value}}{2}$$

Compute the midrange for the data in Problem 19. Is the midrange resistant? Explain.

**Technology Step-By-Step**
Determining the Mean and Median

**TI-83 Plus**
*Step 1:* Enter the raw data in L1 by pressing STAT and selecting 1:Edit.

*Step 2:* Press STAT, highlight the CALC menu, and select 1:1-Var Stats.

*Step 3:* With 1-Var Stats appearing on the HOME screen, press 2ⁿᵈ 1 to insert L1 on the HOME screen. Press ENTER.

**MINITAB**
*Step 1:* Enter the data in C1.

*Step 2:* Select the **Stat** menu, highlight **Basic Statistics** and then highlight **Display Descriptive Statistics**.

*Step 3:* In the **Variables** window, enter C1. Click OK.

**Excel**
*Step 1:* Enter the data in column A.

*Step 2:* Select the **Tools** menu and highlight **Data Analysis** ...

*Step 3:* In the Data Analysis window, highlight **Descriptive Statistics** and click OK.

*Step 4:* With the cursor in the **Input Range** window, use the mouse to highlight the data in column A.

*Step 5:* Select the **Summary statistics** option and click OK.

## 3.2 Measures of Dispersion

**Objectives**
1. Compute the range of a variable from raw data
2. Compute the variance of a variable from raw data
3. Compute the standard deviation of a variable from raw data
4. Use the Empirical Rule

In Section 3.1, we discussed measures of central tendency. The purpose of these measures is to describe the typical value of a variable or the center of the distribution. However, in addition to measuring the central tendency of a variable, we would also like to know the amount of dispersion in the variable. By dispersion, we mean the degree to which the data are "spread out". An example should help to explain why measures of central tendency are not sufficient in describing a distribution.

▶ EXAMPLE 1 **Comparing Two Sets of Data**

*Problem:* The data in Table 7 represent the IQ scores of a random sample of 100 students from two different universities. For each university, compute the arithmetic mean IQ score and draw a histogram, using a lower class limit of 50 for the first class and a class width of 10.

| TABLE 7 | | | | | | | | | | | | | | | | | | | |
|---|---|---|---|---|---|---|---|---|---|---|---|---|---|---|---|---|---|---|---|
| **University A** | | | | | | | | | | **University B** | | | | | | | | | |
| 73 | 103 | 91 | 93 | 136 | 108 | 92 | 104 | 90 | 78 | 86 | 91 | 107 | 94 | 105 | 107 | 89 | 96 | 102 | 96 |
| 108 | 93 | 91 | 78 | 81 | 130 | 82 | 86 | 111 | 93 | 92 | 109 | 103 | 106 | 98 | 95 | 97 | 95 | 109 | 109 |
| 102 | 111 | 125 | 107 | 80 | 90 | 122 | 101 | 82 | 115 | 93 | 91 | 92 | 91 | 117 | 108 | 89 | 95 | 103 | 109 |
| 103 | 110 | 84 | 115 | 85 | 83 | 131 | 90 | 103 | 106 | 110 | 88 | 97 | 119 | 90 | 99 | 96 | 104 | 98 | 95 |
| 71 | 69 | 97 | 130 | 91 | 62 | 85 | 94 | 110 | 85 | 87 | 105 | 111 | 87 | 103 | 92 | 103 | 107 | 106 | 97 |
| 102 | 109 | 105 | 97 | 104 | 94 | 92 | 83 | 94 | 114 | 107 | 108 | 89 | 96 | 107 | 107 | 96 | 95 | 117 | 97 |
| 107 | 94 | 112 | 113 | 115 | 106 | 97 | 106 | 85 | 99 | 98 | 89 | 104 | 99 | 99 | 87 | 91 | 105 | 109 | 108 |
| 102 | 109 | 76 | 94 | 103 | 112 | 107 | 101 | 91 | 107 | 116 | 107 | 90 | 98 | 98 | 92 | 119 | 96 | 118 | 98 |
| 107 | 110 | 106 | 103 | 93 | 110 | 125 | 101 | 91 | 119 | 97 | 106 | 114 | 87 | 107 | 96 | 93 | 99 | 89 | 94 |
| 118 | 85 | 127 | 141 | 129 | 60 | 115 | 80 | 111 | 79 | 104 | 88 | 99 | 97 | 106 | 107 | 112 | 97 | 94 | 106 |

*Approach:* Use statistical software to compute the arithmetic mean and draw a histogram for each university.

*Solution:* We enter the data into Minitab and determine that the arithmetic mean IQ score of both universities is 100.0. Figure 11 shows the histograms.  ◀◀

Figure 11

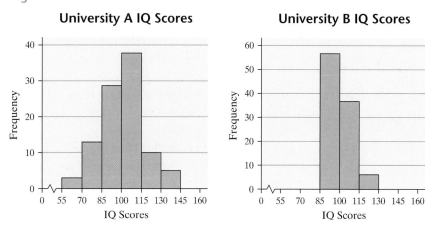

We notice that both universities have the same mean IQ, but the histograms indicate the IQs from University A are more spread out—that is, more dispersed. While an IQ of 100.0 is "typical" for both universities, it appears to be a more reliable description of the typical student from University B than that of University A. That is, a higher proportion of students have IQ scores within, say, 15 points of the mean of 100.0 from University B than from University A.

Our goal in this section is to discuss numerical measures of dispersion so that we can quantify the spread of data. In this section, we discuss three numerical measures for describing the dispersion or spread of data: the range, variance, and standard deviation. In Section 3.4, we will discuss another measure of dispersion, the *interquartile range* (IQR).

## ① The Range

The simplest measure of dispersion is the range. In order to compute the range, the data must be quantitative.

*Definition*

> The **range**, **R**, of a variable is the difference between the largest data value and the smallest data value. That is,
>
> Range = $R$ = Largest Data Value − Smallest Data Value

▶ **EXAMPLE 2   Determining the Range of a Set of Data**

*Problem:* The data in Table 8 represent the number of home runs hit by all teams in the American League in 2001. Compute the range.

*Approach:* The range is found by computing the difference between the largest and smallest data values. In other words, we determine the difference between the highest and lowest numbers.

*Solution:* The Texas Rangers hit the most home runs, 246, while the Tampa Bay Devil Rays hit the fewest home runs, 121. The range, *R*, is

$$R = 246 - 121 = 125$$

| TABLE 8 | |
| --- | --- |
| **Team** | **Home Runs** |
| 1. Anaheim Angels | 158 |
| 2. Baltimore Orioles | 136 |
| 3. Boston Red Sox | 198 |
| 4. Chicago White Sox | 214 |
| 5. Cleveland Indians | 212 |
| 6. Detroit Tigers | 139 |
| 7. Kansas City Royals | 152 |
| 8. Minnesota Twins | 164 |
| 9. New York Yankees | 203 |
| 10. Oakland Athletics | 199 |
| 11. Seattle Mariners | 169 |
| 12. Tampa Bay Devil Rays | 121 |
| 13. Texas Rangers | 246 |
| 14. Toronto Blue Jays | 195 |

*Source:* Major League Baseball

**In Your Own Words**

The range is not resistant.

All the teams in the American League hit between 121 and 246 home runs. The difference between the best home-run-hitting team and the worst home-run-hitting team in the American League is 125 home runs. ◄◄

**NW** *Now compute the range of the data in Problem 3.*

Notice the range is affected by extreme values in the data set, so the range is not resistant. If the Texas Rangers had hit 500 home runs in 2001, the range would be $R = 500 - 121 = 379$. In addition, the range is computed using only two values in the data set (the largest and smallest). The variance and the standard deviation, on the other hand, use all the data values in the computations.

## Variance

Just as there is a population mean and sample mean, we have a population variance and a sample variance. Variance is based upon the difference between each observation and the arithmetic mean; that is, it is based upon the **deviation about the mean**. For a population, the deviation about the mean for the $i$th observation is $x_i - \mu$. For a sample, the deviation about the mean for the $i$th observation is $x_i - \overline{x}$. The further an observation is from the mean, the larger the absolute deviation.

Because the mean represents the "center of gravity," the sum of all deviations about the mean must equal zero. That is,

$$\Sigma(x_i - \mu) = 0 \quad \text{and} \quad \Sigma(x_i - \overline{x}) = 0$$

In other words, observations larger than the mean are offset by observations smaller than the mean.

**Definition**

The **population variance** of a variable is the sum of the squared deviations about the population mean divided by the number of observations in the population, $N$. That is, it is the arithmetic mean of the squared deviations about the population mean. The population variance is symbolically represented by $\sigma^2$ (lower case Greek sigma squared).

$$\sigma^2 = \frac{\Sigma(x_i - \mu)^2}{N} = \frac{(x_1 - \mu)^2 + (x_2 - \mu)^2 + \cdots + (x_N - \mu)^2}{N} \ast \quad \textbf{(1)}$$

where $x_1, x_2, \ldots, x_N$ are the $N$ observations in the population and $\mu$ is the population mean.

NOTE: In using Formula (1), do not round until the last computation. Use as many decimal places as allowed by your calculator in order to avoid round-off errors.

*An algebraically equivalent formula for computing the population variance is

$$\sigma^2 = \frac{\Sigma x_i^2 - \dfrac{(\Sigma x_i)^2}{N}}{N},$$ where $\Sigma x_i^2$ means to square each observation and then sum these squared values, and $(\Sigma x_i)^2$ means to add up all the observations and then square the sum.

► **EXAMPLE 3**   **Computing a Population Variance**

*Problem:* Use the data in Table 8 to compute the population variance of the home runs hit by American League teams in 2001.

*Approach:*

**Step 1:** Create a table with the population data (number of home runs) in the first column. In the second column, enter the population mean: $\mu = 179$.
**Step 2:** Compute the deviation about the mean for each of the data values. That is, compute $x_i - \mu$ for each data value. Enter these values in column 3.
**Step 3:** Square the values in column 3 and enter the results in column 4.
**Step 4:** Sum the squared deviations in column 4 and divide this result by the size of the population, $N$.

*Solution:*

**Step 1:** See Table 9. Column 1 lists the observations in the data set, and column 2 contains the population mean.
**Step 2:** Compute the deviations about the mean, $x_i - \mu$, for each observation as shown in column 3. For example, the deviation about the mean for the Anaheim Angels is $158 - 179 = -21$.
**Step 3:** Column 4 shows the squared deviations about the mean.

**In Your Own Words**

Follow Steps 1–4 to compute the population variance "by hand."

| TABLE 9 | | | |
|---|---|---|---|
| Number of Home Runs | Population Mean, $\mu$ | Deviation about the Mean, $x_i - \mu$ | Squared Deviations about the Mean, $(x_i - \mu)^2$ |
| 158 | 179 | $158 - 179 = -21$ | $(-21)^2 = 441$ |
| 136 | 179 | $-43$ | 1849 |
| 198 | 179 | 19 | 361 |
| 214 | 179 | 35 | 1225 |
| 212 | 179 | 33 | 1089 |
| 139 | 179 | $-40$ | 1600 |
| 152 | 179 | $-27$ | 729 |
| 164 | 179 | $-15$ | 225 |
| 203 | 179 | 24 | 576 |
| 199 | 179 | 20 | 400 |
| 169 | 179 | $-10$ | 100 |
| 121 | 179 | $-58$ | 3364 |
| 246 | 179 | 67 | 4489 |
| 195 | 179 | 16 | 256 |
| | | $\Sigma(x_i - \mu) = 0$ | $\Sigma(x_i - \mu)^2 = 16{,}704$ |

**Step 4:** We sum the entries in column 4 to obtain the numerator of Formula (1), $\Sigma(x_i - \mu)^2$. We compute the population variance by dividing the sum of the entries in column 4 by the number of teams, 14:

$$\sigma^2 = \frac{\Sigma(x_i - \mu)^2}{N} = \frac{16{,}704}{14} = 1{,}193.1 \qquad ◄◄$$

Notice the unit of measure of the variance in Example 3 is home runs squared. This unit of measure results from squaring the deviations about the mean. Because home runs squared does not have any real meaning, the use of the variance for any practical purpose is limited.

The sample variance is computed using sample data.

*Definition*

The **sample variance**, $s^2$, is computed by determining the sum of the squared deviations about the sample mean and dividing this result by $n - 1$. The formula for the sample variance from a sample of size $n$ is

$$s^2 = \frac{\sum(x_i - \bar{x})^2}{n - 1} = \frac{(x_1 - \bar{x})^2 + (x_2 - \bar{x})^2 + \cdots + (x_n - \bar{x})^2}{n - 1} \quad ^* \qquad \textbf{(2)}$$

where $x_1, x_2, \ldots, x_n$ are the $n$ observations in the sample and $\bar{x}$ is the sample mean.

**Caution**

When computing the sample variance, be sure to divide by $n - 1$, not $n$.

Notice that the sample variance is obtained by dividing by $n - 1$. If we divided by $n$, as we would expect, the sample variance would consistently underestimate the population variance. Whenever a statistic consistently overestimates or underestimates a parameter, it is called **biased**. To obtain an unbiased estimate of the population variance, divide the sum of the squared deviations about the mean by $n - 1$.

To help understand the idea of a biased estimator, consider the following situation: Suppose you work for a carnival in which you must guess a person's age. After 20 people come to your booth, you notice that you have a tendency to underestimate people's age. (You guess too low.) What would you do about this? In all likelihood, you would adjust your guesses higher so that you don't underestimate any more. In other words, before the adjustment, your guesses were biased. To remove the bias, you increase your guess. That's what dividing by $n - 1$ in the sample variance formula accomplishes. Dividing by $n$ results in an underestimate, so we divide by a smaller number to increase our "guess."

Although a proof that establishes why we divide by $n - 1$ is beyond the scope of the text, we can provide an explanation that has intuitive appeal. We already know that the sum of the deviations about the mean, $\sum(x_i - \bar{x})$ must equal zero. Therefore, if the sample mean is known and the first $n - 1$ observations are known, then the $n$th observation must be the value that causes the sum of the deviations to equal zero. For example, suppose $\bar{x} = 4$ based upon a sample of size three. In addition, suppose $x_1 = 2$ and $x_2 = 3$, then we can determine $x_3$.

$$\frac{x_1 + x_2 + x_3}{3} = \bar{x}$$

$$\frac{2 + 3 + x_3}{3} = 4 \qquad x_1 = 2, x_2 = 3, \bar{x} = 4$$

$$5 + x_3 = 12$$

$$x_3 = 7$$

**In Your Own Words**

We have $n - 1$ degrees of freedom in the computation of $s^2$ because an unknown parameter, $\mu$, is estimated with $\bar{x}$. For each parameter estimated, we lose one degree of freedom.

Therefore, the first $n - 1$ observations have freedom to be whatever value they wish, but the $n$th value has no freedom; it must be whatever value forces the sum of the deviations about the mean to equal zero. In this regard, we say that there are $n - 1$ **degrees of freedom**.

Again, you should notice that Greek letters are used for parameters while Roman letters are used for statistics. Do not use rounded values of the sample mean in Formula (2).

---

*An algebraically equivalent formula for computing the sample variance is

$$s^2 = \frac{\sum x_i^2 - \dfrac{(\sum x_i)^2}{n}}{n - 1},$$ where $\sum x_i^2$ means to square each observation and then sum these squared values, whereas $(\sum x_i)^2$ means to add up all the observations and then square the sum.

▶ **EXAMPLE 4**   Computing a Sample Variance

*Problem:* Compute the sample variance of the sample obtained in Example 1(b) on page 85 from Section 3.1.

*Approach:* We follow the same approach that we used to compute the population variance, but this time using the sample mean instead of the population mean.

*Solution:* The Anaheim Angels (158 home runs), Baltimore Orioles (136 home runs), Detroit Tigers (139 home runs), Tampa Bay Devil Rays (121 home runs), and Seattle Mariners (169 home runs) are in our sample.

*Step 1:* We create a table just as we did in the computation of the population variance. Column 1 in Table 10 lists the data in the sample and column 2 contains the unrounded sample mean, $\bar{x} = 144.6$.

*Step 2:* Compute the deviations about the mean, $x_i - \bar{x}$, for each observation as shown in column 3.

*Step 3:* Column 4 in Table 10 shows the squared deviations about the mean.

| TABLE 10 | | | |
|---|---|---|---|
| Number of Home Runs | Sample Mean, $\bar{x}$ | Deviation about the Mean, $x_i - \bar{x}$ | Squared Deviations about the Mean, $(x_i - \bar{x})^2$ |
| 158 | 144.6 | $158 - 144.6 = 13.4$ | $13.4^2 = 179.56$ |
| 136 | 144.6 | $-8.6$ | 73.96 |
| 139 | 144.6 | $-5.6$ | 31.36 |
| 121 | 144.6 | $-23.6$ | 556.96 |
| 169 | 144.6 | 24.4 | 595.36 |
| | | $\Sigma(x_i - \bar{x}) = 0$ | $\Sigma(x_i - \bar{x})^2 = 1437.2$ |

*Step 4:* We sum the entries in column 4 and obtain $\Sigma(x_i - \bar{x})^2 = 1437.2$. We compute the sample variance by dividing the sum of the entries in column 4 by the number of individuals in the sample less one, $5 - 1 = 4$:

$$s^2 = \frac{\Sigma(x_i - \bar{x})^2}{n - 1} = \frac{1437.2}{5 - 1} = 359.3 \qquad ◀◀$$

Notice that the sample variance obtained for this sample is an underestimate of the population variance. This discrepancy does not violate our definition of an unbiased estimator, however. A biased estimator is one that *consistently* under- or overestimates.

 **EXPLORATION**   Using the results from the Exploration in Section 3.1 on page 86, compute two additional sample variances. Conclude that different samples lead to different sample variances. ◀◀

③ **Standard Deviation**

The standard deviation and the arithmetic mean are the most popular methods for numerically describing the distribution of a variable. This is because these two measures are used for most types of statistical inference.

*Definitions*   The **population standard deviation**, $\sigma$, is obtained by taking the square root of the population variance. That is,

$$\sigma = \sqrt{\sigma^2}$$

The **sample standard deviation**, $s$, is obtained by taking the square root of the sample variance. That is,

$$s = \sqrt{s^2}$$

▶ **EXAMPLE 5**    **Obtaining the Population and Sample Standard Deviation**

*Problem:* Use the results obtained in Examples 3 and 4 to compute the population and sample standard deviation number of home runs hit in 2001 by teams in the American League.

*Approach:* The population standard deviation is the square root of the population variance. The sample standard deviation is the square root of the sample variance.

*Solution:* The population standard deviation is

$$\sigma = \sqrt{\sigma^2} = \sqrt{\frac{\Sigma(x_i - \mu)^2}{N}} = \sqrt{\frac{16{,}704}{14}} = 34.5 \text{ home runs}$$

The sample standard deviation for the sample obtained in Example 1 from Section 3.1 is

$$s = \sqrt{s^2} = \sqrt{\frac{\Sigma(x_i - \bar{x})^2}{n - 1}} = \sqrt{\frac{1437.2}{5 - 1}} = 19.0 \text{ home runs} \quad ◀◀$$

**Caution**

Never use the rounded variance to compute the standard deviation.

To avoid round-off error, never use the rounded value of the variance to compute the standard deviation.

**EXPLORATION**    Compute the two sample standard deviations using the two sample variances obtained in the Exploration on page 105. Conclude that different samples result in different sample standard deviations. ◀◀

**NW** *Now Work Problem 11.*

**Using Technology:** Statistical spreadsheets and many calculators have the ability to compute the variance and standard deviation. Figure 12(a) shows the population standard deviation obtained in Example 5, and Figure 12(b) shows the sample standard deviation obtained in Example 5 using a TI-83 Plus graphing calculator.

**Figure 12**

Population standard deviation →
```
1-Var Stats
 x̄=179
 Σx=2506
 Σx²=465278
 Sx=35.8458237
 σx=34.54190002
↓n=14
```

Sample standard deviation →
```
1-Var Stats
 x̄=144.6
 Σx=723
 Σx²=105983
 Sx=18.95521037
 σx=16.95405556
↓n=5
```

(a)                              (b)

## Interpretations of the Standard Deviation

The standard deviation is used in conjunction with the mean in order to numerically describe distributions. The mean measures the center of the distribution, while the standard deviation measures the spread of the distribution. So how does the value of the standard deviation relate to the dispersion of the distribution? If we are comparing two populations, then

**the larger the standard deviation, the more dispersion the distribution has**. This rule of thumb is true provided that the variable of interest from the two populations has the same unit of measure. The units of measure must be the same so that we are comparing "apples" with "apples." For example, a standard deviation of $100 is not the same as 100 Japanese yen, because $1 is equivalent to about 120 yen. This means a standard deviation of $100 is substantially higher than a standard deviation of 100 yen.

▶ **EXAMPLE 6**   **Comparing the Variance and Standard Deviation of Two Data Sets**

*Problem:* Refer to the data in Example 1 on page 100. Use the standard deviation to determine whether University A or University B has more dispersion in the IQ scores of its students.

*Approach:* We will use Minitab to compute the standard deviation of IQ for each university. The university with the higher standard deviation will be the university with more dispersion in IQ scores. Recall that on the basis of the histograms, it was apparent that University A had more dispersion. Therefore, we would expect University A to have a higher sample standard deviation.

*Solution:* We enter the data into Minitab and compute the descriptive statistics. See Figure 13.

**Figure 13**   **Descriptive statistics**

| Variable | N | Mean | Median | TrMean | StDev | SEMean |
|---|---|---|---|---|---|---|
| UnivA | 100 | 100.00 | 102.00 | 99.97 | 16.08 | 1.61 |
| UnivB | 100 | 100.00 | 98.000 | 99.73 | 8.35 | 0.83 |

| Variable | Minimum | Maximum | Q1 | Q3 |
|---|---|---|---|---|
| UnivA | 60.00 | 141.00 | 90.00 | 110.00 |
| UnivB | 86.00 | 119.00 | 94.00 | 107.00 |

The sample standard deviation is larger for University A (16.1) than University B (8.4).* Therefore, University A has IQ scores that are more dispersed.   ◀◀

④ **The Empirical Rule**

If data have a distribution that is bell shaped, the following rule can be used to approximate the percentage of data that will lie within $k$ standard deviations of the mean.

**The Empirical Rule**

If a distribution is roughly bell shaped, then
- Approximately 68% of the data will lie within one standard deviation of the mean. That is, approximately 68% of the data lie between $\mu - 1\sigma$ and $\mu + 1\sigma$.
- Approximately 95% of the data will lie within two standard deviations of the mean. That is, approximately 95% of the data lie between $\mu - 2\sigma$ and $\mu + 2\sigma$.
- Approximately 99.7% of the data will lie within three standard deviations of the mean. That is, approximately 99.7% of the data lie between $\mu - 3\sigma$ and $\mu + 3\sigma$.

*Don't forget that we agreed to round the mean and standard deviation to one more decimal place than the original data.

**9. Rats in Space** In the experiment "Regulation of Erythropoiesis during Spaceflight," led by NASA scientist Dr. Robert D. Lange, rats were sent to space on Spacelab Life Sciences I to measure the effect of space travel on weight. The following data represent the weights (in grams) of rats in an Animal Enclosure Module (AEM) on the flight return day for both a ground (control) group and a flight (experimental) group. Compute the range and sample standard deviation for both the control and the experimental groups. Does there appear to be a difference in the variability of weights? What might account for any difference?

| Control Group | | Flight Group | |
|---|---|---|---|
| 305 | 330 | 330 | 320 |
| 320 | 316 | 310 | 326 |
| 345 | | 383 | |

*Source:* NASA Life Sciences Data Archive

**10. Reaction Time** In an experiment conducted on-line at the University of Mississippi, study participants are asked to react to a stimulus. In one experiment, the participant must press a key upon seeing a blue screen. The time (in seconds) to press the key is measured. The same person is then asked to press a key upon seeing a red screen, again with the time to react measured. The results for six study participants are listed in the table. Compute the range and sample standard deviation reaction time for both blue and red. Does there appear to be a difference in the variability of reaction time? What might account for any difference?

| Participant Number | Reaction Time to Blue | Reaction Time to Red |
|---|---|---|
| 1 | 0.582 | 0.408 |
| 2 | 0.481 | 0.407 |
| 3 | 0.841 | 0.542 |
| 4 | 0.267 | 0.402 |
| 5 | 0.685 | 0.456 |
| 6 | 0.45 | 0.533 |

*Source:* PsychExperiments at the University of Mississippi (http://www.olemiss.edu/psychexps/)

**11. Pulse Rates** The following data represent the pulse rates (beats per minute) of the nine students enrolled in a section of Sullivan's course in introductory statistics. Treat the nine students as a population.

(a) Compute the population variance and population standard deviation.

| Student | Pulse |
|---|---|
| Perpectual Bempah | 76 |
| Megan Brooks | 60 |
| Jeff Honeycutt | 60 |
| Clarice Jefferson | 81 |
| Crystal Kurtenbach | 72 |
| Janette Lantka | 80 |
| Kevin McCarthy | 80 |
| Tammy Ohm | 68 |
| Kathy Wojdyla | 73 |

(b) Determine two simple random samples of size three, and compute the sample variance and sample standard deviation of each sample.

(c) Which samples underestimate the population standard deviation? Which overestimate the population standard deviation?

**12. Travel Time** The following data represent the travel time (in minutes) to school for the nine students enrolled in Sullivan's College Algebra course. Treat the nine students as a population.

| Student | Travel Time | Student | Travel Time |
|---|---|---|---|
| Amanda | 39 | Scot | 45 |
| Amber | 21 | Erica | 11 |
| Tim | 9 | Tiffany | 12 |
| Mike | 32 | Glenn | 39 |
| Nicole | 30 | | |

(a) Compute the population variance and population standard deviation.

(b) Determine three simple random samples of size four, and compute the sample variance and sample standard deviation of each sample.

(c) Which samples underestimate the population standard deviation? Which overestimate the population standard deviation?

*In Problems 13–20, compute the range, sample variance, and sample standard deviation.*

**13. Waiting for a Table** The following data represent the number of customers waiting for a table at 6:00 P.M. on Saturday for 40 consecutive Saturdays at Bobak's Restaurant:

| | | | | | | | |
|---|---|---|---|---|---|---|---|
| 11 | 6 | 5 | 8 | 11 | 6 | 3 | 7 |
| 4 | 6 | 5 | 4 | 13 | 14 | 9 | 11 |
| 13 | 8 | 10 | 10 | 9 | 9 | 6 | 5 |
| 10 | 8 | 8 | 8 | 7 | 7 | 3 | 8 |
| 7 | 8 | 9 | 6 | 10 | 11 | 4 | 8 |

**14. Highway Repair** The following data represent the number of potholes on 50 randomly selected 1-mile stretches of highway in the city of Chicago:

| | | | | | |
|---|---|---|---|---|---|
| 2 | 7 | 4 | 7 | 2 | 7 |
| 2 | 2 | 2 | 3 | 4 | 3 |
| 1 | 2 | 3 | 2 | 1 | 4 |
| 2 | 2 | 5 | 2 | 3 | 4 |
| 4 | 1 | 7 | 10 | 3 | 5 |
| 4 | 3 | 3 | 2 | 2 | |
| 1 | 6 | 5 | 7 | 9 | |
| 2 | 2 | 2 | 1 | 5 | |
| 3 | 5 | 1 | 3 | 5 | |

**15. Tensile Strength** Tensile strength is the maximum stress at which one can be reasonably certain failure will not occur. The following data represent tensile strength (in thousands of pounds per square inch) of a composite material to be used in an aircraft.

| | | | | |
|---|---|---|---|---|
| 203.41 | 185.97 | 184.41 | 160.44 | 174.63 |
| 209.58 | 190.67 | 200.73 | 180.95 | 185.34 |
| 213.35 | 207.88 | 206.51 | 201.95 | 205.59 |
| 218.56 | 210.80 | 209.84 | 204.60 | 212.00 |
| 242.76 | 231.46 | 212.15 | 219.51 | 225.25 |

*Source:* Vangel, Mark G., "New Methods for One-sided Tolerance Limits for a One-way Balanced Random-Effects ANOVA Model," *Technometrics*; May, 1992, Vol. 34, Issue 2, pp. 176–185.

**16. Miles on a Cavalier** A random sample of 36 three-year-old Chevy Cavaliers was obtained in the Miami, Florida area and the number of miles on the car was recorded. The data are listed in the following table.

| | | | | |
|---|---|---|---|---|
| 34,122 | 34,511 | 17,595 | 30,500 | 31,883 |
| 39,416 | 34,500 | 44,224 | 35,936 | 32,923 |
| 28,281 | 41,620 | 26,455 | 44,448 | 40,614 |
| 37,021 | 30,283 | 27,368 | 41,043 | 42,398 |
| 30,747 | 15,499 | 37,904 | 30,739 | |
| 35,439 | 26,051 | 23,221 | 38,995 | |
| 17,685 | 32,305 | 38,041 | 41,194 | |
| 29,307 | 39,884 | 35,295 | 29,289 | |

*Source:* cars.com

**17. Serum HDL** Dr. Paul Oswiecmiski randomly selects 40 of his 20–29-year-old patients and obtains the following data regarding their serum HDL cholesterol:

| | | | | | | | |
|---|---|---|---|---|---|---|---|
| 70 | 56 | 48 | 48 | 53 | 52 | 66 | 48 |
| 36 | 49 | 28 | 35 | 58 | 62 | 45 | 60 |
| 38 | 73 | 45 | 51 | 56 | 51 | 46 | 39 |
| 56 | 32 | 44 | 60 | 51 | 44 | 63 | 50 |
| 46 | 69 | 53 | 70 | 33 | 54 | 55 | 52 |

**18. Volume of Philip Morris Stock** The following data represent the number of millions of shares of Philip Morris stock sold in a random sample of 35 trading days in 2000:

| | | | | |
|---|---|---|---|---|
| 3.98 | 8.90 | 10.40 | 7.52 | 13.84 |
| 14.04 | 9.14 | 7.96 | 11.40 | 13.30 |
| 5.29 | 9.69 | 8.94 | 12.10 | 13.25 |
| 4.62 | 10.85 | 24.52 | 14.62 | 16.10 |
| 4.24 | 8.54 | 11.59 | 6.75 | 27.54 |
| 16.06 | 17.62 | 7.17 | 6.13 | 25.32 |
| 7.71 | 10.14 | 4.28 | 7.24 | 10.88 |

*Source:* http://finance.yahoo.com

**19. M&Ms** The following data represent the weights (in grams) of a simple random sample of 50 M&M plain candies:

| | | | | | | |
|---|---|---|---|---|---|---|
| 0.87 | 0.88 | 0.82 | 0.90 | 0.90 | 0.84 | 0.84 |
| 0.91 | 0.94 | 0.86 | 0.86 | 0.86 | 0.88 | 0.87 |
| 0.89 | 0.91 | 0.86 | 0.87 | 0.93 | 0.88 | |
| 0.83 | 0.94 | 0.87 | 0.93 | 0.91 | 0.85 | |
| 0.91 | 0.91 | 0.86 | 0.89 | 0.87 | 0.93 | |
| 0.88 | 0.88 | 0.89 | 0.79 | 0.82 | 0.83 | |
| 0.90 | 0.88 | 0.84 | 0.93 | 0.76 | 0.90 | |
| 0.88 | 0.92 | 0.85 | 0.79 | 0.84 | 0.86 | |

*Source:* Michael Sullivan

**20. Old Faithful** The following data represent the length of eruption (in seconds) for a random sample of eruptions of the Old Faithful geyser in California:

| | | | | | | |
|---|---|---|---|---|---|---|
| 108 | 108 | 99 | 105 | 103 | 103 | 94 |
| 102 | 99 | 106 | 90 | 104 | 110 | 110 |
| 103 | 109 | 109 | 111 | 101 | 101 | |
| 110 | 102 | 105 | 110 | 106 | 104 | |
| 104 | 100 | 103 | 102 | 120 | 90 | |
| 113 | 116 | 95 | 105 | 103 | 101 | |
| 100 | 101 | 107 | 110 | 92 | 108 | |

*Source:* Ladonna Hansen, Park Curator

21. **Rates of Return of Stocks** Stocks may be categorized by industry. The following data represent the five-year rates of return for a simple random sample of financial stocks and energy stocks for the period ending December 28, 2001:

**Financial Stocks**

| | | | | |
|---|---|---|---|---|
| 16.12 | 18.61 | 4.04 | 13.93 | 17.85 |
| 3.34 | 8.39 | 21.32 | 34.22 | 16.56 |
| 28.51 | 13.29 | 11.03 | 14.61 | 6.67 |
| 2.98 | 14.06 | 28.90 | 11.26 | 9.24 |
| 10.97 | 7.13 | 5.32 | 0.63 | 13.43 |
| 17.10 | 22.21 | 6.18 | 5.80 | 1.39 |
| 0.83 | 5.63 | 8.39 | 8.02 | 1.23 |
| 3.32 | 9.42 | 2.98 | 13.62 | 6.36 |

*Source:* Morningstar.com

**Energy Stocks**

| | | | | |
|---|---|---|---|---|
| 2.49 | 9.70 | 21.67 | 8.31 | 5.07 |
| 7.70 | 23.72 | 5.59 | 6.09 | 9.69 |
| 1.82 | 12.68 | 6.07 | 11.09 | 6.35 |
| 2.00 | 14.22 | 13.17 | 6.29 | 6.25 |
| 7.39 | 9.09 | 30.39 | 11.25 | 7.17 |
| 9.48 | 9.23 | 9.70 | 5.41 | 10.29 |
| 6.11 | 7.38 | 11.60 | 12.10 | 13.38 |

*Source:* Morningstar.com

   (a) Compute the arithmetic mean and the median for each industry. Which sector has the higher arithmetic mean rate of return? Which sector has the higher median rate of return?
   (b) Compute the standard deviation for each industry. In finance, the standard deviation rate of return is called **risk**. Which sector is riskier?

22. **American League versus National League** The following data represent the earned-run average of the top 37 pitchers in both the American League and the National League during the 2001 season:

**American League**

| | | | | |
|---|---|---|---|---|
| 3.05 | 3.15 | 3.16 | 3.29 | 3.37 |
| 3.43 | 3.45 | 3.49 | 3.51 | 3.59 |
| 3.60 | 3.65 | 3.77 | 3.90 | 3.94 |
| 3.99 | 4.05 | 4.08 | 4.09 | 4.09 |
| 4.09 | 4.25 | 4.32 | 4.36 | 4.37 |
| 4.39 | 4.42 | 4.45 | 4.45 | 4.50 |
| 4.76 | 4.93 | 5.02 | 5.08 | 5.17 |
| 5.54 | 5.82 | | | |

**National League**

| | | | | |
|---|---|---|---|---|
| 2.49 | 2.98 | 3.04 | 3.05 | 3.09 |
| 3.16 | 3.29 | 3.31 | 3.36 | 3.40 |
| 3.42 | 3.50 | 3.57 | 3.57 | 3.69 |
| 3.70 | 3.80 | 4.03 | 4.05 | 4.05 |
| 4.19 | 4.31 | 4.33 | 4.34 | 4.42 |
| 4.42 | 4.45 | 4.46 | 4.46 | 4.47 |
| 4.48 | 4.49 | 4.63 | 4.79 | 4.81 |
| 4.85 | 4.90 | | | |

   (a) Compute the arithmetic mean and the median for each league. Which league has the higher arithmetic mean earned-run average? Which league has the higher median earned-run average?
   (b) Compute the standard deviation for each league. Which league has more dispersion?

23. **Golf Scores** The data to the right represent the golf scores of Michael and Kevin for their most recent 15 rounds of golf:

   (a) Compute the arithmetic mean score of each golfer.
   (b) Compute the median score of each golfer.
   (c) Compute the mode score of each golfer.
   (d) Compute the range of each golfer.
   (e) Compute the sample standard deviation of each golfer.
   (f) Which player is more consistent? Why?

**Michael's Scores**

| | | | | |
|---|---|---|---|---|
| 83 | 88 | 81 | 75 | 79 |
| 80 | 83 | 85 | 82 | 80 |
| 77 | 78 | 82 | 83 | 81 |

**Kevin's Scores**

| | | | | |
|---|---|---|---|---|
| 90 | 73 | 73 | 80 | 85 |
| 89 | 78 | 83 | 75 | 79 |
| 88 | 84 | 86 | 82 | 73 |

**24. SAT Scores** The following data represent SAT Math scores for 15 randomly selected students in two different high school districts:

| District 115 | | | | | District 231 | | | | |
|---|---|---|---|---|---|---|---|---|---|
| 377 | 412 | 530 | 594 | 561 | 409 | 445 | 519 | 559 | 538 |
| 415 | 565 | 464 | 523 | 529 | 447 | 540 | 478 | 514 | 518 |
| 537 | 459 | 468 | 532 | 495 | 523 | 474 | 480 | 520 | 497 |

(a) Compute the arithmetic mean SAT Math score of each district.
(b) Compute the median SAT Math score of each district.
(c) Compute the mode SAT Math score of each district.
(d) Compute the range of each district.
(e) Compute the sample standard deviation of each district.
(f) Which district has less dispersion in the scores? Why?

---

**25. The Empirical Rule** A random sample of 50 gas stations in Cook County, Illinois, resulted in a mean price per gallon of $1.60 and a standard deviation of $0.07.

(a) A histogram of the data indicates that the data follow a bell-shaped distribution. Use the Empirical Rule to determine the percentage of gas stations that have prices within three standard deviations of the mean.
(b) Determine the percentage of gas stations that have prices between $1.46 and $1.74, according to the Empirical Rule.

**26. The Empirical Rule** The weight, in grams, of both kidneys based upon a random sample of 30 normal men aged 40–49 resulted in a mean of 325 grams, with a standard deviation of 30 grams.

(a) A histogram of the data indicates that the data follow a bell-shaped distribution. Use the Empirical Rule to determine the percentage of kidneys that weigh within three standard deviations of the mean.
(b) Determine the percentage of kidneys that weigh between 265 grams and 385 grams, according to the Empirical Rule.

**27. The Empirical Rule** Use the results of Problem 17 in this section, Problem 17 in Section 3.1, and Problem 21 in Section 2.2.

(a) On the basis of the histogram drawn in Problem 21 in Section 2.2, comment on the appropriateness of using the Empirical Rule.
(b) Use the Empirical Rule to approximate the percentage of patients with serum HDL within three standard deviations of the mean.
(c) Use the Empirical Rule to approximate the percentage of patients with serum HDL between 29.3 and 72.9. *Hint*: $\bar{x} = 51.1$, $s = 10.9$.
(d) Determine the actual percentage of patients with serum HDL between 29.3 and 72.9. Compare the results to those obtained from the Empirical Rule.

**28. The Empirical Rule** Use the results of Problem 19 in this section, Problem 19 in Section 3.1 and Problem 23 in Section 2.2.

(a) On the basis of the histogram drawn in Problem 23 in Section 2.2, comment on the appropriateness of using the Empirical Rule.
(b) Use the Empirical Rule to approximate the percentage of M&Ms within three standard deviations of the mean.
(c) Use the Empirical Rule to approximate the percentage of M&Ms that weigh between 0.794 and 0.954 grams. *Hint*: $\bar{x} = 0.874$, $s = 0.040$.
(d) Determine the actual percentage of M&Ms that weigh between 0.794 miles and 0.954 grams. Compare the results to those obtained from the Empirical Rule.

**29.** Refer to the tensile strength data from Problem 15. Suppose the data value entered as 185.97 was accidentally recorded as 815.97. Recompute the range and standard deviation for the corrected data set. How did this change affect the statistics? Are the data more or less dispersed? Are these statistics resistant?

**30.** Refer to the data from Problem 16. Suppose the data value entered as 17,685 should have been recorded as 117,685. Recompute the range, and standard deviation for the corrected data set. How did this change affect the statistics? Are the data more or less dispersed? Are these statistics resistant?

**31.** Use a sample of five test scores of 65, 70, 71, 75, and 95 to answer the following questions:

(a) Compute the sample standard deviation.
(b) Suppose the professor decides to curve the exam by adding 4 points to each person's score. Compute the sample standard deviation of the adjusted scores.
(c) Compare the unadjusted test score's standard deviation to the curved test score's standard deviation. What effect did adding 4 to each score have on the sample standard deviation?
(d) Multiply each exam score by 2 and determine the sample standard deviation of the new test scores.
(e) What effect did multiplying the scores by 2 have on the sample standard deviation?

**32.** Which histogram depicts a higher standard deviation? Justify your answer.

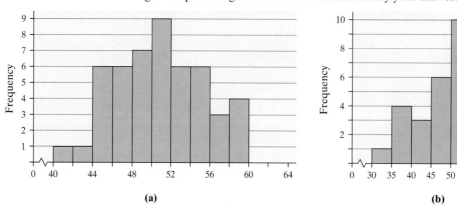

(a)           (b)

**33.** Match the histograms to the summary statistics given.

|       | Mean | Median | Standard Deviation |
|-------|------|--------|--------------------|
| (I)   | 53   | 53     | 1.3                |
| (II)  | 60   | 60     | 11                 |
| (III) | 53   | 53     | 9                  |
| (IV)  | 53   | 53     | 0.12               |

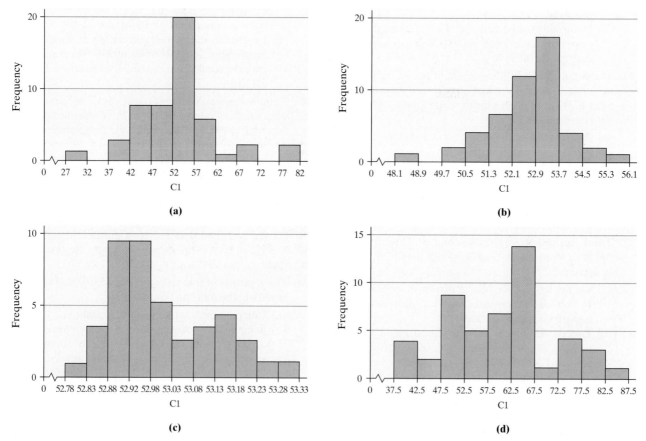

(a)           (b)

(c)           (d)

**34.** Compute the sample standard deviation of the following test scores: 78, 78, 78, 78. What can be said about a data set in which all the values are identical?

## Technology Step-by-Step
### Determining the Range, Variance, and Standard Deviation

The same steps followed to obtain the measures of central tendency from raw data can be used to obtain the measures of dispersion.

## 3.3   Measures of Central Tendency and Dispersion from Grouped Data

**Preparing for This Section**   Before getting started, review the following:

✓ Summarizing discrete data in tables (Section 2.2, pp. 55–56)

✓ Summarizing continuous data in tables (Section 2.2, pp. 56–58)

**Objectives**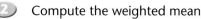

**1** Approximate the mean of a variable from grouped data

**2** Compute the weighted mean

**3** Approximate the variance and standard deviation of a variable from grouped data

We have discussed how to compute descriptive statistics from raw data, but many times the data that we have access to have already been summarized in frequency distributions (grouped data). While we cannot obtain exact values of the mean or standard deviation without raw data, these measures can be approximated using the techniques discussed in this section.

**1** **The Arithmetic Mean from Grouped Data**

Before we can provide a method for computing the mean from grouped data, we must learn how to obtain the class midpoint of a class.

*Definition*   The **class midpoint** is found by adding the lower class limit and upper class limit of a class and dividing the result by 2. That is,

$$\text{Class midpoint} = \frac{\text{Lower class limit } + \text{ Upper Class limit}}{2}$$

Since raw data cannot be retrieved from a frequency table, we assume that, within each class, all the data values are equal to the class midpoint. We then multiply the class midpoint by the frequency. This product is expected to be close to the sum of the data that lie within the class. We repeat the process for each class and sum the results. This sum approximates the sum of all the data.

*Definition*   **The Approximate Mean of a Variable from a Frequency Distribution**

Population Mean

$$\mu = \frac{\sum x_i f_i}{\sum f_i} = \frac{x_1 f_1 + x_2 f_2 + \cdots + x_n f_n}{f_1 + f_2 + \cdots + f_n}$$

Sample Mean

$$\bar{x} = \frac{\sum x_i f_i}{\sum f_i} = \frac{x_1 f_1 + x_2 f_2 + \cdots + x_n f_n}{f_1 + f_2 + \cdots + f_n} \tag{1}$$

where   $x_i$ is the midpoint or value of the $i$th class

$f_i$ is the frequency of the $i$th class

$n$ is the number of classes

In Formula (1), $x_1f_1$ approximates the sum of all the data values in the first class, $x_2f_2$ approximates the sum of all the data values in the second class, and so on. Notice that the formulas for the population mean and sample mean are identical, just as they were for computing the mean from raw data.

▶ EXAMPLE 1    **Approximating the Mean for Continuous Quantitative Data from the Frequency Distribution**

**TABLE 11**

| Class (three-year rate of return) | Frequency |
| --- | --- |
| 10.0–14.9 | 7 |
| 15.0–19.9 | 11 |
| 20.0–24.9 | 8 |
| 25.0–29.9 | 6 |
| 30.0–34.9 | 3 |
| 35.0–39.9 | 3 |
| 40.0–44.9 | 0 |
| 45.0–49.9 | 2 |

*Problem:* The frequency distribution in Table 11 represents the three-year rate of return of a random sample of 40 small-capitalization growth mutual funds. Approximate the mean three-year rate of return.

*Approach:* We perform the following steps to approximate the mean:

**Step 1:** Determine the class midpoint of each class. Recall that the class midpoint is found by adding the lower class limit and the upper class limit and dividing the result by 2.

**Step 2:** Compute the sum of the frequencies, $\Sigma f_i$.

**Step 3:** Multiply the class midpoint by the frequency to obtain $x_if_i$ for each class.

**Step 4:** Compute $\Sigma x_if_i$.

**Step 5:** Substitute into Formula (1) to obtain the mean from grouped data.

*Solution:*

**Step 1:** The class midpoint of the first class is $\dfrac{10.0 + 14.9}{2} = 12.45$, so $x_1 = 12.45$. The remaining class midpoints are listed in column 2 of Table 12.

**Step 2:** We add the frequencies in column 3 to obtain $\Sigma f_i = 7 + 11 + \ldots + 2 = 40$.

**Step 3:** Compute the values of $x_if_i$ by multiplying each class midpoint by the corresponding frequency and obtain the results shown in column 4 of Table 12.

**Step 4:** We add the values in column 4 of Table 12 to obtain $\Sigma x_if_i = 928$.

**TABLE 12**

| Class (three-year rate of return) | Class Midpoint, $x_i$ | Frequency, $f_i$ | $x_if_i$ |
| --- | --- | --- | --- |
| 10.0–14.9 | 12.45 | 7 | $(12.45)(7) = 87.15$ |
| 15.0–19.9 | 17.45 | 11 | 191.95 |
| 20.0–24.9 | 22.45 | 8 | 179.6 |
| 25.0–29.9 | 27.45 | 6 | 164.7 |
| 30.0–34.9 | 32.45 | 3 | 97.35 |
| 35.0–39.9 | 37.45 | 3 | 112.35 |
| 40.0–44.9 | 42.45 | 0 | 0 |
| 45.0–49.9 | 47.45 | 2 | 94.9 |
| | | $\Sigma f_i = 40$ | $\Sigma x_if_i = 928$ |

**Step 5:** Substituting into Formula (1), we obtain

$$\bar{x} = \frac{\Sigma x_if_i}{\Sigma f_i} = \frac{928}{40} \approx 23.2$$

The approximate mean three-year rate of return is 23.2 percent.    ◀◀

**EXPLORATION**   Compute the mean three-year rate of return from the raw data listed in Example 3 on page 57 from Section 2.2. Conclude that the mean from the frequency distribution does not necessarily equal the mean from the raw data.   ◄◄

**Note**   To compute the mean from a frequency distribution where the data are discrete, treat each category of data as the class midpoint. For discrete data the mean from grouped data will equal the mean from raw data.   ◄

**NW**   *Now compute the mean of the frequency distribution in Problem 3.*

## ② Weighted Mean

Sometimes, certain data values have a higher importance or weight associated with them. In this case, we compute the *weighted mean*. For example, your grade point average is a weighted mean, with the weights equal to the number of credit hours in the course. The value of the variable is equal to the grade converted to a point value.

*Definition*

The **weighted mean**, $\bar{x}_w$, of a variable is found by multiplying the values of the variable by their corresponding weight and dividing the result by the sum of the weights. It can be expressed using the formula

$$\bar{x}_w = \frac{\sum w_i x_i}{\sum w_i} = \frac{w_1 x_1 + w_2 x_2 + \cdots + w_n x_n}{w_1 + w_2 + \cdots + w_n} \qquad (2)$$

where   $w_i$ is the weight of the $i$th observation

$x_i$ is the value of the $i$th observation

▶ **EXAMPLE 2**   **Computing the Weighted Mean**

*Problem:* Marissa just completed her first semester in college. She earned an "A" in her four-hour statistics course, a "B" in her three-hour sociology course, an "A" in her three-hour psychology course, a "C" in her five-hour computer programming course and an "A" in her one-hour drama course. Determine Marissa's grade point average.

*Approach:* We must assign point values to each grade. Let an "A" equal four points, a "B" equal three points and a "C" equal two points. The number of credit hours for each course determines its weight. So a five-hour course gets a weight of five, a four-hour course gets a weight of four, and so on. We multiply the weight of each course by the points earned in the course, sum these products, and divide the sum by the number of credit hours.

*Solution:*

$$\text{GPA} = \bar{x}_w = \frac{\sum w_i x_i}{\sum w_i} = \frac{4(4) + 3(3) + 3(4) + 5(2) + 1(4)}{4 + 3 + 3 + 5 + 1} = \frac{51}{16} = 3.19$$

**NW**   *Now Work Problem 9.*

Marissa's grade point average for her first semester is 3.19.   ◄◄

③ ## Approximating the Variance and Standard Deviation from Grouped Data

The procedure for approximating the variance and standard deviation from grouped data is similar to that of finding the mean from grouped data. Again, because we do not have access to the original data, the variance is approximate.

*Definition*

> **The Approximate Variance of a Variable from a Frequency Distribution:**
>
> Population Variance        Sample Variance
>
> $$\sigma^2 = \frac{\sum (x_i - \mu)^2 f_i}{\sum f_i} \quad * \qquad s^2 = \frac{\sum (x_i - \overline{x})^2 f_i}{(\sum f_i) - 1} \qquad (3)$$
>
> where  $x_i$ is the midpoint or value of the $i$th class
>
> $f_i$ is the frequency of the $i$th class

We approximate the standard deviation by taking the square root of the variance.

▶ **EXAMPLE 3**   **Approximating the Variance and Standard Deviation from a Frequency Distribution**

*Problem:* The frequency distribution in Table 11 on page 116 summarizes the three-year rate of return of a random sample of 40 small-capitalization growth mutual funds. Approximate the variance and standard deviation three-year rate of return.

*Approach:* We will use the sample variance Formula (3).

*Step 1:* Create a table with the class in the first column, the class midpoint in the second column, the frequency in the third column, and the unrounded mean in the fourth column.

*Step 2:* Compute the deviation about the mean, $x_i - \overline{x}$, for each class where $x_i$ is the class midpoint of the $i$th class and $\overline{x}$ is the sample mean. Enter the results in column 5.

*Step 3:* Square the deviation about the mean and multiply this result by the frequency to obtain $(x_i - \overline{x})^2 f_i$. Enter the results in column 6.

*Step 4:* Add the entries in columns 3 and 6 to obtain $\sum f_i$ and $\sum (x_i - \overline{x})^2 f_i$.

*Step 5:* Substitute the values obtained in Step 4 into Formula (3) to obtain an approximate value for the sample variance.

*Solution:*

*Step 1:* We create Table 13. Column 1 contains the classes. Column 2 contains the class midpoint of each class. Column 3 contains the frequency of each class. Column 4 contains the unrounded sample mean obtained in Example 1.

*Step 2:* Column 5 of Table 13 contains the deviation about the mean, $x_i - \overline{x}$, for each class.

*Step 3:* Column 6 contains the values of the squared deviation about the mean multiplied by the frequency $(x_i - \overline{x})^2 f_i$.

*Step 4:* We add up the entries in columns 3 and 6 and obtain $\sum f_i = 40$ and $\sum (x_i - \overline{x})^2 f_i = 3327.5$.

---

*An algebraically equivalent formula for the population variance is $\dfrac{\sum x_i^2 f_i - \dfrac{(\sum x_i f_i)^2}{\sum f_i}}{\sum f_i}$.

| Class (three-year rate of return) | Class Midpoint, $x_i$ | Frequency, $f_i$ | $\bar{x}$ | $x_i - \bar{x}$ | $(x_i - \bar{x})^2 f_i$ |
|---|---|---|---|---|---|
| 10.0–14.9 | 12.45 | 7 | 23.2 | −10.75 | 808.9375 |
| 15.0–19.9 | 17.45 | 11 | 23.2 | −5.75 | 363.6875 |
| 20.0–24.9 | 22.45 | 8 | 23.2 | −0.75 | 4.5 |
| 25.0–29.9 | 27.45 | 6 | 23.2 | 4.25 | 108.375 |
| 30.0–34.9 | 32.45 | 3 | 23.2 | 9.25 | 256.6875 |
| 35.0–39.9 | 37.45 | 3 | 23.2 | 14.25 | 609.1875 |
| 40.0–44.9 | 42.45 | 0 | 23.2 | 19.25 | 0 |
| 45.0–49.9 | 47.45 | 2 | 23.2 | 24.25 | 1176.125 |
| | | $\Sigma f_i = 40$ | | | $\Sigma(x_i - \bar{x})^2 f_i = 3327.5$ |

**TABLE 13**

**Step 5:** Substitute these values into Formula (3) to obtain an approximate value for the sample variance.

$$s^2 = \frac{\Sigma(x_i - \bar{x})^2 f_i}{(\Sigma f_i) - 1} = \frac{3327.5}{39} \approx 85.32$$

Take the square root of the unrounded estimate of the sample variance to obtain an approximation of the sample standard deviation.

$$s = \sqrt{s^2} = \sqrt{\frac{3327.5}{39}} \approx 9.24 \text{ percent}$$

We approximate the sample standard deviation three-year rate of return to be 9.24 percent.   ◂◂

**Using Technology:** Statistical software can be used to compute means and standard deviations from grouped data, but it is easier to use a graphing calculator with advanced statistical features. Figure 16 shows the results from Examples 1 and 3 using a TI-83 Plus graphing calculator.

Figure 16

Approximate mean

Approximate sample standard deviation

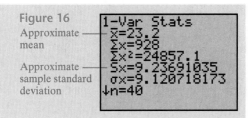

```
1-Var Stats
 x̄=23.2
 Σx=928
 Σx²=24857.1
 Sx=9.23691035
 σx=9.120718173
↓n=40
```

**EXPLORATION**   Compute the sample standard deviation three-year rate of return from the raw data listed in Example 3 on page 57 from Section 2.2. Conclude that the standard deviation from the frequency distribution does not necessarily equal the standard deviation from the raw data.   ◂◂

**NW**   *Now compute the standard deviation from the frequency distribution in Problem 3.*

## 3.3 Assess Your Understanding

### Concepts and Vocabulary

1. Explain the role of the class midpoint in the formulas to approximate the arithmetic mean and the standard deviation.

### Exercises

1. **Health Insurance** The following data represent the number of people aged 25–64 covered by health insurance in 1998. Approximate the arithmetic mean and standard deviation age.

| Age | Number (in millions) |
|-----|----------------------|
| 25–34 | 38.5 |
| 35–44 | 44.7 |
| 45–54 | 35.2 |
| 55–64 | 22.9 |

*Source:* United States Census Bureau, Current Population Reports

2. **Earthquakes** The following data represent the magnitude of earthquakes in the United States in 2001. Approximate the arithmetic mean and standard deviation magnitude.

| Magnitude | Number |
|-----------|--------|
| 0–0.9 | 374 |
| 1.0–1.9 | 2 |
| 2.0–2.9 | 604 |
| 3.0–3.9 | 760 |
| 4.0–4.9 | 268 |
| 5.0–5.9 | 37 |
| 6.0–6.9 | 6 |
| 7.0–7.9 | 1 |

*Source:* National Earthquake Information Center

3. **Meteorology** The following data represent the high-temperature distribution for the month of August in Chicago since 1872.

| Temperature | Days |
|-------------|------|
| 50–59 | 1 |
| 60–69 | 308 |
| 70–79 | 1,519 |
| 80–89 | 1,626 |
| 90–99 | 503 |
| 100–109 | 11 |

*Source:* NOAA

(a) Approximate the arithmetic mean and standard deviation temperature.

(b) Draw a frequency histogram of the data to verify that the data are bell shaped.

(c) According to the Empirical Rule, 95% of days in the month of August will be between what two temperatures?

4. **Multiple Births** The following data represent the number of live multiple delivery births (three or more babies) in 1999 for women 15–49 years old.

| Age | Number of Multiple Births |
|-----|---------------------------|
| 15–19 | 66 |
| 20–24 | 411 |
| 25–29 | 1,653 |
| 30–34 | 2,926 |
| 35–39 | 1,813 |
| 40–44 | 356 |
| 45–49 | 96 |

*Source:* National Vital Statistics Report, Vol. 49, No. 1; April 17, 2001

(a) Approximate the arithmetic mean and standard deviation age.

(b) Draw a frequency histogram of the data to verify that the data are bell shaped.

(c) According to the Empirical Rule, 95% of mothers of multiple births will be between what two ages?

5. **Tensile Strength** Use the first frequency distribution obtained in Problem 19 in Section 2.2 to approximate the arithmetic mean and standard deviation tensile strength. Compare these results to the values obtained in Problem 15 in Sections 3.1 and 3.2.

6. **Miles on a Cavalier** Use the first frequency distribution obtained in Problem 20 in Section 2.2 to approximate the arithmetic mean and standard deviation miles. Compare these results to the values obtained in Problem 16 in Sections 3.1 and 3.2.

7. **Serum HDL** Use the first frequency distribution obtained in Problem 21 in Section 2.2 to approximate the arithmetic mean and standard deviation serum HDL. Compare these results to the values obtained in Problem 17 in Sections 3.1 and 3.2.

8. **Volume of Philip Morris Stock** Use the first frequency distribution obtained in Problem 22 in Section 2.2 to approximate the arithmetic mean and standard deviation number of shares traded. Compare these results to the values obtained in Problem 18 in Sections 3.1 and 3.2.

9. **Grade Point Average** Marissa has just completed her second semester in college. She earned a "B" in her five-hour calculus course, an "A" in her three-hour social

work course, an "A" in her four-hour biology course and a "C" in her three-hour American literature course. Assuming that an "A" equals four points, a "B" equals three points, and a "C" equals two points, determine Marissa's grade point average.

10. **Computing Class Average** In Marissa's calculus course, attendance counts for 5% of the grade, quizzes count for 10% of the grade, exams count for 60% of the grade, and the final exam counts for 25% of the grade. Marissa had a 100% average for attendance, 93% for quizzes, 86% for exams, and 85% on the final. Determine Marissa's course average.

11. **Mixed Chocolates** Michael and Kevin want to buy chocolates; however, they can't agree on whether they want chocolate-covered almonds, chocolate-covered peanuts, or chocolate-covered raisins. They agree to create a mix. They bought 4 pounds of chocolate-covered almonds at $3.50 per pound, 3 pounds of chocolate-covered peanuts for $2.75 per pound, and 2 pounds of chocolate-covered raisins for $2.25 per pound. Determine the cost per pound of the mix.

12. **Nut Mix** Michael and Kevin return to the candy store, but this time they want to purchase nuts. They can't decide between peanuts, cashews or almonds. They again agree to create a mix. They bought 2.5 pounds of peanuts for $1.30 per pound, 4 pounds of cashews for $4.50 per pound and 2 pounds of almonds for $3.75 per pound. Determine the price per pound of the mix.

13. **Population** The following data represent the male and female population by age of the United States for residents under 100 years old in 1999:

| Age | Male Resident Population (in thousands) | Female Resident Population (in thousands) |
|---|---|---|
| 0–9 | 19,881 | 18,994 |
| 10–19 | 20,174 | 19,143 |
| 20–29 | 18,264 | 18,038 |
| 30–39 | 21,001 | 21,305 |
| 40–49 | 20,550 | 21,091 |
| 50–59 | 14,187 | 15,148 |
| 60–69 | 9,312 | 10,669 |
| 70–79 | 6,926 | 9,191 |
| 80–89 | 2,664 | 4,788 |
| 90–99 | 384 | 1,112 |

*Source:* U.S. Census Bureau

(a) Approximate the population mean and standard deviation age for males.
(b) Approximate the population arithmetic mean and standard deviation age for females.
(c) Which gender has the higher mean age?
(d) Which gender has more dispersion in age?

14. **Age of Mother** The following data represent the age of the mother at childbirth for 1980 and 1997.

| Age of Mother | Number of Births in 1980 (in thousands) | Number of Births in 1997 (in thousands) |
|---|---|---|
| 10–14 | 1.1 | 1.2 |
| 15–19 | 53.0 | 52.9 |
| 20–24 | 115.1 | 110.9 |
| 25–29 | 112.9 | 114.3 |
| 30–34 | 61.9 | 85.4 |
| 35–39 | 19.8 | 36.0 |
| 40–44 | 3.9 | 6.9 |
| 45–49 | 0.2 | 0.3 |

*Source:* U.S. National Center for Health Statistics

(a) Approximate the population mean and standard deviation age of mothers in 1980.
(b) Approximate the population mean and standard deviation age of mothers in 1997.
(c) Which year has the higher mean age?
(d) Which year has more dispersion in age?

## Technology Step-by-Step
Determining the Mean and Standard Deviation from Grouped Data

*TI-83 Plus*  **Step 1:** Enter the class midpoint in L1 and the frequency or relative frequency in L2 by pressing STAT and selecting 1:Edit.

**Step 2:** Press STAT, highlight the CALC menu and select 1:1-Var Stats

**Step 3:** With 1-Var Stats appearing on the HOME screen, press 2nd 1 to insert L1 on the HOME screen. Then press the comma and press 2nd 2 to insert L2 on the HOME screen. So, the HOME screen should have the following:

```
1-Var Stats L1, L2
```

Press ENTER to obtain the mean and standard deviation.

**Objectives** ❶ Determine and interpret z-scores

❷ Determine and interpret percentiles

❸ Determine and interpret quartiles

❹ Check a set of data for outliers

In Section 3.1, we were able to find measures of central tendency. Measures of central tendency are meant to describe the "typical" data value. Section 3.2 discussed measures of dispersion, which describe the amount of "spread" in a set of data. In this section, we discuss measures of position—that is, we wish to describe the relative position of a certain data value within the entire set of data.

❶ **Z-scores**

During the 2000 baseball season, the city of New York sponsored a "subway" series. Both of its teams (the Yankees and Mets) were in the World Series. Which team was the better home-run-hitting team? In 2000, the New York Yankees hit 205 home runs while the New York Mets hit 198. A quick comparison might lead one to think the Yankees are a better home-run-hitting team than the Mets. However, this comparison would be unfair because the two teams play in different leagues—the Yankees play in the American League, where a designated hitter bats for the pitcher, whereas the Mets play in the National League, where the pitcher must bat (pitchers are typically poor hitters). In order to compare the two teams' home-run production, we need to determine their relative standings in their respective leagues. This can be accomplished using a *z-score*.

**Definition**

The **z-score** represents the number of standard deviations that a data value is from the mean. It is obtained by subtracting the mean from the data value and dividing this result by the standard deviation. There is both a population z-score and a sample z-score; their formulas follow:

$$\text{Population Z-Score} \qquad \text{Sample Z-Score}$$

$$z = \frac{x - \mu}{\sigma} \qquad\qquad z = \frac{x - \overline{x}}{s} \qquad \textbf{(1)}$$

The z-score is unitless; it has mean 0 and standard deviation 1.

**In Your Own Words**

Z-scores provide a way to compare "apples" to "oranges" by converting variables with different centers and/or spreads to variables with the same center (0) and spread (1).

If a data value is larger than the mean, the z-score will be positive. If a data value is smaller than the mean, the z-score will be negative. If the data value equals the mean, the z-score will be zero. Z-scores measure the number of standard deviations above or below the mean. For example, a z-score of 1.24 would be interpreted as "the data value is 1.24 standard deviations above the mean." A z-score of $-2.31$ would be interpreted as "the data value is 2.31 standard deviations below the mean".

We are now prepared to determine whether the Yankees or Mets had a better year in home-run production.

▶ **EXAMPLE 1** Comparing z-scores

*Problem:* Determine whether the New York Yankees or New York Mets had a relatively better home-run-hitting season. The Yankees hit 205 home runs and play in the American League, where the mean number of home runs was $\mu = 192$. The standard deviation was $\sigma = 36.0$. The Mets hit 198 home runs and play in the National League, where the mean number of home runs was $\mu = 186.9$. The standard deviation was $\sigma = 28.7$.

*Approach:* To determine which team had the relatively better home-run-hitting season, we compute each team's $z$-score. The team with the higher $z$-score had the better season. Because we know the values of the population parameters, we will compute the population $z$-score.

*Solution:* First, we compute the $z$-score for the Yankees:

$$z\text{-score} = \frac{x - \mu}{\sigma} = \frac{205 - 192}{36.0} = 0.36*$$

Next, we compute the $z$-score for the Mets:

$$z\text{-score} = \frac{x - \mu}{\sigma} = \frac{198 - 186.9}{28.7} = 0.39$$

So, the Yankees had home-run production 0.36 standard deviation above the mean, while the Mets had home-run production 0.39 standard deviation above the mean. Therefore, the Mets had a relatively better year at hitting home runs. ◄◄

**EXPLORATION** Compute the $z$-score for all teams in the American League in 2001, using the data from Table 1 in Section 3.1 on page 85. Compute the mean and standard deviation of the $z$-scores. Conclude that the mean of $z$-scores is 0 and the standard deviation is 1. ◄◄

**NW** *Now Work Problem 1.*

## ❷ Percentiles

Recall that the median divides the lower 50% of a set of data from the upper 50% of the set. In general, the **$k$th percentile**, denoted $P_k$, of a set of data divides the lower $k$% of a data set from the upper $(100 - k)$% of the set. Percentiles divide a data set that is written in ascending order into 100 parts, so that there are 99 possible percentiles that can be computed. For example, $P_1$ divides the bottom 1% of the data from the top 99% of the data, while $P_{99}$ divides the lower 99% of the data from the top 1% of the data. Figure 17 displays the 99 possible percentiles.

Figure 17

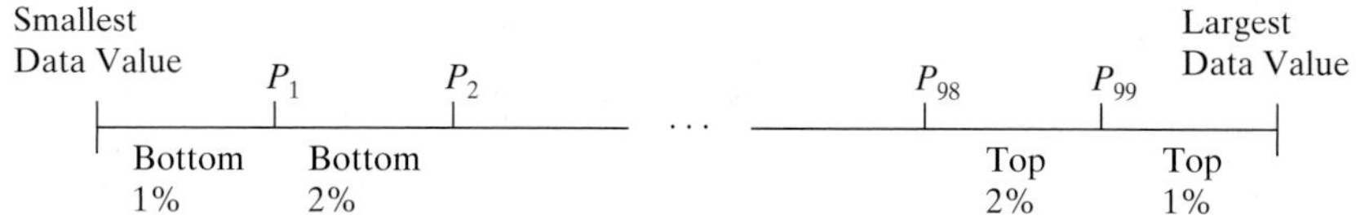

Smallest Data Value · · · Largest Data Value

$P_1$  $P_2$  $P_{98}$  $P_{99}$

Bottom 1%  Bottom 2%  Top 2%  Top 1%

If a data value lies at the $40^{\text{th}}$ percentile, then approximately 40% of the data is less than this value and approximately 60% of the data is higher than this value.

Percentiles are often used in order to give the relative standing of a data value. Many standardized exams, such as the SAT college entrance exam, use percentiles to provide students with an understanding of how they scored on the exam in relation to all other students who took the exam. For example, in 2000, an SAT verbal score of 580 was at the $75^{\text{th}}$ percentile. This means approximately 75% of the scores are below 580 and 25% are above 580. Pediatricians use percentiles to describe the progress of a newborn baby's weight gain relative to other newborn babies. A three-to-five-month-old male child who weighs 14.3 pounds would be at the $15^{\text{th}}$ percentile.

*Z-scores are typically rounded to two decimal places.

The following steps can be used to compute the $k$th percentile:

**Determining the $k$th Percentile, $P_k$**

**Step 1:** Arrange the data in ascending order.

**Step 2:** Compute an index $i$, using the formula

$$i = \left(\frac{k}{100}\right)n \qquad (2)$$

where $k$ is the percentile of the data value and $n$ is the number of individuals in the data set.

**Step 3: (a)** If $i$ is not an integer, round up* to the next highest integer. Locate the $i$th value of the data set written in ascending order. This number represents the $k$th percentile.

**(b)** If $i$ is an integer, the $k$th percentile is the arithmetic mean of the $i$th and $(i + 1)$st data value.

**Caution**

Don't forget to write the data in ascending order before finding the percentile.

An example should clarify the procedure.

▶ **EXAMPLE 2** **Determining the Percentile of a Data Value**

*Problem:* In order to qualify for "exempt" status on the PGA tour, a golfer must be in the top 130 money winners for the previous season. The data in Table 14 represent the earnings of the top 130 golfers on the PGA tour in 2000. Find the earnings that correspond to the 85th percentile.

| TABLE 14 | | | | | | |
|---|---|---|---|---|---|---|
| 1. 339,242 | 20. 425,624 | 39. 519,740 | 58. 638,422 | 77. 877,390 | 96. 1,263,585 | 115. 1,968,685 |
| 2. 346,569 | 21. 459,812 | 40. 527,741 | 59. 649,674 | 78. 889,153 | 97. 1,320,278 | 116. 2,002,068 |
| 3. 379,349 | 22. 460,024 | 41. 528,959 | 60. 660,707 | 79. 889,381 | 98. 1,368,888 | 117. 2,023,465 |
| 4. 387,716 | 23. 461,981 | 42. 537,105 | 61. 669,709 | 80. 896,098 | 99. 1,384,508 | 118. 2,025,781 |
| 5. 388,341 | 24. 464,480 | 43. 538,706 | 62. 673,387 | 81. 947,118 | 100. 1,543,818 | 119. 2,068,499 |
| 6. 391,075 | 25. 466,345 | 44. 548,070 | 63. 700,738 | 82. 963,974 | 101. 1,550,592 | 120. 2,099,943 |
| 7. 393,059 | 26. 466,712 | 45. 552,795 | 64. 724,580 | 83. 964,346 | 102. 1,557,720 | 121. 2,169,727 |
| 8. 393,316 | 27. 467,431 | 46. 564,918 | 65. 728,635 | 84. 990,215 | 103. 1,563,115 | 122. 2,337,765 |
| 9. 397,610 | 28. 469,590 | 47. 580,510 | 66. 731,925 | 85. 999,460 | 104. 1,597,139 | 123. 2,413,345 |
| 10. 398,393 | 29. 482,028 | 48. 583,605 | 67. 741,995 | 86. 1,004,827 | 105. 1,604,952 | 124. 2,462,846 |
| 11. 402,017 | 30. 482,744 | 49. 590,109 | 68. 747,312 | 87. 1,040,244 | 106. 1,631,695 | 125. 2,547,829 |
| 12. 403,982 | 31. 485,589 | 50. 597,021 | 69. 753,709 | 88. 1,048,166 | 107. 1,642,221 | 126. 2,573,835 |
| 13. 406,591 | 32. 493,906 | 51. 604,199 | 70. 762,979 | 89. 1,054,338 | 108. 1,702,317 | 127. 3,061,444 |
| 14. 414,123 | 33. 494,307 | 52. 608,535 | 71. 774,249 | 90. 1,063,456 | 109. 1,747,643 | 128. 3,469,405 |
| 15. 414,509 | 34. 495,975 | 53. 610,432 | 72. 784,754 | 91. 1,096,131 | 110. 1,804,433 | 129. 4,746,457 |
| 16. 415,430 | 35. 498,749 | 54. 611,209 | 73. 827,691 | 92. 1,138,749 | 111. 1,819,323 | 130. 9,188,321 |
| 17. 417,646 | 36. 507,308 | 55. 612,882 | 74. 838,054 | 93. 1,142,789 | 112. 1,842,221 | |
| 18. 418,780 | 37. 511,414 | 56. 617,242 | 75. 854,822 | 94. 1,207,104 | 113. 1,932,280 | |
| 19. 424,309 | 38. 514,193 | 57. 631,752 | 76. 867,372 | 95. 1,262,535 | 114. 1,940,519 | |

*Source:* Yahoo! Sports

*Approach:* We will follow the steps given above.

*Solution:*

**Step 1:** The data provided in Table 14 are already listed in ascending order.

*Rounding up is different from rounding off. Rounding up 15.2 results in 16; rounding off 15.2 results in 15.

***Step 2:*** To find the 85th percentile, $P_{85}$, we compute the index $i$ with $k = 85$ and $n = 130$:

$$i = \left(\frac{85}{100}\right)130 = 110.5$$

***Step 3:*** Round 110.5 up to 111. The 111th observation in the data set is $1,819,323, so $P_{85} = $1,819,323. Approximately 85% of the players in the top 130 earned less than $1,819,323, and approximately 15% of the players in the top 130 earned more than $1,819,323.   ◄◄

**NW**  *Now Work Problems 7(a)–(c).*

Often, we are interested in knowing the percentile to which a specific data value corresponds. The $k$th percentile of a data value, $x$, from a data set that contains $n$ values is computed by using the following steps:

---

**Finding the Percentile that Corresponds to a Data Value**

***Step 1:*** Arrange the data in ascending order.

***Step 2:*** Use the following formula to determine the percentile of the score, $x$:

$$\text{Percentile of } x = \frac{\text{Number of data values less than } x}{n} \times 100 \quad \textbf{(3)}$$

Round this number to the nearest integer.

---

► **EXAMPLE 3**   **Finding the Percentile of a Specific Data Value**

*Problem:* In order to qualify for the last tournament of the year, the Tour Championship, a player must be among the top 30 money winners. Find the percentile that corresponds to the $30^{\text{th}}$ player on the top 130 money list, Carlos Franco, with earnings $1,550,592.

*Approach:* We will follow the steps given above.

*Solution:*

***Step 1:*** The data provided in Table 14 are in ascending order.

***Step 2:*** There are 100 players in the top 130 who earned less than $1,550,592. Therefore,

$$\text{Percentile rank of Carlos Franco} = \frac{100}{130} \cdot 100 = 77$$

Carlos Franco's earnings are at the $77^{\text{th}}$ percentile. This means approximately 77% of the players in the top 130 earned less than Carlos and approximately 23% of players in the top 130 earned more than Carlos.   ◄◄

**NW**  *Now Work Problem 7(d).*

## ③ Quartiles

The most common percentiles are quartiles. **Quartiles** divide data sets into fourths, or four equal parts. The first quartile, denoted $Q_1$, divides the bottom 25% of the data from the top 75%. Therefore, the first quartile is equivalent

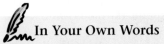
to the 25<sup>th</sup> percentile. The second quartile divides the bottom 50% of the data from the top 50% of the data, so the second quartile is equivalent to the 50<sup>th</sup> percentile, which is equivalent to the median. Finally, the third quartile divides the bottom 75% of the data from the top 25% of the data, so that the third quartile is equivalent to the 75<sup>th</sup> percentile. Figure 18 illustrates the concept of quartiles.

**Figure 18**

▶ **EXAMPLE 4** **Finding the Quartiles of a Data Set**

*Problem:* Find the first, second, and third quartiles for the earnings of top 130 PGA golfers listed in Table 14.

*Approach:*

**Step 1:** The first quartile, $Q_1$, is the 25<sup>th</sup> percentile, $P_{25}$. We let $k = 25$ in Formula (2) to obtain the index, $i$.
**Step 2:** The second quartile, $Q_2$, is the 50<sup>th</sup> percentile, $P_{50}$. We let $k = 50$ in Formula (2) to obtain the index, $i$.
**Step 3:** The third quartile, $Q_3$, is the 75<sup>th</sup> percentile, $P_{75}$. We let $k = 75$ in Formula (2) to obtain the index, $i$.

*Solution:*

**Step 1:** The index for the first quartile, $Q_1$, is

$$i = \left(\frac{25}{100}\right)130 = 32.5$$

Round this value up to 33. The 33<sup>rd</sup> observation will be the first quartile. So we have $Q_1 = P_{25} = \$494{,}307$.
**Step 2:** The index for the second quartile, $Q_2$, is

$$i = \left(\frac{50}{100}\right)130 = 65$$

Because the index is an integer, the second quartile will be the arithmetic mean of the 65<sup>th</sup> and 66<sup>th</sup> data values.

$$Q_2 = P_{50} = M = \frac{\$728{,}635 + \$731{,}925}{2} = \$730{,}280$$

**Step 3:** The index for the third quartile, $Q_3$, is

$$i = \left(\frac{75}{100}\right)130 = 97.5$$

Round this value up to 98. The 98<sup>th</sup> observation will be the third quartile. So, we have $Q_3 = P_{75} = \$1{,}368{,}888$.  ◀◀

**NW** *Now Work Problem 9(b).*

### Outliers

Whenever performing any type of data analysis, we should always check for extreme observations in the data set. Extreme observations are referred to as **outliers**. Whenever outliers are encountered, their origin must be investigated. They can occur because of error in the measurement of a variable, during data entry, or from errors in sampling. For example, in the 2000 presidential election, a precinct in New Mexico accidentally recorded 610 absentee ballots for Al Gore as 110. Workers in the Gore camp discovered the data-entry error through an analysis of vote totals.

Sometimes, extreme observations are in the nature of the population. For example, suppose we wanted to estimate the mean price of a European car. We might take a random sample of size 5 from the population of all European automobiles. If our sample included a Ferrari 360 Spider (approximately $170,000), it probably would be an outlier, because this car costs much more than the typical European automobile. This type of outlier would be considered *unusual* because it is not a "typical" value from the data set.

We can check for outliers with quartiles. The following steps can be followed in order to check for outliers by using quartiles.

> **Caution**
>
> Outliers distort both the arithmetic mean and the standard deviation, since neither is resistant. Because these measures often form the basis for most statistical inference, any conclusions drawn from a set of data that contains outliers can be flawed.

---

**Checking for Outliers by Using Quartiles**

*Step 1:* Determine the first and third quartiles of the data.

*Step 2:* Compute the interquartile range. The **interquartile range** or **IQR** is the difference between the third and first quartile. That is,

$$IQR = Q_3 - Q_1$$

*Step 3:* Determine the fences. **Fences** serve as cutoff points for determining outliers.

$$\text{Lower Fence} = Q_1 - 1.5(IQR)$$

$$\text{Upper Fence} = Q_3 + 1.5(IQR)$$

*Step 4:* If a data value is less than the lower fence or greater than the upper fence, then it is considered an outlier.

---

▶ **EXAMPLE 5**   **Checking for Outliers**

*Problem:* Check the data that represent the earnings of the top 130 PGA golfers in 2000 for outliers.

*Approach:* We follow the preceding steps. Any data value that is less than the lower fence or greater than the upper fence will be considered an outlier.

*Solution:* We consider the top 130 PGA golfers to be a population. Figure 19* shows the output provided by Minitab. Notice that the mean is much larger than the median, so the data is skewed right.

Figure 19   **Descriptive Statistics**

| Variable | N | Mean | Median | Tr Mean | StDev | SE Mean |
|---|---|---|---|---|---|---|
| Earnings | 130 | 1070682 | 730280 | 935458 | 1020972 | 89545 |

| Variable | Min | Max | Q1 | Q3 |
|---|---|---|---|---|
| Earnings | 339242 | 9188321 | 494207 | 1372793 |

*A comparison of the quartiles reported by Minitab and those obtained by hand in Example 4 reveals a discrepancy. Minitab uses a sophisticated approach to computing the quartiles and can be considered more accurate. Be aware that different software use different techniques for finding quartiles and, therefore, the results can differ.

Minitab's quartiles are more accurate than the ones obtained by hand in Example 4, so we will use $Q_1 = \$494,207$, and $Q_3 = \$1,372,793$; so IQR $= \$1,372,793 - \$494,207 = \$878,586$. Substitute these values into the formulas for the lower and upper fence:

$$\text{Lower Fence} = Q_1 - 1.5(\text{IQR}) = \$494,207 - 1.5(\$878,586) = -\$823,672$$

$$\text{Upper Fence} = Q_3 + 1.5(\text{IQR}) = \$1,372,793 + 1.5(\$878,586) = \$2,690,672$$

Clearly, there are no outliers resulting from the value of the lower fence; however, we do have outliers resulting from the upper fence: $9,188,321; $4,746,457; $3,469,405; and $3,061,444. These earnings correspond to Tiger Woods, Phil Mickelson, Ernie Els, and Hal Sutton, respectively.  ◄◄

**NW** *Now Work*
*Problem 9(c) and (d).*

Figure 20 shows a histogram of the data. We can easily identify the outliers corresponding to Tiger Woods and Phil Mickelson in Figure 20; however, the outliers for the other two players are less obvious.

Figure 20

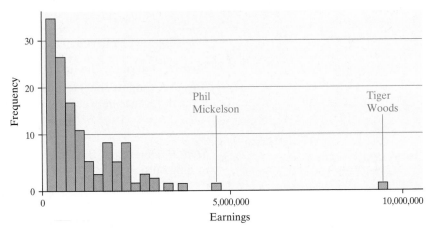

**PGA Earnings, 2000**

## 3.4 Assess Your Understanding

### Concepts and Vocabulary

1. Write a paragraph that explains the meaning of percentiles.
2. Suppose you received the highest score on an exam. Your friend scored the second-highest score, yet you both were in the 99th percentile. How can this be?
3. Morningstar is a mutual fund rating agency. They rank a fund's performance by using from one to five stars. A one-star mutual fund is in the bottom 20% of its investment class; a five-star mutual fund is in the top 20% of its investment class. Interpret the meaning of a four-star mutual fund.
4. When outliers are discovered, should they always be removed from the data set before further analysis?
5. Mensa is an organization designed for people of high intelligence. One qualifies for Mensa if one's intelligence is measured at or above the 98th percentile. Explain what this means.

### Exercises

1. **ACT versus SAT** You are registering for classes during **NW** your first semester at college. The advisor asks to see either your ACT math score or your SAT math score so she can enroll you in the appropriate math class. You scored 25 on the ACT math portion and 650 on the SAT math score. In the math portion of the ACT exam, the mean score is 20.7, with a standard deviation of 5, while the SAT math portion has a mean score of 514, with a standard deviation of 113. Which test score should you provide to the advisor, assuming that you wish to enroll in the highest-level course possible? Why?

2. **ACT versus SAT** You are filling out an application for college. The application requests either your ACT or SAT I score. You scored 26 on the ACT composite and 650 on the SAT I. On the ACT exam, the composite mean score is 21 with a standard deviation of 5, while the SAT I has a mean score of 514 with a standard deviation of 113. Which test score should you provide on the application? Why?

3. **Birth Weights** In 1998, babies born after a gestation period of 32–35 weeks had a mean weight of 2600 grams

and a standard deviation of 680 grams. In the same year, babies born after a gestation period of 40 weeks had a mean weight of 3500 grams and a standard deviation of 480 grams. Suppose a 34-week-gestation-period baby weighs 2400 grams and a 40-week-gestation-period baby weighs 3300 grams. Which baby weighs less relative to the gestation period?

4. **Birth Weights** In 1998, babies born after a gestation period of 32–35 weeks had a mean weight of 2600 grams and a standard deviation of 680 grams. In the same year, babies born after a gestation period of 40 weeks had a mean weight of 3500 grams and a standard deviation of 480 grams. Suppose a 34-week-gestation-period baby weighs 3000 grams and a 40-week-gestation-period baby weighs 4000 grams. Which baby weighs more relative to the gestation period?

5. **Men versus Women** The average 20–29-year-old man is 69.9 inches tall, with a standard deviation of 3.0 inches, while the average 20–29-year-old woman is 64.6 inches tall, with a standard deviation of 2.8 inches. Who is relatively taller, a 75-inch man or a 70-inch woman?

6. **Men versus Women** The average 20–29-year-old man is 69.9 inches tall, with a standard deviation of 3.0 inches, while the average 20–29-year-old woman is 64.6 inches tall, with a standard deviation of 2.8 inches. Who is relatively taller, a 67-inch man or a 62-inch woman?

7. **PGA Earnings** Use the data in Table 14 regarding the earnings of PGA tour players in 2000 to answer the following questions.
(a) Find and interpret the 40th percentile.
(b) Find and interpret the 95th percentile
(c) Find and interpret the 10th percentile.
(d) What is the percentile rank of $459,812?
(e) What is the percentile rank of $1,563,115?

8. **PGA Earnings** Use the data in Table 14 regarding the earnings of PGA tour players in 2000 to answer the following questions.
(a) Find and interpret the 30th percentile.
(b) Find and interpret the 90th percentile.
(c) Find and interpret the 5th percentile.
(d) What is the percentile rank of $507,308?
(e) What is the percentile rank of $1,804,433?

9. **April Showers** The following data represent the number of inches of rain in Chicago, Illinois, during the month of April for 20 randomly selected years.

| 2.47 | 3.97 | 3.94 | 4.11 |
| 1.14 | 4.02 | 3.41 | 1.85 |
| 5.22 | 0.97 | 6.14 | 2.34 |
| 3.48 | 4.77 | 2.78 | 4.00 |
| 6.28 | 5.50 | 7.69 | 5.79 |

*Source:* NOAA, Climate Diagnostics Center

(a) Compute the z-score corresponding to the rainfall in 1971, 0.97 inches. Interpret this result.
(b) Determine the quartiles.
(c) Compute the interquartile range, IQR.
(d) Determine the lower and upper fences. Are there any outliers, according to this criterion?

10. **Hemoglobin in Cats** The following data represent the hemoglobin (in g/dL) for 20 randomly selected cats.

| 7.8 | 9.6 | 13.4 | 9.5 | 5.7 |
| 8.7 | 8.9 | 12.9 | 10.3 | 9.6 |
| 7.7 | 11.2 | 9.9 | 10.0 | 11.7 |
| 13.0 | 11.0 | 9.4 | 10.6 | 10.7 |

*Source:* Joliet Junior College Veterinarian Technology Program

(a) Compute the z-score corresponding to the hemoglobin of Blackie, 7.8 g/dL. Interpret this result.
(b) Determine the quartiles.
(c) Compute the interquartile range, IQR.
(d) Determine the lower and upper fences. Are there any outliers, according to this criterion?

11. **Red Blood Cell Count** The following data represent the red blood cell count (in M/μL) for 20 randomly selected dogs.

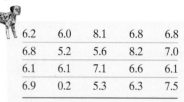

| 6.2 | 6.0 | 8.1 | 6.8 | 6.8 |
| 6.8 | 5.2 | 5.6 | 8.2 | 7.0 |
| 6.1 | 6.1 | 7.1 | 6.6 | 6.1 |
| 6.9 | 0.2 | 5.3 | 6.3 | 7.5 |

*Source:* Laura Blackburn, Student, Joliet Junior College

(a) Compute the z-score corresponding to the red blood cell count of Sampson, 0.2 M/μL. Interpret this result.
(b) Determine the quartiles.
(c) Compute the interquartile range, IQR.
(d) Determine the lower and upper fences. Are there any outliers, according to this criterion?

12. **Acid Rain** A biologist measures the pH level of the rain in Pierce County, Washington for a random sample of 19 rain dates and obtains the following data.

| 5.08 | 4.66 | 4.70 | 4.87 |
| 4.78 | 5.00 | 4.50 | 4.73 |
| 4.79 | 4.65 | 4.91 | 5.07 |
| 5.03 | 4.78 | 4.77 | 4.60 |
| 4.73 | 5.05 | 4.70 | |

*Source:* National Atmospheric Deposition Program

(a) Compute the $z$-score corresponding to 5.08. Interpret this result.

(b) Determine the quartiles.

(c) Compute the interquartile range, IQR.

(d) Determine the lower and upper fences. Are there any outliers, according to this criterion?

**13. Concentration of Dissolved Organic Carbon** The following data represent the concentration of organic carbon (mg/l) collected from organic soil.

| 22.74 | 29.8 | 27.1 | 16.51 | 6.51 |
|---|---|---|---|---|
| 8.81 | 5.29 | 20.46 | 14.9 | 33.67 |
| 30.91 | 14.86 | 15.91 | 15.35 | 9.72 |
| 19.8 | 14.86 | 8.09 | 17.9 | 18.3 |
| 5.2 | 11.9 | 14 | 7.4 | 17.5 |
| 10.3 | 11.4 | 5.3 | 15.72 | 20.46 |
| 16.87 | 15.42 | 22.49 | | |

*Source:* Lisa Emili, Ph.D. Candidate, University of Waterloo

(a) Compute the $z$-score corresponding to 20.46. Interpret this result.

(b) Determine the quartiles.

(c) Compute the interquartile range, IQR.

(d) Determine the lower and upper fences. Are there any outliers, according to this criterion?

**14. Concentration of Dissolved Organic Carbon** The following data represent the concentration of organic carbon (mg/l) collected from mineral soil.

| 8.5 | 3.91 | 9.29 | 21 | 10.89 |
|---|---|---|---|---|
| 10.3 | 11.56 | 7 | 3.99 | 3.79 |
| 5.5 | 4.71 | 7.66 | 11.72 | 11.8 |
| 8.05 | 10.72 | 21.82 | 22.62 | 10.74 |
| 3.02 | 7.45 | 11.33 | 7.11 | 9.6 |
| 12.57 | 12.89 | 9.81 | 17.99 | 21.4 |
| 8.37 | 7.92 | 17.9 | 7.31 | 16.92 |
| 4.6 | 8.5 | 4.8 | 4.9 | 9.1 |
| 7.9 | 11.72 | 4.85 | 11.97 | 7.85 |
| 9.11 | 8.79 | | | |

*Source:* Lisa Emili, Ph.D. Candidate, University of Waterloo

(a) Compute the $z$-score corresponding to 17.99. Interpret this result.

(b) Determine the quartiles.

(c) Compute the interquartile range, IQR.

(d) Determine the lower and upper fences. Are there any outliers, according to this criterion?

**15. Rates of Return of Stocks** Stocks can be categorized by industry. The following data represent the five-year rates of return for a simple random sample of financial stocks and energy stocks ending December 28, 2001.

(a) Determine whether there are any outliers in either industry.

(b) Draw histograms for these data, and label the outliers if any.

| Financial Stocks | | | | |
|---|---|---|---|---|
| 16.12 | 18.61 | 4.04 | 13.93 | 17.85 |
| 3.34 | 8.39 | 21.32 | 34.22 | 16.56 |
| 28.51 | 13.29 | 11.03 | 14.61 | 6.67 |
| 2.98 | 14.06 | 28.90 | 11.26 | 9.24 |
| 10.97 | 7.13 | 5.32 | 0.63 | 13.43 |
| 17.10 | 22.21 | 6.18 | 5.80 | 1.39 |
| 0.83 | 5.63 | 8.39 | 8.02 | 1.23 |
| 3.32 | 9.42 | 2.98 | 13.62 | 6.36 |

*Source:* Morningstar.com

| Energy Stocks | | | | |
|---|---|---|---|---|
| 2.49 | 9.70 | 21.67 | 8.31 | 5.07 |
| 7.70 | 23.72 | 5.59 | 6.09 | 9.69 |
| 1.82 | 12.68 | 6.07 | 11.09 | 6.35 |
| 2.00 | 14.22 | 13.17 | 6.29 | 6.25 |
| 7.39 | 9.09 | 30.39 | 11.25 | 7.17 |
| 9.48 | 9.23 | 9.70 | 5.41 | 10.29 |
| 6.11 | 7.38 | 11.60 | 12.10 | 13.38 |

*Source:* Morningstar.com

**16. American League versus National League** The following data represent the earned-run averages of the top 37 pitchers in the American League and the top 37 in the National League during the 2001 season.

| American League | | | | |
|---|---|---|---|---|
| 3.05 | 3.15 | 3.16 | 3.29 | 3.37 |
| 3.43 | 3.45 | 3.49 | 3.51 | 3.59 |
| 3.60 | 3.65 | 3.77 | 3.90 | 3.94 |
| 3.99 | 4.05 | 4.08 | 4.09 | 4.09 |
| 4.09 | 4.25 | 4.32 | 4.36 | 4.37 |
| 4.39 | 4.42 | 4.45 | 4.45 | 4.50 |
| 4.76 | 4.93 | 5.02 | 5.08 | 5.17 |
| 5.54 | 5.82 | | | |

| National League | | | | |
|---|---|---|---|---|
| 2.49 | 2.98 | 3.04 | 3.05 | 3.09 |
| 3.16 | 3.29 | 3.31 | 3.36 | 3.40 |
| 3.42 | 3.50 | 3.57 | 3.57 | 3.69 |
| 3.70 | 3.80 | 4.03 | 4.05 | 4.05 |
| 4.19 | 4.31 | 4.33 | 4.34 | 4.42 |
| 4.42 | 4.45 | 4.46 | 4.46 | 4.47 |
| 4.48 | 4.49 | 4.63 | 4.79 | 4.81 |
| 4.85 | 4.90 | | | |

(a) Determine whether there are any outliers in either league.

(b) Draw histograms for these data, and label the outliers if any.

17. **Student Survey of Income** A survey of 50 randomly se-

    lected full-time Joliet Junior College students was con-
    ducted during the Fall 2000 semester; in the survey, the
    students were asked to disclose their weekly income
    from employment. If the student did not work, $0 was
    entered. The results of the survey are as follows:

    | | | | | |
    |---|---|---|---|---|
    | 0 | 262 | 0 | 635 | 0 |
    | 244 | 521 | 476 | 100 | 650 |
    | 12777 | 567 | 310 | 527 | 0 |
    | 83 | 159 | 0 | 547 | 188 |
    | 719 | 0 | 367 | 316 | 0 |
    | 479 | 0 | 82 | 579 | 289 |
    | 375 | 347 | 331 | 281 | 628 |
    | 0 | 203 | 149 | 0 | 403 |
    | 0 | 454 | 67 | 389 | 0 |
    | 671 | 95 | 736 | 300 | 181 |

    (a) Check the data set for outliers.
    (b) Provide an explanation for the outliers.

18. **Student Survey of Entertainment Spending** A survey
    of 40 randomly selected full-time Joliet Junior College
    students was conducted in the Fall 2000 semester; in the

survey, the students were asked to disclose their weekly
spending on entertainment. The results of the survey
are as follows:

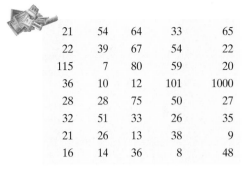

| | | | | |
|---|---|---|---|---|
| 21 | 54 | 64 | 33 | 65 |
| 22 | 39 | 67 | 54 | 22 |
| 115 | 7 | 80 | 59 | 20 |
| 36 | 10 | 12 | 101 | 1000 |
| 28 | 28 | 75 | 50 | 27 |
| 32 | 51 | 33 | 26 | 35 |
| 21 | 26 | 13 | 38 | 9 |
| 16 | 14 | 36 | 8 | 48 |

(a) Check the data set for outliers.
(b) Provide an explanation for the outliers.

19. **Pulse Rate** Use the results of Problem 11 in Sections
    3.1 and 3.2 to compute the $z$-scores for all the students.
    Compute the mean and standard deviation of these
    $z$-scores.

20. **Travel Time** Use the results of Problem 12 in Sections
    3.1 and 3.2 to compute the $z$-scores for all the students.
    Compute the mean and standard deviation of these
    $z$-scores.

---

## Technology Step-by-Step
Determining Percentiles

| | |
|---|---|
| ***TI-83 Plus*** | The TI-83 Plus can only compute the quartiles. Follow the same Steps as were given to compute the mean and median from raw data. |
| ***MINITAB*** | Minitab can only compute the quartiles. Follow the same Steps as were given to compute the mean and median from raw data. |
| ***Excel*** | ***Step 1:*** Enter the raw data into column A. |
| | ***Step 2:*** With the data analysis tool pak enabled, select the **Tools** menu and highlight **Data Analysis**.... |
| | ***Step 3:*** Select **Rank and Percentile** from the Data Analysis window. |
| | ***Step 4:*** With the cursor in the **Input Range** cell, highlight the data. Press OK. |

---

## 3.5   The Five-Number Summary; Boxplots

**Objectives**   ① Compute the five-number summary
②  Draw and interpret boxplots

Some aspects of statistical analysis attempt to verify a conjecture by means
of observational studies or designed experiments. In other words, a theory is
conjectured, and then data is collected in order to support the theory. For
example, a dietitian might conjecture that exercise will lower an individual's
cholesterol. The dietitian would carefully design an experiment that randomly divides study participants into two groups—the control group, and
the experimental (treatment) group. She would impose the treatment (exercise) on the two groups and then measure the effect on the response variable, cholesterol levels.

However, there is another aspect of statistics that looks at data in order to spot any interesting results that might be concluded from the data. In other words, rather than develop a theory and use data to support or disprove the theory, a researcher would start with data and look for a theory. This area of statistics is referred to as **exploratory data analysis (EDA)**. The idea behind exploratory data analysis is to draw graphs of data and obtain measures of central tendency and spread, in order to form some conjectures regarding the data.

The methods of exploratory data analysis were developed by John Tukey (1915–2000). A complete presentation of the materials found in this chapter can be found in his text *Exploratory Data Analysis* (Addison-Wesley, 1977).

## ① The Five-Number Summary

Rather than numerically describing a distribution via the mean and standard deviation, exploratory data analysis summarizes a distribution by using measures that are resistant to extreme observations. Recall that the median is a measure of central tendency that divides the lower 50% from the upper 50% of the data. This particular measure of central tendency is resistant to extreme values and hence is the preferred measure of central tendency when data is skewed to either the right or the left. The three measures of dispersion presented in Section 3.2 are not resistant to extreme values. However, the interquartile range, $Q_3 - Q_1$ is resistant; it measures the spread of the data by determining the difference between the 25th and 75th percentiles. It is interpreted as the range of the middle 50% of the data. However, the median, $Q_1$ and $Q_3$ do not provide information about the tails of the distribution of the data. To get this information, we need to know the smallest and largest values in the data set. The **five-number summary** of a set of data consists of the smallest data value, $Q_1$, the median, $Q_3$, and the largest data value. Symbolically, the five-number summary is presented as follows.

The Five-Number Summary

MINIMUM   $Q_1$   $M$   $Q_3$   MAXIMUM

▶ EXAMPLE 1   Obtaining the Five-Number Summary

*Problem:* Use the data in Table 12 on page 57 from Section 2.2 to obtain the five-number summary of the three-year rates of return of 40 small-capitalization growth mutual funds.

*Approach:* The five-number summary requires that we determine the minimum data value, $Q_1$ (the 25th percentile), $M$ (the median), $Q_3$ (the 75th percentile), and the maximum data value. We will need to arrange the data in ascending order, then use the procedures introduced in Section 3.4 to obtain $Q_1$, $M$, and $Q_3$.

*Solution:* The data in ascending order are as follows:

10.8, 10.9, 11.6, 12.7, 12.8, 14.7, 14.7, 15.8, 16.6, 16.7, 17.4, 18.2, 18.4, 18.4, 18.4, 18.5, 19.2, 19.6, 21.3, 21.9, 22.2, 22.6, 23.3, 23.7, 23.7, 24.1, 25.5, 27.0, 27.4, 28.5, 29.1, 29.6, 31.1, 32.0, 32.1, 35.9, 37.0, 38.1, 45.9, 47.7

The smallest number in the data set (i.e., the lowest return) is 10.8. The largest number in the data set is 47.7. The first quartile, $Q_1$, is 17.05. The median, $M$, is 22.05. The third quartile, $Q_3$, is 28.8. The five-number summary is

10.8   17.05   22.05   28.8   47.7   ◀◀

Using Technology: Many calculators and statistical software packages have the ability to present the five-number summary. For example, Figure 21 presents the five-number summary for the data in Example 1, obtained from a TI-83 Plus graphing calculator.

Figure 21

```
1-Var Stats
↑n=40
minX=10.8
Q₁=17.05
Med=22.05
Q₃=28.8
maxX=47.7
```

## ② Boxplots

The five-number summary can be used to create another graph, called the *boxplot*.

**Drawing a Boxplot**

*Step 1:* Determine the lower and upper fences:

$$\text{Lower fence} = Q_1 - 1.5(\text{IQR})$$

$$\text{Upper fence} = Q_3 + 1.5(\text{IQR})$$

*Step 2:* Draw vertical lines at $Q_1$, $M$, and $Q_3$. Enclose these vertical lines in a box.

*Step 3:* Label the lower and upper fences.

*Step 4:* Draw a line from $Q_1$ to the smallest data value that is larger than the lower fence. Draw a line from $Q_3$ to the largest data value that is smaller than the upper fence.

*Step 5:* Any data values less than the lower fence or greater than the upper fence are outliers and are marked with an asterisk (*).

▶ **EXAMPLE 2**   **Constructing a Boxplot**

*Problem:* Use the results from Example 1 to a construct a boxplot of the three-year rates of return of small-capitalization growth mutual funds.

*Approach:* Follow the steps presented above.

*Solution:* From the results of Example 1, we know that $Q_1 = 17.05$, $M = 22.05$, $Q_3 = 28.8$. Therefore, the interquartile range = $IQR = Q_3 - Q_1 = 28.8 - 17.05 = 11.75$. The difference between the 75[th] percentile and 25[th] percentile is a rate of return of 11.75 percent.

*Step 1:* We compute the lower and upper fences:

$$\text{Lower fence} = Q_1 - 1.5(\text{IQR}) = 17.05 - 1.5(11.75) = -0.575$$

$$\text{Upper fence} = Q_3 + 1.5(\text{IQR}) = 28.8 + 1.5(11.75) = 46.425$$

*Step 2:* Draw a horizontal number line with a scale that will accommodate our graph. Draw vertical lines at $Q_1 = 17.05$, $M = 22.05$, and $Q_3 = 28.8$. Enclose these lines in a box. See Figure 22(a).

*Step 3:* We label the lower and upper fence with brackets ([ and ]). See Figure 22(b).

*Step 4:* The smallest data value that is larger than $-0.575$ (the lower fence) is 10.8. The largest data value that is smaller than 46.425 (the upper fence) is 45.9. We draw horizontal lines from $Q_1$ to 10.8 and from $Q_3$ to 45.9. See Figure 22(c).

*Step 5:* Label any values less than $-0.575$ (the lower fence) or greater than 46.425 (the upper fence) as outliers. We label 47.7 as an outlier. See Figure 22(d).

Figure 22

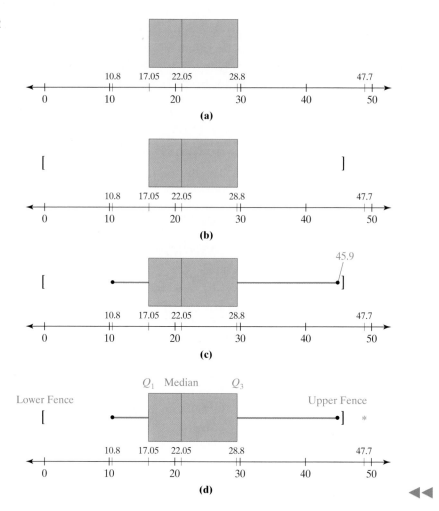

Boxplots can be interpreted the same way that histograms can be interpreted. In fact, we can describe the shape of the distribution using the boxplot.

> **Distribution Shape Based upon Boxplot**
> 1. If the median is near the center of the box and each of the horizontal lines is of approximately equal length, then the distribution is roughly symmetric.
> 2. If the median is to the left of the center of the box or the right line is substantially longer than the left line, the distribution is skewed right.
> 3. If the median is to the right of the center of the box or the left line is substantially longer than the right line, the distribution is skewed left.

**Caution**

Identifying the shape of a distribution from a boxplot (or from a histogram, for that matter) is subjective. When identifying the shape of a distribution from a graph, be sure to support your opinion.

Figure 23 on page 135 provides examples of boxplots that are (a) symmetric, (b) skewed right, and (c) skewed left, along with the corresponding histograms.

The boxplot in Figure 22(d) suggests that the distribution is skewed right, since the right line is longer than the left and the median is in the left of the center of the box. (Refer to Figure 9 on page 59 in Section 2.2 to see the histogram of these data.)

**NW** *Now Work Problem 9.*

Figure 23

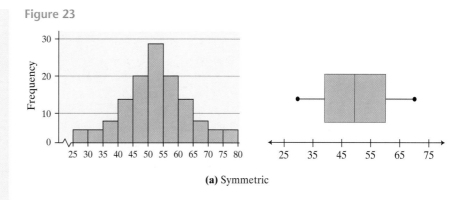

**(a)** Symmetric

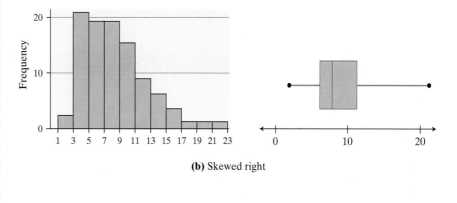

**(b)** Skewed right

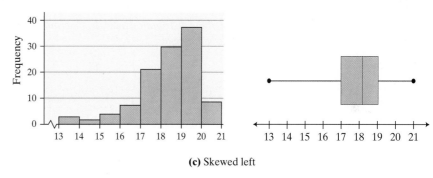

**(c)** Skewed left

### Historical Note

John Tukey was born on July 16, 1915 in New Bedford, Massachusetts. His parents graduated numbers 1 and 2 from Bates College and were elected "the couple most likely to give birth to a genius." In 1936, Tukey graduated from Brown University with an undergraduate degree in chemistry. He went on to earn a master's degree in chemistry at Brown. In 1939, Tukey earned his doctorate in mathematics from Princeton. He remained at Princeton and in 1965 became the founding chair of the Department of Statistics. Among his many accomplishments, Tukey is credited with coining the terms *software* and *bit*. In the early 1970s, he discussed the negative effects of aerosol cans on the ozone layer. Tukey recommended that the 1990 Census be adjusted by means of statistical formulas. John Tukey died in New Brunswick, New Jersey, July 26, 2000.

▶ **EXAMPLE 3**   Comparing Two Distributions by Using Boxplots

*The Problem:* In the Spacelab Life Sciences 2, fourteen male rats were sent to space. Upon their return, the red blood cell mass (in milliliters) of the rats was determined. A control group of fourteen male rats was held under the same conditions (except for spaceflight) as the space rats, and their red blood cell mass was also determined when the space rats returned. The project was led by Dr. Paul X. Callahan. The data in Table 15 were obtained. Construct boxplots for red blood cell mass for the flight group and control group. Does it appear that the flight to space affected the red blood cell mass of the rats?

*Approach:* When comparing two data sets, we draw the boxplots on the same horizontal number line to make comparison easy. Therefore, we draw boxplots for flight and control, using the same number line.

*Solution:* See Figure 24. From the boxplots, it appears that the spaceflight has reduced the red blood cell mass of the rats.

**TABLE 15**

| Flight | | Control | |
|--------|--------|---------|--------|
| 8.59 | 8.64 | 8.65 | 6.99 |
| 6.87 | 7.89 | 7.62 | 7.44 |
| 7.00 | 8.80 | 7.33 | 8.58 |
| 6.39 | 7.54 | 7.14 | 9.14 |
| 7.43 | 7.21 | 8.40 | 9.66 |
| 9.79 | 6.85 | 8.55 | 8.70 |
| 9.30 | 8.03 | 9.88 | 9.94 |

*Source:* NASA Life Sciences Data Archive

Figure 24

**Flight versus Control**

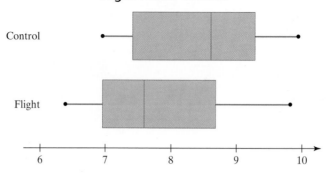

## 3.5 Assess Your Understanding

### Concepts and Vocabulary

1. Explain the circumstances under which the median and interquartile range would be better measures of central tendency and dispersion than the mean and standard deviation.

### Exercises

#### • Skill Building

*In Problems 1 and 2, (a) identify the shape of the distribution, and (b) determine the five-number summary. Assume that each of the numbers in the five-number summary is an integer.*

1.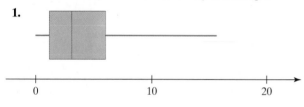

2.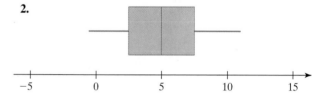

#### • Applying the Concepts

*In Problems 3–6, find the five-number summary, and construct a boxplot for the data in the indicated problem. Comment on the shape of the distribution.*

3. **Age at Inauguration** The following data represent the age of United States presidents on their respective inauguration days.

| | | | | | | |
|---|---|---|---|---|---|---|
| 57 | 61 | 57 | 57 | 58 | 57 | 61 |
| 54 | 68 | 51 | 49 | 64 | 50 | 48 |
| 65 | 52 | 56 | 46 | 54 | 49 | 50 |
| 47 | 55 | 55 | 54 | 42 | 51 | 56 |
| 55 | 51 | 54 | 51 | 60 | 62 | 43 |
| 55 | 56 | 61 | 52 | 69 | 64 | 46 |
| 54 | | | | | | |

4. **Divorce Rate** The following data represent percentage of divorces (as a percentage of marriages in 1996) for selected countries. (*Note*: The United States is at 49%. The highest is Belarus, at 68%, and the lowest is Macedonia, at 5%.)

| | | | | | | |
|---|---|---|---|---|---|---|
| 68 | 65 | 64 | 63 | 63 | 61 | 56 |
| 56 | 55 | 53 | 52 | 49 | 46 | 45 |
| 43 | 43 | 41 | 41 | 40 | 39 | 39 |
| 38 | 35 | 34 | 28 | 26 | 26 | 25 |
| 24 | 21 | 19 | 18 | 18 | 18 | 17 |
| 15 | 15 | 13 | 13 | 12 | 12 | 12 |
| 7 | 6 | 5 | | | | |

***Source:*** Time Almanac 2000

**5. Grams of Fat in McDonald's Breakfast** The following data represent the number of grams of fat in breakfast meals offered at McDonald's.

| 12 | 23 | 28 | 2 | 31 | 37 | 34 |
|----|----|----|----|----|----|----|
| 15 | 23 | 38 | 31 | 16 | 11 | 8 |
| 8 | 17 | 20 | | | | |

*Source:* McDonald's Corporation

**6. Miles per Gallon** The following data represent the number of miles per gallon on the highway of domestic cars for the 2001 model year.

| 30 | 30 | 30 | 29 | 24 | 27 | 26 |
|----|----|----|----|----|----|----|
| 26 | 30 | 34 | 28 | 32 | 32 | 32 |
| 37 | 28 | 26 | 26 | 27 | 26 | 28 |
| 35 | 20 | 25 | 38 | 38 | 30 | 28 |
| 28 | 25 | 25 | 34 | 24 | 28 | 32 |
| 28 | 28 | 35 | 23 | 29 | 30 | 30 |
| 29 | 33 | 40 | 33 | 40 | | |

*Source: Road and Track* magazine

**7. Tensile Strength** Tensile strength is the maximum stress at which one can be reasonably certain that failure will not occur. The following data represent tensile strength (in thousands of pounds per square inch) of a composite material to be used in an aircraft.

| 203.41 | 185.97 | 184.41 | 160.44 | 174.63 |
|--------|--------|--------|--------|--------|
| 209.58 | 190.67 | 200.73 | 180.95 | 185.34 |
| 213.35 | 207.88 | 206.51 | 201.95 | 205.59 |
| 218.56 | 210.80 | 209.84 | 204.60 | 212.00 |
| 242.76 | 231.46 | 212.15 | 219.51 | 225.25 |

*Source:* Vangel, Mark G., "New Methods for One-sided Tolerance Limits for a One-way Balanced Random-Effects ANOVA Model," *Technometrics*; May, 1992, Vol. 34, Issue 2, pp. 176–185.

(a) Compute the five-number summary.
(b) Draw a boxplot of the data.
(c) Determine the shape of the distribution from the boxplot. Refer to the histogram drawn in Problem 19 in Section 2.2 to test your answer.

**8. Miles on a Cavalier** A random sample of 36 three-year-old Chevy Cavaliers was obtained in the Miami, Florida area, and the number of miles on each car was recorded.

(a) Compute the five-number summary.
(b) Draw a boxplot of the data.
(c) Determine the shape of the distribution from the boxplot. Refer to the histogram drawn in Problem 20 in Section 2.2 to test your answer.

| 34,122 | 34,511 | 17,595 | 30,500 | 31,883 |
|--------|--------|--------|--------|--------|
| 39,416 | 34,500 | 44,224 | 35,936 | 32,923 |
| 28,281 | 41,620 | 26,455 | 44,448 | 40,614 |
| 37,021 | 30,283 | 27,368 | 41,043 | 42,398 |
| 30,747 | 15,499 | 37,904 | 30,739 | |
| 35,439 | 26,051 | 23,221 | 38,995 | |
| 17,685 | 32,305 | 38,041 | 41,194 | |
| 29,307 | 39,884 | 35,295 | 29,289 | |

*Source:* cars.com

**9. Serum HDL** Dr. Paul Oswiecmiski randomly selects forty of his 20–29-year-old patients and obtains the following data regarding their serum HDL cholesterol:

| 70 | 56 | 48 | 48 | 53 | 52 | 66 | 48 |
|----|----|----|----|----|----|----|----|
| 36 | 49 | 28 | 35 | 58 | 62 | 45 | 60 |
| 38 | 73 | 45 | 51 | 56 | 51 | 46 | 39 |
| 56 | 32 | 44 | 60 | 51 | 44 | 63 | 50 |
| 46 | 69 | 53 | 70 | 33 | 54 | 55 | 52 |

(a) Compute the five-number summary.
(b) Draw a boxplot of the data.
(c) Determine the shape of the distribution from the boxplot. Refer to the histogram drawn in Problem 21 in Section 2.2 to test your answer.

**10. Volume of Philip Morris Stock** The following data represent a random sample of the number of millions of shares of Philip Morris stock traded for 35 days in 2000.

| 3.98 | 8.90 | 10.40 | 7.52 | 13.84 |
|------|------|-------|------|-------|
| 14.04 | 9.14 | 7.96 | 11.40 | 13.30 |
| 5.29 | 9.69 | 8.94 | 12.10 | 13.25 |
| 4.62 | 10.85 | 24.52 | 14.62 | 16.10 |
| 4.24 | 8.54 | 11.59 | 6.75 | 27.54 |
| 16.06 | 17.62 | 7.17 | 6.13 | 25.32 |
| 7.71 | 10.14 | 4.28 | 7.24 | 10.88 |

*Source:* http://finance.yahoo.com

(a) Compute the five-number summary.
(b) Draw a boxplot of the data.
(c) Determine the shape of the distribution from the boxplot. Refer to the histogram drawn in Problem 22 in Section 2.2 to test your answer.

**11. M&Ms** The following data represent the weights (in grams) of a simple random sample of 50 M&M plain candies.

| 0.87 | 0.88 | 0.82 | 0.90 | 0.90 | 0.84 | 0.84 |
| 0.91 | 0.94 | 0.86 | 0.86 | 0.86 | 0.88 | 0.87 |
| 0.89 | 0.91 | 0.86 | 0.87 | 0.93 | 0.88 | |
| 0.83 | 0.94 | 0.87 | 0.93 | 0.91 | 0.85 | |
| 0.91 | 0.91 | 0.86 | 0.89 | 0.87 | 0.93 | |
| 0.88 | 0.88 | 0.89 | 0.79 | 0.82 | 0.83 | |
| 0.90 | 0.88 | 0.84 | 0.93 | 0.76 | 0.90 | |
| 0.88 | 0.92 | 0.85 | 0.79 | 0.84 | 0.86 | |

*Source:* Michael Sullivan

(a) Compute the five-number summary.
(b) Draw a boxplot of the data.
(c) Determine the shape of the distribution from the boxplot. Refer to the histogram drawn in Problem 23 in Section 2.2 to test your answer.

**12. Old Faithful** The following data represent the length of eruption (in seconds) for a random sample of eruptions of the Old Faithful geyser in California.

| 108 | 108 | 99 | 105 | 103 | 103 | 94 |
| 102 | 99 | 106 | 90 | 104 | 110 | 110 |
| 103 | 109 | 109 | 111 | 101 | 101 | |
| 110 | 102 | 105 | 110 | 106 | 104 | |
| 104 | 100 | 103 | 102 | 120 | 90 | |
| 113 | 116 | 95 | 105 | 103 | 101 | |
| 100 | 101 | 107 | 110 | 92 | 108 | |

*Source:* Ladonna Hansen, Park Curator

(a) Compute the five-number summary.
(b) Draw a boxplot of the data.
(c) Determine the shape of the distribution from the boxplot. Refer to the histogram drawn in Problem 24 in Section 2.2 to test your answer.

---

*In Problems 13–16, compare the two data sets by determining the five-number summary and constructing boxplots on the same scale.*

**13. Vehicle Death Rates** The following data represent the death rates for a sample of passenger minivans and sport utility vehicles (SUVs). The death rates are expressed as deaths per 1 million registered vehicles. For example, the Nissan Quest had 18 deaths for each 1 million vehicles registered. Which vehicles appear to be safer, according to the boxplots?

| Van | Death Rate | SUV | Death Rate |
|---|---|---|---|
| Nissan Quest | 18 | Chevy Suburban | 44 |
| Dodge Grand Caravan | 22 | Ford Explorer (4 door) | 103 |
| Chevy Astro | 25 | Isuzu Rodeo | 151 |
| Ford Windstar | 33 | Chevy S10 Blazer | 195 |
| Honda Odyssey | 41 | Jeep Grand Cherokee | 78 |
| Chrysler Town and Country | 45 | Ford Explorer (2 door) | 231 |
| Mercury Villager | 46 | Jeep Cherokee | 74 |
| Dodge Caravan | 48 | Ford Expedition | 39 |
| GMC Safari | 81 | Chevy Tahoe | 45 |
| | | Toyota 4Runner | 125 |

*Source:* Insurance Institute for Highway Safety

**14. Automobile Batteries** The following data represent the number of cold cranking amps of group size 24 and group size 35 batteries. The cold cranking amps number measures the amps produced by the battery at 0° Fahrenheit. Which type of battery would you prefer?

| Group Size 24 Batteries | | | Group Size 35 Batteries | | |
|---|---|---|---|---|---|
| 800 | 600 | 675 | 525 | 620 | 550 |
| 600 | 525 | 700 | 560 | 675 | 550 |
| 500 | 660 | 550 | 530 | 570 | 640 |
| 585 | 675 | | 525 | 640 | 640 |

*Source: Consumer Reports,* October 2000

15. **Thermocouples** A thermocouple is a temperature sensor. One use of a thermocouple is to measure the temperature inside a furnace. A company has three thermocouples, but one is suspected of being broken. They wrap the three thermocouples together, place them in an oven and record the temperature. They repeat this process 20 times and obtain the following data. Which thermocouple appears to be broken? Which thermocouple provides the most consistent measurement?

| Thermocouple 1 | | Thermocouple 2 | | Thermocouple 3 | |
|---|---|---|---|---|---|
| 326.06 | 326.20 | 323.59 | 323.55 | 326.03 | 325.97 |
| 326.09 | 326.00 | 323.63 | 323.76 | 326.06 | 326.20 |
| 326.07 | 325.97 | 323.62 | 323.55 | 326.03 | 325.95 |
| 326.08 | 326.20 | 323.64 | 323.74 | 326.06 | 326.18 |
| 326.05 | 326.07 | 323.64 | 323.66 | 326.02 | 326.04 |
| 326.05 | 326.11 | 323.60 | 323.66 | 326.02 | 326.08 |
| 326.03 | 326.00 | 323.62 | 323.57 | 326.01 | 325.98 |
| 326.08 | 326.20 | 323.64 | 323.75 | 326.01 | 326.16 |
| 326.00 | 326.13 | 323.58 | 323.70 | 325.99 | 326.12 |
| 326.16 | 326.12 | 323.70 | 323.68 | 326.13 | 326.08 |

*Source:* Christenson, Ronald and Blackwood, Larry, "Tests for Precision and Accuracy of Multiple Measuring Devices" *Technometrics*, Nov. 93, Vol 35, Issue 4, pp. 411–421.

16. **Homicide Rates** The following data represent the homicide rates (homicides per 100,000 residents) for 14–24 year olds in all states (except Florida and Kansas) in 1999. A researcher wanted to know whether the homicide rates for states west of the Mississippi differed from those for states east of the Mississippi. Which part of the country appears to have the higher homicide rate?

| West | | | East | | |
|---|---|---|---|---|---|
| 16.84 | 3.49 | 8.60 | 6.49 | 10.56 | 1.22 |
| 13.98 | 2.49 | 11.98 | 13.51 | 18.25 | 10.94 |
| 9.08 | 2.07 | 3.22 | 7.92 | 5.04 | 6.02 |
| 5.01 | 19.77 | 10.49 | 6.99 | 23.96 | 11.29 |
| 2.44 | 15.51 | 2.58 | 2.75 | 4.09 | 11.92 |
| 3.96 | 11.91 | 5.65 | 14.20 | 13.81 | 0.00 |
| 4.84 | 4.52 | 4.39 | 21.99 | 15.88 | 8.93 |
| 13.45 | 11.91 | | 13.20 | 15.87 | 4.94 |

*Source:* United States Department of Justice

## Technology Step-by-Step
Drawing Boxplots Using Technology

**TI-83 Plus**   **Step 1:** Enter the raw data into L1.

**Step 2:** Press $2^{nd}$ Y = and select 1:Plot 1.

**Step 3:** Turn the plots ON. Use the cursor to highlight the boxplot icon. Your screen should look as follows:

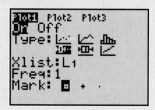

**Step 4:** Press ZOOM and select 9: ZoomStat.

**MINITAB**   **Step 1:** Enter the raw data into column C1.

**Step 2:** Select the **Graph** menu and highlight **Boxplot**...

**Step 3:** Select the data to be graphed. If you want the boxplot to be horizontal rather than vertical, select the OPTIONS button and transpose X and Y. Click OK.

**Excel**   **Step 1:** Start the PHStat Add-in.

**Step 2:** Enter the raw data into column A.

**Step 3:** Select the **PHStat** menu and highlight **Box-and-Whisker Plot**... With the cursor in the "Data Variable Cell Range" cell, highlight the data in column A.

**Step 4:** Click OK.

## CHAPTER 3 REVIEW

### Summary

This chapter concentrated on describing distributions numerically. Measures of central tendency are used to indicate the "typical" value in a distribution. Three measures of central tendency were discussed. The arithmetic mean measures the center of gravity of the distribution; the median separates the bottom 50% of the data from the top 50%. Both of these measures require that the data be quantitative. The mode measures the most frequent observation. The data can be either quantitative or qualitative in order to compute the mode. The median is resistant to extreme values, while the mean is not. A comparison between the median and mean can help determine the shape of the distribution.

Measures of dispersion describe how "spread out" the data are. The range is the difference between the highest and lowest data value. The variance measures the average squared deviation about the mean. The standard deviation is the square root of the variance. The mean and standard deviation are used in many types of statistical inference.

The mean, median, and mode can be approximated from grouped data. The variance and standard deviation can also be approximated from grouped data.

We can determine the relative position of an observation in a data set by using z-scores and percentiles. Z-scores determine the number of standard deviations from the mean that an observation is, whereas percentiles determine the percent of observations that lie above and below an observation. The interquartile range can be used to identify potential outliers. Any potential outlier must be investigated to determine whether it was the result of a data entry error, of some other error in the data collection process, or of an unusual value in the data set.

Exploratory data analysis is used to investigate data without any preconceived notions as to the distribution of the data. The five-numbers summary provides an idea about the center and spread of a data set, through the median and the interquartile range. The length of the tails in the distribution can be determined from the smallest and the largest data values. The five-number summary is used to construct boxplots. Boxplots can be used to describe the shape of the distribution.

## Formulas

**Population Mean**

$$\mu = \frac{\sum x_i}{N}$$

**Sample Mean**

$$\overline{x} = \frac{\sum x_i}{n}$$

**Population Variance**

$$\sigma^2 = \frac{\sum(x_i - \mu)^2}{N}$$

**Sample Variance**

$$s^2 = \frac{\sum(x_i - \overline{x})^2}{n - 1}$$

**Population Standard Deviation**

$$\sigma = \sqrt{\sigma^2}$$

**Sample Standard Deviation**

$$s = \sqrt{s^2}$$

**Range** = Largest Data Value − Smallest Data Value

**Weighted Mean**

$$\overline{x}_w = \frac{\sum w_i x_i}{\sum w_i}$$

**Population Mean from Grouped Data**

$$\mu = \frac{\sum x_i f_i}{\sum f_i}$$

**Sample Mean from Grouped Data**

$$\overline{x} = \frac{\sum x_i f_i}{\sum f_i}$$

**Population Variance from Grouped Data**

$$\sigma^2 = \frac{\sum(x_i - \mu)^2 f_i}{\sum f_i}$$

**Approximating the Sample Variance from Grouped Data**

$$s^2 = \frac{\sum(x_i - \overline{x})^2 f_i}{(\sum f_i) - 1}$$

**Population z-score**

$$z = \frac{x - \mu}{\sigma}$$

**Sample z-score**

$$z = \frac{x - \overline{x}}{s}$$

**Percentile of** $x = \dfrac{\text{Number of data values less than } x}{n} \cdot 100$

**Interquartile Range**

$$\text{IQR} = Q_3 - Q_1$$

**Lower and Upper Fences**

Lower Fence $= Q_1 - 1.5(\text{IQR})$

Upper Fence $= Q_3 + 1.5(\text{IQR})$

## Vocabulary

| | | |
|---|---|---|
| Parameter (p. 84) | Population variance (p. 102) | Z-score (p. 122) |
| Statistic (p. 84) | Sample variance (p. 104) | $k$th percentile (p. 123) |
| Arithmetic mean (p. 85) | Biased (p. 104) | Quartiles (p. 125) |
| Median (p. 87) | Degrees of freedom (p. 104) | Interquartile range (p. 127) |
| Mode (p. 89) | Population standard deviation (p. 105) | Outlier (p. 127) |
| Resistant (p. 90) | Sample standard deviation (p. 105) | Fences (p. 127) |
| Range (p. 101) | Class midpoint (p. 115) | Exploratory data analysis (p. 132) |
| Deviation about the mean (p. 102) | Weighted mean (p. 117) | Five-number summary (p. 132) |

## Objectives

| Section | You should be able to . . . | Review Exercises |
|---|---|---|
| **3.1** | 1 Determine the arithmetic mean of a variable from raw data (p. 84) | 1–8 (a) |
| | 2 Determine the median of a variable from raw data (p. 87) | 1–8 (a) |
| | 3 Determine the mode of a variable from raw data (p. 89) | 5–6 |
| | 4 Use the mean and median to help identify the shape of a distribution (p. 90) | 15, 16 (a)–(c) |
| **3.2** | 1 Compute the range of a variable from raw data (p. 101) | 1–8 (b) |
| | 2 Compute the variance of a variable from raw data (p. 102) | 1–8 (b), 15(d), 16(d) |
| | 3 Compute the standard deviation of a variable from raw data (p. 105) | 1–8(b), 15(d), 16(d) |
| | 4 Use the Empirical Rule (p. 107) | 9, 10 (c)–(e) |
| **3.3** | 1 Approximate the mean of a variable from grouped data (p. 115) | 11, 12 (a) |
| | 2 Compute the weighted mean (p. 117) | 13, 14 |
| | 3 Approximate the variance and standard deviation of a variable from grouped data (p. 118) | 11, 12 (b) |
| **3.4** | 1 Determine and interpret z-scores (p. 122) | 17, 18 |
| | 2 Determine and interpret percentiles (p. 123) | 19(a)–(e) |
| | 3 Determine and interpret quartiles (p. 125) | 19(f) |
| | 4 Check a set of data for outliers (p. 127) | 15, 16 (h) |
| **3.5** | 1 Compute the five-number summary (p. 132) | 15, 16 (e) |
| | 2 Draw and interpret boxplots (p. 133) | 15(f), 16(f), 19(g), 20 |

## Review Exercises

**1. Muzzle Velocity** The following data represent the muzzle velocity (in meters per second) of rounds fired from a 155-mm gun.

| | | | | |
|---|---|---|---|---|
| 793.8 | 793.1 | 792.4 | 794.0 | 791.4 |
| 792.4 | 791.7 | 792.3 | 789.6 | 794.4 |

*Source:* Christenson, Ronald and Blackwood, Larry; "Tests for Precision and Accuracy of Multiple Measuring Devices," *Technometrics*, Nov. 93, Vol. 35, Issue 4, pp. 411–421.

(a) Compute the sample mean and median muzzle velocity.
(b) Compute the range, sample variance, and sample standard deviation.

**2. Pulse Rates** The following data represent the pulse rate of 8 randomly selected females after stepping up and down on a 6 inch platform for 3 minutes. Pulse is measured in beats per minute.

| | | | | |
|---|---|---|---|---|
| 136 | 169 | 120 | 128 | 129 |
| 143 | 115 | 146 | 96 | 86 |

*Source:* Michael McCraith, Joliet Junior College

(a) Compute the sample mean and median pulse.
(b) Compute the range, sample variance, and sample standard deviation.

**3. Price of Chevy Cavaliers** The following data represent the sales price for nine two-year-old Chevrolet Cavaliers.

| | | | | |
|---|---|---|---|---|
| 16,495 | 15,300 | 13,995 | 13,995 | 12,995 |
| 11,990 | 10,995 | 10,948 | 10,900 | |

*Source:* cars.com

(a) Compute the sample mean and median price.
(b) Compute the range, and sample standard deviation.
(c) Redo (a) and (b) as if the data value 16,495 and incorrectly been entered as 61,495. How does this change affect the mean? The median? The range? The standard deviation? Which of these values is resistant?

**4. Home Sales** The following data represent the closing prices (in U.S. dollars) of 15 randomly selected homes sold in Joliet, Illinois, in December 1999.

| | | | |
|---|---|---|---|
| 138,820 | 140,794 | 136,833 | 157,216 |
| 169,541 | 153,146 | 115,000 | 149,380 |
| 135,512 | 99,000 | 124,757 | 136,529 |
| 149,143 | 136,924 | 128,429 | |

*Source:* Transamerica Intellitech

(a) Compute the sample mean and median sale price.
(b) Compute the range, and sample standard deviation.

**5. Chief Justices** The following data represent the ages of chief justices of the United States Supreme Court when they were appointed.

| Justice | Age |
|---|---|
| John Jay | 44 |
| John Rutledge | 56 |
| Oliver Ellsworth | 51 |
| John Marshall | 46 |
| Roger B. Taney | 59 |
| Salmon P. Chase | 56 |
| Morrison R. Waite | 58 |
| Melville W. Fuller | 55 |
| Edward D. White | 65 |
| William H. Taft | 64 |
| Charles E. Hughes | 68 |
| Harlan F. Stone | 69 |
| Frederick M. Vinson | 56 |
| Earl Warren | 62 |
| Warren E. Burger | 62 |
| William H. Rehnquist | 62 |

*Source:* Information Please Almanac

(a) Compute the population mean, median, and mode ages.
(b) Compute the range and population standard deviation ages.
(c) Obtain two simple random samples of size four, and compute the sample mean and sample standard deviation ages.

**6. National Champion Trees** The following data represent the height (in feet) of the 10 National Champion trees, as selected by *American Forests*.

| Tree | Height (in feet) |
|---|---|
| American Beech | 115 |
| Black Willow | 76 |
| Coast Douglas-Fir | 329 |
| Coast Redwood | 313 |
| Giant Sequoia | 275 |
| Loblolly Pine | 148 |
| Pinyon Pine | 69 |
| Sugar Maple | 65 |
| Sugar Pine | 232 |
| White Oak | 96 |

*Source:* World Almanac, 2000

(a) Compute the population mean, median, and mode.
(b) Compute the range and population standard deviation.
(c) Obtain two simple random samples of size four, and compute the sample mean and sample standard deviation.

7. **Family Size** By random sample, 36 married couples who had been married seven years were asked the number of children they had. The results of the survey follow:

| 0 | 0 | 3 | 1 | 2 | 3 |
| 3 | 4 | 3 | 3 | 0 | 3 |
| 1 | 2 | 1 | 3 | 0 | 3 |
| 4 | 2 | 3 | 2 | 2 | 4 |
| 2 | 1 | 3 | 4 | 1 | 3 |
| 0 | 3 | 3 | 3 | 2 | 1 |

(a) Compute the sample mean and the median number of children.
(b) Compute the range and the sample standard deviation.

8. **Waiting in Line** The following data represent the number of cars that arrived at a McDonald's drive-through between 11:50 A.M. and 12:00 noon each Wednesday for the past 30 weeks:

| 1 | 3 | 2 | 8 | 6 |
| 6 | 6 | 3 | 3 | 1 |
| 5 | 6 | 3 | 3 | 1 |
| 4 | 9 | 5 | 3 | 5 |
| 2 | 6 | 7 | 5 | 8 |
| 7 | 8 | 3 | 2 | 3 |

(a) Compute the sample mean and the median number of cars.
(b) Compute the range and the sample standard deviation waiting time.

9. **The Empirical Rule** Lightbulbs from a sample of 200 lightbulbs have a mean life of 600 hours and a standard deviation of 53 hours.
(a) Suppose a histogram of the data indicates the sample data follow a bell-shaped distribution. According to the Empirical Rule, 99.7% of lightbulbs will have life times between _____ and _____ hours.
(b) Assuming the distribution of the data is bell shaped, determine the percentage of lightbulbs that will have a life between 494 and 706 hours.
(c) If the company that manufactures the lightbulb guarantees to replace any bulb that doesn't last at least 441 hours, what percent of lightbulbs can the firm expect to have to replace, according to the Empirical Rule?

10. **The Empirical Rule** In a random sample of 250 toner cartridges, the mean number of pages a toner cartridge can print is 4,302 and the standard deviation is 340.
(a) Suppose a histogram of the data indicates that the sample data follow a bell-shaped distribution. According to the Empirical Rule, 99.7% of toner cartridges will print between _____ and _____ pages.
(b) Assuming that the distribution of the data is bell shaped, determine the percentage of toner cartridges whose print total is between 3,622 and 4,982 pages.
(c) If the company that manufactures the toner cartridges guarantees to replace any cartridge that doesn't print at least 3,622 pages, what percent of cartridges can the firm expect to be responsible for replacing, according to the Empirical Rule?

11. **Vehicle Fatalities** The frequency distribution listed in the table represents the number of drivers in fatal crashes in 1996, by age, for males aged 20 to 84 years old.

| Age | Number of Drivers |
| --- | --- |
| 20–24 | 6,148 |
| 25–29 | 5,073 |
| 30–34 | 4,834 |
| 35–39 | 4,414 |
| 40–44 | 3,563 |
| 45–49 | 2,935 |
| 50–54 | 2,164 |
| 55–59 | 1,655 |
| 60–64 | 1,398 |
| 65–69 | 1,154 |
| 70–74 | 1,055 |
| 75–79 | 894 |
| 80–84 | 684 |

*Source:* National Highway Traffic Safety Administration

(a) Approximate the mean age of a male involved in a traffic fatality.
(b) Approximate the standard deviation age of a male involved in a traffic fatality.

12. **Vehicle Fatalities** The frequency distribution shown in the table represents the number of drivers in fatal crashes in 1996, by age, for females aged 20 to 84 years old.

| Age | Number of Drivers |
| --- | --- |
| 20–24 | 1,747 |
| 25–29 | 1,558 |
| 30–34 | 1,561 |
| 35–39 | 1,503 |
| 40–44 | 1,180 |
| 45–49 | 957 |
| 50–54 | 752 |
| 55–59 | 522 |
| 60–64 | 498 |
| 65–69 | 491 |
| 70–74 | 550 |
| 75–79 | 485 |
| 80–84 | 314 |

(a) Approximate the mean age of a female involved in a traffic fatality.
(b) Approximate the standard deviation age of a female involved in a traffic fatality.
(c) Compare the results to those obtained in Problem 11. How do you think an insurance company might use this information?

13. **Weighted Mean** Michael has just completed his first semester in college. He earned an "A" in his 5-hour calculus course, a "B" in his 4-hour chemistry course, an "A" in his 3-hour speech course, and a "C" in his 3-hour psychology course. Assuming an "A" equals 4 points, a "B" equals 3 points, and a "C" equals 2 points, determine Michael's grade-point average if grades are weighted by class hours.

14. **Weighted Mean** Yolanda wishes to develop a new type of meat loaf to sell at her restaurant. She decides to combine 2 pounds of ground sirloin (cost $2.70 per pound), 1 pound of ground turkey ($1.30 per pound), and $\frac{1}{2}$ pound of ground pork (cost $1.80 per pound). What is the cost per pound of the meat loaf?

15. **Mets versus Yankees** The following data represent the salaries of the 25 players on the playoff rosters of the 2000 New York Mets and the 2000 New York Yankees.

| New York Mets | | New York Yankees | |
|---|---|---|---|
| **Player** | **Salary** | **Player** | **Salary** |
| Mike Piazza | 12,121,428 | Bernie Williams | 12,357,143 |
| Robin Ventura | 8,000,000 | David Cone | 12,000,000 |
| Al Leiter | 7,750,000 | Derek Jeter | 10,000,000 |
| Mike Hampton | 5,750,000 | Mariano Rivera | 7,250,000 |
| Bobby J. Jones | 5,366,667 | Andy Pettitte | 7,000,000 |
| Edgardo Alfonzo | 4,375,000 | David Justice | 7,000,000 |
| Rick Reed | 4,375,000 | Paul O'Neill | 6,500,000 |
| Todd Zeile | 4,333,333 | Roger Clemens | 6,350,000 |
| Darryl Hamilton | 3,633,333 | Chuck Knoblauch | 6,000,000 |
| Armando Benitez | 3,437,500 | Scott Brosius | 5,250,000 |
| John Franco | 3,350,000 | Tino Martinez | 4,800,000 |
| Mike Bordick | 3,025,000 | Denny Neagle | 4,750,000 |
| Dennis Cook | 2,200,000 | Jose Vizcaino | 3,500,000 |
| Turk Wendell | 2,050,014 | Jose Canseco | 3,000,000 |
| Lenny Harris | 1,100,000 | Mike Stanton | 2,400,000 |
| Rick White | 610,000 | Orlando Hernandez | 1,950,000 |
| Kurt Abbott | 500,000 | Jeff Nelson | 1,916,666 |
| Todd Pratt | 500,000 | Glenallen Hill | 1,500,000 |
| Matt Franco | 462,500 | Jorge Posada | 1,250,000 |
| Glendon Rusch | 270,000 | Jason Grimsley | 750,000 |
| Bubba Trammell | 253,000 | Luis Sojo | 450,000 |
| Joe McEwing | 250,000 | Chris Turner | 375,000 |
| Benny Agbayani | 220,000 | Clay Bellinger | 206,650 |
| Jay Payton | 215,000 | Dwight Gooden | 200,000 |
| Timo Perez | 200,000 | Luis Polonia | 200,000 |

*Source: USA Today* research by Hal Bodley

(a) Compute the population mean salary for each team.
(b) Compute the median salary for each team.
(c) Given the results of (a) and (b), decide whether the distribution is symmetric, skewed right, or skewed left.
(d) Compute the population standard deviation salary for each team. Which team has its salaries more dispersed?
(e) Compute the five-number summary for each team.
(f) On the same graph, draw boxplots for the two teams. Annotate the graph with some general remarks comparing the team salaries.
(g) Describe the shape of the distribution of each team, as illustrated by the boxplots. Does this confirm the result obtained in part (c)?
(h) Determine whether the data for each team contain outliers.

16. **Air-Traffic Control Errors** An air-traffic control error is said to occur when planes come too close to one another. The following data represent the number of air-traffic control errors for a random sample of regions around the United States for fiscal years 1996 and 2000.

| Center | Errors, 2000 | Errors, 1996 | Center | Errors, 2000 | Errors, 1996 |
|---|---|---|---|---|---|
| Washington | 102 | 24 | Cleveland | 74 | 32 |
| New York ARTCC | 71 | 44 | Chicago ARTCC | 70 | 26 |
| Indianapolis | 54 | 39 | Atlanta | 40 | 36 |
| Memphis | 38 | 21 | Dallas–Fort Worth | 34 | 23 |
| Los Angeles | 33 | 19 | Denver | 33 | 11 |
| Jacksonville, FL | 30 | 27 | Kansas City, MO | 28 | 20 |
| New York TRACON | 27 | 33 | Albuquerque | 25 | 21 |
| Miami | 21 | 15 | Boston | 21 | 13 |
| Northern California | 18 | 30 | Houston | 18 | 7 |
| Oakland | 17 | 20 | Minneapolis | 15 | 13 |
| Chicago TRACON | 14 | 13 | Salt Lake City | 12 | 8 |
| Oakland Bay | 8 | 6 | Miami | 8 | 2 |
| Washington Dulles | 7 | 2 | | | |

*Source:* Associated Press

(a) Compute the sample mean number of errors in 1996 and 2000.
(b) Compute the median number of errors in 1996 and 2000.
(c) Given the results of (a) and (b), decide whether the distribution of "number of errors in 1996" is symmetric, skewed right, or skewed left. What about "number of errors in 2000"?
(d) Compute the sample standard deviation number of errors in 1996 and 2000. Which year is more dispersed?
(e) Compute the five-number summary for each year.
(f) On the same graph, draw boxplots for the two years. Add some general remarks comparing each year's errors.
(g) Describe the shape of the distribution of each year, as shown in the boxplots. Does this confirm the result obtained in part (c)?
(h) Determine whether the data for each year contain outliers.
(i) Explain why comparing 1996 data to 2000 data may be misleading.

17. **Cholesterol** According to the National Center for Health Statistics, the mean serum total cholesterol for females 20 years and older is 206, with a standard deviation of 44.7. The mean serum total cholesterol for males 20 years and older is 202, with a standard deviation of 41.0. Who has a relatively higher serum total cholesterol: a female whose serum total cholesterol is 230 or a male whose serum total cholesterol is 230? Why?

18. **Weights of Males versus Females** According to the National Center for Health Statistics, the mean weight of a 20–29-year-old female is 141.7 pounds, with a standard deviation of 27.2 pounds. The mean weight of a 20–29-year-old male is 172.1 pounds, with a standard deviation of 36.5 pounds. Who is relatively heavier: a 20–29-year-old female who weighs 150 pounds or a 20–29-year-old male who weighs 185 pounds? Why?

19. **Infant Mortality Rate** The data on pages 146–147 represent the infant mortality rate (deaths per 1000 infants) for 226 countries throughout the world. The data have already been sorted in ascending order.
(a) Find the 20th percentile.
(b) Find the 95th percentile.
(c) Find the 99th percentile.
(d) What is the percentile rank of the United States?
(e) What is the percentile rank of Nepal?
(f) Find the quartiles.
(g) Draw a boxplot. Are there any outliers?

| Country | Infant Mortality Rate | Country | Infant Mortality Rate | Country | Infant Mortality Rate |
|---|---|---|---|---|---|
| 1. Sweden | 3.49 | 56. Chile | 9.6 | 109. Saint Helena | 23.23 |
| 2. Iceland | 3.58 | 57. Poland | 9.61 | 110. Oman | 23.28 |
| 3. Singapore | 3.65 | 58. Virgin Islands | 9.64 | 111. Tuvalu | 23.3 |
| 4. Finland | 3.82 | 59. Puerto Rico | 9.71 | 112. North Korea | 24.29 |
| 5. Japan | 3.91 | 60. Guadeloupe | 9.77 | 113. Colombia | 24.7 |
| 6. Norway | 3.98 | 61. Bermuda | 9.82 | 114. Cook Islands | 24.7 |
| 7. Andorra | 4.08 | 62. Cayman Islands | 10.44 | 115. Suriname | 25.06 |
| 8. Netherlands | 4.42 | 63. American Samoa | 10.63 | 116. Bosnia/Herzegovina | 25.17 |
| 9. Macau | 4.49 | 64. Nauru | 10.9 | 117. Solomon Islands | 25.26 |
| 10. Austria | 4.5 | 65. Montenegro | 10.97 | 118. Anguilla | 25.44 |
| 11. France | 4.51 | 66. Costa Rica | 11.49 | 119. Trinidad and Tobago | 25.76 |
| 12. Switzerland | 4.53 | 67. Kuwait | 11.55 | 120. Belize | 25.97 |
| 13. Slovenia | 4.56 | 68. Netherlands | 11.74 | 121. Gaza Strip | 25.97 |
| 14. Belgium | 4.76 | 69. Antilles Barbados | 12.37 | 122. Venezuela | 26.17 |
| 15. Germany | 4.77 | 70. Estonia | 12.92 | 123. Mexico | 26.19 |
| 16. Luxembourg | 4.83 | 71. Macedonia | 13.35 | 124. China | 28.92 |
| 17. Spain | 4.99 | 72. French Guiana | 13.99 | 125. El Salvador | 29.22 |
| 18. Australia | 5.04 | 73. Fiji | 14.45 | 126. Lebanon | 29.3 |
| 19. Guernsey | 5.07 | 74. Tonga | 14.45 | 127. Philippines | 29.52 |
| 20. Liechtenstein | 5.07 | 75. Jamaica | 14.61 | 128. Iran | 30.02 |
| 21. Canada | 5.08 | 76. Belarus | 14.63 | 129. Libya | 30.08 |
| 22. Denmark | 5.11 | 77. Grenada | 14.63 | 130. Tunisia | 30.09 |
| 23. Gibraltar | 5.6 | 78. Lithuania | 14.67 | 131. Paraguay | 30.81 |
| 24. Ireland | 5.62 | 79. Brunei | 14.84 | 132. Vietnam | 31.13 |
| 25. Czech Republic | 5.63 | 80. Bulgaria | 15.13 | 133. Honduras | 31.29 |
| 26. United Kingdom | 5.63 | 81. Uruguay | 15.14 | 134. Thailand | 31.48 |
| 27. Jersey | 5.71 | 82. Saint Lucia | 15.64 | 135. Samoa | 32.75 |
| 28. Northern Mariana Islands | 5.79 | 83. Latvia | 15.71 | 136. Micronesia | 33.48 |
| 29. Italy | 5.92 | 84. Sri Lanka | 16.51 | 137. Nicaragua | 34.79 |
| 30. Monaco | 5.92 | 85. Saint Kitts and Nevis | 16.72 | 138. Syria | 34.86 |
| 31. Hong Kong | 5.93 | 86. Bahamas | 16.99 | 139. Ecuador | 35.13 |
| 32. Malta | 5.94 | 87. Saint Vincent and the Grenadines | 17.06 | 140. Dominican Republic | 35.93 |
| 33. Portugal | 6.05 | 88. Palau | 17.12 | 141. Brazil | 38.04 |
| 34. San Marino | 6.33 | 89. Dominica | 17.13 | 142. Guyana | 39.07 |
| 35. New Zealand | 6.39 | 90. United Arab Emirates | 17.17 | 143. Peru | 40.6 |
| 36. Aruba | 6.51 | 91. Mauritius | 17.73 | 144. Marshall Islands | 40.95 |
| 37. Greece | 6.51 | 92. Seychelles | 17.74 | 145. Mongolia | 41.22 |
| 38. Man, Isle of | 6.54 | 93. Greenland | 18.26 | 146. Albania | 41.33 |
| 39. United States | 6.82 | 94. Argentina | 18.31 | 147. Armenia | 41.48 |
| 40. Guam | 6.83 | 95. Turks and Caicos Islands | 18.66 | 148. Algeria | 41.97 |
| 41. Faroe Islands | 6.94 | 96. Romania | 19.84 | 149. Indonesia | 42.21 |
| 42. Taiwan | 7.06 | 97. Serbia | 20.13 | 150. Moldova | 43.32 |
| 43. Croatia | 7.35 | 98. Russia | 20.33 | 151. Guatemala | 47.03 |
| 44. Cuba | 7.51 | 99. Bahrain | 20.48 | 152. Turkey | 48.9 |
| 45. South Korea | 7.85 | 100. Panama | 20.8 | 153. Morocco | 49.72 |
| 46. Israel | 7.9 | 101. Malaysia | 20.96 | 154. São Tomé and Principe | 50.41 |
| 47. Martinique | 7.97 | 102. Maldives | 20.96 | 155. Saudi Arabia | 52.9 |
| 48. Cyprus | 8.07 | 103. British Virgin Islands | 21.05 | 156. Georgia | 52.94 |
| 49. New Calendonia | 8.57 | 104. Jordan | 21.11 | 157. Cape Verde | 54.58 |
| 50. Saint Pierre and Miquelon | 8.61 | 105. Ukraine | 21.67 | 158. Kiribati | 55.36 |
| 51. Reunion | 8.67 | 106. Qatar | 22.14 | 159. Ghana | 57.43 |
| 52. Montserrat | 9.1 | 107. West Bank | 22.33 | 160. Senegal | 58.08 |
| 53. Hungary | 9.15 | 108. Antigua and Barbuda | 23.05 | 161. South Africa | 58.88 |
| 54. Slovakia | 9.18 | | | 162. Kazakhstan | 59.39 |
| 55. French Polynesia | 9.3 | | | 163. Papua New Guinea | 59.89 |
| | | | | 164. Bolivia | 60.44 |
| | | | | 165. Botswana | 61.68 |

| Country | Infant Mortality Rate | Country | Infant Mortality Rate | Country | Infant Mortality Rate |
|---|---|---|---|---|---|
| 166. Zimbabwe | 62.25 | 188. Mauritania | 78.15 | 208. Djibouti | 103.32 |
| 167. Egypt | 62.32 | 189. Gambia, The | 79.29 | 209. Central African Republic | 106.69 |
| 168. Iraq | 62.49 | 190. Tanzania | 80.97 | 210. Burkina Faso | 108.53 |
| 169. Vanuatu | 62.52 | 191. Pakistan | 82.49 | 211. Swaziland | 108.95 |
| 170. India | 64.9 | 192. Lesotho | 82.97 | 212. Bhutan | 110.99 |
| 171. Cambodia | 66.82 | 193. Azerbaijan | 83.41 | 213. Guinea-Bissau | 112.25 |
| 172. Kenya | 68.74 | 194. Madagascar | 85.26 | 214. Tajikistan | 117.42 |
| 173. Sudan | 70.21 | 195. Comoros | 86.33 | 215. Rwanda | 120.06 |
| 174. Yemen | 70.28 | 196. Benin | 90.84 | 216. Malawi | 122.28 |
| 175. Cameroon | 70.87 | 197. Zambia | 92.38 | 217. Mali | 123.25 |
| 176. Namibia | 70.88 | 198. Uganda | 93.25 | 218. Niger | 124.9 |
| 177. Mayotte | 71.31 | 199. Laos | 94.8 | 219. Somalia | 125.77 |
| 178. Burundi | 71.5 | 200. Equatorial Guinea | 94.83 | 220. Guinea | 130.98 |
| 179. Togo | 71.55 | 201. Cote d'Ivoire | 95.06 | 221. Western Sahara | 133.59 |
| 180. Bangladesh | 71.66 | 202. Gabon | 96.3 | 222. Liberia | 134.63 |
| 181. Uzbekistan | 72.13 | 203. Chad | 96.66 | 223. Mozambique | 139.86 |
| 182. Turkmenistan | 73.3 | 204. Haiti | 97.1 | 224. Sierra Leone | 148.66 |
| 183. Nigeria | 74.18 | 205. Ethiopia | 101.29 | 225. Afghanistan | 149.28 |
| 184. Burma | 75.3 | 206. Congo, Republic of the | 101.55 | 226. Angola | 195.78 |
| 185. Nepal | 75.93 | 207. Congo, Democratic Republic of the | 101.71 | | |
| 186. Eritrea | 76.66 | | | | |
| 187. Kyrgyzstan | 77.08 | | | | |

**20. Crime Rate** Answer the accompanying questions regarding the boxplot, which illustrates crime-rate data for the 50 United States:

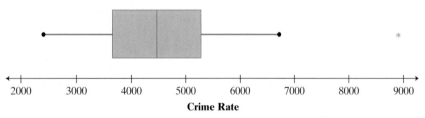

(a) Approximately, what is the median crime rate in the United States?
(b) Approximately, what is the 25th percentile crime rate in the United States?
(c) Are there any outliers? If so, identify them.
(d) What is the lowest crime rate?

**Chapter 3 Projects located at www.prenhall.com/sullivanstats**

# Describing the Relation between Two Variables

## Outline

**For additional study help, go to**
www.prenhall.com/sullivanstats

    Materials include

- Projects
  - ■ Case Study: The Mystery of Global Temperature Change
  - ■ Decisions: What Car Should I Buy?
  - ■ Consumer Reports Project
- Self-Graded Quizzes
- "Preparing for This Section" Quizzes
- STATLETs
- PowerPoint Downloads
- Step-by-Step Technology Guide
- Graphing Calculator Help

## Putting It All Together

In Chapters 2 and 3 we examined data in which a single variable was measured for each individual in the study (**univariate data**), such as the three-year rate of return (the variable) for various mutual funds (the individual). We obtained descriptive measures for the variable that were both graphical and numerical.

However, much research is designed in order to describe the relation that may exist between two variables. For example, a researcher may be interested in the relationship between per capita gross domestic product (GDP)

and the life expectancy of residents of a country (per capita GDP can be thought of as the average income of residents of the country). Here, the individual is the country and the two variables are per capita GDP and life expectancy. This type of data is referred to as *bivariate data*. **Bivariate data** is data in which two variables are measured on an individual. In order to describe the relation between the two variables, we first graphically represent the data and then obtain some numerical descriptions of the data, just as we did when analyzing univariate data.

## 4.1  Scatter Diagrams; Correlation

**Preparing for This Section**  Before getting started, review the following:

✓ Mean (Section 3.1, pp. 84–87)

✓ Standard deviation (Section 3.2, pp. 105–106)

✓ Z-scores (Section 3.4, pp. 122–123)

**Objectives**  Draw scatter diagrams

 Interpret scatter diagrams

 Understand the properties of the linear correlation coefficient

 Compute and interpret the linear correlation coefficient

Before we can graphically represent bivariate data, a fundamental question must be asked. Am I interested in using the value of one variable to predict the value of the other variable? For example, it seems reasonable to think that as the level of per capita gross domestic product (GDP) within a country increases, the life expectancy of the country's residents also increases. Therefore, we might use per capita GDP to predict life expectancy. We call life expectancy the *response* (or *dependent*) *variable* and per capita GDP the *predictor* (or *independent*) *variable*.

*Definition*

> The **response variable** is the variable whose value can be explained by, or is determined by, the value of the **predictor variable**.

It is important to recognize that if the data used in the study are observational, then the relation determined between the two variables is not a causal relation. We cannot say that changes in the level of the predictor variable *cause* changes in the level of the response variable. In fact, it may be that the two are related through some *lurking variable*. A **lurking variable** is a variable that is related to either the response or predictor variable or both, but is excluded from the analysis. For example, air-conditioning bills can be used to predict lemonade sales: As air-conditioning bills rise, the sales of lemonade rise. This relation does not mean that high air-conditioning bills cause high lemonade sales, because both high air-conditioning bills and high lemonade sales are associated with high summer temperatures. Therefore, summer temperature is a lurking variable.

**Caution**

If bivariate data are observational, then the predictor and response are not causally related.

###  Scatter Diagrams

The first step in identifying the type of relation, if any, that might exist between two variables is to draw a picture. Bivariate data can be represented graphically through a *scatter diagram*.

*Definition*

> A **scatter diagram** is a graph that shows the relationship between two quantitative variables measured on the same individual. Each individual in the data set is represented by a point in the scatter diagram. The predictor variable is plotted on the horizontal axis and the response variable is plotted on the vertical axis. Do not connect the points when drawing a scatter diagram.

▶ **EXAMPLE 1**  Drawing a Scatter Diagram

*Problem:*  The per capita gross domestic product (GDP) of a country is the average income of a resident of the country. A researcher would like to know if per capita GDP could be used to predict life expectancy. The data in

Table 1 represent the per capita GDP (in thousands of U.S. dollars) for randomly selected countries in Western Europe and the life expectancy of residents of the country. Draw a scatter diagram of the data.

| | Per Capita GDP (000s) | Life Expectancy | Country | Per Capita GDP (000s) | Life Expectancy |
|---|---|---|---|---|---|
| Country | | | | | |
| Austria | 21.4 | 77.48 | Ireland | 18.6 | 76.39 |
| Belgium | 23.2 | 77.53 | Italy | 21.5 | 78.51 |
| Finland | 20.0 | 77.32 | Netherlands | 22.0 | 78.15 |
| France | 22.7 | 78.63 | Switzerland | 23.8 | 78.99 |
| Germany | 20.8 | 77.17 | United Kingdom | 21.2 | 77.37 |

TABLE 1

*Source:* Time Almanac 2000

*Approach:* Because the researcher wants to use per capita GDP to predict life expectancy, per capita GDP will be the predictor variable (horizontal axis) and life expectancy will be the response variable (vertical axis). We plot the ordered pairs (21.4, 77.48), (23.2, 77.53), and so on, in a rectangular coordinate system.

*Solution:* The scatter diagram is shown in Figure 1.

Figure 1

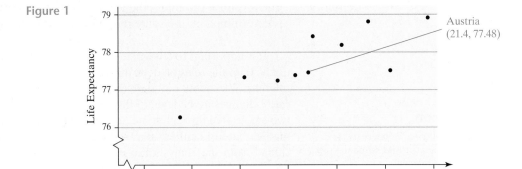

It would appear from the graph that as per capita GDP rises, the life expectancy of the country's residents rise as well. ◀◀

It is not always clear which variable should be considered the response variable and which should be considered the predictor variable. For example, does high school GPA predict a student's SAT score or can the SAT score be used to predict GPA? The researcher must determine which variable plays the role of predictor variable based upon the questions he or she wants answered. For example, if the researcher is interested in predicting SAT scores on the basis of high school GPA, then high school GPA will play the role of predictor variable.

**NW** *Now Work Problems 11(a) and 11(b).*

### ② Interpreting Scatter Diagrams

Scatter diagrams show the type of relation that exists between two variables. Our goal in interpreting scatter diagrams will be to distinguish scatter diagrams that imply a linear relation from those which imply a nonlinear relation and those which imply no relation. Figure 2 displays various scatter diagrams and the type of relation implied.

Figure 2

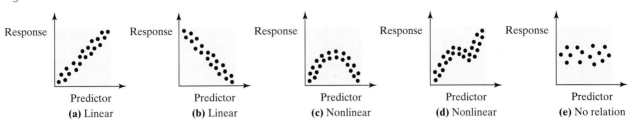

(a) Linear    (b) Linear    (c) Nonlinear    (d) Nonlinear    (e) No relation

As we compare Figure 2(a) with Figure 2(b), we notice a distinct difference. In Figure 2(a), the data follow a linear pattern that slants upward to the right, while the data in Figure 2(b) follow a linear pattern that slants downward to the right.

*Definitions*

Two variables that are linearly related are said to be **positively associated** when above-average values of one variable are associated with above-average values of the corresponding variable. That is, two variables are positively associated if, whenever the values of the predictor variable increase, the values of the response variable also increase.

Two variables that are linearly related are said to be **negatively associated** when above-average values of one variable are associated with below-average values of the corresponding variable. That is, two variables are negatively associated if, whenever the values of the predictor variable increase, the values of the response variable decrease.

**In Your Own Words**

If two variables that are linearly related are positively associated, then as one goes up the other also goes up. If two variables that are linearly related are negatively associated, then as one goes up the other goes down.

So the scatter diagram from Figure 1 implies that per capita GDP is positively associated with life expectancy.

**NW** *Now Work Problem 1.*

**3**  **Correlation**

It is dangerous to use only a scatter diagram to conclude whether two variables might follow a linear relation. Suppose we redraw the scatter diagram in Figure 1 using a different scale. See Figure 3.

Figure 3

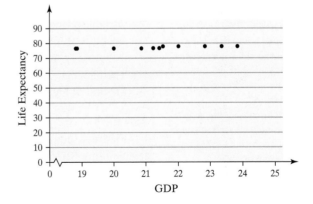

**Caution**

The horizontal or vertical scale of a scatter diagram can be set so that the scatter diagram misleads a reader.

From Figure 3, we might conclude that per capita GDP and life expectancy are not related. The moral of the story is this: Just as we can manipulate the scale of graphs of univariate data, we can also manipulate the scale of the graphs of bivariate data, thereby encouraging incorrect conclusions. Therefore, numerical summaries of bivariate data should be used in conjunction with graphs in order to determine the type of relation, if any, that exists between two variables.

*Definition*

The **linear correlation coefficient** or **Pearson product moment correlation coefficient** is a measure of the strength of linear relation between two quantitative variables. We use the Greek letter $\rho$ (rho) to represent the population correlation coefficient and $r$ to represent the sample correlation coefficient. We shall present only the formula for the sample correlation coefficient.

**Sample Correlation Coefficient**

$$r = \frac{\sum \left( \dfrac{x_i - \overline{x}}{s_x} \right)\left( \dfrac{y_i - \overline{y}}{s_y} \right)}{n - 1} \quad * \tag{1}$$

where    $\overline{x}$ is the sample mean of the predictor variable

$s_x$ is the sample standard deviation of the predictor variable

$\overline{y}$ is the sample mean of the response variable

$s_y$ is the sample standard deviation of the response variable

$n$ is the number of individuals in the sample

The Pearson linear correlation coefficient is named in honor of Karl Pearson (1857–1936).

**Historical Note**

**Properties of the Linear Correlation Coefficient**

1. The linear correlation coefficient is always between $-1$ and 1, inclusive. That is, $-1 \leq r \leq 1$.

2. If $r = +1$, there is a perfect positive linear relation between the two variables. See Figure 4(a).

3. If $r = -1$, there is a perfect negative linear relation between the two variables. See Figure 4(d).

4. The closer $r$ is to $+1$, the stronger is the evidence of positive association between the two variables. See Figures 4(b) and 4(c).

5. The closer $r$ is to $-1$, the stronger is the evidence of negative association between the two variables. See Figures 4(e) and 4(f).

6. If $r$ is close to 0, there is evidence of no *linear* relation between the two variables. Because the linear correlation coefficient is a measure of strength of linear relation, $r$ close to 0 does not imply no relation, just no linear relation. See Figures 4(g) and 4(h).

7. The linear correlation coefficient is a unitless measure of association. So the unit of measure for $x$ and $y$ plays no role in the interpretation of $r$.

Figure 4 demonstrates these properties through various scatter diagrams, along with their linear correlation coefficients.

*An equivalent formula for the linear correlation coefficient is

$$r = \frac{\sum x_i y_i - \dfrac{\sum x_i \sum y_i}{n}}{\sqrt{\left( \sum x_i^2 - \dfrac{\left( \sum x_i \right)^2}{n} \right)}\sqrt{\left( \sum y_i^2 - \dfrac{\left( \sum y_i \right)^2}{n} \right)}} = \frac{S_{xy}}{\sqrt{S_{xx}}\sqrt{S_{yy}}}$$

Figure 4

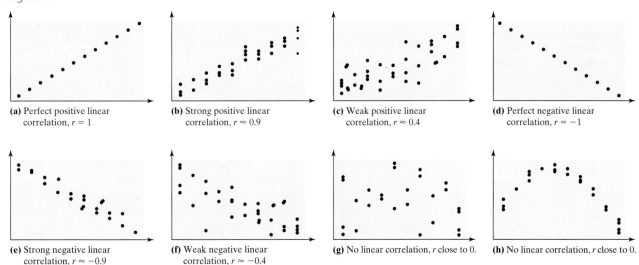

**(a)** Perfect positive linear
correlation, $r = 1$

**(b)** Strong positive linear
correlation, $r \approx 0.9$

**(c)** Weak positive linear
correlation, $r \approx 0.4$

**(d)** Perfect negative linear
correlation, $r = -1$

**(e)** Strong negative linear
correlation, $r \approx -0.9$

**(f)** Weak negative linear
correlation, $r \approx -0.4$

**(g)** No linear correlation, $r$ close to 0.

**(h)** No linear correlation, $r$ close to 0.

**Caution**

A linear correlation coefficient close
to 0 does not imply no relation, just
no linear relation. For example,
although the scatter diagram drawn
in Figure 4(h) indicates that the
predictor and response variables are
related, the linear correlation
coefficient of this data is close to 0.

In looking carefully at Formula (1), we should notice that the numerator of the formula is the product of $z$-scores for the predictor ($x$) and response ($y$) variables. A positive linear correlation coefficient means that the sum of the product of the $z$-scores for $x$ and $y$ must be positive. Under what circumstances does this occur? Figure 5 shows a scatter diagram that implies a positive association between $x$ and $y$. The vertical dashed line represents the value of $\bar{x}$ and the horizontal dashed line represents the value of $\bar{y}$. These two dashed lines divide our scatter diagram into 4 quadrants, labeled I, II, III, and IV.

Notice that a majority of the data lie in quadrants I and III. If a certain $x$-value is above its mean, $\bar{x}$, then the corresponding $y$-value will be above its mean, $\bar{y}$. If a certain $x$-value is below its mean, $\bar{x}$, then the corresponding $y$-value will be below its mean, $\bar{y}$. Therefore, for data in quadrant I, we would have $\left( \dfrac{x_i - \bar{x}}{s_x} \right)$ positive and $\left( \dfrac{y_i - \bar{y}}{s_y} \right)$ positive, so that their product is positive. For data in quadrant III, we would have $\left( \dfrac{x_i - \bar{x}}{s_x} \right)$ negative and $\left( \dfrac{y_i - \bar{y}}{s_y} \right)$ negative, so that their product is positive. The sum of these products will be positive, and therefore we have a positive linear correlation coefficient. A similar argument can be made for negative correlation.

Figure 5

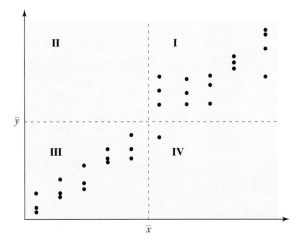

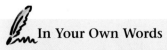

**In Your Own Words**

The correlation coefficient describes the strength and the direction of the linear relationship between a predictor and a response variable.

Now suppose the data are equally dispersed in the four quadrants. Then the negative products (resulting from data in quadrants II and IV) will offset the positive products (resulting from data in quadrants I and III). The result is a linear correlation coefficient close to 0.

 *Now Work Problem 5.*

④ ▶ EXAMPLE 2   **Computing and Interpreting the Correlation Coefficient**

*Problem:* In Table 2, columns 2 and 3 represent the per capita gross domestic product (in thousands of U.S. dollars) and the average life expectancy of the population, for randomly selected countries in Western Europe. Compute and interpret the linear correlation coefficient.

*Approach:* We treat per capita GDP as the predictor variable, $x$, and life expectancy as the response variable, $y$.

*Step 1:* Compute $\bar{x}$, $s_x$, $\bar{y}$, and $s_y$.

*Step 2:* Determine $\dfrac{x_i - \bar{x}}{s_x}$ and $\dfrac{y_i - \bar{y}}{s_y}$.

*Step 3:* Compute $\left(\dfrac{x_i - \bar{x}}{s_x}\right)\left(\dfrac{y_i - \bar{y}}{s_y}\right)$.

*Step 4:* Determine $\sum\left(\dfrac{x_i - \bar{x}}{s_x}\right)\left(\dfrac{y_i - \bar{y}}{s_y}\right)$ and substitute this value into Formula (1).

*Solution:*

*Step 1:* We compute $\bar{x}$, $s_x$, $\bar{y}$, and $s_y$:

$$\bar{x} = 21.52,\ s_x = 1.531738301,\ \bar{y} = 77.754,\ \text{and}\ s_y = 0.794847295$$

Notice we do not round these values. This is to avoid round-off error when using Formula (1).

*Step 2:* We determine $\dfrac{x_i - \bar{x}}{s_x}$ and $\dfrac{y_i - \bar{y}}{s_y}$ in Columns 4 and 5 in Table 2.

| TABLE 2 | | | | | |
|---|---|---|---|---|---|
| Country | Per Capita GDP, $x_i$ | Life Expectancy, $y_i$ | $\dfrac{x_i - \bar{x}}{s_x}$ | $\dfrac{y_i - \bar{y}}{s_y}$ | $\left(\dfrac{x_i - \bar{x}}{s_x}\right)\left(\dfrac{y_i - \bar{y}}{s_y}\right)$ |
| Austria | 21.4 | 77.48 | $\dfrac{21.4 - 21.52}{1.531738301}$ $= -0.0783424$ | $\dfrac{77.48 - 77.754}{0.794847295}$ $= -0.34472$ | $(-0.0783424)(-0.34472)$ $= 0.027006$ |
| Belgium | 23.2 | 77.53 | 1.0967931 | -0.28182 | -0.30909 |
| Finland | 20.0 | 77.32 | -0.9923366 | -0.54602 | 0.541832 |
| France | 22.7 | 78.63 | 0.7703666 | 1.102098 | 0.84902 |
| Germany | 20.8 | 77.17 | -0.4700542 | -0.73473 | 0.345364 |
| Ireland | 18.6 | 76.39 | -1.9063309 | -1.71605 | 3.271365 |
| Italy | 21.5 | 78.51 | -0.0130571 | 0.951126 | -0.01242 |
| Netherlands | 22.0 | 78.15 | 0.3133695 | 0.498209 | 0.156123 |
| Switzerland | 23.8 | 78.99 | 1.4885049 | 1.555016 | 2.314648 |
| United Kingdom | 21.2 | 77.37 | -0.2089130 | -0.48311 | 0.100928 |

$$\sum\left(\frac{x_i - \bar{x}}{s_x}\right)\left(\frac{y_i - \bar{y}}{s_y}\right) = 7.284776$$

***Step 3:*** We multiply the entries in columns 4 and 5 to obtain the entries in column 6.

***Step 4:*** We add up the entries in column 6 to obtain $\sum\left(\dfrac{x_i - \bar{x}}{s_x}\right)\left(\dfrac{y_i - \bar{y}}{s_y}\right) = 7.284776$. Substitute this value into Formula (1) to obtain the correlation coefficient.

$$r = \frac{\sum\left(\dfrac{x_i - \bar{x}}{s_x}\right)\left(\dfrac{y_i - \bar{y}}{s_y}\right)}{n - 1} = \frac{7.284776}{10 - 1} = 0.809$$

The linear correlation between per capita GDP and life expectancy is 0.809, indicating a strong positive association between the two variables. The higher the per capita GDP of a country, the longer its residents can expect to live.

◄◄

Notice in Example 2 that we carry many decimal places in the computation of the correlation coefficient, in order to avoid rounding error. Also, compare the signs of the entries in columns 4 and 5. Notice that negative values in column 4 correspond with negative values in column 5 and that positive values in column 4 correspond with positive values in column 5 (except for Belgium and Italy). This means that above-average values of $x$ are associated with above-average values of $y$ and below-average values of $x$ are associated with below-average values of $y$. This is why the linear correlation coefficient is positive.

**Using Technology:** Many calculators and statistical software have the ability to compute the correlation between two quantitative variables. Figure 6 shows the correlation coefficient computed in Example 2 using Excel. Notice Excel provides a **correlation matrix**, which means it will compute the correlation between every pair of columns and display the correlation in the bottom triangle of the matrix.

**Figure 6**

|  | Per Capita GDP | Life Expectancy |
|---|---|---|
| Per Capita GDP | 1 |  |
| Life Expectancy | **0.809419506** | 1 |

 *Now Work Problems 11(c) and 11(d).*

Recall from Chapter 1 we stated that there are two types of studies—observational and experimental. When data analysis is performed upon observational data, a researcher cannot make claims regarding causation. It is vital that this point be understood, as it pertains to the linear correlation coefficient.

The data used in Example 2 are observational data. Therefore, we cannot claim that a high per capita GDP causes a higher life expectancy. It may be that a lurking variable is related to both per capita GDP and a higher life expectancy. Recall that a lurking variable is a variable that is correlated to both the response and predictor variables. For example, high per capita GDP and life expectancy may be related through a lurking variable such as the quality of health care. If data are obtained through a controlled experiment, then a linear correlation coefficient that implies a strong positive or negative association also implies causation.

**Caution!**

A linear correlation coefficient that implies a strong positive or negative association that is computed using observational data does not imply causation.

## 4.1 Assess Your Understanding

### Concepts and Vocabulary

1. Describe the difference between univariate and bivariate data.
2. Explain what is meant by a lurking variable. Provide an example.
3. What does it mean to say that two variables are positively associated?
4. What does it mean to say that the linear correlation coefficient between two variables equals one? What would the scatter diagram look like?
5. What does it mean if $r = 0$?
6. Is the linear correlation coefficient a resistant measure? Support your answer.
7. Explain what is wrong with the following statement: "We have concluded that there is a high correlation between male drivers and rates of automobile accidents."
8. Write a statement that explains the concept of correlation. Include a discussion of the role that $x_i - \bar{x}$ and $y_i - \bar{y}$ play in the computation.

### Exercises

- **Skill Building**

*In Problems 1–4, determine whether the scatter diagram indicates that a linear relation may exist between the two variables. If the relation is linear, determine whether it is indicative of positive or negative association between the variables.*

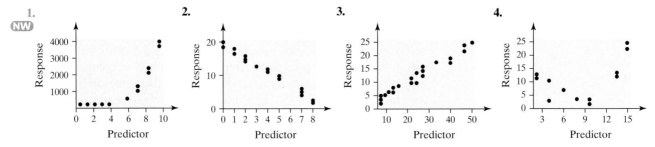

5. Match the linear correlation coefficient to the scatter diagram. The scales on the $x$- and $y$-axis are the same for each scatter diagram.

(a) $r = 0.787$    (b) $r = 0.523$    (c) $r = 0.810$    (d) $r = 0.946$

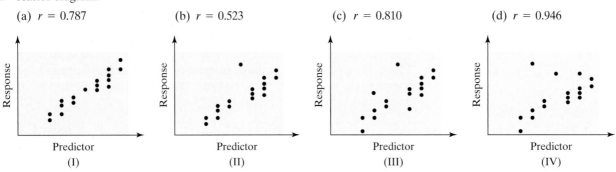

6. Match the linear correlation coefficient to the scatter diagram. The scales on the $x$- and $y$-axis are the same for each scatter diagram.

(a) $r = -0.969$    (b) $r = -0.049$    (c) $r = -1$    (d) $r = -0.992$

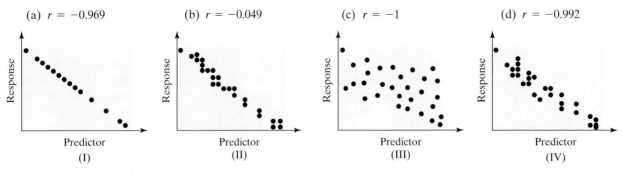

*In Problems 7–10, (a) draw a scatter diagram of the data, (b) by hand, compute the correlation coefficient, (c) comment on the type of relation that appears to exist between x and y.*

**7.**

| x | 2 | 4 | 8 | 8 | 9 |
|---|---|---|---|---|---|
| y | 1.4 | 1.8 | 2.1 | 2.3 | 2.6 |

**8.**

| x | 2 | 3 | 5 | 6 | 6 |
|---|---|---|---|---|---|
| y | 5.7 | 5.2 | 2.8 | 1.9 | 2.2 |

**9.**

| $x$ | 1.2 | 1.8 | 2.3 | 3.5 | 4.1 |
|---|---|---|---|---|---|
| $y$ | 8.4 | 7 | 7.3 | 4.5 | 2.4 |

**10.**

| $x$ | 0 | 0.5 | 1.4 | 2.1 | 3.9 | 4.6 |
|---|---|---|---|---|---|---|
| $y$ | 0.8 | 1.3 | 1.9 | 2.5 | 5.0 | 6.8 |

## • Applying the Concepts

**11. Height versus Head Circumference** A pediatrician
NW wants to determine the relation that may exist between
a child's height and head circumference. She randomly
selects 11 three-year-old children from her practice,
measures their height and head circumference, and obtains the data shown in the table.

| Height (inches) | Head Circumference (inches) | Height (inches) | Head Circumference (inches) |
|---|---|---|---|
| 27.75 | 17.5 | 26.5 | 17.3 |
| 24.5 | 17.1 | 27 | 17.5 |
| 25.5 | 17.1 | 26.75 | 17.3 |
| 26 | 17.3 | 26.75 | 17.5 |
| 25 | 16.9 | 27.5 | 17.5 |
| 27.75 | 17.6 | | |

*Source:* Denise Slucki, Student at Joliet Junior College

(a) If the pediatrician wants to use height to predict
head circumference, determine which variable is
the predictor variable and which is the response
variable.
(b) Draw a scatter diagram.
(c) Compute the linear correlation coefficient between
the height and head circumference of a child.
(d) Comment on the type of relation that appears to
exist between the height and head circumference of
a child on the basis of the scatter diagram and linear
correlation coefficient.

**12. Round-cut Diamonds** A gemologist is interested in determining the relation that exists between the size of a
diamond (carats) and the price. The following data represent the weight and price of D color, VS1 clarity,
round-cut diamonds.

| Carats | Price | Carats | Price |
|---|---|---|---|
| 0.66 | $3282 | 0.80 | $4378 |
| 0.75 | $3950 | 0.90 | $5682 |
| 0.70 | $3543 | 0.91 | $6426 |
| 0.71 | $3788 | 1.18 | $9362 |
| 0.77 | $4108 | | |

*Source:* diamonds.com

(a) The researcher wants to use the number of carats to
predict price. Determine which variable is the predictor variable and which is the response variable.
(b) Draw a scatter diagram.
(c) Compute the linear correlation coefficient between
the number of carats and the price of a diamond.
(d) Comment on the type of relation that appears to
exist between the number of carats and the price,

based upon the scatter diagram and linear correlation coefficient.
(e) Remove the diamond that weighs 1.18 carats from
the data set and recompute the linear correlation coefficient between the number of carats and price of
a diamond. What effect did the removal of the data
value have on the linear correlation coefficient?
Provide a justification for this result.

**13. Gestation Period versus Life Expectancy** A researcher
would like to know if the gestation period of an animal
can be used to predict the life expectancy. She collects
the following data.

| Animal | Gestation (or Incubation) Period (days) | Life Expectancy (years) |
|---|---|---|
| Cat | 63 | 11 |
| Chicken | 22 | 7.5 |
| Dog | 63 | 11 |
| Duck | 28 | 10 |
| Goat | 151 | 12 |
| Lion | 108 | 10 |
| Parakeet | 18 | 8 |
| Pig | 115 | 10 |
| Rabbit | 31 | 7 |
| Squirrel | 44 | 9 |

*Source:* Time Almanac 2000

(a) Suppose a researcher wants to use the gestation period of an animal to predict its life expectancy. Determine which variable is the predictor variable and
which is the response variable.
(b) Draw a scatter diagram.
(c) Compute the linear correlation coefficient between
gestation period and life expectancy.
(d) Comment on the type of relation that appears to
exist between gestation period and life expectancy
based upon the scatter diagram and linear correlation coefficient.
(e) Remove the goat from the data set and recompute
the linear correlation coefficient between the gestation period and life expectancy. What effect did the
removal of the data value have on the linear correlation coefficient? Provide a justification for this result.

**14. Tar and Nicotine** Every year the Federal Trade Commission must report tar and nicotine levels in cigarettes
to the Congress. It obtains the tar and nicotine levels of
over 1200 brands of cigarettes. A sample from those reported to Congress is given in the table on page 158.

| Brand | Tar (mg) | Nicotine (mg) |
|---|---|---|
| Barclay 100 | 5 | 0.4 |
| Benson and Hedges King | 16 | 1.1 |
| Camel Regular | 24 | 1.7 |
| Chesterfield King | 24 | 1.4 |
| Doral | 8 | 0.5 |
| Kent Golden Lights | 9 | 0.8 |
| Kool Menthol | 9 | 0.8 |
| Lucky Strike | 24 | 1.5 |
| Marlboro Gold | 15 | 1.2 |
| Newport Menthol | 18 | 1.3 |
| Salem Menthol | 17 | 1.3 |
| Virginia Slims Ultra Light | 5 | 0.5 |
| Winston Light | 10 | 0.8 |

*Source:* Federal Trade Commission

(a) Suppose a researcher wants to use the level of tar to predict the nicotine level in a cigarette. Determine which variable is the predictor variable and which is the response variable.
(b) Draw a scatter diagram.
(c) Compute the linear correlation coefficient between the amount of tar and nicotine.
(d) Comment on the type of relation that appears to exist between the amount of tar and nicotine, based upon the scatter diagram and linear correlation coefficient.

**15. Bone Length** Research performed at NASA and led by Dr. Emily R. Morey-Holton measured the lengths of the right humerus and right tibia in 11 rats that were sent to space on Spacelab Life Sciences 2. The following data were collected.

| Right Humerus (mm) | Right Tibia (mm) | Right Humerus (mm) | Right Tibia (mm) |
|---|---|---|---|
| 24.8 | 36.05 | 25.9 | 37.38 |
| 24.59 | 35.57 | 26.11 | 37.96 |
| 24.59 | 35.57 | 26.63 | 37.46 |
| 24.29 | 34.58 | 26.31 | 37.75 |
| 23.81 | 34.2 | 26.84 | 38.5 |
| 24.87 | 34.73 | | |

*Source:* NASA Life Sciences Data Archive

(a) Draw a scatter diagram treating the length of the right humerus as the predictor variable and the length of the right tibia as the response variable.
(b) Compute the linear correlation coefficient between the length of the right humerus and the length of the right tibia.

(c) Comment on the type of relation that appears to exist between the length of the right humerus and the length of the right tibia based upon the scatter diagram and the linear correlation coefficient.
(d) Convert the data to inches (1 mm = 0.03937 inch) and recompute the linear correlation coefficient. What effect did the conversion from millimeters to inches have on the linear correlation coefficient?

**16. Weight of a Car versus Miles per Gallon** An engineer wanted to determine how the weight of a car affected the gas mileage. The following data represent the weight of various domestic cars and their miles per gallon in the city for the 2001 model year.

| Car | Weight (pounds) | Miles Per Gallon |
|---|---|---|
| Buick LeSabre | 3565 | 19 |
| Buick Regal | 3440 | 20 |
| Cadillac Seville | 3970 | 17 |
| Chevrolet Camaro | 3305 | 19 |
| Chevrolet Monte Carlo | 3340 | 20 |
| Chrysler PT Cruiser | 3200 | 20 |
| Chrysler Sebring Sedan | 3230 | 19 |
| Dodge Neon | 2560 | 28 |
| Ford Focus | 2520 | 28 |
| Ford Mustang | 3065 | 20 |
| Lincoln LS | 3600 | 18 |
| Mercury Sable | 3300 | 19 |
| Oldsmobile Aurora | 3625 | 19 |
| Pontiac Bonneville | 3590 | 19 |
| Pontiac Sunfire | 2605 | 23 |
| Saturn Coupe | 2370 | 28 |

*Source: Road and Track* magazine

(a) Determine which variable is the likely predictor variable and which is the likely response variable.
(b) Draw a scatter diagram.
(c) Compute the linear correlation coefficient between the weight of a car and its miles per gallon in the city.
(d) Comment on the type of relation that appears to exist between the weight of a car and its miles per gallon in the city based upon the scatter diagram and the linear correlation coefficient.

**17. General Electric versus the S&P 500** An investor wants to determine whether the rate of return of the S&P 500 can be used to predict the rate of return of the stock of General Electric (GE). The following data represent the rate of return of GE stock for 13 months versus the rate of return of the Standard & Poor's Index of 500 stocks.

| Month | Rate of Return of S&P 500 | Rate of Return of GE |
|---|---|---|
| Sep-99 | −2.9 | 5.6 |
| Oct-99 | 6.3 | 14.3 |
| Nov-99 | 1.9 | −4.0 |
| Dec-99 | 5.8 | 18.9 |
| Jan-00 | −5.0 | −13.4 |
| Feb-00 | −2.0 | −1.2 |
| Mar-00 | 10.0 | 17.6 |
| Apr-00 | −3.1 | 1.0 |
| May-00 | −2.2 | 0.4 |
| Jun-00 | 2.4 | 0.7 |
| Jul-00 | −1.6 | −2.5 |
| Aug-00 | 6.1 | 13.4 |
| Sep-00 | −5.3 | −1.2 |

*Source:* Yahoo! Finance

(a) Draw a scatter diagram treating the rate of return of the S&P 500 as the predictor variable.
(b) Compute the linear correlation coefficient between the rate of return of GE stock and the rate of return of the S&P 500.
(c) Comment on the type of relation that appears to exist between the rate of return of GE stock and the rate of return of the S&P 500, on the basis of the scatter diagram and the linear correlation coefficient.

18. **Fat-free Mass versus Energy Expenditure** In an effort to measure the dependence of energy expenditure on body build, researchers used underwater weighing techniques to determine the fat-free body mass for seven men. In addition, they measured the total 24-hour energy expenditure during inactivity. The results are shown in the table.

| Fat-free Mass (kg) | Energy Expenditure (Kcal) |
|---|---|
| 49.3 | 1894 |
| 59.3 | 2050 |
| 68.3 | 2353 |
| 48.1 | 1838 |
| 57.6 | 1948 |
| 78.1 | 2528 |
| 76.1 | 2568 |

*Source:* Webb, P. Energy expenditure and fat-free mass in men and women. *American Journal of Clinical Nutrition*, **34**, (1981):1816–1826.

(a) Determine which variable is the likely predictor variable and which is the likely response variable.
(b) Draw a scatter diagram.
(c) Compute the linear correlation coefficient between fat-free mass and energy expenditure.
(d) Comment on the type of relation that appears to exist between fat-free mass and energy expenditure on the basis of the scatter diagram and the linear correlation coefficient.

19. **Age versus HDL Cholesterol** A doctor wanted to determine whether there was a relation between a male's age and his HDL (so-called "good") cholesterol. He randomly selected 17 of his patients and determined their HDL cholesterol. He obtained the following data.

| Age | HDL Cholesterol | Age | HDL Cholesterol |
|---|---|---|---|
| 38 | 57 | 38 | 44 |
| 42 | 54 | 66 | 62 |
| 46 | 34 | 30 | 53 |
| 32 | 56 | 51 | 36 |
| 55 | 35 | 27 | 45 |
| 52 | 40 | 52 | 38 |
| 61 | 42 | 49 | 55 |
| 61 | 38 | 39 | 28 |
| 26 | 47 | | |

*Source:* Data based upon information obtained from the National Center for Health Statistics

(a) Draw a scatter diagram of the data treating age as the predictor variable. What type of relation, if any, appears to exist between age and HDL cholesterol?
(b) Compute the linear correlation coefficient between age and HDL cholesterol.
(c) Comment on the type of relation that appears to exist between age and HDL cholesterol on the basis of the scatter diagram and the linear correlation coefficient.

20. **Intensity of a Lightbulb** Cathy is conducting an experiment to measure the relation between a lightbulb's intensity and the distance from the light source. She measures a 100-watt lightbulb's intensity 1 meter from the bulb and at 0.1-meter intervals up to 2 meters from the bulb and obtains the following data.

| Distance (meters), x | Intensity, y | Distance (meters), x | Intensity, y |
|---|---|---|---|
| 1.0 | 0.29645 | 1.6 | 0.11450 |
| 1.1 | 0.25215 | 1.7 | 0.10243 |
| 1.2 | 0.20547 | 1.8 | 0.09231 |
| 1.3 | 0.17462 | 1.9 | 0.08321 |
| 1.4 | 0.15342 | 2.0 | 0.07342 |
| 1.5 | 0.13521 | | |

(a) Draw a scatter diagram of the data, treating distance as the predictor variable.
(b) Do you think that it is appropriate to compute the linear correlation between distance and intensity? Why?

21. **Does Size Matter?** Researchers Willerman, Schultz, Rutledge, and Bigler wondered whether the size of a person's brain was related to the individual's mental capacity. They selected a sample of right-handed introductory psychology students who had Scholastic Aptitude Test Scores higher than 1350. The subjects were administered the Wechsler (1981) Adult Intelligence Scale-Revised in order to obtain their IQ scores. The MRI Scans were performed at the

same facility for the subjects. The scans consisted of 18 horizontal MR images. The computer counted all pixels with nonzero gray scale in each of the 18 images and the total count served as an index for brain size.

| Gender | MRI Count | IQ | Gender | MRI Count | IQ |
|--------|-----------|-----|--------|-----------|-----|
| Female | 816,932 | 133 | Male | 949,395 | 140 |
| Female | 951,545 | 137 | Male | 1,001,121 | 140 |
| Female | 991,305 | 138 | Male | 1,038,437 | 139 |
| Female | 833,868 | 132 | Male | 965,353 | 133 |
| Female | 856,472 | 140 | Male | 955,466 | 133 |
| Female | 852,244 | 132 | Male | 1,079,549 | 141 |
| Female | 790,619 | 135 | Male | 924,059 | 135 |
| Female | 866,662 | 130 | Male | 955,003 | 139 |
| Female | 857,782 | 133 | Male | 935,494 | 141 |
| Female | 948,066 | 133 | Male | 949,589 | 144 |

*Source:* Willerman, L., Schultz, R., Rutledge, J. N., and Bigler, E. (1991), "In Vivo Brain Size and Intelligence," *Intelligence*, 15, 223–228.

(a) Draw a scatter diagram, treating MRI Count as the predictor variable and IQ as the response variable. Comment on what you see.
(b) Compute the linear correlation coefficient between MRI Count and IQ. Do you think that MRI Count and IQ are linearly related?
(c) A lurking variable in the analysis is gender. Draw a scatter diagram, treating MRI Count as the predictor variable and IQ as the response variable, but use a different plotting symbol for each gender. For example, use a circle for males and a triangle for females. What do you notice?
(d) Compute the linear correlation coefficient between MRI Count and IQ for females. Compute the linear correlation coefficient between MRI Count and IQ for males. Do you still believe that MRI Count and IQ are linearly related? What is the moral?

22. **Male versus Female Drivers** The following data represent the number of licensed drivers in various age groups and the number of accidents within the age group by gender.

| Age Group | Number of Male Licensed Drivers (000s) | Number of Crashes Involving a Male (000s) | Number of Female Licensed Drivers (000s) | Number of Crashes Involving a Female (000s) |
|-----------|----------------------------------------|-------------------------------------------|-------------------------------------------|---------------------------------------------|
| 16 | 816 | 244 | 764 | 178 |
| 17 | 1198 | 233 | 1115 | 175 |
| 18 | 1342 | 243 | 1212 | 164 |
| 19 | 1454 | 229 | 1333 | 145 |
| 20–24 | 7866 | 951 | 7394 | 618 |
| 25–29 | 9356 | 899 | 8946 | 595 |
| 30–34 | 10,121 | 875 | 9871 | 571 |
| 35–39 | 10,521 | 901 | 10,439 | 566 |
| 40–44 | 9776 | 692 | 9752 | 455 |
| 45–49 | 8754 | 667 | 8710 | 390 |
| 50–54 | 6840 | 390 | 6763 | 247 |
| 55–59 | 5341 | 290 | 5258 | 165 |
| 60–64 | 4565 | 218 | 4486 | 133 |
| 65–69 | 4234 | 191 | 4231 | 121 |
| 70–74 | 3604 | 167 | 3749 | 104 |
| 75–79 | 2563 | 118 | 2716 | 77 |
| 80–84 | 1400 | 61 | 1516 | 45 |
| ≥ 85 | 767 | 34 | 767 | 20 |

*Source:* National Highway and Traffic Safety Institute

(a) On the same graph, draw a scatter diagram for both males and females. Be sure to use a different plotting symbol for each group. For example, use a square (□) for males and a plus sign (+) for females. Treat number of licensed drivers as the predictor variable.

(b) Based on the scatter diagrams, do you think that insurance companies are justified in charging different insurance rates for males and females? Why?

(c) Compute the linear correlation coefficient between number of licensed drivers and number of crashes for males.

(d) Compute the linear correlation coefficient between number of licensed drivers and number of crashes for females.

(e) Which gender has the stronger linear relation between number of licensed drivers and number of crashes. Why?

---

**23. Gestation Period versus Life Expectancy** Suppose we add humans to the data in Problem 13. Humans have a gestation period of 268 days and a life expectancy of 76.5 years.

(a) Redraw the scatter diagram with humans included.

(b) Recompute the linear correlation coefficient with humans included.

(c) Compare the results of (a) and (b) with the results of Problem 13. Provide a statement that explains the results.

**24. Weight of a Car versus Miles per Gallon** Suppose we add the Dodge Viper to the data in Problem 16. A Dodge Viper weighs 3425 pounds and gets 11 miles per gallon.

(a) Redraw the scatter diagram with the Viper included.

(b) Recompute the linear correlation coefficient with the Viper included.

(c) Compare the results of (a) and (b) with the results of Problem 16. Why are the results reasonable?

**25.** Consider the following set of data:

| x | 2.2 | 3.7 | 3.9 | 4.1 | 2.6 | 4.1 | 2.9 | 4.7 |
|---|-----|-----|-----|-----|-----|-----|-----|-----|
| y | 3.9 | 4.0 | 1.4 | 2.8 | 1.5 | 3.3 | 3.6 | 4.9 |

(a) Draw a scatter diagram of the data and compute the linear correlation coefficient.

(b) Draw a scatter diagram of the data and compute the linear correlation coefficient with the additional data point (10.4, 9.3). Comment on the effect of the additional data point. Conclude that correlations should always be reported with scatter diagrams.

**26.** On the basis of the accompanying scatter diagram, explain what is wrong with the following statement: "Because the linear correlation coefficient between them is 0.012, there is no relation between age and median income."

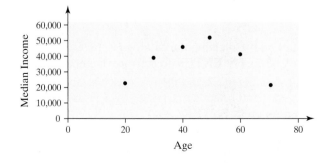

**27.** For each of the following statements, explain whether you think the variables will have positive correlation, negative correlation, or no correlation. Support your opinion.

(a) Number of children in the household under the age of 3 and expenditures on diapers

(b) Interest rates on car loans and number of cars sold

(c) Number of hours per week on the treadmill and cholesterol level

(d) Price of a Big Mac and number of McDonald's french fries sold in a week

(e) Shoe size and IQ

**28.** Consider the following data set:

| x | 5 | 6 | 7 | 7 | 8 | 8 | 8 | 8 |
|---|-----|-----|-----|-----|-----|-----|-----|-----|
| y | 4.2 | 5 | 5.2 | 5.9 | 6 | 6.2 | 6.1 | 6.9 |
| x | 9 | 9 | 10 | 10 | 11 | 11 | 12 | 12 |
| y | 7.2 | 8 | 8.3 | 7.4 | 8.4 | 7.8 | 8.5 | 9.5 |

(a) Draw a scatter diagram with the x-axis starting at 0 and ending at 30, the y-axis starting at 0 and ending at 20.

(b) Compute the linear correlation coefficient.

(c) Now multiply both x and y by 2.

(d) Draw a scatter diagram of the new data with the x-axis starting at 0 and ending at 30, the y-axis starting at 0 and ending at 20. Compare the scatter diagrams.

(e) Compute the linear correlation coefficient.

(f) Conclude that multiplying each value in the data set does not affect the correlation between the variables. Explain why this is the case.

**29.** Consider the following four data sets:

| Data Set 1 | | Data Set 2 | | Data Set 3 | | Data Set 4 | |
|---|---|---|---|---|---|---|---|
| x | y | x | y | x | y | x | y |
| 10 | 8.04 | 10 | 9.14 | 10 | 7.46 | 8 | 6.58 |
| 8 | 6.95 | 8 | 8.14 | 8 | 6.77 | 8 | 5.76 |
| 13 | 7.58 | 13 | 8.74 | 13 | 12.74 | 8 | 7.71 |
| 9 | 8.81 | 9 | 8.77 | 9 | 7.11 | 8 | 8.84 |
| 11 | 8.33 | 11 | 9.26 | 11 | 7.81 | 8 | 8.47 |
| 14 | 9.96 | 14 | 8.10 | 14 | 8.84 | 8 | 7.04 |
| 6 | 7.24 | 6 | 6.13 | 6 | 6.08 | 8 | 5.25 |
| 4 | 4.26 | 4 | 3.10 | 4 | 5.39 | 8 | 5.56 |
| 12 | 10.84 | 12 | 9.13 | 12 | 8.15 | 8 | 7.91 |
| 7 | 4.82 | 7 | 7.26 | 7 | 6.42 | 8 | 6.89 |
| 5 | 5.68 | 5 | 4.47 | 5 | 5.73 | 19 | 12.50 |

*Source:* Anscombe, Frank, "Graphs in statistical analysis," *American Statistician* 27 (1973):17–21.

(a) Compute the linear correlation coefficient for each data set.
(b) Draw a scatter diagram for each data set. Conclude that linear correlation coefficients and scatter diagrams must be used together in any statistical analysis of bivariate data.

---

**30.** In a study published in the Journal of the American Medical Association (May 16, 2001), researchers found that breast-feeding may help to prevent obesity in kids. In an interview, the head investigator stated, "It's not clear whether breast milk has obesity-preventing properties or the women who are breast-feeding are less likely to have fat kids because they are less likely to be fat themselves and may be more health-conscious". Using this researcher's statement, explain what might be wrong with the conclusion that breast-feeding prevents obesity. Identify some lurking variables in the study.

**31.** Compute the linear correlation coefficient between the points (3, 8) and (5, 12). Explain why the result is reasonable.

**32.** Compute the linear correlation coefficient between the points (2, 4) and (6, 1). Explain why the result is reasonable.

**33.** Explain how you might use the linear correlation coefficient to identify lurking variables in an observational study.

---

## Technology Step-by-Step
### Drawing Scatter Diagrams and Determining the Correlation Coefficient

**TI-83 Plus** — **Scatter Diagrams**

**Step 1:** Enter the predictor variable in L1 and the response variable in L2.

**Step 2:** Press $2^{nd}$ Y = to bring up the StatPlot menu. Select 1: Plot1.

**Step 3:** Turn Plot 1 ON by putting the cursor on the ON button and pressing ENTER.

**Step 4:** Highlight scatter diagram icon (see the figure) and press ENTER. Be sure that Xlist is L1 and Ylist is L2.

**Step 5:** Press ZOOM and select 9: ZoomStat.

**Correlation Coefficient**

**Step 1:** Turn the diagnostics on by selecting the catalog ($2^{nd}$ ∅). Scroll down and select DiagnosticOn. Hit ENTER to activate diagnostics.

**Step 2:** With the predictor variable in L1 and the response variable in L2, press STAT, highlight CALC and select 4: LinReg (ax + b). With LinReg on the HOME screen press ENTER.

**MINITAB** — **Scatter Diagrams**

**Step 1:** Enter the predictor variable in C1 and the response variable in C2. You may want to name the variables.

**Step 2:** Select the **Graph** menu and highlight **Plot**...

*Step 3:* With the cursor in the Y column, select the response variable. With the cursor in the X column, select the predictor variable. Click OK.

**Correlation Coefficient**

*Step 1:* With the predictor variable in C1 and the response variable in C2, select the **Stat** menu, highlight **Basic Statistics**. Highlight **Correlation**.

*Step 2:* Select the variables whose correlation you wish to determine and click OK.

*Excel*  **Scatter Diagrams**

*Step 1:* Enter the predictor variable in column A and the response variable in column B. Select the Chart Wizard icon.

*Step 2:* Follow the instructions in the Chart Wizard.

**Correlation Coefficient**

*Step 1:* Be sure the Data Analysis Tool Pak is activated, by selecting the **Tools** menu and highlighting **Add-Ins** ... Check the box for the Analysis ToolPak and select OK.

*Step 2:* Select **Tools** and highlight **Data Analysis** ... Highlight **Correlation** and select OK.

*Step 3:* With the cursor in the Input Range, highlight the data. Select OK.

# 4.2 Least-Squares Regression

**Preparing for This Section**   Before getting started, review the following:

✓ Lines (Section C.4, available on CD)

**Objectives**    Find the least-squares regression line

 Interpret the slope and the *y*-intercept of the least-squares regression line

 Predict the value of the response variable based upon the least-squares regression line

 Determine residuals based upon the least-squares regression line

⑤ Compute the sum of squared residuals

Once the scatter diagram and linear correlation coefficient indicate that a linear relation exists between two variables, we proceed to find a linear equation that describes the relation between the two variables. One way to obtain a line that describes the relation would be to select two points from the data that appear to provide a good "fit" and find the equation of the line through these points.

▶ **EXAMPLE 1**   **Finding an Equation that Describes Linearly Related Data**

*Problem:* The data in Table 3 represent the per capita gross domestic product of randomly selected countries in Western Europe and the life expectancy of the residents. We determined that these data are linearly related in the last section.

**(a)** Find a linear equation that relates per capita GDP, *x* (the predictor variable), and life expectancy, *y* (the response variable), by selecting two points and finding the equation of the line containing the points.

**(b)** Graph the equation on the scatter diagram.

**(c)** Use the equation to predict the life expectancy of a resident of Spain where per capita GDP is $16.4 thousand.

| | TABLE 3 | | |
|---|---|---|---|
| Country | Per Capita GDP, x | Life Expectancy, y | (x, y) |
| Austria | 21.4 | 77.48 | (21.4, 77.48) |
| Belgium | 23.2 | 77.53 | (23.2, 77.53) |
| Finland | 20.0 | 77.32 | (20.0, 77.32) |
| France | 22.7 | 78.63 | (22.7, 78.63) |
| Germany | 20.8 | 77.17 | (20.8, 77.17) |
| Ireland | 18.6 | 76.39 | (18.6, 76.39) |
| Italy | 21.5 | 78.51 | (21.5, 78.51) |
| Netherlands | 22.0 | 78.15 | (22.0, 78.15) |
| Switzerland | 23.8 | 78.99 | (23.8, 78.99) |
| United Kingdom | 21.2 | 77.37 | (21.2, 77.37) |

*Source: Time Almanac*, 2000

**In Your Own Words**

A "good fit" means that the line drawn appears to describe the relation between the two variables well.

*Approach:* To answer Part (a), we perform the following steps:

*Step 1:* Select two points from Table 3 such that a line drawn through the points appears to give a "good fit." Call the points $(x_1, y_1)$ and $(x_2, y_2)$. Refer to Figure 1 on page 150 for the scatter diagram.

*Step 2:* Find the slope of the line containing these two points using $m = \dfrac{y_2 - y_1}{x_2 - x_1}$.

*Step 3:* Use the point-slope formula, $y - y_1 = m(x - x_1)$, to find the line through the points selected in Step 1. Express the line in the form $y = mx + b$, where $m$ is the slope and $b$ is the $y$-intercept.

For part (b), draw a line through the points selected in Step 1 of part (a). Finally, for part (c), we let $x = 16.4$ in the equation found in part (a).

*Solution:*

**(a)** *Step 1:* We will select Ireland, $(x_1, y_1) = (18.6, 76.39)$, and Switzerland, $(x_2, y_2) = (23.8, 78.99)$, because a line drawn through these two points seems to give a good fit.

*Step 2:* $m = \dfrac{y_2 - y_1}{x_2 - x_1} = \dfrac{78.99 - 76.39}{23.8 - 18.6} = \dfrac{2.6}{5.2} = 0.5*$

*Step 3:* We use the point-slope form of a line to find the equation of the line.

$$y - y_1 = m(x - x_1)$$

$$y - 76.39 = 0.5(x - 18.6) \qquad m = 0.5, x_1 = 18.6, y_1 = 76.39$$

$$y - 76.39 = 0.5x - 9.3$$

$$y = 0.5x + 67.09 \tag{1}$$

The slope of the line is 0.5 and the $y$-intercept is 67.09.

**(b)** Figure 7 shows the scatter diagram along with the line drawn through the points (18.6, 76.39) and (23.8, 78.99).

*Unless otherwise noted, we will round to four decimal places. As always, do not round until the last computation.

Figure 7

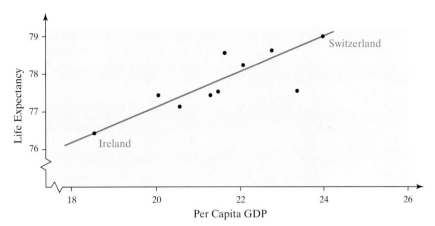

**(c)** We let $x = 16.4$ in equation (1) to predict the life expectancy of a resident of Spain.

$$y = 0.5(16.4) + 67.09$$
$$= 75.29$$

We predict that a resident of Spain can expect to live 75.29 years. In fact, in Spain, the life expectancy is 77.71 years, so we underestimated the life expectancy of a resident of Spain.   ◄◄

**In Your Own Words**

The residual represents how close our prediction comes to actual observation. The smaller the residual, the better is the prediction.

**NW**  *Now Work Problems 3(a), 3(b), and 3(c).*

**①  Least-Squares Regression**

Although the line that we found in Example 1 appears to describe the relation between per capita GDP and life expectancy well, is there a line that fits the data better? Is there a line that fits the data "best"?

Consider Figure 8. Each $y$-coordinate on the line corresponds to a predicted life expectancy for a given level of per capita GDP. For example, the per capita GDP of Italy is $21.5 thousand, so the predicted life expectancy is $0.5(21.5) + 67.09 = 77.84$ years. The observed life expectancy in Italy is 78.51 years. The difference between the observed value of $y$ and the predicted value of $y$ is the error or **residual**. For Italy, the residual is

$$\text{Residual} = \text{observed } y - \text{predicted } y$$
$$= 78.51 - 77.84$$
$$= 0.67 \text{ years}$$

The residual for Italy is labeled in Figure 8.

**Historical Note**

Adrien Marie Legendre was born on September 18, 1752. He was born into a wealthy family and educated in mathematics and physics at the College Mazarin in Paris. From 1775 to 1780, he taught at Ecole Militaire. On March 30, 1783, Legendre was appointed an adjoin't in the Académie des Sciences. On May 13, 1791, he became a member of the committee of the Académie des Sciences and was charged with the task of standardizing weights and measures. The committee worked to compute the length of the meter. During the revolution, Legendre lost his small fortune. In 1794, Legendre published Eléments de géométrie, which was the leading elementary text in geometry for around 100 years. In 1806, Legendre published a book on orbits, in which he developed the theory of least squares. He died on January 10, 1833.

Figure 8

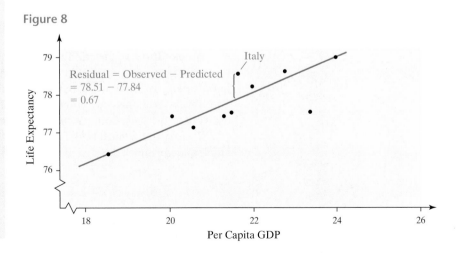

The line that "best" describes the relation between two variables is the one that makes the residuals "as small as possible." The most popular technique for making the residuals "as small as possible" is the *method of least squares*, discovered by Adrien Marie Legendre.

*Definition*

## Least-Squares Regression Criterion

The **least-squares regression line** is the one that minimizes the sum of the squared errors. It is the line that minimizes the square of the vertical distance between observed values of $y$ and those predicted by the line, $\hat{y}$ (read "$y$-hat"). We represent this as

$$\text{Minimize } \sum \text{residuals}^2$$

The advantage of the least-squares criterion is that it allows for statistical inference on the predicted value and slope (Chapter 11). Another advantage of the least-squares criterion is explained by Legendre in his text *Nouvelles méthods pour la détermination des orbites des comètes*, published in 1805.

> Of all the principles that can be proposed for this purpose, I think there is none more general, more exact, or easier to apply, than that which we have used in this work; it consists of making the sum of squares of the errors a *minimum*. By this method, a kind of equilibrium is established among the errors which, since it prevents the extremes from dominating, is appropriate for revealing the state of the system which most nearly approaches the truth.

*Theorem*

### Equation of the Least-Squares Regression Line

The equation of the least-squares regression line is given by

$$\hat{y} = b_1 x + b_0$$

where

$$b_1 = r \cdot \frac{s_y}{s_x} \text{ is the \textbf{slope} of the least-squares regression line*} \qquad (2)$$

and

$$b_0 = \bar{y} - b_1 \bar{x} \text{ is the \textbf{y-intercept} of the least-squares regression line} \quad (3)$$

**Note:** $\bar{x}$ is the sample mean and $s_x$ is the sample standard deviation of the predictor variable $x$; $\bar{y}$ is the sample mean and $s_y$ is the sample standard deviation of the response variable $y$.

The notation $\hat{y}$ is used in the least-squares regression line to serve as a reminder that it is a predicted value of $y$ for a given value of $x$. An interesting property of the least-squares regression line, $\hat{y} = b_1 x + b_0$, is that the line always contains the point $(\bar{x}, \bar{y})$. This property can be useful in the drawing of the least-squares regression line by hand.

Since $s_y$ and $s_x$ must both be positive, the sign of the linear correlation coefficient determines the sign of the slope of the least-squares regression line.

*An equivalent formula is $b_1 = \dfrac{S_{xy}}{S_{xx}} = \dfrac{\sum x_i y_i - \dfrac{(\sum x_i)(\sum y_i)}{n}}{\sum x_i^2 - \dfrac{(\sum x_i)^2}{n}}$

▶ **EXAMPLE 2**   **Finding the Least-Squares Regression Line**

*Problem:* For the data in Table 3 on page 164,

(a) Find the least-squares regression line.
(b) Interpret the slope and intercept.
(c) Predict the life expectancy of a resident of Italy where per capita GDP is $21.5 thousand.
(d) Compute the residual for Italy.
(e) Draw the least-squares regression line on the scatter diagram of the data.

*Approach:*

(a) From Example 2 in Section 4.1, we have the following: $r = 0.8094$, $\bar{x} = 21.52$, $s_x = 1.531738$, $\bar{y} = 77.754$ and $s_y = 0.794847$. We substitute these values into Formula (2) to find the slope of the least-squares regression line. We use Formula (3) to find the intercept of the least-squares regression.

(b) The slope represents the change in life expectancy divided by the change in per capita GDP, so we interpret the slope as the additional life expectancy if per capita GDP rises by $1 thousand. The intercept represents the value of the response variable when the predictor variable is 0.

(c) Substitute $x = 21.5$ into the least-squares regression line found in part (a) to find $\hat{y}$.

(d) The residual is the difference between the observed $y$ and the predicted $y$. That is, residual $= y - \hat{y}$.

(e) To draw the least-squares regression line, select two values of $x$ and use the equation to find the predicted values of $y$. Plot these points on the scatter diagram and draw a line through the points.

*Solution:*

(a) Substituting $r = 0.8094$, $s_x = 1.531738$, and $s_y = 0.794847$ into Formula (2), we obtain

$$b_1 = r \cdot \frac{s_y}{s_x} = 0.8094 \cdot \frac{0.794847}{1.531738} = 0.4200*$$

We have that $\bar{x} = 21.52$ and $\bar{y} = 77.754$. Substituting these values into Formula (3), we obtain

$$b_0 = \bar{y} - b_1 \bar{x} = 77.754 - 0.4200(21.52) = 68.7156$$

The least-squares regression line is

$$\hat{y} = 0.4200x + 68.7156$$

(b) The slope of the least-squares regression line is 0.4200. If per capita GDP increases by $1 thousand, life expectancy increases by 0.42 years. The intercept is 68.7156 years. The intercept is found by letting $x = 0$, so the intercept is the life expectancy in a country where per capita GDP is $0. It is probably unreasonable to interpret the intercept since countries with per capita GDP of $0 do not occur in our data set.

(c) We let $x = 21.5$ in the equation $y = 0.4200x + 68.7156$ to predict the life expectancy in Italy.

$$\hat{y} = 0.4200(21.5) + 68.7156$$

$$= 77.75 \text{ years}$$

We predict that the life expectancy in Italy is 77.75 years.

*Throughout the text, we will round the slope and y-intercept values to four decimal places.

**Historical Note**

**(d)** The actual life expectancy in Italy is 78.51 years. The residual is

$$\text{residual} = \text{observed } y - \text{predicted } y$$
$$= y - \hat{y}$$
$$= 78.51 - 77.75$$
$$= 0.76 \text{ years}$$

We underestimated life expectancy in Italy by 0.76 years.

**(e)** Figure 9 shows the graph of the least-squares regression line drawn on the scatter diagram with the residual labeled.

**Figure 9**

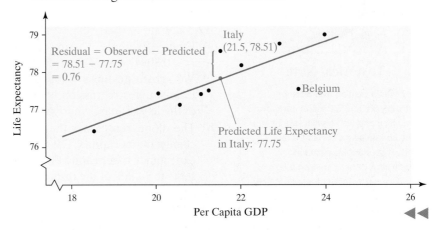

Notice that an underestimate results in a positive residual, while an overestimate will result in a negative residual.

We can think of any point on the least-squares regression line as the mean value of the response variable for a given value of the predictor variable. For example, the mean life expectancy for a country that has per capita GDP of $21.5 thousand is 77.75 years. Because the life expectancy in Italy is 78.51 years, the Italians have a life expectancy that is higher than the mean life expectancy of countries that have per capita GDP of $21.5 thousand. Perhaps it's the red wine!

Look back at Figure 9. Which country would result in the worst prediction? The worst prediction would be the country whose residual differs the most from zero (i.e., has the furthest vertical distance from the line of best fit). From Figure 9, it would seem that Belgium (per capita GDP: $23.2 thousand, life expectancy: 77.53) would result in the worst prediction. The predicted life expectancy for Belgium is

$$\hat{y} = 0.4200(23.2) + 68.7156$$
$$= 78.46 \text{ years}$$

The residual for Belgium is then

$$\text{Residual} = \text{observed } y - \text{predicted } y$$
$$= y - \hat{y}$$
$$= 77.53 - 78.46$$
$$= -0.93 \text{ years}$$

We overestimated the life expectancy in Belgium by 0.93 years. Which country results in the best prediction?

## The Scope of the Model

It is important to note that the *y*-intercept will have meaning only if the following two conditions are met:

**1.** A value of 0 for the predictor variable makes sense.

**2.** There are observed values of the predictor variable near 0.

**In Your Own Words**

An underestimate means the residual is positive; an overestimate means the residual is negative. A residual of zero means the prediction is right on!

## Caution

Never use the least-squares regression line to make predictions for values of the predictor variable that are much larger or much smaller than those observed.

The second condition is especially important because statisticians do not use the regression model to make predictions **outside the scope of the model**. In other words, statisticians do not recommend using the regression model to make predictions for values of the predictor variable that are much larger or much smaller than those observed because we cannot be certain of the behavior of the data for which we have no observations.

For example, it would be inappropriate to use the line we determined in Example 2 to predict life expectancy in a country where per capita GDP is $40 thousand. The highest observed per capita GDP in our data set is $23.8 thousand and we cannot be certain that the linear relation between per capita GDP and life expectancy will continue. It is likely that at high levels of per capita GDP, life expectancy will level off. See Figure 10.

Figure 10

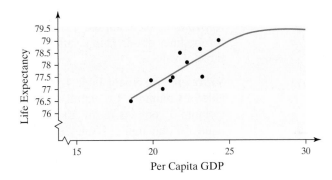

**Using Technology:** While we have presented the procedure for computing the least-squares regression line by hand, in practice a calculator or statistical spreadsheet program is used to determine the least-squares regression line. Figure 11(a) shows the output obtained from a TI-83 Plus graphing calculator, and Figure 11(b) shows the output obtained from Minitab with the slope and y-intercept highlighted. Figure 11(c) shows partial output from Excel with the slope and intercept highlighted.

Figure 11

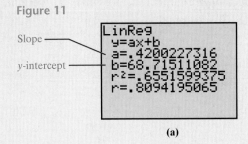

(a)

**Regression**

The regression equation is
y = 68.7 + 0.420 x

| Predictor | Coef | StDev | T | P |
|---|---|---|---|---|
| Constant | 68.715 | 2.324 | 29.57 | 0.000 |
| x | 0.4200 | 0.1077 | 3.90 | 0.005 |

S = 0.4951    R-Sq = 65.5%    R-Sq(adj) = 61.2%

(b)

| | Coefficients | Standard Error | t Stat | P-value |
|---|---|---|---|---|
| Intercept | 68.71511082 | 2.323769861 | 29.57053 | 1.85E-09 |
| X Variable 1 | 0.420022732 | 0.107736532 | 3.89861 | 0.004554 |

(c)

**NW** *Now Work Problems 3(d) and 3(e).*

**Least-Squares Criterion Revisited**

Recall that the least-squares regression line is the line that minimizes the sum of the squared residuals. This means that the sum of the squared residuals, $\sum$ residuals$^2$, for the least-squares line will be smaller than for any other line that may describe the relation between the two variables. In particular, the sum of the squared residuals for the line obtained in Example 2 using the method of least-squares will be smaller than the sum of the squared residuals for the line obtained in Example 1. It is worthwhile to verify this result.

▶ EXAMPLE 3    **Comparing the Sum of Squared Residuals**

*Problem:* Compare the sum of squared residuals for the lines obtained in Examples 1 and 2.

*Approach:* We compute $\sum$ residuals$^2$ using the predicted values of $y$, $\hat{y}$, for the lines obtained in Examples 1 and 2. This is best done by creating a table of values.

*Solution:* We create Table 4, which contains the value of the predictor variable in Column 1. Column 2 contains the corresponding response variable. Column 3 contains the predicted value of the response variable using the equation obtained in Example 1, $\hat{y} = 0.5x + 67.09$. In Column 4, we compute the residuals for each observation, residual $=$ observed $y$–predicted $y$. For example, the first residual using the equation found in Example 1 is observed $y$– predicted $y = 77.48$–$77.79 = -0.31$. Column 5 contains the squares of the residuals obtained in Column 4. Column 6 contains the predicted value of the response variable using the least-squares regression equation obtained in Example 2: $\hat{y} = 0.4200x + 68.7156$. Column 7 represents the residuals for each observation and Column 8 represents the squared residuals.

| | | $y = 0.5x +$ 67.09 | | | $\hat{y} = 0.4200x +$ 68.7156 | | |
|---|---|---|---|---|---|---|---|
| Per Capita GDP | Life Expectancy | EXAMPLE 1 | residual | residual$^2$ | EXAMPLE 2 | residual | residual$^2$ |
| 21.4 | 77.48 | 0.5(21.4) + 67.09 = 77.79 | 77.48 − 77.79 = −0.31 | $(-.31)^2$ = 0.0961 | 0.4200(21.4) + 68.7156 = 77.70 | 77.48 − 77.70 = −0.22 | $(-0.22)^2$ = 0.0484 |
| 23.2 | 77.53 | 78.69 | −1.16 | 1.3456 | 78.46 | −0.93 | 0.8649 |
| 20 | 77.32 | 77.09 | 0.23 | 0.0529 | 77.12 | 0.20 | 0.04 |
| 22.7 | 78.63 | 78.44 | 0.19 | 0.0361 | 78.25 | 0.38 | 0.1444 |
| 20.8 | 77.17 | 77.49 | −0.32 | 0.1024 | 77.45 | −0.28 | 0.0784 |
| 18.6 | 76.39 | 76.39 | 0 | 0 | 76.53 | −0.14 | 0.0196 |
| 21.5 | 78.51 | 77.84 | 0.67 | 0.4489 | 77.75 | 0.76 | 0.5776 |
| 22 | 78.15 | 78.09 | 0.06 | 0.0036 | 77.96 | 0.19 | 0.0361 |
| 23.8 | 78.99 | 78.99 | 0 | 0 | 78.71 | 0.28 | 0.0784 |
| 21.2 | 77.37 | 77.69 | −0.32 | 0.1024 | 77.62 | −0.25 | 0.0625 |
| | | | $\sum$ residual = −0.96 | $\sum$ residual$^2$ = 2.188 | | $\sum$ residual = −0.01* | $\sum$ residual$^2$ = 1.9503 |

TABLE 4

The sum of the squared residuals for the line found in Example 1 is 2.188, while the sum of the squared residuals for the least-squares regression line is 1.9503. Again, any line that describes the relation between per capita GDP and life expectancy will have a sum of squared residuals that is greater than 1.9503.

◀◀

**NW** *Now Work Problems 3(f) and 3(g)*

* The sum of the residuals for the least-squares regression line does not sum to zero due to rounding error.

We draw the graphs of the two lines obtained in Examples 1 and 2 on the same scatter diagram in Figure 12 to help the reader visualize the difference.

Figure 12

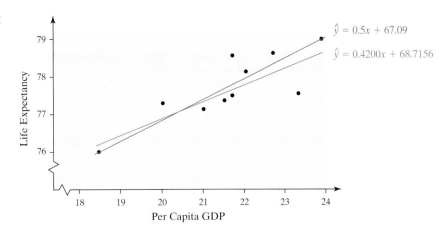

Look back at Table 4. Notice that the sum of the residuals based on the line found in Example 1 (Column 4) does not add up to zero. Because the sum of the residuals adds up to $-0.96$, the model is consistently overestimating life expectancy. In least-squares regression models, the fact that the sum of the residuals is always zero means that the overestimates are offset by underestimates. That is, the predicted value is an **unbiased estimator** of the actual observation.

## 4.2 Assess Your Understanding

### Concepts and Vocabulary

1. Explain the least-squares regression criterion.
2. What is a residual? What does it mean when a residual is positive?
3. Explain the phrase "outside the scope of the model." Why is it dangerous to make predictions "outside the scope of the model"?
4. If the linear correlation between two variables is negative, what can be said about the slope of the regression line?
5. What is meant by an unbiased estimator as it pertains to least-squares regression?
6. In your own words, explain the meaning of Legendre's quote given on page 166.

### Exercises

#### • Skill Building

1. For the data set

| $x$ | 0 | 2 | 3 | 5 | 6 | 6 |
|-----|-----|-----|-----|-----|-----|-----|
| $y$ | 5.8 | 5.7 | 5.2 | 2.8 | 1.9 | 2.2 |

(a) Draw a scatter diagram. Comment on the type of relation that appears to exist between $x$ and $y$.
(b) Given that $\bar{x} = 3.667$, $s_x = 2.42212$, $\bar{y} = 3.933$, $s_y = 1.8239152$, and $r = -0.9476938$, determine the least-squares regression line.
(c) Graph the least-squares regression line on the scatter diagram drawn in part (a).

2. For the data set

| $x$ | 2 | 4 | 8 | 8 | 9 |
|-----|-----|-----|-----|-----|-----|
| $y$ | 1.4 | 1.8 | 2.1 | 2.3 | 2.6 |

(a) Draw a scatter diagram. Comment on the type of relation that appears to exist between $x$ and $y$.
(b) Given that $\bar{x} = 6.2$, $s_x = 3.03315$, $\bar{y} = 2.04$, $s_y = 0.461519$, and $r = 0.957241$, determine the least-squares regression line.
(c) Graph the least-squares regression line on the scatter diagram drawn in part (a).

(a) Find the least-squares regression line treating the length of the right humerus as the predictor variable and the length of the right tibia as the response variable.

(b) Interpret the slope and intercept, if appropriate.

(c) Determine the residual if the length of the right humerus is 26.11 mm and the actual length of the right tibia is 37.96 mm. Is the length of this tibia above average or below average?

(d) Draw the least-squares regression line on the scatter diagram and label the residual.

(e) Suppose one of the rats sent to space received a broken right tibia due to a severe landing. The length of the right humerus is determined to be 25.31 mm. Use the least-squares regression line to estimate the length of the right tibia.

14. **Weight of a Car versus Miles per Gallon** An engineer wanted to determine how the weight of a car affected the gas mileage. The following data represent the weight of various domestic cars and their miles per gallon in the city for the 2001 model year.

| Car | Weight (pounds) | Miles Per Gallon |
|---|---|---|
| Buick LeSabre | 3565 | 19 |
| Buick Regal | 3440 | 20 |
| Cadillac Seville | 3970 | 17 |
| Chevrolet Camaro | 3305 | 19 |
| Chevrolet Monte Carlo | 3340 | 20 |
| Chrysler PT Cruiser | 3200 | 20 |
| Chrysler Sebring Sedan | 3230 | 19 |
| Dodge Neon | 2560 | 28 |
| Ford Focus | 2520 | 28 |
| Ford Mustang | 3065 | 20 |
| Lincoln LS | 3600 | 18 |
| Mercury Sable | 3300 | 19 |
| Oldsmobile Aurora | 3625 | 19 |
| Pontiac Bonneville | 3590 | 19 |
| Pontiac Sunfire | 2605 | 23 |
| Saturn Coupe | 2370 | 28 |

*Source: Road and Track magazine*

(a) Find the least-squares regression line treating weight as the predictor variable and miles per gallon as the response variable.

(b) Interpret the slope and intercept, if appropriate.

(c) Predict the miles per gallon of an Oldsmobile Aurora and compute the residual. Is the miles per gallon of an Aurora above average or below average for cars of this weight?

(d) Draw the least-squares regression line on the scatter diagram of the data and label the residual.

(e) Would it be reasonable to use the least-squares regression line to predict the miles per gallon of a Honda Insight—a hybrid gas and electric car? Why?

15. **General Electric versus the S&P 500** The following data represent the rate of return of GE stock for 13 months versus the rate of return of the Standard & Poor's Index of 500 stocks.

| Month | Rate of Return of S&P 500 | Rate of Return of GE |
|---|---|---|
| Sep-99 | −2.9 | 5.6 |
| Oct-99 | 6.3 | 14.3 |
| Nov-99 | 1.9 | −4.0 |
| Dec-99 | 5.8 | 18.9 |
| Jan-00 | −5.0 | −13.4 |
| Feb-00 | −2.0 | −1.2 |
| Mar-00 | 10.0 | 17.6 |
| Apr-00 | −3.1 | 1.0 |
| May-00 | −2.2 | 0.4 |
| Jun-00 | 2.4 | 0.7 |
| Jul-00 | −1.6 | −2.5 |
| Aug-00 | 6.1 | 13.4 |
| Sep-00 | −5.3 | −1.2 |

*Source: Yahoo! Finance*

(a) Find the least-squares regression line treating the rate of return of the S&P 500 as the predictor variable and the rate of return of General Electric as the response variable.

(b) Interpret the slope and y-intercept, if appropriate. The slope of the line is referred to as the **beta** of the stock. Stock analysts use beta to determine how a stock will perform relative to the market. For example, a beta of 1.1 means that in a month in which the S&P 500 increases 1%, the stock will increase 1.1%. So stocks with betas greater than 1 have more price movement than the overall market; they are, therefore, considered riskier than those stocks with betas less than 1.

(c) Predict the rate of return of General Electric if the rate of return of the S&P 500 is 8.9 percent.

16. **Fat-free Mass versus Energy Expenditure** In an effort to measure the dependence of energy expenditure on body build, researchers used underwater weighing techniques to determine the fat-free body mass for seven men. In addition, they measured the total 24-hour energy expenditure during inactivity. The results are shown in the table.

| Fat-free Mass (kg) | Energy Expenditure (Kcal) |
|---|---|
| 49.3 | 1894 |
| 59.3 | 2050 |
| 68.3 | 2353 |
| 48.1 | 1838 |
| 57.6 | 1948 |
| 78.1 | 2528 |
| 76.1 | 2568 |

*Source: Webb, P. (1981) Energy expenditure and fat-free mass in men and women. American Journal of Clinical Nutrition, 34, 1816–1826*

(a) Find the least-squares regression line treating fat-free mass as the predictor variable and energy expenditure as the response variable.
(b) Interpret the slope and intercept, if appropriate.
(c) Predict the energy expenditure where fat-free mass is 52.4 kg.

17. **Does Size Matter?** Researchers Willerman, Schultz, Rutledge, and Bigler wondered whether the size of a person's brain was related to the individual's mental capacity. They selected a sample of right-handed Anglo introductory psychology students who had Scholastic Aptitude Test Scores higher than 1350. The subjects were administered the Wechsler (1981) Adult Intelligence Scale-Revised in order to obtain their IQ scores. The MRI Scans, performed at the same facility, consisted of 18 horizontal MR images. The computer counted all pixels with nonzero gray scale in each of the 18 images and the total count served as an index for brain size. The resulting data are presented in the table below.

(a) Find the least-squares regression line treating MRI Count as the predictor variable and IQ as the response variable.
(b) What do you notice about the value of the slope? Why does this result seem reasonable based upon the scatter diagram and linear correlation coefficient obtained in Problem 21 of Section 4.1?

| Gender | MRI Count | IQ | Gender | MRI Count | IQ |
|--------|-----------|----|--------|-----------|----|
| Female | 816,932 | 133 | Male | 949,395 | 140 |
| Female | 951,545 | 137 | Male | 1,001,121 | 140 |
| Female | 991,305 | 138 | Male | 1,038,437 | 139 |
| Female | 833,868 | 132 | Male | 965,353 | 133 |
| Female | 856,472 | 140 | Male | 955,466 | 133 |
| Female | 852,244 | 132 | Male | 1,079,549 | 141 |
| Female | 790,619 | 135 | Male | 924,059 | 135 |
| Female | 866,662 | 130 | Male | 955,003 | 139 |
| Female | 857,782 | 133 | Male | 935,494 | 141 |
| Female | 948,066 | 133 | Male | 949,589 | 144 |

*Source:* Willerman, L., Schultz, R., Rutledge, J. N., and Bigler, E. (1991), "In Vivo Brain Size and Intelligence," *Intelligence*, 15, 223–228.

(c) When there is no relation between the predictor and response variable, we use the mean value of the response variable, $\bar{y}$, to predict. Predict the IQ of an individual whose MRI Count is 1,000,000. Predict the IQ of an individual whose MRI Count is 830,000.

18. **Male versus Female Drivers** The following data represent the number of licensed drivers in various age groups and the number of accidents within the age group by gender.

| Age Group | Number of Male Licensed Drivers (000s) | Number of Crashes Involving a Male (000s) | Number of Female Licensed Drivers (000s) | Number of Crashes Involving a Female (000s) |
|-----------|------|------|------|------|
| 16 | 816 | 244 | 764 | 178 |
| 17 | 1198 | 233 | 1115 | 175 |
| 18 | 1342 | 243 | 1212 | 164 |
| 19 | 1454 | 229 | 1333 | 145 |
| 20–24 | 7866 | 951 | 7394 | 618 |
| 25–29 | 9356 | 899 | 8946 | 595 |
| 30–34 | 10,121 | 875 | 9871 | 571 |
| 35–39 | 10,521 | 901 | 10,439 | 566 |
| 40–44 | 9776 | 692 | 9752 | 455 |
| 45–49 | 8754 | 667 | 8710 | 390 |
| 50–54 | 6840 | 390 | 6763 | 247 |
| 55–59 | 5341 | 290 | 5258 | 165 |
| 60–64 | 4565 | 218 | 4486 | 133 |
| 65–69 | 4234 | 191 | 4231 | 121 |
| 70–74 | 3604 | 167 | 3749 | 104 |
| 75–79 | 2563 | 118 | 2716 | 77 |
| 80–84 | 1400 | 61 | 1516 | 45 |
| ≥ 85 | 767 | 34 | 767 | 20 |

*Source:* National Highway and Traffic Safety Institute

(a) Find the least-squares regression line for males, treating number of licensed drivers as the predictor variable and number of crashes as the response variable. Repeat this procedure for females.

(b) Interpret the slope of the least-squares regression line for each gender, if appropriate. How might an insurance company use this information?

(c) Predict the number of accidents for males if there were 8700 thousand licensed drivers. Predict the number of accidents for females if there were 8700 thousand licensed drivers.

---

## Technology Step-by-Step
Determining the Least-Squares Regression Line

**TI-83 Plus** Use the same steps that were followed to obtain the correlation coefficient.

**MINITAB** *Step 1:* With the predictor variable in C1 and the response variable in C2, select the **Stat** menu and highlight **Regression**. Highlight **Regression**. . . .

*Step 2:* Select the predictor and response variables and click OK.

**Excel** *Step 1:* Be sure the Data Analysis Tool Pak is activated by selecting the **Tools** menu and highlighting **Add-Ins** . . . Check the box for the Analysis ToolPak and select OK.

*Step 2:* Enter the predictor variable in column A and the response variable in column B.

*Step 3:* Select the **Tools** menu and highlight **Data Analysis** . . .

*Step 4:* Select the **Regression** option.

*Step 5:* With the cursor in the Y-range cell, highlight the column that contains the response variable. With the cursor in the X-range cell, highlight the column that contains the predictor variable. Press OK.

---

## 4.3 The Coefficient of Determination

**Objective** ① Compute and interpret the coefficient of determination

In Section 4.2, we discussed the procedure for obtaining the least-squares regression line. In this section, we discuss another numerical measure of the strength of relation that exists between two quantitative variables.

### ① The Coefficient of Determination

Consider the per capita GDP versus life expectancy data introduced in Section 4.1. If we were asked to predict the life expectancy of a randomly selected Western European, what would a good guess be? Our best guess might be the average life expectancy of all Western Europeans. Since we don't know this value, we would use the average life expectancy from the sample data given in Table 1 on page 150, $\bar{y} = 77.75$.

Now suppose we were told this particular European lives in France where the per capita GDP is \$22.7 thousand. We could use the least-squares regression line to adjust our "guess" to $\hat{y} = 0.4200(22.7) + 68.7156 = 78.25$ years. Knowing the linear relation that exists between per capita GDP and life expectancy allows us to increase our estimate of the life expectancy of the individual. In statistical terms, we say that some of the variation in life expectancy is explained by the linear relation between per capita GDP and life expectancy.

A descriptive measure of the percentage of variation in life expectancy that is explained by the least-squares regression line is the *coefficient of determination*.

**Definition** The **coefficient of determination**, $R^2$, measures the percentage of total variation in the response variable that is explained by the least-squares regression line.

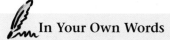

**In Your Own Words**

The coefficient of determination is a measure of how well the least-squares regression line describes the relation between the predictor and response variable. The closer $R^2$ is to 1, the better the line describes how changes in the predictor affect the value of the response.

The coefficient of determination is a number between 0 and 1, inclusive. That is, $0 \leq R^2 \leq 1$. If $R^2 = 0$, the least-squares regression line has no explanatory value; if $R^2 = 1$, the least-squares regression line explains 100% of the variation in the response variable.

Consider Figure 13, where a horizontal line is drawn at $\overline{y} = 77.75$. This value represents the predicted life expectancy of a Western European without any knowledge of her or his country's per capita GDP. Armed with the additional information that the individual is from France, we increased our "guess" to 78.25 years. The difference between the predicted life expectancy of 78.25 years and the predicted life expectancy of 77.75 years is due to the fact that the individual is from France. In other words, the difference between the prediction of $\hat{y} = 78.25$ and $\overline{y} = 77.75$ is explained by the linear relation between per capita GDP and life expectancy. The actual life expectancy in France is 78.63 years. The difference between our predicted value, $\hat{y} = 78.25$, and the actual value, $y = 78.63$, is due to factors other than the per capita GDP in France (perhaps the French have a healthy life style) and random error. In general, these differences are called **deviations**.

**Figure 13**

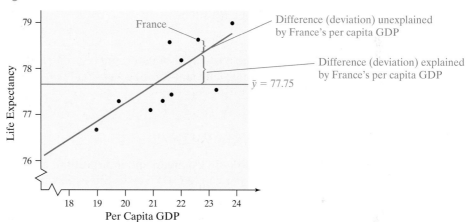

**In Your Own Words**

The word "deviations" comes from "deviate." "To deviate" means "to stray."

The deviation between the observed value of the response variable, $y$, and the mean value of the response variable, $\overline{y}$, is called the **total deviation**, so total deviation $= y - \overline{y}$. The deviation between the predicted value of the response variable, $\hat{y}$, and the mean value of the response variable, $\overline{y}$, is called the **explained deviation**, so explained deviation $= \hat{y} - \overline{y}$. Finally, the deviation between the observed value of the response variable, $y$, and the predicted value of the response variable, $\hat{y}$, is called the **unexplained deviation**, so unexplained deviation $= y - \hat{y}$. See Figure 14.

**Figure 14**

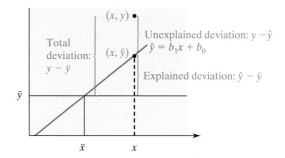

From the figure, it should be clear that

$$y - \overline{y} = (y - \hat{y}) + (\hat{y} - \overline{y})$$

or

Total deviation = unexplained deviation + explained deviation.

Although beyond the scope of this text, it can be shown that

$$\text{Total variation} = \text{unexplained variation} + \text{explained variation}$$

Dividing both sides by total variation, we obtain

$$1 = \frac{\text{unexplained variation}}{\text{total variation}} + \frac{\text{explained variation}}{\text{total variation}}$$

Subtracting $\dfrac{\text{unexplained variation}}{\text{total variation}}$ from both sides, we obtain

$$R^2 = \frac{\text{explained variation}}{\text{total variation}} = 1 - \frac{\text{unexplained variation}}{\text{total variation}}$$

Unexplained variation is found by summing the squares of the residuals, $\sum \text{residuals}^2$. So, the smaller the sum of squared residuals, the smaller the unexplained variation, and therefore, the larger $R^2$ will be. Therefore, the closer the observed $y$'s are to the regression line (the predicted $y$'s), the larger $R^2$ will be.

The coefficient of determination, $R^2$, is the square of the linear correlation coefficient for the least-squares regression model.

▶ **EXAMPLE 1** **Computing the Coefficient of Determination, $R^2$**

*Problem:* Compute and interpret the coefficient of determination, $R^2$, for the per capita GDP versus life-expectancy data shown in Table 1 on page 150 in Section 4.1.

*Approach:* To compute $R^2$, we square the linear correlation coefficient, $r$, found in Example 2 from Section 4.1 on page 154.

*Solution:* $R^2 = r^2 = 0.8094^2 = 0.6551 = 65.51\%$.

*Interpretation:* 65.51% of the variation in life expectancy is explained by the least-squares regression line, while 34.49% of the variation in life expectancy is explained by other factors. ◀◀

**Caution**

Squaring the linear correlation coefficient to obtain the coefficient of determination works only for the least-squares linear regression model

$$\hat{y} = b_0 + b_1 x.$$

The method does not work in general.

To help reinforce the concept of the coefficient of determination, consider the three data sets in Table 5.

| TABLE 5 | | | | | |
|---|---|---|---|---|---|
| **DATA SET A** | | **DATA SET B** | | **DATA SET C** | |
| x | y | x | y | x | y |
| 3.6 | 8.9 | 3.1 | 8.9 | 2.8 | 8.9 |
| 8.3 | 15.0 | 9.4 | 15.0 | 8.1 | 15.0 |
| 0.5 | 4.8 | 1.2 | 4.8 | 3.0 | 4.8 |
| 1.4 | 6.0 | 1.0 | 6.0 | 8.3 | 6.0 |
| 8.2 | 14.9 | 9.0 | 14.9 | 8.2 | 14.9 |
| 5.9 | 11.9 | 5.0 | 11.9 | 1.4 | 11.9 |
| 4.3 | 9.8 | 3.4 | 9.8 | 1.0 | 9.8 |
| 8.3 | 15.0 | 7.4 | 15.0 | 7.9 | 15.0 |
| 0.3 | 4.7 | 0.1 | 4.7 | 5.9 | 4.7 |
| 6.8 | 13.0 | 7.5 | 13.0 | 5.0 | 13.0 |

Figure 15(a) represents the scatter diagram of data set A, Figure 15(b) represents the scatter diagram of data set B, and Figure 15(c) represents the scatter diagram of data set C.

**Figure 15**

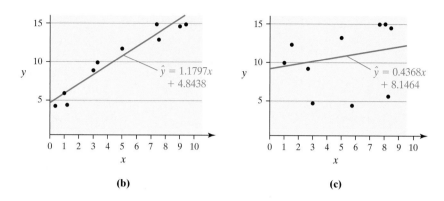

(a)                              (b)                              (c)

Notice that the $y$-values in each of the three data sets are the same. The variance of $y$ is 17.49. If we look at the scatter diagram in Figure 15(a), we notice that 100% of the variability in $y$ can be explained by the least-squares regression line, because all of the data lie on a straight line. In Figure 15(b), a high percentage of the variability in $y$ can be explained by the least-squares regression line because the data have a strong linear relation. (Higher $x$ values are associated with higher $y$ values.) Finally, in Figure 15(c), a low percentage of the variability in $y$ is explained by the least-squares regression line. (If $x$ increases, we cannot easily predict the change in $y$.) If we compute the coefficient of determination, $R^2$, for the three data sets in Table 5, we obtain the following results:

Coefficient of Determination for Data Set A: 100%

Coefficient of Determination for Data Set B: 94.7%

Coefficient of Determination for Data Set C: 9.4%

Notice that as the explanatory ability of the line decreases, so does the coefficient of determination, $R^2$.

**Figure 16**

SUMMARY OUTPUT

| Regression Statistics | |
|---|---|
| Multiple R | 0.80942 |
| R Square | 0.65516 |
| Adjusted R Square | 0.612055 |
| Standard Error | 0.495073 |
| Observations | 10 |

**Using Technology:** Statistical spreadsheets and graphing calculators with advanced statistical features have the ability to compute the coefficient of determination. Figure 16 displays partial output from Excel using the per capita GDP data with the value of $R^2$ highlighted. Notice the results differ slightly from the value of $R^2$ we obtained in Example 1. The difference is due to rounding error.

**NW** *Now Work Problems 1 and 7(a).*

## 4.3 Assess Your Understanding

### Concepts and Vocabulary

**1.** Suppose it is determined that $R^2 = 0.75$ when a linear regression is performed. Interpret this result.

**2.** Explain what is meant by total deviation, explained deviation, and unexplained deviation.

## Exercises

### • Skill Building

1. Match the coefficient of determination to the scatter diagram. The scales on the horizontal and vertical axis are the same
   for each scatter diagram.

(a) $R^2 = 0.58$    (b) $R^2 = 0.90$    (c) $R^2 = 1$    (d) $R^2 = 0.12$

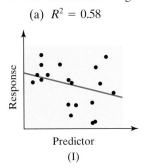
Predictor
(I)

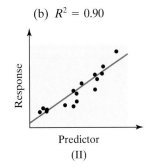
Predictor
(II)

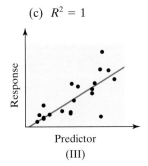
Predictor
(III)

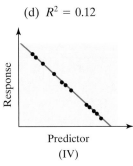
Predictor
(IV)

2. Use the linear correlation coefficient given to determine the coefficient of determination, $R^2$. Interpret each $R^2$.

(a) $r = -0.32$    (b) $r = 0.13$    (c) $r = 0.40$    (d) $r = 0.93$

---

### • Applying the Concepts

3. **The Other Old Faithful** Perhaps you are familiar with the famous Old Faithful geyser in Yellowstone National Park. There is another Old Faithful geyser located in Calistoga in California's Napa Valley. The following data represent the time between eruptions and the length of eruption for 11 randomly selected eruptions.

| Time Between Eruptions, x | Length of Erup-tion, y | Time Between Eruptions, x | Length of Erup-tion, y |
|---|---|---|---|
| 12.17 | 1.88 | 11.70 | 1.82 |
| 11.63 | 1.77 | 12.27 | 1.93 |
| 12.03 | 1.83 | 11.60 | 1.77 |
| 12.15 | 1.83 | 11.72 | 1.83 |
| 11.30 | 1.70 | | |

*Source:* Ladonna Hansen, Park Curator

The coefficient of determination is determined to be 83.0%. Interpret this result.

4. **Concrete** As concrete cures, it gains strength. The following data represent the 7-day and 28-day strength (in pounds per square inch) of a certain type of concrete.

| 7-day Strength, x | 28-day Strength, y | 7-day Strength, x | 28-day Strength, y |
|---|---|---|---|
| 2300 | 4070 | 2480 | 4120 |
| 3390 | 5220 | 3380 | 5020 |
| 2430 | 4640 | 2660 | 4890 |
| 2890 | 4620 | 2620 | 4190 |
| 3330 | 4850 | 3340 | 4630 |

The coefficient of determination, $R^2$, is determined to be 57.5%. Interpret this result.

5. **Calories versus Sugar** The following data represent the number of calories per serving and the number of grams of sugar per serving for a random sample of high-fiber cereals.

| Calories, x | Sugar, y | Calories, x | Sugar, y |
|---|---|---|---|
| 200 | 18 | 210 | 23 |
| 210 | 23 | 210 | 16 |
| 170 | 17 | 210 | 17 |
| 190 | 20 | 190 | 12 |
| 200 | 18 | 190 | 11 |
| 180 | 19 | 200 | 11 |

*Source: Consumer Reports,* October, 1999

(a) A scatter diagram with the least-squares regression line is shown. Do you think that calories and sugar content are linearly related? Why?

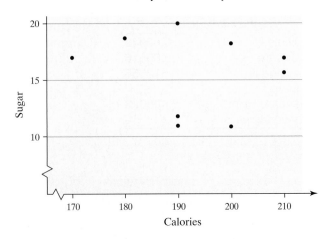

(b) The coefficient of determination, $R^2$, for these data is 6.8%. Interpret this result.

**6. Consumption versus Income** The following data represent the per capita disposable income (income after taxes) and per capita consumption in constant 1996 dollars in the United States for 1993–2001.

| Year | Per Capita Disposable Income, x | Per Capita Consumption, y |
|------|--------------------------------|---------------------------|
| 1993 | 20,233 | 18,796 |
| 1994 | 20,504 | 19,253 |
| 1995 | 20,795 | 19,630 |
| 1996 | 21,069 | 20,058 |
| 1997 | 21,464 | 20,563 |
| 1998 | 22,354 | 21,303 |
| 1999 | 22,641 | 22,098 |
| 2000 | 23,148 | 22,917 |
| 2001 | 23,691 | 23,312 |

*Source:* Bureau of Economic Analysis

(a) A scatter diagram of the data is shown. Do you think that per capita disposable income and per capita consumption are linearly related? Why or why not?

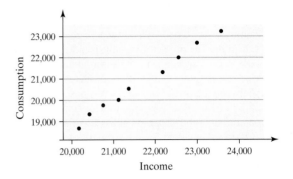

(b) The coefficient of determination, $R^2$, for the data is 99.0%. Interpret this result. Does this support your conclusion from part (a)? Why?

---

*Problems 7–14 use the results from Problems 11–18 in Section 4.1 and Problems 9–16 in Section 4.2.*

**7. Height versus Head Circumference** Use the results from Problem 11 in Section 4.1 and Problem 9 in Section 4.2 to
(a) compute the coefficient of determination, $R^2$.
(b) interpret the coefficient of determination.

**8. Round-Cut Diamonds** Use the results from Problem 12 in Section 4.1 and Problem 10 in Section 4.2 to
(a) compute the coefficient of determination, $R^2$.
(b) interpret the coefficient of determination.

**9. Gestation Period versus Life Expectancy** Use the results from Problem 13 in Section 4.1 and Problem 11 in Section 4.2 to
(a) compute the coefficient of determination, $R^2$.
(b) interpret the coefficient of determination.

**10. Tar and Nicotine** Use the results from Problem 14 in Section 4.1 and Problem 12 in Section 4.2 to
(a) compute the coefficient of determination, $R^2$.
(b) interpret the coefficient of determination and comment on the adequacy of the linear model.

**11. Bone Length** Use the results from Problem 15 in Section 4.1 and Problem 13 in Section 4.2 to
(a) compute the coefficient of determination, $R^2$.
(b) interpret the coefficient of determination and comment on the adequacy of the linear model.

**12. Weight of a Car versus Miles per Gallon** Use the results from Problem 16 in Section 4.1 and Problem 14 in Section 4.2 to
(a) compute the coefficient of determination, $R^2$.
(b) interpret the coefficient of determination.

**13. General Electric versus the S&P 500** Use the results from Problem 17 in Section 4.1 and Problem 15 in Section 4.2 to
(a) compute the coefficient of determination, $R^2$.
(b) interpret the coefficient of determination.

**14. Fat-Free Mass versus Energy Expenditure** Use the results from Problem 18 in Section 4.1 and Problem 16 in Section 4.2 to
(a) compute the coefficient of determination, $R^2$.
(b) interpret the coefficient of determination.

---

## Technology Step-by-Step
Determining $R^2$

**TI-83 Plus**  **The Coefficient of Determination, $R^2$**
Use the same steps that were followed to obtain the correlation coefficient to obtain $R^2$. Diagnostics must be on.

**MINITAB**  **The Coefficient of Determination, $R^2$**
**Step 1:** This is provided in the standard regression output.

**Excel**  **The Coefficient of Determination, $R^2$**
**Step 1:** This is provided in the standard regression output.

# CHAPTER 4 REVIEW

## Summary

In this chapter, we introduced techniques that allow us to describe the relation between two quantitative variables. The first step in identifying the type of relation that might exist is to draw a scatter diagram. The predictor variable is plotted on the horizontal axis, the corresponding response variable on the vertical axis. The scatter diagram can be used to discover whether the relation between the predictor and the response variable is linear. In addition, for linear relations, we can judge whether the linear relation shows positive or negative association.

A numerical measure for the strength of linear relation between two variables is the linear correlation coefficient. It is a number between $-1$ and $1$, inclusive. Values of the correlation coefficient near $-1$ are indicative of a negative linear relation between the two variables; values of the correlation coefficient near $+1$ indicate a positive linear relation between the two variables. If the correlation coefficient is near $0$, then there is little *linear* relation between the two variables.

Once a linear relation between the two variables has been discovered, we describe the relation by finding the least-squares regression line. This line best describes the linear relation between the predictor and the response variables. We can use the least-squares regression line to predict a value of the response variable for a given value of the predictor variable.

The coefficient of determination, $R^2$, measures the percent of variation in the response variable that is explained by the least-squares regression line. It is a measure between $0$ and $1$ inclusive. The closer $R^2$ is to $1$, the more explanatory value the line has.

One item worth mentioning again is that a researcher should never claim causation between two variables in a study unless the data are experimental. Observational data allow us to say that two variables might be associated, but we cannot claim causation.

## Formulas

### Correlation Coefficient

$$r = \frac{\sum \left( \dfrac{x_i - \bar{x}}{s_x} \right)\left( \dfrac{y_i - \bar{y}}{s_y} \right)}{n - 1}$$

### Equation of the Least-Squares Regression Line

The equation of the least-squares regression line is given by

$$\hat{y} = b_1 x + b_0$$

where

$\hat{y}$ is the predicted value of the response variable,

$b_1 = r \cdot \dfrac{s_y}{s_x}$ is the slope of the least-squares regression

line, and

$b_0 = \bar{y} - b_1 \bar{x}$ is the intercept of the least-squares regression line.

### Coefficient of Determination, $R^2$

$$R^2 = \frac{\text{variation explained by predictor variable}}{\text{total variation}}$$

$$= 1 - \frac{\text{unexplained variation}}{\text{total variation}}$$

$$= r^2 \text{ for the least-squares regression model } \hat{y} = b_0 + b_1 x$$

## Vocabulary

| | | |
|---|---|---|
| Bivariate data (p. 148) | Positively associated (p. 151) | Outside the scope of the model (p. 169) |
| Response variable (p. 149) | Negatively associated (p. 151) | Coefficient of determination (p. 176) |
| Predictor variable (p. 149) | Linear correlation coefficient (p. 152) | Total deviation (p. 177) |
| Lurking variable (p. 149) | Residuals (p. 165) | Explained deviation (p. 177) |
| Scatter diagram (p. 149) | Least-squares regression line (p. 166) | Unexplained deviation (p. 177) |

## Objectives

| Section | You should be able to ... | Review Exercises |
|---|---|---|
| **4.1** | 1 Draw scatter diagrams (p. 149) | 1–4, 9, 10, 20, 33–35(a) |
| | 2 Interpret scatter diagrams (p. 150) | 1–4(c) |
| | 3 Understand the properties of the linear correlation coefficient (p. 151) | 36 |
| | 4 Compute and interpret the linear correlation coefficient (p. 154) | 1–4(c), 9(d), 10(d) |
| **4.2** | 1 Find the least-squares regression line (p. 165) | 5–8(a) |
| | 2 Interpret the slope and y-intercept of the least-squares regression line (p. 167) | 5–8(c) |
| | 3 Predict the value of the response variable, based upon the least-squares regression line (p. 167) | 5–8(d) |
| | 4 Determine residuals, based upon the least-squares regression line (p. 168) | 5–8(e) |
| | 5 Compute the sum of squared residuals (p. 170) | 9, 10 (f) and (g) |
| **4.3** | 1 Compute and interpret the coefficient of determination (p. 176) | 11–14 |

# Review Exercises

1. **Engine Displacement versus Fuel Economy** The following data represent the size of a car's engine (in liters) versus its miles per gallon in the city for various 2001 domestic automobiles.
   (a) Draw a scatter diagram treating engine displacement as the predictor variable and miles per gallon as the response variable.
   (b) Compute the linear correlation coefficient between engine displacement and miles per gallon.
   (c) Based upon the scatter diagram and the linear correlation coefficient, comment on the type of relation that appears to exist between the two variables.

| Car | Engine Displacement (in liters) | City Miles per Gallon |
|---|---|---|
| Buick Century | 3.1 | 20 |
| Buick LeSabre | 3.8 | 19 |
| Cadillac DeVille | 4.6 | 16 |
| Chevrolet Camaro | 3.8 | 19 |
| Chevrolet Cavalier | 2.2 | 24 |
| Chevrolet Malibu | 3.1 | 23 |
| Chrysler LHS | 3.5 | 18 |
| Dodge Intrepid | 2.7 | 19 |
| Ford Crown Victoria | 4.6 | 17 |
| Ford Focus | 2.0 | 28 |
| Ford Mustang | 3.8 | 20 |
| Oldsmobile Aurora | 3.5 | 19 |
| Pontiac Grand Am | 2.4 | 22 |
| Pontiac Sunfire | 2.2 | 23 |
| Saturn Coupe | 1.9 | 28 |

*Source: Road and Track magazine*

2. **Temperature versus Cricket Chirps** Crickets make a chirping noise by sliding their wings rapidly over each other. Perhaps you have noticed that the number of chirps seems to increase with the temperature. The following data list the temperature (in Fahrenheit) and the number of chirps per second for the striped ground cricket.

| Temperature | Chirps per Second | Temperature | Chirps per Second |
|---|---|---|---|
| 88.6 | 20.0 | 71.6 | 16.0 |
| 93.3 | 19.8 | 84.3 | 18.4 |
| 80.6 | 17.1 | 75.2 | 15.5 |
| 69.7 | 14.7 | 82.0 | 17.1 |
| 69.4 | 15.4 | 83.3 | 16.2 |
| 79.6 | 15.0 | 82.6 | 17.2 |
| 80.6 | 16.0 | 83.5 | 17.0 |
| 76.3 | 14.4 | | |

*Source: The Songs of Insects, Pierce, George W., Cambridge, Mass.: Harvard University Press, 1949, pp. 12–21*

   (a) Draw a scatter diagram treating temperature as the predictor variable and chirps per second as the response variable.
   (b) Compute the linear correlation coefficient between temperature and chirps per second.
   (c) Based upon the scatter diagram and the linear correlation coefficient, comment on the type of relation that appears to exist between the two variables.

3. **Apartments** The following data represent the square footage and rents for apartments in North Chicago and in the western suburbs of Chicago.

| Location | Square Footage | Rent Per Month |
|---|---|---|
| North Chicago | 800 | 1320 |
| North Chicago | 820 | 1507 |
| North Chicago | 1050 | 1895 |
| North Chicago | 1415 | 2195 |
| North Chicago | 936 | 1555 |
| North Chicago | 800 | 1615 |
| North Chicago | 891 | 1125 |
| North Chicago | 970 | 1660 |
| North Chicago | 950 | 1600 |
| North Chicago | 1250 | 1900 |
| North Chicago | 1050 | 1830 |
| West Suburbs | 1161 | 1265 |
| West Suburbs | 1000 | 1050 |
| West Suburbs | 1034 | 915 |
| West Suburbs | 910 | 1003 |
| West Suburbs | 934 | 920 |
| West Suburbs | 860 | 890 |
| West Suburbs | 1056 | 1035 |
| West Suburbs | 1000 | 1190 |
| West Suburbs | 784 | 730 |
| West Suburbs | 962 | 1005 |
| West Suburbs | 897 | 940 |

*Source: apartments.com*

   (a) On the same graph, draw a scatter diagram for both North Chicago and West Suburban apartments treating square footage as the predictor variable. Be sure to use a different plotting symbol for each group.
   (b) Compute the linear correlation coefficient between square footage and rent for each location.
   (c) Given the scatter diagram and the linear correlation coefficient, comment on the type of relation that appears to exist between the two variables for each group.
   (d) Does location appear to be a factor in rent?

**4. Boys versus Girls** The following data represent the height (in inches) of boys and girls between the ages of 2 and 10 years. Data are based upon results obtained from the National Center for Health Statistics.

| Age | Boy Height | Girl Height | Age | Boy Height | Girl Height |
|---|---|---|---|---|---|
| 2 | 36.1 | 39.0 | 6 | 49.8 | 43.7 |
| 2 | 34.2 | 38.6 | 7 | 43.2 | 50.5 |
| 2 | 31.1 | 33.6 | 7 | 47.9 | 47.7 |
| 3 | 36.3 | 41.3 | 8 | 51.4 | 44 |
| 3 | 39.5 | 40.9 | 8 | 48.3 | 62.1 |
| 4 | 41.5 | 43.2 | 8 | 50.9 | 44.8 |
| 4 | 38.6 | 39.8 | 9 | 52.2 | 50.9 |
| 5 | 45.6 | 50.5 | 9 | 51.3 | 55.6 |
| 5 | 44.8 | 38.3 | 10 | 55.6 | 61.4 |
| 5 | 44.6 | 43.9 | 10 | 59.5 | 50.8 |

(a) On the same graph, draw a scatter diagram for both boys and girls treating age as the predictor variable. Be sure to use a different plotting symbol for each gender.
(b) Compute the linear correlation coefficient between age and height for each gender.
(c) Based upon the scatter diagram and the linear correlation coefficient, comment on the type of relation that appears to exist between the two variables for each gender.
(d) Does gender appear to be a factor in determining height?

**5.** Using the data and results from Problem 1, do the following:
(a) Find the least-squares regression line, treating engine displacement as the predictor variable and miles per gallon as the response variable.
(b) Draw the least-squares regression line on the scatter diagram.
(c) Interpret the slope and intercept, if appropriate.
(d) Predict the miles per gallon of a Ford Mustang whose engine displacement is 3.8 liters.

(e) Compute the residual of the prediction found in part (d).
(f) Is the miles per gallon above or below average for a Ford Mustang?

**6.** Using the data and results from Problem 2, do the following:
(a) Find the least-squares regression line, treating temperature as the predictor variable and chirps per second as the response variable.
(b) Draw the least-squares regression line on the scatter diagram.
(c) Interpret the slope and intercept, if appropriate.
(d) Predict the chirps per second if it is 83.3°F.
(e) Compute the residual of the prediction found in part (d).
(f) Were chirps per second above or below average at 83.3°F?

**7.** Using the "West Suburbs" data and results from Problem 3, do the following:
(a) Find the least-squares regression line, treating square footage as the predictor variable and rent as the response variable.
(b) Draw the least-squares regression line on the scatter diagram.
(c) Interpret the slope and intercept, if appropriate.
(d) Predict the rent of a 934-square-foot apartment.
(e) Compute the residual of the prediction found in part (d).
(f) Is this apartment's rent above or below average?

**8.** Using the "boy height" data and results from Problem 4, do the following:
(a) Find the least-squares regression line, treating age as the predictor variable and height as the response variable.
(b) Draw the least-squares regression line on the scatter diagram.
(c) Interpret the slope and intercept, if appropriate.
(d) Predict the height of a six-year-old boy.
(e) Compute the residual of the prediction found in part (d).
(f) Is this boy's height above or below average?

*In Problems 9 and 10, do the following:*

*(a) Draw a scatter diagram treating x as the predictor variable and y as the response variable.*
*(b) Select two points from the scatter diagram and find the equation of the line containing the points selected.*
*(c) Graph the line found in part (b) on the scatter diagram.*
*(d) Determine the least-squares regression line.*
*(e) Graph the least-squares regression line on the scatter diagram.*
*(f) Compute the sum of the squared residuals for the line found in part (b).*
*(g) Compute the sum of the squared residuals for the least-squares regression line found in part (d).*
*(h) Comment on the "fit" of the line found in part (b) versus the least-squares regression line found in part (d).*

**9.**

| x | 3 | 4 | 6 | 7 | 9 |
|---|---|---|---|---|---|
| y | 2.1 | 4.2 | 7.2 | 8.1 | 10.6 |

**10.**

| x | 10 | 14 | 17 | 18 | 21 |
|---|---|---|---|---|---|
| y | 105 | 94 | 82 | 76 | 63 |

**11.** Use the results from Problems 1 and 5 to compute and interpret : $R^2$.

**12.** Use the results from Problems 2 and 6 to compute and interpret $R^2$.

**13.** Use the results from Problems 3 and 7 to compute and interpret $R^2$.

**14.** Use the results from Problems 4 and 8 to compute and interpret $R^2$.

**Chapter 4 Projects located at www.prenhall.com/sullivanstats**

# Probability

## Outline

 **For additional study help, go to**
www.prenhall.com/sullivanstats

> **Materials include**
>
> • Projects
>   ■ Case Study: The Case of the Body in the Bag
>   ■ Decisions: Sports Probabilities
>   ■ Consumer Reports Project
> • Self-Graded Quizzes
> • "Preparing for This Section" Quizzes
> • STATLETs
> • PowerPoint Downloads
> • Step-by-Step Technology Guide
> • Graphing Calculator Help

## Putting It All Together

In Chapter 1, we learned the methods of collecting data. In Chapters 2 and 3, we learned how to summarize raw data through tables, graphs, and numbers. As far as the statistical process goes, we have discussed the collecting, organizing and summarizing part of the process.

Before we can proceed with the analysis of data, we introduce probability, which forms the basis of inferential statistics. Why? Well, we can think of the probability of an occurrence as the likelihood of observing that occurrence. If something has a high likelihood of occurring, we assign it a high probability (close to 1). If something has a small chance of occurring, we assign it a low probability (close to 0). For example, in rolling a single die, it is unlikely that we would roll five straight sixes, so we assign this result a low probability. In fact, the probability of rolling five straight sixes is 0.0001286. So if we were playing a game that entailed throwing a single die, and one of the players threw five sixes in a row, we would consider the player to be lucky (or a cheater) because it is such an unusual occurrence. Statisticians use probability in the same way. If something occurs that has a low probability, we investigate to find out "what's up."

## 5.1 Probability of Simple Events

**Preparing for This Section**    Before getting started, review the following:

✓ Relative frequency (Section 2.1, pp. 41–42)

**Objectives**    **1** Understand the properties of probabilities
**2** Compute and interpret probabilities using the classical method
**3** Compute and interpret probabilities using the empirical method
**4** Use simulation to obtain probabilities
**5** Understand subjective probabilities

**In Your Own Words**

Probability describes how likely it is that some event will occur. If we can look at the proportion of times an event has occurred over a long period of time (or over a large number of trials), we can be more certain of the likelihood of its occurrence.

**Probability** is a measure of the likelihood of a random phenomenon or chance behavior. Probability describes the long-term proportion with which a certain **outcome** will occur in situations with short-term uncertainty. The long-term predictability of chance behavior is best understood through a simple experiment. Flip a coin 100 times and compute the proportion of heads observed after each toss of the coin. Suppose the first flip is tails, so the proportion of heads is 0/1; the second flip is heads, so the proportion of heads is 1/2; the third flip is heads, so the proportion of heads is 2/3, and so on. Plot the proportion of heads versus the number of flips and obtain the graph in Figure 1(a). We repeat this experiment with the results shown in Figure 1(b).

Look at the graphs in Figures 1(a) and (b). Notice that in the short term (fewer flips of the coin), the proportion of heads is different and unpredictable for each experiment. As the number of flips of the coin increases, however, both graphs tend toward a proportion of 0.5. This is the basic premise of probability. Probability deals with experiments that yield random short-term results or outcomes, yet reveal long-term predictability. **The long-term proportion with which a certain outcome is observed is the probability of that outcome.** Therefore, we say that the probability of observing a head is $\frac{1}{2}$ or 50% because as the number of repetitions of the experiment increases (as we flip the coin more and more), the proportion of heads tends toward $\frac{1}{2}$. This phenomenon is referred to as the *Law of Large Numbers*.

**Figure 1**

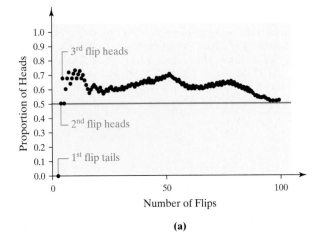

(a)

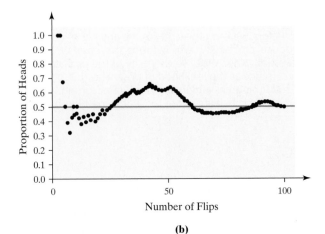

(b)

**The Law of Large Numbers**

As the number of repetitions of a probability experiment increases, the proportion with which a certain outcome is observed gets closer to the probability of the outcome.

The Law of Large Numbers is illustrated in Figure 1. For a few flips of the coin, the probability fluctuates wildly around 0.5, but as the number of flips increases, the proportion of heads settles down near 0.5. Jakob Bernoulli (a major contributor to the field of probability) felt that The Law of Large Numbers was common sense. This is evident by the following quote from his text *Ars Conjectandi*: "For even the most stupid of men, by some instinct of nature, by himself and without any instruction, is convinced that the more observations have been made, the less danger there is of wandering from one's goal."

In probability, an **experiment** is any process that can be repeated in which the results are uncertain. Probability experiments do not always produce the same results or outcome, so the result of any single trial of the experiment is not known ahead of time. However, the results of the experiment over many trials do produce regular patterns that enable us to predict with remarkable accuracy. For example, an insurance company cannot know ahead of time whether a specific 16-year-old driver will be involved in an accident over the course of a year. However, based upon historical records, the company can be fairly certain that about three out of every ten 16-year-old male drivers will be involved in a traffic accident during the course of a year. Therefore, of the 816,000 16-year-old drivers (816,000 repetitions of the experiment), the insurance company is fairly confident that about 30%, or 244,800, of the 816,000 drivers will be involved in an accident. This prediction forms the basis for establishing insurance rates for any particular 16-year-old male driver.

We now introduce some terminology that is required in the study of computing probabilities.

*Definitions*

**In Your Own Words**

A simple event is the result of one trial of a probability experiment. The sample space is a list of all possible results of a probability experiment.

A **simple event** is any single outcome from a probability experiment. Each simple event is denoted $e_i$.

The **sample space, $S$,** of a probability experiment is the collection of all possible simple events.

An **event** is any collection of outcomes from a probability experiment. An event may consist of one or more simple events. Events are denoted using capital letters such as $E$.

We present an example to illustrate the definitions.

▶ **EXAMPLE 1**   Identifying Events and the Sample Space of a Probability Experiment

*Problem:* A probability experiment consists of rolling a single "fair" die.*

**(a)** Identify the simple events of the probability experiment.
**(b)** Determine the sample space.
**(c)** Define the event $E$ = "roll an even number."

*A fair die is one in which each possible outcome is equally likely. For example, rolling a 2 is just as likely as rolling a 5. We contrast this with a "loaded" die in which a certain outcome is more likely. For example, if rolling a 1 is more likely than rolling a 2, 3, 4, 5, or 6, then the die is loaded.

*Approach:* The simple events are the possible outcomes that can occur based upon one trial of the experiment. The sample space is a list of all possible simple events. So we need to make a list of all possible ways the die can land.

*Solution:*

**(a)** The simple events from rolling a single fair die are $e_1 =$ "rolling a one" $= \{1\}$, $e_2 =$ "rolling a two" $= \{2\}$, $e_3 =$ "rolling a three" $= \{3\}$, $e_4 =$ "rolling a four" $= \{4\}$, $e_5 =$ "rolling a five" $= \{5\}$, and $e_6 =$ "rolling a six" $= \{6\}$.

**(b)** Combining these simple events, we form the sample space, $S = \{1, 2, 3, 4, 5, 6\}$. There are 6 simple events in the sample space.

**(c)** The event $E =$ "roll an even number" $= \{2, 4, 6\}$. ◄◄

## ① Properties of Probabilities

We define the **probability of an event**, denoted $P(E)$, as the likelihood of that event occurring. Probabilities have some properties that must be satisfied.

---

**Properties of Probabilities**

**1.** The probability of any event $E$, $P(E)$, must be between 0 and 1 inclusive. That is, $0 \le P(E) \le 1$.

**2.** If an event is **impossible**, the probability of the event is 0.

**3.** If an event is a **certainty**, the probability of the event is 1.

**4.** If $S = \{e_1, e_2, \ldots, e_n\}$, then $P(e_1) + P(e_2) + \ldots + P(e_n) = 1$.

---

- Property 1 states that a probability less than 0 or greater than 1 is not possible. Therefore, probabilities of 1.13 or $-0.32$ are not possible.
- Property 2 states that impossible events have a probability of 0. For example, in the United States of America it is impossible for Thanksgiving to be on Monday; therefore, the probability of Thanksgiving occurring on Monday is 0.
- Property 3 states that events that are certain have a probability of 1. Because every Thanksgiving occurs on Thursday, the probability that Thanksgiving occurs on Thursday next year is 1.
- Property 4 states that the sum of the probabilities of all the simple events must be 1.

The closer a probability is to 1, the more likely the event is to occur. The closer a probability is to 0, the less likely the event is to occur. For example, an event with probability 0.8 is more likely to occur than an event with probability 0.75. An event with probability 0.8 will occur about 80 times out of 100 repetitions of the experiment, while an event with probability 0.75 will occur about 75 times out of 100. Be careful of this interpretation. Just because an event has probability of 0.75 does not mean that the event *must* occur 75 times out of 100. It simply means that we *expect* the number of occurrences to be close to 75 in 100 trials of the experiment. The more repetitions of the probability experiment, the closer the proportion with which the event occurs will be to 0.75 (The Law of Large Numbers).

One of the main goals of this section is to learn how probabilities can be used to identify *unusual events*. An **unusual event** is an event that has a low probability of occurring. Typically, an event whose probability is less than 5% is considered unusual, but this "cut-off point" is not set in stone. Ultimately, the researcher must decide the probability that separates unusual

**In Your Own Words**

An unusual event is an event that is not likely to occur.

events from "not so unusual events." The context of the problem dictates this "cut-off point." For example, suppose the probability of being wrongly convicted of a capital crime punishable by death is 3%. The probability is too high in light of the consequences (death for the wrongly convicted). In other words, the probability indicates the event is not unusual (unlikely) enough. We would want this probability to be as close to zero as possible. Now suppose that you are planning a picnic on a day in which there is a 3% chance of rain. In this context, you would consider "rain" an unusual (unlikely) event and proceed with the picnic plans. The point is this: Selecting a probability that separates unusual events from "not so unusual" events is a subjective practice. The probability selected will depend upon the situation. With that said, statisticians typically use "cut-off" points of 1%, 5%, and 10%. For most circumstances, any event that occurs with probability of 5% or less will be considered unusual.

**Caution**

A probability of 0.05 should not always be used to separate unusual events from "not so unusual events."

**NW**   *Now Work Problem 1.*

Next, we introduce three methods for determining the probability of an event: (1) the classical method, (2) the empirical method, and (3) the subjective method.

## 2   The Classical Method

The classical method of computing probabilities requires *equally likely outcomes*. An experiment is said to have **equally likely outcomes** when each simple event has the same probability of occurring. For example, in throwing a fair die once, each of the six simple events in the sample space, $\{1, 2, 3, 4, 5, 6\}$, has an equal chance of occurring. Contrast this situation with a loaded die in which a five or six is twice as likely to occur as a one, two, three, or four.

*Theorem*

**Computing Probability Using the Classical Method**

If an experiment has $n$ equally likely simple events and if the number of ways that an event $E$ can occur is $m$, then the probability of $E$, $P(E)$, is

$$P(E) = \frac{\text{Number of ways that } E \text{ can occur}}{\text{Number of possible outcomes}} = \frac{m}{n} \qquad \textbf{(1)}$$

So, if $S$ is the sample space of this experiment, then

$$P(E) = \frac{N(E)}{N(S)} \qquad \textbf{(2)}$$

where $N(E)$ is the number of simple events in $E$ and $N(S)$ is the number of simple events in the sample space.

▶ **EXAMPLE 2**   **Computing Probabilities Using the Classical Method**

*Problem:*  A pair of fair dice are rolled.

**(a)** Compute the probability of rolling a seven.
**(b)** Compute the probability of rolling "snake eyes"; that is, compute the probability of rolling a two.
**(c)** Comment on the likelihood of rolling a seven versus rolling a two.

*Approach:*  To compute probabilities using the classical method, we count the number of simple events in the sample space and count the number of ways the event can occur.

*Solution:*

(a) In rolling a pair of fair dice, there are 36 equally likely outcomes, as shown in Figure 2.

Figure 2

There are 36 simple events in the sample space, so $N(S) = 36$. The event $E$ = "roll a seven" = $\{(1, 6), (2, 5), (3, 4), (4, 3), (5, 2), (6, 1)\}$ is composed of six simple events, so $N(E) = 6$. Using Formula (2), the probability of rolling a seven is

$$P(E) = P(\text{roll a seven}) = \frac{N(E)}{N(S)} = \frac{6}{36} = \frac{1}{6}$$

(b) The event $F$ = "rolling a two" = $\{(1, 1)\}$ has one simple event, so $N(E) = 1$. Using Formula (2), the probability of rolling a two is

$$P(F) = P(\text{roll a two}) = \frac{N(E)}{N(S)} = \frac{1}{36}$$

(c) Since $P(\text{rolling a seven}) = 6/36$ and $P(\text{rolling a two}) = 1/36$, we are six times as likely to roll a seven as a two. In other words, in 36 rolls of the dice, we would *expect* to observe about 6 sevens and only 1 two.

◄◄

In simple random sampling, each individual has the same chance of being selected. Therefore, we can use the classical method to compute the probability of obtaining a specific sample.

► EXAMPLE 3   Computing Probabilities Using Equally Likely Outcomes

*Problem:* Sophia has three tickets to a concert. Yolanda, Michael, Kevin, and Marissa have all stated they would like to go to the concert with Sophia. To be fair, Sophia decides to randomly select the two people who can go to the concert with her.

(a) Determine the sample space of the experiment. In other words, list all possible simple random samples of size $n = 2$.
(b) Compute the probability of the event "Michael and Kevin attend the concert."
(c) Compute the probability of the event "Marissa attends the concert."

*Approach:* First, we determine the simple events in the sample space by making a table. The probability of each event is the number of simple events in the event divided by the number of simple events in the sample space.

*Solution:*

**(a)** The sample space is listed in Table 1.

| TABLE 1 | | |
|---|---|---|
| Yolanda, Michael | Yolanda, Kevin | Yolanda, Marissa |
| Michael, Marissa | Kevin, Marissa | Michael, Kevin |

**(b)** We have $N(S) = 6$ and there is one way the event "Michael and Kevin attend the concert" can occur. Therefore, the probability that Michael and Kevin attend the concert is 1/6.

**(c)** We have $N(S) = 6$ and there are three ways the event "Marissa attends the concert" can occur. The probability that Marissa will attend is $3/6 = 50\%$. ◀◀

 *Now Work Problems 11 and 27.*

**③  Empirical Probabilities**

When using the classical method, we obtain the probability of an event without actually conducting the probability experiment. Consider Example 2, where we computed the probability of rolling a seven to be 1/6 without actually rolling a pair of dice. Often, however, attempts to count the number of ways an event can occur are difficult or impossible. For example, the probability of winning in the card game Black Jack can be computed using classical methods, but the tedious amount of work that needs to be done to figure out the equally likely events is overwhelming! It may also be the case that the simple events are not equally likely, in which case the classical method is inappropriate. For these reasons, we introduce a second method for determining probabilities that relies on empirical evidence, that is, evidence based upon the outcomes of a probability experiment.

---

**Approximating Probabilities through the Empirical Approach**

The probability of an event $E$ is approximately the number of times event $E$ is observed divided by the number of repetitions of the experiment.

$$P(E) \approx \text{relative frequency of } E = \frac{\text{frequency of } E}{\text{number of trials of experiment}} \quad (3)$$

---

**✍ In Your Own Words**

When probabilities are computed by using the classical approach, they are exact. When probabilities are computed by using the empirical approach, they are approximate. Probabilities computed from population data are exact; probabilities computed from survey (or sample) data are approximate.

Why is the probability obtained using the empirical approach approximate? Because different "runs" of the probability experiment will lead to different outcomes, and therefore different estimates of $P(E)$. Consider flipping a coin 20 times and recording the number of heads. Use the results of the experiment to estimate the probability of obtaining a head. Now repeat the experiment. Because the results of the second "run" of the experiment do not necessarily yield the same results, we cannot say the probability *equals* some proportion, but rather say the probability is *approximately* some value. Of course, as we increase the number of trials of a probability experiment, we have more confidence in our estimate (again, the Law of Large Numbers).

When we survey a random sample of individuals, the probabilities computed from the surveys are approximate. In fact, we can think of a survey as a probability experiment because the results of a survey will be different each time the survey is conducted (because there will be different people in it).

► EXAMPLE 4   Using Relative Frequencies to Approximate Probabilities

*Problem:* The data in Table 2 represent the results of a survey in which 200 people were asked their means of travel to work. The data are based on information obtained from the United States Census Bureau.

| TABLE 2 | |
|---|---|
| **Travel Mean** | **Frequency** |
| Drive Alone | 153 |
| Carpool | 22 |
| Public Transportation | 10 |
| Walk | 5 |
| Other Means | 3 |
| Work at Home | 7 |

**(a)** Approximate the probability that a randomly selected individual carpools to work.

**(b)** Would it be unusual to randomly select an individual who walks to work?

*Approach:* To estimate the probability of each event, we determine the relative frequency of the event.

*Solution:*

**(a)** There are $153 + 22 + \ldots + 7 = 200$ individuals in the survey. The relative frequency of the event "carpool" is $22/200 = 0.11 = 11\%$. We estimate that there is an 11% probability that a randomly selected individual carpools to work.

**(b)** The probability that an individual walks to work is $5/200 = 0.025 = 2.5\%$. It is somewhat unusual for individuals to walk to work.   ◄◄

► EXAMPLE 5   Comparing the Classical Method and Empirical Method

*Problem:* Suppose a survey is conducted in which 50 families with three children are asked to disclose the gender of their children. Based upon the results, it was found that 18 of the families had two boys and one girl.

**(a)** Estimate the probability of having two boys and one girl in a three-child family using the empirical method.

**(b)** Compute the probability of having two boys and one girl in a three-child family using the classical method, assuming boys and girls are equally likely.

*Approach:* To answer part (a), we determine the relative frequency of the event "two boys and one girl." To answer part (b), we must count the number of ways the event "two boys and one girl" can occur and divide this by the number of possible outcomes for this experiment.

*Solution:*

**(a)** The empirical probability of the event $E =$ "two boys and one girl" is

$$P(E) \approx \text{ relative frequency of } E = \frac{18}{50} = 0.36 = 36\%$$

There is about a 36% probability that a family of three children will have two boys and one girl.

**(b)** To determine the sample space, we construct a **tree diagram** to list the equally likely outcomes of the experiment. We draw two branches corresponding to the two possible outcomes (boy or girl) for the first repetition

**Historical Note**

Girolamo Cardano (in English Jerome Cardan) was born in Pavia, Italy, on September 24, 1501. He was an illegitimate child whose father was Fazio Cardano, a lawyer in Milan. Fazio was a part-time mathematician and taught Girolamo. In 1526, Cardano earned his medical degree; shortly thereafter, his father died. Unable to maintain a medical practice, Cardano spent his inheritance and turned to gambling to help support himself. Cardano developed an understanding of probability which helped him to win. He wrote a booklet on probability, *Liber de Ludo*, which wasn't printed until 1663, 87 years after his death. The booklet is a practical guide to gambling, including cards, dice, and cheating. Eventually Cardano became a lecturer of mathematics at the Piatti Foundation. This position allowed him to practice medicine and develop a favorable reputation as a doctor. In 1545, he published his greatest work, *Ars Magna*.

of the experiment (i.e., the first child). For the second child, we draw four branches: two branches originate from the first boy and two branches originate from the first girl. This is repeated for the third child. (See Figure 3, where B stands for boy and G stands for girl.)

Figure 3

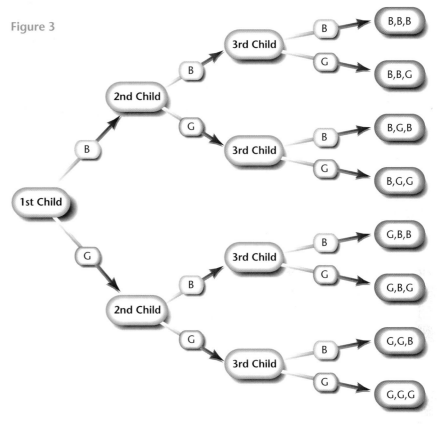

The sample space $S$ of this experiment is found by following each branch to identify all the possible outcomes of the experiment:

$$S = \{\text{BBB, BBG, BGB, BGG, GBB, GBG, GGB, GGG}\}$$

So $N(S) = 8$.

For the event $E =$ "two boys and a girl" $= \{\text{BBG, BGB, GBB}\}$, we have $N(E) = 3$. Since the outcomes are equally likely (for example, BBG is just as likely as BGB), the probability of $E$ is

$$P(E) = \frac{N(E)}{N(S)} = \frac{3}{8} = 0.375 = 37.5\%$$

There is a 37.5% probability that a family of three children will have two boys and one girl.   ◄◄

In comparing the results of Examples 5(a) and 5(b), we notice that the two probabilities are slightly different. Empirical probabilities and classical probabilities often differ in value. However, as the number of repetitions in a probability experiment increase, the empirical probability gets closer to the classical probability. That is, the classical probability is the theoretical relative frequency of an event after a large number of trials of the probability experiment. It is also possible that the two probabilities differ because having a boy or girl are not equally likely events. (Maybe the probability of having a boy is 50.5% and the probability of having a girl is 49.5%.) If this is the case, the empirical probability will not get closer to the classical probability.

**NW**  *Now Work Problem 17.*

 **Simulation**

Suppose we wanted to determine the probability of having a boy. Using classical methods, we assume that having a boy is just as likely as having a girl, so the probability of having a boy is 50%. We could also approximate this probability by looking in the *Statistical Abstract of the United States* under "Vital Statistics" and determining the number of boys and girls born for the most recent year for which data are available. For example, in 1996, there were 1,990,000 boys born and 1,901,000 girls born. Based upon empirical evidence, the probability of a boy is approximately 1,990,000/(1,990,000 + 1,901,000) = 0.511 = 51.1%.

However, instead of obtaining data from existing sources, we could simulate having babies by using a graphing calculator or statistical software to replicate the experiment as many times as we like.

▶ **EXAMPLE 6**   **Simulating Probabilities**

*Problem:*

**(a)** Simulate the experiment of sampling 100 babies.
**(b)** Simulate the experiment of sampling 1000 babies.

*Approach:*  To simulate probabilities, we use a random number generator available in statistical software and most calculators. We assume the simple events "have a boy" and "have a girl" are equally likely.

*Solution:*

**(a)** We use Minitab to perform the simulation. Set the seed in Minitab to any value you wish, say 1204. Use the Integer Distribution* to generate random data that simulate having babies. If we agree to let 0 represent a boy and 1 represent a girl, we can approximate the probability of having a girl by summing the number of 1's (adding up the number of girls) and dividing by the number of repetitions of the experiment, 100. See Figure 4.

Figure 4

| Integer Distribution | ✕ |
|---|---|
| | **G**enerate `100` rows of data |
| | **S**tore in column(s): |
| | `c1` |
| | **M**inimum value: `0` |
| | Ma**x**imum value: `1` |
| Select | |
| Help | **O**K    Cancel |

---

*The Integer Distribution involves a mathematical formula that uses a seed number to generate a sequence of equally likely random integers. Consult the technology manuals for setting the seed and generating sequences of integers.

Using Minitab's Tally command we can determine the number of 0s and 1s that Minitab randomly generated. See Figure 5.

Figure 5

**Summary Statistics for Discrete Variables**

```
C1     Count     Percent
0         48       48.00
1         52       52.00
N=       100
```

Based upon Figure 5, we approximate that there is a 48% probability of having a boy and a 52% probability of having a girl.

**(b)** Again, set the seed to 1204. Figure 6 shows the results of simulating the birth of 1000 babies.

Figure 6     **Summary Statistics for Discrete Variables**

```
C1     Count     Percent
0        501       50.10
1        499       49.90
N=      1000
```

We approximate that there is a 50.1% probability of having a boy and a 49.9% probability of having a girl.

Notice that more repetitions of the experiment (100 babies versus 1000 babies) result in a probability closer to the expected values of $\frac{1}{2}$ for each gender. Again, this result demonstrates the Law of Large Numbers.

**NW** *Now Work Problem 31.*

⑤  ## Subjective Probabilities

**Subjective probabilities** are probabilities obtained based upon an educated guess. For example, three different economists were asked "What is the probability the economy will fall into recession next year?" Each economist provided a different answer. The first economist said the probability is about 25%. The second economist was more gloomy and said the probability is about 50%. Finally, the third economist stated the probability was about 10%. How can three well-trained economists have such different opinions regarding the probability of a recession? Because the probabilities they stated are educated guesses based upon information they currently have available. The differences in the probabilities come from the fact that people interpret information differently!

**Caution!**
Because subjective probabilities are based upon personal judgments, they should be interpreted with extreme skepticism.

## 5.1 Assess Your Understanding

### Concepts and Vocabulary

**1.** Describe the difference between classical and empirical probability.

**2.** What is the probability of an impossible event? Suppose a probability is approximated to be zero based upon empirical results. Does this mean the event is impossible?

**3.** Why should subjective probabilities be interpreted with caution?

**4.** Suppose a coin is flipped and a die is cast. List all simple events for this probability experiment.

**5.** In computing classical probabilities, all simple events must be equally likely. Explain what this means.

**6.** Describe the idea behind using probabilities to identify unusual events.

# Exercises

## • Basic Skills

**1.** Which of the following numbers could be the probability of an event? (NW)

$$0, 0.01, 0.35, -0.4, 1, 1.4$$

**2.** Which of the following numbers could be the probability of an event?

$$1.5, \frac{1}{2}, \frac{3}{4}, \frac{2}{3}, 0, -\frac{1}{4}$$

*For Problems 3–6, let the sample space be* $S = \{1, 2, 3, 4, 5, 6, 7, 8, 9, 10\}$. *Suppose the simple events are equally likely.*

**3.** Compute the probability of the event $E = \{1, 2, 3\}$.

**4.** Compute the probability of the event $F = \{3, 5, 9, 10\}$.

**5.** Compute the probability of the event $E =$ "an even number".

**6.** Compute the probability of the event $F =$ "an odd number".

## • Applying the Concepts

**7. Family of Three** Using the results of Example 5(b) and Figure 3, compute the probability of having three boys in a three-child family.

**8. Family of Three** Using the results of Example 5(b) and Figure 3, compute the probability of having three girls in a three-child family.

**9. Family of Four** Compute the probability of having one boy and three girls in a four-child family assuming boys and girls are equally likely.

**10. Family of Four** Compute the probability of having two girls and two boys in a four-child family assuming boys and girls are equally likely.

**11. Planting Tulips** A bag of 100 tulip bulbs was purchased from a nursery. The bag contains 40 red tulip bulbs, 35 yellow tulip bulbs, and 25 purple tulip bulbs. (NW)

  (a) What is the probability that a randomly selected tulip bulb will be red?

  (b) What is the probability that a randomly selected tulip bulb will be purple?

**12. Golf Balls** The local golf store sells an "onion bag" that contains 80 "experienced" golf balls. Suppose the bag contains 35 Titleists, 25 Maxflis, and 20 Top-Flites.

  (a) What is the probability that a randomly selected golf ball will be a Titleist?

  (b) What is the probability that a randomly selected golf ball will be a Top-Flite?

**13. Roulette** In the game of roulette a wheel consists of 38 slots numbered $0, 00, 1, 2, \ldots, 36$. (See the photo.) To play the game, a metal ball is spun around the wheel and is allowed to fall into one of the numbered slots.

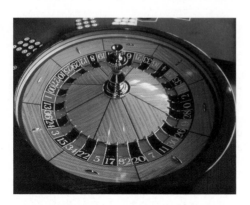

  (a) Determine the sample space.

  (b) Determine the probability that the metal ball falls into the slot marked "8." Interpret this probability.

  (c) Determine the probability that the metal ball lands in an odd slot. Interpret this probability.

**14. A Deck of Cards** A standard deck of cards contains 52 cards. There are 4 suits in the deck: 13 hearts, 13 diamonds, 13 clubs and 13 spades. Each suit has one "ace," one "two," one "three," one "four," one "five," one "six," one "seven," one "eight," one "nine," one "ten," one "jack," one "queen," and one "king." Consider the probability experiment of drawing one card from a deck of cards.

  (a) Determine the sample space of the probability experiment.

  (b) Determine the probability of randomly selecting the king of hearts from the deck of cards. Interpret this probability.

  (c) Determine the probability of randomly selecting any heart from a deck of cards. Interpret this probability.

  (d) Determine the probability of randomly selecting a king from a deck of cards. Interpret this probability.

**15. Birthdays** Exclude leap years from the following calculations:

  (a) Determine the probability that a randomly selected person has a birthday on the 1st day of a month. Interpret this probability.

  (b) Determine the probability that a randomly selected person has a birthday on the 31st day of a month. Interpret this probability.

  (c) Determine the probability that a randomly selected person was born in December. Interpret this probability.

  (d) Determine the probability that a randomly selected person has a birthday on November 8. Interpret this probability.

  (e) If you just met somebody and she asked you to guess her birthday, are you likely to be correct?

  (f) Do you think it is appropriate to use the methods of classical probability to compute the probability that a person is born in December?

**16. Genetics** A gene is composed of two alleles. An allele can be either a dominant allele or a recessive allele. Suppose a husband and wife decide to have a child. Each of these parents has one dominant normal cell allele, $S$, and one recessive normal cell allele, $s$. Therefore, the genotype of each parent is *Ss*. This means that each parent is a carrier of the disease, but does not have the disease. Each parent contributes one allele to his or her offspring with each allele being equally likely.

(a) List the possible genotypes of their offspring.

(b) What is the probability that the offspring will have sickle cell anemia? In other words, what is the probability that the offspring will have genotype *ss*? Interpret this probability.

(c) What is the probability that the offspring will not have sickle cell anemia, but will be a carrier? In other words, what is the probability that the offspring will have one dominant normal cell allele and one recessive sickle cell allele? Interpret this probability.

17. **College Survey** In a national survey conducted by the Centers for Disease Control in order to determine college students' health-risk behaviors, college students were asked: "How often do you wear a seat belt when riding in a car driven by someone else?" The frequencies appear in the following table:

| Response | Frequency |
|---|---|
| Never | 125 |
| Rarely | 324 |
| Sometimes | 552 |
| Most of the time | 1,257 |
| Always | 2,518 |

(a) Approximate the probability that a randomly selected college student never wears a seat belt when riding in a car driven by someone else.

(b) Would you consider it unusual to find somebody who never wears a seat belt when riding in a car driven by someone else? Why?

(c) Approximate the probability that a randomly selected college student sometimes wears a seat belt when riding in a car driven by someone else. Interpret this probability.

18. **College Survey** In a national survey conducted by the Centers for Disease Control in order to determine college students' health-risk behaviors, college students were asked, "How often do you wear a seat belt when driving a car?" The frequencies appear in the following table:

| Response | Frequency |
|---|---|
| Never | 118 |
| Rarely | 249 |
| Sometimes | 345 |
| Most of the time | 716 |
| Always | 3,093 |

(a) Approximate the probability that a randomly selected college student never wears a seat belt when driving a car.

(b) Is it unusual for a college student never to wear a seat belt when driving a car? Why?

(c) Approximate the probability that a randomly selected college student sometimes wears a seat belt when driving a car. Interpret this probability.

19. **Foreign-Born Population** A survey of 300 randomly selected foreign-born residents of the United States was conducted in which the respondents were asked to disclose the region of their birth. The data, based on information obtained from the United States Census Bureau, are as follows:

| Region | Frequency | Region | Frequency |
|---|---|---|---|
| Europe | 46 | Oceania | 2 |
| Asia | 83 | Latin America | 152 |
| Africa | 9 | North America | 8 |

(a) Approximate the probability that a randomly selected foreign-born resident of the United States is from Asia. Interpret this probability.

(b) Approximate the probability that a randomly selected foreign-born resident is from Oceania.

(c) Are foreign-born residents from Oceania unusual?

20. **Larceny Theft** A police officer randomly selected 595 police records of larceny thefts. The following data, based on information obtained from the U.S. Federal Bureau of Investigation, represent the number of offenses for various types of larceny thefts:

(a) Approximate the probability that a randomly selected larceny theft is shoplifting. Interpret this probability.

(b) Approximate the probability that a randomly selected larceny theft is pocket picking.

(c) As far as larceny theft goes, is pocket picking unusual?

| Type of Larceny Theft | Number of Offenses |
|---|---|
| Pocket picking | 5 |
| Purse snatching | 5 |
| Shoplifting | 118 |
| From motor vehicles | 197 |
| Motor vehicle accessories | 77 |
| Bicycles | 43 |
| From buildings | 105 |
| From coin-operated machines | 45 |

**21. Birth Weight** The following data represent the birth weight of all babies born in the United States in 1998:

| Weight (in grams) | Number |
|---|---|
| Less than 500 | 5,950 |
| 500–999 | 22,471 |
| 1,000–1,499 | 28,555 |
| 1,500–1,999 | 58,921 |
| 2,000–2,499 | 182,311 |
| 2,500–2,999 | 649,658 |
| 3,000–3,499 | 1,457,401 |
| 3,500–3,999 | 1,135,572 |
| 4,000–4,499 | 335,087 |
| 4,500–4,999 | 54,809 |
| 5,000 or more | 6,200 |
| **Total** | **3,936,935** |

*Source:* National Vital Statistics Report, Vol. 48, No. 3, March 28, 2000

(a) What is the probability that a randomly selected baby born in 1998 weighs 1000–1499 grams at birth. Interpret this probability.
(b) What is the probability that a randomly selected baby born in 1998 weighs 4000–4499 grams at birth. Interpret this probability.
(c) What is the probability that a randomly selected baby born in 1998 weighs 5000 grams or more. Is this unusual?

**22. Multiple Births** The following data represent the number of live multiple delivery births (three or more babies) in 1999 for women 15–44 years old:

| Age | Number of Multiple Births |
|---|---|
| 15–19 | 66 |
| 20–24 | 411 |
| 25–29 | 1653 |
| 30–34 | 2926 |
| 35–39 | 1813 |
| 40–44 | 956 |

*Source:* National Vital Statistics Report, Vol. 49, No. 1, April 17, 2001

(a) What is the probability that a randomly selected multiple birth in 1999 had a mother 30–34 years old. Interpret this probability.
(b) What is the probability that a randomly selected multiple birth in 1999 had a mother 40–44 years old. Interpret this probability.
(c) What is the probability that a randomly selected multiple birth in 1999 had a mother 15–19 years old. Is a multiple birth where the mother is 15–19 years old unusual?

**23. Tornadoes** Of the 28,538 tornadoes that occurred between 1916 and 1985 in the United States, 3262 of them occurred between the hours of 5:00 P.M. and 6:00 P.M. If a tornado occurs, approximate the probability it occurs between 5:00 P.M. and 6:00 P.M. Is it unusual for torna-

does to occur between 5:00 P.M. and 6:00 P.M.? Source: Dr. T. Fujita, *U.S. Tornadoes*, Part I.

**24. Tornadoes** Of the 28,538 tornadoes that occurred between 1916 and 1985 in the United States, 324 of them occurred between the hours of 5:00 A.M. and 6:00 A.M. If a tornado occurs, approximate the probability that it occurs between 5:00 A.M. and 6:00 A.M. Is it unusual for tornadoes to occur between 5:00 A.M. and 6:00 A.M.? Source: Dr. T. Fujita, *U.S. Tornadoes*, Part I.

**25. Chicago Weather** In Chicago, during the month of June, 11.5 days are partly cloudy, 7.3 days are clear, and 11.2 days are cloudy. Approximate the probability that a randomly selected day is clear. Interpret this probability.

**26. Chicago Weather** In Chicago, during the month of July, 12.3 days are partly cloudy, 8.2 days are clear, and 10.5 days are cloudy. Approximate the probability that a randomly selected day is clear. Interpret this probability.

**27. Randomly Selecting Cabinet Members** During the presidency of Andrew Jackson there were six cabinet positions: secretary of state, secretary of the treasury, secretary of war, attorney general, postmaster general, and secretary of the navy. Suppose two members of the cabinet are to be randomly selected to represent the United States at the funeral of a foreign diplomat.

(a) List the sample space.
(b) What is the probability the two representatives will be the secretary of state and the secretary of the treasury?
(c) What is the probability one of the two representatives will be the secretary of state?

**28. Presidential Elections Prior to 1804** Prior to the presidential election of 1804, the individual running for president who received the most votes was declared president, while the individual receiving the second most votes was declared vice president. In 1792, the individuals running in the presidential election were (1) George Washington, (2) John Adams, (3) George Clinton, (4) Thomas Jefferson, and (5) Aaron Burr.

(a) List sample space for president and vice president. Remember order matters here, so (Washington, Adams) is different from (Adams, Washington).
(b) Suppose the election were purely random. What is the probability that Washington would be president and Adams vice president?
(c) Suppose the election were purely random. What is the probability that Washington would be selected as president?

**29. Mark McGwire** On September 8, 1998, Mark McGwire hit his 62nd home run of the season. Of the 62 home runs he hit, 26 went to left field, 21 went to left center, 12 went to center field, 3 went to right center field and 0 went to right field.

(a) What is the probability that a randomly selected home run was hit to left field? Interpret this probability.
(b) What is the probability that a randomly selected home run was hit to right field?
(c) Is it impossible for Mark McGwire to hit a homer to right field?

**30. Rolling a Die**

(a) Roll a single die 50 times, recording the result of each throw of the die. Use the results to approximate the probability of throwing a three.

(b) Roll a single die 100 times, recording the result of each throw of the die. Use the results to approximate the probability of throwing a three.

(c) Compare the results of (a) and (b) to the classical probability of rolling a three.

**31. Simulation** Use a graphing calculator or statistical software to simulate rolling a six-sided die 100 times, using the Integer Distribution with numbers one through six.

(a) Use the results of the simulation to compute the probability of obtaining a one.

(b) Repeat the simulation. Compute the probability of rolling a one.

(c) Simulate rolling a six-sided die 500 times. Compute the probability of rolling a one.

(d) Which simulation resulted in the closest estimate to the probability that would be obtained using the classical method?

**32. Blood Types** A person can have one of four blood types: A, B, AB, or O. If people are randomly selected, is the probability they have blood type A equal to $\frac{1}{4}$? Why?

**33. Rolling Dice** When rolling a pair of dice, there are 11 possible outcomes: "rolling a 2", "rolling a 3", … "rolling a 12." Is the probability of "rolling a 2" equal to 1/11? Why?

**34. Classifying Probability** Determine whether the following probabilities are computed using classical methods, empirical methods, or subjective methods.

(a) The probability of having eight girls in an eight-child family is 0.390625%.

(b) On the basis of a study of families with eight children, the probability of a family having eight girls is 0.54%.

(c) According to a sports analyst, the probability that the Chicago Bears will win their next game is about 30%.

(d) On the basis of clinical trials, the probability of efficacy of a new drug is 75%.

**35. Checking for Loaded Dice** You suspect a pair of dice to be loaded and conduct a probability experiment by rolling each die 200 times. The outcome of the experiment is listed in the following table:

| Value of Die | Frequency | Value of Die | Frequency |
|---|---|---|---|
| 1 | 105 | 4 | 49 |
| 2 | 47 | 5 | 51 |
| 3 | 44 | 6 | 104 |

Do you think the die is loaded? Why?

**36.** Conduct a survey in your school by randomly asking 50 students whether they drive to school. Based upon the results of the survey, approximate the probability that a randomly selected student drives to school.

**37. Median Income** In 1998, the median income of families in the United States was $46,737. What is the probability that a randomly selected family will have an income greater than $46,737?

**38. Education** In 1998, 17% of Americans aged 25 years old or older did not have a high-school diploma. What is the probability that a randomly selected American aged 25 years old or older does not have a high school diploma?

---

## Technology Step-by-Step
Simulation

**TI-83 Plus**  **Step 1:** Set the seed by entering any number on the HOME screen. Press the STO ⟹ button, press the MATH button, highlight the PRB menu and highlight 1:rand and hit ENTER. With the cursor on the HOME screen, hit ENTER.

**Step 2:** Press the MATH button and highlight the PRB menu. Highlight 5:randInt( and hit ENTER.

**Step 3:** After the randInt( on the HOME screen type 1,n, number of repetitions of experiment) where $n$ is the number of equally likely simple events. For example, to simulate rolling a single die 50 times, we type

```
randInt(1, 6, 50)
```

**Step 4:** Press the STO ⟹ button and then 2ⁿᵈ 1, and hit ENTER to store the data in L1.

**Step 5:** Draw a histogram of the data using the simple events as classes. TRACE to obtain outcomes.

**MINITAB**  **Step 1:** Set the seed by selecting the **Calc** menu and highlighting **Set Base** … Insert any seed you wish into the cell and click OK.

**(b)** There are 13.6 million United States residents 18 years old or older who are widowed. The probability that a randomly selected United States resident 18 years old or older is widowed is 13.6/197.4 = 0.069 = 6.9%.

**(c)** The events "widowed" and "divorced" are mutually exclusive. Do you see why? We use the Addition Rule for Mutually Exclusive Events:

$$P(\text{widowed or divorced}) = P(\text{widowed}) + P(\text{divorced}) = \frac{13.6}{197.4} + \frac{19.4}{197.4} = \frac{33}{197.4} = 0.167 = 16.7\%$$

**(d)** The events "male" and "widowed" are not mutually exclusive. In fact, there are 2.6 million males that are widowed in the United States. Therefore, we use the Addition Rule Formula (1) to compute $P(\text{male or widowed})$:

$$P(\text{male or widowed}) = P(\text{male}) + P(\text{widowed}) - P(\text{male and widowed})$$

 *Now Work Problem 35.*

$$= \frac{95}{197.4} + \frac{13.6}{197.4} - \frac{2.6}{197.4} = \frac{106}{197.4} = 0.537 = 53.7\%$$   ◄◄

## ② Complements

Suppose the probability of an event $E$ is known and we would like to determine the probability that $E$ does not occur. This can easily be accomplished using the idea of *complements*.

*Definition*

**Complement of an Event**

Let $S$ denote the sample space of a probability experiment and let $E$ denote an event. The **complement of $E$**, denoted $\overline{E}$, is all simple events in the sample space $S$ that are not simple events in the event $E$.

**In Your Own Words**

The Complement Rule is used when you know the probability that some event will occur and you want to know the opposite: the chance it will not occur.

Because $E$ and $\overline{E}$ are mutually exclusive,

$$P(E \text{ or } \overline{E}) = P(E) + P(\overline{E}) = P(S) = 1$$

Subtracting $P(E)$ from both sides, we obtain

$$P(\overline{E}) = 1 - P(E)$$

We have the following result.

*Theorem*

**Complement Rule**

If $E$ represents any event and $\overline{E}$ represents the complement of $E$, then
$$P(\overline{E}) = 1 - P(E)$$

Figure 11 illustrates the Complement Rule using a Venn diagram.

**Figure 11**

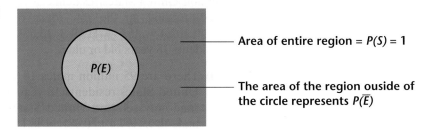

Area of entire region = $P(S) = 1$

The area of the region ouside of the circle represents $P(\overline{E})$

$P(E)$

► EXAMPLE 4 **Computing Probabilities Using Complements**

*Problem:* According to the National Gambling Impact Study Commision, 52% of Americans have played state lotteries. What is the probability that a randomly selected American has not played a state lottery?

*Approach:* Not playing a state lottery is the complement of playing a state lottery. We compute the probability using the Complement Rule.

*Solution:*

$$P(\text{not played state lottery}) = 1 - P(\text{played state lottery}) = 1 - 0.52 = 0.48$$

There is a 48% probability of randomly selecting an American that has not played a state lottery. ◄◄

► EXAMPLE 5 **Computing Probabilities Using Complements**

*Problem:* The data in Table 4 represent the income distribution of households in the United States in 2000.

| TABLE 4 | | | |
|---|---|---|---|
| Annual Income | Number (in thousands) | Annual Income | Number (in thousands) |
| Less than $10,000 | 10,023 | $50,000 to $74,999 | 20,018 |
| $10,000 to $14,999 | 6,995 | $75,000 to $99,999 | 10,480 |
| $15,000 to $24,999 | 13,994 | $100,000 to $149,999 | 8,125 |
| $25,000 to $34,999 | 13,491 | $150,000 to $199,999 | 2,337 |
| $35,000 to $49,999 | 17,032 | $200,000 or more | 2,239 |

*Source:* U.S. Bureau of the Census

**(a)** Compute the probability that a randomly selected household earned $200,000 or more in 2000.

**(b)** Compute the probability that a randomly selected household earned less than $200,000 in 2000.

**(c)** Compute the probability that a randomly selected household earned at least $10,000 in 2000.

*Approach:* The probabilities will be determined by finding the relative frequency of each event. We need to find the total number of households in the United States in 2000.

*Solution:*

**(a)** There were a total of $10{,}023 + 6{,}995 + \ldots + 2{,}239 = 104{,}734$ thousand households in the United States in 2000 and 2,239 thousand of them earned $200,000 or more. The probability that a randomly selected household in the United States earned $200,000 or more in 2000 is $2{,}239/104{,}734 = 0.021 = 2.1\%$.

**(b)** We could compute the probability of randomly selecting a household that earned less than $200,000 in 2000 by adding the relative frequencies of each category less than $200,000, but it is easier to use complements. The complement of earning less than $200,000 is earning $200,000 or more. Therefore,

$$P(\text{less than } \$200{,}000) = 1 - P(\$200{,}000 \text{ or more}) = 1 - 0.021 = 0.979 = 97.9\%$$

There is a 97.9% probability of randomly selecting a household that earned less than $200,000 in 2000.

(c) The phrase "at least" means greater than or equal to. The complement of "at least \$10,000" is "less than \$10,000." 10,023 thousand households earned less than \$10,000. The probability of randomly selecting a household that earned at least \$10,000 is

$$P(\text{at least } \$10{,}000) = 1 - P(\text{less than } \$10{,}000) = 1 - \frac{10{,}023}{104{,}734} = 0.904 = 90.4\%$$

**NW** *Now Work Problem 31.*

There is a 90.4% probability of randomly selecting a household that earned \$10,000 or more. ◄◄

## 5.2 Assess Your Understanding

### Concepts and Vocabulary

1. What does it mean when two events are mutually exclusive?
2. Why is $P(E \text{ and } F)$ subtracted from $P(E) + P(F)$ in the Addition Rule when $E$ and $F$ are not mutually exclusive?
3. What does it mean when two events are complements?

### Exercises

#### • Basic Skills

*In Problems 1–8, a probability experiment is conducted in which the sample space of the experiment is* $S = \{1, 2, 3, 4, 5, 6, 7, 8, 9, 10, 11, 12\}$*. Let event* $E = \{2, 3, 4, 5, 6, 7\}$*, event* $F = \{5, 6, 7, 8, 9\}$*, event* $G = \{9, 10, 11, 12\}$*, and event* $H = \{2, 3, 4\}$*.*

1. List the simple events in "$E$ and $F$." Are $E$ and $F$ mutually exclusive?
2. List the simple events in "$F$ and $G$." Are $F$ and $G$ mutually exclusive?
3. List the simple events in $F$ or $G$. Now find $P(F \text{ or } G)$ by counting the number of simple events in $F$ or $G$. Determine $P(F \text{ or } G)$ using the Addition Rule.
4. List the simple events in $E$ or $H$. Now find $P(E \text{ or } H)$ by counting the number of simple events in $E$ or $H$. Determine $P(E \text{ or } H)$ using the Addition Rule.

5. List the simple events in $E$ and $G$. Are $E$ and $G$ mutually exclusive?
6. List the simple events in $F$ and $H$. Are $F$ and $H$ mutually exclusive?
7. List the simple events in $\overline{E}$. Find $P(\overline{E})$.
8. List the simple events in $\overline{F}$. Find $P(\overline{F})$.

*In Problems 9–14, find the probability of the indicated event if* $P(A) = 0.25$ *and* $P(B) = 0.45$*.*

9. $P(A \text{ or } B)$ if $P(A \text{ and } B) = 0.15$
10. $P(A \text{ and } B)$ if $P(A \text{ or } B) = 0.6$
11. $P(A \text{ or } B)$ if $A$ and $B$ are mutually exclusive
12. $P(A \text{ and } B)$ if $A$ and $B$ are mutually exclusive
13. $P(\overline{A})$
14. $P(\overline{B})$
15. If $P(A) = 0.60$, $P(A \text{ or } B) = 0.85$, and $P(A \text{ and } B) = 0.05$, find $P(B)$.
16. If $P(B) = 0.30$, $P(A \text{ or } B) = 0.65$, and $P(A \text{ and } B) = 0.15$, find $P(A)$.

#### • Applying the Concepts

*In Problems 17–20, a golf ball is selected at random from a golf bag. If the golf bag contains 9 Titleists, 8 Maxflis, and 3 Top-Flites, find the probability of each event.*

17. The golf ball is a Titleist or Maxfli.
18. The golf ball is a Maxfli or Top-Flite.
19. The golf ball is not a Titleist.
20. The golf ball is not a Top-Flite.
21. **Family of Three** Compute the probability of having one or two boys in a three-child family.
22. **Family of Three** Compute the probability of having two or three girls in a three-child family.

23. **A Deck of Cards** A standard deck of cards contains 52 **NW** cards as shown in Figure 10. One card is randomly selected from the deck.
    (a) Compute the probability of randomly selecting a heart or club from a deck of cards.
    (b) Compute the probability of randomly selecting a heart or club or diamond from a deck of cards.
    (c) Compute the probability of randomly selecting an ace or heart from a deck of cards.

24. **A Deck of Cards** A standard deck of cards contains 52 cards as shown in Figure 10. One card is randomly selected from the deck.
    (a) Compute the probability of randomly selecting a two or three from a deck of cards.
    (b) Compute the probability of randomly selecting a two or three or four from a deck of cards.
    (c) Compute the probability of randomly selecting a two or club from a deck of cards.

25. **Birthdays** Exclude leap years from the following calculations.
    (a) Compute the probability that a randomly selected person does not have a birthday on November 8.
    (b) Compute the probability that a randomly selected person does not have a birthday on the 1st day of a month.
    (c) Compute the probability that a randomly selected person does not have a birthday on the 31st day of a month.
    (d) Compute the probability that a randomly selected person was not born in December.

26. **Roulette** In the game of roulette a wheel consists of 38 slots numbered 0, 00, 1, 2, … 36. The odd-numbered slots are red, while the even-numbered slots are black. The numbers 0 and 00 are green. To play the game, a metal ball is spun around the wheel and is allowed to fall into one of the numbered slots.
    (a) What is the probability that the metal ball lands on green or red?
    (b) What is the probability that the metal ball does not land on green?

27. **College Survey** In a national survey conducted by the Centers for Disease Control in order to determine college students' health-risk behaviors, college students were asked: "How often do you wear a seat belt when riding in a car driven by someone else?" The frequencies appear in the following table:

| Response | Frequency |
|---|---|
| Never | 125 |
| Rarely | 324 |
| Sometimes | 552 |
| Most of the time | 1,257 |
| Always | 2,518 |

    (a) Approximate the probability that a randomly selected college student never or rarely wears a seat belt when riding in a car driven by someone else.
    (b) Approximate the probability that a randomly selected college student does not always wear a seat belt when riding in a car driven by someone else.

28. **College Survey** In a national survey conducted by the Centers for Disease Control in order to determine college students' health-risk behaviors, college students were asked: "How often do you wear a seat belt when driving a car?" The frequencies are displayed in the following table:

| Response | Frequency |
|---|---|
| Never | 118 |
| Rarely | 249 |
| Sometimes | 345 |
| Most of the time | 716 |
| Always | 3,093 |

    (a) Approximate the probability that a randomly selected college student never or rarely wears a seat belt when driving a car.
    (b) Approximate the probability that a randomly selected college student does not always wear a seat belt when driving a car.

29. **Foreign-Born Population** A survey of 300 randomly selected foreign-born residents of the United States was conducted in which the respondents were asked to disclose the region of their birth. The data, based on information obtained from the United States Census Bureau, are as follows:

| Region | Frequency | Region | Frequency |
|---|---|---|---|
| Europe | 46 | Oceania | 2 |
| Asia | 83 | Latin America | 152 |
| Africa | 9 | North America | 8 |

    (a) Approximate the probability that a randomly selected foreign-born resident of the United States is from Europe or Asia.
    (b) Approximate the probability that a randomly selected foreign-born resident of the United States is from Europe, Asia, or Latin America.
    (c) Approximate the probability that a randomly selected foreign-born resident of the United States is not from Europe.

30. **Larceny Theft** A police officer randomly selected 595 police records of larceny thefts. The following data, based on information obtained from the U.S. Federal Bureau of Investigation, represent the number of offenses for various types of larceny thefts.

| Type of Larceny Theft | Number of Offenses |
|---|---|
| Pocket picking | 5 |
| Purse snatching | 5 |
| Shoplifting | 118 |
| From motor vehicles | 197 |
| Motor vehicle accessories | 77 |
| Bicycles | 43 |
| From buildings | 105 |
| From coin-operated machines | 45 |

    (a) Approximate the probability that a randomly selected larceny theft is shoplifting or purse snatching.
    (b) Approximate the probability that a randomly selected larceny theft is shoplifting or purse snatching or pocket picking.
    (c) Approximate the probability that a randomly selected larceny theft is not pocket picking.

31. **Birth Weight** The following data represent the birth
    NW  weight of all babies born in the United States in 1998:

| Weight (in grams) | Number | Weight (in grams) | Number |
|---|---|---|---|
| Less than 500 | 5,950 | 3,000–3,499 | 1,457,401 |
| 500–999 | 22,471 | 3,500–3,999 | 1,135,572 |
| 1,000–1,499 | 28,555 | 4,000–4,499 | 335,087 |
| 1,500–1,999 | 58,921 | 4,500–4,999 | 54,809 |
| 2,000–2,499 | 182,311 | 5,000 or more | 6,200 |
| 2,500–2,999 | 649,658 | **Total** | **3,936,935** |

*Source:* National Vital Statistics Report, Vol. 48, No. 3,
March 28, 2000

(a) Determine the probability that a randomly selected
baby born in 1998 weighs less than 1000 grams at
birth.

(b) Determine the probability that a randomly selected
baby born in 1998 weighs 1000 grams or more at
birth.

(c) Determine the probability that a randomly selected
baby born in 1998 weighs less than 5000 grams at
birth.

(d) Determine the probability that a randomly selected
baby born in 1998 weighs at least 500 grams at
birth.

32. **Multiple Births** The following data represent the number of live multiple delivery births (three or more babies) in 1999 for women 15–44 years old:

| Age | Number of Multiple Births |
|---|---|
| 15–19 | 66 |
| 20–24 | 411 |
| 25–29 | 1653 |
| 30–34 | 2926 |
| 35–39 | 1813 |
| 40–44 | 956 |

*Source:* National Vital Statistics Report,
Vol. 49, No. 1, April 17, 2001

(a) Determine the probability that a randomly selected
multiple birth in 1999 for women 15–44 years old
had a mother 30–39 years old.

(b) Determine the probability that a randomly selected
multiple birth in 1999 for women 15–44 years old
had a mother that was not 30–39 years old.

(c) Determine the probability that a randomly selected
multiple birth in 1999 for women 15–44 years old
had a mother who was less than 40 years old.

(d) Determine the probability that a randomly selected
multiple birth in 1999 for women 15–44 years old
had a mother who was at least 20 years old.

33. **Tornadoes** Of the 28,538 tornadoes that occurred between 1916 and 1985 in the United States, 3262 of them
occurred between the hours of 5:00 P.M. and 6:00 P.M.. If

a tornado occurs, approximate the probability it does
not occur between 5:00 P.M. and 6:00 P.M.. *Source:* Dr. T.
Fujita, U.S. Tornadoes Part I.

34. **Tornadoes** Of the 28,538 tornadoes that occurred between 1916 and 1985 in the United States, 324 of them
occurred between the hours of 5:00 A.M. and 6:00 A.M..
If a tornado occurs, approximate the probability it does
not occur between 5:00 A.M. and 6:00 A.M.. *Source:* Dr. T.
Fujita, U.S. Tornadoes Part I.

35. **Medicaid** The following data, in thousands, represent
    NW  the age of persons receiving Medicaid and their poverty
level:

| | | AGE | | |
|---|---|---|---|---|
| | <18 | 18–44 | 45–64 | >65 |
| **Below Poverty Level** | 8,550 | 4,356 | 1,520 | 958 |
| **Above Poverty Level** | 5,884 | 3,741 | 1,756 | 1,943 |

*Source:* U.S. Census Bureau

(a) Determine the probability that a randomly selected
individual receiving Medicaid has income below the
poverty line.

(b) Determine the probability that a randomly selected
individual receiving Medicaid is less than 18 years
old.

(c) Determine the probability that a randomly selected
individual receiving Medicaid has income below the
poverty line and is less than 18 years old.

(d) Determine the probability that a randomly selected
individual receiving Medicaid has income below the
poverty line or is less than 18 years old.

36. **SAT Verbal Scores** The following data represent the
scores received on the 2000 SAT I: Reasoning Test—
Verbal, by gender:

| | Male | Female |
|---|---|---|
| 200–249 | 6,755 | 7,512 |
| 250–299 | 9,594 | 10,816 |
| 300–349 | 25,630 | 30,058 |
| 350–399 | 50,042 | 60,139 |
| 400–449 | 78,276 | 95,780 |
| 450–499 | 98,989 | 117,976 |
| 500–549 | 102,423 | 120,183 |
| 550–599 | 84,524 | 96,381 |
| 600–649 | 62,205 | 68,515 |
| 650–699 | 36,202 | 39,843 |
| 700–749 | 18,044 | 18,716 |
| 750–800 | 10,647 | 11,028 |
| **TOTAL** | **583,331** | **676,947** |

*Source:* College Board

(a) If a test taker is randomly selected, what is the probability that the test taker is female?
(b) If a test taker is randomly selected, what is the probability that the test taker scored 600–649?
(c) If a test taker is randomly selected, what is the probability that the test taker is female and scored 600–649?
(d) If a test taker is randomly selected, what is the probability that the test taker is female or scored 600–649?

37. **Driver Fatalities** The data to the right represent the number of driver fatalities in 1996 by age group for male and female drivers:

(a) Determine the probability that a randomly selected driver fatality involves a male driver.
(b) Determine the probability that a randomly selected driver fatality involves a 20–24-year-old.
(c) Determine the probability that a randomly selected driver fatality involves a 20–24-year-old male driver.
(d) Determine the probability that a randomly selected driver fatality involves a 20–24-year-old or a male driver.

| | Male | Female |
|---|---|---|
| 16–19 | 1922 | 762 |
| 20–24 | 2814 | 699 |
| 25–29 | 2126 | 617 |
| 30–34 | 1986 | 627 |
| 35–39 | 1743 | 604 |
| 40–44 | 1437 | 485 |
| 45–49 | 1169 | 391 |
| 50–54 | 859 | 347 |
| 55–59 | 684 | 260 |
| 60–64 | 637 | 270 |
| 65–69 | 614 | 268 |
| 70–74 | 598 | 358 |
| 75–79 | 554 | 323 |
| 80–84 | 474 | 230 |
| 85 or older | 329 | 146 |
| **TOTAL** | **17,946** | **6,387** |

*Source:* National Highway Traffic Institute

38. **Marital Status** The following data represent the marital status of Americans 25 years old or older and their level of education in 1999:

| | Did Not Graduate High School | High School Graduate | Some College | College Graduate |
|---|---|---|---|---|
| Never Married | 4,012 | 7,790 | 6,565 | 7,685 |
| Married, Spouse Present | 15,122 | 36,076 | 27,003 | 29,811 |
| Married, Spouse Absent | 1,877 | 2,006 | 1,435 | 1,489 |
| Separated | 1,094 | 1,515 | 1,214 | 598 |
| Widowed | 5,000 | 4,757 | 2,264 | 1,456 |
| Divorced | 2,951 | 7,047 | 5,649 | 3,766 |

*Source:* U.S. Census Bureau

(a) Determine the probability that a randomly selected person has never married.
(b) Determine the probability that a randomly selected person is a high school graduate.
(c) Determine the probability that a randomly selected person has never married and is a high school graduate.
(d) Determine the probability that a randomly selected person has never married or is a high school graduate.

## 5.3 The Multiplication Rule

**Objectives**  Compute probabilities using the Multiplication Rule

Compute probabilities using the Multiplication Rule for Independent Events

Compute "at least" probabilities

 In Section 5.2 we studied the probability of event *E* or *F* occurring in a single trial of an experiment. In this section, we concentrate on methods for computing the probability of events *E* and *F*.

▶ **EXAMPLE 1** **Illustrating the Multiplication Rule**

*Problem:* Two members from a five-member committee are to be randomly selected to serve as chairperson and secretary. The first person selected is the chairperson and the second person selected is the secretary. The members of the committee are Bob, Faye, Elena, Melody, and Dave. Determine the probability that Elena is chairperson and Dave is secretary.

*Approach:* The first person randomly selected is chairperson. Because an individual cannot serve as both chairperson and secretary, we are sampling without replacement.* We list the possibilities for chairperson and secretary in a tree diagram. We count the number of ways Elena can be chairperson and Dave can be secretary and divide this result by the number of possible outcomes of the experiment.

*Solution:* We list the possibilities for chairperson and secretary in the tree diagram in Figure 12 on page 211. The first trial of the experiment involves selecting the chairperson. There are five branches on the tree diagram corresponding to the five possible choices for chairperson. For each choice of chairperson, there are four remaining choices for secretary. Therefore, four branches are drawn from each choice of chairperson. From the diagram, we see there are twenty ways the chairperson and secretary can be selected. There is one way the chairperson can be Elena and the secretary can be Dave. Therefore, the probability that Elena is chairperson and Dave is secretary is 1/20. ◀◀

The method used to obtain the probability in Example 1 was rather tedious. Imagine the amount of work involved had there been 20 members on the original committee! Fortunately, the *Multiplication Rule* can be used to obtain probabilities of the form *E* and *F*. Before we present the Multiplication Rule, we discuss the concept of *conditional probability*.

**Conditional Probability**

The notation $P(F \mid E)$ is read "the probability of event *F* given event *E*." It is the probability of an event *F* occurring given the occurrence of the event *E*.

For example, we might compute the probability of selecting Dave to serve on the committee as secretary given that Elena has already been selected as chairperson. Under this scenario, we concentrate on the branch of the tree diagram in which Elena is chairperson. Look again at Figure 12. If we let $E =$ "Elena is chairperson" and $F =$ "Dave is secretary," then $P(F \mid E) = P(\text{Dave is secretary given that Elena is chairperson}) = \frac{1}{4}$ since there is one way for Dave to be secretary and four possible outcomes after Elena was chosen as the chairperson.

### In Your Own Words

A tree diagram is a great way to list outcomes in a probability experiment.

---

*Recall from Chapter 1 that sampling without replacement means that once an individual is selected, the individual is removed from the sample space and cannot be selected again.

Figure 12

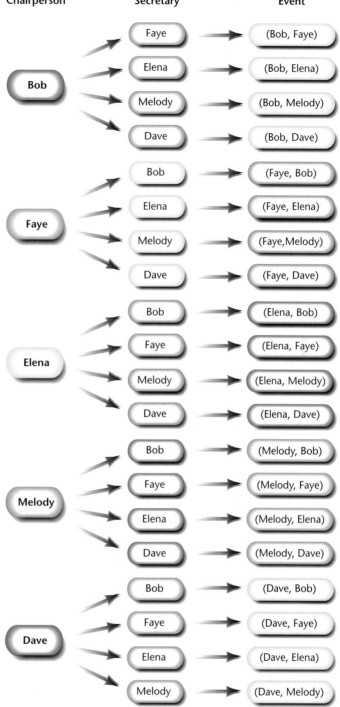

| Chairperson | Secretary | Event |

We now present the Multiplication Rule.

**The Multiplication Rule**

The probability that two events $E$ and $F$ both occur is

$$P(E \text{ and } F) = P(E) \cdot P(F \mid E)$$

In words, the probability of $E$ and $F$ is the probability of event $E$ occurring times the probability of event $F$ occurring given the occurrence of event $E$.

**In Your Own Words**

The Addition Rule is used for "or" probabilities. The Multiplication Rule is for "and" probabilities.

Let's redo Example 1 using the Multiplication Rule.

▶ EXAMPLE 2   Computing Probabilities Using the Multiplication Rule

*Problem:* Redo Example 1 using the Multiplication Rule.

*Approach:* We first compute the probability that Elena is selected as chairperson. We multiply this result by the probability Dave is secretary given Elena is chairperson.

*Solution:* We define event $E$ = "Elena is chairperson" and event $F$ = "Dave is secretary." The first trial of the experiment is randomly selecting a chairperson. Since there are five members on the committee, then $P(\text{Elena is chairperson}) = P(E) = 1/5$. The second trial of the experiment is randomly selecting a secretary given that Elena has already been selected as chairperson. Therefore, $P(\text{"Dave is secretary" given that "Elena is chairperson"}) = P(F\,|\,E) = \frac{1}{4}$ since there is one way Dave can be selected and there are four different outcomes of the experiment (since Elena is not available to serve as secretary). Using the Multiplication Rule we find

$$P(\text{Elena is chairperson and Dave is secretary}) = P(E \text{ and } F) = P(E) \cdot P(F\,|\,E)$$
$$= \frac{1}{5} \cdot \frac{1}{4} = \frac{1}{20}$$

This is the same result as Example 1, and we obtained it without drawing a tree diagram.   ◀◀

**NW** *Now Work Problem 7.*

The Multiplication Rule applies to more than just a sequence of events (such as selecting a chair and then selecting a secretary). Consider the next example.

▶ EXAMPLE 3   Using the Multiplication Rule

*Problem:* The probability that a police officer is hiding behind the billboard next to the highway is 0.08. The probability that a randomly selected driver is speeding, given that a police officer is behind the billboard, is 0.2. What is the probability that a randomly selected driver is speeding and there is a police officer hiding behind the billboard next to the highway?

*Approach:* Let $E$ represent the event "police officer hiding behind the billboard" and let $F$ represent the event "driver is speeding." We use the Multiplication Rule to compute $P(E \text{ and } F)$.

*Solution:* $P(\text{driver is speeding and a police officer is hiding behind the billboard}) = P(E \text{ and } F) = P(E) \cdot P(F|E) = 0.08(0.2) = 0.016 = 1.6\%$. There is a 1.6% probability that a driver is speeding and there is a police officer hiding behind the billboard.   ◀◀

**NW** *Now Work Problem 17.*

**In Your Own Words**

In determining whether two events are independent, ask yourself whether the probability of one event is affected by the other event. For example, what is the probability that a 20-to-29-year-old male has high cholesterol? What is the probability that a 20-to-29-year-old male has high cholesterol, given that he eats fast food four times a week? Does the fact that the individual eats fast food four times a week change the probability that he has high cholesterol? If yes, the events are not independent.

In certain experiments, the probability of one event is not affected by some other event. In this case, we say the two events are *independent*.

*Definition*   Two events $E$ and $F$ are **independent** if the occurrence of event $E$ in a probability experiment does not affect the probability of event $F$. Two events are **dependent** if the occurrence of event $E$ in a probability experiment affects the probability of event $F$.

To contrast independent and dependent events, consider an experiment in which an urn contains three red chips and five blue chips. Suppose two chips are drawn without replacement and the first chip drawn is red. The probability that the second chip is blue is 5/7 because there are now seven chips left in the urn and five of them are blue. However, if the two chips are drawn with replacement (we put the red chip back after selecting it), the probability that the second chip drawn is blue is 5/8, the same as the probability of selecting a blue chip on the first trial. In sampling without replacement, the events "the first chip is red" and "the second chip is blue" are dependent because selecting a red chip on the first draw affects the probability that the second draw will result in a blue chip. In sampling with replacement, the events "the first chip is red" and "the second chip is blue" are independent because the occurrence of the first chip being red does not affect the probability of the second chip being blue.

In Example 1, the events "Elena is chairperson" and "Dave is secretary" are dependent because the probability of Dave being selected as secretary *before* Elena is chosen as chairperson is 1/5. However, once Elena is selected as chairperson, the probability Dave is secretary is $\frac{1}{4}$.

We now formalize the definition of independence.

*Definition*

**Independent Events**

Two events $E$ and $F$ are independent if and only if $P(F \mid E) = P(F)$ or $P(E \mid F) = P(E)$.

This means that if events $E$ and $F$ are independent, then knowledge of the event $E$ has no effect on the probability of the event $F$ occurring and vice-versa.

▶ EXAMPLE 4    **Distinguishing Independent and Dependent Events**

**Caution**

Independence is different from mutually exclusive. For example, the event "not a cloud in the sky" and the event "raining" are mutually exclusive. However, the two events are not independent. $P(\text{rain}) \neq P(\text{rain} \mid \text{not a cloud in the sky})$.

According to the National Center for Health Statistics, the probability that a randomly selected mother will have a low-birth-weight baby is 0.078. The probability that a randomly selected mother will have a low-birth-weight baby given that the mother smokes is 0.121. Since $P(\text{low birth weight}) \neq P(\text{low birth weight} \mid \text{mother smokes})$, the events "low birth weight" and "mother smokes" are not independent. In fact, if a mother smokes there is an increased probability of a low-birth-weight baby.  ◀◀

 *Now Work Problem 3.*

 If events $E$ and $F$ are independent, then we can modify the Multiplication Rule.

**Multiplication Rule for Independent Events**

If $E$ and $F$ are independent events, the probability that $E$ and $F$ both occur is

$$P(E \text{ and } F) = P(E) \cdot P(F)$$

In words, the probability of $E$ and $F$ is the probability of event $E$ times the probability of event $F$.

▶ EXAMPLE 5    **Computing Probabilities of Independent Events**

*Problem:* In the game of roulette, there is a wheel with slots numbered 0, 00, and 1 through 36. A metal ball is allowed to roll around a wheel until it falls into one of the numbered slots. You decide to play the game and place a bet on the number 17. What is the probability the ball will land in the slot numbered 17 two times in a row?

*Approach:* There are 38 simple events in the sample space corresponding to the 38 possible outcomes of the experiment. We use the classical method of computing probabilities because the simple events are equally likely. In addition, we use the Multiplication Rule for Independent Events. The events "17 on first trial" and "17 on second trial" are independent because the ball does not remember it landed on 17 on the first trial, so this cannot affect the probability of landing on 17 on the second trial.

*Solution:* Because there are 38 possible outcomes to the experiment, the probability of the ball landing on 17 is 1/38. Because the events "17 on first trial" and "17 on second trial" are independent, we have

$P$(ball lands in slot 17 in the first game and ball lands in slot 17 in the second game)

$= P$(ball lands in slot 17 in the first game) $\cdot P$(ball lands in slot 17 in the second game)

$= \dfrac{1}{38} \cdot \dfrac{1}{38} = \dfrac{1}{1{,}444} \approx 0.0006925$

It is very unlikely that the ball will land on 17 twice in a row. We'd expect the ball to land on 17 twice in a row about 7 times in 10,000 trials.    ◄◄

We can extend the Multiplication Rule for three or more independent events.

---

**Multiplication Rule for *n* Independent Events**

If events $E, F, G, \ldots$ are independent, then

$$P(E \text{ and } F \text{ and } G \text{ and } \cdots) = P(E) \cdot P(F) \cdot P(G) \cdots$$

---

► **EXAMPLE 6    Life Expectancy**

*Problem:* The probability that a randomly selected male 24 years old will survive the year is 99.85% according to the National Vital Statistics Report, Vol. 47, No. 28. What is the probability that three randomly selected 24-year-old males will survive the year?

*Approach:* We can safely assume that the outcomes of the probability experiment are independent because there is no indication that the survival of one of the males affects the survival of the other two. For example, if two of the males lived in the same house, then a house fire could kill both males and we lose independence. (Knowledge that one of the males died in a house fire certainly would affect the probability that the other died.) By randomly selecting the males, we minimize the chances that they are related in any way.

*Solution:*

$P$(All three males survive) $= P$(first survives and second survives and third survives)

$= P$(first survives) $\cdot P$(second survives) $\cdot P$(third survives)

$= (0.9985)(0.9985)(0.9985)$

$= 0.9955$

There is a 99.55% probability that all three males survive the year.    ◄◄

  *Now Work Problem 19(a) and 19(b).*

Whenever a small random sample is taken from a large population, it is reasonable to compute probabilities of events assuming independence. Consider the following example.

▶ EXAMPLE 7    Sickle Cell Anemia

*Problem:* In a survey of 10,000 African-Americans, it was determined that 27 had sickle cell anemia.

**(a)** Suppose we randomly select one of the 10,000 African-Americans surveyed. What is the probability that he or she will have sickle cell anemia?
**(b)** If two individuals from this group are randomly selected, what is the probability that both have sickle cell anemia?
**(c)** Compute the probability of randomly selecting two individuals from this group who have sickle cell anemia, assuming independence.

*Approach:* We will let the event $E$ = "sickle cell anemia," so that $P(E)$ = number of African-Americans who have sickle cell anemia divided by the number in the survey. To answer part (b), we let $E_1$ = "first person has sickle cell anemia" and $E_2$ = "second person has sickle cell anemia," and then we compute $P(E_1 \text{ and } E_2) = P(E_1) \cdot P(E_2 \mid E_1)$. To answer part (c), we use the Multiplication Rule for Independent Events.

*Solution:*

**(a)** If one individual is selected, $P(E) = 27/10,000 = 0.0027 = 0.27\%$.
**(b)** Using the Multiplication Rule, we have

$$P(E_1 \text{ and } E_2) = P(E_1) \cdot P(E_2 \mid E_1) = \frac{27}{10000} \cdot \frac{26}{9999} \approx 0.00000702$$

Notice $P(E_2 \mid E_1) = \frac{26}{9999}$ because we are sampling without replacement, so after event $E_1$ occurs there is one less person with sickle cell anemia and one less person in the sample space.

**(c)** The assumption of independence means that the outcome of the first trial of the experiment does not impact the probability of the second trial. (It's like sampling with replacement.) Therefore, we assume $P(E_1) = P(E_2) = 27/10,000$. Then

$$P(E_1 \text{ and } E_2) = P(E_1) \cdot P(E_2) = \frac{27}{10,000} \cdot \frac{27}{10,000} \approx 0.00000729 \quad ◀◀$$

The probabilities in Examples 7(b) and 7(c) are extremely close in value. Based upon these results, we infer the following principle:

> If small random samples are taken from large populations without replacement, it is reasonable to assume independence of the events. As a general rule of thumb, if the sample size is less than 5% of the population size, then we treat the events as independent.

For example, in Example 7, we can compute the probability of randomly selecting two African-Americans that have sickle cell anemia using independence because the sample size is less than 5% of the population size $\left(\frac{2}{10,000} = 0.0002 = 0.02\%\right)$.

 *Now Work Problem 29.*

We now present an example in which we compute "at least" probabilities. These probabilities utilize the Complement Rule. The phrase "at least" means greater than or equal to. For example, a person must be at least 17 years old to see an "R" rated movie.

▶ EXAMPLE 8    **Computing "At Least" Probabilities**

*Problem:* Compute the probability that at least 1 male out of 1000 aged 24 years will die during the course of the year if the probability that a randomly selected 24-year-old male survives the year is 0.9985.

*Approach:* The phrase "at least" means greater than or equal, so we wish to know the probability that 1 or 2 or 3 or ... or 1000 males will die during the year. These events are mutually exclusive, so

$$P(1 \text{ or } 2 \text{ or } 3 \text{ or} \ldots \text{or } 1000 \text{ die}) = P(1 \text{ dies}) + P(2 \text{ die}) + P(3 \text{ die}) + \ldots + P(1000 \text{ die}).$$

Clearly, computing these probabilities would be very time consuming. However, we should notice that the complement of "at least one dying" is "none die." We shall use the Complement Rule to compute the probability.

*Solution:*

$$
\begin{aligned}
P(\text{at least one dies}) &= 1 - P(\text{none die}) \\
&= 1 - P(\text{first survives and second survives and} \ldots \text{and thousandth survives}) \\
&= 1 - P(\text{first survives}) \cdot P(\text{second survives}) \cdot \ldots \cdot P(\text{thousandth survives}) \quad \text{independent events} \\
&= 1 - (0.9985)^{1000} \\
&= 1 - 0.2229 \\
&= 0.7771 \\
&= 77.71\%
\end{aligned}
$$

There is a 77.71% probability that at least one 24-year-old male out of 1000 will die during the course of the year.    ◀◀

 *Now Work Problem 19(c).*

You may have noticed that "or" probabilities utilize the Addition Rule, while "and" probabilities utilize the Multiplication Rule. Accordingly, "or" probabilities imply addition, while "and" probabilities imply multiplication.

## 5.3 Assess Your Understanding

### Concepts and Vocabulary

**1.** The word "and" in probability implies that we use the ____ Rule. The word "or" in probability implies that we use the ____ Rule.

**2.** Describe the concept of independence.

**3.** State two methods for determining whether two events are independent.

**4.** Describe the difference between mutually exclusive events and independent events.

**5.** Comment on the difference between sampling with replacement and sampling without replacement.

**6.** Describe the circumstances under which sampling without replacement can be assumed to be independent.

### Exercises

• **Basic Skills**

**1.** Determine whether the events $E$ and $F$ are independent or dependent. Justify your answer.

(a) *E:* It rains on June 30
    *F:* It is cloudy on June 30
(b) *E:* Your car has a flat tire
    *F:* The price of gasoline increases overnight

(c) *E:* You live at least 80 years
    *F:* You smoke a pack of cigarettes every day of your life

2. Determine whether the events *A* and *B* are independent or dependent. Justify your answer.

   (a) *A:* You earn an "A" on an exam
       *B:* You study for an exam
   (b) *A:* You are late for work
       *B:* Your car has a flat tire
   (c) *A:* You earn more than $50,000 per year
       *B:* You are born in the month of July

3. According to the United States Census Bureau, the probability a randomly selected individual in the United States earns more than $75,000 per year is 18.4%. The probability a randomly selected individual in the United States earns more than $75,000 per year given that the individual has earned a bachelor's degree is 35.0%. Are the events "earn more than $75,000 per year" and "earned a bachelor's degree" independent?

4. Are events *E* and *F* independent if $P(F) = 0.95$; $P(F | E) = 0.95$?

5. Consider an experiment in which a single card is drawn from a deck. Let *E* be the event "draw a face card." Let *F* be the event "draw a club."

   (a) What is $P(F)$?
   (b) What is $P(F | E)$?
   (c) Are events *E* and *F* independent? Why?

6. Consider an experiment in which a single card is drawn from a deck.

   (a) What is the probability of the event *E* = "draw an ace"?
   (b) Let *F* be the event "draw a club." What is $P(E | F)$?
   (c) Are events *E* and *F* independent? Why?

---

• **Applying the Concepts**

7. **Acceptance Sample** Suppose you just received a shipment of six televisions. Two of the televisions are defective. If two televisions are randomly selected, use a tree diagram to compute the probability both televisions work. Then compute the probability using the Multiplication Rule.

8. **Committee** A committee consists of four women and three men. The committee will randomly select two people to attend a conference in Hawaii. Use a tree diagram to find the probability that both are women. Then compute the probability using the Multiplication Rule.

9. **Drawing Cards** Suppose two cards are randomly selected from a standard 52-card deck.

   (a) What is the probability that the first card is a king and the second card is a king if the sampling is done without replacement?
   (b) What is the probability that the first card is a king and the second card is a king if the sampling is done with replacement?

10. **Drawing Cards** Suppose two cards are randomly selected from a standard 52-card deck.

    (a) What is the probability that the first card is a club and the second card is a club if the sampling is done without replacement?
    (b) What is the probability that the first card is a club and the second card is a club if the sampling is done with replacement?

11. **Board Work** This past semester, I had a small business calculus section. The students in the class were Mike, Neta, Jinita, Kristin and Dave. Suppose I randomly select two people to go to the board to do homework. What is the probability that Dave is the first person chosen to go to the board and Neta is the second?

12. **Party** My wife has organized a monthly neighborhood party. There are five people involved in the group: Yolanda (my wife), Lorrie, Laura, Kim, and Anne Marie. They decide to randomly select the first and second home that will host the party. What is the probability that my wife hosts the first party and Lorrie hosts the second? NOTE: Once a home has hosted, it cannot host again until all other homes have hosted.

13. **Playing a CD on the "Random" Setting** Suppose a compact disk (CD) you just purchased has 13 tracks. After listening to the CD, you decide that you like 5 of the songs. With the random feature on your CD player, each of the 13 songs is played once in random order. Find the probability that among the first two songs played,

    (a) You like both of them. Would this be unusual?
    (b) You like neither of them.
    (c) You like exactly one of them.
    (d) Redo (a)–(c) if a song can be replayed before all 13 songs are played (if, for example, track 2 can play twice in a row).

14. **Packaging Error** Due to a manufacturing error, three cans of regular soda were accidentally filled with diet soda and placed into a 12-pack. Suppose that two cans are randomly selected from the case.

    (a) Determine the probability that both contain diet soda.
    (b) Determine the probability that both contain regular soda. Would this be unusual?
    (c) Determine the probability that exactly one is diet and one is regular.

15. **Planting Tulips** A bag of 30 tulip bulbs was purchased from a nursery. The bag contains 12 red tulip bulbs, 10 yellow tulip bulbs, and 8 purple tulip bulbs.

    (a) What is the probability that two randomly selected tulip bulbs will both be red?
    (b) What is the probability that the first bulb selected is red and the second yellow?
    (c) What is the probability that the first bulb selected is yellow and the second is red?
    (d) What is the probability that one bulb is red and the other yellow?

16. **Golf Balls** The local golf store sells an "onion bag" that contains 35 "experienced" golf balls. Suppose the bag contains 20 Titleists, 8 Maxflis, and 7 Top-Flites.

    (a) What is the probability that two randomly selected golf balls will both be Titleists?
    (b) What is the probability that the first ball selected will be a Titleist and the second will be a Maxfli?

(c) What is the probability that the first ball selected will be a Maxfli and the second will be a Titleist?

(d) What is the probability that one golf ball is a Titleist and the other is a Maxfli?

17. **Smokers** According to the National Center for Health Statistics, there is a 23.4% probability that a randomly selected resident of the United States aged 25 years or older is a smoker. In addition, there is a 21.7% probability that a randomly selected resident of the United States aged 25 years or older is female, given that he or she smokes. What is the probability that a randomly selected resident of the United States aged 25 years or older is female and smokes? Would it be unusual to randomly select a resident of the United States aged 25 years or older who is female and smokes?

18. **Multiple Jobs** According to the United States Bureau of Labor Statistics, there is a 5.84% probability that a randomly selected employed individual has more than one job (a multiple job holder). In addition, there is a 52.6% probability that a randomly selected employed individual is male given that he has more than one job. What is the probability that a randomly selected employed individual is a multiple job holder and male? Would it be unusual to randomly select an employed individual who is a multiple job holder and male?

19. **Life Expectancy** The probability that a randomly selected 40-year-old male will live to be 41 years old is 0.99718 according to the National Vital Statistics Report, Vol. 48, No. 18.

(a) What is the probability that two randomly selected 40-year-old males will live to be 41 years old?

(b) What is the probability that five randomly selected 40-year-old males will live to be 41 years old?

(c) What is the probability that at least one of five randomly selected 40-year-old males will not live to be 41 years old? Would it be unusual that at least one of five randomly selected 40-year-old males will not live to be 41 years old?

20. **Life Expectancy** The probability that a randomly selected 40-year-old female will live to be 41 years old is 0.99856 according to the National Vital Statistics Report, Vol. 48, No. 18.

(a) What is the probability that two randomly selected 40-year-old females will live to be 41 years old?

(b) What is the probability that five randomly selected 40-year-old females will live to be 41 years old?

(c) What is the probability that at least one of five randomly selected 40-year-old females will not live to be 41 years old? Would it be unusual that at least one of five randomly selected 40-year-old females will not live to be 41 years old?

21. **Blood Types** Blood types can be classified as either $Rh^+$ or $Rh^-$. According to the Information Please Almanac, 99% of the Chinese population has $Rh^+$ blood.

(a) What is the probability that two randomly selected Chinese have $Rh^+$ blood?

(b) What is the probability that six randomly selected Chinese have $Rh^+$ blood?

(c) What is the probability that at least one of six randomly selected Chinese has $Rh^-$ blood? Would it be unusual that at least one of six randomly selected Chinese has $Rh^-$ blood?

22. **Quality Control** Suppose a company selects two people who work independently inspecting two-by-four timbers. Their job is to identify low-quality timbers. Suppose the probability that an inspector does not identify a low-quality timber is 20%.

(a) What is the probability that two inspectors do not identify a low-quality timber?

(b) How many inspectors should be hired in order to keep the probability of not identifying a low-quality timber below 1%?

23. **The Birthday Problem** Determine the probability that at least two people in a room of ten people share the same birthday, ignoring leap years, by answering the following questions:

(a) Compute the probability that 10 people have different birthdays. (*Hint*: The first person's birthday can occur 365 ways, the second person's birthday can occur 364 ways because he or she can't have the same birthday as the first person, the third person's birthday can occur 363 ways because he or she can't have the same birthday as the first or second person, and so on.)

(b) The complement of "10 people have different birthdays" is "At least 2 share a birthday." Use this information to compute the probability that at least 2 people out of 10 share the same birthday.

24. **The Birthday Problem** Using the procedure given in Problem 23, compute the probability that at least 2 people in a room of 23 people share the same birthday.

25. **A Flush** A flush in the card game of poker occurs if a player gets five cards that are all the same suit (clubs, diamonds, hearts, or spades). Answer the following questions to obtain the probability of being dealt a flush in five cards.

(a) We initially concentrate on one suit, say clubs. There are 13 clubs in a deck. Compute $P(\text{five clubs}) = P(\text{first card is clubs and second card is clubs and third card is clubs and fourth card is clubs and fifth card is clubs})$.

(b) A flush can occur if we get five clubs or five diamonds or five hearts or five spades. Compute $P(\text{five clubs or five diamonds or five hearts or five spades})$. Note the events are mutually exclusive.

26. **A Royal Flush** A royal flush in the game of poker occurs if the player gets the cards Ten, Jack, Queen, King, and Ace all in the same suit. Use the results of Problem 25 to compute the probability of being dealt a royal flush.

27. **Cold Streaks** Players in sports are said to have "hot streaks" and "cold streaks." For example, a batter in baseball might be considered to be in a "slump" or "cold streak" if he has made 10 outs in 10 consecutive at-bats. Suppose a hitter successfully reaches base 30% of the time he comes to the plate.

(a) Find the probability that the hitter makes 10 outs in 10 consecutive at-bats, assuming that at-bats are independent events. (*Hint*: The hitter makes an out 70% of the time.)

(b) Are "cold streaks" unusual?

28. **Hot Streaks** Wilt Chamberlain holds the National Basketball Association record for most consecutive field

goals in a game with 18. Over his career, Wilt made 54% of his field goal attempts.

(a) Find the probability of Wilt Chamberlain making 18 consecutive field goals, assuming that field goal attempts are independent events.

(b) Is this record likely to be broken?

29. **Independence in Small Samples from Large Populations** Suppose a computer chip company has just shipped 10,000 computer chips to a computer company. Unfortunately, 50 of the chips are defective.

(a) Compute the probability that two randomly selected chips are defective using conditional probability.

(b) There are 50 defective chips out of 10,000 shipped. The probability that the first chip randomly selected is defective is 50/10,000 = 0.005 = 0.5%. Compute the probability that two randomly selected chips are defective under the assumption of independent events. Compare your results to part (a). Conclude that when small samples are taken from large populations without replacement, the assumption of independence does not significantly affect the probability.

30. **Independence in Small Samples from Large Populations** Suppose a poll is being conducted in the Village of Lemont. The pollster identifies her target population as all residents of Lemont 18 years old or older. The list of this population has 6,494 people on it.

(a) Compute the probability that the first resident selected to participate in the poll is Roger Cummings and the second is Rick Whittingham. [These people are my neighbors in Lemont.]

(b) The probability that any particular resident of Lemont is the first person picked is 1/6494. Compute the probability that the two people selected are Roger and Rick assuming independence. Compare your results to part (a). Conclude that when small samples are taken from large populations without replacement, the assumption of independence does not significantly affect the probability.

31. **Playing Five-Card Stud** In the game of five-card stud, one card is dealt face down to each player and the remaining four cards are dealt face up. After two cards are dealt (one down and one up), the players bet. Players continue to bet after each additional card is dealt. Suppose three cards have been dealt to each of the five players at the table. You currently have three clubs in your hand, so you will attempt to get a flush (all cards in the same suit). Of the cards dealt, there are two clubs showing in other player's hands.

(a) How many clubs are in a standard 52-card deck?

(b) How many cards remain in the deck or are not known by you? Of this amount, how many are clubs?

(c) What is the probability you get dealt a club on the next card?

(d) What is the probability you get dealt two clubs in a row?

(e) Should you stay in the game?

32. Refer back to the tree diagram in Figure 12.

(a) What is the probability of Dave's being chairperson?

(b) What is the probability of Dave's being secretary, given that he has been selected as chairperson?

## 5.4    Conditional Probability

**Objectives**  Compute conditional probabilities

 Use the Multiplication Rule to check for independent events

Oftentimes, we have knowledge that affects the probability of an event. For example, in Example 4 from Section 5.3, we saw that the probability that a randomly selected mother gives birth to a low-birth-weight baby is 0.078. However, if we know that the mother is a smoker, the probability increases to 0.121.

 We use conditional probabilities in the Multiplication Rule as follows:

$$P(E \text{ and } F) = P(E) \cdot P(F|E)$$

Solving this expression for $P(F|E)$, we obtain the *conditional probability rule*.

**Conditional Probability Rule**

If $E$ and $F$ are any two events, then

$$P(F|E) = \frac{P(E \text{ and } F)}{P(E)} = \frac{N(E \text{ and } F)}{N(E)} \qquad (1)$$

The probability of event $F$ occurring given the occurrence of event $E$ is found by dividing the probability of $E$ and $F$ by the probability of $E$. Or, the probability of event $F$ occurring given the occurrence of event $E$ is found by dividing the number of simple events in $E$ and $F$ by the number of simple events in $E$.

▶ **EXAMPLE 1**   **Computing a Conditional Probability**

*Problem:* The data in Table 5 represent the marital status and gender of the residents of the United States aged 18 years old or older in 1998.

|  | Males (in millions) | Females (in millions) | Totals (in millions) |
|---|---|---|---|
| **TABLE 5** | | | |
| Never Married | 25.5 | 21.0 | 46.5 |
| Married | 58.6 | 59.3 | 117.9 |
| Widowed | 2.6 | 11.0 | 13.6 |
| Divorced | 8.3 | 11.1 | 19.4 |
| Totals (in millions) | 95.0 | 102.4 | 197.4 |

*Source:* U.S. Census Bureau, Current Population Reports

**(a)** Compute the probability that a randomly selected male has never married.

**(b)** Compute the probability that a randomly selected individual who has never married is male.

*Approach:*

**(a)** We are given that the randomly selected person is male, so we concentrate on the "male" column. There are 95 million males and 25.5 million people who are male and never married, so $N(\text{male}) = 95$ million and $N(\text{male and never married}) = 25.5$ million. Compute the probability via the Conditional Probability Rule.

**(b)** We are given that the randomly selected person has never married, so we concentrate on the "never married" row. There are 46.5 million people who have never married and 25.5 million people who are male and have never married, so $N(\text{never married}) = 46.5$ million and $N(\text{male and never married}) = 25.5$ million. Compute the probability via the Conditional Probability Rule.

*Solution:*

**(a)** Substituting into Formula (1), we obtain

$$P(\text{never married} \mid \text{male}) = \frac{N(\text{never married and male})}{N(\text{male})} = \frac{25.5}{95} \approx 0.268$$

There is a 26.8% probability the randomly selected individual has never married, given that he is male.

**(b)** Substituting into Formula (1), we obtain

$$P(\text{male} \mid \text{never married}) = \frac{N(\text{male and never married})}{N(\text{never married})} = \frac{25.5}{46.5} \approx 0.548$$

There is a 54.8% probability the randomly selected individual is male, given that he or she has never married. ◀◀

What is the difference between the results of Example 1(a) and (b)? In Example 1(a), we found that 26.8% of males have never married, while in Example 1(b), we found that 54.8% of individuals who have never married are male. Do you see the difference?

**NW** *Now Work Problem 7.*

▶ **EXAMPLE 2** **Birth Weights of Preterm Babies**

*Problem:* In 1998, 11.47% of all births were preterm. (The gestation period of the pregnancy was less than 37 weeks.) Also in 1998, for 0.23% of all births, the baby was preterm and weighed 8 pounds, 13 ounces or more. What is the probability that a randomly selected baby weighs 8 pounds, 13 ounces or more, given that the baby was preterm?

*Approach:* We want to know the probability that the baby weighs 8 pounds, 13 ounces or more given the baby was preterm. We know that 0.23% of all babies weighed 8 pounds, 13 ounces or more and were preterm, so $P(\text{Weighs 8 pounds, 13 ounces or more and preterm}) = 0.23\%$. We also know that 11.47% of all births were preterm, so $P(\text{preterm}) = 11.47\%$. We compute the probability by dividing the probability that a baby will weigh 8 pounds, 13 ounces or more *and* be preterm by the probability that a baby will be preterm.

*Solution:* $P(\text{Weighs 8 pounds, 13 ounces or more} \mid \text{preterm})$

$$= \frac{P(\text{Weighs 8 pounds, 13 ounces or more and preterm})}{P(\text{preterm})}$$

$$= \frac{0.23\%}{11.47\%} = \frac{0.0023}{0.1147} \approx 0.0201 = 2.01\%$$

$$\approx 2.01\%$$

There is a 2.01% probability that a randomly selected baby will weigh 8 pounds, 13 ounces or more, given the baby is preterm. It is unusual for preterm babies to weigh 8 pounds, 13 ounces or more. ◀◀

**NW** *Now Work Problem 11.*

**②** **Checking for Independence**

In Section 5.3, we said two events $E$ and $F$ are independent if $P(E|F) = P(E)$ or $P(F|E) = P(F)$. An alternative method for checking for independence is through the Multiplication Rule.

> **Checking for Independence**
>
> If $P(E \text{ and } F) = P(E) \cdot P(F)$, then events $E$ and $F$ are independent.

▶ **EXAMPLE 3** **Checking for Independence**

*Problem:* Table 5 on page 220 represents the gender and marital status of the residents of the United States aged 18 years old and older. Determine whether the events "male" and "never married" are independent.

**Caution**

The method of checking for independence presented here works only for population data. To determine whether sample data are independent requires procedures introduced in Section 11.2.

*Approach:* In order for the events to be independent, $P(\text{male}|\text{never married}) = P(\text{male})$ or $P(\text{never married}|\text{male}) = P(\text{never married})$. Alternatively, we could check for independence by determining whether $P(\text{male and never married}) = P(\text{male}) \cdot P(\text{never married})$.

*Solution:* From Example 1(b), we found that $P(\text{male}|\text{never married}) = 0.548$. In Example 3(a) of Section 5.2, we computed $P(\text{male}) = 95/197.4 \approx 0.481$. Because $P(\text{male}|\text{never married}) \neq P(\text{male})$, the events are dependent. In fact, knowing that individuals have never married increases the probability that they are male. Another method for checking for independence would be to determine whether $P(\text{male and never married}) = P(\text{male}) \cdot P(\text{never married})$. From Table 5, we find that $P(\text{male and never married}) = 25.5/197.4 \approx 0.129$. We also find that

$$P(\text{male}) \cdot P(\text{never married}) = \frac{95}{197.4} \cdot \frac{46.5}{197.4} \approx 0.113.$$ Because $P(\text{male}$ and never married$) \neq P(\text{male}) \cdot P(\text{never married})$, the events "male" and "never married" are dependent. ◄◄

**NW** *Now Work Problem 15.*

**Mutually Exclusive Events versus Independent Events** It is important that we understand that mutually exclusive events and independent events are different concepts. Recall that two events are mutually exclusive if they have no simple events in common. That is, two events are mutually exclusive if the occurrence of one of the events precludes the other event from occurring. However, two events are independent if knowledge of one event does not affect the probability of the other. Consider a simple experiment in rolling a single die. Let $E =$ "roll an even number" and let $F =$ "roll an odd number." Clearly, $E$ and $F$ are mutually exclusive, but they are not independent, because $P(E) = 1/2$ and $P(E|F) = 0$.

**Caution**

Two events that are mutually exclusive are not independent.

## 5.4 Assess Your Understanding

### Concepts and Vocabulary

1. State two methods for discovering whether two events are independent.

2. Describe the difference between mutually exclusive events and independent events.

### Exercises

- **Basic Skills**

1. Suppose that $E$ and $F$ are two events and that $P(E \text{ and } F) = 0.6$ and $P(E) = 0.8$. What is $P(F|E)$?

2. Suppose that $A$ and $B$ are two events and that $P(A \text{ and } B) = 0.21$ and $P(A) = 0.4$. What is $P(B|A)$?

3. Suppose that $E$ and $F$ are two events and that $N(E \text{ and } F) = 420$ and $N(E) = 740$. What is $P(F|E)$?

4. Suppose that $A$ and $B$ are two events and that $N(A \text{ and } B) = 380$ and $N(A) = 925$. What is $P(B|A)$?

5. Suppose that $E$ and $F$ are two events and that $P(E \text{ and } F) = 0.24$, $P(E) = 0.4$, and $P(F) = 0.6$. Are $E$ and $F$ independent? Why?

6. Suppose that $A$ and $B$ are two events and that $P(A \text{ and } B) = 0.3$, $P(A) = 0.5$, and $P(B) = 0.8$. Are $A$ and $B$ independent? Why?

- **Applying the Concepts**

7. **Medicaid** The data to the right, in thousands, represent the age of persons receiving Medicaid and their poverty level.

   (a) What is the probability that a randomly selected individual who is less than 18 years old is below the poverty level?

   (b) What is the probability that a randomly selected individual who is below the poverty level is less than 18 years old?

| | AGE | | | |
|---|---|---|---|---|
| | <18 | 18–44 | 45–64 | >65 |
| **Below Poverty Level** | 8,550 | 4,356 | 1,520 | 958 |
| **Above Poverty Level** | 5,884 | 3,741 | 1,756 | 1,943 |

*Source:* U.S. Census Bureau

**8. SAT Verbal Scores** The following data represent the scores received on the 2000 SAT I: Reasoning Test—Verbal, by gender:

| | Male | Female | | Male | Female |
|---|---|---|---|---|---|
| 200–249 | 6,755 | 7,512 | **550–599** | 84,524 | 96,381 |
| 250–299 | 9,594 | 10,816 | **600–649** | 62,205 | 68,515 |
| 300–349 | 25,630 | 30,058 | **650–699** | 36,202 | 39,843 |
| 350–399 | 50,042 | 60,139 | **700–749** | 18,044 | 18,716 |
| 400–449 | 78,276 | 95,780 | **750–800** | 10,647 | 11,028 |
| 450–499 | 98,989 | 117,976 | **TOTAL** | 583,331 | 676,947 |
| 500–549 | 102,423 | 120,183 | | | |

*Source:* College Board

(a) What is the probability that a randomly selected female scored 750–800 on the 2000 SAT I: Reasoning Test—Verbal?

(b) What is the probability that a randomly selected individual who scored 750–800 on the 2000 SAT I: Reasoning Test—Verbal is female?

**9. Driver Fatalities** The following data represent the number of driver fatalities in 1996, by age group, for male and female drivers:

| | Male | Female | | Male | Female |
|---|---|---|---|---|---|
| **16–19** | 1922 | 762 | **60–64** | 637 | 270 |
| **20–24** | 2814 | 699 | **65–69** | 614 | 268 |
| **25–29** | 2126 | 617 | **70–74** | 598 | 358 |
| **30–34** | 1986 | 627 | **75–79** | 554 | 323 |
| **35–39** | 1743 | 604 | **80–84** | 474 | 230 |
| **40–44** | 1437 | 485 | **85 or older** | 329 | 146 |
| **45–49** | 1169 | 391 | **TOTAL** | 17,946 | 6,387 |
| **50–54** | 859 | 347 | | | |
| **55–59** | 684 | 260 | | | |

*Source:* National Highway Traffic Institute

(a) What is the probability that a randomly selected driver fatality who was male was 20–24 years old?

(b) What is the probability that a randomly selected driver fatality who was 20–24 years old was male?

(c) Suppose you are a police officer called to the scene of a traffic accident with a fatality. The dispatcher states the victim is 20–24 years old, but the gender is not known. Is the victim more likely to be male or female? Why?

**10. Marital Status** The following data represent the marital status of Americans 25 years old or older and their level of education in 1999:

| | Did Not Graduate High School | High School Graduate | Some College | College Graduate |
|---|---|---|---|---|
| **Never Married** | 4,012 | 7,790 | 6,565 | 7,685 |
| **Married, Spouse Present** | 15,122 | 36,076 | 27,003 | 29,811 |
| **Married, Spouse Absent** | 1,877 | 2,006 | 1,435 | 1,489 |
| **Separated** | 1,094 | 1,515 | 1,214 | 598 |
| **Widowed** | 5,000 | 4,757 | 2,264 | 1,456 |
| **Divorced** | 2,951 | 7,047 | 5,649 | 3,766 |

*Source:* U.S. Census Bureau

(a) What is the probability that a randomly selected individual who has never married is a high school graduate?

(b) What is the probability that a randomly selected individual who is a high school graduate has never married?

**11. Rainy Days** For the month of June in the city of Chicago, 37% of the days are cloudy. Also in the month of June in the city of Chicago, 21% of the days are cloudy and rainy. What is the probability that a randomly selected day in June will be rainy if it is cloudy?

**12. Cause of Death** According to the U.S. National Center for Health Statistics, in 1997, 0.2% of deaths in the United States were of 25–34-year-olds whose cause of death was cancer. In addition, 1.97% of all decedants were 25–34 years old. What is the probability that a randomly selected death is the result of cancer if the individual is known to have been 25–34 years old?

**13. High School Dropout** According to the U.S. Census Bureau, 9.1% of 16–17-year-olds drop out of high school. In addition, 5.8% of White 16–17-year-olds drop out of high school. What is the probability a randomly selected dropout is White, given that he or she is 16–17 years old?

**14. Income by Region** According to the U.S Census Bureau, 19.1% of the U.S. households are in the Northeast. In addition, 4.4% of American households earn $75,000 per year or more and are located in the Northeast. Determine the probability that a randomly selected American household earns more than $75,000 per year, given that the house is located in the Northeast.

15. **Medicaid** Refer to the table in Problem 7. Are the events " < 18 years old" and "below poverty level" independent?

16. **SAT Verbal Scores** Refer to the table in Problem 8. Are the events "female" and "scored 750–800" independent?

17. **Driver Fatalities** Refer to the table in Problem 9. Are the events "male" and "20–24 years old" independent?

18. **Marital Status** Refer to the table in Problem 10. Are the events "never married" and "high school graduate" independent?

19. **Household Income** In 1997, 18.4% of all households earned more than $75,000 per year. Also in 1997, 4.2% of all households earned more than $75,000 per year and lived in the Northeast. What is the probability that a randomly selected household lives in the Northeast, given that the household earns more than $75,000 per year?

## 5.5 Counting Techniques

**Objectives**  Solve counting problems using the Multiplication Principle
 Solve counting problems using permutations
 Solve counting problems using combinations
 Compute probabilities involving permutations and combinations

 Counting plays a major role in many diverse areas, including probability. In this section, we look at special types of counting problems and develop general formulas for solving them.

We begin with an example that will demonstrate a general counting principle.

▶ **EXAMPLE 1** Counting the Number of Possible Meals

*Problem:* The fixed-price dinner at Mabenka Restaurant provides the following choices:

Appetizer: soup or salad
Entrée: baked chicken, broiled beef patty, baby beef liver, or roast beef au jus
Dessert: ice cream or cheese cake

How many different meals can be ordered?

*Approach:* Ordering such a meal requires three separate decisions:

**Choose an Appetizer**    **Choose an Entrée**    **Choose a Dessert**
2 choices              4 choices             2 choices

We will draw a tree diagram that lists the possible meals that can be ordered.

*Solution:* Look at the tree diagram in Figure 13 on page 225. We see that, for each choice of appetizer, there are 4 choices of entrée—and that, for each of these $2 \cdot 4 = 8$ choices, there are 2 choices for dessert. A total of

$$2 \cdot 4 \cdot 2 = 16$$

different meals can be ordered. ◀◀

Example 1 illustrates a general counting principle.

Figure 13

Appetizer      Entree      Dessert

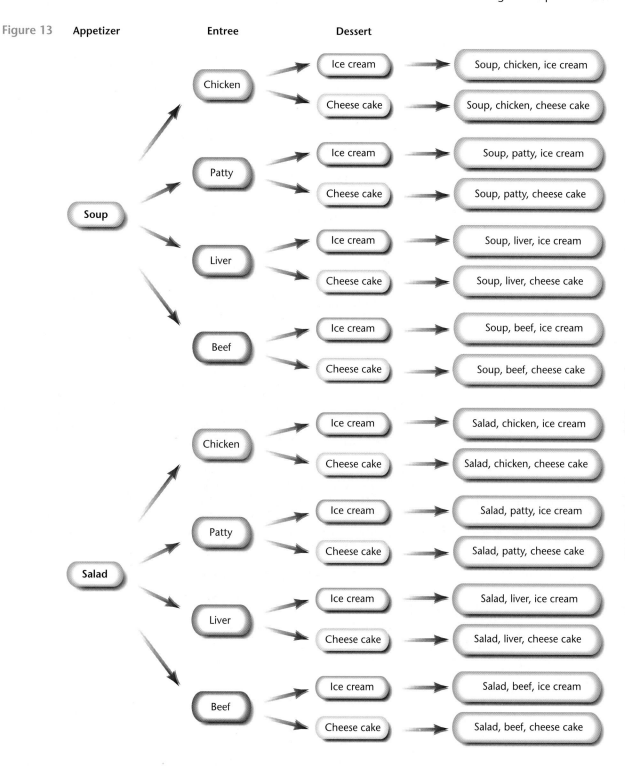

*Theorem*     **Multiplication Principle of Counting**

If a task consists of a sequence of choices in which there are $p$ selections for the first choice, $q$ selections for the second choice, $r$ selections for the third choice, and so on, then the task of making these selections can be done in

$$p \cdot q \cdot r \cdot \ldots$$

different ways.

▶ EXAMPLE 2    Counting Airport Codes

*Problem:* The International Airline Transportation Association (IATA) assigns three-letter codes to represent airport locations. For example, the airport code for Fort Lauderdale is FLL. How many different airport codes are possible?

*Approach:* We are choosing 3 letters from 26 letters and arranging them in order. We notice that repetition of letters is allowed. We use the Multiplication Principle of Counting, recognizing there are 26 ways to choose the first letter, 26 ways to choose the second letter, and 26 ways to choose the third letter.

*Solution:* By the Multiplication Principle, there are

$$26 \cdot 26 \cdot 26 = 17,576$$

different airport codes possible.                                              ◀◀

   In Example 2, we were allowed to repeat a letter. For example, a valid airport code is FLL (Ft. Lauderdale International Airport), in which the letter L appears twice. In the next example, repetition is not allowed.

▶ EXAMPLE 3    Counting without Repetition

*Problem:* Three members from a 14-member committee are to be randomly selected to serve as chair, vice chair, and secretary. The first person selected is the chair, the second person selected vice chair, and the third secretary. How many different committee structures are possible?

*Approach:* The task consists of making three selections. The first selection requires choosing from 14 members. Because a member cannot serve in more than one capacity, the second selection requires choosing from 13 members. The third selection requires choosing from 12 members. (Do you see why?) We use the Multiplication Principle to determine the number of possible committees.

*Solution:* By the Multiplication Principle, there are

$$14 \cdot 13 \cdot 12 = 2184$$

different committee structures possible.                                       ◀◀

**NW**  *Now Work Problem 27.*

## The Factorial Symbol

We now introduce a special symbol that can assist us in representing certain types of counting problems.

*Definition*    If $n \geq 0$ is an integer, the **factorial symbol $n!$** is defined as follows:

$$0! = 1 \qquad 1! = 1$$
$$n! = n(n-1) \cdot \ldots \cdot 3 \cdot 2 \cdot 1$$

   For example, $2! = 2 \cdot 1 = 2, 3! = 3 \cdot 2 \cdot 1 = 6, 4! = 4 \cdot 3 \cdot 2 \cdot 1 = 24$, and so on. Table 6 lists the values of $n!$ for $0 \leq n \leq 6$.

| TABLE 6 | | | | | | |
|---|---|---|---|---|---|---|
| $n$ | 0 | 1 | 2 | 3 | 4 | 5 | 6 |
| $n!$ | 1 | 1 | 2 | 6 | 24 | 120 | 720 |

Using Technology: Your calculator has a factorial key. Use it to see how fast factorials increase in value. Find the value of 69!. What happens when you try to find 70!? In fact, 70! is larger than $10^{100}$ (a *googol*), the largest number most calculators can display.

▶ EXAMPLE 4   The Traveling Salesman

*Problem:* You have just been hired as a book representative for Prentice Hall. On your first day, you must travel to seven schools to introduce yourself. How many different routes are possible?

*Approach:* The seven schools are different. Let's call the schools *A, B, C, D, E, F,* and *G*. School *A* can be visited first, second, third, fourth, fifth, sixth or seventh. So, we have seven choices for school *A*. We would then have six choices for school *B*, five choices for school *C* and so on. We can use the factorial to find our solution.

 *Now Work*
*Problems 1 and 29.*

*Solution:* There are $7 \cdot 6 \cdot 5 \cdot 4 \cdot 3 \cdot 2 \cdot 1 = 7! = 5040$ different routes possible.   ◀◀

**2**   **Permutations**

Examples 3 and 4 illustrate a type of counting problem referred to as a *permutation*.

*Definition*

A **permutation** is an ordered arrangement in which *r* objects are chosen from *n* distinct (different) objects and repetition is not allowed. The symbol $_nP_r$ represents the number of permutations of *r* objects selected from *n* objects.

So we could represent the solution to the question posed in Example 3 as

$$_nP_r = {}_{14}P_3 = 14 \cdot 13 \cdot 12 = 2184$$

and the solution to Example 4 could be represented as

$$_7P_7 = 7 \cdot 6 \cdot 5 \cdot 4 \cdot 3 \cdot 2 \cdot 1 = 5040$$

To arrive at a formula for $_nP_r$, we note that there are *n* choices for the first selection, $n - 1$ choices for the second selection, $n - 2$ choices for the third selection, ..., and $n - (r - 1)$ choices for the *r*th selection. By the Multiplication Principle, we have

$$
\begin{array}{ccccccccc}
 & & 1^{st} & & 2^{nd} & & 3^{rd} & & r^{th} \\
_nP_r & = & n & \cdot & (n-1) & \cdot & (n-2) & \cdot \ldots \cdot & [n-(r-1)] \\
 & = & n & \cdot & (n-1) & \cdot & (n-2) & \cdot \ldots \cdot & (n-r+1)
\end{array}
$$

This formula for $_nP_r$ can be written in **factorial notation**:

$$_nP_r = n \cdot (n-1) \cdot (n-2) \cdot \ldots \cdot (n-r+1)$$

$$= n \cdot (n-1) \cdot (n-2) \cdot \ldots \cdot (n-r+1) \cdot \frac{(n-r) \cdot \ldots \cdot 3 \cdot 2 \cdot 1}{(n-r) \cdot \ldots \cdot 3 \cdot 2 \cdot 1} = \frac{n!}{(n-r)!}$$

We have the following result.

*Theorem*

**Number of Permutations of $n$ Distinct Objects Taken $r$ at a Time**

The number of arrangements of $r$ objects chosen from $n$ objects, in which

1. the $n$ objects are distinct
2. once an object is used it cannot be repeated, and
3. order is important,

is given by the formula

$$_nP_r = \frac{n!}{(n-r)!} \qquad \textbf{(1)}$$

▶ **EXAMPLE 5** **Computing Permutations**

*Problem:* Evaluate: **(a)** $_7P_5$ **(b)** $_8P_2$ **(c)** $_5P_5$

*Approach:* To answer (a), we use Formula (1) with $n = 7$ and $r = 5$. To answer (b), we use Formula (1) with $n = 8$ and $r = 2$. To answer (c), we use Formula (1) with $n = 5$ and $r = 5$.

*Solution:*

**(a)** $_7P_5 = \dfrac{7!}{(7-5)!} = \dfrac{7!}{2!} = \dfrac{7 \cdot 6 \cdot 5 \cdot 4 \cdot 3 \cdot 2!}{2!} = \underbrace{7 \cdot 6 \cdot 5 \cdot 4 \cdot 3}_{5 \text{ factors}} = 2{,}520$

**(b)** $_8P_2 = \dfrac{8!}{(8-2)!} = \dfrac{8!}{6!} = \dfrac{8 \cdot 7 \cdot 6!}{6!} = \underbrace{8 \cdot 7}_{2 \text{ factors}} = 56$

**(c)** $_5P_5 = \dfrac{5!}{(5-5)!} = \dfrac{5!}{0!} = 5! = 5 \cdot 4 \cdot 3 \cdot 2 \cdot 1 = 120$ ◀◀

**Using Technology:** We can use a graphing calculator to compute permutations. Figure 14 shows the result of doing Example 5(a) on a TI-83 Plus calculator.

**NW** *Now Work Problem 7.*

**Figure 14**

```
7 nPr 5
                2520
```

Example 5(c) illustrates a general result.

*Theorem*

$$_rP_r = r!$$

We need to recognize that, in a permutation, order matters. That is, if we wanted to permute the letters $ABC$ by selecting them 3 at a time, the following arrangements are all different:

$$ABC, ACB, BAC, BCA, CAB, CBA$$

▶ **EXAMPLE 6** **Betting on the Trifecta**

*Problem:* In how many ways can horses in a 10-horse race finish first, second, and third?

*Approach:* The 10 horses are distinct. Once a horse crosses the finish line, that horse will not cross the finish line again, and, in a race, order is important. We have a permutation of 10 objects taken 3 at a time.

*Solution:* The top three horses can finish a 10-horse race in

$$_{10}P_3 = \frac{10!}{(10-3)!} = \frac{10!}{7!} = \frac{10 \cdot 9 \cdot 8 \cdot 7!}{7!} = \underbrace{10 \cdot 9 \cdot 8}_{3 \text{ factors}} = 720 \text{ ways} \blacktriangleleft\blacktriangleleft$$

**NW**  *Now Work Problem 41.*

### ③ Combinations

In a permutation, order is important; for example, the arrangements *ABC, ACB, BAC, BCA, CAB*, and *CBA* are considered different arrangements of the letters *A, B*, and *C*. In many situations, though, order is unimportant. If order is unimportant, the six arrangements of the letters *A, B*, and *C* given above are not different. That is, we do not distinguish *ABC* from *BAC*, for example. In the card game of poker, the order in which the cards are received does not matter; the *combination* of the cards is what matters.

**Definition**

> A **combination** is a collection, without regard to order, of *n* distinct objects without repetitions. The symbol $_nC_r$ represents the number of combinations of *n* distinct objects taken *r* at a time.

▶ **EXAMPLE 7**   **Listing Combinations**

*Problem:* Roger, Rick, Randy and Jay are going to play golf. They will randomly select teams of two players each. List all possible team combinations. That is, list all the combinations of the four people Roger, Rick, Randy, and Jay taken two at a time. What is $_4C_2$?

*Approach:* We list the possible teams. We note that order is unimportant, so {Roger, Rick} is the same as {Rick, Roger}.

*Solution:* The list of all such teams (combinations) is

Roger, Rick; Roger, Randy; Roger, Jay; Rick, Randy; Rick, Jay; Randy, Jay

Thus,

$$_4C_2 = 6$$

There are six ways of forming teams of two from a group of four players.
$\blacktriangleleft\blacktriangleleft$

We can find a formula for $_nC_r$ by noting that the only difference between a permutation and a combination is that we disregard order in combinations. To determine $_nC_r$, we eliminate from the formula for $_nP_r$ the number of permutations that were rearrangements of a given set of *r* objects. In Example 7, for example, selecting {Roger, Rick} was the same as selecting {Rick, Roger}, so there were 2! = 2 rearrangements of the two objects. This can be determined from the formula for $_nP_r$ by calculating $_rP_r = r!$. So, if we divide $_nP_r$ by *r*!, we will have the desired formula for $_nC_r$:

$$_nC_r = \frac{_nP_r}{r!} = \frac{n!}{r!(n-r)!}$$

We have proven the following result.

*Theorem* | **Number of Combinations of *n* Distinct Objects Taken *r* at a Time**

The number of different arrangements of *n* objects using $r \leq n$ of them, in which

**1.** the *n* objects are distinct

**2.** once an object is used, it cannot be repeated, and

**3.** order is not important

is given by the formula

$$_nC_r = \frac{n!}{r!(n-r)!} \tag{2}$$

Using Formula (2) to solve the problem presented in Example 7, we obtain

$$_4C_2 = \frac{4!}{2!(4-2)!} = \frac{4!}{2!2!} = \frac{4 \cdot 3 \cdot 2!}{2 \cdot 1 \cdot 2!} = \frac{12}{2} = 6$$

▶ **EXAMPLE 8** **Using Formula (2)**

*Problem:* Use Formula (2) to find the value of each expression.

**(a)** $_4C_1$

**(b)** $_6C_4$

**(c)** $_6C_2$

*Approach:* We will use Formula (2).

*Solution:*

**(a)** $_4C_1 = \frac{4!}{1!(4-1)!} = \frac{4!}{1! \cdot 3!} = \frac{4 \cdot 3!}{1 \cdot 3!} = 4$

**(b)** $_6C_4 = \frac{6!}{4!(6-4)!} = \frac{6!}{4! \cdot 2!} = \frac{6 \cdot 5 \cdot 4!}{4! \cdot 2 \cdot 1} = \frac{30}{2} = 15$

**(c)** $_6C_2 = \frac{6!}{2!(6-2)!} = \frac{6!}{2!4!} = \frac{6 \cdot 5 \cdot 4!}{2 \cdot 1 \cdot 4!} = \frac{30}{2} = 15$   ◀◀

**Using Technology:** We can use a graphing calculator to compute combinations. Figure 15 shows the result of doing Example 8(b) an a TI-83+ graphing calculator.

Figure 15

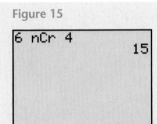

Notice in Example 8 that $_6C_4 = _6C_2$. This result can be generalized.

*Theorem* | $$_nC_r = _nC_{n-r}$$

**NW** *Now Work Problem 15.*

▶ **EXAMPLE 9** **Simple Random Samples**

*Problem:* How many different simple random samples of size 4 can be obtained from a population whose size is 20?

*Approach:* The 20 individuals in the population are distinct. In addition, the order in which an individual is selected to be in the sample is unimportant. Thus, the number of simple random samples of size 4 from a population of size 20 is a combination of 20 objects taken 4 at a time.

*Solution:* Use Formula (2) with $n = 20$ and $r = 4$:

$$_{20}C_4 = \frac{20!}{4!(20-4)!} = \frac{20!}{4!16!} = \frac{20 \cdot 19 \cdot 18 \cdot 17 \cdot 16!}{4 \cdot 3 \cdot 2 \cdot 1 \cdot 16!} = \frac{116,280}{24} = 4,845$$

There are 4,845 different simple random samples of size 4 from a population whose size is 20.  ◄◄

 *Now Work Problem 47.*

## Summary

To summarize the differences between combinations and the various types of permutations, we present Table 7.

| TABLE 7 | | |
|---|---|---|
| | **Description** | **Formula** |
| **Combination** | The selection of $r$ objects from a set of $n$ different objects where the order in which the objects is selected does not matter (so $AB$ is the same as $BA$) and an object cannot be selected more than once (repetition is not allowed) | $_nC_r = \dfrac{n!}{r!(n-r)!}$ |
| **Permutation of Distinct Items with Replacement** | The selection of $r$ objects from a set of $n$ different objects where the order in which the objects are selected matters (so $AB$ is different from $BA$) and an object may be selected more than once (repetition is allowed) | $n^r$ |
| **Permutation of Distinct Items without Replacement** | The selection of $r$ objects from a set of $n$ different objects where the order in which the objects are selected matters (so $AB$ is different from $BA$) and an object cannot be selected more than once (repetition is not allowed) | $_nP_r = \dfrac{n!}{(n-r)!}$ |

**④ Probabilities Involving Permutations and Combinations**

The counting techniques presented in this section can be used to determine probabilities of certain events by utilizing the classical method of computing probabilities. Recall that this method stated the probability of an event $E$ is the number of ways event $E$ can occur divided by the number of different possible outcomes of the experiment.

► EXAMPLE 10   Winning the Lottery

*Problem:* In the Illinois Lottery, an urn contains balls numbered 1–54. From this urn, 6 balls are randomly chosen without replacement. For a \$1 bet, a player chooses two sets of 6 numbers. In order to win, all six numbers must match those chosen from the urn. The order in which the balls are selected does not matter. What is the probability of winning Lotto?

*Approach:* The probability of winning is given by the number of ways a ticket could win, divided by the size of the sample space. Each ticket has two sets of six numbers, so there are two chances (for the two sets of numbers) of winning for each ticket. The size of the sample space $S$ is the number of ways that 6 objects can be selected from 54 objects without replacement and without regard to order, so that $N(S) = {}_{54}C_6$.

*Solution:* The size of the sample space is

$$N(S) = {}_{54}C_6 = \frac{54!}{6! \cdot (54-6)!} = \frac{54 \cdot 53 \cdot 52 \cdot 51 \cdot 50 \cdot 49 \cdot 48!}{6! \cdot 48!} = 25,827,165$$

Each ticket has two sets of 6 numbers, so a player has two chances of winning for each \$1. If $E$ is the event "winning ticket," then $N(E) = 2$. The probability of $E$ is

$$P(E) = \frac{2}{25{,}827{,}165} \approx 0.000000077$$

There is about a 1 in 13,000,000 chance of winning the Illinois Lottery! ◄◄

▶ **EXAMPLE 11**   **Probabilities Involving Combinations**

*Problem:* A shipment of 120 fasteners that contains 4 defective fasteners was sent to a manufacturing plant. The quality-control manager at the manufacturing plant randomly selects five fasteners and inspects them. What is the probability that exactly one of the fasteners is defective?

*Approach:* The probability that exactly one fastener is defective is found by calculating the number of ways of selecting exactly 1 defective fastener in 5 fasteners and then dividing this result by the number of ways of selecting 5 fasteners from 120 fasteners. To choose exactly 1 defective in the 5 requires choosing one defective from the four defectives and 4 nondefectives from the 116 nondefectives. The order in which the fasteners are selected does not matter, so we use combinations.

*Solution:* The number of ways of choosing 1 defective fastener from 4 defective fasteners is $_4C_1$. The number of ways of choosing four nondefective fasteners from 116 nondefectives is $_{116}C_4$. Using the Multiplication Principle, we find that the number of ways of choosing 1 defective and 4 nondefective fasteners is

$$(_4C_1) \cdot (_{116}C_4) = 4 \cdot 7{,}160{,}245 = 28{,}640{,}980$$

The number of ways of selecting 5 fasteners from 120 fasteners is $_{120}C_5 = 190{,}578{,}024$. The probability of selecting exactly 1 defective fastener is

$$P(\text{one defective fastener}) = \frac{(_4C_1)(_{116}C_4)}{_{120}C_5} = \frac{4 \cdot 7{,}160{,}245}{190{,}578{,}024} \approx 0.1503 = 15.03\%$$

There is a 15.03% probability of randomly selecting exactly one defective fastener. ◄◄

**NW** *Now Work Problem 53.*

## 5.5 Assess Your Understanding

### Concepts and Vocabulary

**1.** Explain the difference between a combination and a permutation.

### Exercises

• **Skill Building**

*In Problems 1–6, find the value of each factorial.*

**1.** 5!          **2.** 7!          **3.** 10!          **4.** 12!          **5.** 0!          **6.** 1!

*In Problems 7–14, find the value of each permutation.*

**7.** $_6P_2$          **8.** $_7P_2$          **9.** $_4P_4$          **10.** $_7P_7$

**11.** $_5P_0$          **12.** $_4P_0$          **13.** $_8P_3$          **14.** $_9P_4$

*In Problems 15–22, find the value of each combination.*

**15.** $_8C_3$
(NW)

**16.** $_9C_2$

**17.** $_{10}C_2$

**18.** $_{12}C_3$

**19.** $_{52}C_1$

**20.** $_{40}C_{40}$

**21.** $_{48}C_3$

**22.** $_{30}C_4$

---

**23.** List all the permutations of five objects *a*, *b*, *c*, *d*, and *e* taken two at a time without repetition. What is $_5P_2$?

**24.** List all the permutations of four objects *a*, *b*, *c*, and *d* taken two at a time without repetition. What is $_4P_2$?

**25.** List all the combinations of five objects *a*, *b*, *c*, *d*, and *e* taken two at a time. What is $_5C_2$?

**26.** List all the combinations of four objects *a*, *b*, *c*, and *d* taken two at a time. What is $_4C_2$?

• **Applying the Concepts**

**27.** **Clothing Options** A man has six shirts and four ties. As-
(NW) suming that they all match, how many different shirt-and-tie combinations can he wear?

**28.** **Clothing Options** A woman has five blouses and three skirts. Assuming that they all match, how many different outfits can she wear?

**29.** **Arranging Songs on a CD** Suppose Dan is going to
(NW) burn a compact disk (CD) that will contain 12 songs. In how many ways can Dan arrange the 12 songs on the CD?

**30.** **Arranging Students** In how many ways can 15 students be lined up?

**31.** **Traveling Salesman** A salesman must travel to 8 cities in order to promote a new marketing campaign. How many different trips are possible, if any route between cities is possible?

**32.** **Randomly Playing Songs** A certain compact disk play-er randomly plays each of 10 songs on a CD. Once a song is played, it is not repeated until all the songs on the CD have been played. In how many different ways can the CD player play the 10 songs?

**33.** **Stocks on the NYSE** Companies whose stocks are list-ed on the New York Stock Exchange (NYSE) have their company name represented by either one, two, or three letters (repetition of letters is allowed). What is the maximum number of companies that can be listed on the New York Stock Exchange?

**34.** **Stocks on the NASDAQ** Companies whose stocks are listed on the NASDAQ stock exchange have their company name represented by either four or five let-ters (repetition of letters is allowed). What is the max-imum number of companies that can be listed on the NASDAQ?

**35.** **Garage Door Code** Outside of my home, there is a key-pad that can be used to open the garage if the correct four-digit code is entered.
(a) How many codes are possible?
(b) What is the probability of entering the correct code on the first try, assuming that I don't remember the code?

**36.** **Social Security Numbers** A Social Security number is used to identify each resident of the United States uniquely. The number is of the form xxx–xx–xxxx, where each x is a digit from 0 to 9.
(a) How many Social Security numbers can be formed?
(b) What is the probability of correctly guessing the So-cial Security number of the President of the United States?

**37.** **Usernames** Suppose a local area network requires eight letters for user names. Lower- and uppercase let-ters are considered the same. How many user names are possible for the local area network?

**38.** **Passwords** Suppose a local area network requires eight characters for a password. The first character must be a letter, but the remaining seven characters can be either a letter or a digit (0 through 9). Lower- and uppercase letters are considered the same. How many passwords are possible for the local area network?

**39.** **Combination Locks** A combination lock has 50 num-bers on it. To open it, you turn counterclockwise to a number, then rotate clockwise to a second number, and then counterclockwise to the third number.
(a) How many different lock combinations are there?
(b) What is the probability of guessing a lock combina-tion on the first try?

**40.** **Forming License Plate Numbers** How many different license plate numbers can be made by using one letter followed by five digits selected from the digits 0 through 9?

**41.** **INDY 500** Suppose 40 cars start at the Indianapolis
(NW) 500. In how many ways can the top three cars finish the race?

**42.** **Betting on the Perfecta** In how many ways can the top two horses finish in a 10-horse race?

**43.** **Forming a Committee** Four members from a 20-person committee are to be selected randomly, to serve as chairperson, vice chairperson, secretary, and treasurer. The first person selected is the chairperson, the second person selected is the vice chairperson, the third is the secretary, and the fourth is the treasurer. How many dif-ferent leadership structures are possible?

**44.** **Forming a Committee** Four members from a 50-person committee are to be selected randomly, to serve as chairperson, vice chairperson, secretary, and treasurer. The first person selected is the chairperson, the second person selected is the vice chairperson, the third is the secretary, and the fourth is the treasurer. How many dif-ferent leadership structures are possible?

**45.** **Lottery** Suppose a lottery exists where balls numbered 1–25 are placed in an urn. To win, you must match the four balls chosen in the correct order. How many possi-ble outcomes are there for this game?

46. **Forming a Committee** In the United States Senate, there are 21 members on the Committee on Banking, Housing, and Urban Affairs. Nine of these 21 members are selected to be on the Subcommittee on Economic Policy. How many different committee structures are possible for this subcommittee?

47. **Simple Random Sample** How many different simple random samples of size 5 can be obtained from a population whose size is 50?
(NW)

48. **Simple Random Sample** How many different simple random samples of size 7 can be obtained from a population whose size is 100?

49. **Children** A family has six children. If this family has exactly two boys, how many different birth and gender orders are possible?

50. **Children** A family has eight children. If this family has exactly three boys, how many different birth and gender orders are possible?

51. **Little Lotto** In the Illinois Lottery game "Little Lotto," an urn contains balls numbered 1–30. From this urn, 5 balls are chosen randomly, without replacement. For a $1 bet, a player chooses one set of 5 numbers. In order to win, all five numbers must match those chosen from the urn. The order in which the balls are selected does not matter. What is the probability of winning Little Lotto with one ticket?

52. **The Big Game** In the Big Game, an urn contains balls numbered 1–50, and a second urn contains balls numbered 1–36. From the first urn, 5 balls are chosen randomly, without replacement. From the second urn, 1 ball is chosen randomly. For a $1 bet, a player chooses one set of 5 numbers to match the balls selected from the first urn and one number to match the ball selected from the second urn. In order to win, all 6 numbers must match, that is, the player must match the first 5 balls selected from the first urn *and* the single ball selected from the second urn. What is the probability of winning the Big Game with a single ticket?

53. **Selecting a Jury** The grade appeal process at a university requires that a jury be structured by selecting five individuals randomly from a pool of eight students and ten faculty.
(NW)
    (a) What is the probability of selecting a jury of all students?
    (b) What is the probability of selecting a jury of all faculty?
    (c) What is the probability of selecting a jury of two students and three faculty?

54. **Selecting a Committee** Suppose there are 55 Democrats and 45 Republicans in the United States Senate. A committee of seven senators is to be formed by selecting members of the Senate randomly.
    (a) What is the probability that the committee is composed of all Democrats?
    (b) What is the probability that the committee is composed of all Republicans?
    (c) What is the probability that the committee is composed of three Democrats and four Republicans?

55. **Acceptance Sampling** Suppose a shipment of 120 electronic components contains 4 defective components. In order to determine whether the shipment should be accepted, a quality-control engineer randomly selects 4 of the components and tests them. If 1 or more of the components is defective, the shipment is rejected. What is the probability the shipment is rejected?

56. **In the Dark** A box containing twelve 40-watt lightbulbs and eighteen 60-watt lightbulbs is stored in your basement. Unfortunately, the box is stored in the dark and you need two 60-watt bulbs. What is the probability, of randomly selecting two 60-watt bulbs from the box?

57. **Randomly Playing Songs** Suppose a compact disk (CD) you just purchased has 13 tracks. After listening to the CD, you decide that you like 5 of the songs. The random feature on your CD player will play each of the 13 songs once in a random order. Find the probability that, among the first 4 songs played,
    (a) you like 2 of them;
    (b) you like 3 of them;
    (c) you like all 4 of them.

58. **Packaging Error** Through a manufacturing error, three cans marked "regular soda" were accidentally filled with diet soda and placed into a 12-pack. Suppose that three cans are randomly selected from the 12-pack.
    (a) Determine the probability that exactly two contain diet soda.
    (b) Determine the probability that exactly one contains diet soda.
    (c) Determine the probability that all three contain diet soda.

59. **Three of a Kind** Suppose you are dealt 5 cards from a standard 52-card deck. Determine the probability of being dealt three of a kind (i.e., three aces, three kings, etc.) by answering the following questions:
    (a) How many ways can 5 cards be selected from a 52-card deck?
    (b) Each deck contains 4 two's, 4 three's, and so on. How many ways can three of the same card be selected from the deck?
    (c) The remaining 2 cards must be different from the 3 chosen and from each other. For example, if we drew three kings, the 4th card cannot be a king. After selecting the three of a kind, there are 12 of the same rank of card remaining in the deck that can be chosen. For example, if we have three aces, then we can choose two's, three's, and so on. Of the 12 ranks remaining, we choose 2 of them and there are 4 cards in each rank. How many ways can we select the remaining 2 cards?
    (d) Use the Multiplication Rule to compute the probability of obtaining three of a kind. That is, what is the probability of selecting three of a kind and two cards that are not like?

60. **Two of a Kind** Follow the outline presented in Problem 59 to determine the probability of being dealt exactly a pair.

61. **Acceptance Sampling** Suppose you have just received a shipment of 20 modems. Although you don't know this,

3 of the modems are defective. To determine whether you will accept the shipment, you randomly select 4 modems and test them. If all 4 modems work, you accept the shipment; otherwise, the shipment is rejected. What is the probability of accepting the shipment?

62. **Acceptance Sampling** Suppose you have just received a shipment of 100 televisions. Although you don't know this, 6 are defective. To determine whether you will accept the shipment, you randomly select 5 televisions and test them. If all 5 televisions work, you accept the shipment; otherwise, the shipment is rejected. What is the probability of accepting the shipment?

---

## Technology Step-by-Step
Factorials, Permutations, and Combinations

**TI-83 Plus**  **Factorials**

**Step 1:** To compute 7!, type 7 on the HOME screen.

**Step 2:** Press MATH, then highlight PRB, and then highlight 4 : ! Press ENTER. With 7! on the HOME screen, press ENTER again.

**Permutations and Combinations**

**Step 1:** To compute $_7P_3$, type 7 on the HOME screen.

**Step 2:** Press MATH, then highlight PRB, and then highlight 2 : $_nP_r$ and press ENTER.

**Step 3:** Type 3 on the HOME screen, and press ENTER. NOTE: To compute $_7C_3$, select 3 : $_nC_r$ instead of 2: $_nP_r$.

**Excel** To do combinations or factorials, select the *fx* button. Highlight Math & Trig in the Function category. Select FACT in the function name to obtain a factorial, and fill in the appropriate cells. Select COMBIN in the function name to obtain a combination, and fill in the appropriate cells. To do permutations, select the *fx* button. Highlight Statistical in the Function category. Select PERMUT in the function name, and fill in the appropriate cells.

---

# CHAPTER 5 REVIEW

## Summary

In this chapter, we introduced the concept of probability. Probability is a measure of the likelihood of a random phenomenon or chance behavior. Because we are measuring a random phenomenon, there is short-term uncertainty. However, this short-term uncertainty gives rise to long-term predictability.

Probabilities are numbers between zero and one, inclusive. The closer a probability is to one, the more likely the event is to occur. If an event has a probability zero, it is said to be impossible; events with a probability one are said to be certain.

We introduced three methods for computing probabilities: (1) the classical method, (2) the empirical (or relative frequency) method, and (3) subjective probabilities. Classical probabilities require the events in the experiment to be equally likely. We count the number of ways an event can occur and divide this by the number of possible outcomes of the experiment. Empirical probabilities rely on the relative frequency with which an event occurs. Therefore, empirical probabilities actually require that an experiment be performed, whereas classical probability does not. Subjective probabilities are probabilities based upon the opinion of the individual providing the probability. They are simply educated guesses about the likelihood of an event's occurring.

Compound probabilities are probabilities in which we are interested in the occurrence of multiple events. For example, we might be interested in the probability that either event $E$ or event $F$ happens. The Addition Rule is used to compute the probability of $E$ or $F$; the Multiplication Rule is used to compute the probability of both $E$ and $F$. Two events are mutually exclusive if they do not have any simple events in common. Two events $E$ and $F$ are independent if the probability of event $E$ occurring does not change if event $F$ is known to have occurred. The complement of an event $E$, denoted $\overline{E}$, is all the simple events in the sample space that are not in $E$.

Finally, we introduced counting methods. The Multiplication Rule is used to count the number of ways a sequence of events can occur. Permutations are used to count the number of ways $r$ distinct items can be arranged from a set of $n$ items without replacement. Combinations are used to count the number of ways $r$ distinct items can be selected from a set of $n$ items without replacement and without regard to order. These counting techniques can be used to calculate probabilities via the classical method.

## Formulas

**Classical Probability**

$$P(E) = \frac{\text{number of ways that } E \text{ can occur}}{\text{number of possible outcomes}} = \frac{N(E)}{N(S)}$$

**Empirical Probability**

$$P(E) \approx \frac{\text{frequency of } E}{\text{number of trials of experiment}}$$

**Addition Rule**

$P(E \text{ or } F) = P(E) + P(F) - P(E \text{ and } F)$

**Addition Rule for Mutually Exclusive Events**

$P(E \text{ or } F) = P(E) + P(F)$

**Probabilities of Complements**

$P(\overline{E}) = 1 - P(E)$

**Multiplication Rule**

$P(E \text{ and } F) = P(E) \cdot P(F|E)$

**Multiplication Rule for Independent Events**

$P(E \text{ and } F) = P(E) \cdot P(F)$

**Multiplication Rule for $n$ Independent Events**

$P(E \text{ and } F \text{ and } G \cdots) = P(E) \cdot P(F) \cdot P(G) \cdot \ldots$

**Conditional Probability Rule**

$$P(F|E) = \frac{P(E \text{ and } F)}{P(E)} = \frac{N(E \text{ and } F)}{N(E)}$$

**Factorial**

$n! = n \cdot (n-1) \cdot (n-2) \cdot \ldots \cdot 3 \cdot 2 \cdot 1$

**Permutation**

$$_nP_r = \frac{n!}{(n-r)!}$$

**Combination**

$$_nC_r = \frac{n!}{r!(n-r)!}$$

## Vocabulary

## Objectives

| Section | You should be able to ... | Review Exercises |
|---|---|---|
| 5.1 | 1 Understand the properties of probabilities (p. 188) | 1, 13(d), 19(e) |
|  | 2 Compute and interpret probabilities using the classical method (p. 189) | 2–4, 13(a), 14 |
|  | 3 Compute and interpret probabilities using the empirical method (p. 191) | 15(a), 16, 17(a) and (b), 18(a) and (b), 19(a) and (b), 20(a) and (b) |
|  | 4 Use simulation to obtain probabilities (p. 194) | 33 |
|  | 5 Understand subjective probabilities (p. 195) | 34 |
| 5.2 | 1 Use the Addition Rule (p. 200) | 6, 7, 13(b)–(c), 17(c), 18(d), 19(d), 20(d) |
|  | 2 Compute the probability of an event using complements (p. 204) | 5, 15(b), 16 |
| 5.3 | 1 Compute probabilities using the Multiplication Rule (p. 210) | 11, 15(c), 23, 24 |
|  | 2 Compute probabilities using the Multiplication Rule for Independent Events (p. 213) | 8, 9, 21(a) and (b), 22(a) and (b) |
|  | 3 Compute "at least" probabilities (p. 216) | 15(d), 17(d), 21(c), 22(c) |
| 5.4 | 1 Compute conditional probabilities (p. 219) | 11, 18(e), 18(f), 19(f) |
|  | 2 Use the Multiplication Rule to check for independent events (p. 221) | 18(h), 19(g) |
| 5.5 | 1 Solve counting problems using the Multiplication Principle (p. 224) | 25, 27 |
|  | 2 Solve counting problems using permutations (p. 227) | 26 |
|  | 3 Solve counting problems using combinations (p. 229) | 28, 29 |
|  | 4 Compute probabilities involving permutations and combinations (p. 231) | 30–32 |

## Review Exercises

**1.** (a) Which among the following numbers could be the probability of an event?

0, −0.01, 0.75, 0.41, 1.34

(b) Which among the following numbers could be the probability of an event?

2/5, 1/3, −4/7, 4/3, 6/7

---

*For Problems 2–5, let the sample space be $S = \{$red, green, blue, orange, yellow$\}$. Suppose the simple events are equally likely.*

**2.** Compute the probability of the event $E = \{$yellow$\}$.

**3.** Compute the probability of the event $F = \{$green or orange$\}$.

**4.** Compute the probability of the event $E = \{$red or blue or yellow$\}$.

**5.** Suppose that $E = \{$yellow$\}$. Compute the probability of $\bar{E}$.

**6.** Suppose that $P(E) = 0.76$, $P(F) = 0.45$, and $P(E$ and $F) = 0.32$. What is $P(E$ or $F)$?

**7.** Suppose that $P(E) = 0.36$, $P(F) = 0.12$, and $E$ and $F$ are mutually exclusive. What is $P(E$ or $F)$?

**8.** Suppose that events $E$ and $F$ are independent. In addition, $P(E) = 0.45$ and $P(F) = 0.2$. What is $P(E$ and $F)$?

**9.** Suppose that $P(E) = 0.8$, $P(F) = 0.5$, and $P(E$ and $F) = 0.24$. Are events $E$ and $F$ independent? Why?

**10.** Suppose that $P(E) = 0.59$ and $P(F|E) = 0.45$. What is $P(E$ and $F)$?

**11.** Suppose that $P(E$ and $F) = 0.35$ and $P(F) = 0.7$. What is $P(E|F)$?

**12.** Determine the value of each of the following:

(a) 7!       (d) $_{10}C_3$

(b) 0!       (e) $_9P_2$

(c) $_9C_4$      (f) $_{12}P_4$

**13. Roulette** In the game of roulette, a wheel consists of 38 slots, numbered 0, 00, 1, 2, …, 36. (See the photo in Problem 13 from Section 5.1.) To play the game, a metal ball is spun around the wheel and is allowed to fall into one of the numbered slots. The slots numbered 0 and 00 are green, the odd numbers are red, and the even numbers are black.

(a) Determine the probability that the metal ball falls into a "green" slot. Interpret this probability.

(b) Determine the probability that the metal ball falls into a "green" or a "red" slot. Interpret this probability.

(c) Determine the probability that the metal ball falls into 00 or a "red" slot. Interpret this probability.

(d) Determine the probability that the metal ball falls into the number 31 and a "black" slot simultaneously. What term is used to describe this event?

**14. Craps** Craps is a dice game in which two fair dice are cast. If the roller shoots a "7" or "11" on the first roll, he or she wins. If the roller shoots a "2," "3," or "12" on the first roll, he or she loses.

(a) Compute the probability that the shooter wins on the first roll. Interpret this probability.

(b) Compute the probability that the shooter loses on the first roll. Interpret this probability.

**15. New Year's Eve and Day** There were 192 total traffic fatalities between 6:00 P.M. on Dec. 31, 1996 and 5:59 A.M. on Jan. 1, 1997. Of these 192 fatalities, 129 were alcohol related.

(a) What is the probability that a randomly selected traffic fatality that occurred between 6:00 P.M. on Dec. 31, 1996 and 5:59 A.M. on Jan. 1, 1997 was alcohol related?

(b) What is the probability that a randomly selected traffic fatality that occurred between 6:00 P.M. on Dec. 31, 1996 and 5:59 A.M. on Jan. 1, 1997 was not alcohol related?

(c) What is the probability that two randomly selected traffic fatalities that occurred between 6:00 P.M. on Dec. 31, 1996 and 5:59 A.M. on Jan. 1, 1997 were both alcohol related?

(d) What is the probability that, of a two randomly selected traffic fatalities that occurred between 6:00 P.M. on Dec. 31, 1996 and 5:59 A.M. on Jan. 1, 1997, at least one was alcohol related?

**16. Cyclones** According to the National Hurricane Center, about 11% of tropical cyclones occur in the North Atlantic Ocean. What is the probability that a randomly selected cyclone occurs in the North Atlantic Ocean? What is the probability that a randomly selected cyclone does not occur in the North Atlantic Ocean?

**17. October in Chicago** The following data represent the distribution of daily high temperatures in Chicago in October from 1872 to 1999:

| Temperature | Frequency |
|---|---|
| 20–29 | 4 |
| 30–39 | 40 |
| 40–49 | 436 |
| 50–59 | 1190 |
| 60–69 | 1270 |
| 70–79 | 794 |
| 80–89 | 238 |
| 90–99 | 8 |

*Source:* Chicago Tribune

(a) What is the probability that a randomly selected October day from 1872 to 1999 in Chicago has a high in the 80s?

(b) What is the probability that a randomly selected October day from 1872 to 1999 in Chicago has a high in the 20s? Is it unusual to have a high temperature in the 20s in October in Chicago?

(c) What is the probability that a randomly selected October day from 1872 to 1999 in Chicago has a high in the 80s or the 90s?

(d) What is the probability that a randomly selected October day from 1872 to 1999 in Chicago has a high of at least 30° F?

18. **Square Footage of Housing** The following data represent the square footage of housing in the United States in 1997, by geographic region:

| | Region | | | |
|---|---|---|---|---|
| | North-east | Mid-west | South | West |
| <500 sq. ft. | 82 | 132 | 356 | 207 |
| 500–749 sq. ft. | 227 | 535 | 1127 | 398 |
| 750–999 sq. ft. | 531 | 1329 | 2647 | 1079 |
| 1,000–1,499 sq. ft. | 1443 | 3386 | 7045 | 3558 |
| 1,500–1,999 sq. ft. | 1828 | 3387 | 5367 | 3488 |
| 2,000–2,499 sq. ft. | 1992 | 2949 | 3281 | 1973 |
| 2,500–2,999 sq. ft. | 1303 | 1758 | 1633 | 967 |
| 3,000–3,999 sq. ft. | 1244 | 1651 | 1501 | 815 |
| 4,000 or more sq. ft. | 801 | 890 | 903 | 429 |

*Source:* U.S. Census Bureau, Current Housing Reports, Series H150/97, American Housing Survey

(a) What is the probability that a randomly selected home in 1997 is located in the Northeast?
(b) What is the probability that a randomly selected home in 1997 has 2000–2499 square feet?
(c) What is the probability that a randomly selected home in 1997 is located in the Northeast and has 2000–2499 square feet?
(d) What is the probability that a randomly selected home in 1997 has 2000–2499 square feet or is located in the Northeast?
(e) What is the probability that a randomly selected home in 1997 has 2000–2499 square feet, given that it is located in the Northeast?
(f) What is the probability that a randomly selected home in 1997 located in the Midwest has 4000 square feet or more?
(g) What is the probability that a randomly selected home in 1997 located in the Midwest has less than 4000 square feet?
(h) Are the events "Midwest" and "2000–2499 square feet" independent? Why?

19. **Gestation Period versus Weight** The following data represent the birth weights (in grams) of babies born in 1998, along with the period of gestation:
(a) What is the probability that a randomly selected baby born in 1998 was postterm?
(b) What is the probability that a randomly selected baby born in 1998 weighed 3000–3499 grams?
(c) What is the probability that a randomly selected baby born in 1998 weighed 3000–3499 grams and was postterm?
(d) What is the probability that a randomly selected baby born in 1998 weighed 3000–3499 grams or was postterm?
(e) What is the probability that a randomly selected baby born in 1998 weighed less than 500 grams and was postterm? Is this event impossible?

| Birth Weight (in grams) | Period of Gestation | | |
|---|---|---|---|
| | Preterm | Term | Postterm |
| Less than 500 | 2,365 | 6 | 0 |
| 500–999 | 7,753 | 45 | 4 |
| 1,000–1,499 | 7,868 | 387 | 74 |
| 1,500–1,999 | 12,550 | 2,419 | 239 |
| 2,000–2,499 | 22,395 | 21,103 | 1,440 |
| 2,500–2,999 | 25,941 | 105,757 | 8,303 |
| 3,000–3,499 | 18,503 | 192,488 | 18,259 |
| 3,500–3,999 | 6,722 | 105,325 | 12,045 |
| 4,000–4,499 | 1,137 | 23,615 | 2,992 |
| 4,500–4,999 | 187 | 3,441 | 507 |
| Greater than 5,000 | 38 | 486 | 51 |

*Source:* National Vital Statistics Report, Vol. 48, No. 3, March 28, 2000

(f) What is the probability that a randomly selected baby born in 1998 weighed 3000–3499 grams, given the baby was postterm?
(g) Are the events "postterm baby" and "weighs 3000–3499 grams" independent? Why?

20. **SAT Math Scores** The following data represent the scores received on the 2000 SAT I: Reasoning Test—Math, by gender:

| SAT Score | Gender | |
|---|---|---|
| | Male | Female |
| 200–249 | 3,497 | 5,752 |
| 250–299 | 7,773 | 13,263 |
| 300–349 | 20,124 | 35,052 |
| 350–399 | 37,524 | 62,879 |
| 400–449 | 65,945 | 102,487 |
| 450–499 | 87,624 | 121,762 |
| 500–549 | 93,742 | 112,520 |
| 550–599 | 91,264 | 94,200 |
| 600–649 | 73,272 | 64,686 |
| 650–699 | 53,880 | 38,989 |
| 700–749 | 31,534 | 17,935 |
| 750–800 | 17,152 | 7,422 |

*Source:* College Board

(a) If a test taker is randomly selected, what is the probability that the test taker is female?
(b) If a test taker is randomly selected, what is the probability that the test taker scored 600–649?
(c) If a test taker is randomly selected, what is the probability that the test taker is female and scored 600–649?
(d) If a test taker is randomly selected, what is the probability that the test taker either is female or scored 600–649?

21. **Home-Court Advantage** During the 2000 season, the Los Angeles Sparks of the WNBA won 97.5% of their

home games. Assume the outcomes of basketball games are independent, and answer the following questions:
(a) What is the probability that the Sparks will win two home games in a row?
(b) What is the probability that the Sparks will win five home games in a row?
(c) What is the probability that the Sparks will lose at least one of their next five home games?

22. **Home Ice** During the 2000–2001 season, the Pittsburgh Penguins of the National Hockey League won about 43% of their home games. Assume that the outcomes of the hockey games are independent, and answer the following questions:
(a) What is the probability that the Penguins will win two home games in a row?
(b) What is the probability that the Penguins will win seven home games in a row?
(c) What is the probability that the Penguins will lose at least one of their next seven home games?

23. **Acceptance Sampling** Suppose you just received a shipment of 10 DVD players. One of the DVD players is defective. You will accept the shipment if two randomly selected DVD players work. What is the probability that you will accept the shipment?

24. **Drawing Cards** Suppose you draw three cards from a standard 52-card deck. What is the probability that all three cards are aces?

25. **Forming License Plates** A license plate is designed so that the first two characters are letters and the last four characters are digits (0 through 9). How many different license plates can be formed?

26. **Choosing a Seat** If four students enter a classroom that has 10 vacant seats, in how many ways can they be seated?

27. **Jumble** In the game of Jumble, the letters of a word are scrambled. The player must form the correct word. In a recent game in a local newspaper, the Jumble "word" was LINCEY. How many different arrangements are there of the letters in this "word"?

28. **Simple Random Sampling** How many different simple random samples of size 8 can be obtained from a population whose size is 55?

29. **Forming Committees** The United States Senate Appropriations Committee has 29 members. Suppose that a subcommittee is to be formed by randomly selecting 5 of the members of the Appropriations Committee. How many different committees could be formed?

30. **Arizona's Fantasy 5** In one of Arizona's lotteries, balls are numbered 1–35. Five balls are selected randomly, without replacement. The order in which the balls are selected does not matter. To win, your numbers must match the five selected. Determine your probability of winning Arizona's Fantasy 5 with one ticket.

31. **Pennsylvania's Cash 5** In one of Pennsylvania's lotteries, balls are numbered 1–39. Five balls are selected randomly, without replacement. The order in which the balls are selected does not matter. To win, your numbers must match the five selected. Determine your probability of winning Pennsylvania's Cash 5 with one ticket.

32. **Packaging Error** Because of a mistake in packaging, a case of 12 bottles of red wine contained 5 Merlot and 7 Cabernet, each without labels. All the bottles look alike and have an equal probability of being chosen. Three bottles are randomly selected.
(a) What is the probability that all 3 are Merlot?
(b) What is the probability that exactly 2 are Merlot?
(c) What is the probability that none is a Merlot?

33. **Simulation** Use a graphing calculator or statistical software to simulate the playing of the game of roulette, using the Integer Distribution with numbers 1 through 38. Repeat the simulation 100 times. Let the number 37 represent 0 and the number 38 represent 00. Use the results of the simulation to answer the following questions.
(a) What is the probability that the ball lands in the slot marked 7?
(b) What is the probability that the ball lands either in the slot marked 0 or in the one marked 00?

34. Explain what is meant by a subjective probability. List some examples of subjective probabilities.

**Chapter 5 Projects located at www.prenhall.com/sullivanstats**

CHAPTER **6**

# The Binomial Probability Distribution

## Outline

**For additional study help, go to**
www.prenhall.com/sullivanstats

**Materials include**

- Projects
  - Case Study: The Voyage of the St. Andrew
  - Decisions: Should We Convict?
  - Consumer Reports Project
- Self-Graded Quizzes
- "Preparing for This Section" Quizzes
- STATLETs
- PowerPoint Downloads
- Step-by-Step Technology Guide
- Graphing Calculator Help

## Putting It All Together

Chapter 5 presented methods for determining probabilities of an event. We introduced two ways of assigning a probability to an event: (1) the classical method and (2) the empirical method. The classical method uses counting techniques and equally likely outcomes to determine probabilities, while the empirical method requires that the probability experiment actually be conducted. We then approximate the probability by dividing the number of times we observed the event by the number of times the experiment was conducted.

In this chapter, we introduce the concept of a *random variable* and *discrete probability distributions*. In Section 6.1, we discuss the properties of discrete probability distributions. Section 6.2 presents a specific discrete probability distribution: the binomial probability distribution.

## 6.1 Probability Distributions

**Preparing for This Section**
Before getting started, review the following:

✓ Discrete versus continuous variables (Section 1.1, pp. 5–6)

✓ Relative frequency histograms for discrete data (Section 2.2, p. 56)

✓ Mean (Section 3.1, pp. 84–87)

✓ Standard deviation (Section 3.2, pp. 105–106)

✓ Mean from grouped data (Section 3.3, pp. 115–117)

✓ Standard deviation from grouped data (Section 3.3, pp. 118–119)

**Objectives**

 Distinguish between discrete and continuous random variables

 Construct and identify probability distributions

 Construct probability histograms

 Compute and interpret the mean of a discrete random variable

 Compute the variance and standard deviation of a discrete random variable

 Compute and interpret the expected value of a discrete random variable

 **Random Variables**

In Chapter 5, we presented the concept of an experiment and the outcomes of an experiment. When experiments are conducted in a way such that the outcome is a numerical result, we say the outcome is a *random variable*.

**Definition**

A **random variable** is a numerical measure of the outcome of a probability experiment, so its value is determined by chance. Random variables are denoted using letters such as $X$.

We will follow the practice of using a capital letter to identify the random variable and a small letter to list the possible values of the random variable, that is, the sample space of the experiment. For example, if an experiment is conducted in which a single die is cast, then $X$ represents the number of pips showing on the die and the possible values of $X$ are $x = 1, 2, 3, 4, 5$, or $6$. As another example, suppose an experiment is conducted in which the time between arrivals of cars at a drive through is measured. The random variable $T$ might describe the time between arrivals, so the sample space of the experiment is $t > 0$.

There are two types of random variables, *discrete* and *continuous*.

**Definitions**

A **discrete random variable** is a random variable that has either a finite number of possible values or a countable number of possible values.

A **continuous random variable** is a random variable that has an infinite number of possible values that is not countable.

▶ **EXAMPLE 1** Distinguishing between Discrete and Continuous Random Variables

**(a)** The number of *A*s earned in a section of statistics with 15 students enrolled is a discrete random variable because the value of the random variable results from counting. If we let the random variable $X$ represent the number of *A*s, then the possible values of $X$ are $x = 0, 1, 2, \ldots, 15$.

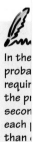

**In Your Own Words**

A probability histogram is constructed the same way as a relative frequency histogram for discrete data. The only difference is that the vertical axis is a probability rather than a relative frequency.

*Solution:* Figure 1 presents the probability histogram.

Figure 1

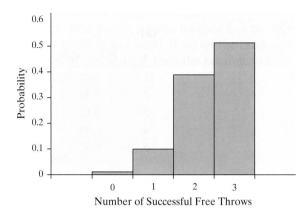

Number of Successful Free Throws

Notice that the area of each rectangle in the probability histogram equals the probability that the random variable assumes the particular value. For example, the area of the rectangle corresponding to the random variable $X = 1$ is $1 \cdot 0.096 = 0.096$ where 1 represents the width of the rectangle and 0.096 represents its height.

Probability histograms help us to determine the shape of the distribution. Recall, we describe distributions as skewed left, skewed right or symmetric. For example, the probability histogram presented in Figure 1 is skewed left.

 *Now Work Problem 13(a) and (b).*

## ④ Mean of a Discrete Random Variable

Remember, when we describe the distribution of a variable, we describe its center, spread, and shape. We now introduce methods for identifying the center and spread of a discrete random variable. We will use the mean to describe the center of a random variable. The variance and standard deviation are used to describe the spread of a random variable.

To help see where the formula for computing the mean of a discrete random variable comes from, consider the following. One semester I had a small statistics class of 10 students. I asked them the age of their car and obtained the following:

$$2, 4, 6, 6, 4, 4, 2, 3, 5, 5$$

What is the mean age of the 10 cars? To answer this question, we let the random variable $X$ represent the age of a car and obtain the probability distribution in Table 2.

| TABLE 2 | |
|---|---|
| **x** | **P(X = x)** |
| 2 | 2/10 = 0.2 |
| 3 | 1/10 = 0.1 |
| 4 | 3/10 = 0.3 |
| 5 | 2/10 = 0.2 |
| 6 | 2/10 = 0.2 |

Now, we compute the mean as follows:

$$\mu = \frac{\sum x_i}{N} = \frac{2 + 4 + 6 + 6 + 4 + 4 + 2 + 3 + 5 + 5}{10}$$

$$= \frac{\overbrace{2 + 2}^{2} + \overbrace{3}^{1} + \overbrace{4 + 4 + 4}^{3} + \overbrace{5 + 5}^{2} + \overbrace{6 + 6}^{2}}{10}$$

$$= \frac{2 \cdot 2 + 3 \cdot 1 + 4 \cdot 3 + 5 \cdot 2 + 6 \cdot 2}{10}$$

$$= 2 \cdot \frac{2}{10} + 3 \cdot \frac{1}{10} + 4 \cdot \frac{3}{10} + 5 \cdot \frac{2}{10} + 6 \cdot \frac{2}{10}$$

$$= 2 \cdot P(X = 2) + 3 \cdot P(X = 3) + 4 \cdot P(X = 4) + 5 \cdot P(X = 5) + 6 \cdot P(X = 6)$$

$$= 2(0.2) + 3(0.1) + 4(0.3) + 5(0.2) + 6(0.2)$$

$$= 4.1$$

Based upon the preceding computations, we conclude that the mean of a discrete probability distribution is found by multiplying each possible value of the random variable by its corresponding probability and adding these products.

### In Your Own Words

To find the mean of a discrete random variable, multiply the value of each random variable by its probability. Then add up these products.

> **The Mean of a Discrete Random Variable**
>
> The mean of a discrete random variable is given by the formula
>
> $$\mu_X = \sum [x \cdot P(X = x)] \qquad (1)$$
>
> where $x$ is the value of the random variable and $P(X = x)$ is the probability of observing the random variable $x$.

▶ **EXAMPLE 5**  **Computing the Mean of a Discrete Random Variable**

*Problem:* Compute the mean of the discrete random variable given in Table 1 from Example 2.

*Approach:* The mean of a discrete random variable is found by multiplying each value of the random variable by its probability and adding up these products.

| TABLE 3 | | |
|---|---|---|
| $x$ | $P(X = x)$ | $x \cdot P(X = x)$ |
| 0 | 0.008 | $0 \cdot 0.008 = 0$ |
| 1 | 0.096 | $1 \cdot 0.096 = 0.096$ |
| 2 | 0.384 | 0.768 |
| 3 | 0.512 | 1.536 |

*Solution:* Refer to Table 3. The first two columns represent the discrete probability distribution. The third column represents $x \cdot P(X = x)$.

We substitute into Formula (1) to find the mean number of free throws made.

$$\mu_X = \sum [x \cdot P(X = x)] = 0(0.008) + 1(0.096) + 2(0.384) + 3(0.512) = 2.4^*$$

◀◀

### Interpretation of the Mean of a Discrete Random Variable

The mean of a discrete random variable can be thought of as the arithmetic mean outcome of the probability experiment if we repeated the experiment many times. Consider the result of Example 5. If we repeated the experiment

*We will follow the practice of rounding the mean, variance, and standard deviation to one more decimal place than the random variable.

of shooting three free throws many times and recorded the number of made free throws, we would expect the average number of made free throws to be around 2.4.

**In Your Own Words**

The mean of a discrete random variable can be thought of as the mean of the outcomes if the experiment is repeated many, many times.

**Interpretation of the Mean of a Discrete Random Variable**

Suppose an experiment is repeated $n$ independent times and the value of the random variable $X$ is recorded. As the number of repetitions of the experiment, $n$, increases, the arithmetic mean value of the $n$ trials will approach $\mu_X$, the mean of the random variable $X$. In other words, let $x_1$ be the value of the random variable $X$ after the first experiment, $x_2$ be the value of the random variable $X$ after the second experiment, and so on. Then

$$\overline{X} = \frac{x_1 + x_2 + \cdots + x_n}{n}$$

The difference between $\overline{X}$ and $\mu_X$ will get closer to 0 as $n$ increases.

▶ EXAMPLE 6 **Illustrating the Interpretation of the Mean of a Discrete Random Variable**

*Problem:* A basketball player historically makes 80% of her free throws. She is asked to shoot three free throws 100 times. Compute the mean number of free throws made.

*Approach:* The player shoots three free throws and the number made is recorded. We repeat this experiment 99 more times, then compute the mean number of free throws made.

*Solution:* The results are presented in Table 4.

| TABLE 4 | | | | | | | | | |
|---|---|---|---|---|---|---|---|---|---|
| 3 | 3 | 2 | 3 | 3 | 3 | 1 | 2 | 3 | 2 |
| 2 | 3 | 3 | 1 | 2 | 2 | 2 | 2 | 2 | 3 |
| 3 | 3 | 2 | 2 | 3 | 2 | 3 | 2 | 2 | 2 |
| 3 | 3 | 2 | 3 | 2 | 3 | 3 | 2 | 3 | 1 |
| 3 | 2 | 2 | 2 | 2 | 0 | 2 | 3 | 1 | 2 |
| 3 | 3 | 2 | 3 | 2 | 3 | 2 | 1 | 3 | 2 |
| 2 | 3 | 3 | 3 | 1 | 3 | 3 | 1 | 3 | 3 |
| 3 | 2 | 2 | 1 | 3 | 2 | 2 | 2 | 3 | 2 |
| 3 | 2 | 2 | 2 | 3 | 3 | 2 | 2 | 3 | 3 |
| 2 | 3 | 2 | 1 | 2 | 3 | 3 | 2 | 3 | 3 |

The first time the experiment was conducted, the player made all three free throws. The second time the experiment was conducted, the player made two out of three free throws. The mean number of free throws made was

$$\overline{X} = \frac{3 + 2 + 3 + \cdots + 3}{100} = 2.35$$

This is close to the mean of 2.4. As the number of repetitions of the experiment increases, we would expect $\overline{X}$ to get even closer to 2.4. ◀◀

Figure 2(a) and Figure 2(b) further demonstrate the interpretation of the mean of a discrete random variable. Figure 2(a) shows the mean number of free throws made versus the number of repetitions of the experiment for the data in Table 4. Figure 2(b) shows the mean number of free throws made versus the number of repetitions of the experiment when the same experiment of shooting three free throws 100 times is conducted a second time. In both plots the player starts off "hot" since the mean number of made free throws is above the theoretical level of 2.4. However, both graphs approach the theoretical mean of 2.4 as the number of repetitions of the experiment increases.

Figure 2

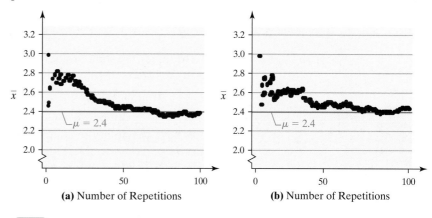

**(a)** Number of Repetitions          **(b)** Number of Repetitions

NW  *Now Work Problem 13(c).*

⑤ **The Variance and Standard Deviation of a Discrete Random Variable**

We now introduce a method for computing the variance and standard deviation of a discrete random variable.

> **Variance and Standard Deviation of a Discrete Random Variable**
>
> The variance of a discrete random variable is given by
>
> $$\sigma_X^2 = \sum [(x - \mu_X)^2 \cdot P(X = x)] \tag{2}$$
>
> where $x$ is the value of the random variable, $\mu_X$ is the mean of the random variable, and $P(X = x)$ is the probability of observing the random variable $x$.
>
> To find the standard deviation of the discrete random variable, take the square root of the variance. That is, $\sigma_X = \sqrt{\sigma_X^2}$.

▶ EXAMPLE 7  **Computing the Variance and Standard Deviation of a Discrete Random Variable**

*Problem:* Find the variance and standard deviation of the discrete random variable given in Table 1 from Example 2.

*Approach:* We know from Example 5 that $\mu_X = 2.4$. We use Formula (2).

*Solution:* Refer to Table 5. The first two columns represent the discrete probability distribution. The third column represents $(x - \mu_X)^2 \cdot P(X = x)$. We sum the entries in the third column.

| TABLE 5 | | |
|---|---|---|
| $x$ | $P(X = x)$ | $(x - \mu_X)^2 \cdot P(X = x)$ |
| 0 | 0.008 | $(0 - 2.4)^2 \cdot 0.008 = 0.04608$ |
| 1 | 0.096 | $(1 - 2.4)^2 \cdot 0.096 = 0.18816$ |
| 2 | 0.384 | $(2 - 2.4)^2 \cdot 0.384 = 0.06144$ |
| 3 | 0.512 | $(3 - 2.4)^2 \cdot 0.512 = 0.18432$ |

$$\sum (x - \mu)^2 \cdot P(X = x) = 0.48$$

The variance of the discrete random variable $X$ is

$$\sigma_X^2 = \sum(x - \mu_X)^2 \cdot P(X = x) = 0.48^*$$

The standard deviation of the discrete random variable is found by taking the square root of the variance.

$$\sigma_X = \sqrt{\sigma_X^2} = \sqrt{0.48} = 0.692820323 \approx 0.7 \quad \blacktriangleleft\blacktriangleleft$$

The larger the standard deviation of the random variable, the more dispersed the observed values of the random variable are from the mean.

**Figure 3**

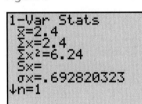

**Using Technology:** Graphing calculators with advanced statistical features have the ability to compute the mean and standard deviation of a discrete random variable. Figure 3 shows the results of Examples 5 and 7 using a TI-83 Plus graphing calculator.

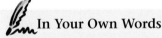 *Now Work Problems 13(d) and 13(e).*

## ⑥ Expected Value

The mean of a random variable is the long-run average outcome of the experiment. In this sense, as the number of trials of the experiment increases, the average result of the experiment gets closer to the mean of the random variable. For this reason, the mean of a random variable is often called the *expected value*.

> **Expected Value**
>
> The **expected value** of a discrete random variable $X$, denoted $E(X)$, is obtained using the formula
>
> $$E(X) = \sum[x \cdot P(X = x)]. \tag{3}$$

*✍ In Your Own Words*

The expected value of a discrete random variable is the mean of the discrete random variable.

The interpretation of expected value is the same as the interpretation of the mean of a discrete random variable.

▶ **EXAMPLE 8** **Finding the Expected Value**

*Problem:* A term life insurance policy will pay a beneficiary a certain sum of money upon the death of the policyholder. These policies have premiums that must be paid annually. Suppose a life insurance company sells a $250,000 one-year term life insurance policy to a 20-year-old male for $350. According to the National Vital Statistics Report, Vol. 47, No. 28, the probability the male will survive the year is 0.99865. Compute the expected value of this policy to the insurance company.

*Approach:* There are two possible outcomes to the experiment: survival or death. Let the random variable $X$ represent the "payout" (money lost or gained), depending upon survival or death of the insured. We assign probabilities to each of these random variables and substitute these values into Formula (3).

---

*If necessary, use the unrounded value of the mean in the computation of the variance in order to avoid rounding errors.

## Historical Note

Christiaan Huygens was born on April 14, 1629, into an influential Dutch family. He studied law and mathematics at the University of Leiden from 1645 to 1647. From 1647 to 1649, he continued to study law and mathematics at the College of Orange at Breda. Among his many great discoveries, Huygens discovered the first moon of Saturn in 1655. The following year, he discovered the shape of the rings of Saturn. While in Paris sharing his discoveries, he learned about probability through the correspondence of Fermat and Pascal. In 1657, Huygens published *De ratiociniis in ludi aleae* (*On Reasoning in Games of Chance*), the first published book on probability theory. In that text, Huygens introduced the idea of mathematical expectation.

*Solution:*

**Step 1:** We have $P$(survives) $= 0.99865$, so that $P$(dies) $= 0.00135$. From the point of view of the insurance company, if the client survives the year, the insurance company makes $350. Therefore, we let $x = \$350$ if the client survives the year. If the client dies during the year, the insurance company must pay $250,000 to the client's beneficiary; however, the company still keeps the $350, so we let $x = \$350 - \$250,000 = -\$249,650$. The value is negative because it is money paid out by the insurance company. We have the probability distribution listed in Table 6.

| TABLE 6 | |
| --- | --- |
| $x$ | $P(X = x)$ |
| $350 (survives) | 0.99865 |
| −$249,650 (dies) | 0.00135 |

**Step 2:** Substituting into Formula (3), we obtain the expected value (from the point of view of the insurance company) of the policy.

$$E(X) = \sum x P(X = x)$$

$$= \$350(0.99865) + (-\$249,650)(0.00135) = \$12.50$$

*Interpretation:* The company expects to make $12.50 for each 20-year-old male client they insure.   ◄◄

The $12.50 profit of the insurance company is a long-term result. Of course, it does not make $12.50 on each person it insures, but rather the average profit per person insured is $12.50.

**NW** *Now Work Problem 21.*

## 6.1 Assess Your Understanding

### Concepts and Vocabulary

1. What is a random variable?
2. What is the difference between a discrete random variable and a continuous random variable? Provide your own examples of each.
3. What are the two requirements for a discrete probability distribution?
4. In your own words, provide an interpretation of the mean of a discrete random variable.
5. Suppose a baseball player historically hits 0.300. (This means that the player averages three hits in every 10 at-bats.) Suppose the player has zero hits in four at-bats in a game, and enters the batter's box for the fifth time, whereupon the announcer declares that the player is "due for a hit." What is the flaw in the announcer's reasoning? If the player had four hits in the last four at-bats, is the player "due to make an out"?
6. In game theory, a game is called a zero-sum game if the expected value of the game is zero. Explain what this means.

## Exercises

### • Basic Skills

*In Problems 1–2, determine whether the random variable is discrete or continuous. In each case, state the possible values of the random variable.*

1. (a) The number of lightbulbs that burn out in the next week in a room of 20 bulbs.
   (b) The time it takes to fly from New York City to Los Angeles.
   (c) The number of hits to a web site in a day.
   (d) The amount of snow in Toronto during the winter.

2. (a) The time it takes for a lightbulb to burn out.
   (b) The weight of a T-bone steak.
   (c) The number of free throw attempts before a shot is made.
   (d) In a random sample of 20 people, the number who are blood type A.

*In Problems 3–8, determine whether the distribution is a probability distribution. If not, state why.*

3.

| x | $P(X = x)$ |
|---|---|
| 0 | 0.2 |
| 1 | 0.2 |
| 2 | 0.2 |
| 3 | 0.2 |
| 4 | 0.2 |

4.

| x | $P(X = x)$ |
|---|---|
| 0 | 0.1 |
| 1 | 0.5 |
| 2 | 0.05 |
| 3 | 0.25 |
| 4 | 0.1 |

5.

| x | $P(X = x)$ |
|---|---|
| 10 | 0.1 |
| 20 | 0.23 |
| 30 | 0.22 |
| 40 | 0.6 |
| 50 | −0.15 |

6.

| x | $P(X = x)$ |
|---|---|
| 10 | 0.5 |
| 20 | 0.1 |
| 30 | 0.3 |
| 40 | −0.2 |
| 50 | 0.3 |

7.

| x | $P(X = x)$ |
|---|---|
| 100 | 0.1 |
| 200 | 0.25 |
| 300 | 0.2 |
| 400 | 0.3 |
| 500 | 0.1 |

8.

| x | $P(X = x)$ |
|---|---|
| 100 | 0.25 |
| 200 | 0.25 |
| 300 | 0.25 |
| 400 | 0.25 |
| 500 | 0.25 |

*In Problems 9 and 10, determine the required value of the missing probability in order to make the distribution a probability distribution.*

9.

| x | $P(X = x)$ |
|---|---|
| 3 | 0.4 |
| 4 | |
| 5 | 0.1 |
| 6 | 0.2 |

10.

| x | $P(X = x)$ |
|---|---|
| 0 | 0.30 |
| 1 | 0.15 |
| 2 | |
| 3 | 0.20 |
| 4 | 0.15 |
| 5 | 0.05 |

### • Applying the Concepts

11. **Parental Involvement** In the following probability distribution, the random variable $X$ represents the number of activities the mother of a K–5th grade student is involved in:

| x | $P(X = x)$ |
|---|---|
| 0 | 0.044 |
| 1 | 0.084 |
| 2 | 0.209 |
| 3 | 0.325 |
| 4 | 0.338 |

*Source:* U.S. National Center for Education Statistics

(a) Verify that this is a probability distribution.
(b) Draw a probability histogram.
(c) Compute and interpret the mean of the random variable $X$.
(d) Compute the variance of the random variable $X$.
(e) Compute the standard deviation of the random variable $X$.
(f) What is the probability that a randomly selected mother is involved in three activities?
(g) What is the probability that a randomly selected mother is involved in three or four activities?

**12. Parental Involvement** In the following probability distribution, the random variable $X$ represents the number of activities the mother of a 6th–8th-grade student is involved in:

| x | P(X = x) |
|---|---|
| 0 | 0.10 |
| 1 | 0.137 |
| 2 | 0.27 |
| 3 | 0.302 |
| 4 | 0.191 |

*Source:* U.S. National Center for Education Statistics

(a) Verify that this is a probability distribution.
(b) Draw a probability histogram.
(c) Compute and interpret the mean of the random variable $X$.
(d) Compute the variance of the random variable $X$.
(e) Compute the standard deviation of the random variable $X$.
(f) What is the probability that a randomly selected mother is involved in three activities?
(g) What is the probability that a randomly selected mother is involved in three or four activities?

**13. Hot Streak** A batter historically gets on base 30% of the time. The following distribution represents the number of at-bats $X$ before he makes an out:

| x | P(X = x) |
|---|---|
| 1 | 0.7000 |
| 2 | 0.2100 |
| 3 | 0.0630 |
| 4 | 0.0189 |
| 5 | 0.0060 |
| 6 | 0.0021 |

(a) Verify that this is a probability distribution.
(b) Draw a probability histogram.
(c) Compute and interpret the mean of the random variable $X$.
(d) Compute the variance of the random variable $X$.
(e) Compute the standard deviation of the random variable $X$.
(f) What is the probability that the batter requires four at-bats before making an out?
(g) What is the probability that the batter requires four or more at-bats before making an out? Would this be unusual?

**14. Waiting in Line** A Wendy's manager performed a study in order to determine the probability distribution for the number of people waiting in line $X$ during lunch. The results were as follows:

| x | P(X = x) | x | P(X = x) | x | P(X = x) |
|---|---|---|---|---|---|
| 0 | 0.011 | 5 | 0.172 | 10 | 0.019 |
| 1 | 0.035 | 6 | 0.132 | 11 | 0.002 |
| 2 | 0.089 | 7 | 0.098 | 12 | 0.006 |
| 3 | 0.150 | 8 | 0.063 | 13 | 0.001 |
| 4 | 0.186 | 9 | 0.035 | 14 | 0.001 |

(a) Verify that this is a probability distribution.
(b) Draw a probability histogram.
(c) Compute and interpret the mean of the random variable $X$.
(d) Compute the variance of the random variable $X$.
(e) Compute the standard deviation of the random variable $X$.
(f) What is the probability that there are 8 people waiting in line for lunch?
(g) What is the probability that there are 10 or more people waiting in line for lunch? Would this be unusual?

---

*In Problems 15–18, (a) construct the probability distribution for the random variable X (Hint: $P(X = x_i) = f_i/N$.), (b) draw the probability histogram, (c) compute and interpret the mean of the probability distribution, and (d) compute the standard deviation of the probability distribution.*

**15. Age of 5–9-Year-Old Boys** The following data represent the number of 5–9-year-old males in the United States in 2000:

| x (age) | Frequency (in thousands) |
|---|---|
| 5 | 2,031 |
| 6 | 2,058 |
| 7 | 2,110 |
| 8 | 2,138 |
| 9 | 2,186 |

*Source:* United States Census Bureau

**16. Age of 5–9-Year-Old Girls** The following data represent the number of 5–9-year-old females in the United States in 2000:

| x (age) | Frequency (in thousands) |
|---|---|
| 5 | 1,934 |
| 6 | 1,961 |
| 7 | 2,008 |
| 8 | 2,041 |
| 9 | 2,081 |

*Source:* United States Census Bureau

**17. Grade School Enrollment** The following data represent the enrollment levels in grades 1–8 in the United States in 1997:

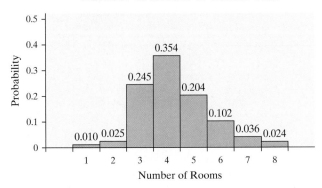

| x (grade level) | Frequency (in thousands) |
| --- | --- |
| 1 | 3,755 |
| 2 | 3,689 |
| 3 | 3,597 |
| 4 | 3,507 |
| 5 | 3,458 |
| 6 | 3,493 |
| 7 | 3,520 |
| 8 | 3,415 |

*Source:* U.S. National Center of Education Statistics

**18. High School Enrollment** The following data represent the enrollment levels in grades 9–12 in the United States in 1997:

| x (grade level) | Frequency (in thousands) |
| --- | --- |
| 9 | 3,819 |
| 10 | 3,377 |
| 11 | 2,972 |
| 12 | 2,673 |

*Source:* U.S. National Center of Education Statistics

**19. Number of Births** The probability histogram below represents the number of live births by a mother 50–54 years old who had a live birth in 1999. The data are from the National Vital Statistics Report, Volume 49, No. 1, April 17, 2001.

(a) What is the probability that a randomly selected 50–54-year-old mother who had a live birth in 1999 has had her fourth live birth?

(b) What is the probability that a randomly selected 50–54-year-old mother who had a live birth in 1999 has had her fourth or fifth live birth?

(c) What is the probability that a randomly selected 50–54-year-old mother who had a live birth in 1999 has had her sixth or more live birth?

**Number of Live Births, 50–54-year-old Mother**

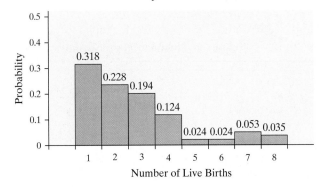

**20. Rental Units** The probability histogram that follows represents the number of rooms in rented housing units in 1997. The data are from the United States Census Bureau's *Current Housing Report.*

**Number of Rooms in Rental Unit**

(a) What is the probability that a randomly selected rental unit has five rooms?

(b) What is the probability that a randomly selected rental unit has five or six rooms?

(c) What is the probability that a randomly selected rental unit has seven or more rooms?

**21. Life Insurance** Suppose a life insurance company sells a $250,000 one-year term life insurance policy to a 20-year-old female for $200. According to the National Vital Statistics Report, Vol. 47, No. 28, the probability that the female survives the year is 0.99952. Compute and interpret the expected value of this policy to the insurance company.

**22. Life Insurance** Suppose a life insurance company sells a $500,000 one-year term life insurance policy to a 50-year-old male for $2,900. According to the National Vital Statistics Report, Vol. 47, No. 28, the probability the male survives the year is 0.99433. Compute and interpret the expected value of this policy to the insurance company.

**23. Roulette** In the game of roulette, a player can place a $5 bet on the number 17 and have a 1/38 probability of winning. If the metal ball lands on 17, the player wins $175, otherwise the casino takes the player's $5. What is the expected value of the game to the player? If you played the game 1,000 times, how much would you expect to lose?

**24. Connecticut Lottery** In the Cash Five Lottery in Connecticut, a player pays $1 for a single ticket with five numbers. Five ping pong balls numbered 1 through 35 are randomly chosen from a bin without replacement. If all five numbers on a player's ticket match the five chosen, the player wins $100,000. The probability of this occurring is 1/324,632. If four numbers match, the player wins $300. This occurs with probability 1/2164. If three numbers match, the player wins $10. This occurs with probability 1/75. Compute and interpret the expected value of the game from the players point of view.

**25. Simulation** Use the probability distribution from Problem 13 and Minitab's DISCRETE command (or some other statistical software) to simulate 100 repetitions of the experiment in which the batter bats until he makes an out. The number of at-bats is recorded. Approximate the mean and standard deviation of the random variable $X$, based upon the simulation. Repeat the simulation by performing 500 repetitions of the experiment. Approximate the mean and standard deviation of the random variable. Compare your results to the theoretical mean and standard deviation. What property is being illustrated?

26. **Simulation** Use the probability distribution from Problem 14 and Minitab's DISCRETE command (or some other statistical software) to simulate 100 repetitions of the experiment. Approximate the mean and standard deviation of the random variable $X$, based upon the simulation. Repeat the simulation by performing 500 repetitions of the experiment. Approximate the mean and standard deviation of the random variable. Compare your results to the theoretical mean and standard deviation. What property is being illustrated?

---

## Technology Step-by-Step
### Finding the Mean and Standard Deviation of a Discrete Random Variable Using Technology

**TI-83 Plus**  **Step 1:** Enter the values of the random variable in L1 and their corresponding probabilities in L2.

**Step 2:** Press STAT, highlight CALC, and select 1: 1-Var Stats.

**Step 3:** With 1-Var Stats on the HOME screen, type L1 followed by a comma, followed by L2 as shown below

```
1-Var Stats L1, L2
```

Hit ENTER.

---

## 6.2   The Binomial Probability Distribution

**Preparing for This Section**   Before getting started, review the following:

✓ Independence (Section 5.3, p. 212)

✓ Combinations (Section 5.5, pp. 229–231)

✓ Multiplication Rule for Independent Events (Section 5.3, pp. 213–214)

✓ Addition Rule (Section 5.2, pp. 200–204)

✓ Complement Rule (Section 5.2, pp. 204–206)

✓ Empirical Rule (Section 3.2, pp. 107–108)

**Objectives**  Determine whether a probability experiment is a binomial experiment

 Compute probabilities of binomial experiments

 Compute the mean and standard deviation of a binomial random variable

 Construct binomial probability histograms

 In Section 6.1, we stated that probability distributions could be presented using tables, graphs or mathematical formulas. Up to this point, we have only presented probability distributions using tables or graphs (probability histograms). In this section, we introduce a specific type of discrete probability distribution that can be presented using a formula, *the binomial probability distribution*.

The binomial probability distribution is a probability distribution that describes probabilities for experiments in which there are two mutually exclusive outcomes. These two outcomes are generally referred to as *success* or *failure*. For example, a basketball player can either make a free throw (success) or miss (failure). A new surgical procedure can result in either life (success) or death (failure).

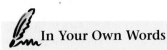

Experiments in which there are only two possible outcomes are referred to as *binomial experiments*, provided that certain criteria are met.

**Criteria for a Binomial Probability Experiment**

An experiment is said to be a **binomial experiment** provided
1. The experiment is performed a fixed number of times. Each repetition of the experiment is called a **trial**.
2. The trials are independent. This means the outcome of one trial will not affect the outcome of the other trials.
3. For each trial, there are two mutually exclusive outcomes: success or failure.
4. The probability of success is fixed for each trial of the experiment.

If a probability experiment satisfies these four requirements, the random variable $X$, the number of successes in $n$ trials, follows the binomial probability distribution. Before introducing the method for computing binomial probabilities, it is worthwhile to introduce some notation.

**Notation Used in the Binomial Probability Distribution**

- There are $n$ independent trials of the experiment.
- Let $p$ denote the probability of success so that $1 - p$ is the probability of failure.
- Let $X$ denote the number of successes in $n$ independent trials of the experiment. So, $0 \le x \le n$.

**Historical Note**

Jacob Bernoulli was born on December 27, 1654 in Basel, Switzerland. He studied philosophy and theology at the urging of his parents. (He resented this.) In 1671, he graduated from the University of Basel with a master's degree in philosophy. In 1676, he received a licentiate in theology. After earning his philosophy degree, Bernoulli traveled to Geneva to tutor. From there, he went to France to study with the great mathematicians of the time. One of Bernoulli's greatest works is *Ars Conjectandi*, published eight years after his death. In this publication, Bernoulli proved the binomial probability formula. To this day, each observed outcome in a binomial probability experiment is called a "Bernoulli" trial.

▶ **EXAMPLE 1   Identifying Binomial Experiments**

*Problem:* Determine which of the following probability experiments qualify as a binomial experiment. For those that are binomial experiments, identify the number of trials, probability of success, probability of failure, and possible values of the random variable $X$.

**(a)** An experiment in which a basketball player who historically makes 80% of his free throws is asked to shoot 3 free throws and the number of made free throws is recorded.

**(b)** The number of people with blood type O-negative based upon a simple random sample of size 10 is recorded. According to the Information Please Almanac, 6% of the human population is blood type O-negative.

**(c)** A probability experiment in which three cards are drawn from a deck without replacement and the number of aces is recorded.

*Approach:* We need to determine whether the four conditions for a binomial experiment are satisfied.

1. The experiment is performed a fixed number of times.
2. The trials are independent.
3. There are only two possible outcomes of the experiment.
4. The probability of success for each trial is constant.

*Solution:*

**(a)** This is a binomial experiment because
   1. There are $n = 3$ trials.
   2. The trials are independent.
   3. There are two possible outcomes: make or miss.

**4.** The probability of success (make) is 0.8 and the probability of failure (miss) is 0.2. They are the same for each trial.

The random variable $X$ is the number of made free throws with $x = 0, 1, 2$, or 3.

**(b)** This is a binomial experiment because
   **1.** There are 10 trials (the 10 randomly selected people).
   **2.** The trials are independent.*
   **3.** There are two possible outcomes: finding a person with blood type O-negative or not.
   **4.** The probability of success is 0.06 and the probability of failure is 0.94.

The random variable $X$ is the number of people with blood type O-negative with $x = 0, 1, 2, 3, \ldots, 10$.

**(c)** This is not a binomial experiment because the trials are not independent. The probability of an ace on the first trial is 4/52. Because we are sampling without replacement, if an ace is selected on the first trial, the probability of an ace on the second trial is 3/51. If an ace is not selected on the first trial, the probability of an ace on the second trial is 4/51. ◄◄

 *Now Work Problem 3.*

It is worth mentioning that the word *success* does not necessarily imply that something positive has occurred. Success simply means that an outcome has occurred that corresponds with $p$, the probability of success. For example, a probability experiment might be to randomly select ten 18-year-old male drivers. We might let $X$ denote the number who have been involved in an accident within the last year. In this case, a success would mean obtaining an 18-year-old male who was involved in an accident. This outcome is certainly not positive, but still represents a success as far as the experiment goes.

**Caution**

The probability of success, $p$, is always associated with the random variable $X$, the number of successes. So if $X$ represents the number of 18-year-olds involved in an accident, then $p$ represents the probability of an 18-year-old being involved in an accident.

## ② Computing Probabilities of Binomial Experiments

We are now prepared to compute probabilities for a binomial random variable $X$.

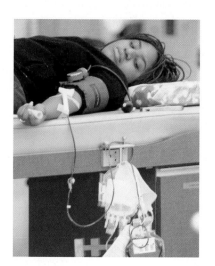

► **EXAMPLE 2** Constructing a Binomial Probability Distribution

*Problem:* According to the *Information Please* almanac, 6% of the human population is blood type O-negative. A simple random sample of size four is taken and the number of people $X$ with blood type O-negative is recorded. Construct a probability distribution for the random variable $X$.

*Approach:* This is a binomial experiment with $n = 4$ trials. We define a success as selecting an individual with blood type O-negative. The probability of success, $p$, is 0.06, and $X$ is the random variable representing the number of successes with $x = 0, 1, 2, 3$, or 4.

*Step 1:* Construct a tree diagram listing the various outcomes of the experiment by listing each outcome as $S$ (success) or $F$ (failure).

*Step 2:* Compute the probabilities for each value of the random variable $X$.

*Step 3:* Construct the probability distribution.

---

*Recall that, in sampling from large populations without replacement, the trials are assumed to be independent, provided that the sample size is small in relation to the size of the population. As a general rule of thumb, if the sample size is less than 5% of the population size, the trials can be assumed to be independent although they are technically dependent.

*Solution:*

**Step 1:** Figure 4 contains a tree diagram listing the 16 possible outcomes of the experiment.

Figure 4

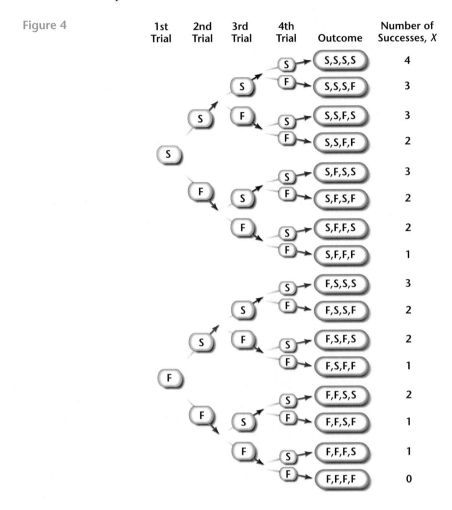

| | 1st Trial | 2nd Trial | 3rd Trial | 4th Trial | Outcome | Number of Successes, X |
|---|---|---|---|---|---|---|

**Step 2:** We now compute the probability for each possible value of the random variable $X$. We start with $P(X = 0)$:

$$P(X = 0) = P(FFFF) = P(F) \cdot P(F) \cdot P(F) \cdot P(F) \quad \text{Multiplication Rule for Independent Events}$$
$$= (0.94)(0.94)(0.94)(0.94)$$
$$= (0.94)^4$$
$$= 0.78075$$

$$P(X = 1) = P(SFFF \text{ or } FSFF \text{ or } FFSF \text{ or } FFFS)$$
$$= P(SFFF) + P(FSFF) + P(FFSF) + P(FFFS) \quad \text{Addition Rule for Mutually Exclusive Events}$$
$$= (0.06)^1(0.94)^3 + (0.06)^1(0.94)^3 + (0.06)^1(0.94)^3 + (0.06)^1(0.94)^3 \quad \text{Multiplication Rule for Independent Events}$$
$$= 4(0.06)^1(0.94)^3$$
$$= 0.19934$$

$$P(X = 2) = P(SSFF \text{ or } SFSF \text{ or } SFFS \text{ or } FSSF \text{ or } FSFS \text{ or } FFSS)$$
$$= P(SSFF) + P(SFSF) + P(SFFS) + P(FSSF) + P(FSFS) + P(FFSS)$$
$$= (0.06)^2(0.94)^2 + (0.06)^2(0.94)^2 + (0.06)^2(0.94)^2 + (0.06)^2(0.94)^2 + (0.06)^2(0.94)^2 + (0.06)^2(0.94)^2$$
$$= 6(0.06)^2(0.94)^2$$
$$= 0.01909$$

We compute $P(X = 3)$ and $P(X = 4)$ similarly and obtain $P(X = 3) = 0.00081$ and $P(X = 4) = 0.00001$. The reader is encouraged to verify these probabilities.

**Step 3:** We use these results and obtain the probability distribution in Table 7.

◀◀

| | TABLE 7 |
|---|---|
| $x$ | $P(X = x)$ |
| 0 | 0.78075 |
| 1 | 0.19934 |
| 2 | 0.01909 |
| 3 | 0.00081 |
| 4 | 0.00001 |

As we look back at the solution in Example 2, we should note some interesting results. Consider the probability of obtaining $X = 1$ success:

$$P(X = 1) = 4(0.06)^1 (0.94)^3$$

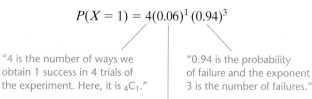

"4 is the number of ways we obtain 1 success in 4 trials of the experiment. Here, it is $_4C_1$."

"0.94 is the probability of failure and the exponent 3 is the number of failures."

"0.06 is the probability of success and the exponent 1 is the number of successes."

The coefficient 4 is the number of ways of obtaining one success in four trials. In general, the coefficient will be $_nC_x$, the number of ways of obtaining $x$ successes in $n$ trials. The second factor in the formula, $(0.06)^1$, is the probability of success, $p$, raised to the number of successes, $x$. The third factor in the formula, $(0.94)^3$, is the probability of failure, $1 - p$, raised to the number of failures, $n - x$. This formula holds in general and we have the *binomial probability distribution function (pdf)*.

*Theorem*

**Binomial Probability Distribution Function**

The probability of obtaining $x$ successes in $n$ independent trials of a binomial experiment where the probability of success is $p$ is given by

$$P(X = x) = {_nC_x}\, p^x (1 - p)^{n-x}   \quad x = 0, 1, 2, \ldots, n \qquad \textbf{(1)}$$

**Caution**

Before using the binomial probability distribution function, be sure the requirements for a binomial experiment are satisfied.

While reading probability problems, pay special attention to key phrases that translate into mathematical symbols. Table 8 lists various phrases and their corresponding mathematical equivalent.

| TABLE 8 | |
|---|---|
| **Phrase** | **Math Symbol** |
| "at least" or "no less than" | $\geq$ |
| "more than" or "greater than" | $>$ |
| "fewer than" or "less than" | $<$ |
| "no more than" or "at most" | $\leq$ |
| "exactly" | $=$ |

▶ **EXAMPLE 3**   **Using the Binomial Probability Distribution Function**

*Problem:* According to Nielsen Media Research, 75% of all United States households have cable television.

**(a)** In a random sample of 15 households, what is the probability that exactly 10 have cable?

**(b)** In a random sample of 15 households, what is the probability that at least 13 have cable?

**(c)** In a random sample of 15 households, what is the probability that fewer than 13 have cable?

*Approach:* This is a binomial experiment with $n = 15$ independent trials with the probability of success, $p$, equal to 0.75. The possible values of the random variable $X$ are $x = 0, 1, 2, \ldots, 15$. We use Formula (1) to compute the probabilities.

*Solution:*

**(a)** $P(X = 10) = {}_{15}C_{10}(0.75)^{10}(1 - 0.75)^{15-10}$     $n = 15, x = 10, p = 0.75$

$$= \frac{15!}{10!(15 - 10)!}(0.75)^{10}(0.25)^5 \quad {}_nC_x = \frac{n!}{x!(n - x)!}$$

$$= 3003(0.0563)(0.00097656)$$

$$= 0.1651$$

*Interpretation:* The probability of getting exactly 10 households out of 15 with cable is 0.1651. In 100 trials of this experiment, we would expect about 17 trials to result in 10 households with cable.

**(b)** The phrase "at least" means greater than or equal to. The values of the random variable $X$ greater than or equal to 13 are $x = 13, 14,$ or 15.

$P(X \geq 13) = P(X = 13 \text{ or } X = 14 \text{ or } X = 15)$

$= P(X = 13) + P(X = 14) + P(X = 15)$     Addition Rule for Mutually Exclusive Events

$= {}_{15}C_{13}(0.75)^{13}(1 - 0.75)^{15-13} + {}_{15}C_{14}(0.75)^{14}(1 - 0.75)^{15-14} + {}_{15}C_{15}(0.75)^{15}(1 - 0.75)^{15-15}$

$= 0.1559 + 0.0668 + 0.0134$

$= 0.2361$

*Interpretation:* There is a 0.2361 probability that in a random sample of 15 households, at least 13 will have cable. In 100 trials of this experiment, we would expect about 24 trials to result in at least 13 households having cable.

**(c)** The values of the random variable $X$ less than 13 are $x = 0, 1, 2, \ldots, 12$. Rather than compute $P(X \leq 12)$ directly by computing $P(X = 0) + P(X = 1) + \ldots + P(X = 12)$, we can use the Complement Rule.

$$P(x < 13) = P(x \leq 12) = 1 - P(x \geq 13) = 1 - 0.2361 = 0.7639$$

*Interpretation:* There is a 0.7639 probability that in a random sample of 15 households, less than 13 will have cable. In 100 trials of this experiment, we would expect about 76 trials to result in fewer than 13 households that have cable. ◄◄

**Using Technology:** Statistical software/spreadsheets and graphing calculators with advanced statistical features have the ability to compute binomial probabilities. Figure 5(a) shows the result of Example 3(a) using Excel's formula wizard. To compute probabilities such as $P(X < 13) = P(X \leq 12)$, it is best to use the **cumulative distribution function** (or **cdf**). The cumulative distribution function computes probabilities less than or equal to a specified value. For example, we can use a TI-83 Plus to compute $P(X \leq 12)$ using the *binomcdf(* command. See Figure 5(b).

Figure 5

```
┌─BINOMDIST──────────────────────────────────────────────┐
│       Number_s │10                    │ [▦] = 10        │
│          Trials │15                    │ [▦] = 15        │
│   Probability_s │0.75                  │ [▦] = 0.75      │
│      Cumulative │False                 │ [▦] = FALSE     │
│                                          = 0.165145981  │
│ Returns the individual term binomial distribution probability. │
│                                                         │
│  Cumulative is a logical value: for the cumulative distribution function, use TRUE; for │
│              the probability mass function, use FALSE.  │
│   [?]       Formula result =0.165145981    [ OK ] [ Cancel ] │
└─────────────────────────────────────────────────────────┘
```

Here's the
probability

```
┌─────────────────────┐
│binomcdf(15,.75,      │
│12)                   │
│         .7639121888  │
└─────────────────────┘
```

**(a)**                                                        **(b)**

**NW** *Now Work Problems 15(a) and 21.*

**③ Mean and Standard Deviation of a Binomial Random Variable**

We discussed finding the mean and standard deviation of a discrete random variable in Section 6.1. These formulas can be used to find the mean and standard deviation of a binomial random variable as well. However, there is a faster method.

> **Mean and Standard Deviation of a Binomial Random Variable**
>
> A binomial experiment with $n$ independent trials and probability of success $p$ will have a mean and standard deviation given by the formulas
>
> $$\mu_X = np \quad \text{and} \quad \sigma_X = \sqrt{np(1-p)} \qquad \text{(2)}$$

▶ **EXAMPLE 4** **Finding the Mean and Standard Deviation of a Binomial Random Variable**

*Problem:* According to Nielsen Media Research, 75% of all United States households have cable television. In a simple random sample of 300 households, determine the mean and standard deviation number of households that will have cable television.

*Approach:* This is a binomial experiment with $n = 300$ and $p = 0.75$. We can use Formula (2) to find the mean and standard deviation, respectively.

*Solution:* $\mu_X = np = 300(0.75) = 225$

and

$$\sigma_X = \sqrt{np(1-p)} = \sqrt{300(0.75)(1-0.75)} = \sqrt{56.25} = 7.5$$

*Interpretation:* We expect that in a random sample of 300 households, 225 will have cable.    ◀◀

**In Your Own Words**

The mean of a binomial random variable equals the number of trials of the experiment times the probability of success. It can be interpreted as the expected number of successes in $n$ trials of the experiment.

**NW** *Now Work Problems 15(b) and15(c).*

**④ The Binomial Probability Histogram**

Constructing binomial probability histograms is no different from constructing probability histograms.

▶ **EXAMPLE 5** **Constructing Binomial Probability Histograms**

*Problem:*

**(a)** Construct a binomial probability histogram with $n = 10$ and $p = 0.2$. Comment on the shape of the distribution.

**(b)** Construct a binomial probability histogram with $n = 10$ and $p = 0.5$. Comment on the shape of the distribution.

**(c)** Construct a binomial probability histogram with $n = 10$ and $p = 0.8$. Comment on the shape of the distribution.

*Approach:* To construct a binomial probability histogram, we will first obtain the probability distribution using Minitab. We then construct the probability histogram of the probability distribution.

*Solution:*

**(a)** We obtain the probability distribution with $n = 10$ and $p = 0.2$. See Table 9. Figure 6 shows the corresponding probability histogram with the mean $\mu_X = 10(0.2) = 2$ labeled. The distribution is skewed right.

| TABLE 9 | |
|---|---|
| **x** | **$P(X = x)$** |
| 0 | 0.107374 |
| 1 | 0.268435 |
| 2 | 0.301990 |
| 3 | 0.201327 |
| 4 | 0.088080 |
| 5 | 0.026424 |
| 6 | 0.005505 |
| 7 | 0.000786 |
| 8 | 0.000074 |
| 9 | 0.000004 |
| 10 | 0.000000 |

Figure 6

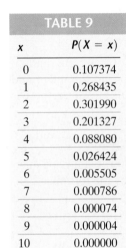

**(b)** We obtain the probability distribution with $n = 10$ and $p = 0.5$. See Table 10. Figure 7 shows the corresponding probability histogram with the mean $\mu_X = 10(0.5) = 5$ labeled. The distribution is symmetric and approximately bell shaped.

| TABLE 10 | |
|---|---|
| **x** | **$P(X = x)$** |
| 0 | 0.000977 |
| 1 | 0.009766 |
| 2 | 0.043945 |
| 3 | 0.117188 |
| 4 | 0.205078 |
| 5 | 0.246094 |
| 6 | 0.205078 |
| 7 | 0.117188 |
| 8 | 0.043945 |
| 9 | 0.009766 |
| 10 | 0.000977 |

Figure 7

**(c)** We obtain the probability distribution with $n = 10$ and $p = 0.8$. See Table 11. Figure 8 shows the corresponding probability histogram with the mean $\mu_X = 10(0.8) = 8$ labeled. The distribution is skewed left.

| TABLE 11 | |
|---|---|
| $x$ | $P(X = x)$ |
| 0 | 0.000000 |
| 1 | 0.000004 |
| 2 | 0.000074 |
| 3 | 0.000786 |
| 4 | 0.005505 |
| 5 | 0.026424 |
| 6 | 0.088080 |
| 7 | 0.201327 |
| 8 | 0.301990 |
| 9 | 0.268435 |
| 10 | 0.107374 |

Figure 8

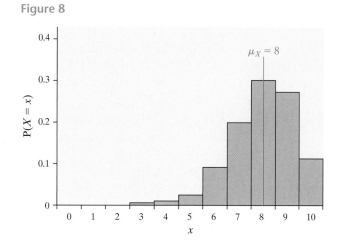

 **NW**  *Now Work Problem 15(d).*

Based upon the results of Example 5, we might conclude that the binomial probability distribution is skewed right if $p < 0.5$, symmetric and approximately bell shaped if $p = 0.5$, and skewed left if $p > 0.5$. Notice that Figure 6 ($p = 0.2$) and Figure 8 ($p = 0.8$) are mirror images of each other.

Remember, the binomial probability distribution depends not only on the parameter $p$, but also upon $n$, the number of trials. What role does $n$ play in the shape of the distribution?

**EXPLORATION**   Construct a binomial probability histogram with $n = 10$ and $p = 0.2$. Comment on the shape of the distribution. Now construct a binomial probability histogram with $n = 30$ and $p = 0.2$. Comment on the shape of the distribution. Finally, construct a binomial probability histogram with $n = 70$ and $p = 0.2$. Comment on the shape of the distribution.

**Result:**   We show the three probability histograms in Figures 9(a), 9(b) and 9(c).

Figure 9(a) is skewed right. Figure 9(b) is slightly skewed right and Figure 9(c) appears bell-shaped. We conclude that as the number of trials $n$ increases, the shape of the distribution becomes more symmetric and bell shaped.

Figure 9

**Binomial Distribution,**
**$n = 10, p = 0.2$**

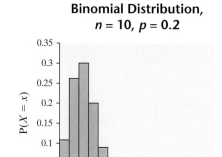

(a)

**Binomial Distribution,**
**$n = 30, p = 0.2$**

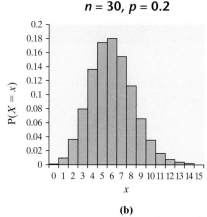

(b)

**Binomial Histogram,**
**$n = 70, p = 0.2$**

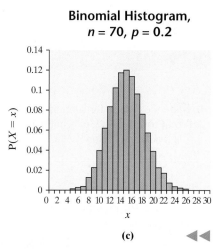

(c)

The conclusion in the Exploration is true in general.

As the number of trials $n$ in a binomial experiment increases, the probability distribution of the random variable $X$ becomes bell shaped. As a general rule of thumb, if $np(1 - p) \geq 10$,* then the probability distribution will be approximately bell shaped.

This result allows us to use the Empirical Rule to identify unusual observations in a binomial experiment. Recall that the Empirical Rule states that in a bell-shaped distribution, about 95% of all observations lie within two standard deviations of the mean. That is, 95% of the observations lie between $\mu - 2\sigma$ and $\mu + 2\sigma$. Any observation that lies outside these intervals may be considered unusual because the observation occurs only 5% of the time.

▶ **EXAMPLE 6**    **Using the Mean, Standard Deviation and Empirical Rule to Check for Unusual Results in a Binomial Experiment**

*Problem:* According to Nielsen Media Research, in 1998, 75% of all United States households had cable television. In a simple random sample of 300 households, 244 had cable. Is this result unusual?

*Approach:* Because $np(1 - p) = 300(0.75)(0.25) = 56.25 \geq 10$, the binomial probability distribution is approximately bell-shaped. Therefore, we can use the Empirical Rule to check for unusual observations. If the observation is less than $\mu - 2\sigma$ or greater than $\mu + 2\sigma$, then we say it is unusual.

*Solution:* From Example 4, we have $\mu = 225$ and $\sigma = 7.5$.

$$\mu - 2\sigma = 225 - 2(7.5) = 225 - 15 = 210$$

and

$$\mu + 2\sigma = 225 + 2(7.5) = 225 + 15 = 240$$

*Interpretation:* Since any value less than 210 or greater than 240 is unusual, 244 is an unusual result. We should attempt to identify reasons for its value. It may be that the percentage of households that have cable has increased since 1998.    ◀◀

**NW**  *Now Work Problem 29.*

*Ramsey, P.P. and P.H. Ramsey, "Evaluating the Normal Approximation to the Binomial Test," Journal of Educational Statistics 13 (1998): 173–182.

## 6.2 Assess Your Understanding

### Concepts and Vocabulary

**1.** State the criteria for a binomial probability experiment.
**2.** What role does $_nC_x$ play in the binomial probability distribution function?

**3.** How can the Empirical Rule be used to identify unusual results from a binomial experiment? When can the Empirical Rule be used?

### Exercises

• **Basic Skills**

*In Problems 1–10, determine which of the following probability experiments represents a binomial experiment. If the probability experiment is not a binomial experiment, state why.*

**1.** A random sample of 15 college seniors is conducted, and the individuals selected are asked to state their ages.

**2.** A random sample of 30 cars in a used car lot is obtained, and their mileage recorded.

**3.** An experimental drug is administered to 100 randomly selected individuals, with the number of individuals responding favorably recorded.

**4.** A poll of 1200 registered voters is conducted in which the respondents are asked whether they believe Congress should reform Social Security.

**5.** Three cards are selected from a standard 52-card deck without replacement. The number of aces selected is recorded.

**6.** Three cards are selected from a standard 52-card deck with replacement. The number of aces selected is recorded.

**7.** A basketball player who makes 80% of her free throws is asked to shoot free throws until she misses. The number of free throw attempts is recorded.

**8.** A baseball player who reaches base safely 30% of the time is allowed to bat until he reaches base safely for the third time. The number of at-bats required is recorded.

**9.** An investor randomly purchases 10 stocks listed on the New York Stock Exchange. Historically, the probability that a stock listed on the NYSE will increase in value over the course of a year is 48%. The number of stocks that increase in value is recorded.

**10.** According to Nielsen Media Research, 75% of all United States households have cable television. In a small town of 40 households, a random sample of 10 households is asked whether they have cable television. The number of households with cable television is recorded.

---

*In Problems 11–14, a binomial probability experiment is conducted with the given parameters. Compute the probability of x successes in the n independent trials of the experiment.*

**11.** $n = 10, p = 0.4, x = 3$

**12.** $n = 15, p = 0.85, x = 12$

**13.** $n = 40, p = 0.99, x = 38$

**14.** $n = 50, p = 0.02, x = 3$

---

*In Problems 15–20, (a) construct a binomial probability distribution with the given parameters; (b) compute the mean and standard deviation of the distribution, using the methods of Section 6.1; (c) compute the mean and standard deviation, using the methods of this section; and (d) draw the probability histogram, comment on its shape, and label the mean on the histogram.*

**15.** $n = 6, p = 0.3$

**16.** $n = 8, p = 0.5$

**17.** $n = 9, p = 0.75$

**18.** $n = 10, p = 0.2$

**19.** $n = 10, p = 0.5$

**20.** $n = 9, p = 0.8$

---

• **Applying the Concepts**

**21.** **On-Time Flights** United Airlines flight 1832 from Chicago to Orlando is on time 80% of the time, according to United Airlines. Suppose 15 flights are randomly selected.
(a) Find the probability that exactly 12 flights are on time.
(b) Find the probability that at least 12 flights are on time.
(c) Find the probability that fewer than 12 flights are on time.
(d) Find the probability that between 10 and 12 flights, inclusive, are on time.

**22.** **Smokers** According to the *Information Please* almanac, 80% of adult smokers started smoking before turning 18 years old. Suppose 10 smokers 18 years old or older are randomly selected.
(a) Find the probability that exactly 8 of them started smoking before 18 years of age.
(b) Find the probability that at least 8 of them started smoking before 18 years of age.
(c) Find the probability that fewer than 8 of them started smoking before 18 years of age.
(d) Find the probability that between 7 and 9 of them, inclusive, started smoking before 18 years of age.

**23.** **Operating Systems** A survey conducted by StatMarket showed that 94% of computer users use Microsoft Windows as their operating system. Suppose 12 computer users are randomly selected.
(a) Find the probability that exactly 10 of them use Microsoft Windows.
(b) Find the probability that 10 or more of them use Microsoft Windows.
(c) Find the probability that 9 or fewer of them use Microsoft Windows.
(d) Find the probability that between 9 and 11 of them, inclusive, use Microsoft Windows.

**24.** **Murder by Firearm** According to *Crime in the United States*, 1998, 65% of murders are committed with a firearm. Suppose 15 murders are randomly selected.
(a) Find the probability that exactly 13 murders are committed with a firearm.
(b) Find the probability that 13 or more murders are committed with a firearm.
(c) Find the probability that 12 or fewer murders are committed with a firearm.
(d) Find the probability that between 10 and 12 murders, inclusive, are committed with a firearm.

25. **Migraine Sufferers** Depakote is a medication whose purpose is to reduce the pain associated with migraine headaches. In clinical trials of Depakote, 2% of the patients in the study experienced weight gain as a side effect. Suppose a random sample of 30 Depakote users is obtained.

    [*Source*: Abbott Laboratories]

    (a) Find the probability that exactly 3 experienced weight gain as a side effect.
    (b) Find the probability that 3 or fewer experienced weight gain as a side effect.
    (c) Find the probability that 4 or more patients experienced weight gain as a side effect.
    (d) Find the probability that between 1 and 4 patients, inclusive, experienced weight gain as a side effect.

26. **Asthma Control** Singulair is a medication whose purpose is to control asthma attacks. In clinical trials of the drug, 18.4% of the patients in the study experienced headaches as a side effect. Suppose a random sample of 30 Singulair users is obtained.

    [*Source*: Merck and Company, Incorporated]

    (a) Find the probability that exactly 5 experienced headaches as a side effect.
    (b) Find the probability that 5 or fewer experienced headaches as a side effect.
    (c) Find the probability that 6 or more patients experienced headaches as a side effect.
    (d) Find the probability that between 6 and 8 patients experienced headaches as a side effect.

27. **Softball** Suppose a softball player reaches base safely 45% of the time. Assuming that trips to the plate are independent events, compute the probability that in the next 10 trips to the plate,

    (a) the player reaches base safely exactly six times.
    (b) the player reaches base safely six or more times.
    (c) the player reaches base safely less than two times.
    (d) the player reaches base safely between three and five times, inclusive.

28. **Basketball** Mark Price holds the record for percentage of free throws made in the National Basketball Association, at 90.4%. Assuming that free throws are independent events, compute the probability that in his next 10 free throws,

    (a) Mark Price makes exactly 8.
    (b) Mark Price makes 8 or more.
    (c) Mark Price makes fewer than 7.
    (d) Mark Price makes between 6 and 8, inclusive.

29. **On-Time Flights** United Airlines flight 1832 from Chicago to Orlando is on time 80% of the time, according to United Airlines.

    (a) Compute the mean and standard deviation of a random variable $X$, the number of on-time flights in 100 trials of the probability experiment.
    (b) Interpret the mean.
    (c) Would it be unusual to observe 90 on-time flights in a random sample of 100 flights from Chicago to Orlando? Why?

30. **Smokers** According to the *Information Please* almanac, 80% of adult smokers started smoking before turning 18 years old.

    (a) Compute the mean and standard deviation of a random variable $X$, the number of smokers who started before 18 in 200 trials of the probability experiment.
    (b) Interpret the mean.
    (c) Would it be unusual to observe 180 smokers who started smoking before turning 18 years old in a random sample of 200 adult smokers? Why?

31. **Operating Systems** According to a survey conducted by StatMarket, 94% of computer users use Microsoft Windows as their operating system.

    (a) Compute the mean and standard deviation of the random variable $X$, the number of Windows users in 200 trials of the probability experiment.
    (b) Interpret the mean.
    (c) Would it be unusual to observe 195 users of Microsoft Windows in a random sample of 200 computer users? Why?

32. **Murder by Firearm** According to *Crime in the United States, 1998*, 65% of murders are committed with a firearm.

    (a) Compute the mean and standard deviation of the random variable $X$, the number of firearm murders in 100 trials of the probability experiment.
    (b) Interpret the mean.
    (c) Would it be unusual to observe 70 murders by firearm in a random sample of 100 murders? Why?

33. **Migraine Sufferers** Depakote is a medication whose purpose is to reduce the pain associated with migraine headaches. In clinical trials and extended studies of Depakote, 2% of the patients in the study experienced weight gain as a side effect.

    (a) Compute the mean and standard deviation of the random variable $X$, the number of patients experiencing weight gain in 600 trials of the probability experiment.
    (b) Interpret the mean.
    (c) Would it be unusual to observe 16 patients who experience weight gain in a random sample of 600 patients who take the medication? Why?

34. **Asthma Control** Singulair is a medication whose purpose is to control asthma attacks. In clinical trials of Singulair, 18.4% of the patients in the study experienced headaches as a side effect.

    (a) Compute the mean and standard deviation of the random variable $X$, the number of patients experiencing headaches in 400 trials of the probability experiment.
    (b) Interpret the mean.
    (c) Would it be unusual to observe 86 patients who experience headaches in a random sample of 400 patients who use this medication? Why?

35. **Males Living at Home** According to the *Information Please* almanac, 59% of males between the ages of 18 and 24 years lived at home in 1998 (unmarried college students living in dorms are counted as living at home). Suppose that a survey is administered at a community college to 20 randomly selected male students between the ages of 18 and 24 years and that 17 of them respond that they live at home.

(a) Based on the sample of 20 students, what proportion of community college males live at home?

(b) Find the probability that 17 or more out of 20 community college male students live at home, assuming that the proportion who live at home is 59%.

(c) What might you conclude from this result?

**36. Females Living at Home** According to the *Information Please* almanac, 48% of females between the ages of 18 and 24 years live at home (unmarried college students living in dorms are counted as living at home). Suppose that a survey is administered at a community college to 20 randomly selected female students between the ages of 18 and 24 years and that 16 of them respond that they live at home.

(a) Based on the sample of 20 students, what proportion of community college females live at home?

(b) Find the probability that 16 or more out of 20 community college female students live at home, assuming that the proportion who live at home is 48%.

(c) What might you conclude from this result?

**37. Boys Are Preferred** In a Gallup poll conducted December 2–4, 2000, 42% of survey respondents said that, if they had only one child, they would prefer the child to be a boy. Suppose you conduct a survey of 15 randomly selected students on your campus and find that 8 of them would prefer a boy.

(a) What proportion of students prefer a boy, according to the sample of 15 students?

(b) Compute the probability that, in a random sample of 15 students, at least 8 would prefer a boy, assuming that the proportion that prefer a boy is 0.42.

(c) Do the results of your survey contradict the results of the Gallup poll? Explain.

**38. Simulation** According to the United States National Center for Health Statistics, there is a 98% probability that a 20-year-old male will survive to age 30.

(a) Using statistical software, such as Minitab, simulate taking 100 random samples of size 30 from this population.

(b) Using the results of the simulation, compute the probability that exactly 29 of the 30 males survive to age 30.

(c) Compute the probability that exactly 29 of the 30 males survive to age 30, using the binomial probability distribution. Compare the results with part (b).

(d) Using the results of the simulation, compute the probability that at most 27 of the 30 males survive to age 30.

(e) Compute the probability that at most 27 of the 30 males survive to age 30, using the binomial probability distribution. Compare the results with part (d).

(f) Compute the mean number of male survivors in the 100 simulations of the probability experiment. Is it close to the expected value?

(g) Compute the standard deviation of the number of male survivors in the 100 simulations of the probability experiment. Compare the result to the theoretical standard deviation of the probability distribution.

(h) Did the simulation yield any unusual results?

**39. Simulation** According to the United States National Center for Health Statistics, there is a 99% probability that a 20-year-old female will survive to age 30.

(a) Using statistical software, such as Minitab, simulate taking 100 random samples of size 30 from this population.

(b) Using the results of the simulation, compute the probability that exactly 29 of the 30 females survive to age 30.

(c) Compute the probability that exactly 29 of the 30 females survive to age 30, using the binomial probability distribution. Compare the results with part (b).

(d) Using the results of the simulation, compute the probability that at most 27 of the 30 females survive to age 30.

(e) Compute the probability that at most 27 of the 30 females survive to age 30, using the binomial probability distribution. Compare the results with part (d).

(f) Compute the mean number of female survivors in the 100 simulations of the probability experiment. Is it close to the expected value?

(g) Compute the standard deviation of the number of female survivors in the 100 simulations of the probability experiment. Compare the result to the theoretical standard deviation of the probability distribution.

(h) Did the simulation yield any unusual results?

**40. Educational Attainment** According to the United States Census Bureau, about 25% of residents of the United States 25 years old or older have completed at least four years of college. Suppose you are performing a study and would like at least 10 people in the study to have completed at least four years of college.

(a) How many residents of the United States do you expect to have to randomly select?

(b) How many residents of the United States do you need to randomly select in order to have a 99% probability that the sample contains at least 10 residents with four or more years of college?

## Technology Step-by-Step
Computing Binomial Probabilities Via Technology

**TI-83 Plus**  Computing $P(X = x)$

*Step 1:* Press $2^{nd}$ VARS to access the probability distribution menu.

*Step 2:* Highlight 0: binompdf( and hit ENTER.

*Step 3:* With binompdf( on the HOME screen, type the number of trials $n$, the probability of success, $p$, and the number of successes, $x$. For example, with $n = 10$, $p = 0.2$, and $x = 4$, type

binompdf(10, 0.2, 4)

Then hit ENTER.

Computing $P(X \le x)$

*Step 1:* Press $2^{nd}$ VARS to access the probability-distribution menu.

*Step 2:* Highlight A: binomcdf( and hit ENTER.

*Step 3:* With binomcdf( on the HOME screen, type the number of trials $n$, the probability of success, $p$, and the number of successes, $x$. For example, with $n = 10$, $p = 0.2$, and $x = 4$, type

binomcdf(10, 0.2, 4)

Then hit ENTER.

**MINITAB**  Computing $P(X = x)$

*Step 1:* Enter the possible values of the random variable $x$ in C1. For example, with $n = 10$, $p = 0.2$, enter $0, 1, 2, \ldots, 10$ into C1.

*Step 2:* Select the **CALC** menu, highlight **Probability Distributions,** then highlight **Binomial** . . . .

*Step 3:* Fill in the window as shown to the left. Click OK.

Computing $P(X \le x)$

Follow the same steps as those for computing $P(X = x)$. In the window that comes up after selecting Binomial Distribution, select Cumulative probability instead of Probability.

**Excel**  Computing $P(X = x)$

*Step 1:* Enter the possible values of the random variable $x$ in column A. For example, with $n = 10$, $p = 0.2$, enter $0, 1, 2, \ldots, 10$ into A.

*Step 2:* With the cursor in cell B1, select the *fx* icon. Highlight Statistical in the Function category window. Highlight BINOMDIST in the Function name window.

**Step 3:** Fill in the window as follows:

```
┌─ BINOMDIST ─────────────────────────────────────────────┐
│        Number_s │A1                          │▤│ = 0      │
│          Trials │10                          │▤│ = 10     │
│    Probability_s │0.2                        │▤│ = 0.2    │
│      Cumulative │False                       │▤│ = FALSE  │
│                                        = 0.107374182      │
│  Returns the individual term binomial distribution probability. │
│                                                            │
│  Cumulative is a logical value: for the cumulative distribution function, use TRUE; for │
│                the probability mass function, use FALSE.   │
│  ┌─┐                                                       │
│  │?│  Formula result =0.107374182     │  OK  │  │ Cancel ││
│  └─┘                                                       │
└────────────────────────────────────────────────────────────┘
```

Click OK.

**Step 4:** Copy the contents in cell B1 to the remaining cells.

**Computing $P(X \leq x)$**

Follow the same steps as those presented for computing $P(X = x)$. In the BINOMDIST window, type "TRUE" in the cumulative cell.

## CHAPTER 6 REVIEW

### Summary

In this chapter, we discussed discrete probability distributions. A random variable represents the numerical measurement of the outcome from a probability experiment. Discrete random variables have either a finite or a countable number of outcomes. The term "countable" means that the values result from counting. Discrete probability distributions must satisfy the following two criteria: (1) All probabilities must be between 0 and 1, inclusive, and (2) the sum of all probabilities must equal 1. Discrete probability distributions can be presented in a table, graph, or mathematical formula. The mean and standard deviation of a probability distribution describe the center and spread of the discrete random variable. The mean of a random variable is also called its expected value.

A probability experiment is considered a binomial experiment if there are $n$ independent trials of the experiment with only two outcomes. The probability of success, $p$, is the same for each trial of the experiment. Special formulas exist for computing the mean and standard deviation of a binomial random variable.

The binomial probability distribution was constructed via classical methods. This means the probabilities are based upon what is expected to happen in a large number of trials of the experiment. If empirical evidence does not agree with the theoretical or expected values, we must investigate the reasons for the disagreement.

### Formulas

**Mean of a Discrete Random Variable**

$$\mu_X = \sum x P(X = x)$$

**Variance of a Discrete Random Variable**

$$\sigma_X^2 = \sum (x - \mu_X)^2 \cdot P(X = x)$$

**Expected Value of a Random Variable $X$**

$$E(X) = \sum x P(X = x)$$

**Binomial Probability Distribution Function**

$$P(X = x) = {}_nC_x\, p^x (1 - p)^{n-x}$$

**Mean of a Binomial Random Variable**

$$\mu_X = np$$

**Standard Deviation of a Binomial Random Variable**

$$\sigma_X = \sqrt{np(1 - p)}$$

### Vocabulary

Random variable (p. 241)
Discrete random variable (p. 241)
Continuous random variable (p. 241)

Probability distribution (p. 242)
Probability histogram (p. 243)
Expected value (p. 248)

Trial (p. 254)
Binomial experiment (p. 254)

## Objectives

## Review Exercises

*In Problems 1 and 2, determine whether the random variable is discrete or continuous. In each case, state the possible values of the random variable.*

1. (a) The number of inches of snow that falls in Buffalo during the winter season.
   (b) The number of days snow accumulates in Buffalo during the winter season.
   (c) The number of golf balls hit into the ocean on the famous 18th hole at Pebble Beach on a randomly selected Sunday.

2. (a) The miles per gallon of gasoline in a 2002 Toyota Sienna.
   (b) The number of children a randomly selected family has.
   (c) The number of goals scored by the Edmonton Oilers in a season.

*In Problems 3 and 4, determine whether the distribution is a probability distribution. If not, state why.*

**3.**

| x | P(X = x) |
|---|---|
| 0 | 0.34 |
| 1 | 0.21 |
| 2 | 0.13 |
| 3 | 0.04 |
| 4 | 0.01 |

**4.**

| x | P(X = x) |
|---|---|
| 0 | 0.40 |
| 1 | 0.31 |
| 2 | 0.23 |
| 3 | 0.04 |
| 4 | 0.02 |

5. **Stanley Cup** The Stanley Cup is a best-of-seven series to determine the champion of the National Hockey League. The following data represent the number of games played, $X$, in the Stanley Cup before a champion was determined, from 1939 to 2000:

| x | Frequency |
|---|---|
| 4 | 20 |
| 5 | 15 |
| 6 | 17 |
| 7 | 10 |

**Source:** *Information Please* Almanac

   (a) Construct a probability distribution for the random variable $X$, the number of games in the Stanley Cup.
   (b) Draw a probability histogram.
   (c) Compute and interpret the mean of the random variable $X$.
   (d) Compute the standard deviation of the random variable $X$.

6. **Property Crime** In 1999, 74% of crime was property crime, according to the National Crime Victimization Survey. Suppose that four crimes are randomly selected. Let the random variable $X$ represent the number of property crimes.
   (a) Construct a probability distribution for the random variable $X$, by constructing a tree diagram.
   (b) Draw a probability histogram.
   (c) Compute and interpret the mean of the random variable $X$.
   (d) Compute the standard deviation of the random variable $X$.

7. **Life Insurance** Suppose a life insurance company sells a $100,000 one-year term life insurance policy to a 35-year-old male for $200. According to the *National Vital Statistics Report*, Vol. 47, No. 28, the probability the male survives the year is 0.99802. Compute and interpret the expected value of this policy to the life insurance company.

8. **Roulette** In the game of roulette, a player can place a $5 bet on red and have a 9/19 probability of winning. If the metal ball lands on red, the player wins $5; otherwise, the casino takes the player's $5. What is the expected value of the game? If the player plays 100 times, how much can the player expect to lose?

*In Problems 9 and 10, determine which of the following probability experiments represents a binomial experiment. If the probability experiment is not a binomial experiment, state why.*

**9.** According to the *Chronicle of Higher Education*, there is a 54% probability that a randomly selected incoming freshman will graduate from college within six years. Suppose 10 incoming male freshmen are randomly selected. After six years, each student is asked whether he or she graduated.

**10.** An experiment is conducted in which a single die is cast until a "3" comes up. The number of throws required is recorded.

**11. High Cholesterol** According to the National Center for Health Statistics, 8% of 20–34-year-old females have high serum cholesterol.
  (a) In a random sample of 10 females 20–34 years old, find the probability that exactly 0 have high serum cholesterol. Interpret this result.
  (b) In a random sample of 10 females 20–34 years old, find the probability that exactly 2 have high serum cholesterol. Interpret this result.
  (c) In a random sample of 10 females 20–34 years old, find the probability that at least 2 have high serum cholesterol. Interpret this result.
  (d) In a random sample of 10 females 20–34 years old, find the probability that exactly 9 will not have high serum cholesterol. Interpret this result.
  (e) In a random sample of 250 females 20–34 years old, what is the expected number with high serum cholesterol? What is the standard deviation?
  (f) If a random sample of 250 females 20–34 years old resulted in 12 of them having high serum cholesterol, would this be unusual? Why?

**12. America Reads** According to a Gallup poll conducted September 10–14, 1999, 56% of Americans 18 years old or older stated they have read at least six books (fiction and nonfiction) within the past year.
  (a) In a random sample of 15 Americans 18 years old or older, find the probability that exactly 10 have read at least six books within the past year. Interpret this result.
  (b) In a random sample of 15 Americans 18 years old or older, find the probability that fewer than 5 have read at least six books within the past year. Interpret this result.
  (c) In a random sample of 15 Americans 18 years old or older, find the probability that at least 5 have read at least six books within the past year. Interpret this result.
  (d) In a random sample of 15 Americans 18 years old or older, find the probability that exactly 12 *have not* read at least six books within the past year. Interpret this result.
  (e) In a random sample of 200 Americans 18 years old or older, what is the expected number who have read at least six books within the past year? What is the standard deviation?
  (f) If a random sample of 200 Americans 18 years old or older resulted in 102 of them having read at least six books within the past year, would this be unusual? Why?

**13. Nielsen Ratings** Nielsen Media Research determines ratings for television programs by placing meters on 5000 televisions throughout the United States. A recent Monday Night Football broadcast resulted in a rating of 7.4, which means 7.4% of households were "tuned in" to the game.
  (a) In a random sample of 20 households, find the probability that exactly 6 are tuned in to Monday Night Football.
  (b) In a random sample of 20 households, find the probability that fewer than 4 were tuned in to Monday Night Football.
  (c) In a random sample of 20 households, find the probability that at least 2 were tuned in to Monday Night Football.
  (d) In a random sample of 20 households, find the probability that exactly 17 were *not* tuned in to Monday Night Football.
  (e) In a random sample of 500 households, what is the expected number who were tuned in to Monday Night Football? What is the standard deviation?
  (f) If a random sample of 500 households results in 50 tuned in to Monday Night Football, would this be unusual? Why?

**14. Quit Smoking** The drug Zyban is meant to create the urge to quit smoking. In clinical trials, 35% of the study's participants experienced insomnia when taking 300 mg of Zyban per day. [*Source:* GlaxoSmithKline]
  (a) In a random sample of 25 users of Zyban, find the probability that exactly 8 will experience insomnia.
  (b) In a random sample of 25 users of Zyban, find the probability that fewer than 4 will experience insomnia.
  (c) In a random sample of 25 users of Zyban, find the probability that at least 5 will experience insomnia.
  (d) In a random sample of 25 users of Zyban, find the probability that exactly 20 will not experience insomnia.
  (e) In a random sample of 1000 users of Zyban, what is the expected number who experience insomnia? What is the standard deviation?
  (f) If a random sample of 1000 users of Zyban results in 330 who experience insomnia, would this be unusual? Why?

*In Problems 15–16, (a) construct a binomial probability distribution with the given parameters, (b) compute the mean and standard deviation of the distribution by using the methods of Section 6.1, (c) compute the mean and standard deviation by using the methods of this section, (d) draw the probability histogram, comment on its shape, and label the mean on the histogram.*

**15.** $n = 5, p = 0.2$

**16.** $n = 8, p = 0.75$

**17.** State the condition required to use the Empirical Rule in order to check for unusual observations in a binomial experiment.

**18.** In sampling without replacement, the assumption of independence required for a binomial experiment is violated. Under what circumstances can we sample without replacement and still use the binomial probability formula to approximate probabilities?

**Chapter 6 Projects located at www.prenhall.com/sullivanstats**

# The Normal Probability Distribution

For additional study help, go to
www.prenhall.com/sullivanstats

> Materials include

- Projects:
  - Case Study: A Tale of Blood Chemistry and Health
  - Decisions: Join the Club
  - Consumer Reports Project
- Self-Graded Quizzes
- "Preparing for This Section" Quizzes
- STATLETs
- PowerPoint Downloads
- Step-by-Step Technology Guide
- Graphing Calculator Help

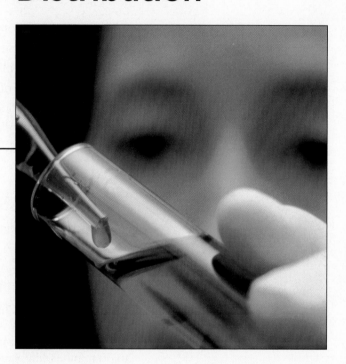

## Putting It All Together

In Chapter 6, we introduced discrete probability distributions and, in particular, the binomial probability distribution. We computed probabilities for this discrete

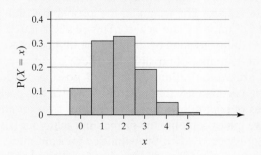

**Binomial Probability Histogram; n = 5, p = 0.35**

distribution using its probability distribution function. However, we could also determine the probability of any discrete random variable from its probability histogram. For example, the figure shows the probability histogram for the binomial random variable $X$ with $n = 5$ and $p = 0.35$.

From the probability histogram, we can see $P(X = 1) \approx 0.31$. Notice that the width of each rectangle in the probability histogram is one. Since the area of a rectangle equals height times width, we can think of $P(X = 1)$ as the area of the rectangle corresponding to $X = 1$. Understanding probability in this fashion makes the transition from computing discrete probabilities to continuous probabilities much easier.

In this chapter, we discuss two continuous distributions, *the uniform distribution* and *the normal distribution*. The greater part of the discussion will center on the normal distribution.

## 7.1   Properties of the Normal Distribution

**Preparing for This Section**   Before getting started, review the following:

✓ Continuous variable (Section 1.1, p. 5)

✓ Requirements for a discrete probability distribution (Section 6.1, p. 242)

✓ $z$-score (Section 3.4, pp. 122–123)

✓ The Empirical Rule (Section 3.2, pp. 107–108)

**Objectives**    ❶ Understand the uniform probability distribution

 ❷ Graph a normal density curve

 ❸ State the properties of the normal curve

❹ Understand the role of area in the normal density function

 ❺ Understand the relation between a normal random variable and a standard normal random variable with regard to area

### ❶ The Uniform Distribution

We illustrate a uniform distribution using an example.

▶ **EXAMPLE 1**   **Illustrating the Uniform Distribution**

Imagine that a friend of yours is always late. Let the random variable $X$ represent the time from when you are supposed to meet your friend until he shows up. Further suppose that your friend could be on time ($x = 0$) or up to 30 minutes late ($x = 30$) with all 1-minute intervals of times between $x = 0$ and $x = 30$ equally likely. That is to say, your friend is just as likely to be from 3 to 4 minutes late as he is to be 25 to 26 minutes late. The random variable $X$ can be any value in the interval from 0 to 30, that is, $0 \leq X \leq 30$. Because any two intervals of equal length between 0 and 30, inclusive, are equally likely, the random variable $X$ is said to follow a **uniform probability distribution**.   ◀◀

When we compute probabilities for discrete random variables, we usually substitute the value of the random variable into a probability distribution function. Things aren't as easy for continuous random variables. Because there are an infinite number of possible outcomes for continuous random variables, the probability of observing a particular value of a continuous random variable is zero. For example, the probability that your friend is exactly 12.9438823 minutes late is zero. This result is based upon the fact that classical probability is found by dividing the number of ways an event can occur by the total number of possibilities. There is one way to observe 12.9438823 and there are an infinite number of possible values between 0 and 30, so we get a probability that is zero. To resolve this problem, we compute probabilities of continuous random variables over an interval of values. For example, we might compute the probability that your friend is between 10 and 15 minutes late. To find probabilities for continuous random variables, we do not use probability distribution functions, but instead we use *probability density functions*.

*Definition*   **Probability Density Function**

A probability density function is an equation used to compute probabilities of continuous random variables that must satisfy the following two properties.

1. The area under the graph of the equation over all possible values of the random variable must equal one.
2. The graph of the equation must be greater than or equal to zero for all possible values of the random variable. That is, the graph of the equation must lie on or above the horizontal axis for all possible values of the random variable.

**In Your Own Words**

To find probabilities for continuous random variables, we do not use probability distribution functions (as we did for discrete random variables). Instead, we use probability density functions.

Property 1 is similar to the property for discrete probability distributions that stated that the sum of the probabilities must add up to one. Property 2 is similar to the property that stated that all probabilities must be greater than or equal to zero.

Figure 1 illustrates the properties for your late friend example. Since all possible values of the random variable between 0 and 30 are equally likely, the graph of the probability density function for uniform random variables is a rectangle. Because the area under the graph of the equation must equal 1 and the random variable is any number between 0 and 30 inclusive, the width of the rectangle is 30. Therefore, the height of the rectangle must be 1/30.*

**Figure 1**
Uniform Density Function

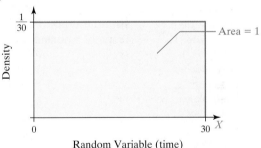

Random Variable (time)

A pressing question remains: How do we use density functions to find probabilities of continuous random variables?

**The area under the graph of a density function over some interval represents the probability of observing a value of the random variable in that interval.**

The following example illustrates this statement.

▶ **EXAMPLE 2**   **Area as a Probability**

*Problem:* Refer to the situation presented in Example 1. What is the probability that your friend will be between 10 and 20 minutes late the next time you meet him?

*Approach:* Figure 1 presented the graph of the density function. We need to find the area under the graph between 10 and 20 minutes.

---

*Area of rectangle = height × width. Rather than using "Probability" as a label on the vertical axis, we use "Density" because the height of the graph does not represent a probability.

## Historical Note

The normal probability distribution is often referred to as the Gaussian distribution in honor of Carl Gauss, the individual thought to have discovered the idea. However, it was actually Abraham de Moivre who first wrote down the equation of the normal distribution. Gauss was born in Brunswick, Germany, on April 30, 1777. Gauss' mathematical prowess was evident early in his life. At age eight he was able to instantly add the first 100 integers. In 1792, Gauss entered the Brunswick Collegium Carolinum and remained there for three years. In 1795, Gauss entered the University of Göttingen. In 1799, Gauss earned his doctorate. The subject of his dissertation was the Fundamental Theorem of Algebra. In 1809, Gauss published a book on the mathematics of planetary orbits. In this book, he further developed the theory of least-squares regression by analyzing the errors. The analysis of these errors led to the discovery that errors follow a normal distribution. Gauss was considered to be "glacially cold" as a person and had troubled relationships with his family. Gauss died February 23, 1855.

*Solution:* Figure 2 presents the graph of the density function with the area we wish to find shaded in green.

Figure 2

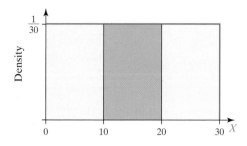

The width of the rectangle is 10 and its height is 1/30; therefore, the area between 10 and 20 is 10(1/30) = 1/3. The probability that your friend is between 10 and 20 minutes late is 1/3. ◀◀

**NW** *Now Work Problem 1.*

We introduced the uniform density function so that we associate probability with area. We are now better prepared to discuss the most popular continuous distribution, the normal distribution.

## ② The Normal Distribution

Many continuous random variables such as IQ scores, birth weights of babies, or serum cholesterol levels have relative frequency histograms that have a shape similar to Figure 3. Relative frequency histograms that are symmetric and bell-shaped (i.e., similar to Figure 3) are said to have the shape of a **normal curve**.

Figure 3

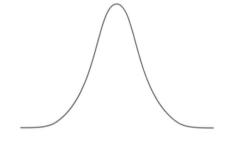

*Definition*

A continuous random variable is **normally distributed** or has a **normal probability distribution** if its relative frequency histogram of the random variable has the shape of a normal curve (bell-shaped and symmetric).

Figure 4 shows a normal curve* demonstrating the role $\mu$ and $\sigma$ play in drawing the curve. The mean, $\mu$, determines the high point of the distribution.

---

*The equation that is used to determine the probability of continuous random variable is called a **probability density function** (or **pdf**). The **normal probability density function** is given by

$y = \dfrac{1}{\sigma\sqrt{2\pi}}e^{-(x-\mu)^2/2\sigma^2}$ where $\mu$ is the mean and $\sigma$ is the standard deviation of the normal random variable. The values $\mu$ and $\sigma$ are the parameters of the distribution.

**Historical Note**

It was Karl Pearson who coined the phrase "normal" curve. He did not do this to imply that a distribution that is not normal is "abnormal," but rather to avoid giving the name of the distribution a proper name, such as Gaussian (as in Carl Friedrich Gauss).

The points at $x = \mu - \sigma$ and $x = \mu + \sigma$ are the *inflection points* on the normal curve. The **inflection points** are the points on the curve where the curvature of the graph changes. To the left of $x = \mu - \sigma$ and to the right of $x = \mu + \sigma$, the curve is drawn upward ( $\bigcup$ or $\bigcup$ ). In between $x = \mu - \sigma$ and $x = \mu + \sigma$, the curve is drawn downward ( $\bigcap$ ).†

Figure 4

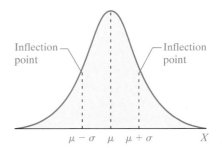

Figure 5 shows how changes in $\mu$ and $\sigma$ change the position or shape of a normal curve. In Figure 5(a), two normal density curves are drawn with the location of the inflection points labeled. One density curve has $\mu = 0, \sigma = 1$, and the other has $\mu = 3, \sigma = 1$. We can see that increasing the mean from 0 to 3 caused the graph to shift three units to the right. In Figure 5(b), two normal density curves are drawn, again with the inflection points labeled. One density curve has $\mu = 0, \sigma = 1$ and the other has $\mu = 0, \sigma = 2$. We can see that increasing the standard deviation from 1 to 2 causes the graph to become flatter and more spread out.

Figure 5

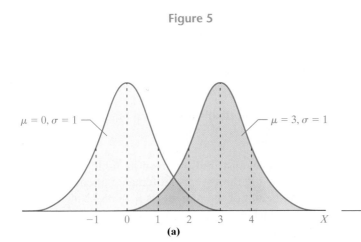
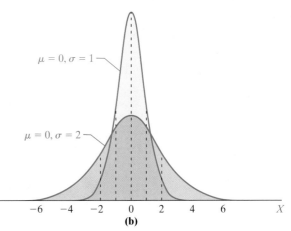

(a)  (b)

NW *Now Work Problem 11.*

†The vertical scale on the graph is purposely omitted. The vertical scale, while important, will not play a role in any of the computations utilizing this curve. We are interested simply in the basic shape of the density curve.

**③** The normal probability density function satisfies all the requirements that are necessary in order to have a legitimate probability distribution. We list the properties of the normal density curve below.

> **Properties of the Normal Density Curve**
>
> 1. It is symmetric about its mean, $\mu$.
> 2. The highest point occurs at $x = \mu$.
> 3. It has inflection points at $\mu - \sigma$ and $\mu + \sigma$.
> 4. The area under the curve is one.
> 5. The area under the curve to the right of $\mu$ equals the area under the curve to the left of $\mu$ equals $1/2$.
> 6. As $x$ increases without bound (gets larger and larger), the graph approaches, but never equals, zero. As $x$ decreases without bound (gets larger and larger in the negative direction) the graph approaches, but never equals, zero.
> 7. The Empirical Rule: Approximately 68% of the area under the normal curve is between $x = \mu - \sigma$ and $x = \mu + \sigma$. Approximately 95% of the area under the normal curve is between $x = \mu - 2\sigma$ and $x = \mu + 2\sigma$. Approximately 99.7% of the area under the normal curve is between $x = \mu - 3\sigma$ and $x = \mu + 3\sigma$. See Figure 6.

**Figure 6**

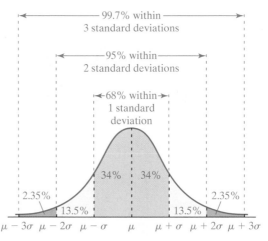

**Normal Distribution**

**④** Let's look at an example of a normally distributed random variable.

▶ **EXAMPLE 3    A Normal Random Variable**

*Problem:* The relative frequency distribution given in Table 1 represents the heights of a pediatrician's 200 three-year-old female patients. The raw data indicate that the mean height of the patients is $\mu = 38.72$ inches with standard deviation $\sigma = 3.17$ inches.

**(a)** Draw a relative frequency histogram of the data. Comment on the shape of the distribution.

**(b)** Draw a normal curve with $\mu = 38.72$ inches and $\sigma = 3.17$ on the relative frequency histogram. Compare the area of the rectangle for heights between 40 and 40.9 inches to the area under the normal curve for heights between 40 and 40.9 inches.

*Approach:*

**(a)** Draw the relative frequency histogram. If the histogram looks like Figure 4 on page 275, we will say that height is approximately normal. We say that height is "approximately normal," rather than "normal" because it is very rare to have a perfectly normal random variable.

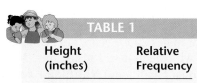

| TABLE 1 | |
|---|---|
| Height (inches) | Relative Frequency |
| 29.0–29.9 | 0.005 |
| 30.0–30.9 | 0.005 |
| 31.0–31.9 | 0.005 |
| 32.0–32.9 | 0.025 |
| 33.0–33.9 | 0.02 |
| 34.0–34.9 | 0.055 |
| 35.0–35.9 | 0.075 |
| 36.0–36.9 | 0.09 |
| 37.0–37.9 | 0.115 |
| 38.0–38.9 | 0.15 |
| 39.0–39.9 | 0.12 |
| 40.0–40.9 | 0.11 |
| 41.0–41.9 | 0.07 |
| 42.0–42.9 | 0.06 |
| 43.0–43.9 | 0.035 |
| 44.0–44.9 | 0.025 |
| 45.0–45.9 | 0.025 |
| 46.0–46.9 | 0.005 |
| 47.0–47.9 | 0.005 |

**Caution**

It is rare for a continuous random variable to be exactly normal. Therefore, we usually say that a random variable is approximately normal if its histogram is bell-shaped and symmetric.

**(b)** Draw the normal curve on the histogram with the high point at $\mu$ and the inflection points at $\mu - \sigma$ and $\mu + \sigma$. Shade the rectangle corresponding to heights between 40 and 40.9 inches and compare the area of the shaded region to that under the normal curve between 40 and 40.9.

*Solution:*

**(a)** Figure 7 shows the relative frequency histogram. The relative frequency histogram is symmetric and bell-shaped. It would appear that the random variable height of three-year-old females is approximately normally distributed.

Figure 7

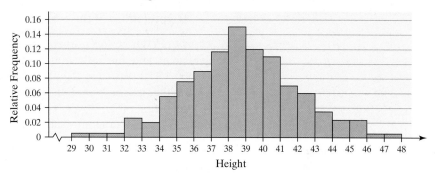

**Heights of Three-Year-Old Females**

**(b)** The normal curve with $\mu = 38.72$ and $\sigma = 3.17$ is superimposed on the relative frequency histogram in Figure 8. The figure demonstrates that the normal curve describes the heights of three-year-old girls fairly well. We conclude that the heights of three-year-old girls are approximately normal with $\mu = 38.72$ and $\sigma = 3.17$. Figure 8 also shows the rectangle corresponding to heights between 40 and 40.9 inches. The area of this rectangle represents the proportion of three-year-old females between 40 and 40.9 inches. Notice that the area of this shaded region is very close to the area under the normal curve for the same region. It is because of this that we can use the area under the normal curve to approximate the proportion of three-year-old females with heights between 40 and 40.9 inches!

Figure 8
Heights of Three-Year-Old Female Patients

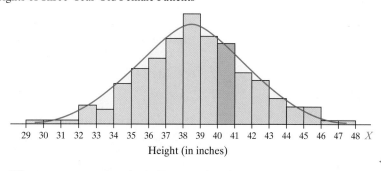

Height (in inches)

◀◀

We now summarize the role area plays in the normal curve.

**The Area under a Normal Curve**

Suppose a random variable $X$ is normally distributed with mean $\mu$ and standard deviation $\sigma$. The area under the normal curve for any range of values of the random variable $X$ represents either

- the proportion of the population with the characteristics described by the range, or

- the probability that a randomly selected individual from the population will have the characteristics described by the range.

▶ **EXAMPLE 4** **Interpreting the Area under a Normal Curve**

*Problem:* The serum total cholesterol for males 20–29 years old is known to be approximately normally distributed with mean $\mu = 180$ and $\sigma = 36.2$ based upon data obtained from the National Health and Nutrition Examination Survey.

(a) Draw a normal curve with the parameters labeled.

(b) An individual with a total cholesterol greater than 200 is considered to have high cholesterol. Shade the region under the normal curve to the right of $X = 200$.

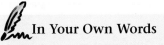
**In Your Own Words**

The area under a normal curve is a proportion or probability.

(c) Suppose the area under the normal curve to the right of $X = 200$ is 0.2903. (You will learn how to find this area in Section 7.3.) Provide two interpretations of this result.

*Approach:* Draw the normal curve with the mean $\mu = 180$ labeled at the high point and the inflection points at $\mu - \sigma = 180 - 36.2 = 143.8$ and $\mu + \sigma = 180 + 36.2 = 216.2$. We then shade the region under the normal curve to the right of $X = 200$. Two interpretations of this shaded region are (1) the proportion of 20–29-year-old males who have high cholesterol and (2) the probability that a randomly selected 20–29-year-old male has high cholesterol.

*Solution:*

(a) Figure 9(a) shows the graph of the normal curve.

Figure 9

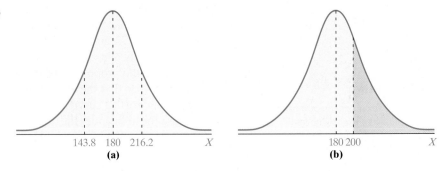

(a)                                             (b)

(b) Figure 9(b) shows the region under the normal curve to the right of $X = 200$ shaded.

(c) The two interpretations for the area of this shaded region are (1) the proportion of 20–29-year-old males that have high cholesterol is 0.2903 and (2) the probability that a randomly selected 20–29-year-old male has high cholesterol is 0.2903. ◄◄

**NW** *Now Work Problems 15 and 19.*

### ⑤ Standardizing a Normal Random Variable

At this point, we know that a random variable $X$ is normally distributed if its relative frequency histogram has the shape of a normal curve. The distribution of the random variable $X$ is normal with mean $\mu$ and standard deviation $\sigma$. The area below the normal curve represents the proportion of the population with a given characteristic or the probability that a randomly selected individual from the population will have a given characteristic.

The question now becomes "How do I find the area under the normal curve?" Finding the area under a curve requires techniques introduced in calculus, which are beyond the scope of this text. An alternative would be to use a series of tables to find areas. However, this would result in an infinite number of tables being created for each possible mean and standard deviation! A solution to the problem lies in the $Z$-score. Recall that the $Z$-score allows us to transform a random variable $X$ with mean $\mu$ and standard deviation $\sigma$ into a random variable $Z$ with mean 0 and standard deviation 1.

---

**Standardizing a Normal Random Variable**

Suppose the random variable $X$ is normally distributed with mean $\mu$ and standard deviation $\sigma$. Then the random variable

$$Z = \frac{X - \mu}{\sigma}$$

is normally distributed with mean $\mu = 0$ and standard deviation $\sigma = 1$. The random variable $Z$ is said to have the **standard normal distribution**.

---

**In Your Own Words**

To find the area under any normal curve, we must first find the Z-score of the normal variable.

This result is powerful indeed! We need only one table of areas corresponding to the standard normal distribution. If a normal random variable has mean different from 0 or standard deviation different from 1, we simply transform the normal random variable into a standard normal random variable $Z$ and use a table to find the area and, therefore, the probability.

We demonstrate the idea behind standardizing a normal variable in the next example.

▶ **EXAMPLE 5**   **Relation between a Normal Random Variable and a Standard Normal Random Variable**

*Problem:* The heights of a pediatrician's 200 three-year-old female patients are known to follow a normal distribution with mean $\mu = 38.72$ and $\sigma = 3.17$. We wish to demonstrate that the area under the normal curve between 35 and 38 inches is equal to the area under the standard normal curve between the $Z$-scores of 35 and 38 inches.

*Approach:*

**Step 1:** Draw a normal curve and shade the area representing the proportion of three-year-old females between 35 and 38 inches tall.

**Step 2:** Standardize the random variable $X = 35$ and $X = 38$ using $Z = \frac{X - \mu}{\sigma}$.

**Step 3:** Draw the standard normal curve with the standardized versions of $X = 35$ and $X = 38$ labeled. Shade the area that represents the proportion of three-year-old females between 35 and 38 inches tall. Comment on the relation between the two shaded regions.

*Solution:*

**Step 1:** Figure 10(a) shows the normal curve with mean $\mu = 38.72$ and $\sigma = 3.17$. The region between $X = 35$ and $X = 38$ is shaded.

**Step 2:** With $\mu = 38.72$ and $\sigma = 3.17$, the standardized version of $X = 35$ is

$$Z = \frac{X - \mu}{\sigma} = \frac{35 - 38.72}{3.17} = -1.17*$$

---

*Recall that we round Z-scores to two decimal places.

The standardized version of $X = 38$ is

$$Z = \frac{X - \mu}{\sigma} = \frac{38 - 38.72}{3.17} = -0.23$$

***Step 3:*** Figure 10(b) shows the standard normal curve with the region between $Z = -1.17$ and $Z = -0.23$ shaded.

**Figure 10**
(a) Normal curve with $\mu = 38.72$ and $\sigma = 3.17$
(b) Standard normal curve

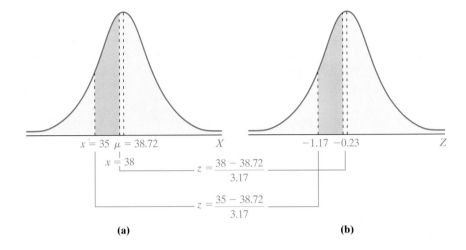

(a)  (b)

The area under the normal curve with $\mu = 38.72$ and $\sigma = 3.17$ bounded to the left by $x = 35$ and bounded to the right by $x = 38$ is equal to the area under the standard normal curve bounded to the left by $z = -1.17$ and bounded to the right by $z = -0.23$.  ◄◄

**NW** *Now Work Problem 21.*

# 7.1 Assess Your Understanding

## Concepts and Vocabulary

1. State the two characteristics of the graph of a probability density function.
2. List the characteristics of a normal density curve.
3. Provide two interpretations of the area under a probability density function.
4. Why do we standardize normal random variables in order to find the area under any normal curve?

5. Discuss the role of $\mu$ and $\sigma$ in constructing the normal density curve.
6. As $\sigma$ increases, the normal density curve becomes more spread out. Knowing the area under the density curve must be 1, what effect does increasing $\sigma$ have on the height of the curve?

## Exercises

### • Skill Building

*Problems 1–4 use the information presented in Examples 1 and 2.*

1. Find the probability that your friend is between 5 and
**NW** 10 minutes late.

2. Find the probability that your friend is between 15 and 25 minutes late.

3. Find the probability that your friend is at least 20 minutes late.

4. Find the probability that your friend is no more than 5 minutes late.

5. **Uniform Distribution** The random number generator on calculators randomly generates a number between 0 and 1. The random variable $X$, the number generated, follows a Uniform Probability Distribution.

(a) Draw the graph of the Uniform Density Function.
(b) What is the probability of generating a number between 0 and 0.2?
(c) What is the probability of generating a number between 0.25 and 0.6?
(d) What is the probability of generating a number greater than 0.95?
(e) Use your calculator or statistical software to randomly generate 200 numbers between 0 and 1. What proportion of the numbers are between 0 and 0.2? Compare the result with part (b).

6. **Uniform Distribution** Suppose the reaction time $X$ (in minutes) of a certain chemical process follows a Uniform Probability Distribution with $5 \leq X \leq 10$.

    (a) Draw the graph of the density curve.
    (b) What is the probability that the reaction time is between 6 and 8 minutes?

    (c) What is the probability that the reaction time is between 5 and 8 minutes?
    (d) What is the probability that the reaction time is less than 6 minutes?

---

*In Problems 7–10, determine whether the histogram indicates that the data may be normally distributed.*

7. **Birth Weights** The following relative frequency histogram represents the birth weights of babies whose term was 36 weeks:

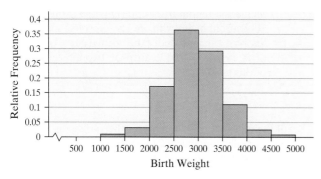

8. **Waiting in Line** The following relative frequency histogram represents the waiting time in line (in minutes) for the Demon Roller Coaster for 2000 randomly selected people on a Saturday afternoon in the summer:

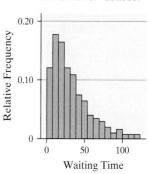

9. **Length of Phone Calls** The following relative frequency histogram represents the length of phone calls on my wife's cell phone during the month of September:

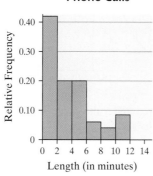

10. **IQ Scores** The following relative frequency histogram represents the IQ scores of a random sample of seventh grade students:

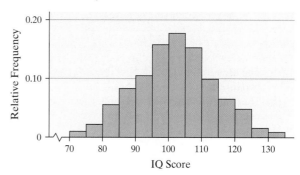

---

*In Problems 11–14, the graph of a normal curve is given. Use the graph to identify the value of $\mu$ and $\sigma$.*

11.    12.    13.    14.

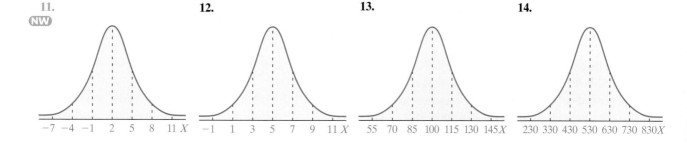

## • Applying the Concepts

**15. SAT Verbal Scores** SAT Verbal scores are known to be normally distributed with mean $\mu = 505$ and standard deviation $\sigma = 110$ based upon data obtained from the College Board.

(a) Draw a normal curve with the parameters labeled.

(b) Shade the region that represents the proportion of test takers who scored less than 395.

(c) Suppose the area under the normal curve to the left of $X = 395$ is 0.1587. Provide two interpretations of this result.

**16. ACT English** ACT English scores are known to be normally distributed with mean $\mu = 20.5$ and standard deviation $\sigma = 5.5$ based upon data obtained from ACT Research.

(a) Draw a normal curve with the parameters labeled.

(b) Shade the region that represents the proportion of test takers who scored more than 27.

(c) Suppose the area under the normal curve to the right of $X = 27$ is 0.1186. Provide two interpretations of this result.

**17. Birth Weights** The birth weight of full-term babies is known to be normally distributed with mean $\mu = 3400$ grams and $\sigma = 505$ grams based upon data obtained from the National Vital Statistics Report, Vol. 48, No. 3.

(a) Draw a normal curve with the parameters labeled.

(b) Shade the region that represents the proportion of full-term babies who weighed more than 4410 grams.

(c) Suppose the area under the normal curve to the right of $X = 4410$ is 0.0228. Provide two interpretations of this result.

**18. Height of 10-Year-Old Males** The height of 10-year-old males is known to be normally distributed with mean $\mu = 55.9$ inches and $\sigma = 5.7$ inches.

(a) Draw a normal curve with the parameters labeled.

(b) Shade the region that represents the proportion of 10-year-old males who are less than 46.5 inches tall.

(c) Suppose the area under the normal curve to the left of $X = 46.5$ is 0.0496. Provide two interpretations of this result.

---

**19. Gestation Period** The length of human pregnancy is normally distributed with $\mu = 266$ days and $\sigma = 16$ days.

(a) The following figure represents the normal curve with $\mu = 266$ days and $\sigma = 16$ days:

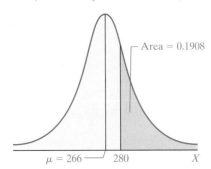

The area to the right of $X = 280$ is 0.1908. Provide two interpretations of this area.

(b) The following figure represents the normal curve with $\mu = 266$ days and $\sigma = 16$ days:

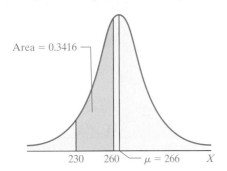

The area between $X = 230$ and $X = 260$ is 0.3416. Provide two interpretations of this area.

**20. Miles per Gallon** Elena conducts an experiment in which she fills up the gas tank on her Toyota Camry 40 times and records the miles per gallon for each fill-up. A histogram

of the miles per gallon indicates the variable is normally distributed with mean of 24.6 miles per gallon and a standard deviation of 3.2 miles per gallon.

(a) The following figure represents the normal curve with $\mu = 24.6$ miles per gallon and $\sigma = 3.2$ miles per gallon:

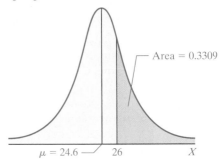

The area under the curve to the right of $X = 26$ is 0.3309. Provide two interpretations of this area.

(b) The following figure represents the normal curve with $\mu = 24.6$ miles per gallon and $\sigma = 3.2$ miles per gallon:

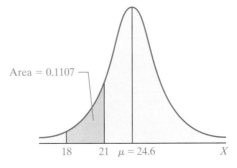

The area under the curve between $X = 18$ and $X = 21$ is 0.1107. Provide two interpretations of this area.

**21.** A random variable $X$ is normally distributed with
$\mu = 10$ and $\sigma = 3$.

(a) Compute $Z_1 = \dfrac{X_1 - \mu}{\sigma}$ for $X_1 = 8$.

(b) Compute $Z_2 = \dfrac{X_2 - \mu}{\sigma}$ for $X_2 = 12$.

(c) The area under the normal curve between $X_1 = 8$ and $X_2 = 12$ is 0.495. What is the area between $Z_1$ and $Z_2$?

**22.** A random variable $X$ is normally distributed with
$\mu = 25$ and $\sigma = 6$.

(a) Compute $Z_1 = \dfrac{X_1 - \mu}{\sigma}$ for $X_1 = 18$.

(b) Compute $Z_2 = \dfrac{X_2 - \mu}{\sigma}$ for $X_2 = 30$.

(c) The area under the normal curve between $X_1 = 18$ and $X_2 = 30$ is 0.6760. What is the area between $Z_1$ and $Z_2$?

---

**23. Hitting a Pitching Wedge**  In the game of golf, "distance control" is just as important as how far a player hits the ball. Suppose Michael went to the driving range with his range finder and hit 75 golf balls with his pitching wedge and measured the distance each ball traveled (in yards). He obtained the data shown to the right.

(a) Use MINITAB or some other statistical software to construct a relative frequency histogram. Comment on the shape of the distribution.

(b) Use MINITAB or some other statistical software to draw the normal density function on the relative frequency histogram.

(c) Do you think the normal density function accurately describes the distance Michael hits a pitching wedge? Why?

| 100 | 97 | 101 | 101 | 103 | 100 | 99 | 100 | 100 |
|---|---|---|---|---|---|---|---|---|
| 104 | 100 | 101 | 98 | 100 | 99 | 99 | 97 | 101 |
| 104 | 99 | 101 | 101 | 101 | 100 | 96 | 99 | 99 |
| 98 | 94 | 98 | 107 | 98 | 100 | 98 | 103 | 100 |
| 98 | 94 | 104 | 104 | 98 | 101 | 99 | 97 | 103 |
| 102 | 101 | 101 | 100 | 95 | 104 | 99 | 102 | 95 |
| 99 | 102 | 103 | 97 | 101 | 102 | 96 | 102 | 99 |
| 96 | 108 | 103 | 100 | 95 | 101 | 103 | 105 | 100 |
| 94 | 99 | 95 | | | | | | |

---

**24. Heights of Five-Year-Old Females**  The data to the right represent the heights (in inches) of eighty randomly selected five-year-old females.

(a) Use Minitab or some other statistical software to construct a relative frequency histogram. Comment on the shape of the distribution.

(b) Use Minitab or some other statistical software to draw the normal density function on the relative frequency histogram.

(c) Do you think the normal density function accurately describes the heights of five-year-old females? Why?

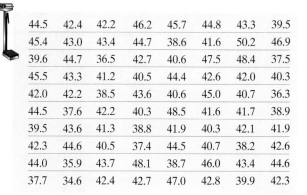

| 44.5 | 42.4 | 42.2 | 46.2 | 45.7 | 44.8 | 43.3 | 39.5 |
|---|---|---|---|---|---|---|---|
| 45.4 | 43.0 | 43.4 | 44.7 | 38.6 | 41.6 | 50.2 | 46.9 |
| 39.6 | 44.7 | 36.5 | 42.7 | 40.6 | 47.5 | 48.4 | 37.5 |
| 45.5 | 43.3 | 41.2 | 40.5 | 44.4 | 42.6 | 42.0 | 40.3 |
| 42.0 | 42.2 | 38.5 | 43.6 | 40.6 | 45.0 | 40.7 | 36.3 |
| 44.5 | 37.6 | 42.2 | 40.3 | 48.5 | 41.6 | 41.7 | 38.9 |
| 39.5 | 43.6 | 41.3 | 38.8 | 41.9 | 40.3 | 42.1 | 41.9 |
| 42.3 | 44.6 | 40.5 | 37.4 | 44.5 | 40.7 | 38.2 | 42.6 |
| 44.0 | 35.9 | 43.7 | 48.1 | 38.7 | 46.0 | 43.4 | 44.6 |
| 37.7 | 34.6 | 42.4 | 42.7 | 47.0 | 42.8 | 39.9 | 42.3 |

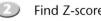 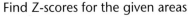

## 7.2    The Standard Normal Distribution

**Preparing for This Section**   Before getting started, review the following:

✓ The Complement Rule (Section 5.2, pp. 204–206)

**Objectives**   ① Find the area under the standard normal curve

② Find Z-scores for the given areas

③ Interpret the area under the standard normal curve as a probability

In Section 7.1, we introduced the normal distribution. We learned that if $X$ is a normally distributed random variable, we can use the area under the normal density function to obtain the proportion of a population, or the probability that a randomly selected individual from the population, has a certain characteristic. To find the area under the normal curve, we first convert the random variable $X$ to a standard normal random variable $Z$ with mean $\mu = 0$ and standard deviation $\sigma = 1$ and find the area under the standard normal curve. This section discusses methods for finding the area under the standard normal curve.

### Properties of the Standard Normal Distribution

**Figure 11**

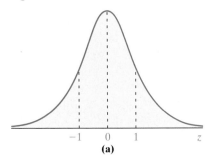

**(a)**

The standard normal distribution has a mean of 0 and a standard deviation of 1. The standard normal curve therefore, will have its high point located at 0 and inflection points located at −1 and +1. We use the random variable $Z$ to represent a standard normal random variable. The graph of the standard normal curve is presented in Figure 11.

Although we stated the properties of normal curves in general in Section 7.1, it is worthwhile to restate them here as they pertain to the standard normal curve.

**Properties of the Standard Normal Curve**

**1.** It is symmetric about its mean, $\mu = 0$.

**2.** Its highest point occurs at $\mu = 0$.

**3.** It has inflection points at $\mu - \sigma = 0 - 1 = -1$ and $\mu + \sigma = 0 + 1 = 1$.

**4.** The area under the curve is 1. This characteristic is required in order to satisfy the requirement that the sum of all probabilities in a legitimate probability distribution equals 1.

**5.** The area under the curve to the right of $\mu = 0$ equals the area under the curve to the left of $\mu = 0$ equals $1/2$.

**6.** As $z$ increases without bound, the graph approaches, but never equals, zero. As $z$ decreases without bound the graph approaches, but never equals, zero.

**7.** The Empirical Rule: Approximately $0.68 = 68\%$ of the area under the standard normal curve is between −1 and 1. Approximately $0.95 = 95\%$ of the area under the standard normal curve is between −2 and 2. Approximately $0.997 = 99.7\%$ of the area under the standard normal curve is between −3 and 3. See Figure 12.

**Figure 12**

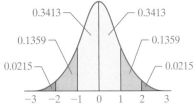

We now discuss the procedure for finding area under the standard normal curve.

① **Finding Area under the Standard Normal Curve**

We discuss two methods for finding area under the standard normal curve. The first method utilizes tables of areas that have been constructed for

Figure 13

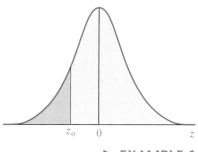

various values of $Z$. The second method involves the use of statistical software or a calculator with advanced statistical features.

Table II, which can be found in the front inside cover of the text or in Appendix A, gives areas under the standard normal curve. The table gives the area under the standard normal curve for values to the left of a specified $Z$-score, $z_0$, as shown in Figure 13.

The shaded region represents the area under the standard normal curve to the left of $Z = z_0$. Whenever finding area under a normal curve, you should sketch a normal curve and shade the area you are finding.

▶ EXAMPLE 1  **Finding Area under the Standard Normal Curve to the Left of a Z-score**

*Problem:* Find the area under the standard normal curve that lies to the left of $Z = 1.68$.

*Approach:*

*Step 1:* Draw a standard normal curve with $Z = 1.68$ labeled and shade the area under the curve to the left of $Z = 1.68$.

*Step 2:* The rows in Table II represent the ones and tenths portion of $Z$, while the columns represent the hundredths portion. To find the area under the curve to the left of $Z = 1.68$, we need to split 1.68 as 1.6 and 0.08. Find the row that represents "1.6" and the column that represents "0.08" in Table II. Identify where the row and column intersect. This value is the area.

Figure 14

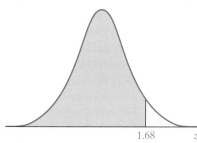

*Solution:*

*Step 1:* Figure 14 shows the graph of the standard normal curve with $Z = 1.68$ labeled. The area left of $Z = 1.68$ is shaded.

*Step 2:* A portion of Table II is presented in Figure 15. We have enclosed the row that represents "1.6" and the column that represents "0.08." The point where the row and column intersect is the area we are seeking. The area to the left of $Z = 1.68$ is 0.9535.

Figure 15

| z | 0.00 | 0.01 | 0.02 | 0.03 | 0.04 | 0.05 | 0.06 | 0.07 | 0.08 | 0.09 |
|---|------|------|------|------|------|------|------|------|------|------|
| 0.0 | 0.5000 | 0.5040 | 0.5080 | 0.5120 | 0.5160 | 0.5199 | 0.5239 | 0.5279 | 0.5319 | 0.5359 |
| 0.1 | 0.5398 | 0.5438 | 0.5478 | 0.5517 | 0.5557 | 0.5596 | 0.5636 | 0.5675 | 0.5714 | 0.5753 |
| 0.2 | 0.5793 | 0.5832 | 0.5871 | 0.5910 | 0.5948 | 0.5987 | 0.6026 | 0.6064 | 0.6103 | 0.6141 |
| 0.3 | 0.6179 | 0.6217 | 0.6255 | 0.6293 | 0.6331 | 0.6368 | 0.6406 | 0.6443 | 0.6480 | 0.6517 |
| 0.4 | 0.6554 | 0.6591 | 0.6628 | 0.6664 | 0.6700 | 0.6736 | 0.6772 | 0.6808 | 0.6844 | 0.6879 |
| 0.5 | 0.6915 | 0.6950 | 0.6985 | 0.7019 | 0.7054 | 0.7088 | 0.7123 | 0.7157 | 0.7190 | 0.7224 |
| 0.6 | 0.7257 | 0.7291 | 0.7324 | 0.7357 | 0.7389 | 0.7422 | 0.7454 | 0.7486 | 0.7517 | 0.7549 |
| 0.7 | 0.7580 | 0.7611 | 0.7642 | 0.7673 | 0.7704 | 0.7734 | 0.7764 | 0.7794 | 0.7823 | 0.7852 |
| 0.8 | 0.7881 | 0.7910 | 0.7939 | 0.7967 | 0.7995 | 0.8023 | 0.8051 | 0.8078 | 0.8106 | 0.8133 |
| 0.9 | 0.8159 | 0.8186 | 0.8212 | 0.8238 | 0.8264 | 0.8289 | 0.8315 | 0.8340 | 0.8365 | 0.8389 |
| 1.0 | 0.8413 | 0.8438 | 0.8461 | 0.8485 | 0.8508 | 0.8531 | 0.8554 | 0.8577 | 0.8599 | 0.8621 |
| 1.1 | 0.8643 | 0.8665 | 0.8686 | 0.8708 | 0.8729 | 0.8749 | 0.8770 | 0.8790 | 0.8810 | 0.8830 |
| 1.2 | 0.8849 | 0.8869 | 0.8888 | 0.8907 | 0.8925 | 0.8944 | 0.8962 | 0.8980 | 0.8997 | 0.9015 |
| 1.3 | 0.9032 | 0.9049 | 0.9066 | 0.9082 | 0.9099 | 0.9115 | 0.9131 | 0.9147 | 0.9162 | 0.9177 |
| 1.4 | 0.9192 | 0.9207 | 0.9222 | 0.9236 | 0.9251 | 0.9265 | 0.9279 | 0.9292 | 0.9306 | 0.9319 |
| 1.5 | 0.9332 | 0.9345 | 0.9357 | 0.9370 | 0.9382 | 0.9394 | 0.9406 | 0.9418 | 0.9429 | 0.9441 |
| 1.6 | 0.9452 | 0.9463 | 0.9474 | 0.9484 | 0.9495 | 0.9505 | 0.9515 | 0.9525 | 0.9535 | 0.9545 |
| 1.7 | 0.9554 | 0.9564 | 0.9573 | 0.9582 | 0.9591 | 0.9599 | 0.9608 | 0.9616 | 0.9625 | 0.9633 |
| 1.8 | 0.9641 | 0.9649 | 0.9656 | 0.9664 | 0.9671 | 0.9678 | 0.9686 | 0.9693 | 0.9699 | 0.9706 |
| 1.9 | 0.9713 | 0.9719 | 0.9726 | 0.9732 | 0.9738 | 0.9744 | 0.9750 | 0.9756 | 0.9761 | 0.9767 |
| 2.0 | 0.9772 | 0.9778 | 0.9783 | 0.9788 | 0.9793 | 0.9798 | 0.9803 | 0.9808 | 0.9812 | 0.9817 |
| 2.1 | 0.9821 | 0.9826 | 0.9830 | 0.9834 | 0.9838 | 0.9842 | 0.9846 | 0.9850 | 0.9854 | 0.9857 |
| 2.2 | 0.9861 | 0.9864 | 0.9868 | 0.9871 | 0.9875 | 0.9878 | 0.9881 | 0.9884 | 0.9887 | 0.9890 |

◀◀

**Using Technology:** The area under the standard normal curve can also be computed using statistical software or graphing calculators with advanced statistical features. Figure 16 shows the results from Example 1 using Minitab. Notice the output is titled "Cumulative Distribution Function." Remember, the word cumulative means "less than or equal to," so Minitab is giving the area under the standard normal curve for $Z$ less than or equal to 1.68.

**Figure 16**

**Cumulative Distribution Function**

```
Normal with mean = 0 and standard deviation = 1.00000

          x          P( X <= x)
     1.6800             0.9535
```

**NW** *Now Work Problem 1.*

Often, rather than being interested in the area under the standard normal curve to the left of $Z = z_0$, we are interested in obtaining the area under the standard normal curve to the right of $Z = z_0$. The idea behind the solution to this type of problem utilizes the fact that the area under the entire standard normal curve is 1 and the Complement Rule. Therefore,

*Theorem*  Area under the normal curve to the right of $z_0$ = 1 − Area to the left of $z_0$.

▶ EXAMPLE 2  **Finding Area under the Standard Normal Curve to the Right of a Z-score**

*In Your Own Words*

area right = 1 − area left

*Problem:* Find the area under the standard normal curve to the right of $Z = -0.46$.

*Approach:*

*Step 1:* Draw a standard normal curve with $Z = -0.46$ labeled and shade the area under the curve to the right of $Z = -0.46$.

*Step 2:* Find the row that represents "−0.4" and the column that represents "0.06" in Table II. Identify where the row and column intersect. This value is the area to the *left* of $Z = -0.46$.

*Step 3:* The area under the standard normal curve to the right of $Z = -0.46$ is 1 minus the area to the left of $Z = -0.46$.

*Solution:*

*Step 1:* Figure 17 shows the graph of the standard normal curve with $Z = -0.46$ labeled. The area to the right of $Z = -0.46$ is shaded.

*Step 2:* A portion of Table II is presented in Figure 18. We have highlighted the row that represents "−0.4" and the column that represents "0.06". The point where the row and column intersect is the area to the left of $Z = -0.46$. The area to the left of $Z = -0.46$ is 0.3228.

*Step 3:* The area under the standard normal curve to the right of $Z = -0.46$ is 1 minus the area to the left of $Z = -0.46$.

**Figure 17**

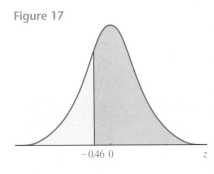

−0.46  0                    z

$$\text{Area right of } -0.46 = 1 - (\text{Area left of } -0.46)$$
$$= 1 - 0.3228$$
$$= 0.6772$$

The area to the right of $Z = -0.46$ is 0.6772.  ◀◀

Figure 18

| z | .00 | .01 | .02 | .03 | .04 | .05 | .06 | .07 | .08 | .09 |
|---|-----|-----|-----|-----|-----|-----|-----|-----|-----|-----|
| −3.4 | 0.0003 | 0.0003 | 0.0003 | 0.0003 | 0.0003 | 0.0003 | 0.0003 | 0.0003 | 0.0003 | 0.0002 |
| −3.3 | 0.0005 | 0.0005 | 0.0005 | 0.0004 | 0.0004 | 0.0004 | 0.0004 | 0.0004 | 0.0004 | 0.0003 |
| −3.2 | 0.0007 | 0.0007 | 0.0006 | 0.0006 | 0.0006 | 0.0006 | 0.0006 | 0.0005 | 0.0005 | 0.0005 |
| −3.1 | 0.0010 | 0.0009 | 0.0009 | 0.0009 | 0.0008 | 0.0008 | 0.0008 | 0.0008 | 0.0007 | 0.0007 |
| −3.0 | 0.0013 | 0.0013 | 0.0013 | 0.0012 | 0.0012 | 0.0011 | 0.0011 | 0.0011 | 0.0010 | 0.0010 |
| −0.5 | 0.3085 | 0.3050 | 0.3015 | 0.2981 | 0.2946 | 0.2912 | 0.2877 | 0.2843 | 0.2810 | 0.2776 |
| −0.4 | 0.3446 | 0.3409 | 0.3372 | 0.3336 | 0.3300 | 0.3264 | 0.3228 | 0.3192 | 0.3156 | 0.3121 |
| −0.3 | 0.3821 | 0.3783 | 0.3745 | 0.3707 | 0.3669 | 0.3632 | 0.3594 | 0.3557 | 0.3520 | 0.3483 |
| −0.2 | 0.4027 | 0.4168 | 0.4129 | 0.4090 | 0.4052 | 0.4013 | 0.3974 | 0.3936 | 0.3897 | 0.3859 |
| −0.1 | 0.4602 | 0.4562 | 0.4522 | 0.4483 | 0.4443 | 0.4404 | 0.4364 | 0.4325 | 0.4286 | 0.4247 |
| −0.0 | 0.5000 | 0.4960 | 0.4920 | 0.4880 | 0.4840 | 0.4801 | 0.4761 | 0.4721 | 0.4681 | 0.4641 |

 *Now Work Problem 3.*

The next example presents a situation in which we are interested in the area between two Z-scores.

▶ **EXAMPLE 3** **Find the Area under the Standard Normal Curve between Two Z-scores**

*Problem:* Find the area under the standard normal curve between $Z = -1.35$ and $Z = 2.01$.

*Approach:*

*Step 1:* Draw a standard normal curve with $Z = -1.35$ and $Z = 2.01$ labeled. Shade the area under the curve between $Z = -1.35$ and $Z = 2.01$.
*Step 2:* Find the area to the left of $Z = -1.35$. Find the area to the left of $Z = 2.01$.
*Step 3:* The area under the standard normal curve between $Z = -1.35$ and $Z = 2.01$ is the area left of $Z = 2.01$ minus the area left of $Z = -1.35$.

Figure 19

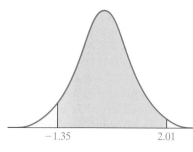

*Solution:*

*Step 1:* Figure 19 shows the standard normal curve with the area between $Z = -1.35$ and $Z = 2.01$ shaded.
*Step 2:* Based upon Table II, the area left of $Z = -1.35$ is 0.0885. The area left of $Z = 2.01$ is 0.9778.
*Step 3:* The area between $Z = -1.35$ and $Z = 2.01$ is

$$(\text{Area between } Z = -1.35 \text{ and } Z = 2.01) = (\text{Area left of } Z = 2.01) - (\text{Area left of } Z = -1.35)$$

$$= 0.9778 - 0.0885$$

$$= 0.8893$$

The area between $Z = -1.35$ and $Z = 2.01$ is 0.8893.   ◀◀

 *Now Work Problem 5.*

We summarize the methods for obtaining area under the standard normal curve in Table 2 on page 288.

### TABLE 2

| Problem | Approach | Solution |
|---|---|---|
| Find the area left of $Z = z_o$ | Shade area left of $Z = z_o$  | Use Table II to find the row and column that correspond to $Z = z_o$. The area is the value where the row and column intersect. Or use technology to find the area. |
| Find the area right of $Z = z_o$ | Shade area right of $Z = z_o$ | Use Table II to find the area left of $Z = z_o$. The area right of $Z = z_o$ is 1 minus the area left of $Z = z_o$. Or use technology to find the area. |
| Find the area between $Z = z_o$ and $Z = z_1$ | Shade the area between $Z = z_o$ and $Z = z_1$ | Use Table II to find the area left of $Z = z_o$ and left of $Z = z_1$. The area between $Z = z_o$ and $Z = z_1$ is (area left of $Z = z_1$) − (area left of $Z = z_o$). Or use technology to find the area. |

**Caution**

State the area under the standard normal curve to the left of $Z = -3.90$ as $< 0.0001$. State the area under the standard normal curve to the left of $Z = 3.90$ as $> 0.9999$.

Because the normal curve extends indefinitely in both directions on the $z$-axis, there is no $Z$-value for which the area under the curve to the left of the $Z$-value is 1. For example, the area to the left of $Z = 10$ is less than 1 even though graphing calculators and statistical software state that the area is 1. This is because there is a limited number of decimal places the machines can compute. We will follow the practice of stating the area left of $Z = -3.90$ or right of $Z = 3.90$ as $< 0.0001$. The area under the standard normal curve left of $Z = 3.90$ or to the right of $Z = -3.90$ will be stated as $> 0.9999$.

### ② Finding Z-scores for Given Areas

Up to this point, we have found areas given the value of a $Z$-score. Often, we are interested in finding a $Z$-score that corresponds to a given area. The procedure to follow is essentially the reverse of the procedure to finding areas given $Z$-scores.

▶ **EXAMPLE 4** **Finding a Z-score from a Specified Area to the Left**

*Problem:* Find the $Z$-score such that the area left of the $Z$-score is 0.32.

*Approach:*

*Step 1:* Draw a standard normal curve with the area and corresponding unknown $Z$-score labeled.

**Step 2:** We look for the area in the table closest to 0.32.

**Step 3:** Find the Z-score that corresponds to the area closest to 0.32.

*Solution:*

**Step 1:** Figure 20 shows the graph of the standard normal curve with the area of 0.32 labeled. We know $z_0$ must be less than 0. Do you know why?

Figure 20

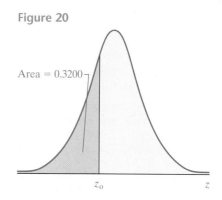

Area = 0.3200

$z_0$                     $z$

**Step 2:** We refer to Table II and look in the body of the table for an area closest to 0.32. The area closest to 0.32 is 0.3192. Figure 21 shows a partial representation of Table II with 0.3192 labeled.

Figure 21

| z | .00 | .01 | .02 | .03 | .04 | .05 | .06 | .07 | .08 | .09 |
|---|-----|-----|-----|-----|-----|-----|-----|-----|-----|-----|
| **−3.4** | 0.0003 | 0.0003 | 0.0003 | 0.0003 | 0.0003 | 0.0003 | 0.0003 | 0.0003 | 0.0003 | 0.0002 |
| **−3.3** | 0.0005 | 0.0005 | 0.0005 | 0.0004 | 0.0004 | 0.0004 | 0.0004 | 0.0004 | 0.0004 | 0.0003 |
| **−3.2** | 0.0007 | 0.0007 | 0.0006 | 0.0006 | 0.0006 | 0.0006 | 0.0006 | 0.0005 | 0.0005 | 0.0005 |
| **−3.1** | 0.0010 | 0.0009 | 0.0009 | 0.0009 | 0.0008 | 0.0008 | 0.0008 | 0.0008 | 0.0007 | 0.0007 |
| **−3.0** | 0.0013 | 0.0013 | 0.0013 | 0.0012 | 0.0012 | 0.0011 | 0.0011 | 0.0011 | 0.0010 | 0.0010 |
| **−0.7** | 0.2420 | 0.2389 | 0.2358 | 0.2327 | 0.2296 | 0.2266 | 0.2236 | 0.2206 | 0.2177 | 0.2148 |
| **−0.6** | 0.2743 | 0.2709 | 0.2676 | 0.2643 | 0.2611 | 0.2578 | 0.2546 | 0.2514 | 0.2483 | 0.2451 |
| **−0.5** | 0.3085 | 0.3050 | 0.3015 | 0.2981 | 0.2946 | 0.2912 | 0.2877 | 0.2843 | 0.2810 | 0.2776 |
| **−0.4** | 0.3446 | 0.3409 | 0.3372 | 0.3336 | 0.3300 | 0.3264 | 0.3228 | 0.3192 | 0.3156 | 0.3121 |
| **−0.3** | 0.3821 | 0.3783 | 0.3745 | 0.3707 | 0.3669 | 0.3632 | 0.3594 | 0.3557 | 0.3520 | 0.3483 |
| **−0.2** | 0.4027 | 0.4168 | 0.4129 | 0.4090 | 0.4052 | 0.4013 | 0.3974 | 0.3936 | 0.3897 | 0.3859 |

Figure 22

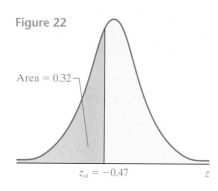

Area = 0.32

$z_0 = -0.47$          $z$

**Step 3:** From reading the table in Figure 21, we can see that the approximate Z-score that corresponds to an area of 0.32 to its left is −0.47. So $z_0 = -0.47$. See Figure 22.   ◄◄

**Using Technology:** Graphing calculators with advanced statistical features and statistical software can be used to find the Z-score corresponding to an area. Figure 23 shows the result of Example 4 using a TI-83 Plus graphing calculator.

Figure 23

```
invNorm(.32,0,1)
          -.4676988012
```

It is useful to remember that if the area to the left of the Z-score is less than 0.5, the Z-score must be less than 0. If the area to the left of the Z-score is greater than 0.5, the Z-score must be greater than 0.

**NW** *Now Work Problem 11.*

The next example deals with situations in which the area to the right of some unknown *Z*-score is given. The solution utilizes the fact that the area under the normal curve is 1.

▶ **EXAMPLE 5** Finding a Z-score from a Specified Area to the Right

*Problem:* Find the *Z*-score such that the area to the right of the *Z*-score is 0.4332.

*Approach:*

*Step 1:* Draw a standard normal curve with the area and corresponding unknown *Z*-score labeled.

*Step 2:* Determine the area to the left of the unknown *Z*-score.

*Step 3:* Look for the area in the table closest to the area determined in Step 2 and record the *Z*-score that corresponds to the closest area.

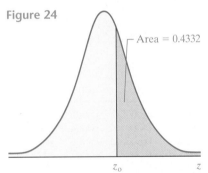

**Caution**

To find a Z-score given the area to the right, you must first determine the area to the left if you are using Table II.

*Solution:*

*Step 1:* Figure 24 shows the standard normal curve with the area and unknown *Z*-score labeled.

*Step 2:* Since the area under the entire normal curve is 1, the area left of the unknown *Z*-score is 1 minus the area right of the unknown *Z*-score. Therefore,

$$\text{area left} = 1 - \text{area right}$$
$$= 1 - 0.4332$$
$$= 0.5668$$

**Figure 24**

*Step 3:* We look in the body of Table II for an area closest to 0.5668. See Figure 25. The area closest to 0.5668 is 0.5675.

**Figure 25**

| z | 0.00 | 0.01 | 0.02 | 0.03 | 0.04 | 0.05 | 0.06 | 0.07 | 0.08 | 0.09 |
|-----|--------|--------|--------|--------|--------|--------|--------|--------|--------|--------|
| 0.0 | 0.5000 | 0.5040 | 0.5080 | 0.5120 | 0.5160 | 0.5199 | 0.5239 | 0.5279 | 0.5319 | 0.5359 |
| 0.1 | 0.5398 | 0.5438 | 0.5478 | 0.5517 | 0.5557 | 0.5596 | 0.5636 | 0.5675 | 0.5714 | 0.5753 |
| 0.2 | 0.5793 | 0.5832 | 0.5871 | 0.5910 | 0.5948 | 0.5987 | 0.6026 | 0.6064 | 0.6103 | 0.6141 |
| 0.3 | 0.6179 | 0.6217 | 0.6255 | 0.6293 | 0.6331 | 0.6368 | 0.6406 | 0.6443 | 0.6480 | 0.6517 |
| 0.4 | 0.6554 | 0.6591 | 0.6628 | 0.6664 | 0.6700 | 0.6736 | 0.6772 | 0.6808 | 0.6844 | 0.6879 |
| 0.5 | 0.6915 | 0.6950 | 0.6985 | 0.7019 | 0.7054 | 0.7088 | 0.7123 | 0.7157 | 0.7190 | 0.7224 |
| 0.6 | 0.7237 | 0.7291 | 0.7324 | 0.7357 | 0.7389 | 0.7422 | 0.7454 | 0.7486 | 0.7517 | 0.7549 |

The approximate *Z*-score that corresponds to an area right of 0.4332 is 0.17. Therefore, $z_0 = 0.17$. See Figure 26.

**Figure 26**

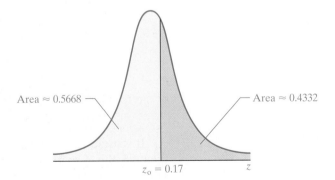

NW *Now Work Problem 15.*

In upcoming chapters, we will often be interested in finding Z-scores that separate the middle area of the standard normal curve from the area in its tails.

▶ **EXAMPLE 6** **Finding the Z-score from an Area in the Middle**

*Problem:* Find the Z-score that divides the middle 90% of the area in the standard normal distribution from the area in the tails.

*Approach:*

*Step 1:* Draw a standard normal curve with the middle 90% = 0.9 of the area separated from the area of 5% = 0.05 in each of the two tails. Label the unknown Z-scores $z_0$ and $z_1$.
*Step 2:* Look in the body of Table II to find the area closest to 0.05.
*Step 3:* Determine the Z-score in the left tail.
*Step 4:* The area to the right of $z_1$ is 0.05. Therefore, the area to the left of $z_1$ is 0.95. Look in Table II for an area of 0.95 and find the corresponding Z-value.

*Solution:*

*Step 1:* Figure 27 shows the standard normal curve with the middle 90% of the area separated from the area in the two tails.

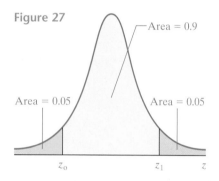

**Figure 27**

Area = 0.9

Area = 0.05    Area = 0.05

$z_0$    $z_1$    $z$

*Step 2:* We look in the body of Table II for an area closest to 0.05. See Figure 28.

**Figure 28**

| z | .00 | .01 | .02 | .03 | .04 | .05 | .06 | .07 | .08 | .09 |
|---|---|---|---|---|---|---|---|---|---|---|
| **−3.4** | 0.0003 | 0.0003 | 0.0003 | 0.0003 | 0.0003 | 0.0003 | 0.0003 | 0.0003 | 0.0003 | 0.0002 |
| **−3.3** | 0.0005 | 0.0005 | 0.0005 | 0.0004 | 0.0004 | 0.0004 | 0.0004 | 0.0004 | 0.0004 | 0.0003 |
| **−3.2** | 0.0007 | 0.0007 | 0.0006 | 0.0006 | 0.0006 | 0.0006 | 0.0006 | 0.0005 | 0.0005 | 0.0005 |
| **−2.0** | 0.0228 | 0.0222 | 0.0217 | 0.0212 | 0.0207 | 0.0202 | 0.0197 | 0.0192 | 0.0188 | 0.0183 |
| **−1.9** | 0.0287 | 0.0281 | 0.0274 | 0.0268 | 0.0262 | 0.0256 | 0.0250 | 0.0244 | 0.0239 | 0.0233 |
| **−1.8** | 0.0359 | 0.0351 | 0.0344 | 0.0336 | 0.0329 | 0.0322 | 0.0314 | 0.0307 | 0.0301 | 0.0294 |
| **−1.7** | 0.0446 | 0.0436 | 0.0427 | 0.0418 | 0.0409 | 0.0401 | 0.0392 | 0.0384 | 0.0375 | 0.0367 |
| **−1.6** | 0.0548 | 0.0537 | 0.0526 | 0.0516 | 0.0505 | 0.0495 | 0.0485 | 0.0475 | 0.0465 | 0.0455 |
| **−1.5** | 0.0668 | 0.0655 | 0.0643 | 0.0630 | 0.0618 | 0.0606 | 0.0594 | 0.0582 | 0.0571 | 0.0559 |

We notice that 0.0495 and 0.0505 are equally close to 0.05. We agree to take the arithmetic mean of the two Z-scores corresponding to the areas.
*Step 3:* The Z-score corresponding to an area of 0.0495 is −1.65. The Z-score corresponding to an area of 0.0505 is −1.64. Therefore, the approximate Z-score corresponding to an area of 0.05 to the left is

$$z_0 = \frac{-1.65 + (-1.64)}{2} = -1.645$$

Figure 29

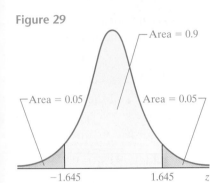

**Step 4:** The area to the right of $z_1$ is 0.05. Therefore, the area left of $z_1 = 1 - 0.05 = 0.95$. In Table II, we find an area of 0.9495 corresponding to $z = 1.64$ and an area of 0.9505 corresponding to $z = 1.65$ in the body. Consequently, the approximate $Z$-score corresponding to an area of 0.05 to the right is

$$z_1 = \frac{1.65 + 1.64}{2} = 1.645$$

See Figure 29.                                                                ◄◄

We could also obtain the solution to Example 6 using symmetry. Because the standard normal curve is symmetric about its mean, 0, the $Z$-score that corresponds to an area to the left of 0.05 will be the additive inverse of the $Z$-score that corresponds to an area to the right of 0.05. Since the area to the left of $Z = -1.645$ is 0.05, the area to the right of $Z = 1.645$ is also 0.05.

**NW** *Now Work Problem 19.*

We are often interested in finding the $Z$-score that has a specified area to the right. For this reason, we have special notation to represent this situation.

*Definition*    The notation $z_\alpha$ (pronounced "z sub alpha") is the $Z$-score such that the area under the standard normal curve to the right of $z_\alpha$ is $\alpha$. Figure 30 illustrates the notation.

Figure 30

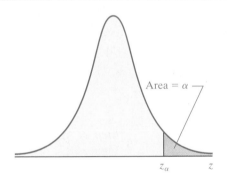

► **EXAMPLE 7**    **Finding the Value of $z_\alpha$**

Figure 31

*Problem:* Find the value of $z_{0.10}$

*Approach:* We wish to find the $Z$-value such that the area under the standard normal curve to the right of the $Z$-value is 0.10.

*Solution:* The area to the right of the unknown $Z$-value is 0.10, so the area to the left of the $Z$-value is $1 - 0.10 = 0.90$. We look in Table II for the area closest to 0.90. The area closest is 0.8997, which corresponds to a $Z$-value of 1.28. Therefore, $z_{0.10} = 1.28$. See Figure 31.                    ◄◄

**NW** *Now Work Problem 23.*

Recall that the area under a normal curve can be interpreted either as a probability or as the proportion of the population with the given characteristic. When interpreting the area under the standard normal curve as a probability, we use the notation introduced in Chapter 6. For example, in Example 6, we found that the area under the standard normal curve to the left of $Z = -1.645$ is 0.05; therefore, the probability of randomly selecting a standard normal random variable that is less than $-1.645$ is 0.05. We write this statement with the notation $P(Z < -1.645) = 0.05$.

We will use the following notation to denote probabilities of a standard normal random variable, $Z$.

---

**Notation for the Probability of a Standard Normal Random Variable**

$P(a < Z < b)$ represents the probability a standard normal random variable is between $a$ and $b$

$P(Z > a)$ represents the probability a standard normal random variable is greater than $a$.

$P(Z < a)$ represents the probability a standard normal random variable is less than $a$.

---

▶ **EXAMPLE 8**

**Finding Probabilities of Standard Normal Random Variables**

*Problem:* Evaluate $P(Z < 1.26)$.

*Approach:*

**Step 1:** Draw a standard normal curve with the area we desire shaded.
**Step 2:** Use Table II to find the area of the shaded region.
**Step 3:** This area represents the probability.

**Figure 32**

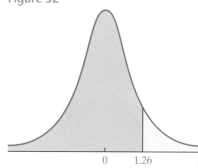

*Solution:*

**Step 1:** Figure 32 shows the standard normal curve with the area to the left of $Z = 1.26$ shaded.
**Step 2:** Using Table II, we find the area under the standard normal curve to the left of $Z = 1.26$ is 0.8962.
**Step 3:** Because the area under the standard normal curve to the left of $Z = 1.26$ is 0.8962, we have $P(Z < 1.26) = 0.8962$.   ◀◀

**NW** *Now Work Problem 29.*

For any continuous random variable, the probability of observing a specific value of the random variable is 0. For example, for a standard normal random variable, $P(Z = a) = 0$ for any value of $a$. This is because there is no area under the standard normal curve associated with a single value, so the probability must be 0. Hence, the following probabilities are equivalent.

$$P(a < Z < b) = P(a \le Z < b) = P(a < Z \le b) = P(a \le Z \le b)$$

For example, $P(Z < 1.26) = P(Z \le 1.26) = 0.8962$.

## 7.2 Assess Your Understanding

## Concepts and Vocabulary

1. State the properties of the standard normal curve.
2. If the area under the standard normal curve to the left of $Z = 1.20$ is 0.8849, what is the area under the standard normal curve to the right of $Z = 1.20$?
3. True or false? The area under the standard normal curve to the left of $Z = 5.30$ is 1. Support your answer.
4. Explain why $P(Z < -1.30) = P(Z \le -1.30)$.

## Exercises

### • Skill Building

*In Problems 1–8, find the indicated areas. For each problem, be sure to draw a standard normal curve and shade the area that is to be found.*

**1.** Determine the area under the standard normal curve that lies to the left of
(a) $Z = -2.45$
(b) $Z = -0.43$
(c) $Z = 1.35$
(d) $Z = 3.49$

**2.** Determine the area under the standard normal curve that lies to the left of
(a) $Z = -3.49$
(b) $Z = -1.99$
(c) $Z = 0.92$
(d) $Z = 2.90$

**3.** Determine the area under the standard normal curve that lies to the right of
(a) $Z = -3.01$
(b) $Z = -1.59$
(c) $Z = 1.78$
(d) $Z = 3.11$

**4.** Determine the area under the standard normal curve that lies to the right of
(a) $Z = -3.49$
(b) $Z = -0.55$
(c) $Z = 2.23$
(d) $Z = 3.45$

**5.** Determine the area under the standard normal curve that lies between
(a) $Z = -2.04$ and $Z = 2.04$
(b) $Z = -0.55$ and $Z = 0$
(c) $Z = -1.04$ and $Z = 2.76$

**6.** Determine the area under the standard normal curve that lies between
(a) $Z = -2.55$ and $Z = 2.55$
(b) $Z = -1.67$ and $Z = 0$
(c) $Z = -3.03$ and $Z = 1.98$

**7.** Determine the area under the standard normal curve
(a) to the left of $Z = -2$ or to the right of $Z = 2$
(b) to the left of $Z = -1.56$ or to the right of $Z = 2.56$
(c) to the left of $Z = -0.24$ or to the right of $Z = 1.20$

**8.** Determine the area under the standard normal curve
(a) to the left of $Z = -2.94$ or to the right of $Z = 2.94$
(b) to the left of $Z = -1.68$ or to the right of $Z = 3.05$
(c) to the left of $Z = -0.88$ or to the right of $Z = 1.23$

---

*In Problems 9 and 10, find the area of the shaded region for each standard normal curve.*

**9.** (a)

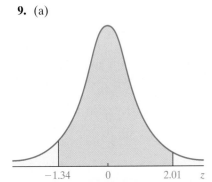

(b)

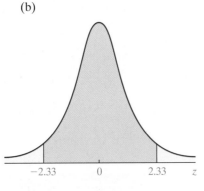

(c)

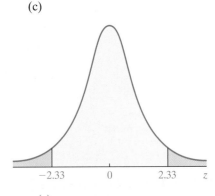

**10.** (a)

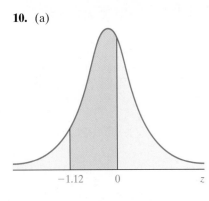

(b)

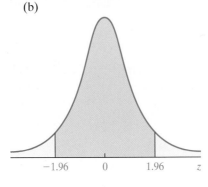

(c)

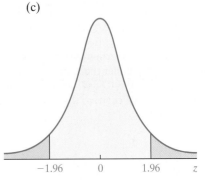

*In Problems 11–22, find the indicated Z-score. Be sure to draw a standard normal curve that depicts the solution.*

11. Find the Z-score such that the area under the standard normal curve to the left is 0.1.

12. Find the Z-score such that the area under the standard normal curve to the left is 0.2.

13. Find the Z-score such that the area under the standard normal curve to the left is 0.98.

14. Find the Z-score such that the area under the standard normal curve to the left is 0.85.

15. Find the Z-score such that the area under the standard normal curve to the right is 0.25.

16. Find the Z-score such that the area under the standard normal curve to the right is 0.35.

17. Find the Z-score such that the area under the standard normal curve to the right is 0.89.

18. Find the Z-score such that the area under the standard normal curve to the right is 0.75.

19. Find the Z-scores that separate the middle 80% of the distribution from the area in the tails of the standard normal distribution.

20. Find the Z-scores that separate the middle 70% of the distribution from the area in the tails of the standard normal distribution.

21. Find the Z-scores that separate the middle 99% of the distribution from the area in the tails of the distribution.

22. Find the Z-scores that separate the middle 94% of the distribution from the area in the tails of the distribution.

*In Problems 23–28, find the value of $z_\alpha$.*

23. $z_{0.05}$     24. $z_{0.35}$     25. $z_{0.01}$     26. $z_{0.02}$     27. $z_{0.20}$     28. $z_{0.15}$

*In Problems 29–40, find the indicated probability of the standard normal random variable Z.*

29. $P(Z < 1.93)$

30. $P(Z < -0.61)$

31. $P(Z > -2.98)$

32. $P(Z > 0.92)$

33. $P(-1.20 \leq Z < 2.34)$

34. $P(1.23 < Z \leq 1.56)$

35. $P(Z \geq 1.84)$

36. $P(Z \geq -0.92)$

37. $P(Z \leq 0.72)$

38. $P(Z \leq -2.69)$

39. $P(Z < -2.56 \text{ or } Z > 1.39)$

40. $P(Z < -0.38 \text{ or } Z > 1.93)$

• **Applying the Concepts**

41. **The Empirical Rule** The Empirical Rule states that about 68% of the data in a bell-shaped distribution lies within 1 standard deviation of the mean. This means about 68% of the data lies between $Z = -1$ and $Z = 1$. Verify this result. Verify that about 95% of the data lies within 2 standard deviations of the mean. Finally, verify that about 99.7% of the data lies within 3 standard deviations of the mean.

42. According to Table II, the area under the standard normal curve to the left of $Z = -1.34$ is 0.0901. Without consulting Table II, determine the area under the standard normal curve to the right of $Z = 1.34$.

43. According to Table II, the area under the standard normal curve to the left of $Z = -2.55$ is 0.0054. Without consulting Table II, determine the area under the standard normal curve to the right of $Z = 2.55$.

44. According to Table II, the area under the standard normal curve between $Z = -1.50$ and $Z = 0$ is 0.4332. Without consulting Table II, determine the area under the standard normal curve between $Z = 0$ and $Z = 1.50$.

45. According to Table II, the area under the standard normal curve between $Z = -1.24$ and $Z = -0.53$ is 0.1906. Without consulting Table II, determine the area under the standard normal curve between $Z = 0.53$ and $Z = 1.24$.

46. (a) Suppose $P(Z < a) = 0.9938$; find $a$.
    (b) Suppose $P(Z \geq a) = 0.4404$; find $a$.
    (c) Suppose $P(-b < Z < b) = 0.8740$; find $b$.

## Technology Step-by-Step
The Standard Normal Distribution

*TI-83 Plus*

**Finding Areas under the Standard Normal Curve**

*Step 1:* From the HOME screen, press $2^{nd}$ VARS to access the DISTRibution menu.

*Step 2:* Select 2:normalcdf(

*Step 3:* With normalcdf( on the HOME screen, type *lowerbound, upperbound, 0, 1*). For example, to find the area left of $Z = 1.26$ under the standard normal curve, type:

$$\text{Normalcdf}(-1E99, 1.26, 0, 1)$$

and hit ENTER.

NOTE: The "E" shown is scientific notation; it is $2^{nd}$, on the keyboard.

**Finding Z-scores Corresponding to an Area**

*Step 1:* From the HOME screen, press $2^{nd}$ VARS to access the DISTRibution menu.

*Step 2:* Select 3:invNorm(

*Step 3:* With invNorm( on the HOME screen, type *"area left", 0, 1*). For example, to find the Z-score such that the area under the normal curve left of the Z-score is 0.68, type:

$$\text{InvNorm}(0.68, 0, 1)$$

and hit ENTER.

*MINITAB*

**Finding Areas under the Standard Normal Curve**

*Step 1:* Minitab will find an area to the left of a specified Z-score. Select the **Calc** menu, highlight **Probability Distributions**, and highlight **Normal** . . . .

*Step 2:* Select **Cumulative Probability**. Set the mean to 0 and the standard deviation to 1. Select **Input Constant**, and enter the specified Z-score. Click OK.

**Finding Z-scores Corresponding to an Area**

*Step 1:* Minitab will find the Z-score for an area to the left of unknown Z-score. Select the **Calc** menu, highlight **Probability Distributions**, and highlight **Normal** . . . .

*Step 2:* Select **Inverse Cumulative Probability**. Set the mean to 0 and the standard deviation to 1. Select **Input Constant**, and enter the specified area. Click OK.

*Excel*

**Finding Areas under the Standard Normal Curve**

*Step 1:* Excel will find the area to the left of a specified Z-score. Select the *fx* button from the tool bar. In **Function Category:**, select "Statistical." In **Function Name:**, select "NormsDist." Click OK.

*Step 2:* Enter the specified Z-score. Click OK.

**Finding Z-scores Corresponding to an Area**

*Step 1:* Excel will find the Z-score for an area to the left of unknown Z-score. Select the *fx* button from the tool bar. In **Function Category:**, select "Statistical." In **Function Name:**, select "NormsInv." Click OK.

*Step 2:* Enter the specified area. Click OK.

## 7.3   Applications of the Normal Distribution

**Preparing for This Section**   Before getting started, review the following:

✓ Percentiles (Section 3.4, pp. 123–125)

**Objectives**  Find and interpret area under a normal curve

 Find the value of a normal random variable

 Suppose that a random variable $X$ is normally distributed with mean $\mu$ and standard deviation $\sigma$. The area below the normal curve represents a proportion or probability. From the discussions in Section 7.1 we know that finding the area under a normal curve requires that we transform a normal random variable $X$ with mean $\mu$ and standard deviation $\sigma$ into a standard normal random variable $Z$ with mean 0 and standard deviation 1. This is accomplished by letting $Z = \dfrac{X - \mu}{\sigma}$ and using Table II to find the area under the standard normal curve. This idea is illustrated in Figure 33.

**Figure 33**

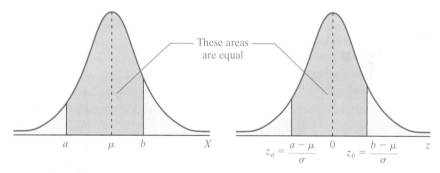

Now that we have the ability to find the area under a standard normal curve, we can find the area under any normal curve. We summarize the procedure below.

---

**Finding the Area under any Normal Curve**

**Step 1:** Draw a normal curve with the desired area shaded.

**Step 2:** Convert the values of $X$ to $Z$-scores, using $Z = \dfrac{X - \mu}{\sigma}$.

**Step 3:** Draw a standard normal curve with the area desired shaded.

**Step 4:** Find the area under the standard normal curve. This is the area under the normal curve drawn in Step 1.

---

▶ **EXAMPLE 1**   **Finding Area under a Normal Curve**

*Problem:* A pediatrician obtains the heights of her 200 three-year-old female patients. The heights are normally distributed with mean 38.72 and standard deviation 3.17. What percent of the three-year-old females have a height less than 35 inches?

*Approach:* Follow the steps listed above.

*Solution:*

**Step 1:** Figure 34 shows the normal curve with the area to the left of 35 shaded.
**Step 2:** We convert $X = 35$ to a standard normal random variable $Z$ with $\mu = 38.72$ and $\sigma = 3.17$.

$$Z = \frac{X - \mu}{\sigma} = \frac{35 - 38.72}{3.17} = -1.17$$

**Step 3:** Figure 35 shows the standard normal curve with the area left of $Z = -1.17$ shaded.

Figure 34

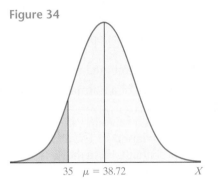

35  $\mu = 38.72$  $X$

Figure 35

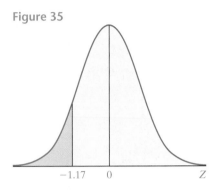

$-1.17$  0  $Z$

The area to the left of $Z = -1.17$ is equal to the area to the left of $X = 35$.
**Step 4:** Using Table II, we find the area to the left of $Z = -1.17$ is 0.121. We conclude that 12.1% of the pediatrician's three-year-old females are less than 35 inches tall.  ◀◀

| TABLE 3 | |
|---|---|
| **Height (inches)** | **Relative Frequency** |
| 29.0–29.9 | 0.005 |
| 30.0–30.9 | 0.005 |
| 31.0–31.9 | 0.005 |
| 32.0–32.9 | 0.025 |
| 33.0–33.9 | 0.02 |
| 34.0–34.9 | 0.055 |
| 35.0–35.9 | 0.075 |
| 36.0–36.9 | 0.09 |
| 37.0–37.9 | 0.115 |
| 38.0–38.9 | 0.15 |
| 39.0–39.9 | 0.12 |
| 40.0–40.9 | 0.11 |
| 41.0–41.9 | 0.07 |
| 42.0–42.9 | 0.06 |
| 43.0–43.9 | 0.035 |
| 44.0–44.9 | 0.025 |
| 45.0–45.9 | 0.025 |
| 46.0–46.9 | 0.005 |
| 47.0–47.9 | 0.005 |

**Using Technology:** Graphing calculators with advanced statistical features and statistical software can be used to compute the area below any normal curve. Figure 36 shows the results of Example 1 using a TI-83 Plus graphing calculator. The results differ from Example 1 because of rounding.

Figure 36

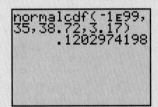

```
normalcdf(-1E99,
35,38.72,3.17)
        .1202974198
```

**NW** *Now Work Problems 11(a) and 11(b).*

According to the results of Example 1, the proportion of three-year-old females shorter than 35 inches is 0.121. If the normal curve is a good model for determining probabilities (or proportions), then about 12.1% of the 200 three-year-olds in Table 1 on page 277 should be shorter than 35 inches. For convenience, the information provided in Table 1 is repeated in Table 3.

From the relative frequency distribution in Table 3, we determine that $0.005 + 0.005 + 0.005 + 0.025 + 0.02 + 0.055 = 0.115 = 11.5\%$ of the three-year olds are less than 35 inches tall. The results based upon the normal curve are in close agreement with the actual results. The normal curve accurately models the heights of three-year-old females.

▶ **EXAMPLE 2**  **Finding the Probability of a Normal Random Variable**

*Problem:* Use the results of Example 1 to compute the probability that a randomly selected three-year-old female is between 35 and 40 inches tall. That is, find $P(35 \leq X \leq 40)$.

*Approach:* We follow the steps listed on page 297.

*Solution:*

**Step 1:** Figure 37 shows the normal curve with the area between $X_1 = 35$ and $X_2 = 40$ shaded.

***Step 2:*** Convert the values of $X_1 = 35$ and $X_2 = 40$ to $Z$-scores.

$$Z_1 = \frac{X_1 - \mu}{\sigma} = \frac{35 - 38.72}{3.17} = -1.17$$

$$Z_2 = \frac{X_2 - \mu}{\sigma} = \frac{40 - 38.72}{3.17} = 0.40$$

***Step 3:*** Figure 38 shows the graph of the standard normal curve with the area between $Z_1 = -1.17$ and $Z_2 = 0.40$ shaded.

Figure 37

Figure 38

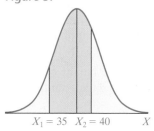

$X_1 = 35$   $X_2 = 40$   $X$

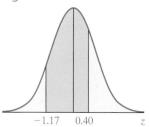

$-1.17$   $0.40$   $z$

***Step 4:*** Using Table II, we find that the area to the left of $Z_2 = 0.40$ is 0.6554 and the area to the left of $Z_1 = -1.17$ is 0.1210. Therefore, the area between $Z_1 = -1.17$ and $Z_2 = 0.40$ is $0.6554 - 0.1210 = 0.5344$. We conclude that the probability a randomly selected three-year-old female is between 35 and 40 inches tall is 0.5344. That is, $P(35 \leq X \leq 40) = P(-1.17 \leq Z \leq 0.40) = 0.5344$. ◄◄

 *Now Work Problem 11(c).*

**In Your Own Words**

The normal probability density function is used to model random variables that appear to be normal (such as girls' heights). A good model is one that yields results that are close to reality.

According to the relative frequency distribution in Table 3, the proportion of the 200 three-year-old females with heights between 35 inches and 40 inches is $0.075 + 0.09 + 0.115 + 0.15 + 0.12 = 0.55 = 55\%$. This is very close to the probability obtained in Example 2!

② **Finding Values of Normal Random Variables**

Often, rather than being interested in the proportion or probability of a normal random variable, we are interested in calculating the value of a normal random variable required for the variable to be at a certain proportion or probability. For example, we might want to know the height of a three-year-old girl at the 20th percentile. This means we want to know the height of a three-year-old girl who is taller than 20% of all three-year-old girls.

**Procedure for Finding the Value of a Normal Random Variable Corresponding to a Specified Proportion or Probability**

***Step 1:*** Draw a normal curve with the area corresponding to the proportion or probability shaded.

***Step 2:*** Use Table II to find the $Z$-score that corresponds to the shaded area.

***Step 3:*** Obtain the normal value from the fact that $X = \mu + Z\sigma$.*

---

$* Z = \dfrac{X - \mu}{\sigma}$    Formula for Standardizing a Random Variable $X$

$Z\sigma = X - \mu$    Multiply both sides by $\sigma$

$X = \mu + Z\sigma$    Add $\mu$ to both sides

▶ **EXAMPLE 3** **Finding the Value of a Normal Random Variable**

*Problem:* The heights of a pediatrician's 200 three-year-old females are normally distributed with mean 38.72 inches and standard deviation 3.17 inches. Find the height of a three-year-old female at the 20ᵗʰ percentile. That is, find the height for a three-year-old female that separates the bottom 20% from the top 80%.

*Approach:* We follow Steps 1–3 given on page 299.

*Solution:*

**Figure 39**

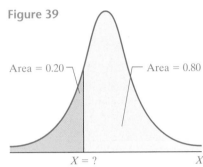

Area = 0.20 ⌐     ⌐ Area = 0.80

X = ?          X

**Step 1:** Figure 39 shows the normal curve with the unknown value of $X$ separating the bottom 20% of the distribution from the top 80% of the distribution.

**Step 2:** From Table II, the area closest to 0.20 is 0.2005. The corresponding $Z$-score is $-0.84$.

**Step 3:** The height of a three-year-old female that separates the bottom 20% of the data from the top 80% is computed as follows:

$$X = \mu + Z\sigma$$
$$= 38.72 + (-0.84)(3.17)$$
$$= 36.1 \text{ inches} \qquad ◀◀$$

**Using Technology:** Graphing calculators with advanced statistical features and statistical software can be used to compute the value of a normal random variable corresponding to a given proportion. Figure 40 shows the results of Example 3 using Minitab.

**Figure 40**

**Inverse Cumulative Distribution Function**

Normal with mean = 38.7200 and standard deviation = 3.17000

```
P ( x <= x)        x
    0.2000      36.0521
```

**NW** *Now Work Problems 21(a) and 21(b).*

▶ **EXAMPLE 4** **Finding the Value of a Normal Random Variable**

*Problem:* The heights of a pediatrician's 200 three-year-old females are normally distributed with mean 38.72 inches and standard deviation 3.17 inches. The pediatrician wishes to determine the heights that separate the middle 98% of the distribution from the bottom and top 1%. In other words, find the 1ˢᵗ and 99ᵗʰ percentiles.

*Approach:* We follow Steps 1–3 given on page 299.

*Solution:*

**Figure 41**

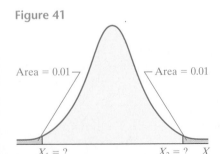

Area = 0.01 ⌐     ⌐ Area = 0.01

$X_1 = ?$          $X_2 = ?$  X

**Step 1:** Figure 41 shows the normal curve with the unknown values of $X$ separating the bottom and top 1% of the distribution from the middle 98% of the distribution.

**Step 2:** First, we will find the $Z$-score that corresponds to the area 0.01 to the left. From Table II, the area closest to 0.01 is 0.0099. The corresponding $Z$-score is $-2.33$. The $Z$-score that corresponds to the area 0.01 to the right will be the $Z$-score that has the area 0.99 to the left. The area closest to 0.99 is 0.9901. The corresponding $Z$-score is 2.33.

**Step 3:** The height of a three-year-old female that separates the bottom 1% of the distribution from the top 99% is

$$X_1 = \mu + Z\sigma$$
$$= 38.72 + (-2.33)(3.17)$$
$$= 31.3 \text{ inches}$$

The height of a three-year-old female that separates the top 1% of the distribution from the bottom 99% is

$$X_2 = \mu + Z\sigma$$
$$= 38.72 + (2.33)(3.17)$$
$$= 46.1 \text{ inches}$$

A three-year-old female whose height is less than 31.3 inches is in the bottom 1% of all three-year-old females and a three-year-old female whose height is more than 46.1 inches is in the top 1% of all three-year-old females. The pediatrician might use this information to identify those patients who have unusual heights. ◄◄

NW  *Now Work Problem 21(c).*

# 7.3 Assess Your Understanding

## Concepts and Vocabulary

**1.** Describe the procedure for finding the area under any normal curve.

**2.** Describe the procedure for finding the score corresponding to a probability.

## Exercises

### • Skill Building

*In Problems 1–10, assume the random variable X is normally distributed with mean $\mu = 50$ and standard deviation $\sigma = 7$. Compute the following probabilities. Be sure to draw a normal curve with the area corresponding to the probability shaded.*

| | | | | |
|---|---|---|---|---|
| **1.** | $P(X > 35)$ | | **6.** | $P(56 < X < 68)$ |
| **2.** | $P(X > 65)$ | | **7.** | $P(55 \leq X \leq 70)$ |
| **3.** | $P(X \leq 45)$ | | **8.** | $P(40 \leq X \leq 49)$ |
| **4.** | $P(X \leq 58)$ | | **9.** | $P(38 < X \leq 55)$ |
| **5.** | $P(40 < X < 65)$ | | **10.** | $P(56 \leq X < 66)$ |

### • Applying the Concepts

**11. Serum Cholesterol** As reported by the U.S. National Center for Health Statistics, the mean serum high-density-lipoprotein (HDL) cholesterol of females 20–29 years old is $\mu = 53$. If serum HDL cholesterol is distributed normally with $\sigma = 13.4$, answer the following questions.

(a) What proportion of 20–29 year old females have a serum HDL cholesterol level below 39?

(b) What proportion of 20–29 year old females have a serum HDL cholesterol above 71?

(c) What proportion of 20–29 year old females have a serum HDL cholesterol between 60 and 75?

(d) What is the probability a randomly selected 20–29-year-old female has a serum HDL cholesterol below 45?

(e) What is the probability a randomly selected 20–29-year-old female has a serum HDL cholesterol between 50 and 60?

(f) Suppose you are a doctor and a 24-year-old female patient of yours has a serum HDL cholesterol of 30. Is this unusual?

**12. Earthquakes** Since 1900, the magnitude of earthquakes that measure 0.1 or higher on the Richter Scale in California is distributed approximately normally, with $\mu = 6.2$ and $\sigma = 0.5$, according to data obtained from the United States Geological Survey.

(a) What is the probability that a randomly selected earthquake in California has a magnitude of 6.0 or higher?

(b) What is the probability that a randomly selected earthquake in California has a magnitude less than 6.4?

(c) What is the probability that a randomly selected earthquake in California has a magnitude between 5.8 and 7.1?

(d) The great San Francisco Earthquake of 1906 had a magnitude of 8.25. Is an earthquake of this magnitude unusual in California?

13. **Stock Market Returns** During the past forty years, the monthly rate of return of stocks has been approximately normally distributed, with $\mu = 0.75$ percent and $\sigma = 4.2$ percent, according to data obtained from Yahoo!Finance.

(a) What is the probability that a randomly selected month has a rate of return of at least 1 percent?

(b) What is the probability that a randomly selected month has a rate of return less than 2 percent?

(c) What proportion of months have a positive rate of return (i.e. a rate of return greater than 0 percent)?

(d) In March, 2000, the monthly rate of return in stocks was 9.7 percent. Is this rate of return unusual? Why?

14. **ACT Math Scores** In 2000, as reported by ACT Research Service, the mean ACT Math score was $\mu = 20.7$. If ACT Math scores are normally distributed with $\sigma = 5$, answer the following questions.

(a) What is the probability that a randomly selected student has an ACT Math score of at least 25?

(b) What is the probability that a randomly selected student has an ACT Math score less than 18?

(c) What is the probability that a randomly selected student has an ACT Math score between 24 and 27?

(d) If a student scores 29 on the ACT Math portion of the exam, what is her percentile rank (i.e. what percent of students scored less than 29)?

(e) If a student scores 17 on the ACT Math portion of the exam, what is her percentile rank?

15. **Heights of Females** As reported by the U.S. National Center for Health Statistics, the mean height of females 20–29 years old is $\mu = 64.1$ inches. If height is normally distributed with $\sigma = 2.8$ inches, answer the following questions.

(a) What percent of 20–29-year-old females are less than 60 inches tall?

(b) What percent of 20–29-year-old females are at least 72 inches tall?

(c) What percent of 20–29-year-old females are between 60 and 70 inches tall?

(d) What is the probability that a randomly selected 20–29-year-old female is between 60 and 70 inches tall?

(e) Would it be unusual for a 20–29-year-old female to be more than $5'10''$?

16. **IQ Scores** Scores on the Stanford–Binet Intelligence Test are normally distributed with mean $\mu = 100$ and standard deviation $\sigma = 16$.

(a) People with IQ scores above 132 are considered to be "gifted." What percent of people are gifted?

(b) What is the probability that a randomly selected individual has an IQ score below 110?

(c) What is the probability that a randomly selected individual has an IQ between 90 and 120?

17. **Blood Pressure** In a study of hypertension and optimal treatment conducted by the Center for Cardiovascular Education, 10,005 patients had a mean systolic blood pressure (the "upper reading") of 161 mm Hg with a standard deviation of 18 mm Hg. Assume systolic blood pressure is normally distributed to answer the following questions.

(a) What percent of the people in the study have systolic blood pressure above 180, severe hypertension?

(b) What percent of the people in the study have systolic blood pressure below 150?

(c) What percent of the people in the study have systolic blood pressure between 140 and 159, mild hypertension?

(d) What is the probability that a randomly selected individual in the study had systolic blood pressure below 143?

(e) What is the probability that a randomly selected individual in the study had systolic blood pressure between 155 and 170?

18. **Gestation Period** The length of human pregnancies is normally distributed with mean $\mu = 266$ days and standard deviation $\sigma = 16$ days.

(a) What percent of pregnancies last more than 270 days?

(b) What percent of pregnancies last less than 250 days?

(c) What percent of pregnancies last between 240 and 280 days?

(d) What is the probability that a randomly selected pregnancy lasts more than 280 days?

(e) What is the probability that a randomly selected pregnancy lasts no more than 245 days?

(f) A "very preterm" baby is one where the gestation period is less than 224 days. What proportion of births are "very preterm"?

19. **Weather in Chicago** The high temperature in Chicago for the month of August is normally distributed with mean $\mu = 80°$ F and standard deviation $\sigma = 8°$ F.

(a) What percent of the days in Chicago in August will have a high temperature above $90°$ F?

(b) What percent of the days in Chicago in August will have a high temperature below $60°$ F?

(c) What percent of the days in Chicago in August will have a high temperature between $85°$ F and $100°$ F?

(d) What is the probability that a randomly selected day in August in Chicago will have a high temperature below $70°$ F?

(e) What is the probability that a randomly selected day in August in Chicago will have a high temperature between $72°$ F and $80°$ F?

20. **Shark Attacks** The number of shark attacks per year in the United States is distributed approximately normal, with $\mu = 31.8$ and $\sigma = 10.0$, according to data obtained from the Florida Museum of Natural History.

(a) What percent of years will have fewer than 30 shark attacks?

(b) What percent of years will have more than 40 shark attacks?

(c) In 2000, there were 51 shark attacks in the United States. Is this an unusually high number of attacks? Why?

21. **Serum Cholesterol** As reported by the U.S. National Center for Health Statistics, the mean serum high-density-lipoprotein (HDL) cholesterol of females 20–29 years old is $\mu = 53$. If serum HDL cholesterol is normally distributed with $\sigma = 13.4$, answer the following questions.

   (a) Determine the $25^{th}$ percentile of serum HDL cholesterol for a 20–29-year-old female. Interpret the result.

   (b) According to the U.S. National Center for Health Statistics, the $25^{th}$ percentile of serum HDL cholesterol for a 20–29-year-old female is 44. How does this compare to the result obtained in part (a)?

   (c) Determine the serum HDL cholesterol levels that make up the middle 80% of serum HDL cholesterol levels.

22. **Earthquakes** Since 1900, the magnitude of earthquakes that measure 0.1 or higher on the Richter Scale in California is distributed approximately normally, with $\mu = 6.2$ and $\sigma = 0.5$, according to data obtained from the United States Geological Survey.

   (a) Determine the $40^{th}$ percentile of the magnitude of earthquakes in California.

   (b) Determine the magnitude of earthquakes that make up the middle 85% of magnitudes.

23. **Stock Market Returns** During the past 40 years, the monthly rate of return of stocks has been distributed approximately normally, with $\mu = 0.75$ percent and $\sigma = 4.2$ percent, according to data obtained from Yahoo!Finance.

   (a) Determine the monthly rate of return that is at the $80^{th}$ percentile.

   (b) Determine the monthly rates of return that make up the middle 95% of returns.

24. **ACT Math Scores** In 2000, as reported by ACT Research Service, the mean ACT Math score was $\mu = 20.7$. If ACT Math scores are normally distributed with $\sigma = 5$, answer the following questions.

   (a) If a student's ACT Math score is better than 80% of the students taking the exam ($80^{th}$ percentile), what was his score?

   (b) A certain college considers only applications in which the ACT Math score is in the top 5% of scores. What score is required to be considered?

25. **Heights of Females** As reported by the U.S. National Center for Health Statistics, the mean height of females 20–29 years old is $\mu = 64.1$ inches. If height is normally distributed with $\sigma = 2.8$ inches, answer the following questions.

   (a) Determine the $40^{th}$ percentile of height for 20–29-year-old females.

(b) Determine the height required to be in the top 2% of all 20–29-year-old females.

(c) A "one-size-fits all" robe is being designed that should fit 95% of 20–29-year-old females; what heights constitute the middle 95% of all 20–29-year-old females?

26. **IQ Scores** Scores on the Stanford–Binet Intelligence Test are normally distributed with mean $\mu = 100$ and standard deviation $\sigma = 16$.

   (a) In order to qualify for Mensa, an organization of people with high IQ scores, one must have an IQ at the $98^{th}$ percentile. What IQ score is required to qualify for Mensa?

   (b) What range of IQ scores make up the middle 50% of the population?

27. **Blood Pressure** In the study of hypertension and optimal treatment mentioned in Problem 17, 10,005 patients had a mean systolic blood pressure of 161 mm Hg with a standard deviation of 18 mm Hg. Assume systolic blood pressure is normally distributed to answer the following questions.

   (a) Determine the $70^{th}$ percentile of systolic blood pressure for this group.

   (b) What range of systolic blood pressures make up the middle 60% of the participants in the study?

28. **Gestation Period** The length of human pregnancies is approximately normally distributed with mean $\mu = 266$ days and standard deviation $\sigma = 16$ days.

   (a) Suppose an unusually long pregnancy is one that is in the top 2%. Determine the length of pregnancy that separates an unusually long pregnancy from one that is not unusually long.

   (b) Determine the length of pregnancy that would be considered typical if we define typical to be the middle 96% of pregnancies.

29. **Weather in Chicago** The high temperature in Chicago for the month of August is distributed approximately normally, with mean $\mu = 80°$ F and standard deviation $\sigma = 8°$ F.

   (a) Determine the high temperature required for a day to be in the bottom 10%.

   (b) If we define an unusually warm day in August as one that is in the top 4%, what temperatures qualify as unusually warm?

   (c) Determine the temperatures that constitute the middle 80% of high temperatures in August.

30. **Shark Attacks** The number of shark attacks per year in the United States is distributed approximately normally, with $\mu = 31.8$ and $\sigma = 10.0$, according to data obtained from Florida Museum of Natural History.

   (a) Determine the number of shark attacks per year that separates the top 2% from the bottom 98%.

   (b) Determine the numbers of shark attacks per year that constitute the middle 80% of shark attacks per year.

**31. SAT Math Scores** The following relative frequency distribution represents the scores on the SAT I: Reasoning Test—Math in 2000:

| Score | Relative Frequency | Score | Relative Frequency |
|-------|--------------------|-------|--------------------|
| 200–249 | 0.0073 | 500–549 | 0.1637 |
| 250–299 | 0.0167 | 550–599 | 0.1472 |
| 300–349 | 0.0438 | 600–649 | 0.1095 |
| 350–399 | 0.0797 | 650–699 | 0.0737 |
| 400–449 | 0.1336 | 700–749 | 0.0393 |
| 450–499 | 0.1661 | 750–800 | 0.0195 |

*Source:* College Board Online

SAT math scores are known to be normally distributed. The mean SAT math score is $\mu = 514$ and the standard deviation is $\sigma = 113$.

(a) Compute the theoretical proportion of test takers in each class by finding the area under the normal curve.
(b) Compare the theoretical proportions to the actual proportions. Are you convinced that the SAT math scores are normally distributed?

**32. SAT Verbal Scores** The following relative frequency distribution represents the scores on the SAT I: Reasoning Test—Verbal in 2000:

| Score | Relative Frequency | Score | Relative Frequency |
|-------|--------------------|-------|--------------------|
| 200–249 | 0.0113 | 500–549 | 0.1766 |
| 250–299 | 0.0162 | 550–599 | 0.1435 |
| 300–349 | 0.0442 | 600–649 | 0.1037 |
| 350–399 | 0.0874 | 650–699 | 0.0603 |
| 400–449 | 0.1381 | 700–749 | 0.0292 |
| 450–499 | 0.1722 | 750–800 | 0.0172 |

*Source:* College Board Online

SAT verbal scores are known to be normally distributed. The mean SAT verbal score is $\mu = 505$ and the standard deviation is $\sigma = 111$.

(a) Compute the theoretical proportion of test takers in each class by finding the area under the normal curve.
(b) Compare the theoretical proportions to the actual proportions. Are you convinced the SAT verbal scores are normally distributed?

**33. Weather in Chicago** The following frequency distribution represents the daily high temperature in Chicago, Nov. 16–30, for the years 1872–1999:

| Temperature | Frequency | Temperature | Frequency |
|-------------|-----------|-------------|-----------|
| 5.0–9.9 | 1 | 40.0–44.9 | 375 |
| 10.0–14.9 | 10 | 45.0–49.9 | 281 |
| 15.0–19.9 | 15 | 50.0–54.9 | 233 |
| 20.0–24.9 | 40 | 55.0–59.9 | 160 |
| 25.0–29.9 | 95 | 60.0–64.9 | 101 |
| 30.0–34.9 | 217 | 65.0–69.9 | 21 |
| 35.0–39.9 | 325 | 70.0–74.9 | 1 |

*Source: Chicago Tribune*, November 27, 2000

(a) Construct a relative frequency distribution.
(b) Draw a relative frequency histogram. Does the distribution of high temperatures appear to be normal?
(c) Compute the mean and standard deviation of (the variable) high temperature.
(d) Use the information obtained in part (c) to compute the theoretical proportion of high temperatures in each class by finding the area under the normal curve.
(e) Are you convinced that high temperatures are normally distributed?

---

## Technology Step-by-Step
The Normal Distribution

**TI-83 Plus** — **Finding Areas under the Normal Curve**

*Step 1:* From the HOME screen, press 2nd VARS to access the DISTRibution menu.

*Step 2:* Select 2:normalcdf(

*Step 3:* With normalcdf( on the HOME screen, type *lowerbound, upperbound, $\mu$, $\sigma$*). For example, to find the area to the left of $X = 35$ under the normal curve with $\mu = 40$ and $\sigma = 10$, type

```
Normalcdf(−1E99, 35, 40, 10)
```

and hit ENTER.
*Note:* The "E" shown is scientific notation; it is 2nd,

**Finding Scores Corresponding to an Area**

*Step 1:* From the HOME screen, press 2$^{nd}$ VARS to access the DISTRibution menu.

*Step 2:* Select 3:invNorm(

*Step 3:* With invNorm( on the HOME screen, type "*area*", $\mu$, $\sigma$). For example, to find the score such that the area under the normal curve to the left of the score is 0.68 with $\mu = 40$ and $\sigma = 10$, type

$$InvNorm(0.68, 40, 10)$$

and hit ENTER.

**MINITAB**   **Finding Areas under the Normal Curve**

*Step 1:* Minitab will find an area to the left of a specified observation. Select the **Calc** menu, highlight **Probability Distributions**, and highlight **Normal**....

*Step 2:* Select **Cumulative Probability**. Enter the mean, $\mu$, and the standard deviation, $\sigma$. Select **Input Constant**, and enter the observation. Click OK.

**Finding Z-scores Corresponding to an Area**

*Step 1:* Minitab will find the score corresponding to a specified area. Select the **Calc** menu, highlight **Probability Distributions**, and highlight **Normal**...

*Step 2:* Select **Inverse Cumulative Probability**. Enter the mean, $\mu$, and the standard deviation, $\sigma$. Select **Input Constant**, and enter the area left of the unknown score. Click OK.

**Excel**   **Finding Areas under the Normal Curve**

*Step 1:* Excel will find an area to the left of a specified observation. Select the *fx* button from the tool bar. In **Function Category**:, select "Statistical." In **Function Name**:, select NormDist. Click OK.

*Step 2:* Enter the specified observation, $\mu$, and $\sigma$, and set **cumulative** to True. Click OK.

**Finding Scores Corresponding to an Area**

*Step 1:* Select the *fx* button from the tool bar. In **Function Category**:, select "Statistical." In **Function Name**:, select NormInv. Click OK.

*Step 2:* Enter the area left of the unknown score, $\mu$, and $\sigma$. Click OK.

---

## 7.4   Assessing Normality

**Preparing for This Section**   Before getting started, review the following:

✓ Shape of a distribution (Section 3.1, pp. 90–93)

**Objectives**    Draw normal probability plots to assess normality

Suppose that we obtain a simple random sample from a population whose distribution is unknown. Many of the statistical tests that we perform on small data sets (sample size less than 30) require that the population from which the sample is drawn be normally distributed. Up to this point, we have said that a random variable $X$ is normally distributed, or at least approximately normal, provided the histogram of the data is symmetric and bell-shaped. This method works well for large data sets, but the shape of a histogram drawn from a small sample of observations does not always accurately represent the shape of the population. For this reason, we need additional methods for assessing the normality of a random variable $X$ when we are looking at a small set of sample data.

## Normal Probability Plots

**In Your Own Words**

Normal probability plots are used to assess normality in small data sets.

A **normal probability plot** plots observed data versus *normal scores*. A **normal score** is the expected $Z$-score of the data value if the distribution of the random variable is normal. The expected $Z$-score of an observed value will depend upon the number of observations in the data set.

To draw a normal probability plot requires the following steps.

> **Drawing a Normal Probability Plot**
>
> **Step 1:** Arrange the data in ascending order.
>
> **Step 2:** Compute $f_i = \dfrac{i - 0.375}{n + 0.25}$,* where $i$ is the index (the $i$th number in the list) and $n$ is the number of observations. This value represents the expected proportion of observations less than or equal to the $i$th data value.
>
> **Step 3:** Find the $Z$-score corresponding to $f_i$ from Table II.
>
> **Step 4:** Plot the observed values on the horizontal axis and the corresponding expected $Z$-scores on the vertical axis.

The idea behind finding the expected $Z$-score is that, if the data come from a population that is normally distributed, we should be able to predict the area to the left of each of the data values. The value of $f_i$ represents the expected area to the left of the $i$th data value when the data come from a population that is normally distributed. For example, $f_1$ is the expected area to the left of the smallest data value, $f_2$ is the expected area to the left of the second-smallest data value, and so on. Figure 42 illustrates the idea.

Once we determine each of the $f_i$, we find the $Z$-scores corresponding to $f_1$, $f_2$, and so on. The smallest observation in the data set will be the smallest expected $Z$-score, and the largest observation in the data set will be the largest expected $Z$-score. In addition, because of the symmetry of the normal curve, the expected $Z$-scores are always paired as positive and negative values.

Normal random variables $X$ and their $Z$-scores are linearly related $[X = \mu + Z\sigma]$, so a plot of observations of normal variables against their expected normal scores will be linear. We conclude the following:

**Figure 42**

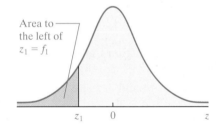

Area to the left of $z_1 = f_1$

> **If sample data are taken from a population that is normally distributed, a normal probability plot of the actual values versus the expected $Z$-scores will be approximately linear.**

Fortunately, graphing calculators with advanced statistical features and statistical software have the ability to create normal probability plots.

Normal probability plots are typically drawn by using graphing calculators or statistical software; however, it is worthwhile to go through an example that demonstrates the procedure so that we can better understand the results supplied by software.

*The derivation of this formula is beyond the scope of this text.

► EXAMPLE 1   **Constructing a Normal Probability Plot**

| TABLE 4 | | | |
|---|---|---|---|
| 15.8 | 18.5 | 22.6 | 27.4 |
| 16.7 | 19.2 | 23.7 | 28.5 |
| 18.2 | 19.5 | 23.7 | 29.1 |
| 18.4 | 21.3 | 25.5 | 29.6 |
| 18.4 | 22.2 | 27.0 | |

*Source:* Morningstar.com

*Problem:* The data in Table 4 represent the three-year rate of return of 19 randomly selected small-capitalization growth mutual funds. Is there evidence to support the belief that the variable "three-year rate of return" is normally distributed?

*Approach:* We follow Steps 1–4 listed on page 306.

*Solution:*

**Step 1:** The first column in Table 5 represents the index $i$. The second column in Table 5 represents the observed values in the data set, written in ascending order.

**Step 2:** The third column in Table 5 represents $f_i = \dfrac{i - 0.375}{n + 0.25}$ for each observation. Remember, this is the expected area under the normal curve to the left of the $i$th observation, under the assumption of normality. For example, $i = 1$ corresponds to the rate of return of 15.8 percent.

$$f_1 = \frac{1 - 0.375}{19 + 0.25} = 0.0325$$

So, the area to the left of 15.8 would be 0.0325 if the sample data come from a population that is normally distributed.

**Step 3:** We use Table II to find the $Z$-scores that correspond to $f_i$. The expected $Z$-scores are listed in the fourth column of Table 5. For example, the area to the left of 15.8 is $f_1 = 0.0325$ if the data come from a population that is normally distributed. Look in Table II for the area closest to $f_1 = 0.0325$. The expected $Z$-score is $-1.85$. Notice that for each negative expected $Z$-score, there is a corresponding positive expected $Z$-score, as a result of the symmetry of the normal curve.

| TABLE 5 | | | |
|---|---|---|---|
| Index, $i$ | Observed Value | $f_i$ | Expected $Z$-score |
| 1 | 15.8 | $\dfrac{1 - 0.375}{19 + 0.25} = 0.0325$ | $-1.85$ |
| 2 | 16.7 | $\dfrac{2 - 0.375}{19 + 0.25} = 0.0844$ | $-1.38$ |
| 3 | 18.2 | 0.1364 | $-1.10$ |
| 4 | 18.4 | 0.1883 | $-0.88$ |
| 5 | 18.4 | 0.2403 | $-0.71$ |
| 6 | 18.5 | 0.2922 | $-0.55$ |
| 7 | 19.2 | 0.3442 | $-0.40$ |
| 8 | 19.5 | 0.3961 | $-0.26$ |
| 9 | 21.3 | 0.4481 | $-0.13$ |
| 10 | 22.2 | 0.5 | 0 |
| 11 | 22.6 | 0.5519 | 0.13 |
| 12 | 23.7 | 0.6039 | 0.26 |
| 13 | 23.7 | 0.6558 | 0.40 |
| 14 | 25.5 | 0.7078 | 0.55 |
| 15 | 27.0 | 0.7597 | 0.71 |
| 16 | 27.4 | 0.8117 | 0.88 |
| 17 | 28.5 | 0.8636 | 1.10 |
| 18 | 29.1 | 0.9156 | 1.38 |
| 19 | 29.6 | 0.9675 | 1.85 |

*Step 4:* We plot the actual observations on the horizontal axis and the expected $Z$-scores on the vertical axis. See Figure 43.

Figure 43
Normal Probability Plot

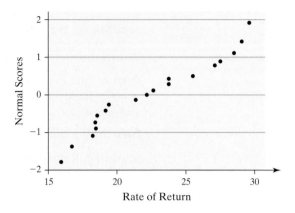

Although the normal probability plot in Figure 43 does show some curvature, it is roughly linear. We conclude that the three-year rate of return of small-capitalization growth mutual funds is approximately normally distributed. ◀◀

Typically, normal probability plots are drawn by using either a graphing calculator with advanced statistical features or statistical software. Certain software, such as Minitab, provides bounds that the data must lie within in order to support the belief that the sample data come from a population that is normally distributed.

▶ **EXAMPLE 2   Assessing Normality Via Statistical Software**

*Problem:* Using Minitab, or some other statistical software, draw a normal probability plot of the data in Table 4 and determine whether there is evidence to support the belief that the sample data come from a population that is normally distributed.

*Approach:* We will construct a normal probability plot using Minitab. Minitab provides curved "bounds" that can be used to assess normality. If the normal probability plot is roughly linear and all the data lie within the bounds provided by the software, then we have reason to believe the data come from a population that is approximately normal.

*Solution:* Figure 44 shows the normal probability plot.

Figure 44                                    Normal Probability Plot for Rate of Return

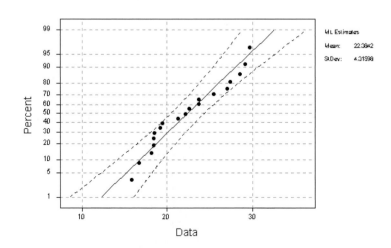

**Caution**

If a normal probability plot of sample data is roughly linear, then it is reasonable to believe that the population is normally distributed. We do not say that the population is normally distributed.

The normal probability plot is roughly linear, and all the data lie within the bounds provided by Minitab. We conclude that the sample data could come from a population that is normally distributed. ◀◀

Figure 45 shows a histogram drawn for the data analyzed in Example 2. Judging from the histogram whether the sample data come from a population that is normally distributed would be very difficult.

Figure 45

Throughout the text, we will provide normal probability plots drawn with Minitab, so that assessing normality is straightforward.

▶ EXAMPLE 3   Assessing Normality

*Problem:* The data in Table 6 represent the time spent waiting in line (in minutes) for the Demon Roller Coaster for 100 randomly selected riders. Is the random variable "waiting time" normally distributed?

| | | | | | | | | | | | | | | | | | |
|---|---|---|---|---|---|---|---|---|---|---|---|---|---|---|---|---|---|
| TABLE 6 | | | | | | | | | | | | | | | | | |
| 7 | 3 | 5 | 107 | 8 | 37 | 16 | 41 | 7 | 25 | 22 | 19 | 1 | 40 | 1 | 29 | 93 |
| 33 | 76 | 14 | 8 | 9 | 45 | 15 | 81 | 94 | 10 | 115 | 18 | 0 | 18 | 11 | 60 | 34 |
| 30 | 6 | 21 | 0 | 86 | 6 | 11 | 1 | 1 | 3 | 9 | 79 | 41 | 2 | 9 | 6 | 19 |
| 4 | 3 | 2 | 7 | 18 | 0 | 93 | 68 | 6 | 94 | 16 | 13 | 24 | 6 | 12 | 121 | 30 |
| 35 | 39 | 9 | 15 | 53 | 9 | 47 | 5 | 55 | 64 | 51 | 80 | 26 | 24 | 12 | 0 | |
| 94 | 18 | 4 | 61 | 38 | 38 | 21 | 61 | 9 | 80 | 18 | 21 | 8 | 14 | 47 | 56 | |

*Approach:* We will use Minitab to draw a normal probability plot. If the normal probability plot is roughly linear and the data lie within the bounds provided by Minitab, conclude that it is reasonable to believe that the sample data come from a population that follows a normal distribution.

*Solution:* Figure 46 shows a normal probability plot of the data drawn using Minitab.

Figure 46

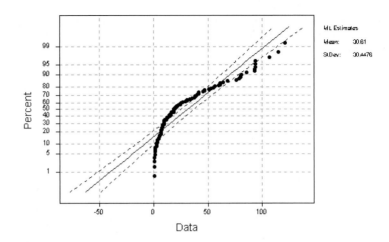

Normal Probability Plot for Wait Time

Clearly, the normal probability plot is not linear. We conclude that the random variable "wait time" is not normally distributed. ◄◄

Figure 47 shows a histogram of the data in Table 6. The histogram indicates that the data are skewed right.

Figure 47

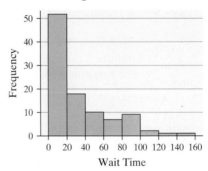

**Histogram of Wait Time**

**NW** *Now Work Problems 1 and 5.*

# 7.4 Assess Your Understanding

## Concepts and Vocabulary

**1.** Explain why normal probability plots should be linear if the data is normally distributed.

**2.** What does $f_i$ represent?

## Exercises

- ### Skill Building

*In Problems 1–6, determine whether the normal probability plot indicates that the sample data could have come from a population that is normally distributed.*

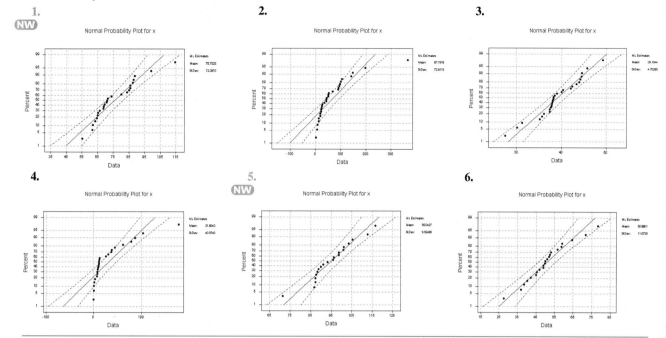

## • Applying the Concepts

*In Problems 7–10, use a normal probability plot to assess whether the sample data could have come from a population that is normally distributed.*

7. **Achievement Scores** A school psychologist conducted an experiment in which 40 third-grade students who were identified as "at-risk" by their teachers were randomly divided into two equal groups of size 20. Group 1 participated in a new school success program; group 2 participated in an ongoing school success program. The following data represent the achievement scores of the students in Group 1 at post-trial assessment.

| | | | | | | |
|---|---|---|---|---|---|---|
| 75 | 110 | 80 | 134 | 102 | 82 | 109 |
| 114 | 83 | 110 | 89 | 89 | 98 | 104 |
| 104 | 99 | 98 | 111 | 104 | 92 | |

8. **Miles on a Cavalier** A random sample of eighteen three-year-old Chevy Cavaliers was obtained in the Miami, FL area, and the number of miles on each car was recorded.

| | | | | |
|---|---|---|---|---|
| 34,122 | 17,685 | 15,499 | 26,455 | 30,500 |
| 39,416 | 29,307 | 26,051 | 27,368 | 35,936 |
| 28,281 | 34,511 | 32,305 | 37,904 | 44,448 |
| 41,194 | 29,289 | 31,883 | | |

*Source:* cars.com

9. **Volume of Philip Morris Stock** The following data represent a random sample of the number of shares of Philip Morris stock traded for 20 days in 2000.

| | | | | |
|---|---|---|---|---|
| 3.98 | 8.90 | 10.40 | 7.52 | 13.84 |
| 5.29 | 9.69 | 8.94 | 12.10 | 13.25 |
| 4.24 | 8.54 | 11.59 | 6.75 | 27.54 |
| 7.71 | 10.14 | 4.28 | 7.24 | 10.88 |

*Source:* http://finance.yahoo.com

10. **Baseball Salaries** The following data represent the baseball salaries of 24 randomly selected players from the 2000 season (data are in thousands, so 2350 means $2,350,000).

| | | | | | |
|---|---|---|---|---|---|
| 2350 | 208 | 260 | 215 | 6000 | 3333 |
| 1400 | 225 | 1000 | 270 | 2750 | 4000 |
| 3950 | 4000 | 269 | 225 | 305 | 200 |
| 2400 | 300 | 962 | 15,714 | 950 | 213 |

*Source:* espn.com

**Technology Step-by-Step**
Normal Probability Plots

**TI-83 Plus**    **Step 1:** Enter the raw data into L1.

**Step 2:** Press $2^{nd}$ Y = to access STAT PLOTS.

**Step 3:** Select 1:Plot1.

**Step 4:** Turn Plot1 ON by highlighting ON and pressing ENTER. Press the down-arrow key. Highlight the *normal probability plot* icon. It is the icon in the lower-right corner under Type:. Press ENTER to select this plot type. The Data List should be set at L1. The data axis should be the *x*-axis.

**Step 5:** Press ZOOM, and select 9:ZoomStat.

**MINITAB**    **Step 1:** Enter the raw data into C1.

**Step 2:** Select the **Graph** menu. Highlight **Probability Plot** ....

**Step 3:** In the Variables cell, enter the column that contains the raw data. Make sure Distribution is set to Normal. Click OK.

**Excel**    **Step 1:** Load the PHStat Add-in.

**Step 2:** Enter the raw data into column A.

**Step 3:** Select the **PHStat** menu. Highlight **Probability Distributions**, then highlight **Normal Probability Plot** ....

**Step 4:** With the cursor in the "Variable Cell Range:" cell, highlight the raw data. Enter a graph title, if desired. Click OK.

---

**7.5** | ## Sampling Distributions; The Central Limit Theorem

**Preparing for This Section**    Before getting started, review the following:

✓ Simple random sampling (Section 1.2, pp. 12–15)

✓ The arithmetic mean (Section 3.1, pp. 84–87)

✓ The standard deviation (Section 3.2, pp. 105–106)

**Objectives**    ① Understand the concept of a sampling distribution

② Compute the mean and standard deviation of a sampling distribution of the mean

③ Compute probabilities of a sample mean obtained from a normal population

④ Compute probabilities of a sample mean utilizing the Central Limit Theorem

 Recall the process of statistics presented in Chapter 1.

**Step 1:** A research question is posed.

**Step 2:** Information (data) is collected so that the question posed can be answered. This information typically is collected through an experiment or survey.

**Step 3:** The information is organized and summarized. This is done through a pictorial representation of the data collected in Step 2 and through numerical summaries, such as the mean and standard deviation.

**Step 4:** Draw conclusions from the data.

The methods for conducting Steps 1 and 2 were discussed in Chapter 1. The methods for conducting Step 3 were discussed in Chapters 2 and 3 .

If the information that is organized in Step 3 is based upon sample data, then we use this data to make inferences about the population. For example, we might compute a sample mean from the information collected in Step 2 and use this information to draw conclusions regarding the population mean.

From our discussions of sampling methods, we know that the values of statistics such as $\bar{x}$ vary from sample to sample (because the individuals in the samples are different). So, using statistics to make inferences regarding a population is subject to variability and, hence, uncertainty. Therefore, we need a way to assess the reliability of inferences made regarding a population. This is where the information learned in Chapters 5 and 6 and in the sections in Chapter 7 preceding this one comes in handy.

Inferential statistics is based upon probability statements. If we know the probability distribution of a random variable, we can obtain probability statements regarding that random variable. For example, suppose IQ scores are normally distributed with mean 100 and standard deviation 16. Given this information, we can make a probability statement regarding the likelihood of randomly selecting an individual with an IQ of 120 or higher.

Because statistics such as $\bar{x}$ vary from sample to sample, they are random variables. As such, statistics have probability distributions associated with them. For example, there is a probability distribution for the sample mean, one for the sample variance, and so on. In order to make probability statements regarding a sample statistic, we need to know the probability distribution of the sample statistic—that is to say, we need to know the shape, center, and spread of the sample statistic's distribution.

## Sampling Distributions

In general, a sampling distribution of a statistic is a probability distribution (such as the normal distribution) for all possible values of the statistic computed from a sample of size $n$. The **sampling distribution of the mean** is a probability distribution of all possible values of the random variable $\bar{x}$ computed from a sample of size $n$ from a population with mean $\mu$ and standard deviation $\sigma$.

The idea behind obtaining the sampling distribution of the mean is as follows:

**Step 1:** Obtain a simple random sample of size $n$.

**Step 2:** Compute the sample mean.

**Step 3:** Assuming that we are sampling from a finite population, repeat Steps 1 and 2 until all simple random samples of size $n$ have been obtained.

We present an example to illustrate the idea behind a sampling distribution.

▶ **EXAMPLE 1**   **Illustrating a Sampling Distribution**

*Problem:* One semester, I had a small statistics class of 7 students. I asked them the age of their cars and obtained the following data:

$$2, 4, 6, 8, 4, 3, 7$$

Construct a sampling distribution of the mean for samples of size $n = 2$. The population mean is $\mu = \dfrac{34}{7} \approx 4.9$ years. What is the probability of obtaining a sample mean between 4 and 6 years, inclusive—that is, what is $P(4 \le \bar{x} \le 6)$?

*Approach:* We follow Steps 1–3 just listed to construct the probability distribution.

*Solution:* There are a total of 7 individuals in the population. We are selecting them two at a time without replacement. Therefore, there are $_7C_2 = 21$ samples of size $n = 2$. We list these 21 samples along with the sample means in Table 7.

| | TABLE 7 | | | | |
|---|---|---|---|---|---|
| Sample | Sample Mean | Sample | Sample Mean | Sample | Sample Mean |
| 2, 4 | 3 | 4, 8 | 6 | 6, 7 | 6.5 |
| 2, 6 | 4 | 4, 4 | 4 | 8, 4 | 6 |
| 2, 8 | 5 | 4, 3 | 3.5 | 8, 3 | 5.5 |
| 2, 4 | 3 | 4, 7 | 5.5 | 8, 7 | 7.5 |
| 2, 3 | 2.5 | 6, 8 | 7 | 4, 3 | 3.5 |
| 2, 7 | 4.5 | 6, 4 | 5 | 4, 7 | 5.5 |
| 4, 6 | 5 | 6, 3 | 4.5 | 3, 7 | 5 |

Table 8 displays the sampling distribution of the sample mean, $\bar{x}$.

| TABLE 8 | | |
|---|---|---|
| Sample Mean | Frequency | Probability |
| 2.5 | 1 | 1/21 |
| 3 | 2 | 2/21 |
| 3.5 | 2 | 2/21 |
| 4 | 2 | 2/21 |
| 4.5 | 2 | 2/21 |
| 5 | 4 | 4/21 |
| 5.5 | 3 | 3/21 |
| 6 | 2 | 2/21 |
| 6.5 | 1 | 1/21 |
| 7 | 1 | 1/21 |
| 7.5 | 1 | 1/21 |

From Table 8, we can see that

$$P(4 \leq \bar{x} \leq 6) = \frac{2}{21} + \frac{2}{21} + \frac{4}{21} + \frac{3}{21} + \frac{2}{21} = \frac{13}{21} = 0.619$$

If we took 10 simple random samples of size 2 from this population, about 6 of them would result in sample means between 4 and 6 years, inclusive. ◄◄

The population mean of the data in Example 1 is $\mu = 4.9$, rounded to one decimal place. Notice that the sample mean with the highest probability is $\bar{x} = 5$. Also notice that the further the sample mean is from 4.9, the lower the probability of obtaining the sample mean. Figure 48 is a probability histogram of the sampling distribution for the sample mean given in Table 8.

Figure 48

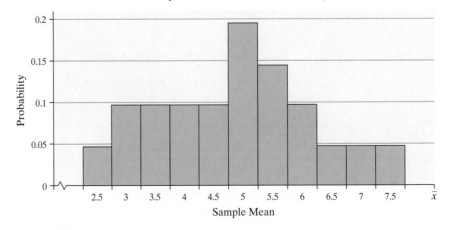

**Probability Distribution of the Sample Mean**

NW Now Work Problem 19.

The point of Example 1 is to help you realize that statistics such as $\bar{x}$ vary and so have distributions associated with them. In practice, a single random sample of size $n$ is obtained from a population. The probability distribution of the sample statistic (or sampling distribution) is determined from statistical theory. We will use simulation to help justify the result that statistical theory provides. We consider two possibilities. In the first case, we sample from a population that is known to be normally distributed. In the second case, we sample from a distribution that is not normally distributed.

▶ EXAMPLE 2    Sampling Distribution of the Sample Mean—Population Normal

*Problem:* The weight of three-year-old female patients is normally distributed with $\mu = 38.72$ pounds and $\sigma = 3.17$ pounds. Approximate the sampling distribution of $\bar{x}$ by taking 100 simple random samples of size $n = 5$.

*Approach:* Use Minitab, Excel, or some other statistical software package to perform the simulation. We will perform the following steps.

*Step 1:* Obtain 100 simple random samples of size $n = 5$ from the population, using simulation.
*Step 2:* Compute the mean of each of the samples.
*Step 3:* Draw a histogram of the sample means.
*Step 4:* Compute the mean and standard deviation of the sample means.

*Solution:*

*Step 1:* We obtain 100 simple random samples of size $n = 5$. All the samples of size $n = 5$ are shown in Table 9.

| Sample | Sample of Size $n = 5$ | | | | | Sample Mean |
|:---:|:---:|:---:|:---:|:---:|:---:|:---:|
| 1 | 36.48 | 39.94 | 42.57 | 39.53 | 33.81 | 38.47 |
| 2 | 43.13 | 37.97 | 42.41 | 39.61 | 43.30 | 41.28 |
| 3 | 41.64 | 39.01 | 37.77 | 38.94 | 41.10 | 39.69 |
| 4 | 40.37 | 43.49 | 37.60 | 40.14 | 38.88 | 40.10 |
| 5 | 38.62 | 33.43 | 45.17 | 42.66 | 39.98 | 39.97 |
| 6 | 38.98 | 41.35 | 36.80 | 43.56 | 39.92 | 40.12 |
| 7 | 42.48 | 37.00 | 35.87 | 39.62 | 38.74 | 38.74 |
| 8 | 39.38 | 37.02 | 41.60 | 40.34 | 37.62 | 39.19 |
| 9 | 42.82 | 45.77 | 35.16 | 42.56 | 39.75 | 41.21 |
| 10 | 36.19 | 35.20 | 37.74 | 40.46 | 37.47 | 37.41 |
| 11 | 36.59 | 41.62 | 42.18 | 39.23 | 39.26 | 39.77 |
| 12 | 38.57 | 42.13 | 45.39 | 38.22 | 46.18 | 42.10 |
| 13 | 38.40 | 39.06 | 43.60 | 31.46 | 37.03 | 37.91 |
| 14 | 34.29 | 47.73 | 37.27 | 41.82 | 33.33 | 38.89 |
| 15 | 42.28 | 43.29 | 37.69 | 37.32 | 40.06 | 40.13 |
| 16 | 34.31 | 43.58 | 40.02 | 41.13 | 42.99 | 40.41 |
| 17 | 38.71 | 39.03 | 39.39 | 42.62 | 38.41 | 39.63 |
| 18 | 38.63 | 39.66 | 39.47 | 41.13 | 38.01 | 39.38 |
| 19 | 39.09 | 33.86 | 37.57 | 41.65 | 35.22 | 37.48 |
| 20 | 40.94 | 37.50 | 38.72 | 41.64 | 35.48 | 38.86 |
| 21 | 38.72 | 35.89 | 37.82 | 35.04 | 37.06 | 36.91 |
| 22 | 39.64 | 36.30 | 35.54 | 40.40 | 38.74 | 38.12 |
| 23 | 38.22 | 38.49 | 33.60 | 40.18 | 39.07 | 37.91 |
| 24 | 40.93 | 40.53 | 37.55 | 37.30 | 37.16 | 38.70 |
| 25 | 33.27 | 38.92 | 37.14 | 39.90 | 33.83 | 36.61 |
| 26 | 39.44 | 37.28 | 35.70 | 41.97 | 36.80 | 38.24 |
| 27 | 38.83 | 41.41 | 38.87 | 39.40 | 37.20 | 39.14 |

### Table 9 (cont'd)

| | | | | | | |
|---|---|---|---|---|---|---|
| 28 | 40.10 | 36.96 | 35.73 | 43.00 | 38.11 | 38.78 |
| 29 | 41.93 | 36.57 | 37.55 | 35.14 | 38.75 | 37.99 |
| 30 | 31.25 | 38.85 | 39.25 | 35.07 | 39.77 | 36.84 |
| 31 | 38.47 | 34.45 | 30.43 | 41.76 | 41.61 | 37.34 |
| 32 | 37.98 | 35.56 | 43.97 | 44.96 | 37.81 | 40.06 |
| 33 | 43.34 | 40.94 | 35.17 | 41.74 | 37.59 | 39.76 |
| 34 | 39.80 | 44.44 | 37.53 | 40.52 | 41.95 | 40.85 |
| 35 | 41.98 | 42.02 | 40.73 | 40.47 | 36.81 | 40.40 |
| 36 | 40.98 | 35.08 | 34.61 | 40.78 | 37.26 | 37.74 |
| 37 | 35.75 | 40.81 | 40.13 | 35.99 | 36.52 | 37.84 |
| 38 | 36.39 | 45.97 | 40.59 | 37.64 | 42.42 | 40.60 |
| 39 | 36.20 | 35.63 | 37.43 | 38.35 | 34.81 | 36.48 |
| 40 | 33.58 | 33.87 | 41.60 | 45.10 | 38.68 | 38.57 |
| 41 | 31.77 | 38.34 | 41.79 | 37.93 | 40.83 | 38.13 |
| 42 | 43.03 | 33.12 | 34.98 | 36.58 | 37.78 | 37.10 |
| 43 | 35.76 | 35.17 | 42.58 | 39.10 | 41.08 | 38.74 |
| 44 | 38.44 | 38.45 | 35.93 | 35.32 | 44.60 | 38.55 |
| 45 | 44.54 | 41.88 | 35.84 | 42.64 | 42.38 | 41.46 |
| 46 | 41.89 | 36.81 | 41.83 | 40.24 | 39.28 | 40.01 |
| 47 | 38.00 | 40.08 | 35.57 | 34.44 | 39.51 | 37.52 |
| 48 | 39.92 | 38.05 | 39.96 | 38.04 | 32.11 | 37.61 |
| 49 | 36.37 | 38.62 | 32.25 | 41.35 | 40.91 | 37.90 |
| 50 | 34.38 | 36.65 | 32.97 | 39.93 | 41.34 | 37.05 |
| 51 | 40.32 | 39.80 | 41.00 | 38.62 | 38.24 | 39.59 |
| 52 | 37.95 | 45.26 | 38.67 | 34.96 | 41.13 | 39.60 |
| 53 | 36.82 | 42.63 | 41.62 | 39.43 | 37.48 | 39.59 |
| 54 | 41.63 | 37.65 | 38.58 | 39.03 | 37.53 | 38.88 |
| 55 | 37.91 | 37.20 | 38.72 | 36.87 | 45.40 | 39.22 |
| 56 | 41.05 | 34.01 | 39.11 | 38.23 | 35.74 | 37.63 |
| 57 | 42.09 | 45.44 | 35.52 | 39.87 | 37.28 | 40.04 |
| 58 | 39.31 | 35.79 | 37.82 | 39.15 | 35.57 | 37.53 |
| 59 | 41.16 | 39.98 | 41.11 | 39.21 | 39.98 | 40.29 |
| 60 | 35.68 | 45.60 | 39.34 | 36.65 | 43.30 | 40.12 |
| 61 | 36.07 | 39.63 | 42.55 | 41.72 | 36.81 | 39.36 |
| 62 | 38.97 | 36.83 | 41.01 | 38.12 | 35.27 | 38.04 |
| 63 | 33.70 | 39.15 | 34.81 | 34.13 | 39.00 | 36.16 |
| 64 | 37.19 | 34.69 | 36.21 | 34.34 | 39.07 | 36.30 |
| 65 | 33.99 | 44.87 | 42.52 | 40.22 | 39.26 | 40.17 |
| 66 | 41.40 | 27.62 | 34.57 | 40.08 | 34.65 | 35.66 |
| 67 | 40.14 | 34.45 | 38.26 | 38.09 | 39.72 | 38.13 |
| 68 | 33.64 | 42.62 | 32.08 | 34.30 | 37.34 | 35.99 |
| 69 | 35.36 | 39.02 | 43.98 | 41.19 | 32.47 | 38.40 |
| 70 | 43.26 | 37.85 | 35.82 | 37.11 | 36.22 | 38.05 |
| 71 | 36.24 | 38.07 | 33.38 | 38.43 | 39.88 | 37.20 |
| 72 | 38.55 | 43.06 | 41.07 | 36.58 | 37.02 | 39.25 |
| 73 | 41.26 | 36.99 | 36.17 | 38.98 | 36.03 | 37.89 |
| 74 | 37.31 | 38.41 | 41.18 | 39.76 | 39.64 | 39.26 |
| 75 | 36.26 | 41.84 | 42.50 | 37.70 | 41.21 | 39.90 |
| 76 | 39.27 | 38.61 | 44.53 | 38.08 | 35.01 | 39.10 |
| 77 | 39.14 | 40.83 | 39.83 | 37.78 | 36.51 | 38.82 |
| 78 | 42.53 | 43.41 | 41.01 | 33.71 | 39.47 | 40.03 |
| 79 | 45.34 | 32.61 | 33.81 | 39.03 | 40.32 | 38.22 |
| 80 | 36.31 | 35.55 | 37.12 | 38.74 | 40.80 | 37.70 |
| 81 | 31.40 | 41.80 | 40.15 | 42.53 | 37.62 | 38.70 |
| 82 | 41.01 | 39.02 | 39.68 | 36.61 | 38.44 | 38.95 |
| 83 | 34.15 | 36.19 | 35.98 | 36.02 | 36.32 | 35.73 |

**Table 9 (cont'd)**

| | | | | | | |
|---|---|---|---|---|---|---|
| 84 | 31.50 | 37.61 | 43.29 | 39.82 | 38.78 | 38.20 |
| 85 | 43.26 | 34.01 | 41.18 | 40.23 | 39.28 | 39.59 |
| 86 | 41.76 | 41.40 | 39.02 | 38.20 | 39.42 | 39.96 |
| 87 | 37.06 | 35.95 | 39.98 | 40.00 | 43.36 | 39.27 |
| 88 | 41.01 | 37.56 | 36.95 | 39.71 | 37.97 | 38.64 |
| 89 | 34.97 | 38.36 | 36.30 | 38.48 | 34.24 | 36.47 |
| 90 | 38.38 | 38.94 | 40.96 | 36.13 | 35.98 | 38.08 |
| 91 | 39.41 | 30.78 | 37.66 | 37.31 | 42.04 | 37.44 |
| 92 | 39.83 | 35.88 | 30.20 | 45.07 | 40.06 | 38.21 |
| 93 | 36.25 | 39.56 | 34.53 | 40.69 | 37.03 | 37.61 |
| 94 | 45.64 | 40.66 | 44.51 | 40.50 | 39.43 | 42.15 |
| 95 | 37.63 | 44.77 | 38.31 | 36.53 | 38.41 | 39.13 |
| 96 | 39.78 | 33.34 | 43.42 | 43.63 | 38.77 | 39.79 |
| 97 | 41.48 | 37.39 | 38.62 | 43.83 | 34.26 | 39.12 |
| 98 | 37.68 | 40.66 | 38.93 | 40.94 | 37.54 | 39.15 |
| 99 | 39.72 | 32.61 | 32.62 | 40.35 | 38.65 | 36.79 |
| 100 | 39.25 | 41.06 | 41.17 | 38.30 | 38.24 | 39.60 |

*Step 2:* We compute the sample means for each of the 100 samples as shown in Table 9.

*Step 3:* We draw a histogram of the 100 sample means. See Figure 49.

Figure 49

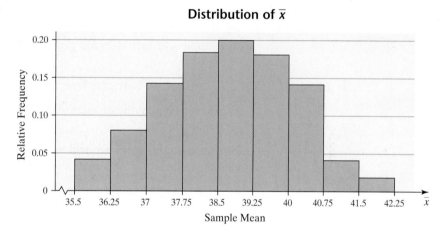

Distribution of $\bar{x}$

*Step 4:* The mean of the 100 sample means is 38.72, and the standard deviation is 1.374. ◄◄

 If we refer to the results obtained in Example 3 from Section 7.1, we see that the histogram of the population data is normal, with mean $\mu = 38.72$ and $\sigma = 3.17$. The histogram in Figure 49 indicates that the distribution of sample means could also be normally distributed. In addition, the mean of the sample means is 38.72, but the standard deviation is only 1.374. We might conclude the following regarding the sampling distribution of $\bar{x}$:

**1.** It is normally distributed.

**2.** It has mean equal to the mean of the population.

**3.** It has standard deviation less than the standard deviation of the population.

A question that one might ask is, "What role does $n$, the sample size, play in the sampling distribution of $\bar{x}$?" Suppose the sample mean is computed for samples of size $n = 1$ through $n = 200$. That is, the sample mean is recomputed each time an additional individual is added to the sample. The sample mean is then plotted against the sample size in Figure 50.

Figure 50

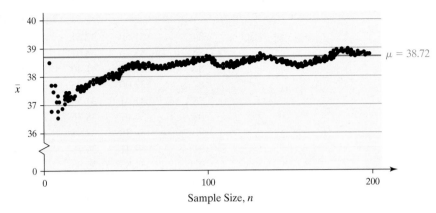

From the graph, we see that, as the sample size $n$ increases, the sample mean gets closer to the population mean. This concept is known as the *Law of Large Numbers*.

**Theorem**

> **The Law of Large Numbers**
>
> As additional observations are added to the sample, the difference between the sample mean, $\bar{x}$, and the population mean $\mu$ approaches zero.

✎ **In Your Own Words**

As the sample size increases, the sample mean gets closer to the population mean.

So, according to the Law of Large Numbers, the more individuals we sample, the closer the sample mean gets to the population mean. This result implies that there is less variability in the distribution of the sample mean as the sample size increases. We demonstrate this result in the next example.

▶ **EXAMPLE 3**   **The Impact of Sample Size on Sampling Variability**

*Problem:* Repeat the problem in Example 2 with sample size $n = 15$ using 200 sample means.

*Approach:* The approach will be identical to that presented in Example 2, except we let $n = 15$ instead of $n = 5$.

*Solution:* Figure 51 shows the histogram of the sample means.

Figure 51

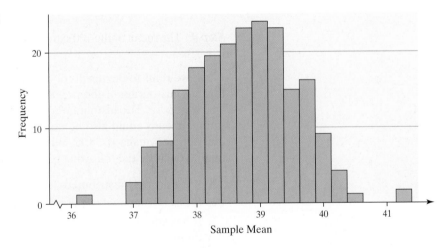

The histogram in Figure 51 is symmetric and bell shaped. This is an indication that the distribution of the sample mean is approximately normally distributed. The mean of the 200 sample means is 38.72 (just as in Example 2); however, the standard deviation is now 0.81. ◀◀

From the results of Examples 2 and 3, we conclude that, as the sample size $n$ increases, the standard deviation of the distribution of $\bar{x}$ decreases. Although the proof is beyond the scope of this text, we should be convinced that the following result is reasonable.

*Theorem*

> **The Mean and Standard Deviation of the Sampling Distribution of $\bar{x}$**
>
> Suppose that a simple random sample of size $n$ is drawn from a population with mean $\mu$ and standard deviation $\sigma$. The sampling distribution of $\bar{x}$ will have mean $\mu_{\bar{x}} = \mu$ and standard deviation $\sigma_{\bar{x}} = \dfrac{\sigma}{\sqrt{n}}$.* The standard deviation of the sampling distribution of $\bar{x}$, $\sigma_{\bar{x}}$, is called the **standard error of the mean**.

![feather pen icon] **In Your Own Words**

Regardless of the distribution of the population, the sampling distribution of $\bar{x}$ will have a mean equal to the mean of the population and a standard deviation equal to the standard deviation of the population divided by the square root of the sample size!

For the population presented in Example 2, if we draw a simple random sample of size $n = 5$, the sampling distribution $\bar{x}$ will have mean $\mu_{\bar{x}} = 38.72$ and standard deviation

$$\sigma_{\bar{x}} = \frac{\sigma}{\sqrt{n}} = \frac{3.17}{\sqrt{5}}$$

**NW** *Now Work Problem 1.*

Now that we now know how to determine the mean and standard deviation for any sampling distribution of $\bar{x}$, we can concentrate on the shape. Refer back to Figures 49 and 51 from Examples 2 and 3. Recall that the population from which the sample was drawn was normal. The shapes of these histograms imply that the sampling distribution of $\bar{x}$ is normal. This leads us to believe that, if the population is normal, then the distribution of the sample mean is also normal.

*Theorem*

> If a random variable $X$ is normally distributed with mean $\mu$ and standard deviation $\sigma$, then the distribution of the sample mean, $\bar{x}$, is normally distributed with mean $\mu_{\bar{x}} = \mu$ and standard deviation $\sigma_{\bar{x}} = \dfrac{\sigma}{\sqrt{n}}$.

For example, the height of three-year-old females is a normal random variable with mean $\mu = 38.72$ and standard deviation $\sigma = 3.17$ inches. The distribution of the sample mean, $\bar{x}$, the mean height of a simple random sample of $n = 5$ three-year-old females, is normal with mean $\mu_{\bar{x}} = 38.72$ and standard deviation $\sigma_{\bar{x}} = \dfrac{3.17}{\sqrt{5}}$. See Figure 52.

Figure 52

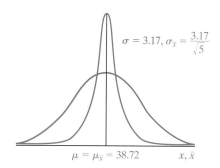

*Technically, $\sigma_{\bar{x}} = \sqrt{\dfrac{N-n}{N-1}} \cdot \dfrac{\sigma}{\sqrt{n}}$ for populations of finite size $N$. However, if the sample size is less than 5% of the population size ($n < 0.05N$), the effect of $\sqrt{\dfrac{N-n}{N-1}}$ (the **finite population correction factor**) can be ignored without affecting the results.

▶ **EXAMPLE 4**  **Describing the Distribution of the Sample Mean**

*Problem:* The height, $X$, of three-year-old females is normally distributed with mean $\mu = 38.72$ inches and standard deviation $\sigma = 3.17$ inches. Compute the probability that a simple random sample of size $n = 10$ results in a sample mean greater than 40 inches. That is, compute $P(\overline{x} > 40)$.

*Approach:* The random variable $X$ is normally distributed, so the sampling distribution of $\overline{x}$ will also be normally distributed. The mean of the sampling distribution is $\mu_{\overline{x}} = \mu$, and its standard deviation is $\sigma_{\overline{x}} = \dfrac{\sigma}{\sqrt{n}}$. We convert the random variable $\overline{x} = 40$ to a $Z$-score and then find the area under the standard normal curve to the right of this $Z$-score.

**Figure 53**

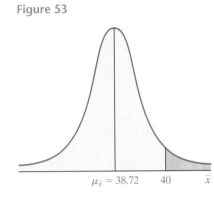

*Solution:* The sample mean is normally distributed with mean $\mu_{\overline{x}} = 38.72$ inches and standard deviation $\sigma_{\overline{x}} = \dfrac{\sigma}{\sqrt{n}} = \dfrac{3.17}{\sqrt{10}} = 1.00$ inch.

Figure 53 displays the normal curve with the area we wish to compute shaded. We convert the random variable $\overline{x} = 40$ to a $Z$-score and obtain

$$Z = \frac{\overline{x} - \mu_{\overline{x}}}{\sigma_{\overline{x}}} = \frac{\overline{x} - \mu_{\overline{x}}}{\dfrac{\sigma}{\sqrt{n}}} = \frac{40 - 38.72}{1.00} = 1.28$$

The area to the right of $Z = 1.28$ is $1 - 0.8997 = 0.1003$.

*Interpretation:* The probability of obtaining a sample mean greater than 40 inches from a population whose mean is 38.72 inches is 0.1003. That is, $P(\overline{x} \geq 40) = 0.1003$. If we took 100 simple random samples of $n = 10$ from this population and if the population mean is 38.72, then about 10 of the samples will have a mean that is 40 or more. ◀◀

**NW** *Now Work Problem 9.*

**4** What if the population from which the sample is drawn is not normal?

▶ **EXAMPLE 5**  **Sampling from a Population That Is Not Normal**

**Figure 54**

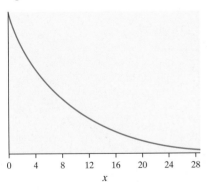

*Problem:* Figure 54 shows the graph of an *exponential density function* with mean and standard deviation equal to 10. The exponential distribution is used to model lifetimes of electronic components and to model the time required to serve a customer or repair a machine.

Clearly, the distribution is not normal. Approximate the sampling distribution of $\overline{x}$ by obtaining, through simulation, 300 random samples of size (a) $n = 3$, (b) $n = 12$, and (c) $n = 30$ from the probability distribution.

*Approach:*

**Step 1:** Use Minitab, Excel, or some other statistical software to obtain 300 random samples for each sample size.

**Step 2:** Compute the sample mean of each of the 300 random samples.

**Step 3:** Draw a histogram of the 300 sample means.

*Solution:*

**Step 1:** Using Minitab, we obtain 300 random samples of size (a) $n = 3$, (b) $n = 12$, and (c) $n = 30$. For example, in the first random sample of size $n = 30$, we obtained the following results:

| 9.2 | 20.0 | 17.0 | 2.4 | 2.6 | 19.9 | 21.2 | 5.7 | 8.1 | 10.8 |
|------|------|------|------|------|------|------|------|------|------|
| 1.2 | 22.3 | 18.4 | 4.2 | 9.9 | 41.8 | 4.2 | 1.2 | 10.8 | 2.1 |
| 11.3 | 17.9 | 28.0 | 12.1 | 3.0 | 0.5 | 4.5 | 14.2 | 5.0 | 11.4 |

**Step 2:** We compute the mean of each of the 300 random samples, using Minitab. For example, the sample mean of the first sample of size $n = 30$ is 11.36.

**Step 3:** Figure 55(a) displays the histogram of $\bar{x}$ that results from simulating 300 random samples of size $n = 3$ from an exponential distribution with $\mu = 10$ and $\sigma = 10$. Figure 55(b) displays the histogram of $\bar{x}$ that results from simulating 300 random samples of size $n = 12$, and Figure 55(c) displays the histogram of $\bar{x}$ that results from simulating 300 random samples of size $n = 30$.

**Figure 55**

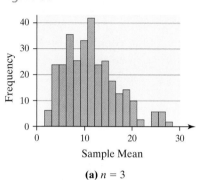

**(a)** $n = 3$

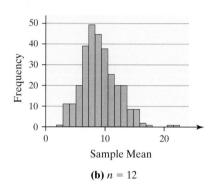

**(b)** $n = 12$

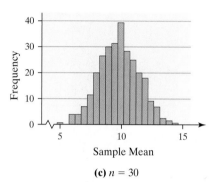

**(c)** $n = 30$

Notice that, as the sample size increases, the distribution of the sample mean becomes more normal, even though the population clearly is not normal!

◀◀

We formally state the results of Example 5 as the *Central Limit Theorem*.

*Theorem*

**The Central Limit Theorem**

Suppose a random variable $X$ has population mean $\mu$ and standard deviation $\sigma$ and that a random sample of size $n$ is taken from this population. Then the sampling distribution of $\bar{x}$ becomes approximately normal as the sample size $n$ increases. The mean of the distribution is $\mu_{\bar{x}} = \mu$ and the standard deviation is $\sigma_{\bar{x}} = \dfrac{\sigma}{\sqrt{n}}$.

*In Your Own Words*

For any population, regardless of its shape, as the sample size increases, the shape of the distribution of the sample mean becomes more "normal."

So, if the random variable $X$ is normally distributed, then the sampling distribution of $\bar{x}$ will be normal. If the sample size is large enough, then the sampling distribution of $\bar{x}$ will be approximately normal, *regardless of the distribution of X*. But how large does the sample size need to be before we can say that the sampling distribution of $\bar{x}$ is approximately normal? Statisticians agree that, if the sample size is 30 or more, the sampling distribution of $\bar{x}$ will be approximately normal.

▶ **EXAMPLE 6**    **Applying the Central Limit Theorem**

*Problem:* According to the United States Department of Agriculture, the mean calorie intake of males 20–39 years old is $\mu = 2716$, with standard deviation $\sigma = 72.8$. Suppose a nutritionist conducts a simple random sample of $n = 35$ males between the ages of 20 and 39 years old and obtains a sample mean $\bar{x} = 2750$. What is the probability that a random sample of 35 males between the ages of 20 and 39 years old would result in a sample mean of 2750 or higher? Are the results of the survey unusual? Why?

**Caution**

The Central Limit Theorem has to do only with the *shape* of the distribution of the sample mean—not with its center and spread! The mean of the distribution of $\bar{x}$ is $\mu$ and the standard deviation of $\bar{x}$ is $\frac{\sigma}{\sqrt{n}}$, regardless of the size of the sample, $n$.

**Historical Note**

Pierre Simon Laplace was born on March 23, 1749 in Normandy, France. Between age 7 and age 16, Laplace attended Benedictine priory school in Beaumont-en-Auge. At age 16, Laplace attended Caen University, where he studied theology. While there, his mathematical talents were discovered, which led him to Paris, where he got a job as professor of mathematics at the École Militaire. While there, he wrote many mathematical papers. In 1773, Laplace was elected to the Académie des Sciences. Laplace was not humble. It is reported that, in 1780, he stated that he was the best mathematician in Paris. In 1799, Laplace published the first two volumes of *Méchanique céleste*, where he discusses methods for calculating the motion of the planets. On April 9, 1810, Laplace presented the Central Limit Theorem to the Academy. Laplace died on March 5, 1827.

*Approach:*

**Step 1:** We recognize that we are computing a probability regarding a sample mean, so we need to know the sampling distribution of $\bar{x}$. Because the population from which the sample is drawn is not known to be normal, the sample size must be greater than or equal to 30 in order to use the results of the Central Limit Theorem.

**Step 2:** Determine the mean and standard deviation of the sampling distribution of $\bar{x}$.

**Step 3:** Convert the sample mean to a $Z$-score.

**Step 4:** Use Table II to find the area under the normal curve.

*Solution:*

**Step 1:** Because the sample size is greater than 30 ($n = 35$), the Central Limit Theorem states that the sampling distribution of $\bar{x}$ is approximately normal.

**Step 2:** The mean of the sampling distribution of $\bar{x}$ will equal the mean of the population, so $\mu_{\bar{x}} = 2716$. The standard deviation of the sampling distribution of $\bar{x}$ will equal the standard deviation of the population divided by the square root of the sample size, so $\sigma_{\bar{x}} = \frac{\sigma}{\sqrt{n}} = \frac{72.8}{\sqrt{35}} = 12.3$.

**Step 3:** We convert $\bar{x} = 2750$ to a $Z$-score.

$$Z = \frac{2750 - 2716}{72.8/\sqrt{35}} = 2.76$$

**Step 4:** We wish to know the probability that a random sample of $n = 35$ from a population whose mean is 2716 results in a sample mean of at least 2750. That is, we wish to know $P(\bar{x} \geq 2750)$. See Figure 56.

**Figure 56**

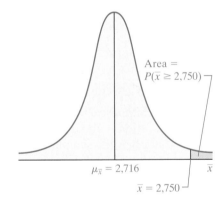

Area = $P(\bar{x} \geq 2,750)$

$\mu_{\bar{x}} = 2,716$

$\bar{x} = 2,750$

This probability is represented by the area under the standard normal curve to the right of $Z = 2.76$.

$$P(\bar{x} \geq 2750) = P(Z \geq 2.76) = 1 - 0.9971 = 0.0029$$

*Interpretation:* The probability that a random sample of 35 males between the ages of 20 and 39 will result in a sample mean of 2750 calories or higher if the population mean is 2716 calories is 0.0029. This means that less than 1 sample in 100 will result in a sample mean of 2750 calories or higher if the population mean is 2716 calories. We can conclude one of two things based upon this result.

1. The mean number of calories for males 20–39 years old is 2716, and we just happened to randomly select 35 individuals who, on average, consume more calories.

2. The mean number of calories consumed by 20–39-year-old males is higher than 2716 calories.

In statistical inference, we are inclined to accept the second possibility as the more reasonable choice. We recognize there is a possibility that our decision is incorrect.                                                          ◀◀

**NW** *Now Work Problem 13.*

## 7.5 Assess Your Understanding

### Concepts and Vocabulary

1. Explain what a sampling distribution is.
2. What are the mean and standard deviation of the sampling distribution of $\overline{x}$, regardless of the distribution of the population from which the sample was drawn?
3. If a random sample of size $n = 20$ is taken from a population, what is required in order to say the sampling distribution of $\overline{x}$ is approximately normal?
4. State the Central Limit Theorem.
5. In order to cut the standard error of the mean in half, the sample size must be increased by a factor of _____.

6. Suppose a simple random sample of size $n = 10$ is obtained from a population that is normally distributed with $\mu = 30$ and $\sigma = 8$. What is the sampling distribution of $\overline{x}$?
7. Suppose a simple random sample of size $n = 40$ is obtained from a population with $\mu = 50$ and $\sigma = 4$. Does the parent population need to be normally distributed in order for the sampling distribution of $\overline{x}$ to be approximately normally distributed? Why? What is the sampling distribution of $\overline{x}$?

### Exercises

#### • Skill Building

*In Problems 1–4, determine $\mu_{\overline{x}}$ and $\sigma_{\overline{x}}$ from the given parameters of the population and the sample size.*

1. $\mu = 50, \sigma = 6, n = 40$
**NW**

2. $\mu = 30, \sigma = 4, n = 34$

3. $\mu = 100, \sigma = 12, n = 20$

4. $\mu = 120, \sigma = 16, n = 10$

5. Suppose a simple random sample of size $n = 40$ is obtained from a population with $\mu = 50$ and $\sigma = 6$.

   (a) Describe the sampling distribution of $\overline{x}$.
   (b) What is $P(\overline{x} > 51)$?
   (c) What is $P(\overline{x} \leq 48)$?
   (d) What is $P(47.5 < \overline{x} < 51.2)$?

6. Suppose a simple random sample of size $n = 34$ is obtained from a population with $\mu = 30$ and $\sigma = 4$.

   (a) Describe the sampling distribution of $\overline{x}$.
   (b) What is $P(\overline{x} > 30.2)$?
   (c) What is $P(\overline{x} \leq 30.6)$?
   (d) What is $P(28.7 \leq \overline{x} < 31.1)$?

7. Suppose a simple random sample of size $n = 10$ is obtained from a population with $\mu = 105$ and $\sigma = 16$.

   (a) What must be true regarding the distribution of the population in order to compute probabilities regarding the sample mean? Describe the sampling distribution of $\overline{x}$.
   (b) Assuming the requirements described in part (a) are satisfied, determine $P(\overline{x} > 103.2)$.
   (c) Assuming the requirements described in part (a) are satisfied, determine $P(\overline{x} < 99.3)$.

8. Suppose a simple random sample of size $n = 20$ is obtained from a population with $\mu = 105$ and $\sigma = 16$.

   (a) What must be true regarding the population in order to compute probabilities regarding the sample mean? Describe the sampling distribution of $\overline{x}$.
   (b) Assuming the requirements described in part (a) are satisfied, determine $P(\overline{x} > 103.2)$.
   (c) Assuming the requirements described in part (a) are satisfied, determine $P(\overline{x} < 99.3)$.
   (d) Compare the results obtained in parts (b) and (c) with the results obtained in parts (b) and (c) in Problem 7. What effect does increasing the sample size have on the probabilities? Why do you think this is the case?

• **Applying the Concepts**

9. **Serum Cholesterol** As reported by the U.S. National
NW Center for Health Statistics, the mean serum high-density-lipoprotein (HDL) cholesterol of females 20–29 years old is $\mu = 53$. If serum HDL cholesterol is normally distributed with $\sigma = 13.4$, answer the following questions.

(a) What is the probability a randomly selected female 20–29 years old will have a serum cholesterol above 60?

(b) What is the probability a random sample of 15 female 20–29-year-olds will have a mean serum cholesterol above 60?

(c) What is the probability that a random sample of 20 female 20–29-year-olds will have a mean serum cholesterol above 60?

(d) What effect does increasing the sample size have on the probability? Provide an explanation for this result.

(e) What might you conclude if a random sample of 20 female 20–29-year-olds had a mean serum cholesterol above 60?

10. **ACT Math Scores** In 2000, as reported by ACT Research Service, the mean ACT Math score was $\mu = 20.7$. If ACT Math scores are normally distributed with $\sigma = 5$, answer the following questions.

(a) What is the probability that a randomly selected student has an ACT Math score less than 18?

(b) What is the probability that a random sample of 10 ACT test takers had a mean math score of 18 or less?

(c) What is the probability that a random sample of 20 ACT test takers had a mean math score of 18 or less?

(d) What might you conclude if a random sample of 20 ACT test takers had a mean math score of 18 or less?

11. **Gestation Period** The length of human pregnancies is approximately normally distributed with mean $\mu = 266$ days and standard deviation $\sigma = 16$ days.

(a) What is the probability a randomly selected pregnancy lasts less than 260 days?

(b) What is the probability that a random sample of 20 pregnancies have a mean gestation period of 260 days or less?

(c) What is the probability that a random sample of 50 pregnancies have a mean gestation period of 260 days or less?

(d) What might you conclude if a random sample of 50 pregnancies resulted in a mean gestation period of 260 days or less?

12. **Weather in Chicago** The high temperature in Chicago for the month of August is approximately normally distributed with mean $\mu = 80°$ F and standard deviation $\sigma = 8°$ F.

(a) What is the probability that a randomly selected day in August has a high temperature less than 78° F?

(b) What is the probability that a random sample of 30 days in August has a mean high temperature less than 78° F?

(c) What is the probability that a random sample of 50 August days has a mean high temperature less than 78° F?

13. **Insect Fragments** The Food and Drug Administration
NW sets Food Defect Action Levels (FDALs) for some of the various foreign substances that inevitably end up in the food we eat and liquids we drink. For example, the FDAL for insect filth in peanut butter is 3 insect fragments (larvae, eggs, body parts, and so on) per 10 grams. A random sample of 50 ten-gram portions of peanut butter is obtained and results in a sample mean of 3.6 insect fragments per ten-gram portion.

(a) Why is the sampling distribution of $\bar{x}$ approximately normal?

(b) What is the mean and standard deviation of the sampling distribution of $\bar{x}$? [*Hint*: This is a Poisson process with $\mu = 3$ and $\sigma = \sqrt{3}$.]

(c) Suppose a simple random sample of $n = 50$ ten-gram samples of peanut butter resulted in a sample mean of 3.6 insect fragments. What is the probability a simple random sample of 50 ten-gram portions results in a mean of at least 3.6 insect fragments? Is this result unusual? What might we conclude?

14. **Burger King's Drive-Through** Suppose cars arrive at Burger King's drive-through at the rate of 20 cars every hour between 12:00 noon and 1:00 PM. A random sample of 40 one-hour time periods between 12:00 noon and 1:00 PM is selected and has 22.1 as the mean number of cars arriving.

(a) Why is the sampling distribution of $\bar{x}$ approximately normal?

(b) What is the mean and standard deviation of the sampling distribution of $\bar{x}$? [*Hint*: This is a Poisson process with $\mu = 20$ and $\sigma = \sqrt{20}$.]

(c) What is the probability that a simple random sample of 40 one-hour time periods results in a mean of at least 22.1 cars? Is this result unusual? What might we conclude?

15. **Winter Temperature** During the daytime in winter when nobody is at home, the mean household temperature is 67.6° F, with standard deviation 4.2° F, for households whose 1997 income was between $10,000 and $24,999, according to the Energy Information Administration. In a random sample of 50 households whose income was between $10,000 and $24,999 and for which nobody is home during the daytime, it was determined the average temperature was 68.3° F. What is the probability that a random sample of 50 households whose income is between $10,000 and $24,999 and for which nobody is home during the daytime results in an average household temperature of 68.3° F or higher? Is this result unusual?

16. **Age of Refrigerator** According to the Energy Information Administration, the mean age of a refrigerator in a home owned by the occupant is 8.5 years, with standard deviation 6.6 years. A random sample of 50 homes owned by the occupant is conducted, and the average age of the refrigerator is found to be 5.7 years. What is the probability that a random sample of 50 homes owned by the occupant results in a mean age of refrigerator of 5.7 years or less?

**17. Household Income** According to the Current Population Survey, the mean household income in the United States in 1997 was \$45,127, with standard deviation \$31,570.

  (a) What is the probability a random sample of 40 homes in Cook County, IL results in a sample mean household income less than \$32,030? What might you conclude based on this result?

  (b) The median household income in the United States in 1997 was \$37,005. Given this information, do you think household income is a normally distributed random variable? Why?

**18. Household Size** According to the Energy Information Administration, the mean household size in the United States in 1997 was 2.6 people, with standard deviation 1.5 people. What is the probability that a random sample of 100 households results in a mean household size of 2.4 people or less?

**19. Sampling Distributions** The following data represent the IQ scores based on the Stanford–Binet test of students in Sullivan's online College Algebra course in the Spring, 2001 semester.

$$98 \quad 106 \quad 104 \quad 120 \quad 100 \quad 114$$

  (a) Compute the population mean, $\mu$.

  (b) List all possible samples with size $n = 2$. There should be $_6C_2 = 15$ samples.

  (c) Construct a sampling distribution for the mean by listing the sample means and their corresponding probabilities.

  (d) Compute the mean of the sampling distribution.

  (e) Compute the probability that the sample mean will be within 5 IQ points of the population mean IQ.

  (f) Repeat parts (b) – (e) using samples of size $n = 3$. Comment on the effect of increasing the sample size.

**20. Sampling Distributions** The following data represent the ages of faculty members hired in the Mathematics Department at Joliet Junior College in the past 3 years.

$$24 \quad 28 \quad 35 \quad 36 \quad 58 \quad 29$$

  (a) Compute the population mean, $\mu$.

  (b) List all possible samples with size $n = 2$. There should be $_6C_2 = 15$ samples.

  (c) Construct a sampling distribution for the mean by listing the sample means and their corresponding probabilities.

  (d) Compute the mean of the sampling distribution.

  (e) Compute the probability that the sample mean will be within 5 years of the population mean age.

  (f) Repeat parts (b)–(e) using samples of size $n = 3$. Comment on the effect of increasing the sample size.

**21. Simulation** Scores on the Stanford–Binet IQ test are normally distributed with $\mu = 100$ and $\sigma = 16$.

  (a) Use Minitab, Excel, or some other statistical software to obtain 500 random samples of size $n = 20$.

  (b) Compute the sample mean of each of the 500 samples.

  (c) Draw a histogram of the 500 sample means. Comment on its shape.

  (d) What do you expect the mean and standard deviation of the sampling distribution of the mean to be?

  (e) Compute the mean and standard deviation of the 500 sample means. Are they close to the expected values?

  (f) Compute the probability that a random sample of 20 people results in a sample mean greater than 108.

  (g) What proportion of the 500 random samples had a sample mean IQ greater than 108? Is this result close to the theoretical value obtained in part (f)?

**22. Simulation** The gestation period of humans is normally distributed with $\mu = 266$ days and $\sigma = 16$ days.

  (a) Use Minitab, Excel, or some other statistical software to obtain 500 random samples of size $n = 15$.

  (b) Compute the sample mean of each of the 500 samples.

  (c) Draw a histogram of the 500 sample means. Comment on its shape.

  (d) What do you expect the mean and standard deviation of the sampling distribution of the mean to be?

  (e) Compute the mean and standard deviation of the 500 sample means. Are they close to the expected values?

  (f) Compute the probability that a random sample of 15 people results in a sample mean greater than 270.

  (g) What proportion of the 500 random samples had a sample mean gestation period greater than 270? Is this result close to the theoretical value obtained in part (f)?

**23. Simulation** The exponential distribution is the distribution of the waiting time $X$ until the occurrence of the first event in a Poisson process with parameter $\beta = 1/\lambda$. The probability density function of the exponential distribution is $y = \dfrac{1}{\beta}e^{-x/\beta}$, where $\beta$ is both the mean and standard deviation of the distribution. Suppose a certain intersection is especially dangerous, and accidents occur there at the rate $\lambda = 0.2$ accidents per day, so that $\beta = 1/0.2 = 5$ days between accidents. This implies we would expect to wait about 5 days before observing the first accident.

  (a) Use Minitab, Excel, or some other statistical software to randomly create 200 observations of the random variable $X$, the number of days between accidents.

  (b) Draw a histogram of the distribution of the random variable $X$. Is the exponential distribution symmetric and bell-shaped? According to this result, what sample size would be required to compute probabilities regarding the mean of the random variable $X$, the number of days between accidents.

  (c) Obtain 500 random samples with size 6 and obtain the sample means for each of the 500 samples.

  (d) Draw a histogram of the sample means. Is the sampling distribution of the mean normally distributed? Is the result what you expected?

  (e) Redo parts (c) and (d) for a sample of size $n = 40$.

**24. Simulation** Let the random variable $X$ represent the sum of the pips on two dice. The probability distribution of $X$ is given below.

| X | 2 | 3 | 4 | 5 | 6 | 7 | 8 | 9 | 10 | 11 | 12 |
|---|---|---|---|---|---|---|---|---|----|----|----|
| $P(X = x)$ | 0.0278 | 0.0556 | 0.0833 | 0.1111 | 0.1389 | 0.1666 | 0.1389 | 0.1111 | 0.0833 | 0.0556 | 0.0278 |

(a) Obtain 500 random samples of size $n = 6$ and obtain the sample means for each of the 500 samples.
(b) Draw a histogram of the sample means. Is the sampling distribution of the mean normally distributed? Did you expect it to be?
(c) Redo parts (a) and (b) with a sample of size $n = 40$.

## 7.6 The Normal Approximation to the Binomial Probability Distribution

**Preparing for This Section**    Before getting started, review the following:

    ✓ Binomial probability distribution (Section 6.2, pp. 253–262)

**Objective**     Approximate binomial probabilities by using the normal curve.

    In Section 6.2, we discussed the binomial probability distribution. A probability experiment is said to be a binomial experiment if the following conditions are met.

> **Criteria for a Binomial Probability Experiment**
>
> A probability experiment is said to be a binomial experiment if all the following are true:
> 1. The experiment is performed $n$ independent times. Each repetition of the experiment is called a **trial**. Independence means that the outcome of one trial will not affect the outcome of the other trials.
> 2. For each trial, there are two mutually exclusive outcomes, success or failure.
> 3. The probability of success, $p$, is the same for each trial of the experiment.

The binomial probability formula can be used to compute probabilities of events in a binomial experiment. When there are a large number of trials of a binomial experiment, the binomial probability formula can be difficult to use. For example, suppose there are 500 trials of a binomial experiment and we wish to compute the probability of 400 or more successes. Using the binomial probability formula would require that we compute the following probabilities:

$$P(X \geq 400) = P(X = 400) + P(X = 401) + \ldots + P(X = 500)$$

This would be time consuming to compute by hand! Fortunately, we have other means for approximating binomial probabilities, provided that certain conditions are met.

Recall the Exploration performed on page 261 in Section 6.2. In that Exploration, we discovered that, as the number of trials, $n$, in a binomial experiment increases, the probability histogram becomes more nearly symmetric and bell shaped. We restate the conclusion below.

*Theorem*    As the number of trials $n$ in a binomial experiment increases, the probability distribution of the random variable $X$ becomes more nearly symmetric and bell-shaped. As a general rule of thumb, if $np(1 - p) \geq 10$, then the probability distribution will be approximately symmetric and bell shaped.

Because of this result, we might be inclined to think that binomial probabilities could be approximated by the area under the normal curve, provided that $np(1 - p) \geq 10$. This intuition would be correct.

*Theorem*

**The Normal Approximation to the Binomial Probability Distribution**

If $np(1 - p) \geq 10$, then the binomial random variable $X$ is approximately normally distributed with mean $\mu_X = np$ and standard deviation $\sigma_X = \sqrt{np(1 - p)}$.

**Historical Note**

The normal approximation to the binomial was discovered by Abraham de Moivre in 1733. With the advance of computing technology, its importance has been diminished.

Figure 57 shows a probability histogram for the binomial random variable $X$ with $n = 40$ and $p = 0.5$ and a normal curve with $\mu_X = np = 40(0.5) = 20$ and standard deviation $\sigma_X = \sqrt{np(1 - p)} = \sqrt{40(0.5)(0.5)} = \sqrt{10}$.

Figure 57

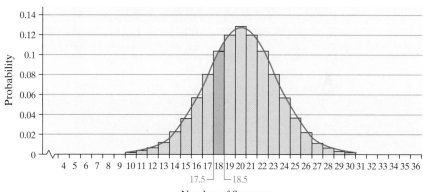

**Binomial Histogram, n = 40, p = 0.5**

We know from Section 6.2 that the area of the rectangle corresponding to $X = 18$ represents $P(X = 18)$. It should also be apparent that the width of each rectangle is one, so the rectangle extends from $X = 17.5$ to $X = 18.5$. The area under the normal curve from $X = 17.5$ to $X = 18.5$ is approximately equal to the area of the rectangle corresponding to $X = 18$. Therefore, the area under the normal curve between $X = 17.5$ and $X = 18.5$ is approximately equal to $P(X = 18)$ where $X$ is a binomial random variable with $n = 40$ and $p = 0.5$. We add and subtract 0.5 from $X = 18$ as a **correction for continuity** because we are using a continuous density function to approximate a discrete probability.

**Caution**

Don't forget about the correction for continuity. It is needed because we are using a continuous density function to approximate the probability of a discrete random variable.

Suppose we want to approximate $P(X \leq 18)$. Figure 58 illustrates the situation.

Figure 58

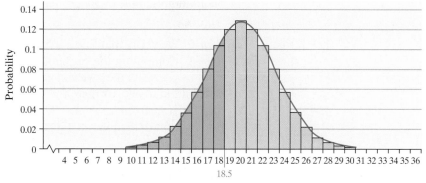

**Binomial Histogram, n = 40, p = 0.5**

In order to approximate $P(X \leq 18)$, we compute the area under the normal curve for $X < 18.5$. Do you see why?

If we wanted to approximate $P(X \geq 18)$, we would compute $P(X \geq 17.5)$. Do you see why? Table 10 summarizes how to use the correction for continuity.

| TABLE 10 | | |
|---|---|---|
| **Exact Probability Using Binomial** | **Approximate Probability Using Normal** | **Graphical Depiction** |
| $P(X = a)$ | $P(a - 0.5 < X < a + 0.5)$ | |
| $P(X \leq a)$ | $P(X < a + 0.5)$ | |
| $P(X \geq a)$ | $P(X > a - 0.5)$ | |
| $P(a \leq X \leq b)$ | $P(a - 0.5 < X < b + 0.5)$ | |

A question remains, however. What do we do if the probability is of the form $P(X > a), P(X < a)$ or $P(a < X < b)$? The solution is to rewrite the inequality in a form with $\leq$ or $\geq$. For example, $P(X > 4) = P(X \geq 5)$ and $P(X < 4) = P(X \leq 3)$ for binomial random variables, because the values of the random variables must be whole numbers.

▶ EXAMPLE 1   The Normal Approximation to a Binomial Random Variable

*Problem:* According to the *Information Please* almanac, 6% of the human population has blood type O-negative. What is the probability that, in a simple random sample of 500, fewer than 25 have blood type O-negative?

*Approach:*

**Step 1:** We verify that this is a binomial experiment.

**Step 2:** Computing the probability by hand would be very tedious. Verify $np(1 - p) \geq 10$; then we will know that the condition for using the normal distribution to approximate the binomial distribution is met.

**Step 3:** Approximate $P(X < 25) = P(X \leq 24)$ by using the normal approximation to the binomial distribution.

*Solution:*

**Step 1:** There are 500 independent trials with each trial having a probability of success equal to 0.06. This is a binomial experiment.

**Figure 59**

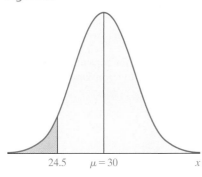

24.5     $\mu = 30$     $x$

**Step 2:** We verify $np(1 - p) \geq 10$:

$$np(1 - p) = 500(0.06)(0.94) = 28.2 \geq 10$$

We can use the normal distribution to approximate the binomial distribution.
**Step 3:** We wish to know the probability that fewer than 25 people in the sample have blood type O-negative—that is, we wish to know $P(X < 25) = P(X \leq 24)$. This is approximately equal to the area under the normal curve to the left of $X = 24.5$, with $\mu = np = 500(0.06) = 30$ and $\sigma = \sqrt{np(1 - p)} = \sqrt{500(0.06)(1 - 0.06)} = \sqrt{28.2} \approx 5.31$. See Figure 59.
We convert $X = 24.5$ to a $Z$-score.

$$Z = \frac{24.5 - 30}{5.31} = -1.04$$

From Table II, we find the area to the left of $Z = -1.04$ is 0.1492. Therefore, the approximate probability that fewer than 25 people will have blood type O-negative is $0.1492 = 14.92\%$.    ◄◄

**Figure 60**

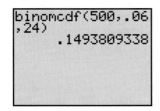

binomcdf(500,.06
,24)
          .1493809338

Using the *binomcdf(* command on a TI-83+ graphing calculator, we find that the exact probability is 0.1494. See Figure 60. The approximate result is close indeed!

**NW** *Now Work Problem 13.*

► **EXAMPLE 2**   **A Normal Approximation to the Binomial**

*Problem:* According to Nielsen Media Research, 75% of all United States households have cable television. Erica conducts a random sample of 1000 households in DuPage County and finds that 800 of them have cable. What might Erica conclude?

*Approach:* This is a binomial experiment with $n = 1000$ and $p = 0.75$. Erica needs to determine the probability of obtaining a random sample of at least 800 households with cable from a sample of size 1000 when assuming 75% of households have cable. Clearly, computing this via the binomial probability formula would be difficult, so Erica will compute the probability using the normal approximation to the binomial, since $np(1 - p) = 1000(0.75)(0.25) = 187.5 \geq 10$. We approximate $P(X \geq 800)$ by computing the area under the standard normal curve to the right of $X = 799.5$ with $\mu = np = 1000(0.75) = 750$ and $\sigma = \sqrt{np(1 - p)} = \sqrt{1000(0.75)(1 - 0.75)} = \sqrt{187.5} \approx 13.693$.

**Figure 61**

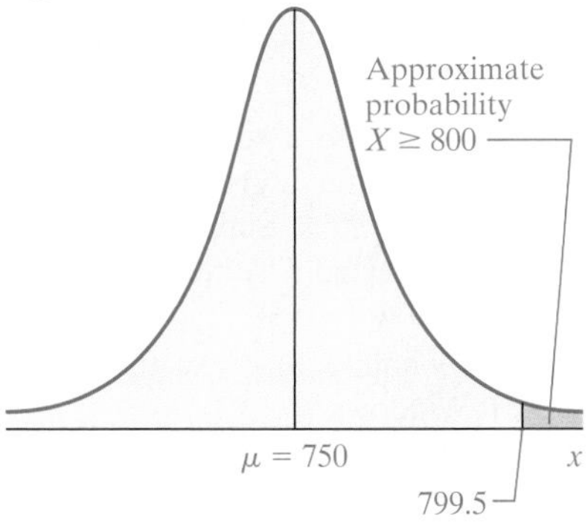

Approximate probability $X \geq 800$

$\mu = 750$     $x$
799.5

*Solution:* Figure 61 shows the area we wish to compute.
We convert $X = 799.5$ to a $Z$-score.

$$Z = \frac{799.5 - 750}{13.693} = 3.61$$

The area under the standard normal curve to the right of $Z = 3.61$ is $1 - 0.9998 = 0.0002$. There is a 0.02% probability of obtaining 800 or more households with cable from a sample of 1000 households, assuming that the percentage of households with cable is 75%. This means that about 2 samples in every 10,000 samples will have 800 or more households with cable. Erica is not inclined to believe that her sample is one of the 2 in 10,000. She would rather believe that the proportion of households in DuPage County with cable is higher than 75%.    ◄◄

**NW** *Now Work Problem 21.*

## 7.6 Assess Your Understanding

### Concepts and Vocabulary

1. List the conditions required for a binomial experiment.
2. Under what circumstances can the normal distribution be used to approximate binomial probabilities?
3. Why must we use a correction for continuity when using the normal distribution to approximate binomial probabilities?

### Exercises

#### • Skill Building

*In Problems 1–8, a discrete random variable is given. Assume the probability of the random variable is going to be approximated via the normal distribution. Describe the area under the normal curve that will be computed. For example, if we wish to compute the probability of finding at least five defective items in a shipment, then we would approximate the probability by computing the area under the normal curve to the right of X = 4.5.*

1. The probability that at least 40 households have a gas stove.
2. The probability of no more than 20 people who want to see Roe versus Wade overturned.
3. The probability that exactly eight defective parts are in the shipment.
4. The probability that exactly 12 students pass the course.
5. The probability that the number of people with blood type O-negative is between 18 and 24 inclusive.
6. The probability that the number of tornadoes that occur in the month of May is between 30 and 40 inclusive.
7. The probability that more than 20 people want to see the marriage tax penalty abolished.
8. The probability that fewer than 40 households have a pet.

*In Problems 9–12, compute $P(X = x)$ via the binomial probability formula. Then, determine whether the normal distribution can be used as an approximation for the binomial distribution. If the normal distribution can be used as an approximation for the binomial distribution, approximate $P(X = x)$ and compare the result to the exact probability.*

9. $n = 60, p = 0.4, X = 20$
10. $n = 80, p = 0.15, X = 18$
11. $n = 75, p = 0.75, X = 60$
12. $n = 85, p = 0.8, X = 70$

#### • Applying the Concepts

13. **On-Time Flights** United Airlines flight 1832 from Chicago to Orlando is on time 80% of the time, according to United Airlines. Suppose 70 flights are randomly selected. Use the normal approximation to the binomial to
    (a) approximate the probability that exactly 60 flights are on time.
    (b) approximate the probability that at least 60 flights are on time.
    (c) approximate the probability that fewer than 50 flights are on time.
    (d) approximate the probability that between 50 and 55 flights, inclusive, are on time.

14. **Smokers** According to *Information Please* almanac, 80% of adult smokers started smoking before they were 18 years old. Suppose 100 smokers 18 years old or older are randomly selected. Use the normal approximation to the binomial to
    (a) approximate the probability that exactly 80 of them started smoking before they were 18 years old.
    (b) approximate the probability that at least 80 of them started smoking before they were 18 years old.
    (c) approximate the probability that fewer than 70 of them started smoking before they were 18 years old.
    (d) approximate the probability that between 70 and 90 of them, inclusive, started smoking before they were 18 years old.

15. **Operating Systems** According to a survey conducted by StatMarket, 94% of computer users use Microsoft Windows as their operating system. Suppose 200 computer users are randomly selected. Use the normal approximation to the binomial to
    (a) approximate the probability that exactly 180 of them use Microsoft Windows.
    (b) approximate the probability that more than 180 of them use Microsoft Windows.
    (c) approximate the probability that no more than 185 of them use Microsoft Windows.
    (d) approximate the probability that between 160 and 180 of them, inclusive, use Microsoft Windows.

16. **Murder by Firearm** According to *Crime in the United States*, 1998, 65% of murders are committed with a firearm. Suppose 150 murders are randomly selected. Use the normal approximation to the binomial to
    (a) approximate the probability that exactly 110 murders are committed with a firearm.

(b) approximate the probability that 100 or more murders are committed with a firearm.

(c) approximate the probability that 90 or fewer murders are committed with a firearm.

(d) approximate the probability that between 100 and 120 murders, inclusive, are committed with a firearm.

**17. Migraine Sufferers** In clinical trials and extended studies of a medication whose purpose is to reduce the pain associated with migraine headaches, 2% of the patients in the study experienced weight gain as a side effect. Suppose a random sample of 600 users of this medication is obtained. Use the normal approximation to the binomial to

(a) approximate the probability that exactly 20 will experience weight gain as a side effect.

(b) approximate the probability that 20 or fewer will experience weight gain as a side effect.

(c) approximate the probability that 22 or more patients will experience weight gain as a side effect.

(d) approximate the probability that between 20 and 30 patients, inclusive, will experience weight gain as a side effect.

**18. Stomach Ulcers** In clinical trials of a drug whose purpose is to treat acid reflux and stomach ulcers, 4.7% of the patients in the study experienced headaches as a side effect. Suppose a random sample of 300 users of this drug is obtained. Use the normal approximation to the binomial to

(a) approximate the probability that exactly 20 experienced headaches as a side effect.

(b) approximate the probability that fewer than 20 experienced headaches as a side effect.

(c) approximate the probability that more than 15 patients experienced headaches as a side effect.

(d) approximate the probability that between 10 and 25 patients, inclusive, experienced headaches as a side effect.

**19. Softball** Suppose a softball player safely reaches base 45% of the time. Assuming at-bats are independent events, use the normal approximation to the binomial to approximate the probability that, in the next 100 at bats,

(a) the player reaches base safely exactly 50 times.

(b) the player reaches base safely 60 or more times.

(c) the player reaches base safely 50 or fewer times.

(d) the player reaches base safely between 60 and 90 times, inclusive.

**20. Basketball** Mark Price holds the record for percentage of free throws made in the National Basketball Association, at 90.4%. Assuming free throws are independent events, use the normal approximation to the binomial to approximate the probability that, in the next 200 free throws,

(a) Mark Price makes exactly 190.

(b) Mark Price makes 180 or more.

(c) Mark Price makes fewer than 180.

(d) Mark Price makes between 180 and 190, inclusive.

**21. Males Living at Home** According to *Information Please* almanac, 59% of males between the ages of 18 and 24 lived at home in 1998. (Unmarried college students living in dorms are counted as living at home.) Suppose a survey is administered at a community college to 200 randomly selected male students between 18 and 24 years old and 135 of them responded they live at home.

(a) Approximate the probability that a survey would result in at least 135 of the respondents living at home under the assumption the true percentage is 59%.

(b) What might you conclude from this result?

**22. Females Living at Home** According to *Information Please* almanac, 48% of females between the ages of 18 and 24 live at home. (Unmarried college students living in dorms are counted as living at home.) Suppose a survey is administered at a community college to 200 randomly selected female students between 18 and 24 years old and 120 of them responded they live at home.

(a) Approximate the probability that a survey would result in at least 120 of the respondents living at home under the assumption the true percentage is 48%.

(b) What might you conclude from this result?

**23. Boys Are Preferred** In a Gallup poll conducted December 2–4, 2000, 42% of survey respondents said that, if they only had one child, they would prefer the child to be a boy. Suppose you conduct a survey of 150 randomly selected students on your campus and find that 80 of them would prefer a boy.

(a) Approximate the probability that, in a random sample of 150 students, at least 80 would prefer a boy, assuming the true percentage is 42%.

(b) Does this result contradict the Gallup poll? Explain.

**24. Liars** According to a *USA Today* "Snapshot," 3% of Americans surveyed lie frequently. Suppose you conduct a survey of 500 college students and find that 20 of them lie frequently.

(a) Compute the probability that, in a random sample of 500 college students, at least 20 lie frequently, assuming the true percentage is 3%.

(b) Does this result contradict the *USA Today* "Snapshot"? Explain.

**25. Time to Graduate** According to a study done by ACT in 1997, 52.8% of students graduate from college in five years or less. Suppose you conduct a survey of 200 recent college graduates and find that 95 of them graduated in five years or less.

(a) Compute the probability that, in a random sample of 200 recent college graduates, 95 or fewer graduated in five years or less, assuming the true percentage is 52.8%.

(b) Does this result contradict the results of the study done by ACT? Explain.

26. **Drunk-Driving Laws** In a Gallup poll conducted July 20–August 3, 2000, 72% of survey respondents said they would favor reducing the drunk-driving limit to 0.08% blood alcohol concentration (BAC). Survey respondents had to be 16 years old or older and a licensed driver. On October 3, 2000, President Bill Clinton signed into law a national BAC of 0.08% required to charge drunk drivers with a crime. Suppose you are a member of SADD (Students Against Drunk Driving)

and conduct a survey of 200 students aged 16 years old or older who were licensed drivers, and 130 of them favored a drunk-driving limit of 0.08% BAC.

(a) Compute the probability that, in a random sample of 200 students, 130 or fewer would favor a drunk-driving law of 0.08% BAC, assuming the true percentage is 72%.

(b) Does this result contradict the results of the Gallup poll? Explain.

# CHAPTER 7 REVIEW

## Summary

In this chapter, we introduced continuous random variables and, in particular, the normal probability density function. A continuous random variable is said to be normally distributed if a histogram of its values is symmetric and bell-shaped. In addition, we can draw normal probability plots that are based upon expected Z-scores. If these normal probability plots are approximately linear, then we say the distribution of the random variable is approximately normal. The area under the normal density function can be used to find theoretical proportions or probabilities for normal random variables. Also, we can find the value of a normal random variable that corresponds to a specific proportion, probability, or percentile.

In Section 7.5, we introduced what is arguably the most important theorem in inferential statistics, the Central Limit

Theorem. We discovered that, if $X$ is a normal random variable or if the sample size is large ($n \geq 30$), the sampling distribution of $\bar{x}$ is approximately normally distributed. The mean of any sampling distribution of $\bar{x}$ is $\mu_{\bar{x}} = \mu$, and its standard deviation is $\sigma_{\bar{x}} = \dfrac{\sigma}{\sqrt{n}}$ (the standard error of the mean). We can compute a probability regarding the random variable $\bar{x}$ by measuring an area under the standard normal curve.

Finally, if $X$ is a binomial random variable with $np(1 - p) \geq 10$, then we can use the area under the normal curve to approximate the probability of a binomial random variable. The parameters of the normal curve are $\mu_X = np$ and $\sigma_X = \sqrt{np(1 - p)}$, where $n$ is the number of trials of the binomial experiment and $p$ is the probability of success.

## Formulas

**Standardizing a Normal Random Variable**

$$Z = \frac{X - \mu}{\sigma} \text{ or } Z = \frac{\bar{x} - \mu}{\dfrac{\sigma}{\sqrt{n}}}$$

**Finding the Score**

$$X = \mu + Z\sigma$$

**Mean of Sampling Distribution of $\bar{x}$**

$$\mu_{\bar{x}} = \mu$$

**Standard Deviation of Sampling Distribution of $\bar{x}$**

$$\sigma_{\bar{x}} = \frac{\sigma}{\sqrt{n}}$$

## Vocabulary

Uniform probability distribution (p. 272)
Probability density function (p. 273)
Normal probability distribution (p. 274)
Normal probability density function (p. 274)
Inflection points (p. 275)

Standard normal distribution (p. 279)
Normal probability plot (p. 306)
Normal scores (p. 306)
Sampling distribution of the mean (p. 313)

The law of large numbers (p. 318)
Standard error of the mean (p. 319)
The Central Limit Theorem (p. 321)
Correction for continuity (p. 327)

## Objectives

| Section | You should be able to . . . | Review Exercises |
|---|---|---|
| **7.1** | 1 Understand the uniform probability distribution (p. 272) | 37 |
| | 2 Graph a normal density curve (p. 274) | 19–22 |
| | 3 State the properties of the normal curve (p. 276) | 38 |
| | 4 Understand the role of area in the normal density function (p. 276) | 1, 2 |
| | 5 Understand the relation between a normal random variable and a standard normal random variable in regards to area (p. 278) | 3, 4 |
| **7.2** | 1 Find the area under the standard normal curve (p. 284) | 5–8 |
| | 2 Find the Z-scores for given areas (p. 288) | 13–18 |
| | 3 Interpret area under standard normal curve as a probability (p. 292) | 9–12 |
| **7.3** | 1 Find and interpret area under a normal curve (p. 297) | 19–22; 23(a)–(c); 24(a)–(c); 25(a)–(c); 26(a)–(d) |
| | 2 Find the value of a normal random variable (p. 299) | 23(d); 24(d), (e); 25(d), (e); 26(e), (f) |
| **7.4** | 1 Draw normal probability plots to assess normality (p. 306) | 29–32 |
| **7.5** | 1 Understand the concept of a sampling distribution (p. 312) | 39 |
| | 2 Compute the mean and standard deviation of a sampling distribution of the mean (p. 317) | 23(e) and (g); 24(f) and (h) |
| | 3 Compute probabilities of a sample mean obtained from a normal population (p. 319) | 23(f), (h); 24(g), (i); 25(f), (g); 26(g) |
| | 4 Compute probabilities of a sample mean, utilizing the Central Limit Theorem (p. 320) | 35, 36 |
| **7.6** | 1 Approximate binomial probabilities, using the normal curve (p. 326) | 27, 28 |

## Review Exercises

**1.**

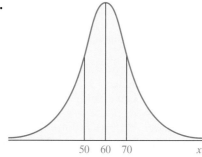

Use the preceding figure to answer the following questions:

(a) What is $\mu$?

(b) What is $\sigma$?

(c) Suppose the area under the normal curve to the right of $X = 75$ is 0.0668. Provide two interpretations for this area.

(d) Suppose the area under the normal curve between $X = 50$ and $X = 75$ is 0.7745. Provide two interpretations for this area.

**2.**

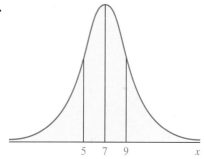

Use the preceding figure to answer the following questions:

(a) What is $\mu$?

(b) What is $\sigma$?

(c) Suppose the area under the normal curve to the left of $X = 10$ is 0.9332. Provide two interpretations for this area.

(d) Suppose the area under the normal curve between $X = 5$ and $X = 8$ is 0.5328. Provide two interpretations for this area.

**3.** A random variable $X$ is normally distributed with $\mu = 20$ and $\sigma = 4$.

(a) Compute $Z_1 = \dfrac{X_1 - \mu}{\sigma}$ for $X_1 = 18$.

(b) Compute $Z_2 = \dfrac{X_2 - \mu}{\sigma}$ for $X_2 = 21$.

(c) The area under the normal curve between $X_1 = 18$ and $X_2 = 21$ is 0.2912. What is the area between $Z_1$ and $Z_2$?

**4.** A random variable $X$ is normally distributed with $\mu = 50$ and $\sigma = 8$.

(a) Compute $Z_1 = \dfrac{X_1 - \mu}{\sigma}$ for $X_1 = 48$.

(b) Compute $Z_2 = \dfrac{X_2 - \mu}{\sigma}$ for $X_2 = 60$.

(c) The area under the normal curve between $X_1 = 48$ and $X_2 = 60$ is 0.4931. What is the area between $Z_1$ and $Z_2$?

*In Problems 5–8, draw a standard normal curve, and shade the area indicated. Then use Table II to find the area under the normal curve.*

**5.** The area left of $Z = -1.04$

**6.** The area right of $Z = 2.04$

**7.** The area between $Z = -0.34$ and $Z = 1.03$

**8.** The area between $Z = 1.93$ and $Z = 3.93$

*In Problems 9–12, find the indicated probability of the standard normal random variable Z.*

**9.** $P(Z < 1.19)$

**10.** $P(Z \geq 1.61)$

**11.** $P(-1.21 < Z \leq 2.28)$

**12.** $P(0.21 < Z < 1.69)$

**13.** Find the $Z$-score such that the area to the left of the $Z$-score is 0.84.

**14.** Find the $Z$-score such that the area right of the $Z$-score is 0.483.

**15.** Find the $Z$-scores that separate the middle 92% of the data from the area in the tails of the standard normal distribution.

**16.** Find the $Z$-scores that separate the middle 88% of the data from the area in the tails of the standard normal distribution.

**17.** Find the value of $z_{0.20}$

**18.** Find the value of $z_{0.04}$

*In Problems 19–22, draw the normal curve with the parameters indicated. Then find the probability of the random variable X. Shade the area that represents the probability.*

**19.** $\mu = 50, \sigma = 6, P(X > 55)$

**20.** $\mu = 30, \sigma = 5, P(X \leq 23)$

**21.** $\mu = 70, \sigma = 10, P(65 < X < 85)$

**22.** $\mu = 20, \sigma = 3, P(22 \leq X \leq 27)$

---

**23. Tire Wear** Suppose Dunlop Tire manufactures tires having the property that the mileage the tire lasts follows a normal distribution with mean 70,000 miles and standard deviation 4400 miles.

(a) What percent of the tires will last at least 75,000 miles?

(b) Suppose Dunlop warrants the tires for 60,000 miles. What percent of the tires will last 60,000 miles or less?

(c) What is the probability that a randomly selected Dunlop tire lasts between 65,000 and 80,000 miles?

(d) Suppose that Dunlop wants to warrant no more than 2% of its tires. What mileage should the company advertise as its warranty mileage?

(e) Suppose that a quality-control engineer obtains a simple random sample of $n = 10$ tires. What is the sampling distribution of $\bar{x}$?

(f) What is the probability that a random sample of 10 tires will result in a sample mean of at least 72,500 miles?

(g) Suppose that a quality-control engineer obtains a simple random sample of $n = 25$ tires. What is the sampling distribution of $\bar{x}$? What effect does increasing the sample size have on $\sigma_{\bar{x}}$?

(h) What is the probability that a random sample of 25 tires will result in a sample mean of at least 72,500 miles?

**24. Talk Time on a Cell Phone** Suppose the "talk time" in digital mode on a Motorola Timeport P8160 is normally distributed with mean 324 minutes and standard deviation 24 minutes.

(a) What proportion of the time will a fully charged battery last at least 300 minutes?

(b) What proportion of the time will a fully charged battery last less than 340 minutes?

(c) Suppose you charge the battery fully. What is the probability it will last between 310 and 350 minutes?

(d) Determine the talk time that is in the top 20%.

(e) Determine the talk time that makes up the middle 90% of talk time.

(f) Suppose that an engineer at Motorola obtains a simple random sample of $n = 15$ batteries. What is the sampling distribution of $\bar{x}$?

(g) What is the probability that a random sample of $n = 15$ batteries results in a mean talk time of at least 330 minutes?

(h) Suppose that an engineer at Motorola obtains a simple random sample of $n = 30$ batteries. What is the sampling distribution of $\bar{x}$? What effect does increasing the sample size have on $\sigma_{\bar{x}}$?

(i) What is the probability that a random sample of $n = 30$ batteries results in a mean talk time of at least 330 minutes?

25. **Serum Cholesterol** As reported by the U.S. National Center for Health Statistics, the mean serum cholesterol of females 16–19 years old is $\mu = 171$. If serum cholesterol is normally distributed with $\sigma = 39.8$, answer the following.
    (a) Determine the proportion of 16–19-year-old females with a serum cholesterol above 180.
    (b) Determine the proportion of 16–19-year-old females with a serum cholesterol between 150 and 200.
    (c) Suppose a 16–19-year-old female is randomly selected. Determine the probability her serum cholesterol is below 140.
    (d) Determine the serum cholesterol that divides the bottom 10% from the top 90% of all serum cholesterol levels of 16–19-year-old females.
    (e) According to the National Center for Health Statistics, the 25th percentile of serum cholesterol for 16–19-year-old females is 145. Is the 25th percentile on the normal curve close to the reported value of 145?
    (f) What is the probability that a random sample of eight females 16–19 years old has a mean serum cholesterol below 167?
    (g) What is the probability that a random sample of 20 females 16–19 years old has a mean serum cholesterol below 167?

26. **Wechsler Intelligence Scale** The Wechsler Intelligence Scale for Children is normally distributed with mean 100 and standard deviation 15.
    (a) What proportion of test takers will score above 125?
    (b) What proportion of test takers will score below 90?
    (c) What proportion of test takers will score between 110 and 140?
    (d) If a child is randomly selected, what is the probability that she scores above 150?
    (e) What intelligence score will place a child in the top 5% of all children?
    (f) If "normal" intelligence is defined as scoring in the middle 95% of all test takers, figure out the scores that differentiate "normal" intelligence from "abnormal" intelligence.

(g) A random sample of 20 children was administered the Wechsler Intelligence Scale for Children, and their mean score was 110. Would it be unusual to obtain a random sample of 20 children whose mean score was 110 or higher?

27. **High Cholesterol** According to the National Center for Health Statistics, 8% of 20–34-year-old females have high serum cholesterol. Suppose you conduct a random sample of 200 20–34-year-old females.
    (a) Verify that the conditions for using the normal distribution to approximate the binomial distribution are met.
    (b) Approximate the probability that exactly 15 have high serum cholesterol. Interpret this result.
    (c) Approximate the probability that more than 20 have high serum cholesterol. Interpret this result.
    (d) Approximate the probability that at least 15 have high serum cholesterol. Interpret this result.
    (e) Approximate the probability that fewer than 25 have high serum cholesterol. Interpret this result.
    (f) Approximate the probability that between 15 and 25, inclusive, have high serum cholesterol. Interpret this result.

28. **America Reads** According to a Gallup poll conducted September 10–14, 1999, 56% of Americans 18 years old or older stated they had read at least 6 books (fiction and nonfiction) within the past year. Suppose you conduct a random sample of 250 Americans 18 years old or older.
    (a) Verify that the conditions for using the normal distribution to approximate the binomial distribution are met.
    (b) Approximate the probability that exactly 125 read at least 6 books within the past year. Interpret this result.
    (c) Approximate the probability that fewer than 120 read at least 6 books within the past year. Interpret this result.
    (d) Approximate the probability that at least 140 read at least 6 books within the past year. Interpret this result.
    (e) Approximate the probability that between 100 and 120, inclusive, read at least 6 books within the past year. Interpret this result.

*In Problems 29 and 30, a normal probability plot of a simple random sample of data from a population whose distribution is unknown was obtained. Given the normal probability plot, is there reason to believe the population is normally distributed?*

**29.**

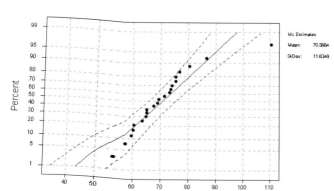

**30.**

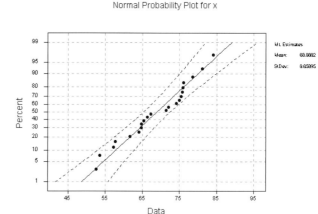

# Confidence Intervals about a Single Parameter

## Outline

**For additional study help, go to**
www.prenhall.com/sullivanstats

Materials include

- Projects:
  - Case Study: The Search for a Fire-Safe Cigarette
  - Decisions: What's Your Major?
  - Consumer Reports Project
- Self-Graded Quizzes
- "Preparing for This Section" Quizzes
- STATLETs
- PowerPoint Downloads
- Step-by-Step Technology Guide
- Graphing Calculator Help

## Putting It All Together

Chapters 1–7 laid the groundwork for the remainder of the text. These chapters dealt with data collection (Chapter 1), descriptive statistics (Chapters 2–4), and probability (Chapters 5–7). We now discuss inferential statistics, the process of using information obtained in a sample and generalizing it to a population. There are two areas of inferential statistics: 1) estimation, in which sample data are used to estimate the value of unknown parameters such as $\mu$ or $\sigma$ and 2) hypothesis testing, in which claims regarding a characteristic of the population are made and sample data are used to test the claim. In this chapter, we will discuss estimation of an unknown parameter and in the next chapter, we will discuss hypothesis testing.

We know from Section 7.5 that $\bar{x}$ is a random variable, and as such, has a distribution associated with it. We call this distribution the sampling distribution of the sample mean. The mean of the distribution of $\bar{x}$ is equal to the mean of the population, $\mu$. The standard deviation of

the distribution of $\bar{x}$ is $\dfrac{\sigma}{\sqrt{n}}$. Finally, we know that the distribution of the sample mean is normal if the population from which the sample was drawn is normal. And the sampling distribution of the sample mean becomes approximately normal as the sample size increases, regardless of the shape of the parent population. This result is known as the Central Limit Theorem. We use these results to help us in the estimation process.

Because the information that is collected from a sample is incomplete (the sample does not contain all the information in the population), we will assign probabilities to our estimates. These probabilities serve as a way of measuring what would happen if we estimated the value of the parameter many times and therefore provide a measure of confidence in our results.

# 8.1 Confidence Intervals about a Population Mean, $\sigma$ Known

**Preparing for This Section**  Before getting started, review the following:

✓ Simple random sampling (Section 1.2, pp. 12–15)

✓ Parameter versus statistic (Section 3.1, p. 84)

✓ Sampling error (Section 1.4, p. 26)

✓ $z_\alpha$ notation (Section 7.2, p. 292)

✓ Central Limit Theorem (Section 7.5, p. 321)

✓ Normal probability plots (Section 7.4, pp. 305–310)

**Objectives**  Compute the point estimate of $\mu$

 Compute confidence intervals about $\mu$ with $\sigma$ known

 Understand the role of margin of error in constructing confidence intervals

 Determine sample size necessary for estimating the population mean

 **Point Estimates**

If we do not have access to population data, then we cannot determine values of population parameters, such as $\mu$ or $\sigma$. Therefore, we use sample data to estimate the values of these parameters. These estimators are called *point estimators.*

*Definition*  A **point estimate** of a parameter is the value of a statistic that estimates the value of the parameter.

**Caution**

Appropriately obtaining individuals to participate in a survey or appropriate design of an experiment is vital for the statistical process to be valid. In other words, if data are carelessly or inappropriately collected, any statistical inference performed on the data is subject to scrutiny. For example, the results of Internet surveys should be looked upon with extreme skepticism.

For example, the sample mean, $\bar{x}$, is a point estimate of the population mean, $\mu$. The sample standard deviation, $s$, is a point estimate of the population standard deviation, $\sigma$.

For many parameters there is more than one point estimate. For example, we could use the sample median or mode as a point estimate of the population mean, $\mu$. So the question becomes, which point estimate do I use if I want to estimate the value of the population mean? The answer is that we use the sample mean. There are three reasons for this:

1. The sample mean, $\bar{x}$, is an unbiased estimator of $\mu$. A statistic is an **unbiased estimator** provided its expected value is equal to the value of the parameter. For example, if we took many samples of size $n$ and computed the sample mean of each sample, then the mean of the sample means would equal $\mu$. In other words, a statistic is unbiased if it does not systematically overestimate or underestimate the value of the parameter it estimates.

**Historical Note**

It was R.A. Fisher who described the criteria of a good statistic.

2. The sample mean provides more **consistent** estimates of the population mean. In other words, the larger your sample, the closer the sample mean gets to the population mean.

3. In repeated samples, a majority of the sample means will be "close" to the value of the population mean. This characteristic of the sample mean is called **efficiency**.

*Theorem*  The sample mean, $\bar{x}$, is the **best point estimate** of the population mean, $\mu$.

▶ **EXAMPLE 1**    **Computing a Point Estimate for $\mu$**

*Problem:* Suppose we were in the market to purchase a used Corvette. We would like to estimate the population mean price of a three-year-old Chevy Corvette.

*Approach:* We estimate the population mean by obtaining a random sample of 15 used Corvettes listed in Table 1. The best point estimate of the population mean is the sample mean.

*Solution:* The sample mean is

$$\bar{x} = \frac{\$47{,}000 + \$32{,}750 + \cdots + \$43{,}785}{15} = \$38{,}247$$

The best point estimate of $\mu$ is $38,247.    ◀◀

 *Now Work Problem 15(a).*

**TABLE 1**

| | | |
|---|---|---|
| $47,000 | $43,108 | $33,995 |
| $32,750 | $33,988 | $43,500 |
| $33,995 | $32,750 | $39,950 |
| $36,900 | $35,995 | $39,998 |
| $37,995 | $37,995 | $43,785 |

*Source:* cars.com

**② Confidence Intervals**

While the sample mean obtained in Example 1 is considered the best point estimate of the population mean, we recognize that there is likely to be some sampling error. Remember that sampling error is the difference between the value of the statistic and the parameter. Sampling error occurs because the sample does not capture all the information that is in the population.

Perhaps the idea that the value of a statistic often does not equal the value of the parameter is best phrased by David Salsburg in this excerpt from his book, *The Lady Tasting Tea*:

> Never will we know if the value of a statistic for a particular set of data is correct. We can only say we used a procedure that produces a statistic that meets these criteria (unbiased, consistent and efficient).

Therefore, rather than reporting the value of the sample mean, it is preferable to report an interval around the sample mean along with a measure of our confidence that this interval contains the population mean. To help understand the idea of this interval, consider the following situation. Suppose that you were asked to guess the mean age of the students in your statistics class. After surveying five of the students, you might guess the mean age is 24 years. This would be a point estimate of $\mu$, the mean age of all students in the class. We could also express our guess by producing a range of ages such as 24 years old give or take 2 years (the *margin of error*). Mathematically, we would write this $24 \pm 2$. Suppose further that you were asked how confident you were that the mean age was between 22 and 26. Your response might be "I am 80% confident that the mean age of students in my statistics class is between 22 and 26 years." If asked to construct an interval in which you were 99% confident, would your margin of error increase or decrease?

In statistics, we construct intervals for a population mean that center around a "guess" as well. The "guess" used in statistics is the sample mean. The margin of error depends upon the level of confidence, the standard deviation of the population, and the sample size.

*Definition*

> A **confidence interval estimate** of a parameter consists of an interval of numbers, along with a measure of the likelihood that the interval contains the unknown parameter. The **level of confidence** in a confidence interval is the proportion of intervals that will contain $\mu$ if a large number of repeated samples are obtained. The level of confidence is denoted $(1 - \alpha) \cdot 100\%$.

For example, a 95% level of confidence ($\alpha = 0.05$) would mean that if 100 confidence intervals were constructed, each based on a different sample from the same population, we would expect 95 of the intervals to contain the population mean. The confidence interval for the population mean depends upon three factors:

1. The point estimate of the population mean,
2. Our level of confidence that the interval contains the population mean,
3. The standard deviation of the sample mean, $\sigma_{\bar{x}} = \dfrac{\sigma}{\sqrt{n}}$.

Because the sample mean is the best point estimate of the population mean, we expect the value of $\bar{x}$ to be close to $\mu$; however, we do not know exactly how close $\bar{x}$ is to the unknown value of $\mu$. The process of constructing a confidence interval relies on the fact that $\bar{x}$ is a random variable that is normally distributed with mean $\mu$ and standard deviation $\dfrac{\sigma}{\sqrt{n}}$, provided the population from which the sample is drawn is normal or the sample size is greater than or equal to 30.

While it is not reasonable to claim to know the population standard deviation without knowing the population mean, for ease of understanding the procedures in constructing confidence intervals, we will assume its value is known in this section. We will drop this assumption in the next section.

Because $\bar{x}$ is normally distributed, we know that 95% of all sample means should lie within 1.96 standard deviations of the population mean, $\mu$, and 2.5% of the the sample means will lie in each tail. See Figure 1.

That is, 95% of all sample means are in the interval

$$\mu - 1.96 \cdot \frac{\sigma}{\sqrt{n}} < \bar{x} < \mu + 1.96 \cdot \frac{\sigma}{\sqrt{n}}$$

With a little algebraic manipulation, we can rewrite this inequality with $\mu$ in the middle and obtain

$$\bar{x} - 1.96 \cdot \frac{\sigma}{\sqrt{n}} < \mu < \bar{x} + 1.96 \cdot \frac{\sigma}{\sqrt{n}}$$

This inequality means that 95% of the sample means will result in confidence intervals that contain the population mean. That is,

$$P\left(\bar{x} - 1.96 \cdot \frac{\sigma}{\sqrt{n}} < \mu < \bar{x} + 1.96 \cdot \frac{\sigma}{\sqrt{n}}\right) = 0.95$$

It is common to write the 95% confidence interval as

$$\bar{x} \pm 1.96 \cdot \frac{\sigma}{\sqrt{n}}$$

From this, we can see the confidence interval is of the form

Point estimate $\pm$ Margin of error

with the point estimate being the sample mean, $\bar{x}$, and the margin of error being $1.96 \cdot \dfrac{\sigma}{\sqrt{n}}$.

> **In Your Own Words**
>
> A confidence interval is a range of numbers, such as 22–30. The level of confidence represents our degree of confidence (expressed as a percent) that the interval contains the population mean.

**Figure 1**

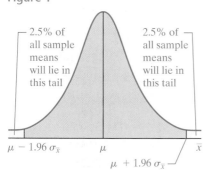

- 2.5% of all sample means will lie in this tail
- 2.5% of all sample means will lie in this tail

$\mu - 1.96\,\sigma_{\bar{x}}$    $\mu$    $\bar{x}$

$\mu + 1.96\,\sigma_{\bar{x}}$

▶ **EXAMPLE 2**   **Constructing 20 95% Confidence Intervals Based on 20 Samples**

*Problem:* It is known that scores on the Stanford–Binet IQ test are normally distributed with $\mu = 100$ and $\sigma = 16$. Use Minitab, Excel, or some other statistical software to simulate obtaining 20 simple random samples of size $n = 15$. Use these 20 different samples to construct 95% confidence intervals for the population mean, $\mu$.

*Approach:*

**Step 1:** We will use Minitab to obtain 20 simple random samples of size 15 from a population that is normally distributed with mean $\mu = 100$ and $\sigma = 16$. We then compute the sample means of each of the 20 samples.

**Step 2:** Construct 95% confidence intervals by computing

$$\overline{x} - 1.96 \cdot \frac{\sigma}{\sqrt{n}} = \overline{x} - 1.96 \cdot \frac{16}{\sqrt{15}} \quad \text{and} \quad \overline{x} + 1.96 \cdot \frac{\sigma}{\sqrt{n}} = \overline{x} + 1.96 \cdot \frac{16}{\sqrt{15}}$$

for each sample mean.

*Solution:*

**Step 1:** Table 2 shows the 20 sample means obtained from Minitab.

**Step 2:** We construct the 20 confidence intervals for each of the 20 sample means and present the results in Table 3.

| TABLE 2 | | | | |
|---|---|---|---|---|
| 104.32 | 93.97 | 108.73 | 104.11 | 100.67 |
| 96.87 | 99.74 | 100.25 | 101.32 | 94.24 |
| 102.23 | 94.32 | 97.66 | 101.44 | 98.19 |
| 107.15 | 100.38 | 95.89 | 104.43 | 102.28 |

| | | TABLE 3 | | |
|---|---|---|---|---|
| Sample | Sample Mean | Margin of Error $1.96 \cdot \dfrac{\sigma}{\sqrt{n}} = 1.96 \cdot \dfrac{16}{\sqrt{15}} = 8.10$ | Lower Bound $\overline{x} - 1.96 \cdot \dfrac{\sigma}{\sqrt{n}}$ | Upper Bound $\overline{x} + 1.96 \cdot \dfrac{\sigma}{\sqrt{n}}$ |
| 1 | 104.32 | 8.10 | $104.32 - 8.10 = 96.22$ | $104.32 + 8.10 = 112.42$ |
| 2 | 93.97 | 8.10 | $93.97 - 8.10 = 85.87$ | $93.97 + 8.10 = 102.07$ |
| 3 | 108.73 | 8.10 | $108.73 - 8.10 = 100.63$ | $108.73 + 8.10 = 116.83$ |
| 4 | 104.11 | 8.10 | 96.01 | 112.21 |
| 5 | 100.67 | 8.10 | 92.57 | 108.77 |
| 6 | 96.87 | 8.10 | 88.77 | 104.97 |
| 7 | 99.74 | 8.10 | 91.64 | 107.84 |
| 8 | 100.25 | 8.10 | 92.15 | 108.35 |
| 9 | 101.32 | 8.10 | 93.22 | 109.42 |
| 10 | 94.24 | 8.10 | 86.14 | 102.34 |
| 11 | 102.23 | 8.10 | 94.13 | 110.33 |
| 12 | 94.32 | 8.10 | 86.22 | 102.42 |
| 13 | 97.66 | 8.10 | 89.56 | 105.76 |
| 14 | 101.44 | 8.10 | 93.34 | 109.54 |
| 15 | 98.19 | 8.10 | 90.09 | 106.29 |
| 16 | 107.15 | 8.10 | 99.05 | 115.25 |
| 17 | 100.38 | 8.10 | 92.28 | 108.48 |
| 18 | 95.89 | 8.10 | 87.79 | 103.99 |
| 19 | 104.43 | 8.10 | 96.33 | 112.53 |
| 20 | 102.28 | 8.10 | 94.18 | 110.38 |

From the results presented in Table 3, we can see that 19 of the 20 (or 95%) samples resulted in 95% confidence intervals that contain the value of the population mean, 100. Notice that sample 3 does not result in a confidence interval that contains the population mean 100 (lower bound = $108.73 - 8.10 = 100.63$; upper bound = $108.73 + 8.10 = 116.83$). ◀◀

We can further see the results of Example 2 through Figure 2, in which we present the sampling distribution of $\overline{x}$, along with the 20 confidence intervals computed in Example 2. Notice that the interval corresponding to $\overline{x} = 108.73$ does not intersect $\mu = 100$.

Figure 2

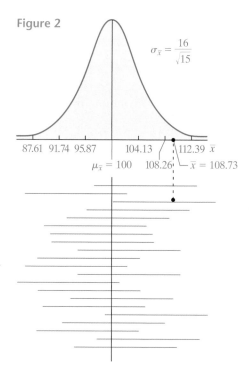

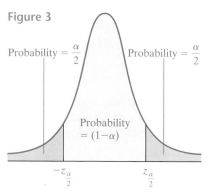

**Note:**  In this particular simulation exactly 95% of the samples resulted in intervals that contained $\mu$. It will not always be the case that exactly 95% of the sample means result in intervals that contain $\mu$ when performing this simulation. It could easily have been the case that all the intervals contained $\mu$ or that 17 out of the 20 contained $\mu$. The interpretation of the confidence interval remains: in a 95% confidence interval, if we obtain many samples of size $n$, the interval constructed will contain the population mean, $\mu$, 95% of the time. Since we obtained only 20 samples, the exact percentage may be slightly different from 95%.  ◀

Of course, we might not always be interested in creating 95% confidence intervals. Therefore, we need a method for constructing any $(1 - \alpha) \cdot 100\%$ confidence interval. If we let $z_{\alpha/2}$ represent the $Z$-score that separates an area of $\alpha/2$ in the right tail from the rest of the distribution, we can construct a confidence interval for any level of confidence. Figure 3 demonstrates the idea. In general, a $(1 - \alpha) \cdot 100\%$ confidence interval will include the population mean, $\mu$, for all sample means that lie within $z_{\alpha/2}$ standard deviations of the population mean, $\mu$. The sample means that are in the tails will not have confidence intervals that include the population mean. The values of $\pm z_{\alpha/2}$ are called the **critical values** of the distribution.

Table 4 shows some of the more common critical values used in the construction of confidence intervals. The reader is encouraged to verify the results presented in the table.

Figure 3

| TABLE 4 | | |
|---|---|---|
| **Level of Confidence, $(1 - \alpha) \cdot 100\%$** | **Area in Each Tail, $\alpha/2$** | **Critical Value, $z_{\alpha/2}$** |
| 90% | 0.05 | 1.645 |
| 95% | 0.025 | 1.96 |
| 99% | 0.005 | 2.575 |

In a 95% confidence interval, $\alpha = 0.05$, so that $z_{\alpha/2} = z_{0.05/2} = z_{0.025} = 1.96$. This means that any sample mean that lies within 1.96 standard deviations of the population mean will result in a confidence interval that contains $\mu$ and any sample mean that is more than 1.96 standard deviations from the population mean will result in a confidence interval that does not contain $\mu$. This is an extremely important point. Whether a confidence interval contains $\mu$ depends solely upon the sample mean, $\overline{x}$. The population mean is a fixed value that either is or is not in the interval, and we do not know which of these possibilities is in fact true for any computed confidence interval. However, if repeated samples are taken, we do know that $(1 - \alpha) \cdot 100\%$ of the confidence intervals will contain the population mean. This result leads to the following interpretation of a confidence interval:

**Interpretation of a Confidence Interval**

A $(1 - \alpha) \cdot 100\%$ confidence interval means that if we obtained many simple random samples of size $n$ from the population whose mean, $\mu$, is unknown, then approximately $(1 - \alpha) \cdot 100\%$ of the intervals will contain $\mu$.

For example, if we constructed a 90% confidence interval with a lower bound of 12 and an upper bound of 18, then we would interpret the interval as follows: "We are 90% confident that the population mean, $\mu$, is between 12 and 18." Be sure that you understand that the level of confidence refers to the method, not the interval. So a 90% confidence interval means that the method will result in an interval that contains the population mean, $\mu$, 90% of the time. It does not mean that there is a 90% probability that $\mu$ lies between 12 and 18.

We are now prepared to present a method for constructing a $(1 - \alpha) \cdot 100\%$ confidence interval about $\mu$.

---

**Constructing a $(1 - \alpha) \cdot 100\%$ Confidence Interval about $\mu$, $\sigma$ Known**

Suppose a simple random sample of size $n$ is taken from a population with unknown mean, $\mu$, and known standard deviation $\sigma$. A $(1 - \alpha) \cdot 100\%$ confidence interval for $\mu$ is given by

$$\text{Lower bound: } \bar{x} - z_{\alpha/2} \cdot \frac{\sigma}{\sqrt{n}} \qquad \text{Upper bound: } \bar{x} + z_{\alpha/2} \cdot \frac{\sigma}{\sqrt{n}}$$

where $z_{\alpha/2}$ is the critical $Z$-value.

**Note:** The size, $n$, of the sample must be greater than or equal to 30 or the population must be normally distributed.

---

When constructing a confidence interval about the population mean, we must verify that the sample either is large ($n \geq 30$) or comes from a population that is normally distributed. Fortunately, the procedures for constructing confidence intervals presented in this section are **robust**. This means that minor departures from normality will not seriously affect the results. Nevertheless, it is important that the requirements for constructing confidence intervals are verified. We shall follow the practice of verifying the normality assumption for small sample sizes by drawing a normal probability plot and checking for outliers by drawing a boxplot.

Because the construction of the confidence interval with $\sigma$ known uses $Z$-scores, it is sometimes referred to as constructing a **Z-interval**.

▶ **EXAMPLE 3** **Constructing a Z-Interval**

*Problem:* Remember that we are in the market to buy a three-year-old Chevy Corvette. In Example 1, we obtained a point estimate of \$38,247 for the population mean. We now wish to construct a 90% confidence interval about the population mean, $\mu$. Assume that the population standard deviation is known to be \$4100.

*Approach:*

**Step 1:** We have a simple random sample of size $n = 15$. We need to verify that the data come from a population that is normally distributed with no outliers. We will verify normality by constructing a normal probability plot. We check for outliers by drawing a boxplot.

**Step 2:** Since we are constructing a 90% confidence interval, $\alpha = 0.10$. Therefore, we need to identify the value of $z_{\alpha/2} = z_{0.10/2} = z_{0.05}$. Recall, this is the $Z$-score such that the area under the standard normal curve to the right of $z_{0.05}$ is 0.05.

**Step 3:** We compute the lower and upper bounds on the interval with $\bar{x} = \$38,247$, $\sigma = \$4100$, and $n = 15$.

**Step 4:** We interpret the result by stating: "We are 90% confident that the mean price of all three-year-old Chevy Corvettes is somewhere between *lower bound* and *upper bound*."

*Solution:*

**Step 1:** A normal probability plot of the data in Table 1 (page 340) is provided in Figure 4(a). A boxplot is presented in Figure 4(b). All the data values lie within the bounds on the normal probability plot, indicating that the data are roughly normal. The boxplot does not display any outliers. The requirements for computing the confidence interval are satisfied.

**Step 2:** We need to determine the value of $z_{0.05}$. We look in Table II for an area equal to 0.95 (remember the table gives areas left of the $Z$-scores). We have $z_{0.05} = 1.645$.

Figure 4        Normal Probability Plot for Price

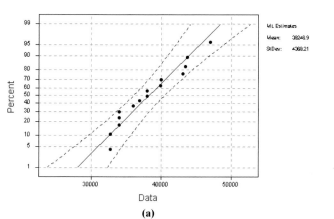

(a)

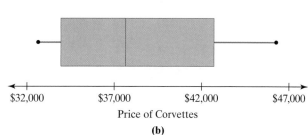

Price of Corvettes

(b)

**Step 3:** We substitute into the formulas for the lower and upper bound of the confidence interval.

$$\text{Lower bound} = \bar{x} - z_{\alpha/2} \cdot \frac{\sigma}{\sqrt{n}} = \$38{,}247 - 1.645 \cdot \frac{\$4100}{\sqrt{15}} = \$38{,}247 - \$1741 = \$36{,}506$$

$$\text{Upper bound} = \bar{x} + z_{\alpha/2} \cdot \frac{\sigma}{\sqrt{n}} = \$38{,}247 + 1.645 \cdot \frac{\$4100}{\sqrt{15}} = \$38{,}247 + \$1741 = \$39{,}988$$

**Step 4:** We are 90% confident that the mean price of all three-year-old Chevy Corvettes is somewhere between \$36,506 and \$39,988.    ◄◄

**Using Technology:** Statistical software and graphing calculators with advanced statistical features have the ability to construct confidence intervals with $\sigma$ known. Figure 5 shows the results using Minitab.

Figure 5

Confidence Intervals

```
The assumed sigma = 4100

Variable            N     Mean    StDev    SeMean      90.0% CI
price              15    38247     4522      1059    (36506, 39988)
```

**NW**  *Now Work Problems 15(b), 15(c), and 15(d).*

### 3  The Role of the Margin of Error

Is there any way that we can reduce the width of the interval? The width of the interval is determined by the *margin of error.*

*Definition*

**The Margin of Error**

The margin of error, $E$, in a $(1 - \alpha) \cdot 100\%$ confidence interval in which $\sigma$ is known is given by

$$E = z_{\alpha/2} \cdot \frac{\sigma}{\sqrt{n}}$$

*In Your Own Words*

The margin of error can be thought of as the "give or take" portion of the statement "The mean age of the class is 23, give or take 3 years."

where $n$ is the sample size.
**Note:**  We require the population from which the sample was drawn be normally distributed or the sample size $n$ be greater than or equal to 30.

As we look at the formula for obtaining the margin of error, we see that its value depends upon three quantities:

**1.** The level of confidence.

**2.** The standard deviation of the population.

**3.** The sample size, $n$.

We cannot control the standard deviation of the population, but we certainly can control the level of confidence and/or the sample size. Let's see how changing these values affect the margin of error.

▶ EXAMPLE 4    **The Role of the Level of Confidence in the Margin of Error**

*Problem:* For the problem of estimating the population mean price of all three-year-old Chevy Corvettes presented in Example 3, determine the effect of increasing the level of confidence on the confidence interval to 99%.

*Approach:* With a 99% level of confidence, we have $\alpha = 0.01$. So, to compute the margin of error, $E$, we simply determine the value of $z_{\alpha/2} = z_{0.01/2} = z_{0.005}$. We then substitute this value into the formula for the margin of error with $\sigma = \$4100$ and $n = 15$.

*Solution:* After consulting Table II, we determine $z_{0.005} = 2.575$. Substituting into the formula for margin of error, we obtain

$$E = z_{\alpha/2} \cdot \frac{\sigma}{\sqrt{n}} = 2.575 \cdot \frac{\$4100}{\sqrt{15}} = \$2726$$

**In Your Own Words**

As the level of confidence increases, the margin of error also increases.

Notice the margin of error has increased from $1741 to $2726 when the level of confidence increases from 90% to 99%. If we want to be more confident the interval contains the population mean, the width of the interval increases. ◀◀

▶ EXAMPLE 5    **The Role of Sample Size in the Margin of Error**

*Problem:* For the problem of estimating the population mean price of all three-year-old Chevy Corvettes presented in Example 3, determine the effect of increasing the sample size to $n = 25$. Leave the level of confidence at 90%.

*Approach:* We compute the margin of error with $n = 25$ instead of $n = 15$.

*Solution:* Substituting $z_{0.05} = 1.645$, $\sigma = \$4100$, and $n = 25$ into the formula for computing margin of error, we obtain

$$E = z_{\alpha/2} \cdot \frac{\sigma}{\sqrt{n}} = 1.645 \cdot \frac{\$4100}{\sqrt{25}} = \$1349$$

By increasing the sample size from $n = 15$ to $n = 25$., the margin of error, $E$, decreases from $1741 to $1349.

**In Your Own Words**

As the sample size increases, the margin of error decreases.

This result should not be very surprising. Remember the Law of Large Numbers states that as the sample size $n$ increases, the sample mean approaches the value of the population mean. This outcome is supported by the smaller margin of error. ◀◀

**NW** *Now Work Problem 11.*

④ **Determining Sample Size**

Suppose we want to know the number of Corvettes that we should sample in order to estimate the price of a three-year-old Corvette within $1500 with 95% confidence. If we solve the formula for the margin of error, $E$, for $n$, we obtain a formula for determining sample size:

$$E = z_{\alpha/2} \cdot \frac{\sigma}{\sqrt{n}}$$

$E\sqrt{n} = z_{\alpha/2} \cdot \sigma$  Multiply both sides by $\sqrt{n}$.

$\sqrt{n} = \dfrac{z_{\alpha/2} \cdot \sigma}{E}$  Divide both sides by the margin of error, $E$.

$n = \left(\dfrac{z_{\alpha/2} \cdot \sigma}{E}\right)^2$  Square both sides.

This gives us a formula for computing the sample size required to obtain a confidence interval for $\mu$ within a prescribed margin of error, $E$, and level of confidence, $(1 - \alpha) \cdot 100\%$.

*Theorem*

**Determining the Sample Size $n$**

The sample size required to estimate the population mean, $\mu$, with a level of confidence $(1 - \alpha) \cdot 100\%$ with a specified margin of error, $E$, is given by

$$n = \left(\frac{z_{\alpha/2} \cdot \sigma}{E}\right)^2$$

where $n$ is rounded up* to the nearest whole number.

When $\sigma$ is unknown, it is common practice to conduct a preliminary survey in order to determine $s$ and use it as an estimate of $\sigma$. When utilizing this approach, the size of the sample should be at least 30.

▶ **EXAMPLE 6**   **Determining Sample Size**

*Problem:* We will once again consider the problem of estimating the population mean price of a three-year-old Chevy Corvette. How large of a sample is required to estimate the mean price within $1000 with 90% confidence?

*Approach:* The sample size required can be obtained using the formula

$$n = \left(\frac{z_{\alpha/2} \cdot \sigma}{E}\right)^2$$

with $z_{\alpha/2} = z_{0.05} = 1.645$, $\sigma = \$4100$, and $E = \$1000$.

*Solution:* We substitute the values of $z$, $\sigma$, and $E$ into the formula for determining sample size and obtain

$$n = \left(\frac{z_{\alpha/2} \cdot \sigma}{E}\right)^2 = \left(\frac{1.645 \cdot \$4100}{\$1000}\right)^2 = 6.7445^2 = 45.488$$

**Caution**

Don't forget to round up when determining sample size.

We round 45.488 up to 46. A sample size of $n = 46$ will result in an interval estimate of the population mean with a margin of error equal to $1000 with 90% confidence. This means that if we obtained 100 samples of size $n = 46$, we would expect about 90 of them to capture the population mean, while 10 would not.  ◀◀

**NW**  *Now Work Problem 27.*

---

*Rounding *up* is different than rounding *off*. We round 5.32 *up* to 6 and *off* to 5.

## Some Final Thoughts

There are many requirements that must be satisfied in the implementation of any type of statistical inference. It is worthwhile to list the requirements for constructing a confidence interval about $\mu$ with $\sigma$ known in a single location for quick reference.

### Requirements in the Construction of a Confidence Interval about $\mu$, $\sigma$ Known

1. The data obtained come from a simple random sample. In Chapter 1, we introduced other sampling techniques, such as stratified, cluster, and systematic. The techniques introduced in this section apply only to samples obtained through simple random sampling. Although methods do exist for constructing confidence intervals when using the other sampling methods, they are beyond the scope of this text. If the data are obtained from a suspect sampling method, such as voluntary response or convenience sampling, no methods exist for constructing confidence intervals. This is because the method of data collection itself is flawed. If data are collected in a flawed manner, any statistical inference performed on the data is useless!

2. The data are obtained from a population that is normally distributed, or the sample size, $n$, is greater than or equal to 30. We can use normal probability plots to help us judge whether the requirement of normality is satisfied. Remember, the techniques introduced in this section are robust. This means that minor departures from requirements will not have a severe effect on the results. However, we need to be aware that the sample mean is not resistant. This means that any outlier(s) in the data will affect the value of the sample mean and therefore affect the confidence interval. If data contain outliers, we should *not* use the methods introduced in this section.

3. The population standard deviation, $\sigma$, is assumed to be known. Clearly, it is unlikely that the population standard deviation is known when the population mean is not. We will drop this assumption in the next section.

## 8.1 Assess Your Understanding

### Concepts and Vocabulary

1. What are the characteristics that an estimator of a parameter should have?
2. The construction of a confidence interval depends upon three factors. What are they?
3. Why does the margin of error increase as the level of confidence increases?
4. Why does the margin of error decrease as the sample size $n$ increases?
5. Suppose a confidence interval has a lower bound of 10 and an upper bound of 18. What is the sample mean used to construct the interval? What is the margin of error?
6. When determining the sample size required to estimate $\mu$, we always round the value of $n$ up to the next whole number. Explain why we do so.

### Exercises

• **Basic Skills**

*In Problems 1–6, a simple random sample of size $n < 30$ has been obtained. From the normal probability plots and boxplots, judge whether a z-interval should be constructed.*

**1.**

**2.**

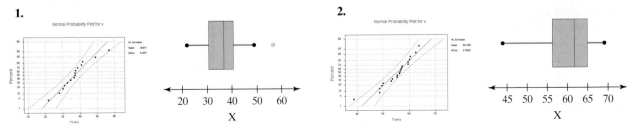

17. **Serum Cholesterol** As reported by the U.S. National Center for Health Statistics, the mean serum high density lipoprotein (HDL) cholesterol of females 20–29 years old is $\mu = 53$. Dr. Paul Oswiecmiski wants to estimate the mean serum HDL cholesterol of his 20–29-year-old female patients. He randomly selects 15 of his 20–29-year-old patients and obtains the data shown. Assume that $\sigma = 13.4$.

| | | | | |
|---|---|---|---|---|
| 65 | 47 | 51 | 54 | 70 |
| 55 | 44 | 48 | 36 | 53 |
| 45 | 34 | 59 | 45 | 54 |

*Source:* Dr. Paul Oswiecmiski

(a) Use the data to compute a point estimate for the population mean serum HDL cholesterol in Dr. Oswiecmiski's patients.

(b) Because the sample size is small, we must verify that serum HDL cholesterol is normally distributed and that the sample does not contain any outliers. The normal probability plot and boxplot are shown below. Are the conditions for constructing a z-interval satisfied?

Normal Probability Plot for Cholesterol

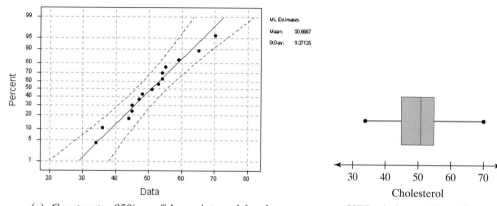

(c) Construct a 95% confidence interval for the mean serum HDL cholesterol for all of Dr. Oswiecmiski's 20–29-year-old female patients. Interpret this interval.

(d) Do Dr. Oswiecmiski's patients appear to have a serum HDL different from that of the general population? Why?

18. **Serum Cholesterol** As reported by the U.S. National Center for Health Statistics, the mean serum high density lipoprotein (HDL) cholesterol of males 20–29 years old is $\mu = 47$. Dr. Paul Oswiecmiski wants to estimate the mean serum HDL cholesterol of his 20–29-year-old male patients. He randomly selects 15 of his 20–29-year-old male patients and obtains the data shown. Assume that $\sigma = 12.5$.

| | | | | |
|---|---|---|---|---|
| 48 | 61 | 46 | 35 | 48 |
| 34 | 64 | 42 | 29 | 71 |
| 64 | 47 | 56 | 27 | 53 |

*Source:* Dr. Paul Oswiecmiski

(a) Use the data to compute a point estimate for the population mean serum HDL cholesterol in Dr. Oswiecmiski's patients.

(b) Because the sample size is small, we must verify that serum HDL cholesterol is normally distributed and that the sample does not contain any outliers. The normal probability plot and boxplot are shown below. Are the conditions for constructing a z-interval satisfied?

Normal Probability Plot for Cholesterol

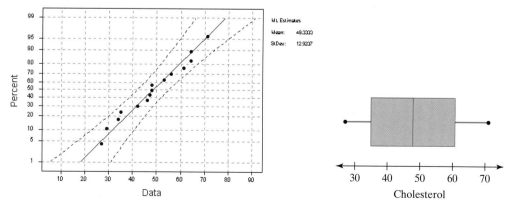

(c) Construct a 95% confidence interval for the mean serum HDL cholesterol for all of Dr. Oswiecmiski's 20–29-year-old male patients. Interpret this interval.

(d) Do Dr. Oswiecmiski's patients appear to have a serum HDL different from that of the general population? Why?

**19. ACT Math Scores** In 2000, as reported by ACT Research Service, the mean ACT Math score was $\mu = 20.7$. Mrs. Teresa Gibson wants to estimate the mean ACT Math score of students in High School District 204. She obtains a simple random sample of 20 students who took the ACT in 2000, looks up their ACT Math scores, and obtains the results shown to the left. Assume that $\sigma = 5$.

| 24 | 23 | 16 | 26 | 25 |
|----|----|----|----|----|
| 22 | 18 | 25 | 26 | 17 |
| 28 | 27 | 23 | 21 | 23 |
| 20 | 25 | 21 | 19 | 30 |

(a) Use the data to compute a point estimate for the population mean ACT Math score in High School District 204.

(b) Because the sample size is small, we must verify that ACT scores are normally distributed and that the sample does not contain any outliers. The normal probability plot and boxplot are shown below. Are the conditions for constructing a $z$-interval satisfied?

Normal Probability Plot for ACT Math

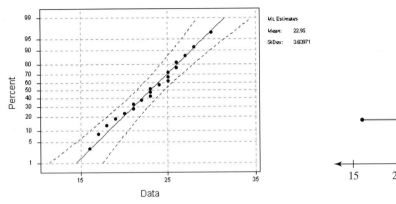

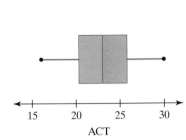

(c) Construct a 92% confidence interval for the mean ACT Math score for all students in District 204 who took the exam. Interpret this interval.

(d) Do Mrs. Gibson's students appear to have a mean ACT Math score different from that of the general population? Why?

**20. ACT Science Scores** In 2000, as reported by ACT Research Service, the mean ACT Science score was $\mu = 21$. Mr. Matt Gibson is the chair of the science department in High School District 203. After speaking with his wife (Problem 19), he decides to estimate the mean ACT Science score of students in his district. He obtains a simple random sample of 20 students who took the ACT in 2000, looks up their ACT Science scores, and obtains the results shown to the left. Assume that $\sigma = 4.5$.

| 25 | 26 | 22 | 24 | 27 |
|----|----|----|----|----|
| 26 | 22 | 18 | 27 | 26 |
| 32 | 18 | 24 | 19 | 23 |
| 20 | 27 | 23 | 25 | 17 |

(a) Use the data to compute a point estimate for the population mean ACT science score in High School District 203.

(b) Because the sample size is small, we must verify that ACT scores are normally distributed and that the sample does not contain any outliers. The normal probability plot and boxplot are shown below. Are the conditions for constructing a $z$-interval satisfied?

Normal Probability Plot for ACT Science

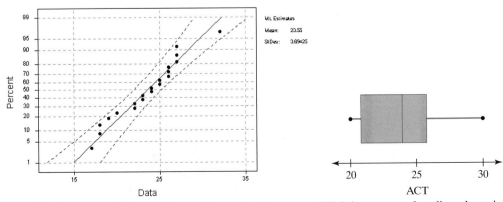

(c) Construct a 94% confidence interval for the mean ACT Science score for all students in District 203 who took the exam. Interpret this interval.

(d) Do Mr. Gibson's students appear to have a mean ACT Science score different from that of the general population? Why?

**21. Gestation Period** The length of human pregnancies is approximately normally distributed with mean $\mu = 266$ days and standard deviation $\sigma = 16$ days. Dr. Margaret Oswiecmiski obtains a simple random sample of 10 of her patients and obtains the following results:

| 279 | 260 | 261 | 266 | 255 |
|---|---|---|---|---|
| 267 | 230 | 266 | 264 | 240 |

(a) Use the data to compute a point estimate for the population mean gestation period.
(b) Support the requirement that gestation period be normally distributed with no outliers by constructing a normal probability plot and boxplot.
(c) Construct a 90% confidence interval for the mean gestation period for all of Dr. Oswiecmiski's patients. Interpret this interval.
(d) Do Dr. Oswiecmiski's patients have a mean gestation period different from 266 days? Why?

**22. Weather in Chicago** The high temperature in Chicago for the month of August is distributed approximately normally, with mean $\mu = 80°$ F and standard deviation $\sigma = 8°$ F. A random sample of 10 days in Chicago during August 2000 results in the following data:

| 77 | 81 | 72 | 86 | 76 |
|---|---|---|---|---|
| 81 | 73 | 92 | 75 | 87 |

(a) Use the data to compute a point estimate for the population mean high temperature in Chicago in August.
(b) Support the requirement that high temperature be normally distributed with no outliers by constructing a normal probability plot and boxplot.
(c) Construct a 95% confidence interval for the mean high temperature in August in Chicago. Interpret this interval.
(d) Is there evidence that August, 2000 was normal?

**23. Volume of Philip Morris Stock** The following data represent a random sample of the number of shares of Philip Morris stock traded for 35 days in 2000:

| 3.98 | 8.90 | 10.40 | 7.52 | 13.84 |
|---|---|---|---|---|
| 14.04 | 9.14 | 7.96 | 11.40 | 13.30 |
| 5.29 | 9.69 | 8.94 | 12.10 | 13.25 |
| 4.62 | 10.85 | 24.52 | 14.62 | 16.10 |
| 4.24 | 8.54 | 11.59 | 6.75 | 27.54 |
| 16.06 | 17.62 | 7.17 | 6.13 | 25.32 |
| 7.71 | 10.14 | 4.28 | 7.24 | 10.88 |

*Source:* http://finance.yahoo.com

The standard deviation of the number of shares traded in 2000 was $\sigma = 4.84$ million shares.

(a) Use the data to compute a point estimate for the population mean number of shares traded per day in 2000.
(b) Construct a 90% confidence interval for the population mean number of shares traded per day in 2000. Interpret the confidence interval.

(c) A second random sample of 35 days in 2000 resulted in the following data:

| 15.22 | 7.58 | 5.16 | 5.76 | 11.78 |
|---|---|---|---|---|
| 10.14 | 5.24 | 21.71 | 8.08 | 9.14 |
| 6.55 | 5.80 | 9.19 | 18.21 | 6.66 |
| 9.31 | 6.58 | 10.49 | 10.54 | 7.95 |
| 6.11 | 5.29 | 7.40 | 3.05 | 2.93 |
| 5.45 | 13.07 | 7.15 | 10.31 | 6.81 |
| 5.57 | 11.59 | 7.89 | 10.91 | 7.28 |

Construct a 90% confidence interval for the population mean number of shares traded per day in 2000. Interpret the confidence interval.
(d) Explain why the confidence intervals obtained in parts (b) and (c) are different.

**24. Volume of Harley Davidson Stock** The following data represent a random sample of the number of shares of Harley Davidson stock traded for 40 days in 2000:

| 0.68 | 0.69 | 0.53 | 0.75 | 0.58 |
|---|---|---|---|---|
| 0.77 | 0.58 | 0.92 | 0.42 | 1.30 |
| 0.63 | 0.68 | 0.46 | 0.48 | 0.63 |
| 0.58 | 0.43 | 0.69 | 1.60 | 0.38 |
| 1.01 | 0.48 | 1.33 | 1.14 | 0.64 |
| 2.86 | 0.48 | 0.35 | 0.30 | 0.74 |
| 0.47 | 0.74 | 1.01 | 1.21 | 1.05 |
| 0.51 | 0.84 | 1.06 | 0.60 | 0.25 |

*Source:* http://finance.yahoo.com

The standard deviation of the number of shares traded in 2000 was $\sigma = 0.94$ million shares.

(a) Use the data to compute a point estimate for the population mean number of shares traded per day in 2000.
(b) Construct a 95% confidence interval for the population mean number of shares traded per day in 2000. Interpret the confidence interval.
(c) A second random sample of 40 days in 2000 resulted in the following data:

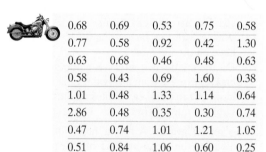

| 1.30 | 0.36 | 0.91 | 0.43 | 0.67 |
|---|---|---|---|---|
| 0.47 | 0.69 | 0.80 | 0.35 | 0.67 |
| 0.82 | 0.82 | 0.84 | 0.61 | 0.49 |
| 0.49 | 1.30 | 0.56 | 0.50 | 0.85 |
| 0.57 | 0.56 | 0.66 | 0.77 | 0.42 |
| 0.56 | 1.62 | 2.86 | 0.79 | 2.95 |
| 0.70 | 0.87 | 1.36 | 0.37 | 0.63 |
| 0.95 | 0.64 | 0.76 | 0.84 | 0.72 |

Construct a 95% confidence interval for the population mean number of shares traded per day in 2000. Interpret the confidence interval.
(d) Explain why the confidence intervals obtained in parts (b) and (c) are different.

**25. Miles on a Saturn** A researcher is interested in approximating the mean number of miles on four-year-old Saturn SC1s. She finds a random sample of 33 such Saturn SC1s in the Chicagoland area and obtains the following results:

| | | | | |
|---|---|---|---|---|
| 45,336 | 90,574 | 42,800 | 84,000 | 57,506 |
| 47,977 | 10,778 | 39,176 | 41,431 | 86,838 |
| 30,114 | 90,100 | 26,560 | 75,312 | 44,411 |
| 29,000 | 30,447 | 25,000 | 49,874 | 76,576 |
| 57,145 | 38,796 | 32,004 | 43,128 | 51,159 |
| 54,000 | 52,181 | 51,305 | 33,867 | 46,178 |
| 51,000 | 39,174 | 59,018 | | |

*Source:* cars.com

(a) Obtain a point estimate of the population mean number of miles on a four-year-old Saturn SC1.
(b) Construct and interpret a 99% confidence interval for the population mean number of miles on a four-year-old Saturn SC1. Assume that $\sigma = 19{,}700$.
(c) Construct and interpret a 95% confidence interval for the population mean number of miles on a four-year-old Saturn SC1. Assume that $\sigma = 19{,}700$.
(d) What effect does decreasing the level of confidence have on the interval?
(e) Do the confidence intervals computed in parts (b) and (c) represent an interval estimate for the population mean number of miles on Saturn SC1s in the United States? Why?

**26. Miles on a Cavalier** A researcher is interested in approximating the mean number of miles on three-year-old Chevy Cavaliers. She finds a random sample of 35 Cavaliers in the Orlando, Florida area and obtains the following results:

| | | | | |
|---|---|---|---|---|
| 37,815 | 20,000 | 57,103 | 46,585 | 24,822 |
| 49,678 | 30,983 | 52,969 | 8,000 | 39,862 |
| 6,000 | 65,192 | 34,285 | 30,906 | 41,841 |
| 39,851 | 43,000 | 74,361 | 52,664 | 33,587 |
| 52,896 | 45,280 | 30,000 | 41,713 | 76,315 |
| 22,442 | 45,301 | 52,899 | 41,526 | 28,381 |
| 55,163 | 51,812 | 36,500 | 31,947 | 16,529 |

*Source:* cars.com

(a) Obtain a point estimate of the population mean number of miles on a three-year-old Cavalier.
(b) Construct and interpret a 99% confidence interval for the population mean number of miles on a three-year-old Cavalier. Assume that $\sigma = 16{,}100$.
(c) Construct and interpret a 95% confidence interval for the population mean number of miles on a three-year-old Cavalier. Assume that $\sigma = 16{,}100$.
(d) What effect does decreasing the level of confidence have on the width of the interval?
(e) Do the confidence intervals computed in parts (b) and (c) represent an interval estimate for the population mean number of miles on Cavaliers in the United States? Why?

**27. Sample Size** Dr. Paul Oswiecmiski wants to estimate the mean serum HDL cholesterol of all 20–29-year-old females. How many subjects would be needed in order to estimate the mean serum HDL cholesterol of all 20–29-year-old females within two points with 99% confidence, assuming that $\sigma = 13.4$? Suppose Dr. Oswiecmiski would be content with 95% confidence. How does the decrease in confidence affect the sample size required?

**28. Sample Size** Dr. Paul Oswiecmiski wants to estimate the mean serum HDL cholesterol of all 20–29-year-old males. How many subjects would be needed in order to estimate the mean serum HDL cholesterol of all 20–29-year-old males within 1.5 points with 90% confidence, assuming that $\sigma = 12.5$? Suppose Dr. Oswiecmiski would prefer 98% confidence. How does the increase in confidence affect the sample size required?

**29. How Often Do You Bathe?** A Gallup poll conducted December 20–21, 1999 asked Americans how many times they bathe during the week. How many subjects would be needed in order to estimate the number of times Americans bathed during the week in 1999 within 0.5 with 95% confidence? Initial survey results indicate that $\sigma = 2.9$.

**30. How Much TV Do You Watch?** A Gallup poll conducted December 20–21, 1999 asked Americans how many hours of TV they watch during the week. How many subjects would be needed in order to estimate the number of hours of TV Americans watched in 1999 during the week within 0.5 hours with 95% confidence? Initial survey results indicate that $\sigma = 1.8$.

**31. Miles on a Saturn** A researcher wishes to estimate the mean number of miles on four-year-old Saturn SC1s.
(a) How many cars should be in a sample in order to estimate the mean number of miles within 1000 miles with 90% confidence, assuming that $\sigma = 19{,}700$?
(b) How many cars should be in a sample in order to estimate the mean number of miles within 500 miles with 90% confidence, assuming that $\sigma = 19{,}700$?
(c) What effect does increasing the required accuracy have on the sample size? Why is this the expected result?

**32. Miles on a Cavalier** A researcher wishes to estimate the mean number of miles on three-year-old Chevy Cavaliers.
(a) How many cars should be in a sample in order to estimate the mean number of miles within 2000 miles with 98% confidence, assuming that $\sigma = 16{,}100$?
(b) How many cars should be in a sample in order to estimate the mean number of miles within 1000 miles with 98% confidence, assuming that $\sigma = 16{,}100$?
(c) What effect does increasing the required accuracy have on the sample size? Why is this the expected result?

**33. Simulation** IQ scores as measured by the Stanford–Binet IQ test are normally distributed with $\mu = 100$ and $\sigma = 16$.

(a) Simulate obtaining 100 samples of size $n = 15$ from this population.

(b) Construct 95% confidence intervals for each of the 100 samples.

(c) How many of the intervals do you expect to include the population mean? How many actually contain the population mean?

**34. Simulation** Suppose the arrival of cars at Burger King's drive-through follows a Poisson process with $\mu = 4$ cars every 10 minutes.

(a) Simulate obtaining 100 samples of size $n = 40$ from this population.

(b) Construct 90% confidence intervals for each of the 100 samples. [*Hint:* $\sigma = \sqrt{\mu}$ in a Poisson process.]

(c) How many of the intervals do you expect to include the population mean? How many actually contain the population mean?

**35. The Effect of Nonnormal Data** The exponential probability distribution is a probability distribution that can be used to model time waiting in line or the lifetime of electronic components. Its density function with $\mu = 5$ is shown in the accompanying figure. Clearly, the distribution is skewed right.

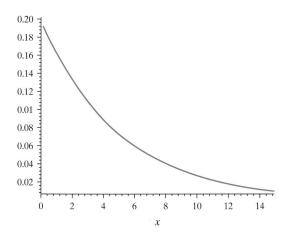

(a) Use Minitab or some other statistical software to generate 100 random samples of size $n = 6$ from a population that follows the exponential probability distribution with $\mu = 5$. (It turns out that $\sigma$ also equals 5.)

(b) Use the 100 samples to determine 100 95% confidence intervals with $\sigma = 5$.

(c) How many of the intervals would we expect to contain $\mu = 5$? How many of the 100 intervals contain $\mu = 5$?

(d) What are the consequences of not having normal data when the sample size is small?

**36. The Effect of Outliers** Suppose the following small data set represents a simple random sample from a population whose mean is 50 and standard deviation is 10:

| 43 | 63 | 53 | 50 | 58 | 44 |
|----|----|----|----|----|----|
| 53 | 53 | 52 | 41 | 50 | 43 |

(a) A normal probability plot indicates the data are normally distributed with no outliers. Compute a 95% confidence interval for this data set, assuming $\sigma = 10$.

(b) Suppose the observation, 41, were inadvertently entered into the computer as 14. Verify that this observation is an outlier.

(c) Construct a 95% confidence interval on the data set with the outlier. What effect does the outlier have on the confidence interval?

(d) Consider the following data set, which represents a simple random sample of size 36 from a population whose mean is 50 and standard deviation is 10:

| 43 | 63 | 53 | 50 | 58 | 44 |
|----|----|----|----|----|----|
| 53 | 53 | 52 | 41 | 50 | 43 |
| 47 | 65 | 38 | 58 | 41 | 52 |
| 49 | 46 | 57 | 50 | 38 | 42 |
| 59 | 54 | 37 | 41 | 33 | 37 |
| 46 | 54 | 42 | 48 | 53 | 41 |

Verify that the sample mean for the large data set is the same as the sample mean for the small data set.

(e) Compute a 95% confidence interval for the large data set, assuming $\sigma = 10$. Compare the results to part (a). What effect does increasing the sample size have on the confidence interval?

(f) Suppose the last observation, 41, were inadvertently entered as 14. Verify that this observation is an outlier.

(g) Compute a 95% confidence interval for the large data set with the outlier, assuming $\sigma = 10$. Compare the results to part (g). What effect does an outlier have on a confidence interval when the data set is large?

**37.** By how many times does the sample size have to be increased to decrease the margin of error by a factor of $\frac{1}{2}$?

**38.** Suppose a certain population, $A$, has standard deviation $\sigma_A = 5$ and a second population, $B$, has standard deviation $\sigma_B = 10$. How many times larger than population $A$'s does population $B$'s sample size need to be, assuming the margin of error and level of confidence are the same? [*Hint:* Compute $n_A / n_B$.]

## Technology Step-by-Step
Confidence Intervals about $\mu$, $\sigma$ Known

**TI-83 Plus**    *Step 1:* If necessary, enter raw data in L1.

*Step 2:* Press STAT, highlight TESTS, and select `7: ZInterval`.

*Step 3:* If the data are raw, highlight DATA—make sure `List1` is set to L1 and `Freq` to 1. If summary statistics are known, highlight STATS and enter the summary statistics. Following $\sigma$:, enter the population standard deviation.

*Step 4:* Enter the confidence level following `C-Level:`.

*Step 5:* Highlight `Calculate`; press ENTER.

**MINITAB**    *Step 1:* Enter raw data in column C1.

*Step 2:* Select the **Stat** menu, highlight **Basic Statistics**, then highlight **1-Sample Z**....

*Step 3:* Enter C1 in the cell marked "Variables". Select Confidence Interval, and enter a confidence level. In the cell marked "Sigma", enter the value of $\sigma$. Click OK.

**Excel**    *Step 1:* If necessary, enter raw data in column A.

*Step 2:* Load the PHStat Add-in.

*Step 3:* Select the **PHStat menu**, highlight **Confidence Intervals** ..., then highlight **Estimate for the mean, sigma known** ....

*Step 4:* Enter the value of $\sigma$ and the confidence level. If the summary statistics are known, click "Sample statistics known" and enter the sample size and sample mean. If summary statistics are unknown, click "Sample statistics unknown". With the cursor in the "Sample cell range" cell, highlight the data in column A. Click OK.

---

## 8.2   Confidence Intervals about a Population Mean, $\sigma$ Unknown

**Preparing for This Section**   Before getting started, review the following:

✓ Simple random sampling (Section 1.2, pp. 12-15)

✓ Parameter versus statistic (Section 3.1, p. 84)

✓ Degrees of freedom (Section 3.2, p. 104)

✓ Central Limit Theorem (Section 7.5, p. 321)

✓ Normal probability plots (Section 7.4, pp. 305–310)

**Objectives**   ① Know the properties of Student's $t$-distribution

② Determine $t$-values

③ Construct confidence intervals about $\mu$, $\sigma$ unknown

④ Determine the method to use to compute an interval about $\mu$

In Section 8.1, we made the assumption that the population standard deviation, $\sigma$, was known, while the population mean, $\mu$, was unknown. In this section, we drop that assumption and compute confidence intervals about $\mu$ with the population standard deviation, $\sigma$, unknown.

### Historical Note

William Sealy Gossett was born on June 13, 1876, in Canterbury, England. Gossett earned a degree in chemistry from New College in Oxford in 1899. He then got a job as a chemist for the Guinness Brewing Company. Gossett, along with other chemists, was asked to find a way to make the best beer at the cheapest cost. This allowed him to concentrate on statistics. In 1904, Gossett wrote a paper on the brewing of beer that included a discussion of standard errors. In July, 1905, Gossett met with Karl Pearson to learn about the theory of standard errors. Over the next few years, he developed his t-distribution. The Guinness Brewery did not allow its employees to publish, so Gossett published his research using the pen name Student. Gossett died October 16, 1937.

## Student's *t*-Distribution

In Section 8.1, we computed confidence intervals about a population mean, $\mu$, on the assumption that the following conditions were met:

1. $\sigma$ was known.
2. The population from which the sample was drawn followed a normal distribution or the sample size $n$ was large ($n \geq 30$).
3. The sample was a simple random sample.

A $(1 - \alpha) \cdot 100\%$ confidence interval was then computed as

$$\bar{x} - z_{\alpha/2} \cdot \frac{\sigma}{\sqrt{n}} \quad \text{and} \quad \bar{x} + z_{\alpha/2} \cdot \frac{\sigma}{\sqrt{n}}$$

If $\sigma$ is unknown, it seems reasonable to replace $\sigma$ with $s$ and proceed with the analysis. However, while it is true that $z = \dfrac{\bar{x} - \mu}{\sigma/\sqrt{n}}$ is normally distributed with mean 0 and standard deviation 1, we cannot replace $\sigma$ with $s$ and say that $z = \dfrac{\bar{x} - \mu}{s/\sqrt{n}}$ is normally distributed with mean 0 and standard deviation 1, because $s$, itself, is a random variable. Instead, $\dfrac{\bar{x} - \mu}{s/\sqrt{n}}$ follows **Student's *t*-distribution**, developed by William Gossett. Gossett was in charge of conducting experiments at the Guinness brewery in order to identify the best barley variety. At the time, the only available distribution was the standard normal distribution, but Gossett always performed experiments with small data sets and he did not know $\sigma$. This led Gossett to determine the sampling distribution of $\dfrac{\bar{x} - \mu}{s/\sqrt{n}}$. Gossett published his findings under the pen name *Student*.

*Theorem*

> **Student's *t*-Distribution**
>
> Suppose a simple random sample of size $n$ is taken from a population. If the population from which the sample is drawn follows a normal distribution, then the distribution of
>
> $$t = \frac{\bar{x} - \mu}{s/\sqrt{n}}$$
>
> follows Student's *t*-distribution with $n - 1$ degrees of freedom,* where $\bar{x}$ is the sample mean and $s$ is the sample standard deviation.

The interpretation of the *t*-statistic is the same as that of the *z*-score. The *t*-statistic represents the number of sample standard errors $\bar{x}$ is from the population mean, $\mu$. It turns out that the shape of the *t*-distribution depends upon the sample size, $n$.

To help see how the *t*-distribution differs from the standard normal (or *z*-) distribution and the role the sample size $n$ plays, we will go through the following exploration.

### EXPLORATION

**(a)** Obtain 1000 simple random samples of size $n = 5$ from a normal population with $\mu = 50$ and $\sigma = 10$.

**(b)** Calculate the sample mean and sample standard deviation for each of the samples.

---

*The reader may wish to review the discussion of degrees of freedom in Section 3.2 on p. 104.

(c) Compute $z = \dfrac{\overline{x} - \mu}{\sigma/\sqrt{n}}$ and $t = \dfrac{\overline{x} - \mu}{s/\sqrt{n}}$ for each sample.

(d) Draw histograms for both $z$ and $t$.

**Results**   We use Minitab to obtain the 1000 simple random samples and compute the 1000 sample means and sample standard deviations. We then compute $z = \dfrac{\overline{x} - \mu}{\sigma/\sqrt{n}} = \dfrac{\overline{x} - 50}{10/\sqrt{5}}$ and $t = \dfrac{\overline{x} - \mu}{s/\sqrt{n}} = \dfrac{\overline{x} - 50}{s/\sqrt{5}}$ for each of the 1000 samples. Figure 6(a) shows the histogram for $z$, and Figure 6(b) shows the histogram for $t$.

Figure 6

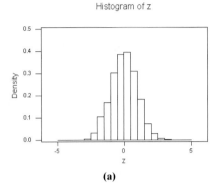

(a)

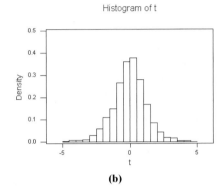

(b)

We notice that the histogram in Figure 6(a) is symmetric and bell shaped, with the center of the distribution at 0 and virtually all the rectangles between $-3$ and 3. In other words, $z$ follows a standard normal distribution. The distribution of $t$ is also symmetric and bell shaped and has its center at 0, but the distribution of $t$ has longer tails (i.e., $t$ is more dispersed), so it is unlikely that $t$ follows a standard normal distribution. The additional spread in the distribution of $t$ can be attributed to the fact that we divide by $\dfrac{s}{\sqrt{n}}$ to find $t$ instead of by $\dfrac{\sigma}{\sqrt{n}}$. Because the sample standard deviation is itself a random variable (rather than a constant such as $\sigma$), we have more dispersion in the distribution of $t$.   ◄◄

We now introduce the properties of the $t$-distribution.

**Properties of the $t$-Distribution**

1. The $t$-distribution is different for different values of $n$, the sample size.
2. The $t$-distribution is centered at 0 and is symmetric about 0.
3. The area under the curve is 1. The area under the curve to the right of 0 equals the area under the curve to the left of 0 equals $1/2$.
4. As $t$ increases without bound, the graph approaches, but never equals, zero. As $t$ decreases without bound, the graph approaches, but never equals, zero.
5. The area in the tails of the $t$-distribution is a little greater than the area in the tails of the standard normal distribution, because we are using $s$ as an estimate of $\sigma$, thereby introducing further variability into the $t$-statistic.
6. As the sample size $n$ increases, the density curve of $t$ gets closer to the standard normal density curve. This result occurs because, as the sample size $n$ increases, the values of $s$ get closer to the values of $\sigma$, by the Law of Large Numbers.

Figure 7

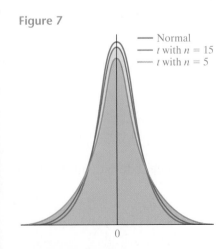

In Figure 7, we show the $t$-distribution for the sample sizes $n = 5$ and $n = 15$. As a point of reference, we have also drawn the standard normal density curve.

**Figure 8**

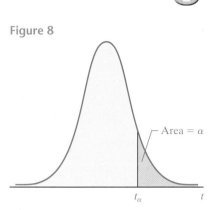

Area = $\alpha$

$t_\alpha$   $t$

**Figure 9**

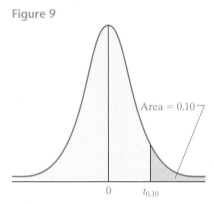

Area = 0.10

0   $t_{0.10}$

**Figure 10**

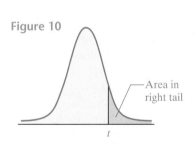

Area in right tail

$t$

(2) Recall that the notation $z_\alpha$ is used to represent the z-score whose area under the normal curve to the right of $z_\alpha$ is $\alpha$. Similarly, we let $t_\alpha$ represent the t-value such that the area under the t-distribution to the right of $t_\alpha$ is $\alpha$. See Figure 8.

The shape of the t-distribution depends upon the sample size, $n$. Therefore, the value of $t_\alpha$ depends not only on $\alpha$, but also on the degrees of freedom, $n - 1$. In Table III in Appendix A, the far left column gives the degrees of freedom. The top row represents the area under the t-distribution to the right of some t-value.

▶ **EXAMPLE 1  Finding t-Values**

*Problem:* Find the t-value such that the area under the t-distribution to the right of the t-value is 0.10, assuming 15 degrees of freedom. That is, find $t_{0.10}$ with 15 degrees of freedom.

*Approach:* We will perform the following steps.

*Step 1:* Draw a t-distribution with the unknown t-value labeled. Shade the area under the curve to the right of the t-value, as in Figure 8.

*Step 2:* Find the row in Table III that corresponds to 15 degrees of freedom and the column that corresponds to an area in the right tail of 0.10. Identify where the row and column intersect. This is the unknown t-value.

*Solution:*

*Step 1:* Figure 9 shows the graph of the t-distribution with 15 degrees of freedom. The unknown value of $t$ is labeled, and the area under the curve to the right of $t$ is shaded.

*Step 2:* A portion of Table III is reproduced in Figure 10. We have enclosed the row that represents 15 degrees of freedom and the column that represents the area 0.10 in the right tail. The point where the row and column intersect is the t-value we are seeking. The value of $t_{0.10}$ with 15 degrees of freedom is 1.341—that is, the area under the t-distribution to the right of $t = 1.341$ with 15 degrees of freedom is 0.10.

| | | | | | Area in Right Tail | | | | | | | |
|---|---|---|---|---|---|---|---|---|---|---|---|---|
| df | 0.25 | 0.20 | 0.15 | 0.10 | 0.05 | 0.025 | 0.02 | 0.01 | 0.005 | 0.0025 | 0.001 | 0.0005 |
| 1 | 1.000 | 1.376 | 1.963 | 3.078 | 6.314 | 12.710 | 15.890 | 31.820 | 63.660 | 127.300 | 318.300 | 636.600 |
| 2 | 0.816 | 1.061 | 1.386 | 1.886 | 2.920 | 4.303 | 4.849 | 6.965 | 9.925 | 14.090 | 22.330 | 31.600 |
| 3 | 0.765 | 0.978 | 1.250 | 1.638 | 2.353 | 3.182 | 3.482 | 4.541 | 5.841 | 7.453 | 10.210 | 12.920 |
| 4 | 0.741 | 0.941 | 1.190 | 1.533 | 2.132 | 2.776 | 2.999 | 3.747 | 4.604 | 5.598 | 7.173 | 8.610 |
| 5 | 0.727 | 0.920 | 1.156 | 1.476 | 2.015 | 2.571 | 2.757 | 3.365 | 4.032 | 4.773 | 5.893 | 6.869 |
| 6 | 0.718 | 0.906 | 1.134 | 1.440 | 1.943 | 2.447 | 2.612 | 3.143 | 3.707 | 4.317 | 5.208 | 5.959 |
| 7 | 0.711 | 0.896 | 1.119 | 1.415 | 1.895 | 2.365 | 2.517 | 2.998 | 3.499 | 4.029 | 4.785 | 5.408 |
| 8 | 0.706 | 0.889 | 1.108 | 1.397 | 1.860 | 2.306 | 2.449 | 2.896 | 3.355 | 3.833 | 4.501 | 5.041 |
| 9 | 0.703 | 0.883 | 1.100 | 1.383 | 1.833 | 2.262 | 2.398 | 2.821 | 3.250 | 3.690 | 4.297 | 4.781 |
| 10 | 0.700 | 0.879 | 1.093 | 1.372 | 1.812 | 2.228 | 2.359 | 2.764 | 3.169 | 3.581 | 4.144 | 4.587 |
| 11 | 0.697 | 0.876 | 1.088 | 1.363 | 1.796 | 2.201 | 2.328 | 2.718 | 3.106 | 3.497 | 4.025 | 4.437 |
| 12 | 0.695 | 0.873 | 1.083 | 1.356 | 1.782 | 2.179 | 2.303 | 2.681 | 3.055 | 3.428 | 3.930 | 4.318 |
| 13 | 0.694 | 0.870 | 1.079 | 1.350 | 1.771 | 2.160 | 2.282 | 2.650 | 3.012 | 3.372 | 3.852 | 4.221 |
| 14 | 0.692 | 0.868 | 1.076 | 1.345 | 1.761 | 2.145 | 2.264 | 2.624 | 2.977 | 3.326 | 3.787 | 4.140 |
| 15 | 0.691 | 0.866 | 1.074 | 1.341 | 1.753 | 2.131 | 2.249 | 2.602 | 2.947 | 3.286 | 3.733 | 4.073 |
| 16 | 0.690 | 0.865 | 1.071 | 1.337 | 1.746 | 2.120 | 2.235 | 2.583 | 2.921 | 3.252 | 3.686 | 4.015 |
| 17 | 0.689 | 0.863 | 1.069 | 1.333 | 1.740 | 2.110 | 2.224 | 2.567 | 2.898 | 3.222 | 3.646 | 3.965 |
| 18 | 0.688 | 0.862 | 1.067 | 1.330 | 1.734 | 2.101 | 2.214 | 2.552 | 2.878 | 3.197 | 3.611 | 3.922 |

◀◀

**Using Technology:** The area under the *t*-distribution can also be computed by graphing calculators with advanced statistical features or by statistical software. Figure 11 shows the results from Minitab. Notice that Minitab asks for the area to the left of the unknown *t*-value because it uses the cumulative distribution. The area under the curve to the right of the unknown *t*-value is 0.10, so the area under the curve to the left must be 0.90.

**Figure 11**

**Inverse Cumulative Distribution Function**

```
Student's t distribution with 15 DF

P( X<= x)                    x
  0.9000                 1.3406
```

**NW** *Now Work Problem 1(a).*

If the degrees of freedom we desire are not available in Table III, we shall follow the practice of choosing the closest number of degrees of freedom available in the table. For example, if we have 43 degrees of freedom, we shall use 40 degrees of freedom from Table III. In addition, the last row of Table III provides the *Z*-values from the standard normal distribution. We shall use these values for situations where the degrees of freedom are substantially more than 1000. This is acceptable because the *t*-distribution starts to behave like the standard normal distribution as *n* increases.

③ **Confidence Intervals**

The construction of confidence intervals about $\mu$ with $\sigma$ unknown follows the exact same logic as the construction of confidence intervals with $\sigma$ known. The only difference is that we use $s$ in place of $\sigma$ and $t$ in place of $z$.

**Constructing a $(1 - \alpha) \cdot 100\%$ Confidence Interval about $\mu$, $\sigma$ Unknown**
Suppose a simple random sample of size $n$ is taken from a population with unknown mean $\mu$ and unknown standard deviation $\sigma$. A $(1 - \alpha) \cdot 100\%$ confidence interval for $\mu$ is given by

$$\text{Lower bound: } \bar{x} - t_{\alpha/2} \cdot \frac{s}{\sqrt{n}} \quad \text{and} \quad \text{Upper bound: } \bar{x} + t_{\alpha/2} \cdot \frac{s}{\sqrt{n}} \quad \text{(1)}$$

where $t_{\alpha/2}$ is computed with $n - 1$ degrees of freedom.

**Note:** The interval is exact when the population is normally distributed; it is approximately correct for nonnormal populations, provided that $n$ is large.

Notice that a confidence interval about $\mu$ with $\sigma$ unknown can be computed for nonnormal populations even though Student's *t*-distribution required the population from which the sample was obtained to be normal. This is because the procedure for constructing the confidence interval is robust—that is, the procedure is accurate even despite moderate departures from the normality assumption. Notice that we said the procedure is accurate for *moderate* departures from normality. If a small data set has outliers, the results are compromised, because neither the sample mean, $\bar{x}$, nor the sample standard deviation, $s$, is resistant to outliers. Sample data should always be inspected for serious departures from normality and for outliers. This is easily done with normal probability plots and boxplots.

Because the interval uses the *t*-distribution, it is often referred to as the **t-interval**.

▶ EXAMPLE 2   **Constructing a Confidence Interval about $\mu$, $\sigma$ Unknown**

*Problem:* Recall that, in Example 3 from Section 8.1, we computed a 90% confidence interval for the mean price of a three-year-old Chevy Corvette. We made the assumption that we knew $\sigma$ to be $4100. Recompute the 90% confidence interval about the population mean with $\sigma$ unknown. The data are displayed in Table 5.

**TABLE 5**

| | | |
|---|---|---|
| $47,000 | $43,108 | $33,995 |
| $32,750 | $33,988 | $43,500 |
| $33,995 | $32,750 | $39,950 |
| $36,900 | $35,995 | $39,998 |
| $37,995 | $37,995 | $43,785 |

*Approach:*

*Step 1:* We draw a normal probability plot and boxplot to verify that there are no serious departures from normality and no outliers.

*Step 2:* Compute $\bar{x}$ and $s$.

*Step 3:* Look up the critical value $t_{\alpha/2}$ with $n - 1$ degrees of freedom.

*Step 4:* Compute the bounds on a $(1 - \alpha) \cdot 100\%$ confidence interval for $\mu$, using the following:

$$\text{Lower bound: } \bar{x} - t_{\alpha/2} \cdot \frac{s}{\sqrt{n}} \qquad \text{Upper bound: } \bar{x} + t_{\alpha/2} \cdot \frac{s}{\sqrt{n}}$$

*Step 5:* We interpret the result by stating: "We are 90% confident that the mean price of all three-year-old Chevy Corvettes is somewhere between *lower bound* and *upper bound*."

*Solution:*

*Step 1:* We have already drawn the normal probability plot and boxplot in Example 3 from Section 8.1. The normal probability plot did not show any serious departures from normality and the boxplot did not indicate any outliers. We can proceed with the construction of the confidence interval.

*Step 2:* We compute $\bar{x}$ and $s$, using a calculator, and discover that $\bar{x} = \$38,247$ and $s = \$4522$.

*Step 3:* Because we wish to determine a 90% confidence interval, we have $\alpha = 0.10$. Therefore, we need to find $t_{0.10/2} = t_{0.05}$ with $15 - 1 = 14$ degrees of freedom. Referring to Table III, we find $t_{0.05} = 1.761$.

*Step 4:* Using Formula (1), we find the lower and upper bounds:

$$\text{Lower Bound: } \bar{x} - t_{\alpha/2} \cdot \frac{s}{\sqrt{n}} = 38,247 - 1.761 \cdot \frac{4522}{\sqrt{15}} = 38,247 - 2056 = 36,191$$

$$\text{Upper Bound: } \bar{x} + t_{\alpha/2} \cdot \frac{s}{\sqrt{n}} = 38,247 + 1.761 \cdot \frac{4522}{\sqrt{15}} = 38,247 + 2056 = 40,303$$

*Step 5:* We are 90% confident that the mean price of all three-year-old Chevy Corvettes is somewhere between $36,191 and $40,303.   ◀◀

**Using Technology:** Statistical software or graphing calculators with advanced statistical features can be used to construct confidence intervals with $\sigma$ unknown. Figure 12 shows the results on a TI-83 Plus graphing calculator.

Figure 12

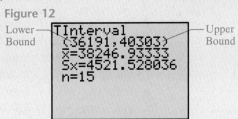

Lower Bound ——— TInterval
(36191,40303)
x̄=38246.93333
Sx=4521.528036
n=15

——— Upper Bound

If we compare the confidence interval with $\sigma$ unknown to the confidence interval with $\sigma$ known (Example 3, Section 8.1), we notice the interval with $\sigma$ unknown is wider. Why do you think this is the case?

**NW** *Now Work Problem 3.*

▶ **EXAMPLE 3**   **The Effect of Outliers**

*Problem:* The management of Disney World wanted to estimate the mean waiting time at the Dumbo ride. They randomly selected 15 riders and measured the amount of time the riders spent waiting in line. The results are in Table 6. Figure 13 shows a normal probability plot and boxplot for the data in the table. Figure 13(a) demonstrates that the sample data could have come from a population that is normally distributed except for the single outlier. Figure 13(b) shows the outlier as well. Determine a 95% confidence interval for the mean waiting time at the Dumbo ride both with and without the outlier in the data set. Comment on the effect the outlier has on the confidence interval.

**Figure 13**   Normal Probability Plot for Wait Time

| TABLE 6 | | | | |
|---|---|---|---|---|
| 30 | 29 | 34 | 41 | 37 |
| 16 | 18 | 30 | 24 | 25 |
| 29 | 16 | 80 | 19 | 30 |

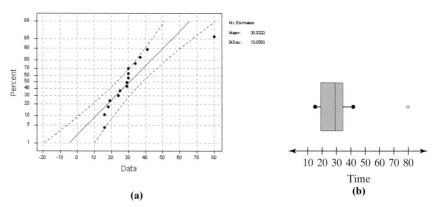

(a)

(b)

*Approach:* We will use Minitab to construct the confidence intervals.

*Solution:* Figure 14(a) shows confidence interval with the outlier included. Figure 14(b) shows the confidence interval with the outlier removed.

**Figure 14**

**One-Sample T: Time(with outlier)**

```
Variable          N     Mean    StDev  SE Mean        95.0% CI
Time             15    30.53    15.59     4.02  (21.90, 39.17)
```
(a)

**One-Sample T: Time(without outlier)**

```
Variable          N     Mean    StDev  SE Mean        95.0% CI
Time             14    27.00     7.75     2.07  (22.53, 31.47)
```
(b)

The 95% confidence interval with the outlier has a lower bound of 21.90 and an upper bound of 39.17. The 95% confidence interval with the outlier removed has a lower bound of 22.53 and an upper bound of 31.47. We notice a few things:

• With the outlier removed, the sample mean decreased, because the sample mean is not resistant.

• With the outlier removed, the sample standard deviation decreased, because the sample standard deviation is not resistant.

• With the outlier removed, the width of the interval decreased from $39.17 - 21.90 = 17.27$ to $31.47 - 22.53 = 8.94$. The confidence interval is nearly twice as wide when the outlier is included. ◀◀

What should we do if the assumptions required to compute a $t$-interval are not met? One option would be to increase the sample size beyond 30 observations. The other option is to use *nonparametric procedures*. **Nonparametric procedures** typically do not require an assumption of normality and the methods are resistant to outliers.

**NW** *Now Work Problem 25.*

### ④ Choosing the Right Method

It can seem difficult to choose the appropriate method for constructing confidence intervals. To help with the decision making, we provide the flowchart in Figure 15.

**Figure 15**

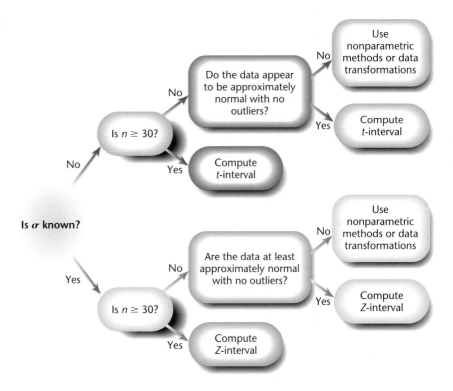

From the flowchart, the first step in choosing which method to use is to determine whether the population standard deviation, $\sigma$, is known. If it is known and the population is distributed at least approximately normally with no outliers, or if the sample size is large, compute the $z$-interval. If $\sigma$ is unknown and the population is distributed at least approximately normally with no outliers, or if the sample size is large, compute the $t$-interval.

**NW** *Now Work Problem 21.*

## 8.2 Assess Your Understanding

### Concepts and Vocabulary

1. Explain the circumstances under which a $z$-interval should be constructed. Under what circumstances should a $t$-interval be constructed? When can neither a $z$- nor a $t$-interval be constructed?

2. Explain why the $t$-distribution has less spread as the number of degrees of freedom increases.

3. The procedures for constructing a $t$-interval are robust. Explain what this means.

4. Consider Figures 6(a) and (b). Why are the $t$-values more dispersed than the $z$-values?

5. Discuss the similarities and differences between the standard normal distribution and the $t$-distribution.

## Exercises

### • Basic Skills

1. (a) Find the $t$-value such that the area in the right tail is 0.10 with 25 degrees of freedom.
   (b) Find the $t$-value such that the area in the right tail is 0.05 with 30 degrees of freedom.
   (c) Find the $t$-value such that the area left of the $t$-value is 0.01 with 18 degrees of freedom. [*Hint:* Use symmetry.]
   (d) Find the critical $t$-value that corresponds to 90% confidence. Assume 20 degrees of freedom.

2. (a) Find the $t$-value such that the area in the right tail is 0.02 with 19 degrees of freedom.
   (b) Find the $t$-value such that the area in the right tail is 0.10 with 32 degrees of freedom.
   (c) Find the $t$-value such that the area left of the $t$-value is 0.05 with 6 degrees of freedom. [*Hint:* Use symmetry.]
   (d) Find the critical $t$-value that corresponds to 95% confidence. Assume 16 degrees of freedom.

3. A simple random sample of size $n$ is drawn from a population that is normally distributed. The sample mean, $\bar{x}$, is found to be 108, and the sample standard deviation, $s$, is found to be 10.
   (a) Construct the 96% confidence interval if the sample size, $n$, is 25.
   (b) Construct the 96% confidence interval if the sample size, $n$, is 10. How does decreasing the sample size affect the margin of error, $E$?
   (c) Construct the 90% confidence interval if the sample size, $n$, is 25. Compare the results to those obtained in part (a). How does decreasing the level of confidence affect the size of the margin of error, $E$?
   (d) Could we have computed the confidence intervals in parts (a)–(c) if the population had not been normally distributed? Why?

4. A simple random sample of size $n$ is drawn from a population that is normally distributed. The sample mean, $\bar{x}$, is found to be 50, and the sample standard deviation, $s$, is found to be 8.
   (a) Construct the 98% confidence interval if the sample size, $n$, is 20.

   (b) Construct the 98% confidence interval if the sample size, $n$, is 15. How does decreasing the sample size affect the margin of error, $E$?
   (c) Construct the 95% confidence interval if the sample size, $n$, is 20. Compare the results to those obtained in part (a). How does decreasing the level of confidence affect the size of the margin of error, $E$?
   (d) Could we have computed the confidence intervals in parts (a)–(c) if the population had not been normally distributed? Why?

5. A simple random sample of size $n$ is drawn. The sample mean, $\bar{x}$, is found to be 18.4, and the sample standard deviation, $s$, is found to be 4.5.
   (a) Construct the 95% confidence interval if the sample size, $n$, is 35.
   (b) Construct the 95% confidence interval if the sample size, $n$, is 50. How does increasing the sample size affect the margin of error, $E$?
   (c) Construct the 99% confidence interval if the sample size, $n$, is 35. Compare the results to those obtained in part (a). How does increasing the level of confidence affect the size of the margin of error, $E$?
   (d) If the sample size is $n = 15$, what conditions must be satisfied in order to compute the confidence interval?

6. A simple random sample of size $n$ is drawn. The sample mean, $\bar{x}$, is found to be 35.1, and the sample standard deviation, $s$, is found to be 8.7.
   (a) Construct the 90% confidence interval if the sample size, $n$, is 40.
   (b) Construct the 90% confidence interval if the sample size, $n$, is 100. How does increasing the sample size affect the margin of error, $E$?
   (c) Construct the 98% confidence interval if the sample size, $n$, is 40. Compare the results to those obtained in part (a). How does increasing the level of confidence affect the size of the margin of error, $E$?
   (d) If the sample size is $n = 18$, what conditions must be satisfied in order to compute the confidence interval?

### • Applying the Concepts

7. **How Often Do You Bathe?** A Gallup poll conducted December 20–21, 1999, asked 1031 randomly selected Americans, "How often do you bathe each week?" Results of the survey indicated that $\bar{x} = 6.9$ and $s = 2.8$. Construct a 99% confidence interval for the mean number of times Americans bathed each week in 1999. Interpret the interval.

8. **How Often Do You Bathe?** A Gallup poll conducted June 29–July 4, 1950, asked 1031 Americans, "How often do you bathe each week?" Results of the survey indicated that $\bar{x} = 3.7$ and $s = 2.3$.
   (a) Construct a 99% confidence interval for the mean number of times Americans bathed each week in 1950. Interpret the interval.

   (b) Compare these results to those of Problem 7. Were Americans bathing more in 1999 than in the 1950s?

9. **How Much TV Do You Watch?** A Gallup poll conducted December 20–21, 1999, asked 1031 Americans, "How much TV do you watch each week?" Results of the survey indicated that $\bar{x} = 3.4$ and $s = 1.8$ hours. Construct a 95% confidence interval for the mean number of hours of TV Americans watched each week in 1999. Interpret the interval.

**10. Rainfall in Chicago** For a simple random sample of 36 years, the mean amount of rainfall in Chicago during the month of April was 3.63 inches, with a standard deviation of 1.63 inches, according to data obtained from the Climate Diagnostics Center of the National Oceanic and Atmospheric Administration. Construct a 95% confidence interval for the mean amount of rainfall in Chicago during the month of April. Interpret the interval.

**11. Concentration of Dissolved Organic Carbon** The following data represent the concentration of organic carbon (mg/l) collected from organic soil:

| 22.74 | 29.8 | 27.1 | 16.51 | 6.51 |
|-------|------|------|-------|------|
| 8.81 | 5.29 | 20.46 | 14.9 | 33.67 |
| 30.91 | 14.86 | 15.91 | 15.35 | 9.72 |
| 19.8 | 14.86 | 8.09 | 17.9 | 18.3 |
| 5.2 | 11.9 | 14 | 7.4 | 17.5 |
| 10.3 | 11.4 | 5.3 | 15.72 | 20.46 |
| 16.87 | 15.42 | 22.49 | | |

*Source:* Lisa Emili, Ph.D. Candidate, University of Waterloo

Construct a 99% confidence interval for the mean concentration of dissolved organic carbon collected from organic soil. Interpret the interval. (*Note:* $\bar{x}$ = 15.92 mg/l and $s$ = 7.38 mg/l.)

**12. Concentration of Dissolved Organic Carbon** The following data represent the concentration of organic carbon (mg/l) collected from mineral soil:

| 8.5 | 3.91 | 9.29 | 21 | 10.89 |
|------|-------|-------|-------|-------|
| 10.3 | 11.56 | 7 | 3.99 | 3.79 |
| 5.5 | 4.71 | 7.66 | 11.72 | 11.8 |
| 8.05 | 10.72 | 21.82 | 22.62 | 10.74 |
| 3.02 | 7.45 | 11.33 | 7.11 | 9.6 |
| 12.57 | 12.89 | 9.81 | 17.99 | 21.4 |
| 8.37 | 7.92 | 17.9 | 7.31 | 16.92 |
| 4.6 | 8.5 | 4.8 | 4.9 | 9.1 |
| 7.9 | 11.72 | 4.85 | 11.97 | 7.85 |
| 9.11 | 8.79 | | | |

*Source:* Lisa Emili, Ph.D. Candidate, University of Waterloo

(a) Construct a 99% confidence interval for the mean concentration of dissolved organic carbon collected from mineral soil. Interpret the interval. (*Note:* $\bar{x}$ = 10.03 mg/l and $s$ = 4.98 mg/l.)

(b) Compare the 99% confidence interval computed for organic soil (Problem 11) to the 99% confidence interval computed for mineral soil. Does there appear to be a difference between the concentration levels for the two soil types?

**13. Wheat Pennies** "Wheat" pennies are pennies that were minted before 1954. The data to the left represent the weights (in grams) of a random sample of 15 wheat pennies:

(a) Because the sample size is small, we must verify that the weights of the pennies are normally distributed and that the sample does not contain any outliers. The normal probability plot and boxplot are shown in the accompanying diagrams. Are the conditions for constructing a *t*-interval satisfied?

| 3.01 | 3.06 | 3.13 | 3.00 | 3.05 |
|------|------|------|------|------|
| 3.07 | 3.02 | 2.98 | 3.04 | 3.02 |
| 3.06 | 3.03 | 3.07 | 3.04 | 2.96 |

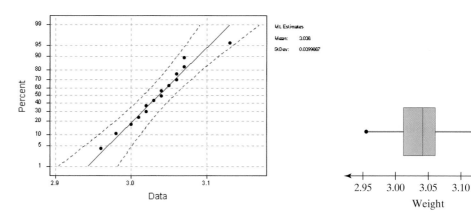

(b) Construct a 90% confidence interval for the mean weight of a "wheat" penny. (*Note:* $\bar{x}$ = 3.036 grams and $s$ = 0.0414 gram.)

(c) What could be done to increase the accuracy of the interval without changing the level of confidence?

**18. Red Blood Cell Mass** The following data represent the red blood cell mass (in millimeters) of 15 rats in a flight group and 15 rats in a ground control group. The red blood cell mass was measured immediately after the flight group returned from space on the Spacelab Life Sciences 2.

| Flight Group | Ground Control Group |
|---|---|
| 8.59 8.64 6.87 7.89 7.00 | 8.65 6.99 7.62 7.33 7.14 |
| 8.80 6.39 7.54 7.43 7.21 | 8.40 8.55 9.88 9.66 7.44 |
| 9.79 6.85 8.23 8.03 9.30 | 8.58 9.14 11.78 8.70 9.94 |

*Source:* NASA Life Sciences Database

(a) Verify that each group's red blood cell mass is normally distributed with no outliers.
(b) Compute a 95% confidence interval for the population mean of each group.
(c) From the confidence intervals, does there appear to be any difference in the red blood cell mass?

---

*In Problems 19–24, construct a 95% z-interval or t-interval about the population mean, μ, whichever is appropriate. If neither can be constructed, state the reason.*

**19. Height of Males** The heights of 20–29-year-old males are known to be normally distributed with $\sigma = 2.9$ inches. A simple random sample of $n = 15$ males 20–29 years old results in the following data:

| | | | | |
|---|---|---|---|---|
| 65.5 | 72.3 | 68.2 | 65.6 | 68.8 |
| 66.7 | 69.6 | 72.6 | 72.9 | 67.5 |
| 71.8 | 73.8 | 70.7 | 67.9 | 73.9 |

**20. Gestation Period** The gestation period of humans is normally distributed with $\sigma = 16$ days. A simple random sample of $n = 12$ live births results in the following data:

| | | | | | |
|---|---|---|---|---|---|
| 266 | 270 | 277 | 278 | 258 | 275 |
| 261 | 260 | 270 | 269 | 252 | 277 |

**21. Officer Friendly** A police officer hides behind a billboard in order to catch speeders. The following data represent the number of minutes he needs to wait before first observing a car that is exceeding the speed limit by more than 10 miles per hour on 10 randomly selected days:

| | | | | |
|---|---|---|---|---|
| 1.0 | 5.4 | 0.8 | 14.1 | 0.5 |
| 0.9 | 3.9 | 0.4 | 1.0 | 3.9 |

**22. Mutual Fund Performance** The following data represent the three-year rate of return for a random sample of 15 large-cap growth mutual funds as of 8/26/2000:

| | | | | |
|---|---|---|---|---|
| 26.8 | 28.5 | 21.3 | 15.6 | 32.5 |
| 45.4 | 36.6 | 21.5 | 28.7 | 25.4 |
| 38.9 | 30.8 | 18.6 | 23.5 | 25.5 |

**23. Pulse** Fifteen randomly selected women were asked to work on the stair master for three minutes. After the three minutes, their pulses were measured and the following data were obtained:

| | | | | |
|---|---|---|---|---|
| 117 | 102 | 98 | 100 | 116 |
| 113 | 91 | 92 | 96 | 136 |
| 134 | 126 | 104 | 113 | 102 |

**24. Law School Salaries** A random sample of recent graduates of law school was conducted in which the graduates were asked to report their starting salary. The data, based upon results reported by the National Association for Law Placement, are as follows:

| | | | | |
|---|---|---|---|---|
| 75,000 | 49,000 | 79,000 | 81,000 | 38,000 |
| 36,500 | 39,000 | 41,500 | 131,000 | 45,500 |
| 92,000 | 62,500 | 68,000 | 37,500 | 39,500 |

**25. The Effect of Outliers** The following data represent the asking price of a simple random sample of homes for sale in Houston, TX in February 2002:

| | | | |
|---|---|---|---|
| 149,900 | 154,900 | 155,500 | 270,000 |
| 259,000 | 270,000 | 166,000 | 229,000 |
| 153,000 | 177,500 | 185,000 | 875,000 |

(a) Construct a boxplot to identify the outlier.
(b) Construct a 99% confidence interval with the outlier included.
(c) Construct a 99% confidence interval with the outlier removed.
(d) Comment on the effect the outlier has on the confidence interval.

**26. The Effect of Outliers** The following data represent the age at which a simple random sample of Chicagoans died in February 2002:

| 67 | 78 | 60 | 73 | 90 | 59 | 68 |
| 86 | 75 | 99 | 80 | 73 | 18 | 39 |

*Source: Chicago Tribune*

(a) Construct a boxplot to identify the outlier.
(b) Construct a 90% confidence interval with the outlier included.
(c) Construct a 90% confidence interval with the outlier removed.
(d) Comment on the effect the outlier has on the confidence interval.

**27. Simulation** IQ scores based upon the Wechsler Intelligence Scale for Children (WIAT) are known to be normally distributed with $\mu = 100$ and $\sigma = 15$.

(a) Simulate obtaining 100 samples of size $n = 15$ from this population.
(b) Obtain the sample mean and standard deviation for each of the 100 samples.
(c) Construct 95% $t$-intervals for each of the 100 samples.
(d) How many of the intervals do you expect to include the population mean? How many actually contain the population mean?

**28. Simulation** Suppose the arrival of cars at Burger King's drive-through follows a Poisson process with $\mu = 4$ cars every 10 minutes.

(a) Simulate obtaining 30 samples of size $n = 35$ from this population.
(b) Obtain the sample mean and standard deviation for each of the 30 samples.
(c) Construct 90% $t$-intervals for each of the 30 samples.
(d) How many of the intervals do you expect to include the population mean? How many actually contain the population mean?

**29. The Effect of Nonnormal Data** The exponential distribution is used to model the time to failure for electronic components and time waiting in line. The figure in Problem 35 from Section 8.1 shows that the density function for an exponential distribution with $\mu = 5$ is skewed right.

(a) Use Minitab or some other statistical software to generate 100 simple random samples of size $n = 6$ from a population that follows an exponential distribution with $\mu = 5$.
(b) Construct 95% confidence intervals for each of the 100 samples. Comment on the margin of error for each sample. What might cause the differences in the margin of error among the various samples?
(c) Use Minitab or some other statistical software to randomly generate 100 simple random samples of size $n = 36$ from a population that follows an exponential distribution with $\mu = 5$.
(d) Construct 95% confidence intervals for each of the 100 samples. Comment on the margin of error for each sample. What might cause the differences in the margin of error among the various samples?

## Technology Step-by-Step
### Confidence Intervals about $\mu$, $\sigma$ Unknown

**TI-83 Plus**   *Step 1:* If necessary, enter raw data in L1.
*Step 2:* Press STAT, highlight TESTS, and select 8: TInterval.
*Step 3:* If the data are raw, highlight DATA—make sure List1 is set to L1 and Freq to 1. If summary statistics are known, highlight STATS and enter the summary statistics.
*Step 4:* Enter the confidence level following C-Level:.
*Step 5:* Highlight Calculate; press ENTER.

**MINITAB**   *Step 1:* Enter raw data in column C1.
*Step 2:* Select the **Stat** menu, highlight **Basic Statistics**, then highlight **1-Sample t** ....
*Step 3:* Enter C1 in the cell marked "Variables." Select Confidence Interval, and enter a confidence level. Click OK.

**Excel**   *Step 1:* If necessary, enter raw data in column A.
*Step 2:* Load the PHStat Add-in.
*Step 3:* Select the **PHStat menu**, highlight **Confidence Intervals** ..., then highlight **Estimate for the mean, sigma unknown** ....
*Step 4:* Enter the confidence level. If the summary statistics are known, click "Sample statistics known" and enter the sample size, sample mean, and sample standard deviation. If summary statistics are unknown, click "Sample statistics unknown." With the cursor in the "Sample cell range" cell, highlight the data in column A. Click OK.

## 8.3   Confidence Intervals about a Population Proportion

**Preparing for This Section**   Before getting started, review the following:

✓ Using the binomial probability distribution and the empirical rule to identify unusual results (Section 6.2, pp. 261–262).

**Objectives**    Obtain a point estimate for the population proportion

② Obtain and interpret a confidence interval for the population proportion

③ Determine the sample size for estimating a population proportion

Probably the most frequently reported confidence interval is one involving the proportion of a population. Researchers are often interested in estimating the proportion of the population that has a certain characteristic. For example, in a poll conducted by the Gallup Organization in early January, 2001, a random sample of 1004 Americans resulted in 13% of the respondents stating that morality/dishonesty is the biggest problem facing America today. The poll had a margin of error of ± 3% with 95% confidence. Based upon discussions in Sections 8.1 and 8.2, we know that this means the pollsters are 95% confident the true proportion of Americans who believe morality/dishonesty is the biggest problem facing America today is somewhere between 10% and 16% (13% ± 3%).

In this section, we will discuss the techniques for estimating the population proportion, *p*. In addition, we present methods for determining the sample size required to estimate the population proportion.

**In Your Own Words**

The symbol ± is read "plus or minus." It means "to add and subtract the quantity following the ± symbol."

① **Point Estimate of the Population Proportion**

Recall that a point estimate is an unbiased estimator of the parameter. The best point estimate is the one that has minimum variance among all unbiased estimators.

**In Your Own Words**

$\hat{p}$ can be thought of as the probability that a randomly selected individual in the sample has a certain characteristic.

> **Point Estimate of a Population Proportion**
>
> Suppose a simple random sample of size *n* is obtained from a population in which each individual either does or does not have a certain characteristic. The best point estimate of *p*, denoted $\hat{p}$ (read "p-hat"), the proportion of the population with a certain characteristic, is given by
>
> $$\hat{p} = \frac{x}{n} \qquad (1)$$
>
> where *x* is the number of individuals in the sample with the specified characteristic.

▶ **EXAMPLE 1**   **Obtaining a Point Estimate of a Population Proportion**

*Problem:* In a poll conducted May 7–10, 2000, by ABC News, a simple random sample of 1068 American adults was asked, "Have you ever been shot at?" Of the 1068 adults, 96 responded yes. Obtain a point estimate for the population proportion of Americans who have been shot at.

*Approach:* The point estimate of the population proportion is $\hat{p} = \dfrac{x}{n}$, where *x* = 96 and *n* = 1068.

*Solution:* Substituting into Formula (1), we obtain $\hat{p} = \dfrac{x}{n} = \dfrac{96}{1068} = 0.090 = 9.0\%$. We estimate that 9.0% of Americans have been shot at. ◀◀

**NW** *Now Work Problem 7(a).*

## ② Confidence Intervals for the Population Proportion

The point estimate obtained in Example 1 is a single estimate of the unknown parameter, $p$. However, just as there is a sampling distribution associated with the sample mean, $\bar{x}$, there is a sampling distribution associated with $\hat{p}$. We recognize it is unlikely that the sample proportion will equal the population proportion (because of sampling and nonsampling error). Therefore, rather than reporting the value of the sample proportion alone, it is preferred to report an interval about the sample proportion.

Recall from Section 6.2 the distribution of $X$, the number of successes in $n$ trials of a binomial experiment, is approximately normal, with mean $\mu = np$ and standard deviation $\sigma = \sqrt{np(1 - p)}$, provided that $np(1 - p) \geq 10$. If we standardize the binomial random variable $X$, we obtain

$$Z = \frac{x - \mu}{\sigma} = \frac{x - np}{\sqrt{np(1 - p)}}$$

Divide the numerator and denominator of the standardized form by $n$, to obtain

$$Z = \frac{x - np}{\sqrt{np(1 - p)}} = \frac{\dfrac{x - np}{n}}{\dfrac{\sqrt{np(1 - p)}}{n}} = \frac{\dfrac{x}{n} - p}{\sqrt{\dfrac{p(1 - p)}{n}}} = \frac{\hat{p} - p}{\sqrt{\dfrac{p(1 - p)}{n}}}$$

and we have the following result:

*Theorem*

**Sampling Distribution of $\hat{p}$**

For a simple random sample of size $n$ such that $n \leq 0.05N$ (that is, the sample size is no more than 5% of the population size), the sampling distribution of $\hat{p}$ is approximately normal with mean $\mu_{\hat{p}} = p$ and standard deviation $\sigma_{\hat{p}} = \sqrt{\dfrac{p(1 - p)}{n}}$, provided that $np(1 - p) \geq 10$.

We use the sampling distribution of $\hat{p}$ to construct a confidence interval about $p$.

**Constructing a $(1 - \alpha) \cdot 100\%$ Confidence Interval for a Population Proportion**

Suppose a simple random sample of size $n$ is taken from a population. A $(1 - \alpha) \cdot 100\%$ confidence interval for $p$ is given by the following quantities:

$$\text{Lower bound: } \hat{p} - z_{\alpha/2} \cdot \sqrt{\frac{\hat{p}(1 - \hat{p})}{n}}$$

$$\text{Upper bound: } \hat{p} + z_{\alpha/2} \cdot \sqrt{\frac{\hat{p}(1 - \hat{p})}{n}} \tag{2}$$

**Note:**  It must be the case that $n\hat{p}(1 - \hat{p}) \geq 10$ to construct this interval.

Notice we use $\hat{p}$ in place of $p$ in the standard deviation. This is because $p$ is unknown and $\hat{p}$ is the best point estimate of $p$.

▶ EXAMPLE 2   Constructing a Confidence Interval for a Population Proportion

*Problem:*  In a poll conducted May 7–10, 2000, by ABC News, a simple random sample of 1068 American adults was asked, "Have you ever been shot at?" Of the 1068 adults, 96 responded yes. Obtain a 95% confidence interval about the population proportion, $p$.

*Approach:*

**Step 1:** Compute the value of $\hat{p}$.
**Step 2:** We can compute a 95% confidence interval about $p$ provided that $n\hat{p}(1 - \hat{p}) \geq 10$.
**Step 3:** Determine the critical value, $z_{\alpha/2}$.
**Step 4:** Determine the lower and upper bounds of the confidence interval.

*Solution:*

**Step 1:** From Example 1, we have that $\hat{p} = 0.090$.
**Step 2:** $n\hat{p}(1 - \hat{p}) = 1068(0.09)(1 - 0.09) = 87.4692 \geq 10$. We can proceed to construct the confidence interval.
**Step 3:** Because we want a 95% confidence interval, we have $\alpha = 0.05$, so $z_{\alpha/2} = z_{0.05/2} = z_{0.025} = 1.96$.
**Step 4:** Substituting into Formula (2) with $n = 1068$, we obtain the lower and upper bounds of the confidence interval:

Lower bound:

$$\hat{p} - z_{\alpha/2} \cdot \sqrt{\frac{\hat{p}(1 - \hat{p})}{n}} = 0.090 - 1.96 \cdot \sqrt{\frac{0.090(1 - 0.090)}{1068}} = 0.090 - 0.017 = 0.073$$

Upper bound:

$$\hat{p} + z_{\alpha/2} \cdot \sqrt{\frac{\hat{p}(1 - \hat{p})}{n}} = 0.090 + 1.96 \cdot \sqrt{\frac{0.090(1 - 0.090)}{1068}} = 0.090 + 0.017 = 0.107$$

Given the results of the survey, we are 95% confident that the proportion of Americans who have been shot at is between 0.073 and 0.107. ◄◄

**Figure 16**

**Using Technology:** Graphing calculators with advanced statistical features and statistical software can be used to construct confidence intervals about the population proportion. Figure 16 shows the results of Example 2 from a TI-83 Plus graphing calculator.

It is important to remember the correct interpretation of a confidence interval. The statement "95% confident" means that, if 1000 samples of size 1068 were taken, about 950 of the intervals would contain the parameter $p$ and about 50 would not. Unfortunately, we cannot know whether the interval we computed in Example 2 is one of the 950 intervals that contains $p$ or one of the 50 that does not contain $p$.

Often, polls will report their results by giving the value of $\hat{p}$ obtained from the sample data along with the margin of error, rather than reporting a confidence interval. In Example 2, ABC News might say "In a survey conducted May 7–10, 2000, 9% of adult Americans stated that they have been shot at. The survey results had a margin of error of 1.7%."

**Caution**

Beware of surveys that do not report a margin of error. Survey results should also report sample size, sampling technique, and the population that was being studied.

**NW** *Now Work Problems 7(b) and (c).*

❸ **Determining Sample Size**

In Section 8.1, we introduced a method for determining the sample size $n$ required to estimate the sample mean within a certain margin of error with a specified level of confidence. This formula was obtained by solving the margin of error, $E = z_{\alpha/2} \cdot \dfrac{\sigma}{\sqrt{n}}$, for $n$. We can follow the same approach to determine sample size when estimating a population proportion. In Formula (2), we notice that the margin of error, $E$, is given by

$E = z_{\alpha/2} \cdot \sqrt{\dfrac{\hat{p}(1 - \hat{p})}{n}}$. We solve the margin of error for $n$ and obtain

**Figure 17**

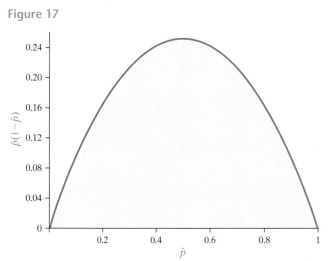

$$n = \hat{p}(1 - \hat{p})\left(\frac{z_{\alpha/2}}{E}\right)^2$$

The problem with this formula is that it depends on $\hat{p}$. Moreover, $\hat{p} = \frac{x}{n}$ depends upon the sample size, $n$, which is what we are trying to determine in the first place! How do we resolve this issue? There are two possibilities: (1) We could use an estimate of $p$ based upon a pilot study or an earlier study, or (2) we could let $\hat{p} = 0.5$ to maximize the value of $\hat{p}(1 - \hat{p})$ (as shown in Figure 17). By maximizing this value, we obtain the largest possible value of $n$ for a given level of confidence and a given margin of error.

The disadvantage of the second option is that it could lead to a larger sample size than is necessary. Because of the time and expense of sampling, it is desirable to avoid too large a sample.

---

**Sample Size for Estimating the Population Proportion $p$**

The sample size required to obtain a $(1 - \alpha) \cdot 100\%$ confidence interval for $p$ with a margin of error $E$ is given by

$$n = \hat{p}(1 - \hat{p})\left(\frac{z_{\alpha/2}}{E}\right)^2 \qquad (3)$$

(rounded up to the next integer) where $\hat{p}$ is a prior estimate of $p$. If a prior estimate of $p$ is unavailable, the sample size required is

$$n = 0.25\left(\frac{z_{\alpha/2}}{E}\right)^2 \qquad (4)$$

rounded up to the next integer.

---

**✒ In Your Own Words**

There are two formulas for determining sample size when estimating the population proportion. Formula (3) requires a prior estimate of $p$; Formula (4) does not.

▶ **EXAMPLE 3**    **Determining Sample Size**

*Problem:* A sociologist wishes to estimate the percentage of the United States population living in poverty. What size sample should be obtained if she wishes the estimate to be within 2 percentage points with 99% confidence if

**(a)** she uses the 1999 estimate of 11.8% obtained from the Current Population Survey.

**(b)** she does not use any prior estimates.

*Approach:* In both cases, we have $E = 0.02$* $(2\% = 0.02)$ and $z_{\alpha/2} = z_{0.01/2} = z_{0.005} = 2.575$. To answer part (a), we let $\hat{p} = 0.118$ in Formula (3). To answer part (b), we use Formula (4).

*Solution:*

**(a)** Substituting $E = 0.02$, $z_{0.005} = 2.575$, and $\hat{p} = 0.118$ into Formula (3), we obtain

$$n = \hat{p}(1 - \hat{p})\left(\frac{z_{\alpha/2}}{E}\right)^2 = 0.118(1 - 0.118)\left(\frac{2.575}{0.02}\right)^2 = 1725.2$$

We round this value up to 1726, so she must survey 1726 randomly selected residents of the United States.

*The margin of error should always be expressed as a decimal when using Formulas (3) and (4).

**(b)** Substituting $E = 0.02$ and $z_{0.005} = 2.575$ into Formula (4), we obtain

$$n = 0.25\left(\frac{z_{\alpha/2}}{E}\right)^2 = 0.25\left(\frac{2.575}{0.02}\right)^2 = 4144.1$$

We round this value up to 4145, so she must survey 4145 randomly selected residents of the United States.  ◄◄

We can see the effect of not having a prior estimate of $p$: The required sample size more than doubled!

**NW** *Now Work Problem 15.*

## 8.3 Assess Your Understanding

### Concepts and Vocabulary

1. What is the best point estimate of a population proportion?
2. What are the requirements that must be satisfied in order to construct a confidence interval about a population proportion?
3. When determining the sample size required in order to obtain an estimate for a population proportion, is the researcher better off using a prior estimate of $p$ or no prior estimate of $p$? Why? List some pros and cons for each scenario.

### Exercises

#### • Basic Skills

*In Problems 1–6, construct a confidence interval of the population proportion at the given level of confidence.*

1. $x = 30, n = 150$, 90% confidence
2. $x = 80, n = 200$, 98% confidence
3. $x = 120, n = 500$, 99% confidence
4. $x = 400, n = 1200$, 95% confidence
5. $x = 860, n = 1100$, 94% confidence
6. $x = 540, n = 900$, 96% confidence

#### • Applying the Concepts

7. **Lipitor** The drug Lipitor is meant to lower cholesterol levels. In a clinical trial of 863 patients who received 10 mg doses of Lipitor daily, 47 reported a headache as a side effect.
   **NW**
   (a) Obtain a point estimate for the population proportion of Lipitor users who will experience a headache as a side effect.
   (b) Verify that the requirements for constructing a confidence interval about $\hat{p}$ are satisfied.
   (c) Construct a 90% confidence interval for the population proportion of Lipitor users who will report a headache as a side effect.
   (d) Interpret the confidence interval.

8. **Pepcid** A study of 74 patients with ulcers was conducted in which they were prescribed 40 mg of Pepcid. After 8 weeks, 58 reported confirmed ulcer healing.
   (a) Obtain a point estimate for the proportion of patients with ulcers receiving Pepcid who will have confirmed ulcer healing.
   (b) Verify that the requirements for constructing a confidence interval about $\hat{p}$ are satisfied.

   (c) Construct a 99% confidence interval for the proportion of patients with ulcers receiving Pepcid who will have confirmed ulcer healing.
   (d) Interpret the confidence interval.

9. **Workplace Violence** In a September 1999 poll conducted by the Gallup Organization, 230 of 1000 randomly selected adults aged 18 years or older stated that they knew someone capable of committing an act of violence in their workplace.
   (a) Obtain a point estimate for the proportion of adults 18 years old or older who know someone capable of committing an act of violence in their workplace.
   (b) Verify that the requirements for constructing a confidence interval about $\hat{p}$ are satisfied.
   (c) Construct a 98% confidence interval for the proportion of adults 18 years old or older who know someone capable of committing an act of violence in their workplace.
   (d) Interpret the confidence interval.

10. **Partial Birth Abortions** In an October 2000 poll conducted by the Gallup Organization, 639 of 1014 randomly selected adults aged 18 years old or older stated they would support a law that would make it illegal to perform a specific abortion procedure conducted in the last six months of pregnancy known as a "partial birth abortion," except in cases necessary to save the life of the mother.

   (a) Obtain a point estimate for the proportion of adults aged 18 years old or older who would support a law that would make it illegal to perform a specific abortion procedure conducted in the last six months of pregnancy known as a "partial birth abortion," except in cases necessary to save the life of the mother.

   (b) Verify that the requirements for constructing a confidence interval about $\hat{p}$ are satisfied.

   (c) Construct a 95% confidence interval for the proportion of adults aged 18 years old or older who would favor such a law.

   (d) Interpret the confidence interval.

11. **Breast-feeding** In a 1995 national survey conducted by the Centers for Disease Control, 5988 of 10,847 women who had a child between 1990 and 1993 indicated having breast-fed the baby.

   (a) Verify that the requirements for constructing a confidence interval about $\hat{p}$ are satisfied.

   (b) Construct a 90% confidence interval for the proportion of women who had a child between 1990 and 1993 who breast-fed their baby. Interpret this interval.

   (c) Construct a 99% confidence interval for the proportion of women who had a child between 1990 and 1993 who breast-fed their baby. Interpret this interval.

   (d) What is the effect of increasing the level of confidence on the width of the interval?

12. **Child Safety Seats** In a Harris Poll conducted February 9, 2000, 1247 of 2208 randomly selected Americans said they judged that state laws governing child safety restraint in vehicles should be strengthened.

   (a) Verify that the requirements for constructing a confidence interval about $\hat{p}$ are satisfied.

   (b) Construct a 92% confidence interval for the proportion of Americans who judge that state laws governing child safety restraint in vehicles should be strengthened. Interpret this interval.

   (c) Construct a 96% confidence interval for the proportion of Americans who judge that state laws governing child safety restraint in vehicles should be strengthened. Interpret this interval.

   (d) What is the effect of increasing the level of confidence on the width of the interval?

13. **Obesity** In a study conducted by the Centers for Disease Control in 1999, 1950 out of 10,485 randomly selected 30–39-year-old Americans were found to be obese.

   (a) Construct a 95% confidence interval, for the proportion of 30–39-year-old Americans who are obese, based upon the 1999 study. Interpret this interval.

   (b) When the same study was performed in 1998, 1937 out of 11,464 randomly selected 30–39-year-old Americans were found to be obese. On the basis of the 1998 study, construct a 95% confidence interval for the proportion of 30–39-year-old Americans who are obese. Interpret this interval.

   (c) Does it appear that the percentage of 30–39-year-old Americans who are obese increased from 1998 to 1999? Why?

14. **Obesity** In a study conducted by the Centers for Disease Control in 1999, 3506 out of 17,020 randomly selected Americans with high school diplomas were found to be obese.

   (a) Construct a 95% confidence interval, for the proportion of Americans with high school diplomas who are obese, based upon the 1999 study. Interpret this interval.

   (b) When the same study was performed on Americans with 4 or more years of college, 2084 out of 14,572 were found to be obese. Construct a 95% confidence interval for the proportion of Americans with 4 or more years of college who are obese. Interpret this interval.

   (c) Does it appear that the percentage of Americans with high school diplomas who are obese is different from the percentage of Americans with 4 or more years of college who are obese? Why?

15. **Child Care** A child psychologist wishes to estimate the percentage of fathers who watch their preschool-aged child when the mother works. What size sample should be obtained if she wishes the estimate to be within 3 percentage points with 99% confidence if

   (a) she uses a 1995 estimate obtained from the U.S. Census Bureau of 18.5%?

   (b) she does not use any prior estimates?

16. **Home Ownership** An urban economist wishes to estimate the percentage of Americans who own their house. What size sample should be obtained if he wishes the estimate to be within 2 percentage points with 90% confidence if

   (a) he uses an estimate of 67.5% from the fourth quarter of 2000 obtained from the U.S. Census Bureau?

   (b) he does not use any prior estimates?

17. **Affirmative Action** A sociologist wishes to conduct a poll in order to estimate the percentage of Americans who judge that affirmative action programs for minorities and women should be continued at some level. What size sample should be obtained if she wishes the estimate to be within 3 percentage points with 95% confidence if

   (a) she uses a 1999 estimate of 80% obtained from a Time/CNN poll?

   (b) she does not use any prior estimates?

18. **Affirmative Action** A sociologist wishes to conduct a poll in order to estimate the percentage of Americans who judge that affirmative action programs should require businesses to hire a specific number or quota of minorities and women. What size sample should be obtained if she wishes the estimate to be within 4 percentage points with 90% confidence if

   (a) she uses a 1999 estimate of 37% obtained from a Time/CNN poll?

   (b) she does not use any prior estimates?

**19. Playing the Horses** A researcher wishes to conduct a poll in order to estimate the percentage of Americans aged 18 years old or older who placed a bet on a horse race last year. What size sample should be obtained if she wishes the estimate to be within 1.5 percentage points with 98% confidence if

(a) she uses a 1998 estimate of 7% obtained from the National Gambling Impact Study Commission?

(b) she does not use any prior estimates?

**20. Credit Card Debt** A school administrator is concerned about the amount of credit card debt college students have. She wishes to conduct a poll in order to estimate the percentage of full-time college students who have credit card debt of $2000 or more. What size sample should be obtained if she wishes the estimate to be within 2.5 percentage points with 94% confidence, if

(a) a pilot study indicates the percentage is 34%?

(b) no prior estimates are used?

**21. The Death Penalty** In a Harris Poll conducted in July, 2000, 64% of the people polled answered yes to the following question: "Do you believe in capital punishment, that is the death penalty, or are you opposed to it?" The margin of error in the poll was $\pm 3\% = \pm 0.03$, and the estimate was made with 95% confidence. How many people were surveyed?

**22. Own a Gun?** In a Harris Poll conducted in May, 2000, 39% of the people polled answered "yes" to the following question: "Do you happen to have in your home or garage any guns or revolvers?" The margin of error in the poll was $\pm 3\% = \pm 0.03$ and the estimate was made with 95% confidence. How many people were surveyed?

**23. Simulation** The following exercise is meant to illustrate the normality of the distribution of the sample proportion, $\hat{p}$.

(a) Using Minitab or some other statistical spreadsheet, randomly generate 200 samples of size 75 from a population with $p = 0.3$. Store the number of successes in a column called $x$.

(b) Determine $\hat{p}$ for each of the 200 samples by computing $x/75$. Store each $\hat{p}$ in a column called *phat*.

(c) Draw a histogram of the 200 estimates of $p$. Comment on the shape of the distribution.

(d) Compute the mean and standard deviation of the sampling distribution of $\hat{p}$ in the simulation.

(e) Compute the theoretical mean and standard deviation of the sampling distribution of $\hat{p}$. Compare the theoretical results to the results of the simulation. Are they close?

(f) Compute the confidence interval for each of the 200 samples. How many of the intervals do you expect to contain $p$? How many actually contain $p$?

**24. The 2000 Presidential Election** The Gallup Organization conducted a poll of 2350 likely voters just prior to the 2000 presidential election. The results of the survey indicated that George W. Bush would receive 48% of the popular vote and Al Gore would receive 46% of the popular vote. The margin of error was reported to be 2%. The Gallup Organization reported that the race was too close to call. Use the concept of a confidence interval to explain what this means.

**25. Finite Population Correction Factor** In this section, we assumed that the sample size was less than 5% of the size of the population. When sampling from a finite population in which $n > 0.05N$ without replacement, the standard deviation of the distribution of $\hat{p}$ is given by

$$s_{\hat{p}} = \sqrt{\frac{\hat{p}(1-\hat{p})}{n-1} \cdot \left(\frac{N-n}{N}\right)}$$

where $N$ is the size of the population. Suppose a survey is conducted at a college having an enrollment of 6,502 students. The student council wants to estimate the percentage of students in favor of establishing a student union. In a random sample of 500 students, it was determined that 410 were in favor of establishing a student union.

(a) Obtain a point estimate of the proportion of students at the college in favor of establishing a student union.

(b) Calculate the standard deviation of the sampling distribution of $\hat{p}$.

(c) Obtain a 95% confidence interval for the proportion of students in favor of establishing a student union. Interpret the interval.

---

**Technology Step-by-Step**
Confidence Intervals about *p*

**TI-83 Plus**   **Step 1:** Press STAT, highlight TESTS, and select `A:1-PropZInt..`

**Step 2:** Enter the values of $x$ and $n$.

**Step 3:** Enter the confidence level following `C-Level:`

**Step 4:** Highlight `Calculate`; press ENTER.

**MINITAB**   **Step 1:** If you have raw data, enter the data in column C1.

**Step 2:** Select the **Stat** menu, highlight **Basic Statistics**, then highlight **1 Proportion** . . . .

**Step 3:** Enter C1 in the cell marked "Sample in Columns" if you have raw data. If you have summary statistics, click "Summarized data" and enter the number of trials, $n$, and the number of successes, $x$.

**Step 4:** Hit the Options button. Select a Confidence Level. Click "Use test based on a normal distribution" (provided that the assumptions stated are satisfied). Click OK twice.

| | |
|---|---|
| ***Excel*** | ***Step 1:*** Load the PHStat Add-in. |
| | ***Step 2:*** Select the **PHStat menu**, highlight **Confidence Intervals** ..., then highlight **Estimate for the proportion** .... |
| | ***Step 3:*** Enter the confidence level. Enter the sample size, $n$, and the number of successes, $x$. Click OK. |

## CHAPTER 8 REVIEW

### Summary

In this chapter, we discussed estimation methods. We estimated the values of the parameters $\mu$ and $p$. We started by estimating the population mean under the assumption that the population standard deviation was known. This assumption allowed us to construct a confidence interval about $\mu$ by utilizing the standard normal distribution. In order to construct the interval, we required either that the population from which the sample was drawn be normal or that the sample size, $n$, be greater than or equal to 30. In addition, the sampling method had to be simple random sampling. With these requirements satisfied, the $(1 - \alpha) \cdot 100\%$ confidence interval about $\mu$ is $\bar{x} \pm z_{\alpha/2} \cdot \dfrac{\sigma}{\sqrt{n}}$. We have $(1 - \alpha) \cdot 100\%$ confidence that the unknown value of $\mu$ lies within the interval.

In Section 8.2, we dropped the assumption that the population standard deviation be known. With $\sigma$ unknown, the sampling distribution of $t = \dfrac{\bar{x} - \mu}{s/\sqrt{n}}$ follows Student's $t$-distribution with $n - 1$ degrees of freedom. We use the

$t$-distribution to construct the confidence interval about $\mu$. In order to construct this interval, either the population from which the sample was drawn must be normal or the sample size must be large. In addition, the sampling method must be simple random sampling. With these requirements satisfied, the $(1 - \alpha) \cdot 100\%$ confidence interval about $\mu$ is $\bar{x} \pm t_{\alpha/2} \cdot \dfrac{s}{\sqrt{n}}$, where $t_{\alpha/2}$ has $n - 1$ degrees of freedom. This means that the procedure results in an interval that contains $\mu$, the population mean, $(1 - \alpha) \cdot 100\%$ of the time.

In Section 8.3, a confidence interval regarding the population proportion, $p$, was constructed. This confidence interval is constructed about the binomial parameter $p$. Provided that the sample is obtained via simple random sampling and that $n\hat{p}(1 - \hat{p}) \geq 10$, the $(1 - \alpha) \cdot 100\%$ confidence interval about $p$ is $\hat{p} \pm z_{\alpha/2} \cdot \sqrt{\dfrac{\hat{p}(1 - \hat{p})}{n}}$. We have $(1 - \alpha) \cdot 100\%$ confidence that the unknown value of $p$ lies within the interval.

### Formulas

**Confidence Intervals**

- A $(1 - \alpha) \cdot 100\%$ confidence interval about $\mu$ with $\sigma$ known is $\bar{x} \pm z_{\alpha/2} \cdot \dfrac{\sigma}{\sqrt{n}}$, provided that the population from which the sample was drawn is normal or that the sample size is large ($n \geq 30$).

- A $(1 - \alpha) \cdot 100\%$ confidence interval about $\mu$ with $\sigma$ unknown is $\bar{x} \pm t_{\alpha/2} \cdot \dfrac{s}{\sqrt{n}}$ where $t_{\alpha/2}$ has $n - 1$ degrees of freedom, provided that the population from which the sample was drawn is normal or that the sample size is large ($n \geq 30$).

- A $(1 - \alpha) \cdot 100\%$ confidence interval about $p$ is $\hat{p} \pm z_{\alpha/2} \cdot \sqrt{\dfrac{\hat{p}(1 - \hat{p})}{n}}$, provided that $n\hat{p}(1 - \hat{p}) \geq 10$.

**Sample Size**

- To estimate the population mean within a margin of error $E$ at a $(1 - \alpha) \cdot 100\%$ level of confidence requires a sample of size $n = \left(\dfrac{z_{\alpha/2} \cdot \sigma}{E}\right)^2$ (rounded up to the next integer).

- To estimate the population proportion within a margin of error $E$ at a $(1 - \alpha) \cdot 100\%$ level of confidence requires a sample of size $n = \hat{p}(1 - \hat{p})\left(\dfrac{z_{\alpha/2}}{E}\right)^2$ (rounded up to the next integer), where $\hat{p}$ is a prior estimate of the population proportion.

- To estimate the population proportion within a margin of error $E$ at a $(1 - \alpha) \cdot 100\%$ level of confidence requires a sample of size $n = 0.25\left(\dfrac{z_{\alpha/2}}{E}\right)^2$ (rounded up to the next integer) when no prior estimate is available.

## Vocabulary

Point estimate (p. 339)
Consistent (p. 339)
Efficiency (p. 339)
Unbiased estimator (p. 339)

Confidence interval estimate (p. 340)
Level of confidence (p. 340)
Critical values (p. 343)
Robust (p. 344)

$z$-interval (p. 344)
Margin of error (p. 345)
Student's $t$-distribution (p. 357)
$t$-interval (p. 360)

## Objectives

| Section | You should be able to ... | Review Exercises |
|---|---|---|
| 8.1 | 1 Compute the point estimate of $\mu$ (p. 339) | 7(b); 8(b); 11(a); 12(a) |
| | 2 Compute confidence intervals about $\mu$ with $\sigma$ known (p. 340) | 3; 5(a)–(c); 6(a)–(c); 7(c),(d); 8(c),(d) |
| | 3 Understand the role of margin of error in constructing confidence intervals (p. 345) | 3; 4 |
| | 4 Find the sample size necessary for estimating the population mean (p. 346) | 5(d); 6(d) |
| 8.2 | 1 Know the properties of Student's $t$-distribution (p. 357) | 17–19 |
| | 2 Determine $t$-values (p. 359) | 1, 2 |
| | 3 Construct confidence intervals about $\mu$, $\sigma$ unknown (p. 360) | 9(b),(c); 10(b),(c); 11(c),(d); 12(c),(d); 13; 14 |
| | 4 Determine the method to use to compute an interval about $\mu$ (p. 363) | 3–14 |
| 8.3 | 1 Obtain a point estimate for the population proportion (p. 370) | 15(a); 16(a) |
| | 2 Obtain and interpret a confidence interval for the population proportion (p. 371) | 15(b); 16(b) |
| | 3 Determine the sample size for estimating a population proportion (p. 372) | 15(c),(d); 16(c),(d) |

## Exercises

*In Problems 1 and 2, find the critical t-value for constructing a confidence interval about a population mean at the given level of confidence for the given sample size, n.*

**1.** 99% confidence; $n = 18$

**2.** 90% confidence; $n = 27$

---

**3.** A simple random sample of size $n$ is drawn from a population that is known to be normally distributed. The sample mean, $\overline{x}$, is determined to be 54.8.

(a) Construct the 90% confidence interval about the population mean if the population standard deviation, $\sigma$, is known to be 10.5 and the sample size, $n$, is 20.

(b) Construct the 90% confidence interval about the population mean if the population standard deviation, $\sigma$, is known to be 10.5 and the sample size, $n$, is 30. How does increasing the sample size affect the width of the interval?

(c) Construct the 99% confidence interval about the population mean if the population standard deviation, $\sigma$, is known to be 10.5 and the sample size, $n$, is 20. Compare the results to those obtained in part (a). How does increasing the level of confidence affect the confidence interval?

**4.** A simple random sample of size $n$ is drawn from a population that is known to be normally distributed. The sample mean, $\overline{x}$, is determined to be 104.3 and the sample standard deviation, $s$, is determined to be 15.9.

(a) Construct the 90% confidence interval about the population mean if the sample size, $n$, is 15.

(b) Construct the 90% confidence interval about the population mean if the sample size, $n$, is 25. How does increasing the sample size affect the width of the interval?

(c) Construct the 95% confidence interval about the population mean if the sample size, $n$, is 15. Compare the results to those obtained in part (a). How does increasing the level of confidence affect the confidence interval?

**5. Tire Wear** Suppose Dunlop wishes to estimate the mean mileage for its SP4000 tire. In a random sample of 35 tires, the sample mean mileage was $\overline{x} = 62,450$ miles.

(a) Why can we say that the sampling distribution of $\overline{x}$ is approximately normal?

(b) Construct a 90% confidence interval for the mean mileage for all SP4000 tires, assuming that $\sigma = 4400$ miles. Interpret this interval.

(c) Construct a 95% confidence interval for the mean mileage for all SP4000 tires, assuming that $\sigma = 4400$ miles. Interpret this interval.

(d) How many tires would Dunlop need in order to estimate the mean mileage for all SP4000 tires within 3000 miles with 99% confidence?

**6. Talk Time on a Cell Phone** Suppose Motorola wishes to estimate the mean "talk time" for its Timeport P8160 digital phone before the battery must be recharged. In a random sample of 40 phones, the sample mean talk time was 315 minutes.

(a) Why can we say that the sampling distribution of $\overline{x}$ is approximately normal?

(b) Construct a 94% confidence interval for the mean "talk time" for all Timeport P8160 digital phones, assuming that $\sigma$ = 24 minutes.

(c) Construct a 98% confidence interval for the mean "talk time" for all Timeport P8160 digital phones, assuming that $\sigma$ = 24 minutes.

(d) How many phones would Motorola need in order to estimate the mean "talk time" for all Timeport P8160 phones within 10 minutes with 95% confidence?

7. **Math Achievement** The following data represent the mathematics achievement test scores for a random sample of 15 students who had just completed high school in Canada and in the United States, according to data obtained from the International Association for the Evaluation of Education Achievement study in 1998:

| Canada | | | | |
|---|---|---|---|---|
| 558 | 556 | 594 | 592 | 458 |
| 600 | 468 | 554 | 614 | 500 |
| 637 | 511 | 557 | 531 | 526 |

| United States | | | | |
|---|---|---|---|---|
| 583 | 402 | 510 | 432 | 500 |
| 531 | 410 | 469 | 459 | 450 |
| 465 | 483 | 553 | 446 | 559 |

(a) Verify that the scores for each country are normally distributed with no outliers.

(b) Obtain a point estimate for the population mean score of each country.

(c) Construct a 99% confidence interval for the population mean achievement score of Canada, assuming that $\sigma$ = 62.6.

(d) Construct a 99% confidence interval for the population mean achievement score of the United States, assuming that $\sigma$ = 71.5.

(e) Does it appear to be the case that Canadians scored higher than Americans on the exam? Why?

8. **Math Achievement** The following data represent the mathematics achievement test scores for a random sample of 15 male and female students who had just completed high school in the United States, according to data obtained from the International Association for the Evaluation of Education Achievement study in 1998:

| Male | | | | |
|---|---|---|---|---|
| 488 | 350 | 547 | 488 | 474 |
| 471 | 443 | 385 | 477 | 452 |
| 418 | 388 | 441 | 463 | 412 |

| Female | | | | |
|---|---|---|---|---|
| 433 | 389 | 520 | 479 | 454 |
| 563 | 411 | 458 | 398 | 337 |
| 418 | 492 | 442 | 494 | 514 |

(a) Verify that the scores for each gender are normally distributed with no outliers.

(b) Obtain a point estimate for the population mean score of each gender.

(c) Construct a 95% confidence interval for the population mean achievement score of males, assuming that $\sigma$ = 64.8.

(d) Construct a 95% confidence interval for the population mean achievement score of females, assuming that $\sigma$ = 56.9.

(e) Does there appear to be any difference between the scores of males and those of females? Why?

9. **Family Size** A random sample of 60 married couples who have been married seven years, were asked the number of children they have. The results of the survey are as follows:

| | | | | | | | | | | | |
|---|---|---|---|---|---|---|---|---|---|---|---|
| 0 | 0 | 0 | 3 | 3 | 3 | 1 | 3 | 2 | 2 | 3 | 1 |
| 3 | 2 | 4 | 0 | 3 | 3 | 3 | 1 | 0 | 2 | 3 | 3 |
| 1 | 4 | 2 | 3 | 1 | 3 | 3 | 5 | 0 | 2 | 3 | 0 |
| 4 | 4 | 2 | 2 | 3 | 2 | 2 | 2 | 2 | 3 | 4 | 3 |
| 2 | 2 | 1 | 4 | 3 | 2 | 4 | 2 | 1 | 2 | 3 | 2 |

*Note:* $\bar{x}$ = 2.27, $s$ = 1.22.

(a) What is the sampling distribution of $\bar{x}$? In other words, what is the shape, mean, and standard deviation of the distribution of $\bar{x}$? Why?

(b) Compute a 95% confidence interval for the mean number of children of all couples who have been married 7 years. Interpret this interval.

(c) Compute a 99% confidence interval for the mean number of children of all couples who have been married 7 years. Interpret this interval.

10. **Waiting in Line** The following data represent the number of cars that arrive at McDonald's drive-through between 11:50 A.M. and 12:00 noon for a random sample of Wednesdays:

*Note:* $\bar{x}$ = 4.08, $s$ = 2.12.

| | | | | | | | | | |
|---|---|---|---|---|---|---|---|---|---|
| 1 | 7 | 3 | 8 | 2 | 3 | 8 | 2 | 6 | 3 |
| 6 | 5 | 6 | 4 | 3 | 4 | 3 | 8 | 1 | 2 |
| 5 | 3 | 6 | 3 | 3 | 4 | 3 | 2 | 1 | 2 |
| 4 | 4 | 9 | 3 | 5 | 2 | 3 | 5 | 5 | 5 |
| 2 | 5 | 6 | 1 | 7 | 1 | 5 | 3 | 8 | 4 |

(a) What is the sampling distribution of $\bar{x}$? In other words, what is the shape, mean and standard deviation of the distribution of $\bar{x}$? Why?

(b) Compute a 90% confidence interval for the mean number of cars waiting in line between 11:50 A.M. and 12:00 noon on Wednesdays. Interpret this interval.

(c) Compute a 95% confidence interval for the mean number of cars waiting in line between 11:50 A.M. and 12:00 noon on Wednesdays. Interpret this interval.

11. **Blood Plasma** In a study of aerobic activity, the blood plasma volume (in liters) of 12 women was measured, and the following data were obtained:

| | | | | | |
|---|---|---|---|---|---|
| 3.15 | 2.99 | 2.77 | 3.12 | 2.45 | 3.85 |
| 2.99 | 3.87 | 4.06 | 2.94 | 3.53 | 3.20 |

*Source: Journal of Applied Physiology 65, 6 (December, 1988), p. 361 from Hogg and Tanis Probability and Statistical Inference 6/e.*

(a) Use the data to compute a point estimate for the population mean and population standard deviation.

(b) Because the sample size is small, we must verify that blood plasma is normally distributed and that the data do not contain any outliers. The figures below show the normal probability plot and boxplot. Are the conditions for constructing a confidence interval about $\mu$ satisfied?

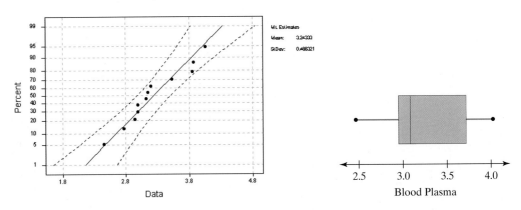

Normal Probability Plot for Blood Plasma

(c) Construct a 95% confidence interval for the mean blood plasma volume for all women. Interpret this interval.

(d) Construct a 99% confidence interval for the mean blood plasma volume for all women. Interpret this interval.

12. **Water Clarity** The campus at Joliet Junior College has a lake. A Secchi dish is used to measure the water clarity of the lake's water by lowering the dish into the water and measuring the distance below water until the dish is no longer visible. The following measurements (in inches) were taken on the lake at various points in time over the course of a year:

| | | | | | |
|---|---|---|---|---|---|
| 82 | 64 | 62 | 66 | 68 | 43 |
| 38 | 26 | 68 | 56 | 54 | 66 |

*Source: Virginia Piekarski, Joliet Junior College*

(a) Use the data to compute a point estimate for the population mean and population standard deviation.

(b) Because the sample size is small, we must verify that the data are normally distributed and do not contain any outliers. The figures below show the normal probability plot and box-plot. Are the conditions for constructing a confidence interval about $\mu$ satisfied?

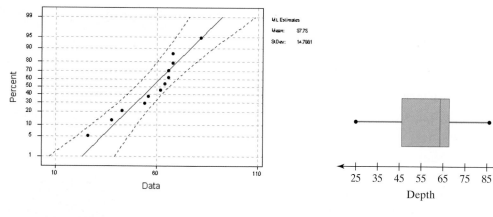

Normal Probability Plot for Depth

(c) Construct a 95% confidence interval for the mean Secchi disk measurement. Interpret this interval.

(d) Construct a 99% confidence interval for the mean Secchi disk measurement. Interpret this interval.

---

13. **Math Achievement** Redo Problems 9(c)–(e), assuming that $\sigma$ is unknown.

14. **Math Achievement** Redo Problems 10(c)–(e), assuming that $\sigma$ is unknown.

15. **Hypertension** In a random sample of 678 adult males 20–34 years of age, it was determined that 58 of them have hypertension (high blood pressure) [from data obtained from the Centers for Disease Control].
(a) Obtain a point estimate for the proportion of adult males 20–34 years of age who have hypertension.
(b) Construct a 95% confidence interval for the proportion of adult males 20–34 years of age who have hypertension. Interpret the confidence interval.
(c) Suppose you wish to conduct your own study to determine the proportion of adult males 20–34 years old who have hypertension. What sample size would be needed for the estimate to be within 3 percentage points with 95% confidence if you use the point estimate obtained in part (a)?
(d) Suppose you wish to conduct your own study to determine the proportion of adult males 20–34 years old who have hypertension. What sample size would be needed for the estimate to be within 3 percentage points with 95% confidence if you don't have a prior estimate?

16. **Carbon Monoxide** From a random sample of 1201 Americans, it was discovered that 1139 of them lived in neighborhoods with acceptable levels of carbon monoxide [from data obtained from the Environmental Protection Agency].

(a) Obtain a point estimate for the proportion of Americans who live in neighborhoods with acceptable levels of carbon monoxide.
(b) Construct a 99% confidence interval for the proportion of Americans who live in neighborhoods with acceptable levels of carbon monoxide.
(c) Suppose you wish to conduct your own study to determine the proportion of Americans who live in neighborhoods with acceptable levels of carbon monoxide. What sample size would be needed for the estimate to be within 1.5 percentage points with 90% confidence if you use the estimate obtained in part (a)?
(d) Suppose you wish to conduct your own study to determine the proportion of Americans who live in neighborhoods with acceptable levels of carbon monoxide. What sample size would be needed for the estimate to be within 1.5 percentage points with 90% confidence if you don't have a prior estimate?

17. The area under the $t$-distribution with 18 degrees of freedom to the right of $t = 1.56$ is 0.0681. What is the area under the $t$-distribution with 18 degrees of freedom to the left of $t = -1.56$? Why?

18. Which is larger, the area under the $t$-distribution with 10 degrees of freedom to the right of $t = 2.32$ or the area under the standard normal distribution to the right of $z = 2.32$? Why?

19. State the properties of Student's $t$-distribution.

**Chapter 8 Projects** located at www.prenhall.com/sullivanstats

# Hypothesis Testing

## Outline

 **For additional study help, go to**
www.prenhall.com/sullivanstats

### Materials include

- Projects:
  - Case Study: Dating Stonehenge
  - Decisions: What Does It Really Weigh?
  - Consumer Reports Project
- Self-Graded Quizzes
- "Preparing for This Section" Quizzes
- STATLETs
- PowerPoint Downloads
- Step-by-Step Technology Guide
- Graphing Calculator Help

## Putting It All Together

In Chapter 8, we mentioned that there are two areas of inferential statistics: (1) estimation and (2) hypothesis testing. We have already discussed procedures for estimating the population mean and the population proportion.

We now focus our attention on hypothesis testing. Hypothesis testing is used to test claims regarding a characteristic of one or more populations. In this chapter, we will test claims regarding a single population parameter. The claims that we test regard the population mean and the population proportion.

## 9.1   The Language of Hypothesis Testing

**Preparing for This Section**   Before getting started, review the following:

  ✓ Simple random sampling (Section 1.2, pp. 12–15)
  ✓ Parameter versus statistic (Section 3.1, p. 84)
  ✓ Table 8 (Section 6.2, p. 257)
  ✓ Central Limit Theorem (Section 7.5, pp. 314–323)

**Objectives**  Determine the null and alternative hypotheses from a claim
 Understand Type I and Type II errors
 Understand the probability of making Type I and Type II errors
 State conclusions to hypothesis tests

Let's begin with an example that introduces the idea behind hypothesis testing.

▶ EXAMPLE 1   **Illustrating Hypothesis Testing**

*Problem:* According to the National Center for Chronic Disease Prevention and Health Promotion, 73.8% of females between the ages of 18 and 29 years exercise. Kathleen believes that more women between the ages of 18 and 29 years are now exercising, so she obtains a simple random sample of 1000 women and finds that 750 of them are exercising. Is this evidence that the percent of women between the ages of 18 and 29 who are exercising has increased? What if Kathleen's sample resulted in 920 women exercising?

*Approach:* Here is the situation Kathleen faces. If 73.8% of 18–29-year-old females exercise, then she would expect 738 of the 1000 respondents in the sample to exercise. The question that Kathleen needs to answer is: "How likely is it to obtain a sample of 750 out of 1000 women exercising from a population when the percentage of women who exercise is 73.8%? How likely is a sample that has 920 women exercising?"

*Solution:* The result of 750 women who exercise is close to what we would expect, so Kathleen is not inclined to believe that the percentage of women exercising has increased. However, the likelihood of obtaining a sample of 920 women who exercise is extremely low if the actual percentage of women who exercise is 73.8%. For the case of obtaining a sample of 920 women who exercise, Kathleen can conclude one of two things: either the proportion of women who exercise is 73.8% and her sample just happens to include a lot of women who exercise or the proportion of women who exercise has increased. Provided the sampling was performed in a correct fashion, Kathleen is more inclined to believe that the percentage of women who exercise has increased.   ◀◀

 Example 1 presents the basic premise behind hypothesis testing: A claim is made, information is collected, and this information is used to test the claim. The steps in conducting a hypothesis test are presented next.

**Steps in Hypothesis Testing**

1. A claim is made.
2. Evidence (sample data) is collected in order to test the claim.
3. The data are analyzed in order to support or refute the claim.

In this section, we introduce the language of hypothesis testing. Sections 9.2–9.4 discuss the formal process of testing a hypothesis.

*Definition*

A **hypothesis** is a statement or claim regarding a characteristic of one or more populations.

**In Your Own Words**

In this chapter, we will focus on hypothesis testing regarding a single population parameter. The characteristic we test regards the population mean or proportion.

In this chapter, we look at hypotheses regarding a single population parameter. The following are examples of claims regarding a characteristic of a single population.

**(A)** According to a Gallup poll conducted in 1995, 74% of Americans felt that men were more aggressive than women. A researcher claims the percentage of Americans that feel men are more aggressive than women is different today (a claim regarding a population proportion).

**(B)** The packaging on a lightbulb states that the bulb will last 500 hours under normal use. A consumer advocate would like to know if the mean lifetime of a bulb is less than 500 hours (a claim regarding the population mean).

**(C)** The standard deviation rate of return for a certain class of mutual funds is 0.08. A mutual fund manager claims the standard deviation rate of return for his fund is less than 0.08 (a claim regarding the population standard deviation).

**Caution**

If population data are available, there is no need for inferential statistics.

We test these types of claims using sample data because it is usually impossible or impractical to gain access to the entire population.

Consider the lightbulb manufacturer mentioned in situation B. The packaging states that the lightbulb lasts 500 hours. The consumer advocate would like to know if the manufacturer is misleading the public, so he wants to know if there is evidence to support the claim that the mean lifetime of the bulb is less than 500 hours. To test the claim, the researcher might obtain a random sample of 49 lightbulbs. Suppose he determines the mean length of time before they burned out is 476 hours. Is this result enough evidence to conclude that the bulb's life is less than 500 hours? If we give the manufacturer the benefit of the doubt, then the mean lifetime of the bulb would be $\mu_{\bar{x}} = 500$. Let's assume that the standard deviation lifetime of the bulb is $\sigma = 42$ hours, so that the standard deviation of $\bar{x}$ is $\sigma_{\bar{x}} = \dfrac{42}{\sqrt{49}} = 6$. In addition, by the Central Limit Theorem we know the distribution of the sample mean is approximately normal (because the sample size is greater than 30). From Figure 1, we can see the sampling distribution of $\bar{x}$ with the observed sample mean of 476 hours labeled.

**Figure 1**

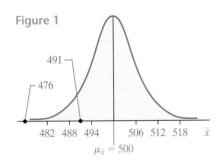

The sample mean of 476 is so far from the claimed mean that we have reason to believe that the manufacturer's claim is probably not true. The consumer advocate would conclude that the population mean is likely some value less than 500 hours.

Suppose that instead of getting a sample mean of 476 hours, the consumer advocate got a sample mean of 491 hours. Would this result be unusual if the bulb does, in fact, have a mean lifetime of 500 hours? Again, we label the observed sample mean on the distribution in Figure 1. This result could very easily occur by chance when the population mean is 500 hours. We have no reason to believe the mean lifetime of the bulb is less than 500 hours.

From the two situations just presented, it is clear from Figure 1 that 476 hours is "too far" from the claimed mean, while 491 hours is "close enough" for us to believe that the population mean could be 500 hours. This raises an interesting question. How far from the presumed mean of 500 hours can our sample mean (evidence) be before we reject the idea that the bulb lasts 500 hours? To answer to this question requires *hypothesis testing*.

*Definition*

**Hypothesis testing** is a procedure, based on sample evidence and probability, used to test claims regarding a characteristic of one or more populations.

Hypothesis testing is based upon two types of hypotheses.

*Definitions*

The **null hypothesis**, denoted $H_o$ (read "H-naught"), is a statement to be tested. The null hypothesis is assumed true until evidence indicates otherwise. In this chapter, it will be a statement regarding the value of a population parameter.

The **alternative hypothesis**, denoted $H_1$ (read "H-one"), is a claim to be tested. We are trying to find evidence for the alternative hypothesis. In this chapter, it will be a claim regarding the value of a population parameter.

**In Your Own Words**

The null hypothesis is a statement of "status quo" or "no difference" and always contains a statement of equality. The null hypothesis is assumed to be true until we have evidence to the contrary. The claim that we seek evidence for always becomes the alternative hypothesis.

For the lightbulb manufacturer the consumer advocate wishes to test the claim that the mean lifetime of the bulb is less than 500 hours. Because we are trying to obtain evidence for this claim, it is expressed as the alternative hypothesis using the notation $H_1: \mu < 500$. The statement made by the manufacturer is that the bulb lasts 500 hours. We express the statement to be tested using the notation $H_o: \mu = 500$.

In this chapter, there are three ways to set up the null and alternative hypotheses.

1. Equal hypothesis versus not equal hypothesis **(two-tailed test)**
   $H_o$: parameter $=$ some value
   $H_1$: parameter $\neq$ some value

2. Equal versus less than **(left-tailed test)**
   $H_o$: parameter $=$ some value
   $H_1$: parameter $<$ some value

3. Equal versus greater than **(right-tailed test)**
   $H_o$: parameter $=$ some value
   $H_1$: parameter $>$ some value

Left- and right-tailed tests are collectively referred to as **one-tailed tests**. Notice that in the left-tailed test the direction of the inequality in the alternative hypothesis points to the left ($<$), while in the right-tailed test the direction of the inequality in the alternative hypothesis points to the right ($>$). Notice that in all three tests the null hypothesis contains a statement of equality. The statement of equality comes from existing information.

Refer to the three claims made on page 384. In Situation A, the null hypothesis would be $H_o: p = 0.74$. This is a statement of "status quo" or "no difference."* It means that American opinions have not changed from 1995. In Situation B, the null hypothesis is $H_o: \mu = 500$. This is a statement of "no difference" between the population mean and the lifetime stated on the label. In Situation C, the null hypothesis is $H_o: \sigma = 0.08$. This is a statement of "no difference" between the population standard deviation rate of return of the manager's mutual fund and all mutual funds.

The wording of the claim, which is dictated by the researcher before any data are collected, determines the structure of the alternative hypothesis (two-tailed, left-tailed, or right-tailed). For example, the label on a can of soda states that the can contains 12 ounces of liquid. A consumer advocate

**In Your Own Words**

Structuring the null and alternative hypothesis:

1. Identify the parameter in the claim.
2. Determine the "status quo" value of the parameter.
3. Identify the claim that we want evidence for to determine the alternative hypothesis.

---

*The Latin phrase *status quo* means "the existing state or condition."

would be concerned only if the mean contents are less than 12 ounces, so the alternative hypothesis would be $H_1: \mu < 12$. However, a quality-control engineer for the soda manufacturer would be concerned if there is too little or too much soda in the can, so the alternative hypothesis would be $H_1: \mu \neq 12$. In both cases, however, the null hypothesis is a statement of no difference between the manufacturer's assertion on the label and the actual mean contents of the can. So the null hypothesis is $H_0: \mu = 12$.

▶ **EXAMPLE 2**  **Forming Hypotheses**

*Problem:* For each of the following claims, determine the null and alternative hypotheses. State whether the test is two-tailed, left-tailed, or right-tailed.

**(a)** The Medco pharmaceutical company has just developed a new antibiotic for children. In competing antibiotics, 2% of children who take the drug experience headaches as a side effect. A researcher for the Food and Drug Administration wishes to test the claim that the percentage of children taking the new antibody who experience headaches as a side effect is more than 2%.

**(b)** The "blue book" value of a used three-year-old Chevy Corvette is $37,500. Grant wishes to test the claim that the mean price of a used three-year-old Chevy Corvette in his neighborhood is different from $37,500.

**(c)** The standard deviation of the contents in a 64-ounce bottle of detergent using an old filling machine was known to be 0.23 ounces. The company just purchased a new filling machine and wants to test the claim that the new filling machine fills the bottles with a standard deviation less than 0.23 ounces.

*Approach:* In each of these cases we must first identify the parameter about which the claim is made and the "status quo." We then assess the direction of the claim (greater than, less than, or not equal to) to help us form the alternative hypothesis.

*Solution:*

**(a)** The claim is regarding a population proportion, $p$. If the new drug is no different from current drugs on the market, then the proportion of individuals taking the new drug who experience a headache will be 0.02, so that the null hypothesis is $H_0: p = 0.02$. The phrase "more than" is represented symbolically as $>$, so the claim is $p > 0.02$. Therefore, the alternative hypothesis is $H_1: p > 0.02$. This is a right-tailed test because the alternative hypothesis contains a $>$ symbol.

**(b)** The claim is regarding a population mean, $\mu$. If the mean price of a three-year-old Corvette in Grant's neighborhood is no different from the blue book price, then the population mean in Grant's neighborhood will be $37,500, so that the null hypothesis is $H_0: \mu = 37,500$. Grant wishes to test the claim that the mean price is different from $37,500, so that the alternative hypothesis is $H_1: \mu \neq 37,500$. This is a two-tailed test because the alternative hypothesis contains a $\neq$ symbol.

**(c)** The claim is regarding a population standard deviation, $\sigma$. If the new machine is no different from the old machine, then the standard deviation amount in the bottles filled by the new machine will be 0.23 ounces, so that the null hypothesis is $H_0: \sigma = 0.23$. The phrase "less than" is represented symbolically by $<$, so the claim is $\sigma < 0.23$ ounces and the alternative hypothesis is $H_1: \sigma < 0.23$. This is a left-tailed test because the alternative hypothesis contains a $<$ symbol.

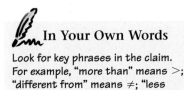
**In Your Own Words**

Look for key phrases in the claim. For example, "more than" means $>$; "different from" means $\neq$; "less than" means $<$; and so on.

**NW** *Now Work Problem 9(a).*
◀◀

**② Type I and Type II Errors**

As stated earlier, we use sample data to determine whether to reject or not reject the null hypothesis. Because the decision to reject or not reject the null hypothesis is based upon incomplete (i.e., sample) information, there is always the possibility of making an incorrect decision. In fact, there are four possible outcomes from hypothesis testing.

**In Your Own Words**

When you are testing a hypothesis, there is always the possibility that your conclusion will be wrong. To make matters worse, you won't know whether you are wrong or not!

---

**Four Outcomes from Hypothesis Testing**

1. We reject $H_o$ when in fact $H_1$ is true. This decision would be correct.
2. We do not reject $H_o$ when in fact $H_o$ is true. This decision would be correct.
3. We reject $H_o$ when in fact $H_o$ is true. This decision would be incorrect. This type of error is called a **Type I error**.
4. We do not reject $H_o$ when in fact $H_1$ is true. This decision would be incorrect. This type of error is called a **Type II error**.

---

Figure 2 illustrates the two types of errors that can be made in hypothesis testing.

**Figure 2**

|  |  | Reality | |
|---|---|---|---|
|  |  | $H_o$ Is True | $H_1$ Is True |
| **Conclusion** | Do Not Reject $H_o$ | Correct Conclusion | Type II Error |
|  | Reject $H_o$ | Type I Error | Correct Conclusion |

We illustrate the idea of Type I and Type II errors by looking at hypothesis testing from the point of view of a criminal trial. In any trial, the defendant is assumed to be innocent. (We give the defendant the benefit of the doubt.) The district attorney must present evidence proving that the defendant is guilty. Because we are seeking evidence for guilt, it becomes the alternative hypothesis. The null hypothesis is innocence, so the hypotheses for a trial would be written:

$H_o$: the defendant is innocent
$H_1$: the defendant is guilty

The trial is the process whereby information (sample data) is obtained. The jury then analyzes the data (the deliberations). Finally, the jury either rejects the null hypothesis and the defendant is considered guilty or does not reject the null hypothesis and declares the defendant not guilty. Note that the defendant is never declared innocent. That is, we never say that the null hypothesis is true. Using this analogy, the two correct decisions would be to conclude that an innocent person is not guilty or conclude that a guilty person is guilty. The two incorrect decisions would be to convict an innocent person (a Type I error) or to let a guilty person go free (a Type II error). It is helpful to think in this way when trying to remember the difference between a Type I and a Type II error.

**In Your Own Words**

A Type I error is like putting an innocent person in jail. A Type II error is like letting a guilty person go free.

▶ **EXAMPLE 3**   Type I and Type II Errors

*Problem:* The Medco pharmaceutical company has just developed a new antibiotic. In competing antibiotics, 2% of children who take the drug experience headaches as a side effect. A researcher for the Food and Drug Administration claims that the percentage of children taking the new antibody who experience a headache as a side effect is more than 2%. To test this claim, we conduct a hypothesis test with $H_0$: $p = 0.02$ and $H_1$: $p > 0.02$. Provide statements explaining what it would mean to make (a) a Type I error and (b) a Type II error.

*Approach:* A Type I error would occur if the null hypothesis is rejected when, in reality, the null hypothesis is true. A Type II error would occur if the null hypothesis is not rejected when, in reality, the alternative hypothesis is true.

*Solution:*

(a) We would make a Type I error if the sample evidence leads us to believe that $p > 0.02$ (i.e., we reject the null hypothesis) when, in fact, the proportion of children who experience a headache is not greater than 0.02.

(b) We would make a Type II error if we do not reject the null hypothesis that the proportion of children experiencing a headache is equal to 0.02 when, in fact, the proportion of children who experience a headache is more than 0.02. For example, the sample evidence led the researcher to believe $p = 0.02$ when in fact the true proportion is $p = 0.023$.   ◀◀

*Now Work Problems 9(b) and 9(c).*

**❸   The Level of Significance**

Recall that we never know whether a confidence interval contains the unknown parameter. In addition, we never know whether the outcome of a hypothesis test results in an error or not. However, just as we place a level of confidence in the construction of a confidence interval, we can determine the probability of making errors. The following notation is commonplace:

$$\alpha = P(\text{Type I error}) = P(\text{rejecting } H_0 \text{ when } H_0 \text{ is true})$$

$$\beta = P(\text{Type II error}) = P(\text{not rejecting } H_0 \text{ when } H_1 \text{ is true})$$

The symbol $\beta$ is the Greek letter beta (pronounced "BAY tah"). The probability of making a Type I error, $\alpha$, is chosen by the researcher *before* the sample data are collected. This probability is referred to as the *level of significance*.

*Definition*   The **level of significance**, $\alpha$, is the probability of making a Type I error.

The choice of the level of significance will depend upon the consequences of making a Type I error. If the consequences are severe, then the level of significance should be small (say, $\alpha = 0.01$). However, if the consequences of making a Type I error are not severe, then a higher level of significance can be chosen (say $\alpha = 0.05$ or $\alpha = 0.10$).

Why is the level of significance not always set at $\alpha = 0.01$? By reducing the probability of making a Type I error, you increase the probability of making a Type II error, $\beta$. Using our court analogy, a jury is instructed that the prosecution must provide proof of guilt "beyond all reasonable doubt." This implies that we are choosing to make $\alpha$ small so that the probability we will send an innocent person to jail is very small. The consequence of the small $\alpha$, however, is a large $\beta$, which means many guilty defendants would go free. For now, we will be content with recognizing the inverse relation between $\alpha$ and $\beta$ (as one goes up the other goes down).

**In Your Own Words**

As the probability of a Type I error increases, the probability of a Type II error decreases, and vice versa.

## Caution

We never "accept" the null hypothesis, because, without having access to the entire population, we don't know the exact value of the parameter stated in the null. Rather, we say that we do not reject the null hypothesis. This is just like the court system. We never declare a defendant innocent, but rather say the defendant is not guilty.

### ④  Writing the Conclusion

Once the decision to reject or not reject the null hypothesis is made, the researcher must state his or her conclusion. It is important to recognize that we never *accept* the null hypothesis. Again, the court system analogy helps to illustrate the idea. The null hypothesis is $H_o$: innocent. When the evidence presented to the jury is not enough to convict beyond all reasonable doubt, the jury comes back with a verdict of not guilty. Notice that the verdict does not state that the null hypothesis of innocence is true; it simply states that there is not enough evidence to support guilt. This is a huge difference. Being told that you are not guilty is very different from being told that you are innocent!

This means that sample evidence can never prove the null hypothesis to be true. When we do not reject the null hypothesis, we are saying that the evidence indicates that the null hypothesis *could* be true.

▶ **EXAMPLE 4**    **Stating the Conclusion**

*Problem:* The Medco pharmaceutical company has just developed a new antibiotic. In competing antibiotics, 2% of children who take the drug experience a headache as a side effect. A researcher for the Food and Drug Administration claims that the percentage of children taking the new antibody who experience a headache as a side effect is more than 2%. From Example 2(a) we know the null hypothesis is $H_o$: $p = 0.02$ and the alternative hypothesis is $H_1$: $p > 0.02$.

**(a)** Suppose the sample evidence indicates that the null hypothesis is rejected. State the conclusion.

**(b)** Suppose the sample evidence indicates that the null hypothesis is not rejected. State the conclusion.

*Approach:* When the null hypothesis is rejected, we say that there is sufficient evidence to support the claim. When the null hypothesis is not rejected, we say that there is not sufficient evidence to support the claim. We never say that the null hypothesis is true!

*Solution:*

**(a)** The claim is that the percentage of children taking the new antibody who experience a headache as a side effect is more than 2%. Because the null hypothesis ($p = 0.02$) is rejected, we conclude there is sufficient evidence to support the claim that the proportion of children who experience a headache as a side effect is more than 0.02.

**(b)** Because the null hypothesis is not rejected, we conclude that there is not sufficient evidence to support the claim that the proportion of children who experience a headache as a side effect is more than 0.02. ◀◀

**NW**  *Now Work Problem 17.*

## 9.1  Assess Your Understanding

### Concepts and Vocabulary

**1.** Explain what it means to make a Type I error. Explain what it means to make a Type II error.

**2.** Suppose the consequences of making a Type I error are severe. Would you choose the level of significance, $\alpha$, to equal 0.01, 0.05 or 0.10? Why?

**3.** What happens to the probability of making a Type II error, $\beta$, as the level of significance, $\alpha$, decreases? Why is this result intuitive?

**4.** If a hypothesis is tested at the $\alpha = 0.05$ level of significance, what is the probability of making a Type I error?

5. The following is a quotation from Sir Ronald A. Fisher, a famous statistician.

"For the logical fallacy of believing that a hypothesis has been proved true, merely because it is not contradicted by the available facts, has no more right to insinuate itself in statistics than in other kinds of scientific reasoning ... It would, therefore, add greatly to the clarity with which the tests of significance are regarded if it were generally understood that tests of significance, when used accurately, are capable of rejecting or invalidating hypotheses, in so far as they are contradicted by the data: but that they are never capable of establishing them as certainly true ... "

In your own words, explain what this quotation means.

## Exercises

### • Basic Skills

*In Problems 1–6, a null and alternative hypothesis are given. Determine whether the hypothesis test is left-tailed, right-tailed, or two-tailed. What parameter is being tested?*

1. $H_0: \mu = 5$
   $H_1: \mu > 5$

2. $H_0: p = 0.2$
   $H_1: p < 0.2$

3. $H_0: \sigma = 4.2$
   $H_1: \sigma \neq 4.2$

4. $H_0: p = 0.76$
   $H_1: p > 0.76$

5. $H_0: \mu = 120$
   $H_1: \mu < 120$

6. $H_0: \sigma = 7.8$
   $H_1: \sigma \neq 7.8$

*For each claim in Problems 7–14, (a) determine the null and alternative hypotheses, (b) explain what it would mean to make a Type I error, and (c) explain what it would mean to make a Type II error.*

7. **Farm Rents** According to the United States Department of Agriculture, the mean farm rent in Indiana was $89.00 per acre in 1995. A researcher for the USDA claims that the mean rent has decreased since then.

8. **Acid Rain** In 2000, the mean pH of rain in Gunnison County, Colorado, was 5.01, according to the National Atmospheric Deposition Program. A researcher claims that the level of acidity in the rain has increased. (This would mean that the pH has decreased.)

9. **Health Insurance** According to the United States Census Bureau, 16.3% of Americans did not have health insurance coverage in 1998. A politician claims that this percentage has decreased since 1998.

10. **Protein** According to the United States Department of Agriculture, 78.2% of females between the ages of 12 and 19 years get the USDA's recommended daily allowance of protein. A nutritionist believes that the percentage of 12-to-19-year-old females at her boarding school who get the USDA's recommended daily allowance of protein is above 78.2%.

11. **Energy Expenditures** According to the United States Energy Information Administration, the mean expenditure for residential energy consumption (electricity, natural gas, etc.) was $1338 in 1997. An economist claims that the mean expenditure for residential energy is different today.

12. **Death Row** According to the U.S. Department of Justice, the mean age of a death row inmate in 1980 was 36.7 years. A district attorney believes the mean age of a death row inmate is different today.

13. **Stock Market Returns** During the past forty years, the standard deviation monthly rate of return for stocks has been 4.2 percent. A stock analyst claims the standard deviation monthly rate of return is higher today.

14. **SAT Verbal Scores** In 2001, the standard deviation SAT verbal score for males was 111. A teacher claims the standard deviation SAT verbal score for male students enrolled at her school is less than 111.

*In Problems 15–26, state the conclusion based upon the results of the test.*

15. For the claim made in Problem 7, suppose the null hypothesis is rejected.

16. For the claim made in Problem 8, suppose the null hypothesis is not rejected.

17. For the claim made in Problem 9, suppose the null hypothesis is not rejected.

18. For the claim made in Problem 10, suppose the null hypothesis is rejected.

19. For the claim made in Problem 11, suppose the null hypothesis is not rejected.

20. For the claim made in Problem 12, suppose the null hypothesis is not rejected.

21. For the claim made in Problem 13, suppose the null hypothesis is rejected.

22. For the claim made in Problem 14, suppose the null hypothesis is not rejected.

23. For the claim made in Problem 7, suppose the null hypothesis is not rejected.

24. For the claim made in Problem 8, suppose the null hypothesis is rejected.

25. For the claim made in Problem 9, suppose the null hypothesis is rejected.

26. For the claim made in Problem 10, suppose the null hypothesis is not rejected.

• **Applying the Concepts**

27. **Potato Consumption** According to the Statistical Abstract of the United States, the mean per capita consumption of potatoes in 1999 was 48.3 pounds. A researcher believes that potato consumption has risen since then.

   (a) Determine the null and alternative hypotheses.
   (b) Suppose sample data indicate that the null hypothesis should be rejected. State the conclusion of the researcher.
   (c) Suppose, in reality, the mean per capita consumption of potatoes is 48.3 pounds. Was a Type I or Type II error committed? If we tested this hypothesis at the $\alpha = 0.05$ level of significance, what is the probability of committing a Type I error?
   (d) If we wanted to decrease the probability of making a Type II error, would we need to increase or decrease the level of significance, $\alpha$?

28. **Test Preparation** The mean score on the SAT I Math exam is 505. A test preparatory company claims that the mean scores of students who take their course is higher than the mean of 505.

   (a) Determine the null and alternative hypotheses.
   (b) Suppose sample data indicate that the null hypothesis should not be rejected. State the conclusion of the company.
   (c) Suppose, in reality, the mean score of students taking the preparatory course is 507. Was a Type I or Type II error committed? If we tested this hypothesis at the $\alpha = 0.01$ level of significance, what is the probability of committing a Type I error?
   (d) If we wanted to decrease the probability of making a Type II error, would we need to increase or decrease the level of significance, $\alpha$?

29. **Traffic Fatalities** In 1998, the proportion of traffic fatalities in which the driver had a blood alcohol content (BAC) of 0.08% or higher was 0.38. After a strong advertising campaign, a member of MADD claims that the proportion of traffic fatalities in which the driver had a BAC of 0.08% or higher has decreased from the 1998 level.

   (a) Determine the null and alternative hypotheses.
   (b) Suppose sample data indicate that the null hypothesis should not be rejected. State the conclusion of the member of MADD.
   (c) Suppose, in reality, the proportion of traffic fatalities in which the driver had a blood alcohol content (BAC) of 0.08% or higher was 0.36. Was a Type I or Type II error committed?

30. **Internet Usage** According to the Statistical Abstract of the United States, in 2000 10.7% of Americans over 65 years of age utilized the Internet. A researcher believes the proportion of Americans over 65 years of age who utilize the Internet is higher than 10.7% today.

   (a) Determine the null and alternative hypotheses.
   (b) Suppose sample data indicate that the null hypothesis should be rejected. State the conclusion of the researcher.
   (c) Suppose, in reality, the percentage of Americans over 65 years of age who utilize the Internet is still 10.7%. Was a Type I or Type II error committed?

31. **Consumer Reports** The following is an excerpt from a *Consumer Reports* article from February, 2001.

   > The *Platinum Gasaver* makes some impressive claims. The device, $188 for two, is guaranteed to increase gas mileage by 22 percent, says the manufacturer, National Fuelsaver. Further, the company quotes "the government" as concluding, "Independent testing shows greater fuel savings with *Gasaver* than the 22 percent claimed by the developer." Readers have told us they want to know more about it.
   >
   > The Environmental Protection Agency (EPA), after its lab tests of the *Platinum Gasaver*, concluded in 1991, "Users of the device would not be expected to realize either an emission or fuel economy benefit." The Federal Trade Commission says, "No government agency endorses gas-saving products for cars."

   Determine the null and alternative hypotheses that the EPA used in order to draw the conclusion stated in the second paragraph.

32. **Prolong Engine Treatment** The manufacturer of Prolong Engine Treatment claims that if you add one 12-ounce bottle of their $20 product, your engine will be protected from excessive wear. In fact, an infomercial claims that a woman drove four hours without oil, thanks to Prolong. *Consumer Reports* magazine tested two engines in which they added Prolong to the motor oil, ran the engines, drained the oil, and then determined the time until the engines seized.

   (a) Determine the null and alternative hypotheses *Consumer Reports* will test.
   (b) Both engines took exactly 13 minutes to seize. What conclusion might *Consumer Reports* draw based upon this evidence?

33. Refer to the claim made in Problem 8. Researchers must choose the level of significance based upon the consequences of making a Type I error. Explain what it would mean to make a Type I error for this hypothesis. Explain what it would mean to make a Type II error. In your opinion, which error is more serious? Why? On the basis of your answer decide on a level of significance, $\alpha$. Be sure to support your opinion.

## 9.2 Testing a Hypothesis about $\mu$, $\sigma$ Known

**Preparing for This Section**  Before getting started, review the following:

✓ Central Limit Theorem (Section 7.5, pp. 314–323)

✓ Computing normal probabilities (Section 7.3, pp. 297–299)

✓ Using probabilities to identify unusual events (Section 5.1, p. 188)

✓ $z_\alpha$ notation (Section 7.2, p. 292)

**Objectives**  Understand the logic of hypothesis testing

 Test a hypothesis about $\mu$ with $\sigma$ known using the classical method

 Test a hypothesis about $\mu$ with $\sigma$ known using the $P$-value approach

Test a hypothesis about $\mu$ with $\sigma$ known using confidence intervals

We are now in a position to present methods for conducting hypothesis tests. In this section, we will present three approaches to testing hypotheses about a population mean, $\mu$, with $\sigma$ known. The first method is often referred to as the classical (or traditional) method, the second method is the more modern $P$-value approach and the third method utilizes confidence intervals. As with constructing confidence intervals, it is not likely that we know the population standard deviation without knowing the population mean, but for ease of understanding the procedures of hypothesis testing, we will assume that the value of the population standard deviation is known in this section.

### Historical Note

Jerzy Neyman was born on April 16, 1894, in Bendery, Russia. In 1921, he moved to Poland. He received his Ph.D. from the University of Warsaw in 1924. He read some of Karl Pearson's works and became interested in statistics; however, Neyman was not impressed with Pearson's mathematical abilities. In 1927, he met Pearson's son, Egon Pearson, who was working on a formal approach to hypothesis testing. It was Neyman who provided the mathematical rigor to their work. Together, they developed the phrases "null hypothesis" and "alternative hypothesis." In 1938, Neyman joined the faculty at the University of California at Berkeley. He died on August 5, 1981.

###  The Logic of Hypothesis Testing

In order to test hypotheses regarding the population mean, we need certain requirements to be satisfied.

- A simple random sample is obtained;
- The population from which the sample is drawn is normally distributed or the sample size is large ($n \geq 30$);
- The population standard deviation, $\sigma$, is known.

If these requirements are met, then the distribution of $\bar{x}$ is normal with mean $\mu$ and standard deviation $\frac{\sigma}{\sqrt{n}}$.

Remember that in hypothesis testing, the null hypothesis must always contain a statement of equality and the null hypothesis is assumed true. In testing a population mean, there are three ways to structure the hypothesis test:

| Two-Tailed | Left-Tailed | Right-Tailed |
|---|---|---|
| $H_0: \mu = \mu_0$ | $H_0: \mu = \mu_0$ | $H_0: \mu = \mu_0$ |
| $H_1: \mu \neq \mu_0$ | $H_1: \mu < \mu_0$ | $H_1: \mu > \mu_0$ |

*Note*: $\mu_0$ is the assumed value of the population mean.

Suppose we work in the quality control department of Ruffles Potato Chips. The quality control manager wants us to verify that the filling machine is calibrated correctly. We wish to determine if the mean amount of chips in a bag is different from 12.5 ounces (the company would be concerned if there are too many or too few chips in the bag). Experience indicates that the population standard deviation is 0.15 ounce. The null and alternative hypotheses are given by

$$H_o: \mu = 12.5 \qquad \text{versus} \qquad H_1: \mu \neq 12.5$$

In order to test the claim, we take a simple random sample of size $n = 40$ bags from the warehouse. If the sample mean is significantly different from 12.5 ounces, we have evidence that the null hypothesis is not true. What do we mean by significantly different? A common criterion is to reject $H_o$ if the sample mean is too many standard deviations away from the assumed (or "status quo") population mean, $\mu_o$. For example, our criterion might be to reject the null hypothesis if the sample mean is more than two standard deviations away from 12.5, the hypothesized mean.

Recall that $Z = \dfrac{\bar{x} - \mu_o}{\sigma/\sqrt{n}}$ represents the number of standard deviations

that $\bar{x}$ is from the population mean, $\mu_o$. Suppose our simple random sample of 40 bags results in sample mean weight of $\bar{x} = 12.43$ ounces. Then, under the assumption that the null hypothesis is true, we have

$$Z = \frac{\bar{x} - \mu_o}{\sigma/\sqrt{n}} = \frac{12.43 - 12.5}{0.15/\sqrt{40}} = -2.95$$

The sample mean is 2.95 standard deviations below the hypothesized mean. Because the sample mean is more than 2 standard deviations (i.e. "too far") from the hypothesized population mean, we will reject the null hypothesis and conclude that there is sufficient evidence to refute the claim that the machine is filling the bag with 12.5 ounces of potato chips. This conclusion would lead the quality control engineer to shut down and recalibrate the filling machine.

Why does this criterion make sense? The area under the standard normal curve to the left of $Z = -2$ is 0.0228 and the area under the standard normal curve to the right of $Z = 2$ is 0.0228. Therefore, the area under the standard normal curve between $Z = -2$ and $Z = 2$ is 0.9544 as shown in Figure 3.

That is, if the null hypothesis is true, then 95.44% of all sample means

should lie within $2\sigma_{\bar{x}} = 2 \cdot \dfrac{\sigma}{\sqrt{n}} = 2 \cdot \dfrac{0.15}{\sqrt{40}} = 0.047$ ounces of the hypothe-

sized mean of 12.5 ounces as shown in Figure 4.

**Caution**

We always test hypotheses assuming that the null hypothesis is true.

**Figure 3**

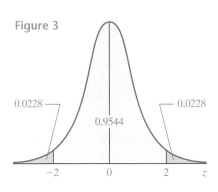

**Figure 4**

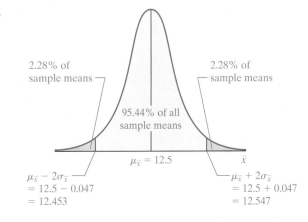

Referring to Figure 4, we see that only 2.28% of the sample means are less than 2 standard deviations below the hypothesized mean and 2.28% of the sample means are more than 2 standard deviations above the hypothesized mean (indicated by the shaded region in green in Figure 4). If the null hypothesis is true, sample means in this range will occur in only $2.28\% + 2.28\% = 4.56\%$ of samples. Recall that we stated in Chapter 5 that events that occur less than 5% of the time are considered unusual. Therefore, if a sample mean lies in the green region, we are inclined to believe that it came from a population whose mean does not equal 12.5 rather than believe that the population mean does equal 12.5 and our sample just happened to result in an unusual outcome (a bunch of under- or over-filled bags).

Figure 5 further illustrates the situation. The sample mean of 12.43 is too far from the assumed population mean of 12.5. Therefore, we reject the null hypothesis that $\mu = 12.5$ and conclude that the sample came from a population with some population mean different from 12.5 ounces, as indicated by the distribution in blue. We don't know what the population mean weight of the bags is, but we have evidence that it is not 12.5 ounces.

Figure 5

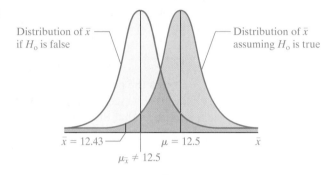

Distribution of $\bar{x}$ if $H_o$ is false

Distribution of $\bar{x}$ assuming $H_o$ is true

$\bar{x} = 12.43$    $\mu = 12.5$    $\bar{x}$

$\mu_{\bar{x}} \neq 12.5$

**In Your Own Words**

If the sample mean is too many standard deviations from the mean stated in the null hypothesis, then we reject the null hypothesis.

Notice that our criterion for rejecting the null will lead to making a Type I error (rejecting the null when the null is true) 2.28% + 2.28% = 4.56% of the time. That is, the probability of making a Type I error is 4.56%.

The previous discussion presents the premise of testing a hypothesis:

**If the sample data result in a statistic that is not likely under the assumption that the null hypothesis is true, we reject the null hypothesis.**

 **The Classical Method of Testing a Hypothesis**

We now formalize the procedure for testing a claim regarding the population mean when the population standard deviation, $\sigma$, is known.

**Hypothesis Test Regarding $\mu$ with $\sigma$ Known**

If a claim is made regarding the population mean with $\sigma$ known, we can use the following steps to test the claim provided:
1. the sample is obtained using simple random sampling,
2. the population from which the sample is drawn is normally distributed or the sample size, $n$, is large ($n \geq 30$).

**Step 1:** A claim is made regarding the population mean. The claim is used to determine the null and alternative hypotheses. Again, the hypotheses can be structured in one of three ways:

| Two-Tailed | Left-Tailed | Right-Tailed |
|---|---|---|
| $H_o: \mu = \mu_o$ | $H_o: \mu = \mu_o$ | $H_o: \mu = \mu_o$ |
| $H_1: \mu \neq \mu_o$ | $H_1: \mu < \mu_o$ | $H_1: \mu > \mu_o$ |

*Note:* $\mu_o$ is the assumed or "status quo" value of the population mean.

**Step 2:** Select a level of significance $\alpha$ based upon the seriousness of making a Type I error. The level of significance is used to determine the *critical value*. The **critical value** represents the maximum number of standard deviations the sample mean can be from $\mu_o$ before the null hypothesis is rejected. For example, the critical value in the left-tailed test is $-z_\alpha$. The shaded region(s) represents the *critical (or rejection) region(s)*.

The **critical region** or **rejection region** is the set of all values such that the null hypothesis is rejected.

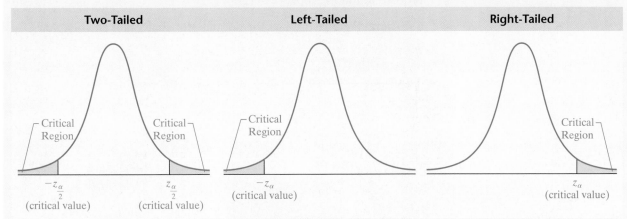

| Two-Tailed | Left-Tailed | Right-Tailed |

**Step 3:** Provided the population from which the sample is drawn is normal or the sample size is large ($n \geq 30$) and the population standard deviation, $\sigma$, is known, the distribution of the sample mean, $\bar{x}$, is normal with mean $\mu_0$ and standard deviation $\frac{\sigma}{\sqrt{n}}$. Therefore,

$$Z = \frac{\bar{x} - \mu_0}{\sigma/\sqrt{n}}$$

represents the number of standard deviations the sample mean is from the assumed mean, $\mu_0$. This value is called the **test statistic**.

**Step 4:** Compare the critical value with the test statistic:

| Two-Tailed | Left-Tailed | Right-Tailed |
|---|---|---|
| If $Z < -z_{\alpha/2}$ or $Z > z_{\alpha/2}$ reject the null hypothesis | If $Z < -z_{\alpha}$ reject the null hypothesis | If $Z > z_{\alpha}$ reject the null hypothesis |

**Step 5:** State the conclusion.

The procedure presented requires that the data come from a population that is normally distributed or that the sample size be large ($n \geq 30$). The procedure is **robust**, which means that minor departures from normality will not adversely affect the results of the test. However, for small samples, if the data have outliers, the procedure should not be used.

For small samples, we will verify that the data come from a population that is normal by constructing normal probability plots (to assess normality) and boxplots (to determine whether there are outliers). If the normal probability plot indicates that the data are not normally distributed or the boxplot reveals outliers, nonparametric tests should be performed.

| TABLE 1 | | |
|---|---|---|
| 8,777 | 19,187 | 12,022 |
| 16,759 | 7,294 | 10,163 |
| 9,021 | 12,682 | 16,747 |
| 16,733 | 15,331 | 5,347 |
| 13,798 | 13,235 | 11,302 |
| 6,929 | 17,576 | 9,455 |
| 16,306 | 8,266 | |

▶ **EXAMPLE 1  The Classical Method of Hypothesis Testing**

*Problem:* According to the U.S. Federal Highway Administration, the mean number of miles driven annually in 1990 was 10,300. Patricia believes that people are driving more today than in 1990. She obtains a simple random sample of 20 people and asks them to disclose the number of miles they drove last year. The results of the survey are presented in Table 1. Assuming $\sigma = 3500$ miles, test Patricia's claim at the $\alpha = 0.1$ level of significance.

*Approach:* Before we can perform the hypothesis test, we must verify that the data come from a population that is approximately normally distributed with no outliers because the sample size, $n$, is less than 30. We will construct a normal probability plot and boxplot to verify these assumptions. We then proceed to follow Steps 1–5.

*Solution:* Figure 6 displays the normal probability plot and boxplot.

Figure 6

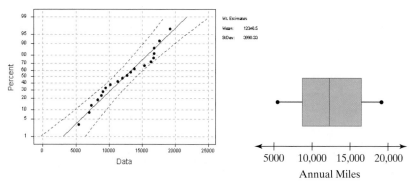

Normal Probability Plot for Annual Miles

The normal probability plot indicates that the data could come from a population that is normal. The boxplot does not show any outliers. We can proceed to perform the hypothesis test.

**Step 1:** Patricia claims people are driving more than 10,300 miles annually. This claim can be written $\mu > 10,300$. This is a right-tailed test and we have

$$H_0: \mu = 10,300 \qquad \text{versus} \qquad H_1: \mu > 10,300$$

Figure 7

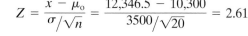

**Step 2:** Because Patricia is performing a right-tailed test, we determine the critical value at the $\alpha = 0.1$ level of significance to be $z_{0.1} = 1.28$. The critical region is displayed in Figure 7.

**Step 3:** Patricia computes the sample mean, $\bar{x}$, to be 12,346.5 miles. The test statistic is

$$Z = \frac{\bar{x} - \mu_0}{\sigma/\sqrt{n}} = \frac{12,346.5 - 10,300}{3500/\sqrt{20}} = 2.61$$

The sample mean of 12,346.5 miles is 2.61 standard deviations above the assumed mean of 10,300 miles.

**Step 4:** Because the test statistic $Z = 2.61$ is more than the critical value of $z_{0.1} = 1.28$, we reject the null hypothesis. That is, the value of the test statistic falls within the critical region so we reject $H_0$. We label this point in Figure 7.

**Step 5:** There is sufficient evidence at the $\alpha = 0.1$ level of significance to support Patricia's claim that the mean number of miles people are driving is greater than 10,300 miles per year.

**NW** *Now Work Problem 9.*

We now present a two-tailed hypothesis test.

▶ **EXAMPLE 2** **The Classical Method of Hypothesis Testing**

**TABLE 2**

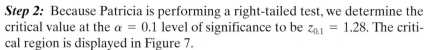

| $47,000 | $32,750 | $35,995 |
| $32,750 | $33,995 | $39,998 |
| $33,995 | $43,500 | $37,995 |
| $43,108 | $39,950 | $37,995 |
| $33,988 | $36,900 | $43,785 |

*Source:* cars.com

*Problem:* Grant is in the market to buy a three-year-old Chevy Corvette. Before shopping for the car, he wants to determine what he should expect to pay. According to the "blue book," the mean price of a three-year-old Chevy Corvette is $37,500. Grant thinks the mean price is different from $37,500 in his neighborhood. After visiting 15 neighborhood dealers online, Grant obtains the data shown in Table 2.

Assuming $\sigma = \$4100$, use these data to test Grant's claim that the mean price of a Corvette is different from $37,500 at the $\alpha = 0.1$ level of significance.

*Approach:* Because the sample size, $n$, is less than 30, we must verify that the data come from a population that is approximately normal with no outliers. We will construct a normal probability plot and boxplot to verify these requirements. We then proceed to follow Steps 1–5.

*Solution:* A normal probability plot and boxplot were drawn for this data in Figure 4 of Section 8.1. They indicate that the price of used Corvettes is approximately normal with no outliers.

*Step 1:* Grant claims the price of a three-year-old Corvette is different from \$37,500. This claim can be written $\mu \neq 37{,}500$ and so we have a two-tailed test. The null and alternative hypotheses are:

$$H_o: \mu = 37{,}500 \qquad \text{versus} \qquad H_1: \mu \neq 37{,}500.$$

*Step 2:* Because Grant is performing a two-tailed test, we determine the critical values at the $\alpha = 0.1$ level of significance to be $-z_{0.1/2} = -z_{0.05} = -1.645$ and $z_{0.1/2} = z_{0.05} = 1.645$. The critical regions are displayed in Figure 8.

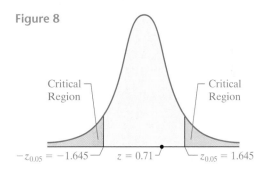

**Figure 8**

Critical Region — Critical Region

$-z_{0.05} = -1.645$ — $z = 0.71$ — $z_{0.05} = 1.645$

**Caution**

In Example 2, we see that the test statistic $Z = 0.71$ is not far enough from the "status quo" value of the population mean, \$37,500. Therefore, we do not have enough evidence to reject the null hypothesis. However, this does not mean we are accepting the null hypothesis that the mean price of a three-year old Corvette is \$37,500. We simply are saying we don't have enough evidence to say it is not \$37,500. Be sure you understand the difference between these two comments.

*Step 3:* Grant computes the sample mean, $\bar{x}$, to be \$38,246.9. The test statistic is

$$Z = \frac{\bar{x} - \mu_o}{\sigma/\sqrt{n}} = \frac{38{,}246.9 - 37{,}500}{4100/\sqrt{15}} = 0.71$$

The sample mean of \$38,246.9 is 0.71 standard deviations above the assumed mean of \$37,500.

*Step 4:* Because the test statistic $Z = 0.71$ is less than $z_{0.05} = 1.645$ and greater than $-z_{0.05} = -1.645$, we do not reject the null hypothesis. That is, the value of the test statistic does not fall within the critical regions. We label this point in Figure 8.

*Step 5:* There is not sufficient evidence at the $\alpha = 0.1$ level of significance to support Grant's claim that the mean price of a three-year-old Corvette is different from \$37,500.   ◄◄

**NW** *Now Work Problem 15.*

### ③ Testing a Hypothesis about $\mu$ with $\sigma$ Known Using *P*-Values

A second approach to testing hypotheses is with *P*-values.

*Definition*

A *P*-value is the probability of observing a sample statistic as extreme or more extreme than the one observed under the assumption that the null hypothesis is true.

Recall our criterion for rejecting the null hypothesis. If the sample mean is too far away from the hypothesized value of the population mean, we reject the null hypothesis. Using the classical approach, "too far away" means too many standard deviations from the mean. The *P*-value is the likelihood or probability that a sample will result in a sample mean such as the one obtained if the null hypothesis is true. A small *P*-value implies that the sample mean is unlikely if the null hypothesis is true and would be considered as evidence against the null hypothesis.

The following procedures can be used to compute *P*-values when testing a hypothesis about a population mean with $\sigma$ known.

## Computing *P*-Values

If a claim is made regarding the population mean with $\sigma$ known, we can use the following steps to compute the *P*-value, provided that
1. the sample is obtained using simple random sampling, and
2. the population from which the sample is drawn is normally distributed or the sample size, $n$, is large ($n \geq 30$).

***Step 1:*** A claim is made regarding the population mean. The claim is used to determine the null and alternative hypotheses. The hypotheses can be structured in one of three ways:

| Two-Tailed | Left-Tailed | Right-Tailed |
|---|---|---|
| $H_0: \mu = \mu_0$ | $H_0: \mu = \mu_0$ | $H_0: \mu = \mu_0$ |
| $H_1: \mu \neq \mu_0$ | $H_1: \mu < \mu_0$ | $H_1: \mu > \mu_0$ |

*Note:* $\mu_0$ is the assumed value of the population mean.

Decide on a level of significance, $\alpha$, depending upon the seriousness of making a Type I error.

***Step 2:*** Compute the test statistic, $Z_0 = \dfrac{\overline{x} - \mu_0}{\sigma / \sqrt{n}}$.

***Step 3:*** Determine the *P*-value:

| Two-Tailed | Left-Tailed | Right-Tailed |
|---|---|---|

$P\text{-value} = P(Z < -|Z_0| \text{ or } Z > |Z_0|)$
$= 2P(Z > |Z_0|)$

$P\text{-value} = P(Z < Z_0)$

$P\text{-value} = P(Z > Z_0)$

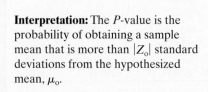
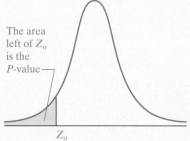
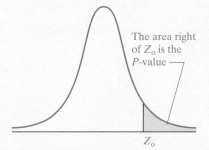

The sum of the area in the tails is the *P*-value

The sum of the area in the tails is the *P*-value

$-|Z_0|$     $|Z_0|$

The area left of $Z_0$ is the *P*-value

$Z_0$

The area right of $Z_0$ is the *P*-value

$Z_0$

**Interpretation:** The *P*-value is the probability of obtaining a sample mean that is more than $|Z_0|$ standard deviations from the hypothesized mean, $\mu_0$.

**Interpretation:** The *P*-value is the probability of obtaining a sample mean of $\overline{x}$ or smaller under the assumption that $H_0$ is true. In other words, it is the probability of obtaining a sample mean that is more than $Z_0$ standard deviations to the left of $\mu_0$.

**Interpretation:** The *P*-value is the probability of obtaining a sample mean of $\overline{x}$ or larger under the assumption that $H_0$ is true. In other words, it is the probability of obtaining a sample mean that is more than $Z_0$ standard deviations to the right of $\mu_0$.

***Step 4:*** Reject the null hypothesis if the *P*-value is less than the level of significance, $\alpha$.

***Step 5:*** State the conclusion.

▶ EXAMPLE 3  **Computing the *P*-Value of a Right-Tailed Test**

*Problem:* Using the data from Example 1, test Patricia's claim using the *P*-value approach at the $\alpha = 0.01$ level of significance.

*Approach:* We will follow the steps just presented. Remember, we've already confirmed in Example 1 that the data come from a population that is approximately normal with no outliers.

*Solution:*

*Step 1:* Patricia claims that people are driving more than 10,300 miles annually. This claim can be written $\mu > 10,300$. We have

$$H_0: \mu = 10,300 \qquad \text{versus} \qquad H_1: \mu > 10,300$$

This is a right-tailed test.

*Step 2:* From Example 1, we have $\bar{x} = 12,346$, $\sigma = 3500$ and $n = 20$. We compute the test statistic, $Z_0$:

$$Z_0 = \frac{\bar{x} - \mu_0}{\sigma / \sqrt{n}} = \frac{12,346.5 - 10,300}{3500 / \sqrt{20}} = 2.61$$

*Step 3:* Because we are performing a right-tailed test, *P*-value $= P(Z > Z_0) = P(Z > 2.61)$. That is, we need to determine the area under the standard normal curve to the right of $Z = 2.61$ as shown in Figure 9.

Figure 9

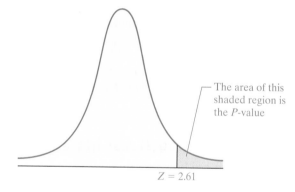

The area of this shaded region is the *P*-value

$Z = 2.61$

Using Table II, we have

$$\text{*P*-value} = P(Z > 2.61) = 1 - P(Z \le 2.61) = 1 - 0.9955 = 0.0045$$

The probability of obtaining a sample mean of 12,346.5 or higher from a population whose mean is 10,300 is 0.0045. So less than 1 sample in 100 will result in a sample mean of 12,346.5 (or higher) from a population whose mean is 10,300 (if the null hypothesis is true).

*Step 4:* If the *P*-value is less than the level of significance, we reject the null hypothesis. Because $0.0045 < 0.01$, we reject the null hypothesis.

*Step 5:* There is evidence to support the claim that the annual number of miles driven is higher than the 1990 level of 10,300 miles at the $\alpha = 0.01$ level of significance.  ◀◀

**NW** *Now Work Problem 19.*

▶ EXAMPLE 4  **Computing the *P*-Value of a Two-Tailed Test**

*Problem:* Use the data in Table 2 from Example 2 to test Grant's claim that the mean price of a three-year-old Corvette is different from $37,500 at the $\alpha = 0.1$ level of significance using the *P*-value approach.

there is not sufficient evidence to support Grant's claim that the mean price of a three-year-old Chevy Corvette is different from \$37,500 at the $\alpha = 0.1$ level of significance. ◀◀

**NW** *Now Work Problem 27.*

One of the more difficult aspects of hypothesis testing is identifying the appropriate procedure to use. In this section, you should notice that all the problems involve testing a population mean where $\sigma$ is known. In the next section, we test hypotheses about a population mean where $\sigma$ is unknown. You are encouraged to carefully read the problems in this section so that it will be easy to recognize when procedures presented in the section are appropriate.

## 9.2 Assess Your Understanding

### Concepts and Vocabulary

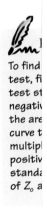

To find
test, fi
test st
negativ
the are
curve t
multipl
positiv
standa
of $Z_o$ a

1. State the requirements that must be satisfied in order to test a claim regarding a population mean with $\sigma$ known.
2. Determine the critical value for a right-tailed test regarding a population mean with $\sigma$ known at the $\alpha = 0.01$ level of significance.
3. Determine the critical value for a two-tailed test regarding a population mean with $\sigma$ known at the $\alpha = 0.05$ level of significance.
4. The procedures for testing a claim regarding a population mean with $\sigma$ known are robust. What does this mean?
5. Explain what a *P*-value is. What is the criterion for rejecting the null hypothesis using the *P*-value approach?

6. Suppose that we are testing the hypotheses $H_0: \mu = \mu_0$ versus $H_1: \mu < \mu_0$ and we find the *P*-value to be 0.23. Explain what this means. Would you reject $H_0$? Why?
7. Suppose that we are testing the hypotheses $H_0: \mu = \mu_0$ versus $H_1: \mu \neq \mu_0$ and we find the *P*-value to be 0.02. Explain what this means. Would you reject $H_0$? Why?
8. Discuss the advantages and disadvantages of using the classical approach to hypothesis testing. Discuss the advantages and disadvantages of using the *P*-value approach to hypothesis testing.

### Exercises

• **Basic Skills**

1. In order to test $H_0: \mu = 50$ versus $H_1: \mu < 50$, a random sample of size $n = 24$ is obtained from a population that is known to be normally distributed with $\sigma = 12$.
   (a) If the sample mean is determined to be $\bar{x} = 47.1$, compute the test statistic.
   (b) If the researcher decides to test this hypothesis at the $\alpha = 0.05$ level of significance, determine the critical value.
   (c) Draw a normal curve that depicts the critical region.
   (d) Will the researcher reject the null hypothesis? Why?

2. In order to test $H_0: \mu = 40$ versus $H_1: \mu > 40$, a random sample of size $n = 25$ is obtained from a population that is known to be normally distributed with $\sigma = 6$.
   (a) If the sample mean is determined to be $\bar{x} = 42.3$, compute the test statistic.
   (b) If the researcher decides to test this hypothesis at the $\alpha = 0.1$ level of significance, determine the critical value.
   (c) Draw a normal curve that depicts the critical region.
   (d) Will the researcher reject the null hypothesis? Why?

3. In order to test $H_0: \mu = 100$ versus $H_1: \mu \neq 100$ a random sample of size $n = 23$ is obtained from a population that is known to be normally distributed with $\sigma = 7$.
   (a) If the sample mean is determined to be $\bar{x} = 104.8$, compute the test statistic.
   (b) If the researcher decides to test this hypothesis at the $\alpha = 0.01$ level of significance, determine the critical values.
   (c) Draw a normal curve that depicts the critical regions.
   (d) Will the researcher reject the null hypothesis? Why?

4. In order to test $H_0: \mu = 80$ versus $H_1: \mu < 80$, a random sample of size $n = 22$ is obtained from a population that is known to be normally distributed with $\sigma = 11$.
   (a) If the sample mean is determined to be $\bar{x} = 76.9$, compute the test statistic.
   (b) If the researcher decides to test this hypothesis at the $\alpha = 0.02$ level of significance, determine the critical value.
   (c) Draw a normal curve that depicts the critical region.
   (d) Will the researcher reject the null hypothesis? Why?

**5.** In order to test $H_o$: $\mu = 20$ versus $H_1$: $\mu < 20$, a random sample of size $n = 18$ is obtained from a population that is known to be normally distributed with $\sigma = 3$.

   (a) If the sample mean is determined to be $\bar{x} = 18.3$, compute and interpret the $P$-value.

   (b) If the researcher decides to test this hypothesis at the $\alpha = 0.05$ level of significance, will the researcher reject the null hypothesis? Why?

**6.** In order to test $H_o$: $\mu = 4.5$ versus $H_1$: $\mu > 4.5$, a random sample of size $n = 13$ is obtained from a population that is known to be normally distributed with $\sigma = 1.2$.

   (a) If the sample mean is determined to be $\bar{x} = 4.9$, compute and interpret the $P$-value.

   (b) If the researcher decides to test this hypothesis at the $\alpha = 0.1$ level of significance, will the researcher reject the null hypothesis? Why?

**7.** In order to test $H_o$: $\mu = 105$ versus $H_1$: $\mu \neq 105$, a random sample of size $n = 35$ is obtained from a population whose standard deviation is known to be $\sigma = 12$.

   (a) Does the population need to be normally distributed in order to compute the $P$-value?

   (b) If the sample mean is determined to be $\bar{x} = 101.2$, compute and interpret the $P$-value.

   (c) If the researcher decides to test this hypothesis at the $\alpha = 0.02$ level of significance, will the researcher reject the null hypothesis? Why?

**8.** In order to test $H_o$: $\mu = 45$ versus $H_1$: $\mu \neq 45$, a random sample of size $n = 40$ is obtained from a population whose standard deviation is known to be $\sigma = 8$.

   (a) Does the population need to be normally distributed in order to compute the $P$-value?

   (b) If the sample mean is determined to be $\bar{x} = 48.3$, compute and interpret the $P$-value.

   (c) If the researcher decides to test this hypothesis at the $\alpha = 0.05$ level of significance, will the researcher reject the null hypothesis? Why?

---

## • Applying the Concepts

**9. Are Mothers Getting Older?** A researcher claims that the average age of a woman before she has her first child is greater than the 1990 mean age of 26.4 years, on the basis of data obtained from the *National Vital Statistics Report*, Vol. 48, No. 14. She obtains a simple random sample of 40 women who gave birth to their first child in 1999 and finds the sample mean age to be 27.1 years. Assume that the population standard deviation is 6.4 years. Test the researcher's claim, using the classical approach at the $\alpha = 0.05$ level of significance.

**10. Miles per Gallon** The Environmental Protection Agency publishes data regarding the miles per gallon of all cars. A researcher claims that fuel additives increase the miles per gallon of cars driven under highway driving conditions. The mean miles per gallon of all large cars manufactured in 1999 without the fuel additive is 25.1, on the basis of data obtained from the EPA. The researcher obtains a simple random sample of 35 large cars manufactured in 1999 and adds a fuel additive. The sample mean is determined to be 26.8 miles per gallon. Assume that the population standard deviation is 3.9. Test the researcher's claim, using the classical approach at the $\alpha = 0.05$ level of significance.

**11. SAT Exam Scores** A school administrator claims that students whose first language learned is not English score worse on the verbal portion of the SAT exam than students whose first language is English. The mean SAT verbal score of students whose first language is English is 516, on the basis of data obtained from the College Board. Suppose a simple random sample of 20 students whose first language learned was not English results in a sample mean SAT verbal score of 458. SAT verbal scores are normally distributed with a population standard deviation of 119.

   (a) Why is it necessary for SAT verbal scores to be normally distributed in order to test the claim using the methods of this section?

   (b) Test the researcher's claim using the classical approach at the $\alpha = 0.10$ level of significance.

**12. SAT Exam Scores** A school administrator claims that students whose first language learned is not English do not score differently on the SAT math portion of the SAT exam from students whose first language is English. The mean SAT math score of students whose first language is English is 516, on the basis of data obtained from the College Board. Suppose a simple random sample of 20 students whose first language learned was not English results in a sample mean SAT math score of 518. SAT math scores are normally distributed with a population standard deviation of 109.

   (a) Why is it necessary for SAT math scores to be normally distributed in order to test the claim using the methods of this section?

   (b) Test the researcher's claim using the classical approach at the $\alpha = 0.10$ level of significance.

13. **Acid Rain** In 1990, the mean pH level of the rain in Pierce County, Washington, was 5.03. A biologist claims that the acidity of rain has increased. (This would mean that the pH level of the rain has decreased.) From a random sample of 19 rain dates in 2000, she obtains the following data:

| | | | |
|---|---|---|---|
| 5.08 | 4.66 | 4.70 | 4.87 |
| 4.78 | 5.00 | 4.50 | 4.73 |
| 4.79 | 4.65 | 4.91 | 5.07 |
| 5.03 | 4.78 | 4.77 | 4.60 |
| 4.73 | 5.05 | 4.70 | |

*Source:* National Atmospheric Deposition Program

(a) Because the sample size is small, she must verify that pH level is normally distributed and the sample does not contain any outliers. The normal probability plot and boxplot are shown below. Are the conditions for testing the hypothesis satisfied?

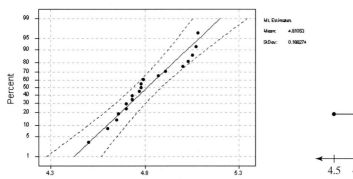

Normal Probability Plot for pH

(b) Test the hypothesis, assuming that $\sigma = 0.2$ at the $\alpha = 0.01$ level of significance.

14. **Age of the Bride** A sociologist claims that the mean age at which women marry in Memphis, Tennessee, is greater than the mean age of 25.0 throughout the United States, on the basis of data from the *Monthly Vital Statistics Report* published by the Centers for Disease Control. Based upon a random sample of 20 recently filed marriage certificates, she obtains the ages shown in the table on the right.

| | | | | |
|---|---|---|---|---|
| 40 | 23 | 30 | 24 | 31 |
| 29 | 28 | 24 | 35 | 34 |
| 24 | 21 | 46 | 29 | 31 |
| 29 | 29 | 21 | 33 | 39 |

*Source:* The Commercial Appeal, June 30, 2001

(a) Because the sample size is small, she must verify that age at marriage is normally distributed and the sample does not contain any outliers. The normal probability plot and boxplot are as shown below. Are the conditions for testing the hypothesis satisfied?

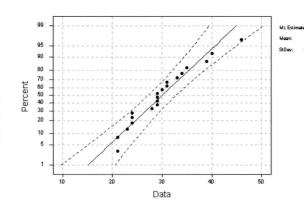

Normal Probability Plot for Age

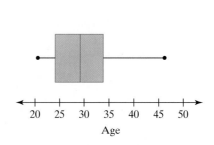

(b) Test the hypothesis, assuming that $\sigma = 6.2$ at the $\alpha = 0.05$ level of significance.

15. **Filling Bottles** A certain brand of apple juice is supposed to have 64 ounces of juice. Because the filling machine is not precise, the exact amount of juice varies from bottle to bottle. The quality control manager wishes to verify that the mean amount of juice in each bottle is 64 ounces, so she can be sure that the machine is not over- or under-filling. She randomly samples 22 bottles of juice and measures the content. She obtains the data on the right.

| | | | | | |
|---|---|---|---|---|---|
| 63.97 | 63.87 | 64.03 | 63.95 | 63.95 | 64.02 |
| 64.01 | 63.90 | 64.00 | 64.01 | 63.92 | 63.94 |
| 63.90 | 64.05 | 63.90 | 64.01 | 63.91 | 63.92 |
| 63.93 | 63.97 | 63.93 | 63.98 | | |

(a) Because the sample size is small, she must verify that the amount of juice is normally distributed and the sample does not contain any outliers. The normal probability plot and boxplot are shown below. Are the conditions for testing the hypothesis satisfied?

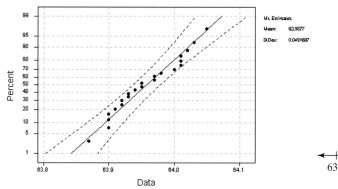

Normal Probability Plot for Ounces

(b) Test the hypothesis, assuming that $\sigma = 0.06$ at the $\alpha = 0.1$ level of significance.
(c) Should the assembly line be shut down so that the machine can be recalibrated?

16. **Price of Gasoline** In February of 2001, an executive of a major oil company claimed that the mean price of regular unleaded gasoline in Cook County, Illinois, was exactly \$1.56. A member of the county board feels that the mean price is different from \$1.56. He randomly sampled 20 gas stations and obtained the data on the right.

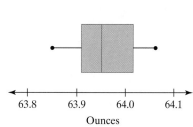

| | | | | |
|---|---|---|---|---|
| 1.52 | 1.61 | 1.55 | 1.58 | 1.66 |
| 1.51 | 1.58 | 1.55 | 1.58 | 1.61 |
| 1.58 | 1.59 | 1.57 | 1.53 | 1.59 |
| 1.56 | 1.61 | 1.57 | 1.65 | 1.53 |

(a) Because the sample size is small, he must verify that the price of gasoline is normally distributed and the sample does not contain any outliers. The normal probability plot and boxplot are shown below. Are the conditions for testing the hypothesis satisfied?

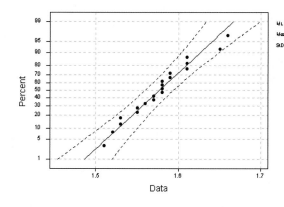

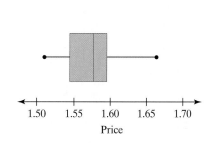

Normal Probability Plot for Price

(b) Test the hypothesis, assuming that $\sigma = 0.05$ at the $\alpha = 0.1$ level of significance.

**17. Are Cars Younger?** Suppose you have just been hired by Ford Motor Company. Management asks you to test the claim that cars are younger today versus 1995. According to the Nationwide Personal Transportation Survey conducted by the U.S. Department of Transportation, the mean age of a car in 1995 was 8.33 years. Based upon a random sample of 18 automobile owners, you obtain the ages as shown in the data on the right.

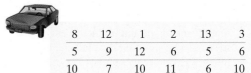

| 8 | 12 | 1 | 2 | 13 | 3 |
|---|----|---|---|----|---|
| 5 | 9 | 12 | 6 | 5 | 6 |
| 10 | 7 | 10 | 11 | 6 | 10 |

(a) Because the sample size is small, you must verify that the data are normally distributed with no outliers by drawing a normal probability plot and boxplot. Based on the following graphs, can you perform a hypothesis test?

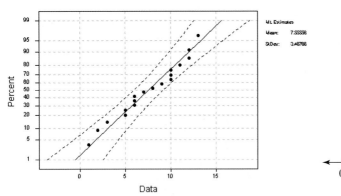

Normal Probability Plot for Age

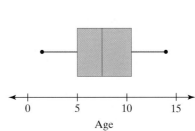

(b) Test the hypothesis, assuming that $\sigma = 3.8$ years at the $\alpha = 0.1$ level of significance.

**18. It's a Hot One!** Recently, a friend of mine claimed that the summer of 2000 in Houston, Texas was hotter than usual. To test his claim, I went to AccuWeather.com and randomly selected 12 days in the summer of 2000. I then recorded the departure from normal, with positive values indicating above-normal temperatures and negative values indicating below-normal temperatures, as shown in the data on the right.

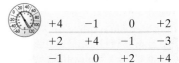

| +4 | −1 | 0 | +2 |
|----|----|---|----|
| +2 | +4 | −1 | −3 |
| −1 | 0 | +2 | +4 |

*Source:* AccuWeather.com

(a) Because the sample size is small, I must verify that the temperature departure is normally distributed and the sample does not contain any outliers. The normal probability plot and boxplot are shown below. Are the conditions for testing the hypothesis satisfied?

Normal Probability Plot for Departure

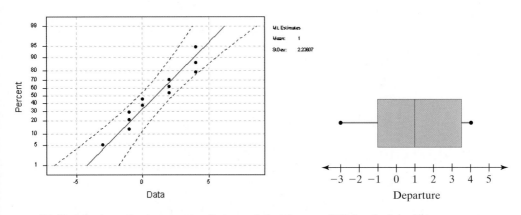

(b) Test the hypothesis, assuming that $\sigma = 1.8$ at the $\alpha = 0.05$ level of significance.

---

**19. Farm Size** In 1990, the average farm size in Kansas was 694 acres, according to data obtained from the U.S. Department of Agriculture. A researcher claims that farm sizes are larger now due to consolidation of farms. She obtains a random sample of 40 farms and determines the mean size to be 731 acres. Assume that $\sigma = 212$ acres. Test the researcher's claim, using the $P$-value approach at the $\alpha = 0.05$ level of significance.

**20. Oil Output** An energy official claims that the oil output per well in the United States has declined from the 1998 level of 11.1 barrels per day. He randomly samples 50 wells throughout the United States and determines the mean output to be 10.7 barrels per day. Assume that $\sigma = 1.3$ barrels. Test the researcher's claim using the *P*-value approach at the $\alpha = 0.05$ level of significance.

21. **Cellular Phone Bills** The mean monthly cellular telephone bill in 1999 was \$40.24, according to the Cellular Telecommunications Industry Association. A researcher at CTIA claims that the average monthly bill has increased since then. She conducts a survey of 49 cellular phone users and determines the mean bill to be \$45.15. Assume that $\sigma = \$21.20$. Test the researcher's claim, using the *P*-value approach at the $\alpha = 0.1$ level of significance.

22. **Length of a Cellular Phone Call** The mean length of a cellular telephone call in 1999 was 2.38 minutes, according to the Cellular Telecommunications Industry Association. A researcher at CTIA claims the average length of a phone call has increased since then. She conducts a survey of 49 cellular phone users and determines the mean length to be 2.48 minutes. Assume that $\sigma = 1.8$ minutes. Test the researcher's claim, using the *P*-value approach at the $\alpha = 0.01$ level of significance.

23. **Volume of Dell Computer Stock** The average daily volume of Dell Computer stock in 2000 was $\mu = 31.8$ million shares, with a standard deviation of $\sigma = 14.8$ million shares, according to Yahoo!Finance. A stock analyst claims that the stock volume in 2001 is different from the 2000 level. Based on a random sample of 35 trading days in 2001, he finds the sample mean to be 39.2 million shares. Test the analyst's claim, using the *P*-value approach at the $\alpha = 0.05$ level of significance.

24. **Volume of Motorola Stock** The average daily volume of Motorola stock in 2000 was $\mu = 11.4$ million shares, with a standard deviation of $\sigma = 8.3$ million shares, according to Yahoo!Finance. A stock analyst claims the stock volume in 2001 is different from the 2000 level. Based on a random sample of 35 trading days in 2001, he finds the sample mean to be 15.0 million shares. Test the analyst's claim, using the *P*-value approach at the $\alpha = 0.05$ level of significance.

25. **Using Confidence Intervals to Test Hypotheses** Test the claim made in Problem 15 by constructing a 95% confidence interval.

26. **Using Confidence Intervals to Test Hypotheses** Test the claim made in Problem 16 by constructing a 95% confidence interval.

27. **Using Confidence Intervals to Test Hypotheses** Test
(NW) the claim made in Problem 23 by constructing a 90% confidence interval.

28. **Using Confidence Intervals to Test Hypotheses** Test the claim made in Problem 24 by constructing a 90% confidence interval.

29. **Statistical Significance versus Practical Significance** A math teacher claims that she has developed a review course that increases the scores of students on the math portion of the SAT exam. Based on data from the College Board, SAT scores are normally distributed with

$\mu = 514$ and $\sigma = 113$. The teacher obtains a random sample of 1800 students, puts them through the review class, and finds that the mean SAT math score of the 1800 students is 518.

(a) State the null and alternative hypotheses.
(b) Test the claim at the $\alpha = 0.10$ level of significance. Is a mean SAT math score of 518 significantly higher than 514?
(c) Do you think that a mean SAT math score of 518 versus 514 will affect the decision of a school admissions administrator? In other words, does the increase in the score have any practical significance?
(d) Test the claim at the $\alpha = 0.10$ level of significance with $n = 400$ students. Assume that the sample mean is still 518. Is a sample mean of 518 significantly more than 514? Conclude that large sample sizes cause *P*-values to shrink substantially, all other things being the same.

30. **Simulation** Simulate drawing 50 simple random samples of size $n = 20$ from a population that is normally distributed with mean 80 and standard deviation 7.

(a) Test the null hypothesis $H_0: \mu = 80$ versus the alternative hypothesis $H_1: \mu \neq 80$.
(b) Suppose we were testing this hypothesis at the $\alpha = 0.1$ level of significance. How many of the 50 samples would you expect to result in a Type I error?
(c) Count the number of samples that lead to a rejection of the null hypothesis. Is it close to the expected value determined in part (b)?
(d) Describe how we know that a rejection of the null hypothesis results in making a Type I error in this situation.

31. **Simulation** Simulate drawing 40 simple random samples of size $n = 35$ from a population that is exponentially distributed with mean 8 and standard deviation $\sqrt{8}$.

(a) Test the null hypothesis $H_0: \mu = 8$ versus the alternative hypothesis $H_1: \mu \neq 8$.
(b) Suppose we were testing this hypothesis at the $\alpha = 0.05$ level of significance. How many of the 40 samples would you expect to result in a Type I error?
(c) Count the number of samples that lead to a rejection of the null hypothesis. Is it close to the expected value determined in part (b)?
(d) Describe how we know that a rejection of the null hypothesis results in making a Type I error in this situation.

32. Suppose a chemical company has developed a catalyst that is meant to reduce reaction time in a chemical process. For a certain chemical process, reaction time is known to be 150 seconds. The researchers conducted an experiment with the catalyst 40 times and measured reaction time. The researchers reported that the catalysts reduced reaction time with a *P*-value of 0.02.

(a) Identify the null and alternative hypotheses.
(b) Explain what this result means. Do you believe that the catalyst is effective?

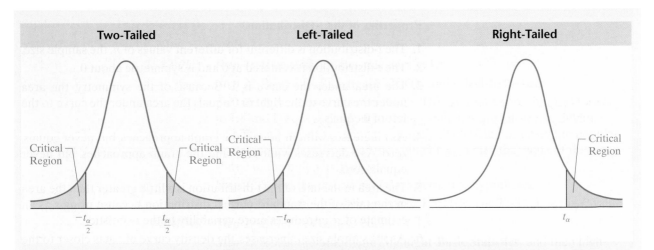

**Step 3:** Compute the test statistic $t = \dfrac{\bar{x} - \mu_0}{s/\sqrt{n}}$, which follows Student's

$t$-distribution with $n - 1$ degrees of freedom.

**Step 4:** Compare the critical value with the test statistic:

| Two-Tailed | Left-Tailed | Right-Tailed |
|---|---|---|
| If $t < -t_{\alpha/2}$ or $t > t_{\alpha/2}$, reject the null hypothesis | If $t < -t_\alpha$, reject the null hypothesis | If $t > t_\alpha$, reject the null hypothesis |

**Step 5:** State the conclusion.

Notice that the procedure just presented requires either that the population from which the sample was drawn be normal or that the sample size be large ($n \geq 30$). The procedure is robust, so minor departures from normality will not adversely affect the results of the test; however, if the data include outliers, the procedure should not be used. Just as we did for hypothesis tests with $\sigma$ known, we will verify this assumption by constructing normal probability plots (to assess normality) and boxplots (to discover whether there are outliers). If the normal probability plot indicates that the data do not come from a normal population or if the boxplot reveals outliers, nonparametric tests should be performed (not discussed in this text).

▶ **EXAMPLE 1** Caffeine Intake of 20–29-Year-Old Females

**TABLE 3**

| | | |
|---|---|---|
| 140.4 | 145.8 | 148.3 |
| 169.8 | 147.9 | 130.0 |
| 161.1 | 164.3 | 130.5 |
| 174.3 | 181.1 | 105.8 |
| 168.1 | 160.3 | 154.1 |
| 117.3 | 103.4 | 127.8 |
| 164.6 | 154.5 | |

*Problem:* In a study conducted by the U.S. Department of Agriculture, it was found that the mean daily caffeine intake of 20–29-year-old females in 1996 was 142.8 milligrams. A nutritionist claims that the mean daily caffeine intake has increased since then. She obtains a simple random sample of 20 females between 20 and 29 years of age and determines their daily caffeine intakes. The results are presented in Table 3. Test the nutritionist's claim at the $\alpha = 0.05$ level of significance.

*Approach:* Before we can perform the hypothesis test, we must verify that the data come from a population that is distributed approximately normally, with no outliers. We will construct a normal probability plot and boxplot to verify these requirements. We then proceed to follow Steps 1–5.

*Solution:* Figure 12 displays the normal probability plot and boxplot.

Figure 12

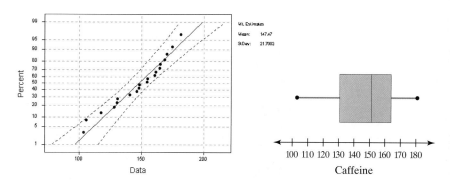

Normal Probability Plot for Caffeine

The normal probability plot indicates that the data come from a population that is approximately normal. The boxplot does not show any outliers. We can proceed to perform the hypothesis test.

*Step 1:* The nutritionist claims that 20–29-year-old females are consuming more than 142.8 mg of caffeine. This claim can be written $\mu > 142.8$. We have

$$H_o: \mu = 142.8 \qquad \text{versus} \qquad H_1: \mu > 142.8.$$

This is a right-tailed test.

*Step 2:* Because the nutritionist is performing a right-tailed test, we determine the critical *t*-value at the $\alpha = 0.05$ level of significance with $n - 1 = 20 - 1 = 19$ degrees of freedom to be $t_{0.05} = 1.729$. The critical region is displayed in Figure 13.

Figure 13

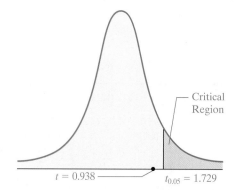

Critical Region

$t = 0.938$                    $t_{0.05} = 1.729$

*Step 3:* The nutritionist computes the sample mean, $\bar{x}$, to be 147.47 mg and the sample standard deviation, $s$, to be 22.26 mg. Because the data are distributed approximately normally, the test statistic is

$$t = \frac{\bar{x} - \mu_o}{s/\sqrt{n}} = \frac{147.47 - 142.8}{22.26/\sqrt{20}} = 0.938$$

The sample mean, 147.47 mg, is 0.938 sample standard deviation above the hypothesized mean, 142.8 mg.

*Step 4:* Because the test statistic $t = 0.938$ is less than the critical value $t_{0.05} = 1.729$, the nutritionist does not reject the null hypothesis. That is, the value of the test statistic does not fall within the critical region, so we do not reject $H_o$. We label this point in Figure 13.

*Step 5:* There is not sufficient evidence to support the nutritionist's claim that the mean number of milligrams of caffeine has increased from the 1996 level of 142.8 mg at the $\alpha = 0.05$ level of significance. ◄◄

**NW** *Now Work Problem 3.*

## Testing a Hypothesis about $\mu$, $\sigma$ Unknown, Using P-Values

We can also test hypotheses about a population mean with the population standard deviation unknown by using P-values. Recall that a P-value is the probability of observing a sample statistic as extreme or more extreme than the one observed, under the assumption that the null hypothesis is true. If the P-value is small, we have evidence against the null hypothesis. Because the t-distribution table (Table III) provides only t-values that correspond to certain areas, we cannot use the table to compute exact P-values.

However, we can use the table to calculate lower and upper bounds on the P-value. In order to find exact P-values, we need to use statistical software or a graphing calculator with advanced statistical features.

**In Your Own Words**

When $\sigma$ is unknown, exact P-values can be found only via technology.

▶ **EXAMPLE 2** Approximating a P-Value

*Problem:* Using the data from Example 1, test the nutritionist's claim, using the P-value approach at the $\alpha = 0.05$ level of significance.

*Approach:* Because this is a right-tailed test, the P-value is the area under the t-distribution with $20 - 1 = 19$ degrees of freedom to the right of the test statistic.

*Solution:* The test statistic is $t = 0.938$. Because we are performing a right-tailed test, P-value $= P(t > t_0) = P(t > 0.938)$, with 19 degrees of freedom—that is, we need to find the area under the t-distribution to the right of $t = 0.938$, as shown in Figure 14.

Using Table III, we find the row that corresponds to 19 degrees of freedom. The test statistic, $t = 0.938$, lies between 0.861 and 1.066. The value of 0.861 has 0.20 as the area under the t-distribution to the right. The area under the t-distribution with 19 degrees of freedom to the right of $t = 1.066$ is 0.15. See Figure 15.

**Figure 14**

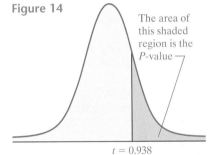

The area of this shaded region is the P-value

$t = 0.938$

**Figure 15**

| df | 0.25 | 0.20 | 0.15 | 0.10 | 0.05 | 0.025 | 0.02 | 0.01 | 0.005 | 0.0025 | 0.001 | 0.0005 |
|---|---|---|---|---|---|---|---|---|---|---|---|---|
| | | | | | | Area in Right Tail | | | | | | |
| 1 | 1.000 | 1.376 | 1.963 | 3.078 | 6.314 | 12.71 | 15.89 | 31.82 | 63.66 | 127.3 | 318.3 | 636.6 |
| 2 | 0.816 | 1.061 | 1.386 | 1.886 | 2.920 | 4.303 | 4.849 | 6.965 | 9.925 | 14.09 | 22.33 | 31.60 |
| 3 | 0.765 | 0.978 | 1.250 | 1.638 | 2.353 | 3.182 | 3.482 | 4.541 | 5.841 | 7.453 | 10.21 | 12.92 |
| 4 | 0.741 | 0.941 | 1.190 | 1.533 | 2.132 | 2.776 | 2.999 | 3.747 | 4.604 | 5.598 | 7.173 | 8.610 |
| 5 | 0.727 | 0.920 | 1.156 | 1.476 | 2.015 | 2.571 | 2.757 | 3.365 | 4.032 | 4.773 | 5.893 | 6.869 |
| 6 | 0.718 | 0.906 | 1.134 | 1.440 | 1.943 | 2.447 | 2.612 | 3.143 | 3.707 | 4.317 | 5.208 | 5.959 |
| 7 | 0.711 | 0.896 | 1.119 | 1.415 | 1.895 | 2.365 | 2.517 | 2.998 | 3.499 | 4.029 | 4.785 | 5.408 |
| 8 | 0.706 | 0.889 | 1.108 | 1.397 | 1.860 | 2.306 | 2.449 | 2.896 | 3.355 | 3.833 | 4.501 | 5.041 |
| 9 | 0.703 | 0.883 | 1.100 | 1.383 | 1.833 | 2.262 | 2.398 | 2.821 | 3.250 | 3.690 | 4.297 | 4.781 |
| 10 | 0.700 | 0.879 | 1.093 | 1.372 | 1.812 | 2.228 | 2.359 | 2.764 | 3.169 | 3.581 | 4.144 | 4.587 |
| 11 | 0.697 | 0.876 | 1.088 | 1.363 | 1.796 | 2.201 | 2.328 | 2.718 | 3.106 | 3.497 | 4.025 | 4.437 |
| 12 | 0.695 | 0.873 | 1.083 | 1.356 | 1.782 | 2.179 | 2.303 | 2.681 | 3.055 | 3.428 | 3.930 | 4.318 |
| 13 | 0.694 | 0.870 | 1.079 | 1.350 | 1.771 | 2.160 | 2.282 | 2.650 | 3.012 | 3.372 | 3.852 | 4.221 |
| 14 | 0.692 | 0.868 | 1.076 | 1.345 | 1.761 | 2.145 | 2.264 | 2.624 | 2.977 | 3.326 | 3.787 | 4.140 |
| 15 | 0.691 | 0.866 | 1.074 | 1.341 | 1.753 | 2.131 | 2.249 | 2.602 | 2.947 | 3.286 | 3.733 | 4.073 |
| 16 | 0.690 | 0.865 | 1.071 | 1.337 | 1.746 | 2.120 | 2.235 | 2.583 | 2.921 | 3.252 | 3.686 | 4.015 |
| 17 | 0.689 | 0.863 | 1.069 | 1.333 | 1.740 | 2.110 | 2.224 | 2.567 | 2.898 | 3.222 | 3.646 | 3.965 |
| 18 | 0.688 | 0.862 | 1.067 | 1.330 | 1.734 | 2.101 | 2.214 | 2.552 | 2.878 | 3.197 | 3.611 | 3.922 |
| 19 | 0.688 | 0.861 | 1.066 | 1.328 | 1.729 | 2.093 | 2.205 | 2.539 | 2.861 | 3.174 | 3.579 | 3.883 |
| 20 | 0.687 | 0.860 | 1.064 | 1.325 | 1.725 | 2.086 | 2.197 | 2.528 | 2.845 | 3.153 | 3.552 | 3.850 |

Because $t = 0.938$ is between 0.861 and 1.066, we know that the P-value is between 0.15 and 0.20. We will present the P-value as follows:

$$0.15 < P\text{-value} < 0.20$$

Because the *P*-value is greater than the level of significance, $\alpha = 0.05$, we do not reject the null hypothesis. There is not sufficient evidence to support the nutritionist's claim.   ◀◀

Suppose we desire a more accurate result. We can use a graphing calculator with advanced statistical features or statistical software to obtain the *P*-value. Figure 16(a) shows the results obtained from a TI-83 Plus graphing calculator by using the calculate option; Figure 16(b) shows the results from using the draw option. The *P*-value is about 0.18.

**Figure 16**   *P*-value ⟶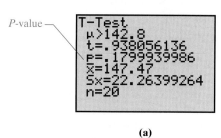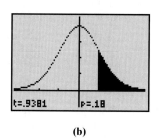

(a)                                          (b)

**NW**  *Now Work Problem 5.*

Sections 9.2 and 9.3 discussed performing hypothesis tests about a population mean. The main criterion for choosing which test to use is whether the population standard deviation, $\sigma$, is known. Provided that the population from which the sample is drawn is normal or that the sample size is large,

- if $\sigma$ is known, use the *z*-test procedures from Section 9.2;
- if $\sigma$ is unknown, use the *t*-test procedures from Section 9.3.

In Section 9.4, we will discuss testing hypotheses about a population proportion.

## 9.3 Assess Your Understanding

### Concepts and Vocabulary

**1.** State the requirements that must be satisfied in order to test a claim about a population mean with $\sigma$ unknown.

**2.** Determine the critical value for a right-tailed test of a population mean with $\sigma$ unknown at the $\alpha = 0.01$ level of significance with 15 degrees of freedom.

**3.** Determine the critical value for a two-tailed test of a population mean with $\sigma$ unknown at the $\alpha = 0.05$ level of significance with 12 degrees of freedom.

**4.** Determine the critical value for a left-tailed test of a population mean with $\sigma$ unknown at the $\alpha = 0.05$ level of significance with 19 degrees of freedom.

### Exercises

- **Basic Skills**

**1.** In order to test $H_0: \mu = 50$ versus $H_1: \mu < 50$, a simple random sample of size $n = 24$ is obtained from a population that is known to be normally distributed.
  - (a) If $\bar{x} = 47.1$ and $s = 10.3$, compute the test statistic.
  - (b) If the researcher decides to test this hypothesis at the $\alpha = 0.05$ level of significance, determine the critical value.
  - (c) Draw a *t*-distribution that depicts the critical region.
  - (d) Will the researcher reject the null hypothesis? Why?

**2.** In order to test $H_0: \mu = 40$ versus $H_1: \mu > 40$, a simple random sample of size $n = 25$ is obtained from a population that is known to be normally distributed.

  - (a) If $\bar{x} = 42.3$ and $s = 4.3$, compute the test statistic.
  - (b) If the researcher decides to test this hypothesis at the $\alpha = 0.1$ level of significance, determine the critical value.
  - (c) Draw a *t*-distribution that depicts the critical region.
  - (d) Will the researcher reject the null hypothesis? Why?

**3.** In order to test $H_0: \mu = 100$ versus $H_1: \mu \neq 100$, a simple random sample of size $n = 23$ is obtained from a population that is known to be normally distributed.
  - (a) If $\bar{x} = 104.8$ and $s = 9.2$, compute the test statistic.
  - (b) If the researcher decides to test this hypothesis at the $\alpha = 0.01$ level of significance, determine the critical values.

(c) Draw a $t$-distribution that depicts the critical region.

(d) Will the researcher reject the null hypothesis? Why?

**4.** In order to test $H_0: \mu = 80$ versus $H_1: \mu < 80$, a simple random sample of size $n = 22$ is obtained from a population that is known to be normally distributed.

(a) If $\bar{x} = 76.9$ and $s = 8.5$, compute the test statistic.

(b) If the researcher decides to test this hypothesis at the $\alpha = 0.02$ level of significance, determine the critical value.

(c) Draw a $t$-distribution that depicts the critical region.

(d) Will the researcher reject the null hypothesis? Why?

**5.** In order to test $H_0: \mu = 20$ versus $H_1: \mu < 20$, a simple random sample of size $n = 18$ is obtained from a population that is known to be normally distributed.

(a) If $\bar{x} = 18.3$ and $s = 4.3$, compute the test statistic.

(b) Draw a $t$-distribution with the area that represents the $P$-value shaded.

(c) Approximate and interpret the $P$-value.

(d) If the researcher decides to test this hypothesis at the $\alpha = 0.05$ level of significance, will the researcher reject the null hypothesis? Why?

**6.** In order to test $H_0: \mu = 4.5$ versus $H_1: \mu > 4.5$, a simple random sample of size $n = 13$ is obtained from a population that is known to be normally distributed.

(a) If $\bar{x} = 4.9$ and $s = 1.3$, compute the test statistic.

(b) Draw a $t$-distribution with the area that represents the $P$-value shaded.

(c) Approximate and interpret the $P$-value.

(d) If the researcher decides to test this hypothesis at the $\alpha = 0.1$ level of significance, will the researcher reject the null hypothesis? Why?

**7.** In order to test $H_0: \mu = 105$ versus $H_1: \mu \neq 105$, a simple random sample of size $n = 35$ is obtained.

(a) Does the population need to be normally distributed in order to test this hypothesis by using the methods presented in this section?

(b) If $\bar{x} = 101.9$ and $s = 5.9$, compute the test statistic.

(c) Draw a $t$-distribution with the area that represents the $P$-value shaded.

(d) Determine and interpret the $P$-value.

(e) If the researcher decides to test this hypothesis at the $\alpha = 0.01$ level of significance, will the researcher reject the null hypothesis? Why?

**8.** In order to test $H_0: \mu = 45$ versus $H_1: \mu \neq 45$, a simple random sample of size $n = 40$ is obtained.

(a) Does the population need to be normally distributed in order to test this hypothesis by using the methods presented in this section?

(b) If $\bar{x} = 48.3$ and $s = 8.5$, compute the test statistic.

(c) Draw a $t$-distribution with the area that represents the $P$-value shaded.

(d) Determine and interpret the $P$-value.

(e) If the researcher decides to test this hypothesis at the $\alpha = 0.01$ level of significance, will the researcher reject the null hypothesis? Why?

• **Applying the Concepts**

**9. Effects of Alcohol on the Brain** In a study published in the *American Journal of Psychiatry* [157:737–744, May 2000], researchers wanted to measure the effect of alcohol on the development of the hippocampal region in adolescents. The hippocampus is the portion of the brain responsible for long-term memory storage. The researchers randomly selected 12 adolescents with adolescent-onset alcohol use disorders. They wanted to test the claim that the hippocampal volumes in the alcoholic adolescents were less than the normal volume of $9.02 \text{ cm}^3$. An analysis of the sample data revealed that the hippocampal volume is approximately normal with $\bar{x} = 8.10$ and $s = 0.7$

(a) Test the researchers' claim at the $\alpha = 0.01$ level of significance, using the classical approach.

(b) Approximate and interpret the $P$-value.

**10. Vitamin A Supplements** In an experiment meant to investigate the effect of Vitamin A supplements on serum retinol levels in low-birth-weight babies, 65 low-birth-weight infants were administered 25,000 IU of vitamin A. ["Effect of Vitamin A Supplementation on Morbidity of Low-Birth-Weight Neonates, *South African Medical Journal*, July 2000, 90(7), pp. 730–736] The sample mean serum retinol level of the infants administered the Vitamin A was determined to be 45.77, with a sample standard deviation equal to 17.07.

(a) Why is it not necessary to verify that the data are normally distributed in order to perform inference on the sample mean?

(b) At the $\alpha = 0.01$ level of significance, test the claim that vitamin A increased the serum retinol level above 12.88 micrograms/dl, the mean level of serum retinol for all low-birth-weight babies, using the classical approach.

(c) Determine and interpret the $P$-value.

**11. Getting Enough Fiber?** A nutritionist claims that the mean daily consumption of fiber for 20–39-year-old males is less than 20 grams per day. (The National Cancer Institute recommends that individuals consume 20–30 grams per day.) In a survey of 457 males who were 20–39 years old, conducted by the U.S. Department of Agriculture, it was found that the mean daily intake of fiber was 19.1 grams, with standard deviation 9.1 grams.

(a) Using the classical approach, decide whether there is enough evidence to support the nutritionist's claim at the $\alpha = 0.01$ level of significance.

(b) Determine and interpret the $P$-value.

12. **Too Much Sodium?** A nutritionist claims that the mean daily consumption of sodium for 20–39-year-old females is more than 2750 mg. (The USDA recommended daily allowance of sodium is 2400 mg.) In a survey conducted by the United States Department of Agriculture of 490 females who were 20–39 years old, it was found that the mean daily intake of sodium was 2919 milligrams, with a standard deviation of 1350 milligrams.

    (a) Using the classical approach, decide whether there is sufficient evidence to support the nutritionist's claim at the $\alpha = 0.05$ level of significance.

    (b) Determine and interpret the $P$-value.

13. **Normal Temperature** Carl Reinhold August Wunderlich said that the mean temperature of humans is 98.6° F. Researchers Philip Mackowiak, Steven Wasserman, and Myron Levine [*JAMA*, Sep 23–30 1992; 268(12):1578–80] felt that the mean temperature of humans is less than 98.6° F. They measured the temperature of 148 healthy adults 1–4 times daily for three days, obtaining 700 measurements. The sample data resulted in a sample mean of 98.2° F and a sample standard deviation of 0.7° F.

    (a) Using the classical approach, judge whether there is evidence to support the researchers' claim at the $\alpha = 0.01$ level of significance.

    (b) Determine and interpret the $P$-value.

14. **Normal Temperature** Carl Reinhold August Wunderlich said that the mean temperature of humans is 98.6° F. Researchers Philip Mackowiak, Steven Wasserman, and Myron Levine [*JAMA*, Sep 23–30 1992; 268(12):1578–80] measured the temperatures of 26 females 1–4 times daily for three days to get a total of 123 measurements. The sample data yielded a sample mean of 98.4° F and a sample standard deviation of 0.7° F.

    (a) Using the classical approach, judge whether there is evidence to support the claim that the normal temperature of women is less than 98.6°F at the $\alpha = 0.01$ level of significance.

    (b) Determine and interpret the $P$-value.

15. **Age of Death-Row Inmates** In 1989, the average age of an inmate on death row was 36.2 years of age, according to data obtained from the U.S. Department of Justice. A sociologist wants to test the claim that the average age of a death-row inmate has changed since then. She randomly selects 32 death-row inmates and finds that their mean age is 38.9, with a standard deviation of 9.6.

    (a) Using the classical approach, test the sociologist's claim at the $\alpha = 0.05$ level of significance.

    (b) Determine and interpret the $P$-value.

16. **Energy Consumption** In 1997, the average household expenditure for energy was $1338, according to data obtained from the U.S. Energy Information Administration. An economist wanted to know whether this amount has changed significantly from its 1997 level. In a random sample of 35 households, he found the mean expenditure (in 1997 dollars) for energy during the most recent year to be $1423, with standard deviation $321.

    (a) Using the classical approach, test the economist's claim that the mean expenditure has changed significantly from the 1997 level at the $\alpha = 0.05$ level of significance.

    (b) Determine and interpret the $P$-value.

17. **Cholesterol** A dietician maintains that the total cholesterol for 40–49-year-old males is high. Anyone with total cholesterol above 200 is considered to have high cholesterol. She conducts a random sample of 40 males between the ages of 40 and 49 years and finds that their mean total cholesterol is 211, with a standard deviation of 39.2 (data obtained from the Centers for Disease Control).

    (a) Using the classical approach, test the dietician's claim that the total cholesterol of males 40–49 years old is more than 200 at the $\alpha = 0.05$ level of significance.

    (b) Determine and interpret the $P$-value.

18. **LDL Cholesterol** A person's low-density lipoprotein (LDL) cholesterol is the so-called "bad" cholesterol, because it contributes to the build-up of plaque in the person's arteries. This plaque restricts blood flow to the heart or brain. If blood flow is restricted to the heart, a heart attack results. If blood flow is restricted to the brain, a stroke results. A dietician claims that the LDL cholesterol for 40–49-year-old females is more than 130. (The American Heart Association states that individuals with an LDL cholesterol higher than 130 are at an increased risk of heart attack or stroke.) In a random sample of 35 females between the ages of 40 and 49 years, it was found that their mean LDL was 147, with a standard deviation of 31.9 (based upon data obtained from the Centers for Disease Control).

    (a) Using the classical approach, test the dietician's claim that the mean LDL cholesterol is more than 130 mg at the $\alpha = 0.01$ level of significance.

    (b) Determine and interpret the $P$-value.

19. **Conforming Golf Balls** The United States Golf Association requires that golf balls have a mean diameter that is 1.68 inches. An engineer for the USGA wishes to discover whether Maxfli XS golf balls have a mean diameter different from 1.68 inches. A random sample of Maxfli XS golf balls was selected; their diameters are shown in the table on the right.

| 1.683 | 1.677 | 1.681 |
|-------|-------|-------|
| 1.685 | 1.678 | 1.686 |
| 1.684 | 1.684 | 1.673 |
| 1.685 | 1.682 | 1.674 |

*Source:* Michael McCraith, Joliet Junior College

(a) Because the sample size is small, he must verify that the diameter is normally distributed and the sample does not contain any outliers. The normal probability plot and boxplot are shown below. Are the conditions for testing the hypothesis satisfied?

Normal Probability Plot for Diameter

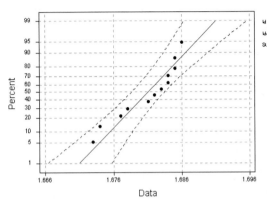

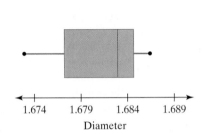

(b) Using the classical approach, test the claim that the golf balls have a mean diameter that is different from 1.68 inches at the $\alpha = 0.05$ level of significance.

(c) Determine and interpret the $P$-value.

20. **Conforming Golf Balls** The USGA requires that golf balls have a weight that is less than 1.62 ounces. An engineer for the USGA wants to test the claim that Maxfli XS golf balls have a mean weight less than 1.62 ounces. He obtains a random sample of 12 Maxfli XS golf balls; their weights are in the table on the right.

| 1.614 | 1.619 | 1.614 |
|-------|-------|-------|
| 1.614 | 1.610 | 1.610 |
| 1.621 | 1.612 | 1.615 |
| 1.621 | 1.602 | 1.617 |

*Source:* Michael McCraith, Joliet Junior College

(a) Because the sample size is small, he must verify that weight is normally distributed and that the sample does not contain any outliers. The normal probability plot and boxplot are shown below. Are the conditions for testing the hypothesis satisfied?

Normal Probability Plot for Weight

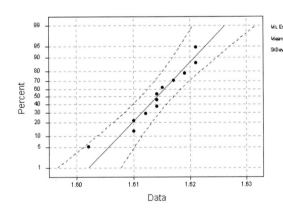

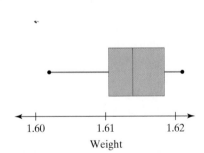

(b) Using the classical approach, decide whether the golf balls meet Maxfli's standard at the $\alpha = 0.1$ level of significance.

(c) Determine and interpret the $P$-value.

**21. Calibrating a pH Meter** An engineer wants to measure the bias in a pH meter. She uses the meter to measure the pH in 14 neutral substances (pH = 7.0) and obtains the data on the right.

| 7.01 | 7.04 | 6.97 | 7.00 | 6.99 | 6.97 | 7.04 |
| 7.04 | 7.01 | 7.00 | 6.99 | 7.04 | 7.07 | 6.97 |

(a) Because the sample size is small, she must verify that pH is normally distributed and the sample does not contain any outliers. The normal probability plot and boxplot are shown below. Are the conditions for testing the hypothesis satisfied?

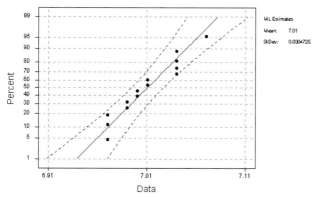

(b) Using the classical approach, is there sufficient evidence to support the claim that the pH meter is not correctly calibrated at the $\alpha = 0.05$ level of significance?

(c) Determine and interpret the $P$-value.

**22. Medication in a Tablet** A drug states that the mean amount of Naproxen Sodium, the active ingredient in reducing pain, in a tablet is 220 mg. A pharmacist randomly selects 17 tablets and measures the amount of Naproxen Sodium in each of them. She obtains the following data:

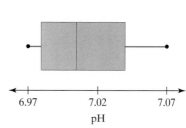

| 219.9 | 219.9 | 224.9 | 218.1 | 224.5 |
| 222.5 | 221.8 | 219.5 | 220.3 | 221.6 |
| 221.3 | 222.9 | 226.3 | 218.5 | 218.1 |
| 222.5 | 218.3 | | | |

(a) Because the sample size is small, she must verify that the amount of Naproxen Sodium in the tablet is normally distributed and the sample does not contain any outliers. The normal probability plot and boxplot are shown below. Are the conditions for testing the hypothesis satisfied?

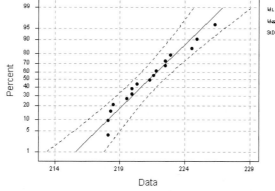

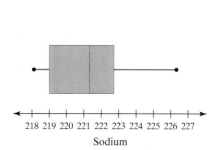

(b) Using the classical approach, is there sufficient evidence to support the claim that the amount of Naproxen Sodium is different from 220 mg at the $\alpha = 0.01$ level of significance?

(c) Determine and interpret the $P$-value.

**23. A Bat's Range** In order to catch flying insects, bats emit high-frequency sounds and then determine the time until they hear an echo. When an insect is detected, the bat can determine the location of the insect by the time it takes the echo to return. Suppose the researchers want to test the claim that the bat's range is more than 35 cm. They collect the following data regarding the detection distance (in centimeters) of 11 bats:

| 62 | 52 | 68 | 23 | 34 | 45 |
|----|----|----|----|----|----|
| 27 | 42 | 83 | 56 | 40 |    |

*Source:* Griffen, Donald R.; Webster, Frederick A.; and Michael, Charles R. "The Echolocation of Flying Insects by Bats." *Animal Behavior,* 8 (1960), p. 148.

(a) Verify that the detection distance of bats is normally distributed, and check for outliers by drawing a normal probability plot and boxplot.
(b) Test the researchers' claim at the $\alpha = 0.05$ level of significance.
(c) Determine and interpret the *P*-value.

**24. P/E Ratio** A stock analyst believes that the price-to-earnings (P/E) ratio of companies listed on the Standard and Poor's 500 (S&P 500) Index is less than its December 1, 2000 level of 22.0, in response to economic uncertainty. The P/E ratio is the price an investor is willing to pay for $1 of earnings. For example, a P/E of 23 means the investor pays $23 for each $1 of earnings. A higher P/E is an indication of investor optimism; lower P/Es are generally assigned to companies with lower earnings growth. In order to test his claim, he randomly samples 14 companies listed on the S&P 500 and calculates their P/E ratios. He obtains the following data:

| Company | P/E Ratio | Company | P/E Ratio |
|---------|-----------|---------|-----------|
| Boeing | 25.1 | Dow Chemical | 13.7 |
| General Motors | 7.8 | Citigroup | 18.5 |
| Halliburton | 35.8 | Merck and Co. | 26.8 |
| Norfolk Southern | 25.6 | Sara Lee | 11.9 |
| Agilent Technologies | 22.5 | Harley-Davidson | 37.4 |
| Old Kent Financial | 20.0 | Circuit City | 14.5 |
| Cendent | 15.0 | Minnesota Mining and Manufacturing | 23.8 |

*Source:* Checkfree Corporation

(a) Verify that P/E ratios are normally distributed, and check for outliers by drawing a normal probability plot and boxplot.
(b) Test the analyst's claim at the $\alpha = 0.05$ level of significance.
(c) Determine and interpret the *P*-value.

**25. Benign Prostatic Hyperplasia** Benign prostatic hyperplasia is a common cause of urinary outflow obstruction in aging males. The efficacy of Cardura (doxazosin mesylate) was measured in clinical trials of 173 patients with benign prostatic hyperplasia. Researchers wanted to discover whether Cardura significantly increased the urinary flow rate. It was found that an average increase of 0.8 mL/sec was obtained. This was said to be significant with a *P*-value less than 0.01. State the null and alternative hypotheses of the researchers and interpret the *P*-value.

**26. Systolic Blood Pressure of Surgical Patients** A nursing student maintained that the mean systolic blood pressure of her male patients on the surgical floor was less than 130 mm Hg. She randomly selected 19 male surgical patients and collected the systolic blood pressures shown on the right.

(a) Because the sample size is small, she must verify that the systolic blood pressure is normally distributed and the sample does not contain any outliers. The normal probability plot and boxplot are shown on page 419. Are the conditions for testing the hypothesis satisfied?

| 116 | 150 | 140 | 148 | 105 |
|-----|-----|-----|-----|-----|
| 118 | 128 | 112 | 124 | 128 |
| 140 | 112 | 126 | 130 | 120 |
| 90 | 134 | 112 | 142 |    |

*Source:* Lora McGuire, Nursing Instructor, Joliet Junior College

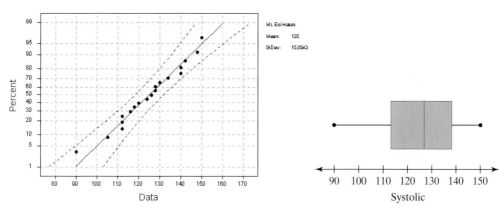

Normal Probability Plot for Systolic

(b) The student enters the data into Minitab and obtains the following results:

### T-Test of the Mean

```
Test of mu = 130.00 vs mu < 130.00
```

| Variable | N | Mean | StDev | SE Mean | T | P |
|---|---|---|---|---|---|---|
| Systolic | 19 | 125 | 15.47 | 3.55 | −1.41 | 0.088 |

What are the null and alternative hypotheses? Identify the *P*-value. Will the nursing student reject the null hypothesis at the $\alpha = 0.05$ level of significance? State her conclusion.

27. **Temperature of Surgical Patients** A nursing student suspects that the mean temperature of surgical patients is above the normal temperature, 98.2°F. (See Problem 13.) She randomly selects 32 surgical patients and obtains the temperatures shown on the right.
   (a) What are the null and alternative hypotheses of the student?
   (b) The student enters the data into Minitab and obtains the following results:

| | | | | | |
|---|---|---|---|---|---|
| 97.4 | 98.6 | 98.2 | 98.2 | 98.4 | 98.6 |
| 99.8 | 97.7 | 97.8 | 98.9 | 97.8 | 97.8 |
| 96.7 | 97.8 | 98.3 | 98.5 | 98.0 | 98.7 |
| 96.8 | 98.6 | 98.4 | 97.4 | 99.1 | 98.7 |
| 97.8 | 99.2 | 99.2 | 98.1 | 98.6 | 98.4 |
| 99.2 | 98.6 | | | | |

*Source:* Lora McGuire, Nursing Instructor, Joliet Junior College

### T-Test of the Mean

```
Test of mu = 98.200 vs mu > 98.200
```

| Variable | N | Mean | StDev | SE Mean | T | P |
|---|---|---|---|---|---|---|
| Temperat | 32 | 98.291 | 0.689 | 0.122 | 0.74 | 0.23 |

What is the *P*-value of the test? State the nursing student's conclusion.

28. **Soybeans** The average yield per acre of soybeans on farms in Iowa in 1999 was 45 bushels, according to data obtained from the U.S. Department of Agriculture. A farmer in Iowa claims the yield is higher this year. He randomly samples 15 acres on his farm and determines the mean yield to be 48.7 bushels, with a standard deviation of 2.48 bushels. He computes the *P*-value to be less than 0.0001 and concludes that the U.S. Department of Agriculture was wrong. Why should his conclusions be looked upon with skepticism?

29. **Simulation** Simulate drawing 40 simple random samples of size $n = 20$ from a population that is normally distribution with mean 50 and standard deviation 10.
   (a) Test the null hypothesis $H_0: \mu = 50$ versus the alternative hypothesis $H_1: \mu \neq 50$ for each of the 40 samples using a *t*-test.
   (b) Suppose we were testing this hypothesis at the $\alpha = 0.05$ level of significance. How many of the 40 samples would you expect to result in a Type I error?
   (c) Count the number of samples that lead to a rejection of the null hypothesis. Is it close to the expected value determined in part (b)?
   (d) Describe why we know a rejection of the null hypothesis results in making a Type I error in this situation.

## Technology Step-by-Step
Hypothesis Tests Regarding $\mu$, $\sigma$ Unknown

**TI-83 Plus**   **Step 1:** If necessary, enter raw data in L1.

**Step 2:** Press STAT, highlight TESTS, and select 2:T-Test.

**Step 3:** If the data are raw, highlight DATA; make sure that List1 is set to L1 and Freq is set to 1. If summary statistics are known, highlight STATS and enter the summary statistics. For the value of $\mu_0$, enter the value of the mean stated in the null hypothesis.

**Step 4:** Select the direction of the alternative hypothesis.

**Step 5:** Highlight **Calculate** and press ENTER. The TI-83 gives the *P*-value.

**MINITAB**   **Step 1:** Enter raw data in column C1.

**Step 2:** Select the **Stat** menu, highlight **Basic Statistics**, then highlight **1-Sample t**....

**Step 3:** Enter C1 in the cell marked "Variables." Select "Test Mean," and enter the value of the mean stated in the null hypothesis. In the cell marked "Alternative," select the direction of the alternative hypothesis. Click OK.

**Excel**   **Step 1:** If necessary, enter raw data in column A.

**Step 2:** Load the PHStat Add-in.

**Step 3:** Select the **PHStat menu**, highlight **One Sample Tests** ..., and then highlight **t Test for the mean, sigma unknown** ....

**Step 4:** Enter the value of the null hypothesis and the level of significance, $\alpha$. If the summary statistics are known, click "Sample statistics known" and enter the sample size, sample mean, and sample standard deviation. If summary statistics are unknown, click "Sample statistics unknown." With the cursor in the "Sample cell range" cell, highlight the data in column A. Click the option corresponding to the desired test (two-tail, upper (right) tail, or lower (left) tail). Click OK.

---

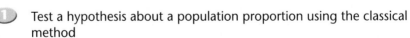

## 9.4   Testing a Hypothesis about a Population Proportion

**Preparing for This Section**   Before getting started, review the following:

   ✓ Binomial probability distribution (Section 6.2, pp. 253–262)

   ✓ Confidence intervals about a population proportion (Section 8.3, pp. 370–372)

**Objectives**   ①   Test a hypothesis about a population proportion using the classical method

②   Test a hypothesis about a population proportion using the *P*-value approach

③   Test a hypothesis about a population proportion on a small sample

①   Recall that the best point estimate of *p*, the proportion of the population with a certain characteristic, is given by

$$\hat{p} = \frac{x}{n}$$

where *x* is the number of individuals in the sample with the specified characteristic and *n* is the sample size. Recall, from Section 8.3, that the sampling distribution of $\hat{p}$ is approximately normal, with mean $\mu_{\hat{p}} = p$ and

**Caution**

When determining the standard error for the sampling distribution of $\hat{p}$, use the assumed value of the population proportion, $p_0$.

standard deviation $\sigma_{\hat{p}} = \sqrt{\dfrac{p(1-p)}{n}}$, provided that $np(1-p) \geq 10.$*

The classical approach to testing a hypothesis about the population proportion, $p$, follows the exact same logic as the testing of hypotheses about a population mean with $\sigma$ known. The only difference is that the **test statistic** is

$$Z = \frac{\hat{p} - p_0}{\sqrt{\dfrac{p_0(1 - p_0)}{n}}}$$

where $p_0$ is the assumed value of the population proportion.

The careful reader will notice that we are using $p_0$ in computing the standard error rather than $\hat{p}$ (as we did in computing confidence intervals about $p$). This is because, when we test a hypothesis, the null hypothesis is always assumed true. Therefore, we are assuming that the population proportion is $p_0$.

---

**Hypothesis Test Regarding a Population Proportion, $p$**

If a claim is made regarding the population proportion, we can use the following steps to test the claim, provided that
1. the sample is obtained via simple random sampling, and
2. $np_0(1 - p_0) \geq 10$ with $n \leq 0.05N$ (the sample size, $n$, is no more than 5% of the population size, $N$).

**Step 1:** A claim is made regarding the population proportion. The claim is used to determine the null and alternative hypotheses. The hypotheses can be structured in one of three ways:

| Two-Tailed | Left-Tailed | Right-Tailed |
|---|---|---|
| $H_0: p = p_0$ | $H_0: p = p_0$ | $H_0: p = p_0$ |
| $H_1: p \neq p_0$ | $H_1: p < p_0$ | $H_1: p > p_0$ |

*Note*: $p_0$ is the assumed value of the population proportion.

**Step 2:** Select a level of significance, $\alpha$, to reflect the seriousness of making a Type I error. The level of significance is used to determine the critical value. The critical value represents the maximum number of standard deviations the sample proportion can be from $p_0$ before the null hypothesis is rejected. For example, the critical value in the left-tailed test is $-z_\alpha$. The shaded region(s) represent(s) the critical region. The critical region is the set of all values such that the null hypothesis is rejected.

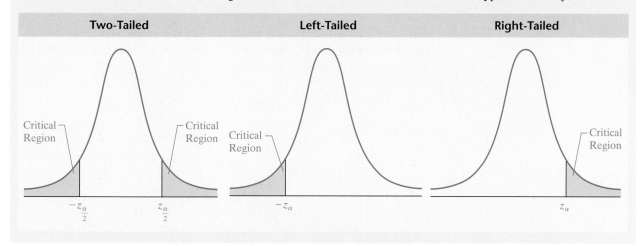

*The sample must be a simple random sample with $n \leq 0.05N$ (i.e., the sample size is no more than 5% of the population size).

**Step 3:** Compute the test statistic:

$$Z = \frac{\hat{p} - p_0}{\sqrt{\dfrac{p_0(1 - p_0)}{n}}}$$

**Step 4:** Compare the critical value with the test statistic.

| Two-Tailed | Left-Tailed | Right-Tailed |
|---|---|---|
| If $Z < -z_{\alpha/2}$ or $Z > z_{\alpha/2}$, reject the null hypothesis. | If $Z < -z_{\alpha}$, reject the null hypothesis. | If $Z > z_{\alpha}$, reject the null hypothesis. |

**Step 5:** State the conclusion.

▶ EXAMPLE 1    **Who Are Thought to Be More Aggressive, Men or Women?**

*Problem:* In 1995, 74% of Americans felt that men were more aggressive than women. In a poll conducted by the Gallup Organization from December 2–4, 2000, a simple random sample of 1026 Americans 18 years old or older resulted in 698 respondents stating that men were more aggressive than women. Is there significant evidence to indicate that the proportion of Americans who believe that men are more aggressive than women has decreased from the level reported in 1995 at the $\alpha = 0.05$ level of significance?

*Approach:* We must verify the requirements to perform the hypothesis test—namely, the sample must be a simple random sample and $np_0(1 - p_0) \geq 10$. In addition, the sample size cannot be more than 5% of the population size. Then, we follow Steps 1–5.

> ⦿
> ⦿
> ⦿ **Caution**
>
> Always verify the requirements before conducting a hypothesis test.

*Solution:* We are testing the claim that the proportion of Americans 18 years old or older who believe that men are more aggressive than women is less than 0.74—that is, $p < 0.74$. The sample is a simple random sample. Also, $np_0(1 - p_0) = (1,026)(0.74)(0.26) = 197.4 > 10$. Because there are over 19 million Americans 18 years old or older, the sample size is less than 5% of the population size. The requirements are satisfied, so we now proceed to follow Steps 1–5.

**Step 1:** The claim is that the proportion of Americans 18 years old or older who believe that men are more aggressive than women is less than 0.74; that is, $p < 0.74$. This is a left-tailed hypothesis with

$$H_0: p = 0.74 \qquad \text{versus} \qquad H_1: p < 0.74$$

**Step 2:** Because this is a left-tailed test, we determine the critical value at the $\alpha = 0.05$ level of significance to be $-z_{0.05} = -1.645$. The critical region is displayed in Figure 17.

Figure 17

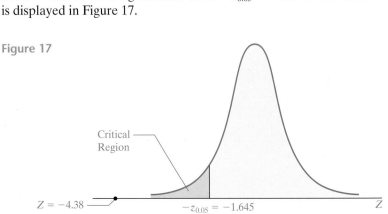

Critical Region

$Z = -4.38$          $-z_{0.05} = -1.645$          $Z$

*Step   3:* The   point   estimate   of   the   population   proportion   is
$\hat{p} = \dfrac{x}{n} = \dfrac{698}{1026} = 0.68$. The test statistic is

$$Z = \frac{\hat{p} - p_o}{\sqrt{\dfrac{p_o(1 - p_o)}{n}}} = \frac{0.68 - 0.74}{\sqrt{\dfrac{0.74(1 - 0.74)}{1026}}} = -4.38$$

The sample proportion of 0.68 is 4.38 standard deviations below the hypothesized proportion of 0.74.

*Step 4:* Because the test statistic $Z = -4.38$ is less than the critical value of $-z_{0.05} = -1.645$, we reject the null hypothesis. That is, the value of the test statistic falls within the critical region, so we reject $H_o$. We label this point in Figure 17.

*Step 5:* There is sufficient evidence to support the claim that the proportion of Americans 18 years old or older who believe that men are more aggressive than women is less than the proportion reported in 1995 at the $\alpha = 0.05$ level of significance.   ◄◄

**NW**  *Now Work Problem 1(a).*

## 2   *P*-values

Recall that a *P*-value is the probability of observing a sample statistic as extreme or more extreme than the one observed, under the assumption that the null hypothesis is true. Our criterion for rejecting the null hypothesis by using *P*-values is as follows:

**Reject the null hypothesis if the *P*-value is less than the level of significance. That is, we reject $H_o$ if *P*-value $< \alpha$.**

The procedures that follow can be used to compute *P*-values when testing a hypothesis about a population proportion.

**Computing *P*-Values**

If a claim is made regarding the population proportion, we can use the following steps to compute the *P*-value, provided that
1. the sample is obtained by using simple random sampling, and
2. $np_o(1 - p_o) \geq 10$ with $n \leq 0.05N$ (the sample size is no more than 5% of the population size).

*Step 1:* A claim is made regarding the population proportion. The claim is used to determine the null and alternative hypotheses. The hypotheses can be structured in one of three ways:

| Two-Tailed | Left-Tailed | Right-Tailed |
|---|---|---|
| $H_o: p = p_o$ | $H_o: p = p_o$ | $H_o: p = p_o$ |
| $H_1: p \neq p_o$ | $H_1: p < p_o$ | $H_1: p > p_o$ |

*Note*: $p_o$ is the assumed value of the population proportion.

*Step 2:* Compute the test statistic $Z_o = \dfrac{\hat{p} - p_o}{\sqrt{\dfrac{p_o(1 - p_o)}{n}}}$.

***Step 3:*** Compute the *P*-value.

| Two-Tailed | Left-Tailed | Right-Tailed |
|---|---|---|

$P$-value $= P(Z < -|Z_o| \text{ or } Z > |Z_o|)$
$= 2P(Z > |Z_o|) = 2P(Z < -|Z_o|)$

$P$-value $= P(Z < Z_o)$

$P$-value $= P(Z > Z_o)$

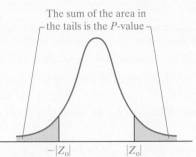

The sum of the area in the tails is the *P*-value

$-|Z_o|$   $|Z_o|$

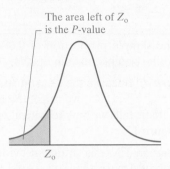

The area left of $Z_o$ is the *P*-value

$Z_o$

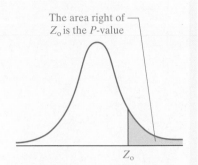

The area right of $Z_o$ is the *P*-value

$Z_o$

**Interpretation:** The *P*-value is the probability of obtaining a sample proportion that is more than $|Z_o|$ standard deviations from the hypothesized proportion, $p_0$.

**Interpretation:** The *P*-value is the probability of obtaining a sample proportion of $\hat{p}$ or smaller, under the assumption that $H_o$ is true. In otherwords, it is the probability of obtaining a sample proportion that is more than $Z_o$ standard deviations to the left of $p_0$.

**Interpretation:** The *P*-value is the probability of obtaining a sample proportion of $\hat{p}$ or larger, under the assumption that $H_o$ is true. In other words, it is the probability of obtaining a sample proportion that is more than $Z_o$ standard deviations to the right of $p_0$.

***Step 4:*** Reject the null hypothesis if the *P*-value is less than the level of significance, $\alpha$.

***Step 5:*** State the conclusion.

▶ **EXAMPLE 2**   **Side Effects of Prevnar**

*Problem:* The drug Prevnar is a vaccine meant to prevent meningitis. (It also helps control ear infections.) It is typically administered to infants. In clinical trials, the vaccine was administered to 710 randomly sampled infants between 12 and 15 months of age. Of the 710 infants, 121 experienced a decrease in appetite. Is there significant evidence to conclude that the proportion of infants who receive Prevnar and experience a decrease in appetite is different from 0.135, the proportion of children who experience a decrease in appetite in competing medications? Test using the *P*-value approach at the $\alpha = 0.01$ level of significance.

*Approach:* We must verify the requirements to perform the hypothesis test—namely, the sample must be a simple random sample with $np_0(1 - p_0) \geq 10$ and the sample size cannot be more than 5% of the population. Then, we follow Steps 1–5, as laid out previously.

*Solution:* We are testing the claim that the proportion of infants who experience a decrease in appetite is different from 0.135—that is, $p \neq 0.135$. The sample is a simple random sample. Also, $np_0(1 - p_0) = (710)(0.135)(1 - 0.135) = 82.9 > 10$. Because there are about 1 million babies between 12 and 15 months of age, the sample size is less than 5% of the population size. The requirements are satisfied, so we now proceed to follow Steps 1–5.

***Step 1:*** We are testing the claim that $p \neq 0.135$. This is a two-tailed test.

$$H_0: p = 0.135 \quad \text{versus} \quad H_1: p \neq 0.135$$

**Step 2:** The point estimate of the population proportion is $\hat{p} = \dfrac{x}{n} = \dfrac{121}{710} = 0.170$. The test statistic is

$$Z_0 = \frac{\hat{p} - p_0}{\sqrt{\dfrac{p_0(1 - p_0)}{n}}} = \frac{0.170 - 0.135}{\sqrt{\dfrac{0.135(1 - 0.135)}{710}}} = 2.73$$

The sample proportion, 0.170, is 2.73 standard deviations above the hypothesized proportion, 0.135.

**Step 3:** Because we are performing a two-tailed test, $P$-value $= 2P(Z < -|Z_0|) = 2P(Z < -2.73)$. That is, we need to determine the area under the standard normal curve to the left of $Z = -2.73$ and to the right of $Z = 2.73$ as shown in Figure 18.

Using Table II, we have

$$P\text{-value} = P(Z < -2.73) + P(Z > 2.73) = 2P(Z < -2.73) = 2(0.0032) = 0.0064$$

The probability of obtaining a sample proportion more than 2.73 standard deviations away from the hypothesized proportion, 0.135, is 0.0064. This means that less than 1 sample in 100 will result in a sample proportion more than 2.73 standard deviations from the hypothesized proportion of 0.135.

**Step 4:** Because the $P$-value is less than the level of significance $(0.0064 < 0.01)$, we reject the null hypothesis.

**Step 5:** The sample evidence supports the claim that the proportion of infants who experienced a decrease in appetite when receiving Prevnar is different from 0.135, the proportion of children who experienced a decrease in appetite in competing medications, at the $\alpha = 0.01$ level of significance. ◄◄

**NW** *Now Work Problem 1(b).*

**Figure 18**

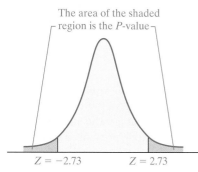

The area of the shaded region is the $P$-value

$Z = -2.73$   $Z = 2.73$

Using Technology: We can obtain the results of Example 2 using a graphing calculator with advanced statistical capabilities or statistical software. Figure 19 shows the results from Minitab.

**Figure 19**

**Test and CI for One Proportion**

```
Test of p = 0.135 vs p not = 0.135

Sample    X    N  Sample p         99.0% CI          Z-Value  P-Value
1        121  710  0.170423  (0.134075, 0.206770)      2.76    0.006
```

### Small-Sample Hypothesis Tests

In order for the sampling distribution of $\hat{p}$ to be approximately normal, we require that $np(1 - p)$ be at least 10. What if this requirement is not satisfied? We can use the binomial probability formula to compute exact $P$-values when the requirements for the sampling distribution of $\hat{p}$ to be approximately normal are not satisfied.

▶ **EXAMPLE 3**   Hypothesis Test for a Population Proportion—Small Sample Size

*Problem:* According to the United States Department of Agriculture, 48.9% of males between 20 and 39 years of age consume the minimum daily requirement of calcium. After an aggressive "got milk" advertising campaign, the USDA conducts a survey of 35 randomly selected males between the ages of 20 and 39 and finds that 21 of them consume the recommended daily allowance of calcium. At the $\alpha = 0.10$ level of significance, is there evidence to conclude that the percentage of males between the ages of 20 and 39 who consume the recommended daily allowance of calcium has increased?

*Approach:* We use the following steps.

*Step 1:* Determine the null and alternative hypotheses.

*Step 2:* Check whether $np_o(1 - p_o)$ is greater than or equal to 10, where $p_o$ is the proportion stated in the null hypothesis. If it is, then the sampling distribution of $\hat{p}$ is approximately normal and we can use the steps presented on pages 423–424. Otherwise, we use Steps 3 and 4, presented next.

*Step 3:* Compute the *P*-value. For right-tailed tests, the *P*-value is the probability of obtaining $x$ or more successes. For left-tailed tests, the *P*-value is the probability of obtaining $x$ or fewer successes.* The *P*-value is always computed with the proportion given in the null hypothesis. Remember, we assume that the null is true until we have evidence to the contrary.

*Step 4:* If the *P*-value is less than the level of significance, $\alpha$, then reject the null hypothesis.

*Solution:*

*Step 1:* The "status quo" or "no change" proportion of 20–39-year-old males who consume the minimum daily requirement of calcium is 0.489. We wish to know whether the advertising campaign helped to increase this proportion. Therefore,

$$H_o: p = 0.489 \quad \text{and} \quad H_1: p > 0.489.$$

*Step 2:* From the null hypothesis, we have $p_o = 0.489$. There were $n = 35$ individuals surveyed, so $np_o(1 - p_o) = 35(0.489)(1 - 0.489) = 8.75$. Because $np_o(1 - p_o) < 10$, the sampling distribution of $\hat{p}$ is not approximately normal.

*Step 3:* Let the random variable $X$ represent the number of individuals who consume the daily requirement of calcium. We have $x = 21$ successes in $n = 35$ trials, so $\hat{p} = \dfrac{21}{35} = 0.6$. We want to judge whether the larger proportion is due to an increase in the population proportion or to sampling error. We obtained $x = 21$ successes in the survey and this is a right-tailed test, so the *P*-value will be $P(X \geq 21)$.

$$P\text{-value} = P(X \geq 21) = 1 - P(X < 21) = 1 - P(X \leq 20).$$

We will compute this *P*-value by using statistical software or a graphing calculator with advanced statistical features, with $n = 35$ and $p = 0.489$. Figure 20 shows the results using a TI-83 Plus graphing calculator.

    The *P*-value is 0.1261. Minitab will compute exact *P*-values via this approach as well.

*Step 4:* The *P*-value is greater than the level of significance $(0.1261 > 0.10)$, so we do not reject $H_o$. There is not significant evidence (at the $\alpha = 0.1$ level of significance) to conclude that the proportion of 20–39-year-old males who consume the recommended daily allowance of calcium has increased.   ◀◀

Figure 20

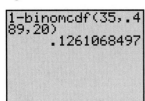

**NW** *Now Work Problem 23.*

---

*We shall not address *P*-values for two-tailed hypothesis tests. For those who are interested, the *P*-value would be 2 times the probability of obtaining $x$ or more successes if $\hat{p} > p$ and 2 times the probability of obtaining $x$ or fewer successes if $\hat{p} < p$.

## 9.4 Assess Your Understanding

### Concepts and Vocabulary

1. State the assumptions required in order to test a hypothesis about a population proportion.

2. A poll that was conducted by CNN, *USA Today*, and Gallup reported the following results: "According to the most recent CNN/USA *Today*/Gallup poll, conducted June 28–July 1, a majority of Americans (52%) approve of the job Bush is doing as president, . . . ." The poll results were obtained by conducting simple random sample of 1014 adults aged 18 years old or older, with a margin of error of $\pm$ 3 percentage points. State what is wrong with the conclusions presented by the pollsters.

### Exercises

#### • Basic Skills

*In Problems 1–6, test the hypothesis, using (a) the Classical Approach and then (b) the P-value approach. Be sure to verify the requirements of the test.*

1. $H_o$: $p = 0.3$ versus $H_1$: $p > 0.3$
   $n = 200$; $x = 75$; $\alpha = 0.05$

2. $H_o$: $p = 0.6$ versus $H_1$: $p < 0.6$
   $n = 250$; $x = 124$; $\alpha = 0.01$

3. $H_o$: $p = 0.55$ versus $H_1$: $p < 0.55$
   $n = 150$; $x = 78$; $\alpha = 0.1$

4. $H_o$: $p = 0.25$ versus $H_1$: $p < 0.25$
   $n = 400$; $x = 96$; $\alpha = 0.1$

5. $H_o$: $p = 0.9$ versus $H_1$: $p \neq 0.9$
   $n = 500$; $x = 440$; $\alpha = 0.05$

6. $H_o$: $p = 0.4$ versus $H_1$: $p \neq 0.4$
   $n = 1000$; $x = 420$; $\alpha = 0.01$

#### • Applying the Concepts

*In Problems 7–22, test the hypothesis, using (a) the Classical Approach and then (b) the P-value approach. Be sure to verify the requirements of the test.*

7. **Gun Control?** In a survey conducted by the Gallup Organization between August 29 and September 5, 2000, 395 of 1012 adults aged 18 years or older said they had a gun in the house. In 1990, 47% of households had a gun. Is there significant evidence to support the claim that the proportion of households that have a gun has decreased since 1990 at the $\alpha = 0.05$ level of significance?

8. **Teenagers and the Death Penalty** In a survey conducted by the Gallup Organization in October of 2000, 320 of 500 teenagers between the ages of 13 and 17 years said that they favor life imprisonment without the possibility of parole for convicted murderers instead of the death penalty. Suppose that it is known that 47% of adults prefer life imprisonment. Is there significant evidence to support the claim that a higher proportion of teenagers support life imprisonment at the $\alpha = 0.1$ level of significance?

9. **Lipitor** The drug Lipitor is meant to reduce total cholesterol and LDL-cholesterol. In clinical trials, 19 out of 863 patients taking 10 mg of Lipitor daily complained of flulike symptoms. Suppose that it is known that 1.9% of patients taking competing drugs complain of flulike symptoms. Is there significant evidence to support the claim that more than 1.9% of Lipitor users experience flulike symptoms as a side effect at the $\alpha = 0.01$ level of significance?

10. **Pepcid** Pepcid is a drug that can be used to heal duodenal ulcers. Suppose the manufacturer of Pepcid claims that more than 80% of patients are healed after taking 40 mg of Pepcid every night for eight weeks. In clinical trials, 148 of 178 patients suffering from duodenal ulcers were healed after eight weeks. Test the manufacturer's claim at the $\alpha = 0.01$ level of significance.

11. **Low Birth Weight** According to the U.S. Census Bureau, 7.1% of all births to nonsmoking mothers are of low birth weight ( < 5 lbs., 8 oz.). An obstetrician wanted to know whether mothers between the ages of 35 and 39 years old gave birth to a higher percentage of low-birth-weight babies. She randomly selected 240 births where the mother was 35–39 years old and found that 22 of them gave birth to low-birth-weight babies. Is there significant evidence to support the claim that mothers between 35 and 39 years old have a higher percentage of low-birth-weight babies at the $\alpha = 0.01$ level of significance?

12. **Exercise** A personal trainer wanted to know whether the proportion of males 30 to 44 years old who do not exercise has decreased from 24.9%, the proportion in 1998 (data obtained from the U.S. National Center for Chronic Disease Prevention and Health Promototion). He randomly selects 150 males between the ages of 30 and 44 years and finds that 28 of them do not exercise. Is there significant evidence that the proportion of males 30 to 44 years old who do not exercise has decreased at the $\alpha = 0.05$ level of significance?

13. **Parents at Home** In 1989, 33% of Americans felt that one parent should stay home to raise the children. In a poll conducted by the Gallup Organization in 2001, 416 of 1015 American adults 18 years of age or older believed that one parent should stay home to raise the children. Is there significant evidence to conclude that the percentage of Americans who believe that one parent should stay home to raise the children has increased at the $\alpha = 0.05$ level of significance?

**14. Own a PC?** According to the Energy Information Administration, 42.1% of households in the United States owned a personal computer in 1998. An economist wants to discover whether this percentage has changed significantly since then. She randomly selects 320 households throughout the United States and finds that 154 of them own a personal computer. Is there sufficient evidence to conclude the percentage of households that own a personal computer is significantly different from the percentage in 1998, at the $\alpha = 0.1$ level of significance?

**15. Free Throws** Shaquille O'Neal makes 43.8% of his free throws. His coach sends him to a free-throw specialist in order to improve his free-throw shooting. After five days with the specialist, his coach randomly selects 80 free throw attempts of Shaq and finds that he makes 39 of them. Is there significant evidence at the $\alpha = 0.05$ level of significance that Shaq's free throwing has improved?

**16. Living Alone?** In 1990, 39% of males aged 15 years of age and older lived alone, according to the U.S. Census Bureau. A sociologist tests whether this percentage is different today by conducting a random sample of 400 males aged 15 years of age and older and finds that 164 are living alone. Is there significant evidence at the $\alpha = 0.10$ level of significance to support belief in a change?

**17. The Environment** In 1990, 58% of Americans 18 years old and older reported they have a great deal of concern regarding air pollution. On February 20, 2001, the Gallup Organization released results of a poll in which 592 of 1004 Americans 18 years old or older stated that they have a great deal of concern regarding the level of air pollution in America. Is there evidence at the $\alpha = 0.05$ level of significance to conclude that the proportion of 2001 Americans having a great deal of concern about the level of air pollution in America is significantly different from the 1990 proportion?

**18. Pathological Gamblers** Pathological gambling is an impulse-control disorder. The American Psychiatry Association lists 10 characteristics that diagnose the disorder in its DSM-IV manual. The National Gambling Impact Study Commission randomly selected 2417 adults and found that 35 were pathological gamblers. Is there evidence to support the claim that more than 1% of the adult population are pathological gamblers at the $\alpha = 0.05$ level of significance?

**19. Teachers Using the Internet** According to the U.S. Department of Education's Office of Educational Technology, 85% of teachers used the Internet to teach in 2000. An educator claims that that percentage has increased from its 2000 level. She randomly samples 340 teachers and discovers that 302 of them have used the Internet in their teaching. Is there significant evidence to support the claim that the percentage of teachers using the Internet in their teaching has increased, at the $\alpha = 0.1$ level of significance?

**20. Does the Internet Affect Watching TV?** A researcher at Neilsen Media Research wanted to determine the impact that the Internet had on households' watching of television. In a study conducted in May, 1999, it was determined that 57.8% of all television households had the television on during "prime time" [*Source*: Neilsen Media Research]. In a random sample of 1025 television and Internet households, 575 had the television on during "prime time." Is there evidence to support the claim that a lower proportion of television and Internet households have the television on during prime time than do television households, at the $\alpha = 0.1$ level of significance?

**21. Talk to the Animals** In a survey conducted by the American Animal Hospital Association, 37% of respondents stated that they talk to their pets on the answering machine or telephone. A veterinarian found this result hard to believe, so he randomly selected 150 pet owners and discovered that 54 of them spoke to their pet on the answering machine or telephone. Test the veterinarian's claim that less than 37% of pet owners speak to their pets on the answering machine or telephone, at the $\alpha = 0.05$ level of significance.

**22. Eating Salad** According to a survey conducted by the Association of Dressings and Sauces (this is an actual association!), 85% of American adults eat salad at least once a week. A nutritionist suspects that the percentage is higher than this. She conducts a survey of 200 American adults and finds that 171 of them eat salad at least once a week. Is there sufficient evidence to support the nutritionist's claim at the $\alpha = 0.10$ level of significance?

**23. Small-Sample Hypothesis Test** In 1997, 4% of mothers smoked more than 21 cigarettes during their pregnancy. An obstetrician believes that the percentage of mothers who smoke 21 cigarettes or more is less than 4% today. She randomly selects 120 pregnant mothers and finds that 3 of them smoked 21 or more cigarettes during pregnancy. Test the researcher's claim at the $\alpha = 0.05$ level of significance.

**24. Small-Sample Hypothesis Test** According to the United States Census Bureau, in 2000, 3.2% of Americans worked at home. An economist believes that the percentage of Americans working at home has increased since then. He randomly selects 150 working Americans and finds that 8 of them work at home. Test the economist's claim at the $\alpha = 0.05$ level of significance.

**25. Small-Sample Hypothesis Test** According to the United States Census Bureau, in 2000, 9.6% of Californians had a travel time to work of more than 60 minutes. An urban economist believes that the percentage of Californians who have a travel time to work of more than 60 minutes has increased since then. She randomly selects 80 working Californians and finds that 13 of them have a travel time to work that is more than 60 minutes. Test the urban economist's claim at the $\alpha = 0.1$ level of significance.

**26. Small-Sample Hypothesis Test** According to the United States Census Bureau, in 2000, 1.5% of males living in Colorado were teachers. A researcher believes that this percentage has decreased since then. She randomly selects 500 males living in Colorado and finds that 3 of them are teachers. Test the researcher's claim at the $\alpha = 0.1$ level of significance.

## Technology Step-by-Step
Hypothesis Tests Regarding a Population Proportion

**TI-83 Plus**  **Step 1:** Press STAT, highlight TESTS, and select 5:1-PropZTest.

**Step 2:** For the value of $p_0$, enter the "status quo" value of the population proportion.

**Step 3:** Enter the number of successes, $x$, and the sample size, $n$.

**Step 4:** Select the direction of the alternative hypothesis.

**Step 5:** Highlight **Calculate** or **Draw**, and press ENTER. The TI-83 gives the $P$-value.

**MINITAB**  **Step 1:** Select the **Stat** menu, highlight **Basic Statistics**, then highlight **1-Proportion**.

**Step 2:** Select "Summarized data."

**Step 3:** Enter the number of trials, $n$, and the number of successes, $x$.

**Step 4:** Click Options. Enter the "status quo" value of the population proportion in the cell "Test proportion." Enter the direction of the alternative hypothesis. If $np_0(1 - p_0) \geq 10$, then check the box marked "Use test and interval based on normal distribution." Click OK twice.

**Excel**  **Step 1:** Load the PHStat Add-in.

**Step 2:** Select the **PHStat menu**, highlight **One Sample Tests . . .** , then highlight **Z Test for proportion**.

**Step 3:** Enter the value of the null hypothesis, the level of significance, $\alpha$, the number of successes, $x$, and the number of trials, $n$. Click the option corresponding to the desired test (two-tail, upper (right) tail, or lower (left) tail). Click OK.

## CHAPTER 9 REVIEW

### Summary

In this chapter, we discussed testing claims by using hypothesis testing. A claim is made regarding a population parameter, which leads to a null and an alternative hypothesis. The null hypothesis is assumed true. We collect sample data, which are used to test the veracity of the claim. Given the sample data, we either reject or do not reject the null hypothesis. In performing a hypothesis test, there is always the possibility of making a Type I error—rejecting the null hypothesis when it is true—or of making a Type II error—not rejecting the null hypothesis when it is false. The probability of making a Type I error is equal to the level of significance, $\alpha$, of the test.

We discussed three types of hypothesis tests in this chapter. First, we performed tests about a population mean with $\sigma$ known. Second, we performed tests about a population mean with $\sigma$ unknown. In both of these cases, we required that either the sample size be large (i.e. $n \geq 30$) or the population be approximately normal with no outliers. For small sample sizes, we verified the normality of the data by using normal probability plots. Boxplots were used to check for outliers.

The third test we performed regarded claims about a population proportion. In order to perform these tests, we required a large sample size, so that $np(1 - p) \geq 10$, yet the sample size could be no more than 5% of the population size. All three hypothesis tests were performed by using classical methods and the $P$-value approach. The $P$-value approach to testing hypotheses has appeal, because the rejection rule is always to reject the null hypothesis if the $P$-value is less than the level of significance, $\alpha$.

In order to assist in the choosing of the appropriate test to perform, we provide the flowchart in Figure 21 on page 430.

**Notes:**
- In testing claims about the population mean, if the sample size is not large or the population is not normally distributed, use the nonparametric methods.
- In testing claims about the population proportion, if $np_0(1 - p_0) < 10$, use the binomial probability formula to test the claim.

Figure 21

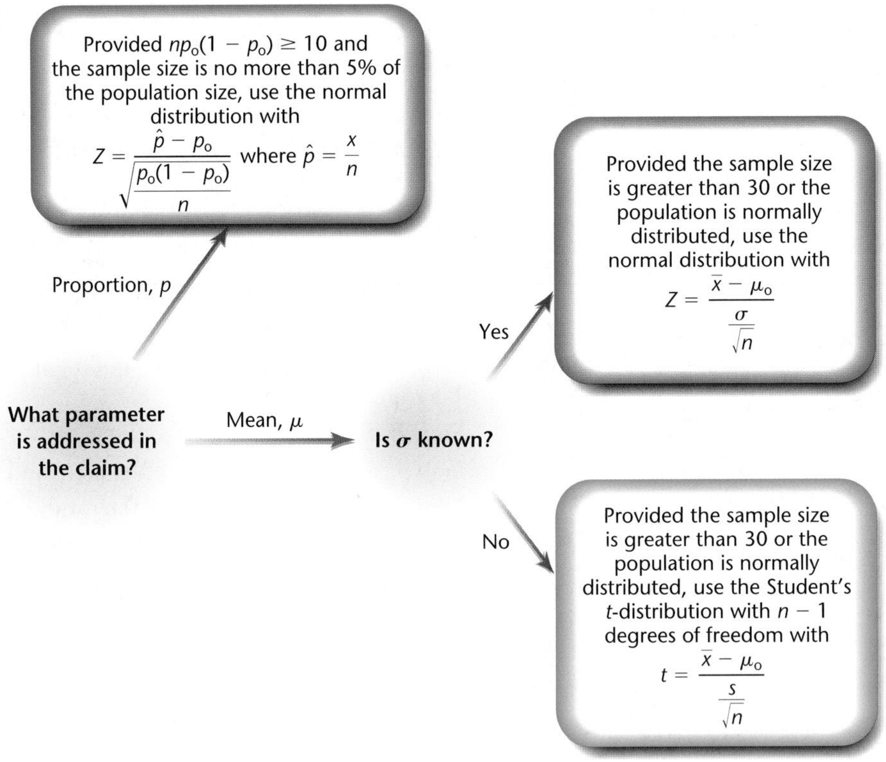

Provided $np_o(1 - p_o) \geq 10$ and the sample size is no more than 5% of the population size, use the normal distribution with

$$Z = \frac{\hat{p} - p_o}{\sqrt{\dfrac{p_o(1 - p_o)}{n}}} \text{ where } \hat{p} = \frac{x}{n}$$

Proportion, $p$

Provided the sample size is greater than 30 or the population is normally distributed, use the normal distribution with

$$Z = \frac{\bar{x} - \mu_o}{\dfrac{\sigma}{\sqrt{n}}}$$

**What parameter is addressed in the claim?**

Mean, $\mu$

**Is $\sigma$ known?**

Yes

No

Provided the sample size is greater than 30 or the population is normally distributed, use the Student's $t$-distribution with $n - 1$ degrees of freedom with

$$t = \frac{\bar{x} - \mu_o}{\dfrac{s}{\sqrt{n}}}$$

## Formulas

### Test Statistics

- $z = \dfrac{\bar{x} - \mu_o}{\sigma/\sqrt{n}}$ follows the standard normal distribution if the population from which the sample was drawn is normal or if the sample size is large ($n \geq 30$).

- $t = \dfrac{\bar{x} - \mu_o}{s/\sqrt{n}}$ follows Student's $t$-distribution with $n - 1$ degrees of freedom if the population from which the sample was drawn is normal or if the sample size is large ($n \geq 30$).

- $z = \dfrac{\hat{p} - p_o}{\sqrt{\dfrac{p_o(1 - p_o)}{n}}}$ follows the standard normal distribution if $np_o(1 - p_o) \geq 10$ and $n \leq 0.05N$.

### Type I/Type II Errors

- $\alpha = P(\text{Type I Error}) = P(\text{rejecting } H_o \text{ when } H_o \text{ is true})$
- $\beta = P(\text{Type II Error}) = P(\text{not rejecting } H_o \text{ when } H_1 \text{ is true})$

## Vocabulary

Hypothesis (p. 384)
Hypothesis testing (p. 385)
Null hypothesis (p. 385)
Alternative hypothesis (p. 385)
Two-tailed test (p. 385)

Left-tailed test (p. 385)
Right-tailed test (p. 385)
Type I error (p. 387)
Type II error (p. 387)
Level of significance (p. 388)

Critical value (p. 394)
Critical region (p. 395)
Test statistic (p. 395)
Robust (p. 395)
$P$-value (p. 397)

## Objectives

| Section | You Should Be Able to ... | Review Exercises |
|---|---|---|
| **9.1** | 1 Determine the null and alternative hypotheses from a claim (p. 383) | 1, 2 |
| | 2 Understand Type I and Type II errors (p. 387) | 1, 2, 15(b); 16(b) |
| | 3 Understand the probability of making Type I and Type II errors (p. 388) | 1–4, 15(c), 16(c) |
| | 4 State conclusions to hypothesis tests (p. 389) | 1, 2 |
| **9.2** | 1 Understand the logic of hypothesis testing (p. 392) | 23 |
| | 2 Test a hypothesis about $\mu$ with $\sigma$ known using the classical method (p. 394) | 5, 6, 13, 14, 17, 18 |
| | 3 Test a hypothesis about $\mu$ with $\sigma$ known using the $P$-value approach (p. 397) | 5, 6, 13, 14, 17, 18 |
| | 4 Test a hypothesis about $\mu$ with $\sigma$ known using confidence intervals (p. 401) | 6(f) |
| **9.3** | 1 Test hypotheses about $\mu$, $\sigma$ unknown, using the classical approach (p. 409) | 7, 8, 11, 12, 15, 16 |
| | 2 Test hypotheses about $\mu$, $\sigma$ unknown, using the $P$-value approach (p. 412) | 7, 8, 11, 12, 15, 16 |
| **9.4** | 1 Test a hypothesis about a population proportion using the classical method (p. 420) | 9, 10, 19, 20 |
| | 2 Test a hypothesis about a population proportion using the $P$-value approach (p. 423) | 9, 10, 19, 20 |
| | 3 Test a hypothesis about a population proportion on a small sample (p. 425) | 21, 22 |

## Review Exercises

*For each claim in Problems 1 and 2, (a) determine the null and alternative hypotheses, (b) explain what it would mean to make a Type I error, (c) explain what it would mean to make a Type II error, (d) state the conclusion that would be drawn if the null hypothesis is not rejected, and (e) state the conclusion that would be reached if the null hypothesis is rejected.*

**1. Charitable Contributions** A government economist maintains that the mean charitable contribution in 2002 was more than $1100 per household.

**2. Online Shopping** According to shop.org, 17.6% of all computer hardware/software purchases in 1999 were made online. A researcher claims the proportion of computer hardware/software purchases in 2000 is more than 0.176.

**3.** Suppose a test is conducted at the $\alpha = 0.05$ level of significance. What is the probability of a Type I error?

**4.** Suppose $\beta$ is computed to be 0.113. What is the probability of a Type II error?

**5.** In order to test $H_0: \mu = 30$ versus $H_1: \mu < 30$, a simple random sample of size $n = 12$ is obtained from a population that is known to be normally distributed with $\sigma = 4.5$.
   (a) If the sample mean is determined to be $\bar{x} = 28.6$, compute the test statistic.
   (b) If the researcher decides to test this hypothesis at the $\alpha = 0.05$ level of significance, determine the critical value.
   (c) Draw a normal curve that depicts the rejection region.
   (d) Will the researcher reject the null hypothesis? Why?
   (e) What is the $P$-value?

**6.** In order to test $H_0: \mu = 65$ versus $H_1: \mu \neq 65$, a simple random sample of size $n = 23$ is obtained from a population that is known to be normally distributed with $\sigma = 12.3$.
   (a) If the sample mean is determined to be $\bar{x} = 70.6$, compute the test statistic.

   (b) If the researcher decides to test this hypothesis at the $\alpha = 0.1$ level of significance, determine the critical values.
   (c) Draw a normal curve that depicts the rejection region.
   (d) Will the researcher reject the null hypothesis? Why?
   (e) What is the $P$-value?
   (f) Test the claim by constructing a 90% confidence interval.

**7.** In order to test $H_0: \mu = 8$ versus $H_1: \mu \neq 8$, a simple random sample of size $n = 15$ is obtained from a population that is known to be normally distributed.
   (a) If $\bar{x} = 7.3$ and $s = 1.8$, compute the test statistic.

   (b) If the researcher decides to test this hypothesis at the $\alpha = 0.02$ level of significance, determine the critical values.
   (c) Draw a $t$-distribution that depicts the rejection region.
   (d) Will the researcher reject the null hypothesis? Why?
   (e) Determine the $P$-value.

**8.** In order to test $H_0: \mu = 3.9$ versus $H_1: \mu < 3.9$, a simple random sample of size $n = 25$ is obtained from a population that is known to be normally distributed.
   (a) If $\bar{x} = 3.5$ and $s = 0.9$, compute the test statistic.
   (b) If the researcher decides to test this hypothesis at the $\alpha = 0.05$ level of significance, determine the critical value.
   (c) Draw a $t$-distribution that depicts the rejection region.
   (d) Will the researcher reject the null hypothesis? Why?
   (e) Determine the $P$-value.

*In Problems 9 and 10, test the hypothesis at the $\alpha = 0.05$ level of significance, using (a) the Classical Approach and (b) the P-value approach. Be sure to verify the requirements of the test.*

9. $H_0: p = 0.6$ versus $H_1: p > 0.6$
   $n = 250; x = 165; \alpha = 0.05$

10. $H_0: p = 0.35$ versus $H_1: p \neq 0.35$
    $n = 420; x = 138; \alpha = 0.01$

11. **Hotel/Motel Costs** According to the American Hotel and Motel Association, the average price of a room was $78.62 per night in 1998. A lodging analyst believes that this value has changed from its 1998 level. He conducts a simple random sample of 50 hotels and motels and determines the mean price to be $80.04 and the standard deviation to be $10.83. Test the analyst's claim at the $\alpha = 0.05$ level of significance.

12. **Net Worth** A household's net worth is defined as all household assets minus any debt. The mean household net worth where the head of household has a high school diploma was $157.8 thousand in 1998, according to the Federal Reserve Bulletin. A government economist claims that the mean household net worth where the head of household has a high school diploma is higher today. She obtains a simple random sample of 100 households and determines the mean net worth to be $165.2 thousand, with a standard deviation of $43.1 thousand (both in 1998 dollars). Test the analyst's claim at the $\alpha = 0.05$ level of signficance.

13. **SAT Math Scores** A mathematics instructor wanted to know whether use of a calculator improves SAT math scores. In 2000, the SAT math scores of students who used a calculator once or twice weekly were normally distributed, with a mean of 474 and a standard deviation 103. In a random sample of 50 students who use a calculator every day, the mean score was 539. Is there significant evidence to support the claim that students who use a calculator "frequently" score better on the SAT math portion than those who use a calculator "infrequently"? Test at the $\alpha = 0.01$ level of significance.

14. **Birth Weight** An obstetrician wants to determine whether a new diet significantly increases the birth weight of babies. In 1998, birth weights of full-term babies (gestation period of 37–41 weeks) were normally distributed, with mean 7.53 pounds and standard deviation 1.15 pounds, according to the National Vital Statistics Report, Vol. 48, No. 3. The obstetrician randomly selects 50 recently pregnant mothers and persuades them to partake of this new diet. The obstetrician then records the birth weights of the babies and obtains a mean of 7.79 pounds. Is there significant evidence to support the claim that the new diet increases the birth weights of newborns at the $\alpha = 0.01$ level of significance?

15. **High Cholesterol** A nutritionist maintains that 20–39-year-old males consume too much cholesterol. The USDA-recommended daily allowance of cholesterol is 300 mg. In a survey conducted by the U.S. Department of Agriculture of 404 20–39-year-old males, it was determined the mean daily cholesterol intake was 326 milligrams, with standard deviation 342 milligrams.
    (a) Is there evidence to support the nutritionist's claim at the $\alpha = 0.05$ level of significance?
    (b) What would it mean for the nutritionist to make a Type I error? A Type II error?
    (c) What is the probability the nutritionist will make a Type I error?

16. **Sodium** A nutritionist claims that 20–39-year-old females consume too much sodium. The USDA-recommended daily allowance of sodium is 2400 mg. In a survey conducted by the U.S. Department of Agriculture of 257 20–39-year-old females, it was determined the mean daily sodium intake was 2919 milligrams and the standard deviation was 906 milligrams.
    (a) Is there evidence to support the nutritionist's claim at the $\alpha = 0.10$ level of significance?
    (b) What would it mean for the nutritionist to make a Type I error? A Type II error?
    (c) What is the probability the nutritionist will make a Type I error?

---

17. **Acid Rain** In 1990, the mean pH level of the rain in Barnstable County, Massachussetts was 4.61. A biologist fears that the acidity of rain has increased (in other words that the pH level of the rain has decreased). She draws a random sample of 25 rain dates in 2000 and obtains the following data:

| | | | | | | |
|------|------|------|------|------|------|------|
| 4.80 | 4.27 | 4.09 | 4.55 | 5.08 | 4.34 | 4.08 |
| 4.36 | 4.82 | 4.70 | 4.40 | 4.73 | 4.62 | 4.48 |
| 4.28 | 4.28 | 4.48 | 4.72 | 4.12 | 4.00 | 4.93 |
| 4.91 | 4.32 | 4.63 | 4.03 | | | |

*Source:* National Atmospheric Deposition Program

(a) Because the sample size is small, she must verify pH level is normally distributed and the sample does not contain any outliers. The normal probability plot and boxplot are shown on page 433. Are the conditions for testing the hypothesis satisfied?

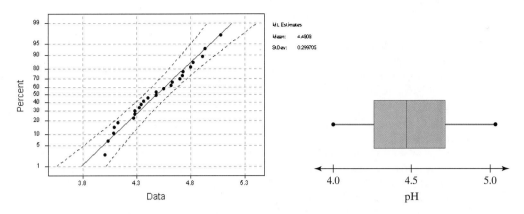

Normal Probability Plot for pH

(b) Test the claim, assuming $\sigma = 0.26$, at the $\alpha = 0.01$ level of significance.
(c) Compute and interpret the $P$-value.

18. **Hemoglobin** A medical researcher maintains that the mean hemoglobin reading of surgical patients is different from 14.0 grams per deciliter. She randomly selects nine surgical patients and obtains the following data:

| 14.6 | 12.8 | 8.9 | 9.0 | 9.9 |
|------|------|-----|-----|-----|
| 10.7 | 13.0 | 12.0 | 13.0 | |

*Source:* Lora McGuire, Nursing
Instructor, Joliet Junior College

(a) Because the sample size is small, she must verify that hemoglobin is normally distributed and the sample does not contain any outliers. The normal probability plot and boxplot are shown below. Are the conditions for testing the hypothesis satisfied?

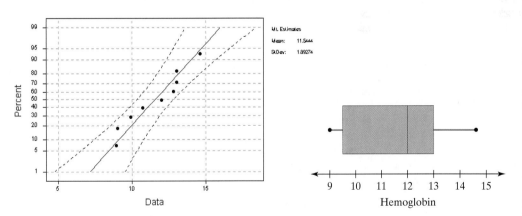

Normal Probability Plot for Hemoglobin

(b) Test the claim, assuming that $\sigma = 2.001$ at the $\alpha = 0.01$ level of significance.
(c) Compute and interpret the $P$-value.

19. **Tuberculosis** According to the Centers for Disease Control, 56% of all tuberculosis cases in 1999 were of foreign-born residents of the United States. A researcher believes that this proportion has increased from its 1999 level. She obtains a simple random sample of 300 tuberculosis cases in the United States and determines that 170 of them are foreign-born. Is there sufficient evidence to support the claim that the percentage of cases of tuberculosis of foreign-born residents has increased at the $\alpha = 0.01$ level of significance?

20. **Phone Purchases** In 1997, 39.4% of females ordered merchandise or services by phone in the last 12 months. A market research analyst feels that the percentage of females ordering merchandise or services by phone has declined from the 1997 level, because of Internet purchases. She obtains a random sample of 500 females and determines that 191 of them have ordered merchandise or services by phone in the last 12 months. Test the researcher's claim that the percentage of females ordering merchandise or services by phone has decreased from its 1997 proportion at the $\alpha = 0.10$ level of significance.

21. **Prestigious Occupation** In 1997, 49% of Americans believed that being a teacher was a prestigious occupation. A researcher wanted to know whether this percentage had risen since then. She surveys 40 randomly selected Americans and finds that 22 of them felt that being a teacher was a prestigious occupation. Is this evidence to support the claim that the proportion of Americans who believe that being a teacher is a prestigious occupation has increased at the $\alpha = 0.05$ level of significance?

22. **Leisure Activities** In 1999, 27% of Americans stated that reading was their favorite leisure-time activity. A researcher wants to know whether this percentage has declined since then. He surveys 35 Americans and finds that 4 of them consider reading to be their favorite leisure-time activity. Is this evidence to support the claim that the proportion of Americans who consider reading to be their favorite leisure-time activity has decreased at the $\alpha = 0.05$ level of significance?

23. In your own words, explain the procedure for testing a hypothesis about a population mean when assuming the population standard deviation is known.

**Chapter 9 Projects located at www.prenhall.com/sullivanstats**

# Inferences on Two Samples

**For additional study help, go to**
www.prenhall.com/sullivanstats

    Materials include

- Projects:
  - Case Study: Control in the Design of an Experiment
  - Decisions: Where Should I Invest?
  - Consumer Reports Project
- Self-Graded Quizzes
- "Preparing for This Section" Quizzes
- STATLETs
- PowerPoint Downloads
- Step-by-Step Technology Guide
- Graphing Calculator Help

## Putting It All Together

In Chapters 8 and 9 we discussed inferences regarding a single population parameter such as $\mu$ or $p$. The inferential methods presented in these chapters will be modified slightly in this chapter so that we can compare two population parameters.

The first two sections of this chapter deal with testing for the difference of two population means. The methods presented in this chapter can be used to determine whether a certain treatment results in significantly different sample statistics. From a design-of-experiments point of view, the methods presented in Section 10.1 are used to handle matched-pairs designs (Section 1.5, page 34) with a quantitative-response variable. For example, we might want to know whether married couples have similar IQs. To test this theory, we could randomly select 20 married couples and determine the difference in their IQs.

Section 10.2 presents inferential methods used to handle completely randomized designs when there is a single

treatment that has two levels (Section 1.5, pages 31–34) and the response variable is quantitative. For example, we might randomly divide 100 volunteers that have a common cold into two groups–a control group and an experimental group. The control group would receive a placebo and the experimental group would receive a predetermined amount of some experimental drug. The response variable might be the time until the cold symptoms go away.

Section 10.3 discusses the difference between two population proportions. Again, we can use a completely randomized design to compare two population proportions. However, rather than having a quantitative response variable, we would have a binomial response variable. That is, either the experimental unit has a characteristic or it does not.

## 10.1 Inference about Two Means: Dependent Samples

**Preparing for This Section**   Before getting started, review the following:

- ✓ Matched-pairs design (Section 1.5, p. 34)
- ✓ Confidence intervals about $\mu$, $\sigma$ unknown (Section 8.2, pp. 360–363)
- ✓ Hypothesis tests about $\mu$, $\sigma$ unknown (Section 9.3, pp. 408–413)
- ✓ Type I and Type II errors (Section 9.1, pp. 387–388)

**Objectives**   **1** Distinguish between independent and dependent sampling

**2** Test claims made regarding matched-pairs data

**3** Construct confidence intervals about the population mean difference of matched-pairs data

**In Your Own Words**

If the individuals in two samples are somehow related (husband–wife, siblings, similar characteristics, or even the same person), then the sampling is dependent.

**1** ### Independent versus Dependent Sampling

In order to perform inference on the difference of two population means, we must first determine whether the data come from an independent or dependent sample. A sampling method is **independent** when the individuals selected for one sample do not dictate which individuals are to be in a second sample. A sampling method is **dependent** when the individuals selected to be in one sample are used to determine the individuals to be in the second sample. For example, if we are conducting a study that compares the IQs of husbands and wives, then once a husband is selected to be in the study, his wife is automatically matched with him. Dependent samples are often referred to as **matched-pairs** samples.

▶ EXAMPLE 1   **Distinguishing between Independent and Dependent Sampling**

*Problem:* For each of the following experiments, determine whether the sampling method is independent or dependent.

**(a)** Researcher Steven J. Sperber, MD, and his associates wanted to determine the effectiveness of a new medication* in the treatment of discomfort associated with the common cold. They randomly divided 430 subjects into two groups: Group 1 received the new medication and Group 2 received a placebo. The goal of the study was to determine whether the mean symptom assessment scores of the individuals receiving the treatment (Group 1) was less than that of the control group (Group 2).

**(b)** In an experiment conducted in a biology class, Professor Andy Neill measured the time required for students to catch a falling meter stick using their dominant hand and nondominant hand for 21 students. The goal of the study was to determine whether the reaction time in an individual's dominant hand is different from the reaction time in the nondominant hand.

---

*The medication was a combination of pseudoephedrine and acetaminophen. The study is published in the *Archives of Family Medicine* 9(2000): 979–985.

*Approach:* We must determine whether the individuals in one group were used to determine the individuals that were in the other group. If so, the sampling method is dependent; if not, the sampling method is independent.

*Solution:*

**(a)** The sampling method is independent because the individuals in the treatment group were not used to determine which individuals are in the control group.

**(b)** The sampling method is dependent because the individuals are related. The measurements for the dominant and nondominant hand are on the same individual.   ◄◄

**NW** *Now Work Problem 1.*

In this section, we will discuss inference on the difference of two means for dependent sampling. Section 10.2 addresses inference when the sampling is independent.

**(2)** ## Hypothesis Tests

The approach to analyzing matched-pairs data is very similar to that presented in constructing a confidence interval or testing a claim regarding a population mean with the population standard deviation unknown. Recall that if the population from which the sample was drawn is normally distributed or the sample size is large ($n \geq 30$), we said that

$$t = \frac{\bar{x} - \mu}{s / \sqrt{n}}$$

*✎ In Your Own Words*

Statistical inference methods on matched-pairs data use the same methods as inference on a single population mean with $\sigma$ unknown, except that the "differences" are analyzed.

follows Student's $t$-distribution with $n - 1$ degrees of freedom.

When analyzing matched-pairs data, we compute the difference in each matched pair and then perform inference on the differenced data using the methods of Section 8.2 or 9.3.

**Testing a Claim Regarding the Difference of Two Means Using a Matched-Pairs Design**

If a claim is made regarding the mean difference of matched-pairs data, we can use the following steps to test the claim, provided that
1. the sample is obtained using simple random sampling;
2. the sample data are matched-pairs;
3. the differences are normally distributed or the sample size, $n$, is large ($n \geq 30$).

*Step 1:* A claim is made regarding the mean difference from matched-pairs data. The claim is used to determine the null and alternative hypotheses. The hypotheses can be structured in one of three ways, where $\mu_d$ is the claimed mean difference of the matched-pairs data.

| Two-Tailed | Left-Tailed | Right-Tailed |
|---|---|---|
| $H_0: \mu_d = 0$ | $H_0: \mu_d = 0$ | $H_0: \mu_d = 0$ |
| $H_1: \mu_d \neq 0$ | $H_1: \mu_d < 0$ | $H_1: \mu_d > 0$ |

**Step 2:** Choose a level of significance $\alpha$ based upon the seriousness of making a Type I error and determine the critical value with $n - 1$ degrees of freedom.

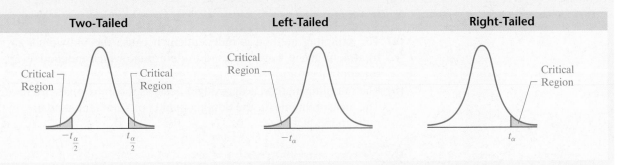

| Two-Tailed | Left-Tailed | Right-Tailed |
|---|---|---|

**Step 3:** Compute the **test statistic** $t = \dfrac{\overline{d} - 0}{s_d \big/ \sqrt{n}} = \dfrac{\overline{d}}{s_d \big/ \sqrt{n}}$ which

approximately follows Student's $t$-distribution with $n - 1$ degrees of freedom. The values of $\overline{d}$ and $s_d$ are the mean and standard deviation of the differenced data.

**Step 4:** Compare the critical value with the test statistic, using the guidelines shown below.

| Two-Tailed | Left-Tailed | Right-Tailed |
|---|---|---|
| If $t < -t_{\alpha/2}$ or $t > t_{\alpha/2}$, reject the null hypothesis | If $t < -t_{\alpha}$, reject the null hypothesis | If $t > t_{\alpha}$, reject the null hypothesis |

**Step 5:** State the conclusion.

The procedures just presented require either that the differenced data be normally distributed or that the sample size be large. The procedure is **robust**, which means that minor departures from normality will not adversely affect the results of the test. If the data have outliers, however, the procedure should not be used. We will verify the assumption that the data come from a population that is normally distributed by constructing normal probability plots. We use boxplots to determine whether there are outliers. If the normal probability plot indicates that the differenced data are not normally distributed or the boxplot reveals outliers, nonparametric tests should be performed.

▶ EXAMPLE 2    Testing a Claim Regarding Matched-Pairs Data

*Problem:* Professor Andy Neill measured the time (in seconds) required to catch a falling meter stick for 12 randomly selected students' dominant hand and nondominant hand. Professor Neill claims that the reaction time in an individual's dominant hand is less than the reaction time in their nondominant hand. Test the claim at the $\alpha = 0.05$ level of significance. The data obtained are presented in Table 1.

*Approach:* This is a matched-pairs sample because the variable is measured on the same subject for both the dominant and nondominant hand. We compute the difference between the dominant time and the nondominant time,

| | TABLE 1 | |
|---|---|---|
| Student | Dominant Hand, $X_1$ | Nondominant Hand, $X_2$ |
| 1 | 0.177 | 0.179 |
| 2 | 0.210 | 0.202 |
| 3 | 0.186 | 0.208 |
| 4 | 0.189 | 0.184 |
| 5 | 0.198 | 0.215 |
| 6 | 0.194 | 0.193 |
| 7 | 0.160 | 0.194 |
| 8 | 0.163 | 0.160 |
| 9 | 0.166 | 0.209 |
| 10 | 0.152 | 0.164 |
| 11 | 0.190 | 0.210 |
| 12 | 0.172 | 0.197 |

*Source:* Professor Andy Neill, Joliet Junior College

$X_1 - X_2$, for each student. If the reaction time in the dominant hand is less than the reaction time in the nondominant hand, we would expect the values of $X_1 - X_2$ to be negative. Before we perform the hypothesis test, we must verify that the differenced data is approximately normally distributed with no outliers because the sample size is small. We will construct a normal probability plot and boxplot of the differenced data to verify these requirements. We then proceed to follow Steps 1–5.

*Solution:* We compute the differences as $d_i = X_1 - X_2 =$ time of dominant hand for $i$th student minus time of nondominant hand for $i$th student. We expect these differences to be negative, so we wish to test the claim that $\mu_d < 0$. Table 2 displays the differences.

| | | TABLE 2 | |
|---|---|---|---|
| Student | Dominant Hand, $X_1$ | Nondominant Hand, $X_2$ | Difference, $d_i$ |
| 1 | 0.177 | 0.179 | $0.177 - 0.179 = -0.002$ |
| 2 | 0.210 | 0.202 | $0.210 - 0.202 = 0.008$ |
| 3 | 0.186 | 0.208 | $-0.022$ |
| 4 | 0.189 | 0.184 | $0.005$ |
| 5 | 0.198 | 0.215 | $-0.017$ |
| 6 | 0.194 | 0.193 | $0.001$ |
| 7 | 0.160 | 0.194 | $-0.034$ |
| 8 | 0.163 | 0.160 | $0.003$ |
| 9 | 0.166 | 0.209 | $-0.043$ |
| 10 | 0.152 | 0.164 | $-0.012$ |
| 11 | 0.190 | 0.210 | $-0.020$ |
| 12 | 0.172 | 0.197 | $-0.025$ |

$$\sum d_i = -0.158$$

We compute the mean and standard deviation of the differences and obtain $\bar{d} = -0.0132$ rounded to four decimal places and $s_d = 0.0164$ rounded to four decimal places. Before we proceed with the test of the hypothesis, we must verify that the data are approximately normal with no outliers. Figure 1 shows the normal probability plot and boxplot of the differenced data.

Figure 1

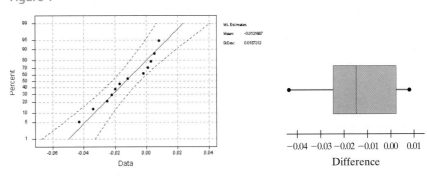

Normal Probability Plot for Difference

## Caution

The way that we define the difference determines the direction of the alternative hypothesis in one-tailed tests. In Example 1, we expect $X_1 < X_2$, so the difference $X_1 - X_2$ is expected to be negative. Therefore, the alternative hypothesis is $H_1$: $\mu_d < 0$, and we have a left-tailed test. However, if we computed the differences as $X_2 - X_1$, then we'd expect the differences to be positive, and we'd have a right-tailed test!

The normal probability plot is roughly linear and the boxplot does not indicate any outliers. We can proceed with the hypothesis test.

**Step 1:** Professor Neill claims that the reaction time in the dominant hand is less than the reaction time in the nondominant hand. We express this claim as $\mu_d < 0$. We have

$$H_o: \mu_d = 0 \qquad \text{versus} \qquad H_1: \mu_d < 0$$

This test is left tailed.

**Step 2:** Because the professor is performing a left-tailed test, we determine the critical $t$-value at the $\alpha = 0.05$ level of significance with $n - 1 = 12 - 1 = 11$ degrees of freedom to be $-t_{0.05} = -1.796$. The critical region is displayed in Figure 2.

**Step 3:** The sample mean difference, $\bar{d}$, is $-0.0132$ second, and the sample standard deviation, $s_d$, is 0.0164 second. Because the data are approximately normally distributed, the test statistic is

$$t = \frac{\bar{d} - 0}{s_d / \sqrt{n}} = \frac{-0.0132 - 0}{0.0164 / \sqrt{12}} = -2.788$$

The sample mean difference of $-0.0132$ second is 2.788 sample standard deviations below the assumed mean difference of 0 seconds.

**Step 4:** Because the test statistic $t = -2.788$ is less than the critical value $-t_{0.05} = -1.796$, Professor Neill rejects the null hypothesis. We label the test statistic in Figure 2.

**Step 5:** There is sufficient evidence at the $\alpha = 0.05$ level of significance to support Professor Neill's claim that the reaction time in the dominant hand is less than the reaction time in the nondominant hand.    ◀◀

Figure 2

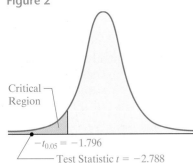

Using Technology:  We can test Professor Neill's claim using statistical software or a graphing calculator with advanced statistical features. For example, using Minitab we obtain the results displayed in Figure 3.

Figure 3

## Paired T-Test and Confidence Interval

Paired T for Dominant − Non-dominant

|          | N  | Mean     | StDev   | SE Mean |
|----------|----|----------|---------|---------|
| Dominant | 12 | 0.17975  | 0.01752 | 0.00506 |
| Non-domi | 12 | 0.19292  | 0.01799 | 0.00519 |
| Difference | 12 | −0.01317 | 0.01643 | 0.00474 |

95% CI for mean difference: (−0.02361, −0.00273)
T-Test of mean difference = 0 (vs < 0): T-Value = −2.78 P-Value = 0.009

> Notice the $P$-value is 0.009. We interpret this by saying that there is a 0.009 probability of obtaining a sample mean difference of $-0.01317$ or less from a population whose mean difference is 0. At the $\alpha = 0.05$ level of significance there is sufficient evidence to reject the null hypothesis because the $P$-value is less than $\alpha$. We conclude that an individual's dominant hand has a faster reaction time than the nondominant hand.

**NW**  *Now Work Problem 11(a).*

**3**  ## Confidence Intervals

We can also create a confidence interval for the mean difference, $\mu_d$, using the sample mean difference, $\overline{d}$, the sample standard deviation difference, $s_d$, the sample size, and $t_{\alpha/2}$. Remember, the format for a confidence interval about a population mean is of the following form:

$$\text{Point estimate} \pm \text{Margin of error}$$

Based on the preceding formula we compute the confidence interval about $\mu_d$ as follows:

---

**Confidence Interval for Matched-Pairs Data**

A $(1 - \alpha) \cdot 100\%$ confidence interval for $\mu_d$ is given by

$$\text{Lower Bound: } \overline{d} - t_{\alpha/2} \cdot \frac{s_d}{\sqrt{n}} \qquad \text{Upper Bound: } \overline{d} + t_{\alpha/2} \cdot \frac{s_d}{\sqrt{n}} \qquad \textbf{(1)}$$

The critical value $t_{\alpha/2}$ is determined using $n - 1$ degrees of freedom. The values of $\overline{d}$ and $s_d$ are the mean and standard deviation of the differenced data.

**Note:**  The interval is exact when the population is normally distributed and approximately correct for nonnormal populations, provided that $n$ is large.

---

▶ **EXAMPLE 3**    **Constructing a Confidence Interval for Matched-Pairs Data**

*Problem:*  Using the data from Table 2, construct a 95% confidence interval estimate of the mean difference, $\mu_d$.

*Approach:*

**Step 1:**  Compute the differenced data. Because the sample size is small, we must verify that the differenced data are approximately normal with no outliers.

**Step 2:**  Compute the sample mean difference, $\overline{d}$, and the sample standard deviation difference, $s_d$.

**Step 3:**  Determine the critical value, $t_{\alpha/2}$, with $\alpha = 0.05$ and $n - 1$ degrees of freedom.

**Step 4:**  Use Formula (1) to determine the lower and upper bounds.

**Step 5:**  Interpret the results.

*Solution:*

**Step 1:**  We computed the differenced data and verified that they are approximately normally distributed with no outliers in Example 2.

**Step 2:**  We computed the sample mean difference, $\overline{d}$, to be $-0.0132$ and sample standard deviation of the difference, $s_d$, to be 0.0164 in Example 2.

**Step 3:**  Using Table III with $\alpha = 0.05$ and $12 - 1 = 11$ degrees of freedom, we find $t_{\alpha/2} = t_{0.025} = 2.201$.

**Step 4:** Substituting into Formula (1), we find

$$\text{Lower Bound: } \overline{d} - t_{\alpha/2} \cdot \frac{s_d}{\sqrt{n}} = -0.0132 - 2.201 \cdot \frac{0.0164}{\sqrt{12}} = -0.0236$$

$$\text{Upper Bound: } \overline{d} + t_{\alpha/2} \cdot \frac{s_d}{\sqrt{n}} = -0.0132 + 2.201 \cdot \frac{0.0164}{\sqrt{12}} = -0.0028$$

**Step 5:** We are 95% confident that the mean difference between the dominant hand's reaction time and the nondominant hand's reaction time is between $-0.0236$ and $-0.0028$ seconds. In other words, we are 95% confident that the dominant hand has a mean reaction time that is somewhere between 0.0028 seconds and 0.0236 seconds faster than the nondominant hand. Notice that the confidence interval does not contain zero. This evidence supports the claim that the reaction time of a person's dominant hand is different from the reaction time of the nondominant hand. ◄◄

We can see that the results of Example 3 agree with the 95% confidence interval determined by Minitab in Figure 3.

 *Now Work Problem 11(b).*

## 10.1 Assess Your Understanding

### Concepts and Vocabulary

1. Explain the difference between independent and dependent sampling.
2. Suppose a researcher claims the mean from population 1 is less than the mean from population 2 in matched-pairs data. How would you define $\mu_d$? How would you determine $d_i$?
3. What are the assumptions that must be satisfied in order to test a claim regarding the difference of two means with dependent sampling?

### Exercises

#### • Skill Building

*In Problems 1–6, determine whether the sampling is dependent or independent.*

1. A sociologist wishes to compare the annual salaries of married couples. She obtains a random sample of 50 married couples in which both spouses work and determines each spouse's annual salary.

2. A researcher wishes to determine the effects of alcohol on people's reaction times to a stimulus. She randomly divides 100 people aged 21 or older into two groups. Group 1 (the treatment group) is asked to drink 3 ounces of alcohol while group 2 drinks a placebo. Both drinks taste the same, so the individuals in the study do not know which group they belong to. Thirty minutes after consuming the drink, the subjects in each group perform a series of tests meant to measure reaction time.

3. An educator wants to determine whether a new curriculum significantly improves standardized test scores for third-grade students. She randomly divides 80 third-graders into two groups. Group 1 (the treatment group) is taught using the new curriculum, while group 2 (the control group) is taught using the traditional curriculum. At the end of the school year, both groups are given the standardized test with the mean scores compared.

4. A psychologist wants to know whether subjects respond faster to a go/no go stimulus or a choice stimulus. With the go/no go stimulus, subjects must respond to a particular stimulus by pressing a button and disregard other stimuli. In the choice stimulus, the subjects respond differently depending on the stimulus. She randomly selects 20 subjects and each subject is presented a series of go/no go stimuli and choice stimuli. The mean reaction time to each stimulus is compared.

5. A study was conducted by researchers designed "to determine the genetic and nongenetic factors to structural brain abnormalities on schizophrenia." The researchers examined the brains of 29 twin patients diagnosed with schizophrenia and compared them with 29 healthy twins. The whole brain volumes of the two groups were compared.
   *Source*: "Volumes of Brain Structures in Twins Discordant for Schizophrenia," William F.C. Baare, et al., *Archives of General Psychiatry* **58**: (2000) 33–40

6. An agricultural researcher wanted to determine whether there were any significant differences in the plowing method used on crop yield. He divided a parcel of land that had uniform soil quality into 30 subplots. He then randomly selected 15 of the plots to be chisele plowed and 15 plots to be fall plowed. He recorded the crop yield at the end of the growing season, to determine whether there was a significant difference in the mean crop yield.

*In Problems 7 and 8, assume that the paired data came from a population that is normally distributed.*

**7.**

| Observation | 1 | 2 | 3 | 4 | 5 | 6 | 7 |
|---|---|---|---|---|---|---|---|
| $X_1$ | 7.6 | 7.6 | 7.4 | 5.7 | 8.3 | 6.6 | 5.6 |
| $X_2$ | 8.1 | 6.6 | 10.7 | 9.4 | 7.8 | 9.0 | 8.5 |

(a) Determine $d_i = X_1 - X_2$ for each pair of data.
(b) Compute $\bar{d}$ and $s_d$.
(c) Test the claim that $\mu_d < 0$ at the $\alpha = 0.05$ level of significance.
(d) Compute a 95% confidence interval about the population mean difference $\mu_d = \mu_1 - \mu_2$.

**8.**

| Observation | 1 | 2 | 3 | 4 | 5 | 6 | 7 | 8 |
|---|---|---|---|---|---|---|---|---|
| $X_1$ | 19.4 | 18.3 | 22.1 | 20.7 | 19.2 | 11.8 | 20.1 | 18.6 |
| $X_2$ | 19.8 | 16.8 | 21.1 | 22.0 | 21.5 | 18.7 | 15.0 | 23.9 |

(a) Determine $d_i = X_1 - X_2$ for each pair of data.
(b) Compute $\bar{d}$ and $s_d$.
(c) Test the claim that $\mu_d \neq 0$ at the $\alpha = 0.01$ level of significance.
(d) Compute a 99% confidence interval about the population mean difference $\mu_d = \mu_1 - \mu_2$.

• **Applying the Concepts**

**9. Muzzle Velocity** The following data represent the muzzle velocity (in feet per second) of rounds fired from a 155 mm gun. For each round, two measurements of the velocity were recorded using two different measuring devices, with the following data obtained:

| Observation | 1 | 2 | 3 | 4 | 5 | 6 | 7 | 8 | 9 | 10 | 11 | 12 |
|---|---|---|---|---|---|---|---|---|---|---|---|---|
| A | 793.8 | 793.1 | 792.4 | 794.0 | 791.4 | 792.4 | 791.7 | 792.3 | 789.6 | 794.4 | 790.9 | 793.5 |
| B | 793.2 | 793.3 | 792.6 | 793.8 | 791.6 | 791.6 | 791.6 | 792.4 | 788.5 | 794.7 | 791.3 | 793.5 |

*Source:* Christenson, Ronald and Blackwood, Larry, "Tests for Precision and Accuracy of Multiple Measuring Devices," *Technometrics*, Nov. 93, Vol 35, Issue 4, pp. 411–421.

(a) Why is this matched-pairs data?
(b) Test the claim that there is a difference in the measurement of the muzzle velocity between device A and device B at the $\alpha = 0.01$ level of significance. *Note:* A normal probability plot and boxplot of the data indicate that the differences are approximately normally distributed with no outliers.
(c) Construct a 99% confidence interval about the population mean difference. Interpret your results.
(d) Draw a boxplot of the differenced data. Does this visual evidence support the results obtained in part (b)?

**10. The Effects of Exercise** A physical therapist wishes to determine whether an exercise program increases flexibility. He measures the flexibility (in inches) of 12 randomly selected subjects both before and after an intensive eight-week training program and obtains the following data:

| Subject | 1 | 2 | 3 | 4 | 5 | 6 | 7 | 8 | 9 | 10 | 11 | 12 |
|---|---|---|---|---|---|---|---|---|---|---|---|---|
| Before | 18.5 | 21.5 | 16.5 | 21 | 20 | 15 | 19.75 | 15.75 | 18 | 22 | 15 | 20.5 |
| After | 19.25 | 21.75 | 16.5 | 20.5 | 22.25 | 16 | 19.5 | 17 | 19.25 | 19.5 | 16.5 | 20 |

*Source:* Michael McCraith, Joliet Junior College

(a) Why is this matched-pairs data?
(b) Test the claim that the flexibility before the exercise program is less than the flexibility after the exercise program at the $\alpha = 0.02$ level of significance. *Note:* A normal probability plot and boxplot of the data indicate that the differences are approximately normally distributed with no outliers.
(c) Construct a 95% confidence interval about the population mean difference. Interpret your results.
(d) Draw a boxplot of the differenced data. Does this visual evidence support the results obtained in part (b)?

11. **Reaction Time** In an experiment conducted on-line at the University of Mississippi, study participants are asked to react to a stimulus. In one experiment, the participant must press a key on seeing a blue screen. Reaction time (in seconds) to press the key is measured. The same person is then asked to press a key on seeing a red screen, again with reaction time measured. The results for six randomly sampled study participants are as follows:

| Participant Number | 1 | 2 | 3 | 4 | 5 | 6 |
|---|---|---|---|---|---|---|
| Reaction Time to Blue | 0.582 | 0.481 | 0.841 | 0.267 | 0.685 | 0.45 |
| Reaction Time to Red | 0.408 | 0.407 | 0.542 | 0.402 | 0.456 | 0.533 |

*Source:* PsychExperiments at the University of Mississippi

(a) Test the claim that the reaction time to the blue stimulus is different from the reaction time to the red stimulus at the $\alpha = 0.01$ level of significance. *Note*: A normal probability plot and boxplot of the data indicate that the differences are approximately normally distributed with no outliers.
(b) Construct a 98% confidence interval about the population mean difference. Interpret your results.
(c) Draw a boxplot of the differenced data. Does this visual evidence support the results obtained in part (a)?

12. **Rat's Hemoglobin.** Hemoglobin helps the red blood cells transport oxygen and remove carbon dioxide. Researchers at NASA wanted to determine the effects of space flight on a rat's hemoglobin. The following data represent the hemoglobin (in grams per deciliter) at lift-off minus 3 days (H-L3) and immediately upon the return (H-R0) for 12 randomly selected rats sent to space on the Spacelab Sciences 1 flight.

| Rat # | 1 | 2 | 3 | 4 | 5 | 6 | 7 | 8 | 9 | 10 | 11 | 12 |
|---|---|---|---|---|---|---|---|---|---|---|---|---|
| H-L3 | 15.2 | 16.1 | 15.3 | 16.4 | 15.7 | 14.7 | 14.3 | 14.5 | 15.2 | 16.1 | 15.1 | 15.8 |
| H-R0 | 15.8 | 16.5 | 16.7 | 15.7 | 16.9 | 13.1 | 16.4 | 16.5 | 16.0 | 16.8 | 17.6 | 16.9 |

*Source:* NASA Life Sciences Data Archive

(a) Test the claim that the hemoglobin levels at lift-off minus 3 days are less than the hemoglobin levels upon return at the $\alpha = 0.05$ level of significance. *Note*: A normal probability plot and boxplot of the data indicate that the differences are approximately normally distributed with no outliers.
(b) Construct a 90% confidence interval about the population mean difference. Interpret your results.
(c) Draw a boxplot of the differenced data. Does this visual evidence support the results obtained in part (a)?

13. **Secchi Disk** A Secchi disk is an eight-inch diameter weighted disk that is painted black and white and attached to a rope. The disk is lowered into water and the depth (in inches) at which it is no longer visible is recorded. The measurement is an indication of water clarity. An environmental biologist is interested in determining whether the water clarity of the lake at Joliet Junior College is improving. She takes measurements at the same location during the course of a year and repeats the measurements on the same dates five years later. She obtains the following results:

| Observation | 1 | 2 | 3 | 4 | 5 | 6 | 7 | 8 |
|---|---|---|---|---|---|---|---|---|
| Date | 5/11 | 6/7 | 6/24 | 7/8 | 7/27 | 8/31 | 9/30 | 10/12 |
| Initial Depth | 38 | 58 | 65 | 74 | 56 | 36 | 56 | 52 |
| Depth 5 Years Later | 52 | 60 | 72 | 72 | 54 | 48 | 58 | 60 |

*Source:* Virginia Piekarski, Joliet Junior College

(a) Test the claim that the clarity of the lake is improving at the $\alpha = 0.05$ level of significance. *Note*: A normal probability plot and boxplot of the data indicate that the differences are approximately normally distributed with no outliers.
(b) Construct a 95% confidence interval about the population mean difference. Interpret your results.

14. **The Effect of Aspirin on Blood Clotting** Blood clotting occurs due to a sequence of chemical reactions. The protein thrombin initiates blood clotting by working with another protein, pro-thrombin. It is common to measure an individual's blood clotting time through prothrombin time—the time between the start of the thrombin–prothrombin reaction and the formation of the clot. Researchers wanted to study the effect of aspirin on prothrombin time. They randomly selected 12 subjects and measured the prothrombin time (in seconds), first without taking aspirin and again three hours after taking two aspirin tablets. They obtained the following data:

| Subject | 1 | 2 | 3 | 4 | 5 | 6 | 7 | 8 | 9 | 10 | 11 | 12 |
|---|---|---|---|---|---|---|---|---|---|---|---|---|
| With Aspirin | 12.3 | 12.0 | 12.0 | 13.0 | 13.0 | 12.5 | 11.3 | 11.8 | 11.5 | 11.0 | 11.0 | 11.3 |
| Without Aspirin | 12.0 | 12.3 | 12.5 | 12.0 | 13.0 | 12.5 | 10.3 | 11.3 | 11.5 | 11.5 | 11.0 | 11.5 |

*Source:* Yochem, Donald and Roach, Darrell. "Aspirin: Effect of Thrombus Formation Time and Prothrombin Time of Human Subjects." *Angiology*, **22** (1971), pp. 70–76.

(a) Test the claim that aspirin affects the time it takes for a clot to form at the $\alpha = 0.05$ level of significance. *Note*: A normal probability plot and boxplot of the data indicate that the differences are approximately normally distributed with no outliers.

(b) Construct a 90% confidence interval about the population mean difference. Interpret your results.

15. **Car Repairs** The Insurance Institute for Highway Safety regularly tests cars for various safety factors. In one such test, the institute tests the bumpers in 5-mph crashes. The following data represent the cost of repair (in dollars) after four different 5-mph crashes on a 2000 Dodge Neon and a 2000 Honda Civic:

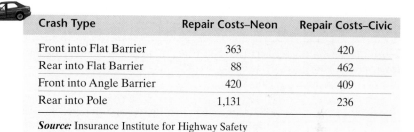

| Crash Type | Repair Costs–Neon | Repair Costs–Civic |
|---|---|---|
| Front into Flat Barrier | 363 | 420 |
| Rear into Flat Barrier | 88 | 462 |
| Front into Angle Barrier | 420 | 409 |
| Rear into Pole | 1,131 | 236 |

*Source:* Insurance Institute for Highway Safety

(a) Test the claim that the repair costs differ at the $\alpha = 0.05$ level of significance. *Note*: A normal probability plot and boxplot of the data indicate that the differences are approximately normally distributed with no outliers.

(b) Construct a 90% confidence interval about the population mean difference. Interpret your results.

16. **Improving SAT Math Scores** A test preparation company claims that its SAT preparation course improves SAT math scores. The company administers the SAT to 12 randomly selected students and determines their scores. The same students then participate in the course. Upon completion, they retake the SAT. The results are presented in the following table:

| Before | 436 | 431 | 270 | 463 | 528 | 377 | 397 | 413 | 525 | 323 | 413 | 292 |
|---|---|---|---|---|---|---|---|---|---|---|---|---|
| After | 443 | 429 | 287 | 501 | 522 | 380 | 402 | 450 | 548 | 349 | 403 | 303 |

(a) Test the claim that the preparatory course improves SAT math scores at the $\alpha = 0.1$ level of significance. *Note*: A normal probability plot and boxplot of the data indicate that the differences are approximately normally distributed with no outliers.

(b) Construct a 95% confidence interval about the population mean difference. Interpret your results.

**17. Car Rentals** The following data represent the daily rental for a compact automobile charged by two car rental companies, Thrifty and Hertz, in 10 locations:

| City | Thrifty | Hertz | City | Thrifty | Hertz |
|------|---------|-------|------|---------|-------|
| Chicago | 21.81 | 18.99 | Seattle | 21.96 | 22.99 |
| Los Angeles | 29.89 | 48.99 | Pittsburgh | 20.90 | 19.99 |
| Houston | 17.90 | 19.99 | Phoenix | 47.75 | 36.99 |
| Orlando | 27.98 | 35.99 | New Orleans | 33.81 | 26.99 |
| Boston | 24.61 | 25.60 | Minneapolis | 33.49 | 20.99 |

*Source:* Yahoo! Travel

(a) Test the claim that Thrifty is less expensive than Hertz at the $\alpha = 0.1$ level of significance. *Note*: A normal probability plot and boxplot of the data indicate that the differences are approximately normally distributed with no outliers.
(b) Construct a 90% confidence interval about the population mean difference. Interpret your results.

**18. Thermocouples** A thermocouple is a temperature sensor used to measure the temperature inside furnaces. A researcher wanted to know whether two thermocouples were measuring the temperature the same. He wrapped the thermocouples together, placed them in an oven and recorded the temperature (in °C) reported by the thermocouple. He did this 9 times and obtained the following data:

| Thermocouple 1 | 326.06 | 326.08 | 326.05 | 326.03 | 326.00 | 326.00 | 325.97 | 326.07 | 326.11 |
|----------------|--------|--------|--------|--------|--------|--------|--------|--------|--------|
| Thermocouple 2 | 326.03 | 326.06 | 326.02 | 326.01 | 325.99 | 325.99 | 325.95 | 326.03 | 326.08 |

*Source:* Christenson, Ronald and Blackwood, Larry, "Tests for Precision and Accuracy of Multiple Measuring Devices," *Technometrics*, Nov. 93, Vol 35, Issue 4, pp. 411–421.

(a) Test the claim that there is a difference in the mean temperature at the $\alpha = 0.05$ level of significance. *Note*: A normal probability plot and boxplot of the data indicate that the differences are approximately normally distributed with no outliers.
(b) Construct a 90% confidence interval about the population mean difference. Interpret your results.

**19. Rat's Red Blood Cell Count** Researchers at NASA wanted to determine the effects of space flight on a rat's red blood cell count (RBC). The following data represent the red blood cell count at lift-off minus three days (RBC-L3) and immediately upon the return (RBC-R0) of 27 rats sent to space on the Spacelab Sciences 1 flight:

| Rat # | RBC–L3 | RBC–R0 | Rat # | RBC–L3 | RBC–R0 | Rat # | RBC–L3 | RBC–R0 | Rat # | RBC–L3 | RBC–R0 |
|-------|--------|--------|-------|--------|--------|-------|--------|--------|-------|--------|--------|
| 112 | 7.53 | 8.16 | 136 | 7.95 | 9.55 | 79 | 7.79 | 8.52 | 156 | 7.27 | 9.11 |
| 145 | 7.79 | 9.15 | 127 | 8.44 | 9.38 | 47 | 8.24 | 8.59 | 128 | 8.23 | 9.81 |
| 15 | 7.84 | 9.09 | 153 | 6.70 | 9.66 | 109 | 7.69 | 7.29 | 99 | 14.50 | 9.86 |
| 142 | 6.86 | 8.42 | 97 | 6.95 | 8.83 | 13 | 7.23 | 8.93 | 90 | 7.73 | 9.90 |
| 45 | 7.93 | 8.96 | 94 | 6.73 | 8.46 | 74 | 7.83 | 8.64 | 82 | 7.31 | 7.64 |
| 150 | 7.48 | 9.25 | 124 | 7.21 | 9.04 | 126 | 8.09 | 9.65 | 55 | 7.84 | 9.54 |
| 162 | 7.94 | 7.40 | 117 | 6.95 | 9.48 | 157 | 8.27 | 8.76 | | | |

*Source:* NASA Life Sciences Data Archive

The researchers used Minitab to test the claim that the red blood cell count three days prior to lift-off was different from the red blood cell count upon their return. The results are as follows:

### Paired T-Test and Confidence Interval
```
Paired T for RBC-L3 - RBC-R0

              N      Mean    StDev    SE Mean
RBC-L3        27     7.864   1.414    0.272
RBC-R0        27     8.929   0.715    0.138
Difference    27    -1.065   1.397    0.269

95% CI for mean difference: (-1.618, -0.512)
T-Test of mean difference = 0 (vs not = 0): T-Value = -3.96 P-Value = 0.001
```

(a) State the null and alternative hypotheses.
(b) What requirements must be satisfied in order to test the claim?
(c) Does it appear that the flight to space affected the red blood cell count of the rats at the $\alpha = 0.05$ level of significance? Why?

**20. Reaction Time Experiment** Researchers at the University of Mississippi wanted to determine the reaction time (in seconds) of students to different stimuli. In the data that follow, the reaction time for subjects was measured after they received a simple stimulus and a go/no go stimulus. The simple stimulus presented an auditory cue and the time from when the cue was given to when the student reacted was measured. The go/no go stimulus required the student to respond to a particular stimulus and not to respond to other stimuli. Again, the reaction time was measured, with the following results:

| Subject Number | Simple | Go/No Go | Subject Number | Simple | Go/No Go | Subject Number | Simple | Go/No Go |
|---|---|---|---|---|---|---|---|---|
| 1 | 0.22 | 0.375 | 7 | 0.255 | 0.442 | 12 | 0.498 | 0.565 |
| 2 | 0.338 | 0.652 | 8 | 0.198 | 0.347 | 13 | 0.262 | 0.402 |
| 3 | 0.266 | 0.467 | 9 | 0.352 | 0.698 | 14 | 0.62 | 0.643 |
| 4 | 0.381 | 0.651 | 10 | 0.259 | 0.488 | 15 | 0.3 | 0.351 |
| 5 | 0.885 | 1.246 | 11 | 0.2 | 0.281 | 16 | 0.424 | 0.38 |
| 6 | 0.25 | 0.654 | | | | | | |

*Source:* PsychExperiments at The University of Mississippi

The researchers used Minitab to test the claim that the simple stimulus had a lower reaction time than the go/no go stimulus. The results of the analysis are as follows:

### Paired T-Test and Confidence Interval
```
Paired T for Simple - Go/No Go

              N      Mean     StDev    SE Mean
Simple        16     0.3567   0.1815   0.0454
Go/No Go      16     0.5401   0.2312   0.0578
Difference    16    -0.1834   0.1308   0.0327

95% CI for mean difference: (-0.2531, -0.1137)
T-Test of mean difference = 0 (vs < 0): T-Value = -5.61 P-Value = 0.000
```

(a) State the null and alternative hypotheses.
(b) What requirements must be satisfied in order to test the claim?
(c) Does it appear that the reaction time to the simple stimulus is less than the reaction time to the go/no go stimulus at the $\alpha = 0.05$ level of significance? Why?

**Technology Step-by-Step**
Two-sample *T*-tests, Dependent Sampling

**TI-83 Plus**    **Hypothesis Tests**

*Step 1:* If necessary, enter raw data in L1 and L2. Let L3 = L1 − L2 (or L2 − L1), depending upon how the alternative hypothesis was defined.

*Step 2:* Press STAT, highlight TESTS, and select `2:T-Test`.

*Step 3:* If the data are raw, highlight `Data` making sure that List is set to L3 with frequency set to 1. If summary statistics are known, highlight `Stats` and enter the summary statistics.

*Step 4:* Highlight the appropriate relation between $\mu_1$ and $\mu_2$ in the alternative hypothesis.

*Step 5:* Highlight Calculate or Draw and press ENTER. Calculate gives the test statistic and *P*-value. Draw will draw the *t*-distribution with the *P*-value shaded.

**Confidence Intervals**

Follow the same steps as those given for hypothesis tests, except select `8: TInterval` Also, select a confidence level (such as 95% = 0.95).

**MINITAB**    *Step 1:* Enter raw data in columns C1 and C2.

*Step 2:* Select the **Stat** menu, highlight **Basic Statistics**, and then highlight **Paired-t . . .**

*Step 3:* Enter C1 in the cell marked "First Sample" and enter C2 in the cell marked "Second Sample." Under OPTIONS, select the direction of the alternative hypothesis and select a confidence level. Click OK.

**Excel**    *Step 1:* Enter raw data in columns A and B.

*Step 2:* Select the **Tools** menu, highlight **Data Analysis . . .**

*Step 3:* Select "*t*-test: Paired Two-Sample for Means." With the cursor in the "Variable 1 Range" cell, highlight the data in column A. With the cursor in the "Variable 2 Range" cell, highlight the data in column B. Enter the hypothesized difference in the means (usually 0) and a value for alpha. Click OK.

---

## 10.2   Inference about Two Means: Independent Samples

**Preparing for This Section**    Before getting started, review the following:

   ✓ Design of experiments (Section 1.5, pp. 31–34)

   ✓ Confidence intervals about $\mu$, $\sigma$ unknown (Section 8.2, pp. 360–363)

   ✓ Hypothesis tests about $\mu$, $\sigma$ unknown (Section 9.3, pp. 408–413)

   ✓ Type I and Type II errors (Section 9.1, pp. 387–388)

**Objectives**  Perform hypothesis tests regarding the difference of two means

Construct confidence intervals regarding the difference of two means

We now turn our attention to inferential methods for comparing means from two independent samples. For example, suppose we wish to know whether a new experimental drug relieves symptoms attributable to the common cold. The response variable might be the time until the cold symptoms go away. If the drug is effective, then the mean time until the cold symptoms go away should be less for individuals taking the drug than for those not taking the drug. If we let $\mu_1$ represent the mean time until cold symptoms go away for the individuals taking the drug, and $\mu_2$ represent the mean time until cold symptoms go away for individuals taking a placebo, the null and alternative hypotheses will be

$$H_o: \mu_1 = \mu_2 \qquad \text{versus} \qquad H_1: \mu_1 < \mu_2$$

or equivalently,

$$H_o: \mu_1 - \mu_2 = 0 \qquad \text{versus} \qquad H_1: \mu_1 - \mu_2 < 0$$

To test this claim, we might randomly divide 500 volunteers who have a common cold into two groups—an experimental group (Group 1) and a control group (Group 2). The control group would receive a placebo and the experimental group would receive a predetermined amount of the experimental drug. Next, determine the time until the cold symptoms go away. Compute $\bar{x}_1$, the sample mean time until cold symptoms go away in the experimental group, and $\bar{x}_2$, the sample mean time until cold symptoms go away in the control group. Now, we would determine whether the difference in the sample means, $\bar{x}_1 - \bar{x}_2$, was significantly different from 0—the assumed difference stated in the null hypothesis. To do this, we need to know the sampling distribution of $\bar{x}_1 - \bar{x}_2$.

*Theorem*

**Sampling Distribution of the Difference of Two Means: Independent Samples with Population Standard Deviations Known**

Suppose a simple random sample of size $n_1$ is taken from a population with mean $\mu_1$ and standard deviation $\sigma_1$. In addition, a simple random sample of size $n_2$ is taken from a second population with mean $\mu_2$ and standard deviation $\sigma_2$. If the two populations are normally distributed or the sample sizes are sufficiently large ($n_1 \geq 30$ and $n_2 \geq 30$), then the sampling distribution of $\bar{x}_1 - \bar{x}_2$ is normally distributed with mean $\mu_1 - \mu_2$ and standard deviation $\sigma_{\bar{x}_1 - \bar{x}_2} = \sqrt{\dfrac{\sigma_1^2}{n_1} + \dfrac{\sigma_2^2}{n_2}}$. The standardized version of $\bar{x}_1 - \bar{x}_2$ is then written as

$$Z = \frac{(\bar{x}_1 - \bar{x}_2) - (\mu_1 - \mu_2)}{\sqrt{\dfrac{\sigma_1^2}{n_1} + \dfrac{\sigma_2^2}{n_2}}}$$

which has the standard normal distribution.

Knowing the sampling distribution of $\bar{x}_1 - \bar{x}_2$, we can use the methods introduced in Sections 8.1 and 9.2 to construct confidence intervals and perform hypothesis tests regarding claims about $\mu_1$ and $\mu_2$ with $\sigma_1$ and $\sigma_2$ known. Unfortunately, it is unreasonable to expect to know information regarding $\sigma_1$ and $\sigma_2$ without knowing information regarding the population means. Therefore, we must develop a sampling distribution for the difference of two means when the population standard deviations are unknown.

The comparison of two means with unequal (and unknown) population variances is known as the "Behrens–Fisher problem." While an exact method for performing inference on the equality of two means with unequal population standard deviations does not exist, an approximate solution is available. The approach that we use is known as **Welch's approximate $t$**, in honor of English statistician Bernard Lewis Welch (1911–1989).

*Theorem*

**Sampling Distribution of the Difference of Two Means: Independent Samples with Population Standard Deviations Unknown (Welch's $t$)**

Suppose a simple random sample of size $n_1$ is taken from a population with unknown mean $\mu_1$ and unknown standard deviation $\sigma_1$. In addition, a simple random sample of size $n_2$ is taken from a second population with unknown mean $\mu_2$ and unknown standard deviation $\sigma_2$. If the two populations are normally distributed or the sample sizes are sufficiently large ($n_1 \geq 30$ and $n_2 \geq 30$), then

$$t = \frac{(\bar{x}_1 - \bar{x}_2) - (\mu_1 - \mu_2)}{\sqrt{\dfrac{s_1^2}{n_1} + \dfrac{s_2^2}{n_2}}} \tag{1}$$

*approximately* follows Student's $t$-distribution with the smaller of $n_1 - 1$ or $n_2 - 1$ degrees of freedom where $\bar{x}_1$ is the sample mean, and $s_1$ is the sample standard deviation from population 1; $\bar{x}_2$ is the sample mean and $s_2$ is the sample standard deviation from population 2.

## ① Inference on the Difference of Two Means with Unknown Population Variances

Now that we know the approximate sampling distribution of $\bar{x}_1 - \bar{x}_2$ with the population standard deviations unknown, we can introduce a procedure that can be used to test claims regarding two population means.

**Hypothesis Test Regarding the Difference of Two Means with $\sigma$ Unknown**

If a claim is made regarding two population means, $\mu_1$ and $\mu_2$, with unknown population standard deviations, we can use the following steps to test the claim, provided that

**1.** the samples are obtained using simple random sampling;

**2.** the samples are independent;

**3.** the populations from which the samples are drawn are normally distributed or the sample sizes are large ($n_1 \geq 30$ and $n_2 \geq 30$).

*Step 1:* A claim is made regarding two population means. The claim is used to determine the null and alternative hypotheses. The hypotheses are structured in one of three ways, as shown below.

| Two-Tailed | Left-Tailed | Right-Tailed |
|---|---|---|
| $H_o: \mu_1 = \mu_2$ | $H_o: \mu_1 = \mu_2$ | $H_o: \mu_1 = \mu_2$ |
| $H_1: \mu_1 \neq \mu_2$ | $H_1: \mu_1 < \mu_2$ | $H_1: \mu_1 > \mu_2$ |

*Note:* $\mu_1$ is the population mean for population 1, and $\mu_2$ is the population mean for population 2.

*Step 2:* Decide on a level of significance $\alpha$ based upon the seriousness of making a Type I error. The critical value is determined using the smaller of $n_1 - 1$ and $n_2 - 1$ degrees of freedom. For example, the critical value in the left-tailed test is $-t_\alpha$, with the smaller of $n_1 - 1$ or $n_2 - 1$ degrees of freedom. The shaded regions represent the critical region as shown:

| Two-Tailed | Left-Tailed | Right-Tailed |
|---|---|---|

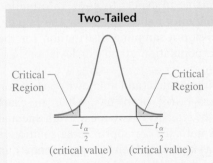

Critical Region — Critical Region

$-t_{\frac{\alpha}{2}}$ (critical value)  $t_{\frac{\alpha}{2}}$ (critical value)

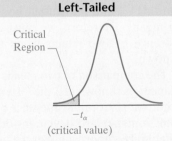

Critical Region

$-t_\alpha$ (critical value)

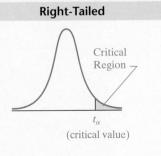

Critical Region

$t_\alpha$ (critical value)

**Step 3:** Compute the **test statistic** $t = \dfrac{(\bar{x}_1 - \bar{x}_2) - (\mu_1 - \mu_2)}{\sqrt{\dfrac{s_1^2}{n_1} + \dfrac{s_2^2}{n_2}}}$, which approximately follows Student's $t$-distribution.

**Step 4:** Compare the critical value with the test statistic, as shown:

| Two-tailed | Left-tailed | Right-tailed |
|---|---|---|
| If $t < -t_{\alpha/2}$ or $t > t_{\alpha/2}$, reject the null hypothesis | If $t < -t_\alpha$, reject the null hypothesis | If $t > t_\alpha$, reject the null hypothesis |

**Step 5:** State the conclusion.

The procedure just presented requires a normal distribution of each population or a large sample size. The procedure is robust, so minor departures from normality will not adversely affect the results of the test. If the data have outliers, however, the procedure should not be used. We will verify these requirements by constructing normal probability plots (to assess normality) and boxplots (to determine whether there are outliers). If the normal probability plot indicates that the data came from a population that is not normally distributed or the boxplot reveals outliers, nonparametric tests should be performed.

► EXAMPLE 1  **Testing a Claim Regarding Two Means**

*Problem:* In the Spacelab Life Sciences 2 payload, 14 male rats were sent to space. Upon their return, the red blood cell mass (in milliliters) of the rats was determined. A control group of 14 male rats was held under the same conditions (except for space flight) as the space rats and their red blood cell mass was also determined when the space rats returned. The project, led by Dr. Paul X. Callahan, resulted in the data listed in Table 3.

| TABLE 3 | | | | | | | |
|---|---|---|---|---|---|---|---|
| Flight | | | | Control | | | |
| 8.59 | 8.64 | 7.43 | 7.21 | 8.65 | 6.99 | 8.40 | 9.66 |
| 6.87 | 7.89 | 9.79 | 6.85 | 7.62 | 7.44 | 8.55 | 8.70 |
| 7.00 | 8.80 | 9.30 | 8.03 | 7.33 | 8.58 | 9.88 | 9.94 |
| 6.39 | 7.54 | | | 7.14 | 9.14 | | |

*Source:* NASA Life Sciences Data Archive

Test the claim that the flight animals have a different red blood cell mass from the control animals at the $\alpha = 0.05$ level of significance.

*Approach:* We verify that each of the samples comes from a population that is approximately normal with no outliers by drawing normal probability plots and boxplots. The boxplots will be drawn on the same graph so that we can visually compare the two populations. We then follow Steps 1–5.

*Solution:* Figure 4 on page 452 shows normal probability plots of the data, which indicate that the data could come from populations that are normal. On the basis of the boxplots, it would seem that there is not much difference in the red blood cell mass of the two populations, although the flight group might have a slightly lower red blood cell mass.

Figure 4

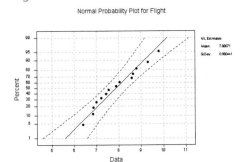

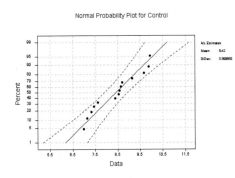

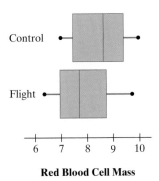

Red Blood Cell Mass

**Step 1:** The claim is that the flight animals have a different red blood cell mass from the control animals. Let $\mu_1$ represent the mean red blood cell mass of the flight animals and $\mu_2$ represent the mean red blood cell mass of the control animals. Then the claim can be expressed as $\mu_1 \neq \mu_2$, and we have the hypotheses

$$H_o: \mu_1 = \mu_2$$

versus

$$H_1: \mu_1 \neq \mu_2$$

which can be written as

$$H_o: \mu_1 - \mu_2 = 0$$

versus

$$H_1: \mu_1 - \mu_2 \neq 0$$

Figure 5

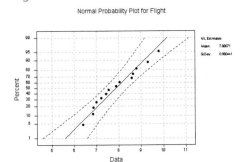

Critical Regions

Test statistic,
$t = -1.437$

$-t_{0.025} = -2.160$     $t_{0.025} = 2.160$

**Step 2:** We have a two-tailed test with $\alpha = 0.05$. Since the sample size of the experimental group and control group are both 14, we have $n_1 - 1 = 14 - 1 = 13$ degrees of freedom. The critical values are $t_{\alpha/2} = t_{0.05/2} = t_{0.025} = 2.160$ and $-t_{0.025} = -2.160$. The critical regions are displayed in Figure 5.

**Step 3:** We compute the sample statistics for each population and obtain the results shown in Table 4. Compute the test statistic as follows:

$$t = \frac{(\bar{x}_1 - \bar{x}_2) - (\mu_1 - \mu_2)}{\sqrt{\dfrac{s_1^2}{n_1} + \dfrac{s_2^2}{n_2}}} = \frac{(7.881 - 8.430) - 0}{\sqrt{\dfrac{1.017^2}{14} + \dfrac{1.005^2}{14}}}$$

$$= \frac{-0.549}{0.3821288115} = -1.437$$

| TABLE 4 | | |
|---|---|---|
| | **Flight Animals** | **Control Animals** |
| **Sample Size** | $n_1 = 14$ | $n_2 = 14$ |
| **Sample Mean** | $\bar{x}_1 = 7.881$ | $\bar{x}_2 = 8.430$ |
| **Sample Standard Deviation** | $s_1 = 1.017$ | $s_2 = 1.005$ |

**Step 4:** We compare the test statistic with the critical values. Because the test statistic, $t = -1.437$, lies between the critical values of $-2.160$ and $2.160$, we do not reject the null hypothesis. That is, the test statistic does not lie within the critical regions, so we do not reject $H_o$. We label this point in Figure 5.

**Step 5:** There is not sufficient evidence at the $\alpha = 0.05$ level of significance to support the claim that there is a significant difference in red blood cell mass between the flight animals and the control animals. ◄◄

The degrees of freedom used to determine the critical value(s) presented in Example 1 are conservative. Results that are more accurate can be obtained by using the following degrees of freedom:

$$df = \frac{\left(\dfrac{s_1^2}{n_1} + \dfrac{s_2^2}{n_2}\right)^2}{\dfrac{\left(\dfrac{s_1^2}{n_1}\right)^2}{n_1 - 1} + \dfrac{\left(\dfrac{s_2^2}{n_2}\right)^2}{n_2 - 1}}. \tag{2}$$

When using Formula (2) to compute degrees of freedom, round down to the nearest integer in order to use Table III. For hand inference, it is recommended that we employ the conservative approach of using the smaller of $n_1 - 1$ or $n_2 - 1$ as the degrees of freedom, to ease computational difficulty. However, computer software will use Formula (2) when computing the degrees of freedom, for increased precision in determining the critical $t$.

**Using Technology:** Statistical spreadsheets and graphing calculators with advanced statistical functions can be used to compare two means. Figure 6 shows the results of Example 1 using a TI-83 Plus graphing calculator. The $P$-value is reported as 0.1627. Because the $P$-value is greater than the level of significance, $\alpha = 0.05$, we do not reject the null hypothesis. There is not sufficient evidence to support the claim that the mean red blood cell mass of flight animals is different from the mean red blood cell mass of control animals.

**⊘ Caution**

*The degrees of freedom in "by hand" solutions will not equal the degrees of freedom in "technology" solutions unless you use Formula (2) to compute degrees of freedom.*

Notice that the degrees of freedom in the technology solution is 25.996 versus 13 in the conservative solution done by hand in Example 1. With the lower degrees of freedom, the critical $t$ is larger (2.160 with 13 degrees of freedom versus 2.056 with approximately 26 degrees of freedom). The larger critical value increases the number of standard deviations the difference in the sample means must be from the hypothesized mean difference before the null hypothesis is rejected. Therefore, in using the smaller of $n_1 - 1$ or $n_2 - 1$ degrees of freedom, we need more substantial evidence to reject the null hypothesis. This requirement decreases the probability of a Type I error (rejecting the null when the null is true) below the actual level of $\alpha$ chosen by the researcher. This is what we mean when we say that the method of using the lesser of $n_1 - 1$ and $n_2 - 1$ as a proxy for degrees of freedom is conservative compared with using Formula (2) on page 452.

**NW** *Now Work Problem 7(a).*

**②** Constructing a confidence interval about the difference of two means is simply an extension of the results presented in Section 8.2.

**Constructing a $(1 - \alpha) \cdot 100\%$ Confidence Interval about the Difference of Two Means**

Suppose a simple random sample of size $n_1$ is taken from a population with unknown mean $\mu_1$ and unknown standard deviation $\sigma_1$. In addition, a simple random sample of size $n_2$ is taken from a population with unknown mean $\mu_2$ and unknown standard deviation $\sigma_2$. If the two populations are normally distributed or the sample sizes are sufficiently large ($n_1 \geq 30$ and $n_2 \geq 30$), then a $(1 - \alpha) \cdot 100\%$ confidence interval about $\mu_1 - \mu_2$ is given by

$$\text{Lower Bound: } (\bar{x}_1 - \bar{x}_2) - t_{\alpha/2} \cdot \sqrt{\frac{s_1^2}{n_1} + \frac{s_2^2}{n_2}}$$

and

$$\text{Upper Bound: } (\bar{x}_1 - \bar{x}_2) + t_{\alpha/2} \cdot \sqrt{\frac{s_1^2}{n_1} + \frac{s_2^2}{n_2}} \qquad (3)$$

where $t_{\alpha/2}$ is computed using the smaller of $n_1 - 1$ or $n_2 - 1$ degrees of freedom or Formula (2).

▶ **EXAMPLE 2** **Constructing a Confidence Interval about the Difference of Two Means**

*Problem:* Construct a 95% confidence interval about $\mu_1 - \mu_2$, using the data presented in Table 3.

*Approach:* A normal probability plot and boxplot (Figure 4) indicate that the data are approximately normal with no outliers. We compute the confidence interval with $\alpha = 0.05$ using Formula (3).

*Solution:* We have already found the sample statistics in Example 1. In addition, we found $t_{\alpha/2} = t_{0.025}$ with 13 degrees of freedom to be 2.160. Substituting into Formula (3), we obtain the following results:

Lower Bound:

$$(\bar{x}_1 - \bar{x}_2) - t_{\alpha/2} \cdot \sqrt{\frac{s_1^2}{n_1} + \frac{s_2^2}{n_2}} = (7.881 - 8.430) - 2.160 \cdot \sqrt{\frac{1.017^2}{14} + \frac{1.005^2}{14}} = -0.549 - 0.825 = -1.374$$

Upper Bound:

$$(\bar{x}_1 - \bar{x}_2) + t_{\alpha/2} \cdot \sqrt{\frac{s_1^2}{n_1} + \frac{s_2^2}{n_2}} = (7.881 - 8.430) + 2.160 \cdot \sqrt{\frac{1.017^2}{14} + \frac{1.005^2}{14}} = -0.549 + 0.825 = 0.276$$

*Interpretation:* We are 95% confident that the mean difference between the red blood cell mass of the flight animals and control animals is between $-1.374$ ml and $0.276$ ml. Because the confidence interval contains zero, there is not sufficient evidence to support the claim that there is a difference in the red blood cell mass of the flight group and the control group. ◄◄

**NW** *Now Work Problem 7(b).*

**Caution**

We would use the pooled two-sample t-test when the two samples come from populations that have the same variance. "Pooling" refers to finding a weighted average of the two sample variances from the independent samples. It is difficult to verify that two sample variances might be equal, so we will always use Welch's t when comparing two means.

## What about the Pooled Two-Sample *t*-tests?

Perhaps you noticed that statistical software and graphing calculators with advanced statistical features provide an option for two types of two-sample *t* tests: one that assumes equal population variances (pooling) and one that does not assume equal population variances. Welch's *t*-statistic does not assume that the population variances are equal and can be used whether the population variances are equal or not. The test that assumes equal population variances is referred to as the *pooled t-statistic*.

The **pooled *t*-statistic** is computed by finding a weighted average of the sample variances and uses this average in the computation of the test statistic. The advantage of this test statistic is that it exactly follows Student's *t*-distribution with $n_1 + n_2 - 2$ degrees of freedom. The disadvantage of the test statistic is that it requires that the population variances be equal. How is this requirement to be verified? While a test for determining the equality of variances does exist (*F*-test, discussed on CD that accompanies the text), the test *requires* that each population be normally distributed. However, the *F*-test is not robust. Any minor departures from normality will make the results of the *F*-test unreliable. It has been recommended by numerous statisticians* that a preliminary *F*-test to check the requirement of equality of variance not be performed. In fact, George Box once said "To make preliminary tests on variances is rather like putting to sea in a rowing boat to find out whether conditions are sufficiently calm for an ocean liner to leave port!"

---

*Moser and Stevens, "Homogeneity of Variance in the Two-Sample Means Test," *The American Statistician*, Vol. 46, No. 1.

Because the formal *F*-test for testing the equality of variances is so volatile, we shall be content in using Welch's *t*. This test is more conservative than the pooled *t*. The price that must be paid for the conservative approach is that the probability of a Type II error is higher in Welch's *t* than in the pooled *t* when the population variances are equal. However, the two tests typically provide the same conclusion, even if the assumption of equal population standard deviations seems reasonable.

## 10.2 Assess Your Understanding

## Concepts and Vocabulary

**1.** What are the requirements that need to be satisfied in order to test a hypothesis regarding the difference of two means with $\sigma$ unknown?

**2.** Explain why using the smaller of $n_1 - 1$ or $n_2 - 1$ degrees of freedom to determine the critical *t* instead of Formula (2) is conservative.

## Exercises

### • Skill Building

*In Problems 1–6, assume that the populations are normally distributed.*

**1.** (a) Test the claim that $\mu_1 \neq \mu_2$ at the $\alpha = 0.05$ level of significance for the given sample data.
   (b) Construct a 95% confidence interval about $\mu_1 - \mu_2$.*

| | Sample for Population 1 | Sample for Population 2 |
|---|---|---|
| $n$ | 15 | 15 |
| $\overline{x}$ | 15.3 | 14.2 |
| $s$ | 3.2 | 3.5 |

**2.** (a) Test the claim that $\mu_1 \neq \mu_2$ at the $\alpha = 0.05$ level of significance for the given sample data.
   (b) Construct a 95% confidence interval about $\mu_1 - \mu_2$.

| | Sample for Population 1 | Sample for Population 2 |
|---|---|---|
| $n$ | 20 | 20 |
| $\overline{x}$ | 111 | 104 |
| $s$ | 8.6 | 9.2 |

**3.** (a) Test the claim that $\mu_1 > \mu_2$ at the $\alpha = 0.1$ level of significance for the given sample data.
   (b) Construct a 90% confidence interval about $\mu_1 - \mu_2$.

| | Sample for Population 1 | Sample for Population 2 |
|---|---|---|
| $n$ | 25 | 18 |
| $\overline{x}$ | 50.2 | 42.0 |
| $s$ | 6.4 | 9.9 |

**4.** (a) Test the claim that $\mu_1 < \mu_2$ at the $\alpha = 0.05$ level of significance for the given sample data.
   (b) Construct a 95% confidence interval about $\mu_1 - \mu_2$.

| | Sample for Population 1 | Sample for Population 2 |
|---|---|---|
| $n$ | 40 | 32 |
| $\overline{x}$ | 94.2 | 115.2 |
| $s$ | 15.9 | 23.0 |

**5.** (a) Test the claim that $\mu_1 < \mu_2$ at the $\alpha = 0.02$ level of significance for the given sample data.
   (b) Construct a 90% confidence interval about $\mu_1 - \mu_2$.

| | Sample for Population 1 | Sample for Population 2 |
|---|---|---|
| $n$ | 32 | 25 |
| $\overline{x}$ | 103.4 | 114.2 |
| $s$ | 12.3 | 13.2 |

**6.** (a) Test the claim that $\mu_1 > \mu_2$ at the $\alpha = 0.05$ level of significance for the given sample data.
   (b) Construct a 95% confidence interval about $\mu_1 - \mu_2$.

| | Sample for Population 1 | Sample for Population 2 |
|---|---|---|
| $n$ | 23 | 13 |
| $\overline{x}$ | 43.1 | 41.0 |
| $s$ | 4.5 | 5.1 |

* The confidence intervals in the back of the text were computed using the smaller of $n_1 - 1$ or $n_2 - 1$ degrees of freedom. These intervals will be wider than those obtained using technology.

## • Applying the Concepts

**7. Treating Acute Bipolar Mania** In a study published in the *Archives of General Psychiatry* entitled "Efficacy of Olanzapine in Acute Bipolar Mania," (Vol. 57, No. 9, pp. 841–849), researchers conducted a randomized, double-blind study to measure the effects of the drug olanzapine on patients diagnosed with bipolar disorder. A total of 115 patients with a DSM-IV diagnosis of bipolar disorder were randomly divided into two groups. Group 1 ($n = 55$) received 5 to 20 mg per day of olanzapine, while Group 2 ($n = 60$) received a placebo. The effectiveness of the drug was measured using the Young–Mania Rating Scale total score with the net improvement in the score recorded. The results are presented in the table below.

| | Treatment Group | Control Group |
|---|---|---|
| $n$ | 55 | 60 |
| Mean improvement | 14.8 | 8.1 |
| Sample standard deviation | 12.5 | 12.7 |

(a) Test the claim that the treatment group experienced a larger mean improvement than the control group at the $\alpha = 0.01$ level of significance.

(b) Construct a 95% confidence interval about $\mu_1 - \mu_2$ and interpret the results.

**8. Treating the Common Cold** Researcher Steven J. Sperber, MD, and his associates wanted to determine the efficacy (effectiveness) of a new medication in the treatment of discomfort associated with the common cold. They published their results in the *Archives of Family Medicines* (2000): 979–985. In their study, they randomly divided 430 subjects into two groups: Group 1 ($n = 216$) received the new medication and Group 2 ($n = 214$) received a placebo. The goal of the study was to determine whether the reduction in the mean symptom assessment score of the individuals receiving the treatment group (Group 1) was more than that of the control group (Group 2). In Group 1, the mean reduction in the symptom assessment scores was 1.30, with a standard deviation of 0.88; in Group 2, participants had a mean reduction in the symptom assessment score of 0.93, with a standard deviation of 0.88.

(a) Test the claim that subjects in Group 1 had a larger reduction in the symptom assessment score than the subjects in Group 2 at the $\alpha = 0.05$ level of significance.

(b) Construct a 95% confidence interval about $\mu_1 - \mu_2$ and interpret the results.

**9. Concrete Strength** An engineer wanted to know whether the strength of two different concrete mix designs differed significantly. He randomly selected 9 cylinders, measuring 6 inches in diameter and 12 inches in height, into which mixture 67-0-301 was poured. After 28 days, he measured the strength (in pounds per square inch) of the cylinder. He also randomly selected 10 cylinders of mixture 67-0-400 and performed the same test. The results are as follows:

| Mixture 67-0-301 | | | Mixture 67-0-400 | | | |
|---|---|---|---|---|---|---|
| 3960 | 4090 | 3100 | 4070 | 4890 | 5020 | 4330 |
| 3830 | 3200 | 3780 | 4640 | 5220 | 4190 | 3730 |
| 4080 | 4040 | 2940 | 4120 | 4620 | | |

(a) Is it reasonable to use Welch's *t*-test? Why? *Note*: Normal probability plots indicate that the data are approximately normal and boxplots indicate that there are no outliers.

(b) Test the claim that mixture 67-0-400 is stronger than mixture 67-0-301 at the $\alpha = 0.05$ level of significance.

(c) Construct a 90% confidence interval about $\mu_{400} - \mu_{301}$ and interpret the results.

(d) Draw boxplots of each data set using the same scale. Does this visual evidence support the results obtained in part (b)?

**10. Measuring Reaction Time** Researchers at the University of Mississippi wanted to determine the whether the reaction time (in seconds) of males differed from that of females to a go/no go stimulus. The researchers randomly selected 20 females and 15 males to participate in the study. The go/no go stimulus required the student to respond to a particular stimulus and not to respond to other stimuli. The data are courtesy of PsychExperiments at the University of Mississippi. The results are as follows:

| Female Students | | | | | | | Male Students | | | | |
|---|---|---|---|---|---|---|---|---|---|---|---|
| 0.588 | 0.652 | 0.442 | 0.293 | 0.380 | 0.434 | 0.613 | 0.375 | 0.256 | 0.427 | 0.373 | 0.224 |
| 0.340 | 0.636 | 0.391 | 0.367 | 0.403 | 0.443 | 0.274 | 0.654 | 0.563 | 0.405 | 0.488 | 0.477 |
| 0.377 | 0.646 | 0.403 | 0.377 | 0.617 | 0.481 | | 0.374 | 0.465 | 0.402 | 0.337 | 0.655 |

(a) Is it reasonable to use Welch's *t*-test? Why? *Note*: Normal probability plots indicate that the data are approximately normal and boxplots indicate that there are no outliers.

(b) Test the claim that there is no difference in the reaction time of males and females at the $\alpha = 0.05$ level of significance.

(c) Construct a 90% confidence interval about $\mu_f - \mu_m$ and interpret the results.

(d) Draw boxplots of each data set using the same scale. Does this visual evidence support the results obtained in part (b)?

**11. Bacteria in Carpeting** Researchers wanted to determine whether carpeted rooms contained more bacteria than uncarpeted rooms. To determine the amount of bacteria in a room, researchers pumped the air from the room over a Petri dish at the rate of one cubic foot per minute for eight carpeted rooms and eight uncarpeted rooms. Colonies of bacteria were allowed to form in the 16 Petri dishes. The results are presented in the table below.

| Carpeted Rooms (Bacteria/cubic foot) | | Uncarpeted Rooms (Bacteria/cubic foot) | |
|---|---|---|---|
| 11.8 | 10.8 | 12.1 | 12.0 |
| 8.2 | 10.1 | 8.3 | 11.1 |
| 7.1 | 14.6 | 3.8 | 10.1 |
| 13.0 | 14.0 | 7.2 | 13.7 |

*Source:* Walter, William G., and Stober, Angie, "Microbial Air Sampling in a Carpeted Hospital." *Journal of Environmental Health,* **30** (1968), p. 405.

(a) Verify that the data are approximately normally distributed with no outliers.
(b) Test the claim that carpeted rooms have more bacteria than uncarpeted rooms at the $\alpha = 0.05$ level of significance.
(c) Construct a 95% confidence interval about $\mu_{carpet} - \mu_{uncarpet}$ and interpret the results.

**12. Visual versus Textual Learners** Researchers wanted to know whether there was a difference in comprehension among students learning a computer program based on the style of the text. They randomly divided 36 students into two groups of 18 each. The researchers verified that the 36 students were similar in terms of educational level, age, and so on. Group 1 individuals learned the software using a visual manual (*multimodal instruction*), while Group 2 individuals learned the software using a textual manual (*unimodal instruction*). The following data represent scores the students received on an exam given to them after they studied from the manuals.

| Visual Manual | | Textual Manual | |
|---|---|---|---|
| 51.08 | 60.35 | 64.55 | 56.54 |
| 57.03 | 76.60 | 57.60 | 39.91 |
| 44.85 | 70.77 | 68.59 | 65.31 |
| 75.21 | 70.15 | 50.75 | 51.95 |
| 56.87 | 47.60 | 49.63 | 49.07 |
| 75.28 | 46.59 | 43.58 | 48.83 |
| 57.07 | 81.23 | 57.40 | 72.40 |
| 80.30 | 67.30 | 49.48 | 42.01 |
| 52.20 | 60.82 | 49.57 | 61.16 |

*Source:* Based on "Multimodal Versus Unimodal Instruction in a Complex Learning Context" by Mark Gellevij, et. al., *Journal of Experimental Education,* 2002, 70(3), pp. 215–239.

(a) Verify that the data are approximately normally distributed with no outliers.
(b) Test the claim that there is a difference in test scores at the $\alpha = 0.05$ level of significance.
(c) Construct a 95% confidence interval about $\mu_{visual} - \mu_{textual}$ and interpret the results.

**13. Cost of Housing** An urban economist wanted to determine whether the mean price of a home in Lemont is less than the mean price of a home in Naperville. A random sample of homes sold in each neighborhood results in the following statistics, where the means and standard deviations are in thousands of dollars:

| Lemont | Naperville |
|---|---|
| $n_1 = 44$ | $n_2 = 32$ |
| $\bar{x}_1 = 253.6$ | $\bar{x}_2 = 310.3$ |
| $s_1 = 106.9$ | $s_2 = 153.9$ |

Source: County's Recorders Office

(a) Test the claim that housing is less expensive in Lemont than in Naperville at the $\alpha = 0.05$ level of significance.
(b) Construct a 90% confidence interval about $\mu_{Lemont} - \mu_{Naperville}$ and interpret the results.
(c) Why do you think it is necessary to have large sample sizes to test this hypothesis?

**14. Debt** Erica Egizio, mathematics instructor at Joliet Junior College, claims that college women have more credit card debt than college men. She conducts a random sample of 38 college men and 32 college women, determines their credit card debt, and obtains the following statistics:

| Men | Women |
|---|---|
| $n_1 = 38$ | $n_2 = 32$ |
| $\bar{x}_1 = 435$ | $\bar{x}_2 = 781$ |
| $s_1 = 1026$ | $s_2 = 1489$ |

(a) Test the claim that college women have more credit card debt than college men at the $\alpha = 0.05$ level of significance.
(b) Construct a 90% confidence interval about $\mu_{women} - \mu_{men}$ and interpret the results.

**15. Hormone Replacement Therapy** Coronary heart disease is the leading cause of death among older women. In observational studies, the number of deaths due to coronary heart disease has been reduced in postmenopausal women who take hormone replacement therapy. Low levels of serum high-density lipoprotein (HDL) cholesterol is considered to be one of the risk factors predictive of death from coronary heart disease. Researchers at the Washington University School of Medicine claimed that serum HDL increases when patients participate in hormone replacement therapy. The researchers randomly divided 59 sedentary women 75 years of age or older into two groups. The 30 patients in Group 1 (the experimental group) received nine months of oral therapy with conjugated estrogens, 0.625 milligrams per day, plus trimonthly medroxyprogesterone acetate, 5 milligrams per day for 13 days. The 29 patients in Group 2 (the control group) received nine months of oral therapy with a placebo. At the conclusion of the treatment, the patient's serum HDL was recorded. The experiment was double-blind. The following results were obtained, where the means and standard deviations are in mg/dL:

| | Experimental Group | Control Group |
|---|---|---|
| Sample size | $n_1 = 30$ | $n_2 = 29$ |
| Mean increase in HDL | $\bar{x}_1 = 8.1$ | $\bar{x}_2 = 2.4$ |
| Sample standard deviation | $s_1 = 10.5$ | $s_2 = 4.3$ |

Source: Ellen F. Binder, et al., "Effects of Hormone Replacement Therapy on Serum Lipids in Elderly Women." *Annals of Internal Medicine* 134 (May 2001): pp. 754–760

(a) What type of experimental design is this? What is the treatment? How many levels does the treatment have?

(b) Test the claim that the treatment group had a larger mean increase in serum HDL levels than the control group at the $\alpha = 0.01$ level of significance (serum HDL is normally distributed).

(c) Construct a 95% confidence interval about $\mu_1 - \mu_2$ and interpret the results.

16. **SAT Test Scores** A researcher wanted to know whether students who do not plan to apply for financial aid score better on the SAT I math test than those who do plan to apply for financial aid. She obtains a random sample of 35 students who do not plan to apply for financial aid and a random sample of 39 students who do plan to apply for financial aid and obtains the following results:

| Do Not Plan to Apply for Financial Aid | Plan to Apply for Financial Aid |
|---|---|
| $n_1 = 35$ | $n_2 = 38$ |
| $\bar{x}_1 = 546.6$ | $\bar{x}_2 = 497.1$ |
| $s_1 = 123.1$ | $s_2 = 119.4$ |

(a) Test the claim that students who do not apply for financial aid score higher on the SAT math I exam than students who do apply for financial aid at the $\alpha = 0.01$ level of significance.

(b) Construct a 98% confidence interval about $\mu_{\text{do not}} - \mu_{\text{do}}$ and interpret the results.

17. **Comparing Step Pulses** A physical therapist wanted to know whether the mean step pulse of men was less than the mean step pulse of women. She randomly selected 51 men and 70 women to participate in the study. Each subject was required to step up and down onto a six-inch platform for three minutes. The pulse of each subject (in beats per minute) was then recorded. After the data were entered into Minitab, the following results were obtained:

### Two Sample T-Test and Confidence Interval

```
Two sample T for Men vs Women

            N    Mean    StDev    SE Mean
Men        51   112.3    11.3        1.6
Women      70   118.3    14.2        1.7

95% CI for mu Men − mu Women: (−10.7, −1.5)
T-Test mu Men = mu Women (vs <):T = −2.61  P = 0.0051  DF = 118
```

(a) State the null and alternative hypotheses.

(b) Identify the $P$-value and state the researcher's conclusion if the level of significance was $\alpha = 0.01$.

(c) What is the 95% confidence interval for the mean difference in pulse rates of men versus women? Interpret this interval.

18. **Comparing Flexibility** A physical therapist claims that women are more flexible than men. She measures the flexibility of 31 randomly selected women and 45 randomly selected men by determining the number of inches subjects could reach while sitting on the floor with their legs straight out and back perpendicular to the ground. The more flexible an individual is, the higher the measured flexibility would be. After entering the data into Minitab, she obtained the following results:

### Two Sample T-Test and Confidence Interval

```
Two sample T for Men vs Women

            N    Mean    StDev    SE Mean
Men        45   18.64    3.29        0.49
Women      31   20.99    2.07        0.37

95% CI for mu Men − mu Women: (−3.58, −1.12)
T-Test mu Men = mu Women (vs <):T = −3.82  P = 0.0001  DF = 73
```

(a) State the null and alternative hypotheses.

(b) Identify the $P$-value and state the researcher's conclusion if the level of significance was $\alpha = 0.01$.

(c) What is the 95% confidence interval for the mean difference in flexibility of men versus women? Interpret this interval.

## Technology Step-by-Step
Two-sample T-tests, Independent Sampling

**TI-83 Plus**  **Hypothesis Tests**

**Step 1:** If necessary, enter raw data in L1 and L2.

**Step 2:** Press STAT, highlight TESTS, and select 4:2-SampTTest...

**Step 3:** If the data are raw, highlight Data, making sure that List1 is set to L1 and List2 is set to L2 with frequencies set to 1. If summary statistics are known, highlight Stats and enter the summary statistics.

**Step 4:** Highlight the appropriate relation between $\mu_1$ and $\mu_2$ in the alternative hypothesis. Set Pooled to NO.

**Step 5:** Highlight Calculate or Draw and press ENTER. Calculate gives the test statistic and $P$-value. Draw will draw the $t$-distribution with the $P$-value shaded.

**Confidence Intervals**

Follow the same steps as those given for hypothesis tests, except select 0:2-SampTInt. Also, select a confidence level (such as 95% = 0.95).

**MINITAB**  **Step 1:** Enter raw data in columns C1 and C2.

**Step 2:** Select the **Stat** menu, highlight **Basic Statistics**, then highlight **2-Sample t . . .**

**Step 3:** Select "Samples in different columns." Enter C1 in the cell marked "First" and enter C2 in the cell marked "Second." Select the direction of the alternative hypothesis and select a confidence level. Click OK.

**Excel**  **Step 1:** Enter raw data in columns A and B.

**Step 2:** Select the **Tools menu**, highlight **Data Analysis . . .**

**Step 3:** Select "$t$-test: Two-Sample Assuming Unequal Variances." With the cursor in the "Variable 1 Range" cell, highlight the data in column A. Enter the hypothesized difference in the means (usually 0) and a value for alpha. Click OK.

---

## 10.3 Inference about Two Population Proportions

**Preparing for This Section**  Before getting started, review the following:

✓ Confidence intervals about a population proportion (Section 8.3, pp. 370–374)

✓ Hypothesis tests about a population proportion (Section 9.4, pp. 420–425)

**Objectives**   Conduct hypothesis tests on the difference between two population proportions

 Construct confidence intervals for the difference between two population proportions

 Determine sample size for the difference between two population proportions

In Sections 8.3 and 9.4, we discussed inference regarding a single population proportion. We will now discuss inferential methods for comparing two population proportions. For example, in clinical trials of the drug Nasonex, a drug that is meant to relieve allergy symptoms, 26% of patients receiving 200 mcg of Nasonex reported a headache as a side effect while 22% of patients receiving a placebo reported a headache as a side effect. Researchers

**Step 3:** Compute the **test statistic** $Z = \dfrac{\hat{p}_1 - \hat{p}_2}{\sqrt{\hat{p}(1-\hat{p})}\sqrt{\dfrac{1}{n_1} + \dfrac{1}{n_2}}}$ where $\hat{p} = \dfrac{x_1 + x_2}{n_1 + n_2}$.

**Step 4:** Compare the critical value with the test statistic.

| Two-Tailed | Left-Tailed | Right-Tailed |
|---|---|---|
| If $Z < -z_{\alpha/2}$ or $Z > z_{\alpha/2}$, reject the null hypothesis | If $Z < -z_{\alpha}$, reject the null hypothesis | If $Z > z_{\alpha}$, reject the null hypothesis |

**Step 5:** State the conclusion.

▶ **EXAMPLE 1**    **Testing a Claim Regarding Two Population Proportions**

*Problem:* In clinical trials of Nasonex, 3774 adult and adolescent allergy patients (patients 12 years and older) were randomly divided into two groups. The patients in Group 1 (experimental group) received 200 mcg of Nasonex, while the patients in Group 2 (control group) received a placebo. Of the 2103 patients in the experimental group, 547 reported headaches as side effect. Of the 1671 patients in the control group, 368 reported headaches as a side effect. Is there significant evidence to support the claim that the proportion of Nasonex users that experienced headaches as a side effect is greater than the proportion in the control group at the $\alpha = 0.05$ level of significance?

*Approach:* We must verify the requirements to perform the hypothesis test. That is, the sample must be a simple random sample and $n_1 \hat{p}_1 (1 - \hat{p}_1) \geq 10$ and $n_2 \hat{p}_2 (1 - \hat{p}_2) \geq 10$. In addition, the sample size cannot be more than 5% of the population. Then we follow the preceding Steps 1–5.

*Solution:* First we verify that the requirements are satisfied:

1. The samples are independently obtained using simple random sampling.

2. We have $x_1 = 547$, $n_1 = 2103$, $x_2 = 368$, and $n_2 = 1671$, so that
$\hat{p}_1 = \dfrac{x_1}{n_1} = \dfrac{547}{2103} = 0.26$ and $\hat{p}_2 = \dfrac{x_2}{n_2} = \dfrac{368}{1671} = 0.22$. Therefore,
$n_1 \hat{p}_1 (1 - \hat{p}_1) = 2103(0.26)(1 - 0.26) = 404.6172 \geq 10$
$n_2 \hat{p}_2 (1 - \hat{p}_2) = 1671(0.22)(1 - 0.22) = 286.7436 \geq 10$

3. There are over 10 million Americans who are 12 years old or older who are allergy sufferers, so the sample sizes are less than 5% of the population size.

All three requirements are satisfied, so we now proceed to follow Steps 1–5.

**Step 1:** The claim is that the proportion of patients taking Nasonex who experience a headache is greater than the proportion of patients taking the placebo who experience a headache. Letting $p_1$ represent the population proportion of patients taking Nasonex who experience a headache and $p_2$ represent the population proportion of patients taking the placebo who experience a headache, we can express the claim as $p_1 > p_2$. This is a right-tailed hypothesis with

$$H_0\colon p_1 = p_2 \quad \text{versus} \quad H_1\colon p_1 > p_2$$

or equivalently,

$$H_0\colon p_1 - p_2 = 0 \quad \text{versus} \quad H_1\colon p_1 - p_2 > 0$$

**Step 2:** Because this is a right-tailed test, we determine the critical value at the $\alpha = 0.05$ level of significance to be $z_{0.05} = 1.645$. The critical region is displayed in Figure 7.

Figure 7

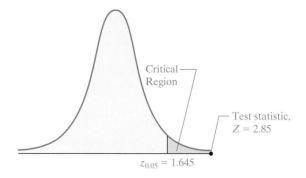

Critical Region

Test statistic, $Z = 2.85$

$z_{0.05} = 1.645$

**Step 3:** From verifying that requirement 2 was satisfied we have that $\hat{p}_1 = \dfrac{x_1}{n_1} = \dfrac{547}{2103} = 0.26$ and $\hat{p}_2 = \dfrac{x_2}{n_2} = \dfrac{368}{1671} = 0.22.$

To find the test statistic, we first must compute the pooled estimate of $p$,

$$\hat{p} = \frac{x_1 + x_2}{n_1 + n_2} = \frac{547 + 368}{2103 + 1671} = 0.242$$

The test statistic is

$$Z = \frac{\hat{p}_1 - \hat{p}_2}{\sqrt{\hat{p}(1 - \hat{p})}\sqrt{\dfrac{1}{n_1} + \dfrac{1}{n_2}}} = \frac{0.26 - 0.22}{\sqrt{0.242(1 - 0.242)}\sqrt{\dfrac{1}{2103} + \dfrac{1}{1671}}} = \frac{0.04}{0.0140357419} = 2.85$$

The sample difference in proportions is 2.85 standard deviations above the assumed difference of 0.

**Step 4:** Because the test statistic $Z = 2.85$ is greater than the critical value of $z_{0.05} = 1.645$, we reject the null hypothesis. That is, the value of the test statistic falls within the critical region so we reject $H_o$. We label this point in Figure 7.

**Step 5:** There is sufficient evidence at the $\alpha = 0.05$ level of significance to support the claim that the proportion of individuals 12 years and older taking 200 mcg of Nasonex who experience headaches is greater than the proportion of individuals 12 years and older taking a placebo who experience headaches. ◄◄

In looking back at the results of Example 1, we notice that the proportion of individuals taking 200 mcg of Nasonex who experience headaches is *statistically significantly* greater than the proportion of individuals 12 years and older taking a placebo who experience headaches. However, we need to ask ourselves a pressing question. Would a mother not give her child an allergy medication because 26% of patients experienced a headache taking the medication versus 22% who experienced a headache taking a placebo? Most mothers would be willing to accept the additional "risk" of a headache in order to relieve her child's allergy symptoms. While the difference of 4% is statistically significant, it does not have any *practical significance*.

*Definition*

**Practical significance** refers to the idea that small differences in statistics can be statistically significant, while not large enough to have any functional value.

**Caution**

In any statistical study, be sure to consider practical significance. Many statistically significant results can be produced simply by increasing the sample size.

When performing significance tests with large samples, small differences between the null hypothesis and the sample statistics will result in a statistically significant difference. Be sure to ask yourself whether the difference has any practical significance.

### *P*-Values

We could also test the hypothesis of Example 1 by using the *P*-value approach. From Step 4, we have the test statistic $Z = 2.85$. The *P*-value is the probability of obtaining a test statistic $Z = 2.85$ or higher (because this is a right-tailed test). Consulting Table II, we find that

$$P\text{-value} = P(Z \geq 2.85) = 1 - P(Z < 2.85) = 1 - 0.9978 = 0.0022$$

Because the *P*-value is less than the level of significance, $\alpha = 0.05$, we reject the null hypothesis.

> **Using Technology:** We could also test the claim presented in Example 1 using statistical software or a graphing calculator with advanced statistical features. Figure 8 shows the results of Example 1, using Minitab.
>
> **Figure 8**
>
> Test and Confidence Interval for Two Proportions
>
> ```
> Sample     X       N   Sample p
> 1         547    2103  0.260105
> 2         368    1671  0.220227
>
> Estimate for p(1) - p(2): 0.0398772
> 95% CI for p(1) - p(2): (0.0125583, 0.0671961)
> Test for p(1) - p(2) = 0 (vs > 0): Z = 2.84  P-Value = 0.002
> ```
>
> Notice that the test statistic is $Z = 2.84$. The discrepancy between the output of Minitab and the solution presented in Example 1 is due simply to rounding. The *P*-value presented by Minitab is 0.002, which agrees with the *P*-value we obtained.

**NW** *Now Work Problem 11(a).*

### ② Confidence Intervals for the Difference between Two Population Proportions

The sampling distribution of the difference of two proportions, $\hat{p}_1 - \hat{p}_2$, can also be used to construct confidence intervals for the difference of two proportions.

> **Constructing a $(1 - \alpha) \cdot 100\%$ Confidence Interval for the Difference between Two Population Proportions**
>
> To construct a $(1 - \alpha) \cdot 100\%$ confidence interval for the difference between two population proportions, the following requirements must be satisfied:
> 1. the samples are obtained independently, using simple random sampling,
> 2. $n_1 \hat{p}_1 (1 - \hat{p}_1) \geq 10$ and $n_2 \hat{p}_2 (1 - \hat{p}_2) \geq 10$,
> 3. $n_1 \leq 0.05 N_1$ and $n_2 \leq 0.05 N_2$ (the sample size is no more than 5% of the population size); this ensures the independence necessary for a binomial experiment.
>
> Provided that these requirements are met, a $(1 - \alpha) \cdot 100\%$ confidence interval for $p_1 - p_2$ is given by
>
> $$\text{Lower Bound: } (\hat{p}_1 - \hat{p}_2) - z_{\alpha/2} \cdot \sqrt{\frac{\hat{p}_1(1 - \hat{p}_1)}{n_1} + \frac{\hat{p}_2(1 - \hat{p}_2)}{n_2}}$$
>
> $$\text{Upper Bound: } (\hat{p}_1 - \hat{p}_2) + z_{\alpha/2} \cdot \sqrt{\frac{\hat{p}_1(1 - \hat{p}_1)}{n_1} + \frac{\hat{p}_2(1 - \hat{p}_2)}{n_2}} \quad \text{(3)}$$

Notice that we do not pool the sample proportions. This is because we are not making any assumptions regarding their equality as we did in hypothesis testing.

▶ **EXAMPLE 2**   **Constructing a Confidence Interval for the Difference between Two Population Proportions**

*Problem:* In clinical trials of Nasonex, 750 randomly selected pediatric patients (ages 3 to 11 years old) were randomly divided into two groups. The patients in Group 1 (experimental group) received 100 mcg of Nasonex, while the patients in Group 2 (control group) received a placebo. Of the 374 patients in the experimental group, 64 reported headaches as side effect. Of the 376 patients in the control group, 68 reported headaches as a side effect. Construct a 90% confidence interval for the difference between the two population proportions, $p_1 - p_2$.

*Approach:* We can compute a 90% confidence interval about $p_1 - p_2$, provided that the requirements stated above are satisfied. We then construct the interval by using Formula (3).

*Solution:*

*Step 1:* We need to verify the requirements for constructing a confidence interval about the difference between two population proportions. (1) The samples were randomly divided into two groups. (2) For the experimental group (Group 1), we have $n_1 = 374$ and $x_1 = 64$, so that $\hat{p}_1 = \dfrac{x_1}{n_1} = \dfrac{64}{374} = 0.171$.

For the control group (Group 2), we have $n_2 = 376$ and $x_2 = 68$, so that $\hat{p}_2 = \dfrac{x_2}{n_2} = \dfrac{68}{376} = 0.181$. Therefore,

$$n_1 \hat{p}_1 (1 - \hat{p}_1) = 374(0.171)(1 - 0.171) = 53.02 \geq 10$$
$$n_2 \hat{p}_2 (1 - \hat{p}_2) = 376(0.181)(1 - 0.181) = 55.74 \geq 10$$

(3) The samples were independently obtained and the sample sizes are less than 5% of the population size. (There are over 20 million children between the ages of 3 and 11 in the United States.)

*Step 2:* Because we want a 90% confidence interval, we have $\alpha = 0.10$, so $z_{\alpha/2} = z_{0.05} = 1.645$.

*Step 3:* Substituting into Formula (3) with $\hat{p}_1 = 0.171$, $n_1 = 374$, $\hat{p}_2 = 0.181$, and $n_2 = 376$, we obtain the lower and upper bounds on the confidence interval:

Lower Bound:

$$(\hat{p}_1 - \hat{p}_2) - z_{\alpha/2} \cdot \sqrt{\frac{\hat{p}_1(1 - \hat{p}_1)}{n_1} + \frac{\hat{p}_2(1 - \hat{p}_2)}{n_2}}$$

$$= (0.171 - 0.181) - 1.645 \cdot \sqrt{\frac{0.171(1 - 0.171)}{374} + \frac{0.181(1 - 0.181)}{376}}$$

$$= -0.010 - 0.046 = -0.056$$

Upper Bound:

$$(\hat{p}_1 - \hat{p}_2) + z_{\alpha/2} \cdot \sqrt{\frac{\hat{p}_1(1 - \hat{p}_1)}{n_1} + \frac{\hat{p}_2(1 - \hat{p}_2)}{n_2}}$$

$$= (0.171 - 0.181) + 1.645 \cdot \sqrt{\frac{0.171(1 - 0.171)}{374} + \frac{0.181(1 - 0.181)}{376}}$$

$$= -0.010 + 0.046 = 0.036$$

Based upon the results of the study, we are 90% confident that the difference between the proportion of headaches in the experimental group and the control group is between $-0.056$ and $0.036$. Because the confidence interval contains 0, there is no evidence to support the claim that the proportion of patients complaining of headaches who receive Nasonex is different from those who do not receive Nasonex at the $\alpha = 0.1$ level of significance. ◄◄

**Figure 9**

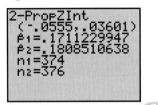

**NW** *Now Work Problem 11(b).*

Using Technology: Graphing calculators with advanced statistical features and statistical spreadsheets can be used to construct confidence intervals about the difference between two population proportions. Figure 9 shows the results using a TI-83 Plus graphing calculator.

## ③ Determining Sample Sizes

In Section 8.3, we introduced a method for determining the sample size $n$ required to estimate a single population proportion within a specified margin of error, $E$, with a specified level of confidence. This formula was obtained by solving the margin of error, $E = z_{\alpha/2} \cdot \sqrt{\dfrac{\hat{p}(1 - \hat{p})}{n}}$, for $n$. We can follow the same approach to determine the sample size when we estimate two population proportions. Notice that the margin of error, $E$, in Formula (3) is given by $E = z_{\alpha/2} \cdot \sqrt{\dfrac{\hat{p}_1(1 - \hat{p}_1)}{n_1} + \dfrac{\hat{p}_2(1 - \hat{p}_2)}{n_2}}$. Assuming that $n_1 = n_2 = n$, we can solve this expression for $n = n_1 = n_2$ and obtain the following result:

*Theorem*

> **Sample Size for Estimating $p_1 - p_2$**
>
> The sample size required to obtain a $(1 - \alpha) \cdot 100\%$ confidence interval with a margin of error, $E$, is given by
>
> $$n = n_1 = n_2 = [\hat{p}_1(1 - \hat{p}_1) + \hat{p}_2(1 - \hat{p}_2)]\left(\frac{z_{\alpha/2}}{E}\right)^2 \qquad \textbf{(4)}$$
>
> rounded up to the next integer if prior estimates of $p_1$ and $p_2$, $\hat{p}_1$ and $\hat{p}_2$, are available. If prior estimates of $p_1$ and $p_2$ are unavailable, the sample size is
>
> $$n = n_1 = n_2 = 0.5\left(\frac{z_{\alpha/2}}{E}\right)^2 \qquad \textbf{(5)}$$
>
> rounded up to the next integer.

► **EXAMPLE 3** Determining Sample Size

*Problem:* A nutritionist wishes to estimate the difference between the proportion of males and females that consume the USDA's recommended daily intake of calcium. What sample size should be obtained if she wishes the estimate to be within 3 percentage points with 95% confidence, assuming that

**(a)** she uses the results of the USDA's 1994–1996 Diet and Health Knowledge Survey, according to which 51.1% of males and 75.2% of females consume the USDA's recommended daily intake of calcium,

**(b)** she does not use any prior estimates.

*Approach:* We have $E = 0.03^*$ and $z_{\alpha/2} = z_{0.05/2} = z_{0.025} = 1.96$. To answer part (a), we let $\hat{p}_1 = 0.511$ (for males) and $\hat{p}_2 = 0.752$ (for females) in Formula (4). To answer part (b), we use Formula (5).

*The margin of error should always be expressed as a decimal when using Formulas (4) and (5).

*Solution:*

**(a)** Substituting $E = 0.03$, $z_{0.025} = 1.96$, $\hat{p}_1 = 0.511$, and $\hat{p}_2 = 0.752$ into Formula (4), we obtain

$$n_1 = n_2 = [\hat{p}_1(1 - \hat{p}_1) + \hat{p}_2(1 - \hat{p}_2)]\left(\frac{z_{\alpha/2}}{E}\right)^2 = [0.511(1 - 0.511) + 0.752(1 - 0.752)]\left(\frac{1.96}{0.03}\right)^2$$

$$= 1862.642444$$

We round this value up to 1863. The nutritionist must survey 1863 randomly selected males and 1863 randomly selected females.

**(b)** Substituting $E = 0.03$ and $z_{0.025} = 1.96$ into Formula (5), we obtain

$$n_1 = n_2 = 0.5\left(\frac{z_{\alpha/2}}{E}\right)^2 = 0.5\left(\frac{1.96}{0.03}\right)^2 = 2134.222222$$

We round this value up to 2135. The nutritionist must survey 2135 randomly selected males and 2135 randomly selected females.   ◄◄

We can see that having prior estimates of the population proportions reduces the number of individuals that need to be surveyed.

 *Now Work Problem 17.*

# 10.3  Assess Your Understanding

## Concepts and Vocabulary

**1.** Explain why we determine a pooled estimate of the population proportion when computing $\sigma_{\hat{p}_1 - \hat{p}_2}$.

**2.** State the requirements that must be satisfied in order to test a claim regarding two population proportions.

**3.** Describe the difference between statistical significance and practical significance.

## Exercises

### • Skill Building

*In Problems 1–4, test the claim at the $\alpha = 0.05$ level of significance by determining (a) the null and alternative hypotheses, (b) the test statistic, (c) the critical value, and (d) the P-value. Assume the samples were obtained independently using simple random sampling.*

**1.** Claim $p_1 > p_2$. Sample data:
$x_1 = 368$, $n_1 = 541$, $x_2 = 351$, $n_2 = 593$

**2.** Claim $p_1 < p_2$. Sample data:
$x_1 = 109$, $n_1 = 475$, $x_2 = 78$, $n_2 = 325$

**3.** Claim $p_1 \neq p_2$. Sample data:
$x_1 = 28$, $n_1 = 254$, $x_2 = 36$, $n_2 = 301$

**4.** Claim $p_1 \neq p_2$. Sample data:
$x_1 = 804$, $n_1 = 874$, $x_2 = 902$, $n_2 = 954$

*In Problems 5–8, construct a confidence interval for $p_1 - p_2$ at the given level of confidence.*

**5.** $x_1 = 368$, $n_1 = 541$, $x_2 = 421$, $n_2 = 593$,
90% confidence

**6.** $x_1 = 109$, $n_1 = 475$, $x_2 = 78$, $n_2 = 325$,
99% confidence

**7.** $x_1 = 28$, $n_1 = 254$, $x_2 = 36$, $n_2 = 301$,
95% confidence

**8.** $x_1 = 804$, $n_1 = 874$, $x_2 = 892$, $n_2 = 954$,
95% confidence

### • Applying the Concepts

**9. Prevnar** The drug Prevnar is a vaccine meant to prevent certain types of bacterial meningitis. It is typically administered to infants starting around two months of age. In randomized, double-blind clinical trials of Prevnar, infants were randomly divided into two groups. Subjects in Group 1 received Prevnar while subjects in Group 2 received a control vaccine. After the first dose, 107 of 710 subjects in the experimental group (Group 1) experienced fever as a side effect. After the first dose, 67 of 611 of the subjects in the control group (Group 2) experienced fever as a side effect.

(a) Test the claim that a higher proportion of subjects in Group 1 experienced fever as a side effect than subjects in Group 2 at the $\alpha = 0.05$ level of significance, using both the classical approach and P-value approach.

(a) Is this dependent or independent sampling? Why?

(b) Test the claim that the weight of wheat pennies is less than the weight of modern pennies at the $\alpha = 0.05$ level of significance. *Note*: A normal probability plot indicates that the data are approximately normal, and a boxplot does not indicate any outliers.

(c) Construct a 95% confidence interval about $\mu_m - \mu_w$. Interpret the result.

(d) Why is the result obtained in part (b) reasonable?

16. **Pulse** A physical therapist wants to check on whether a new exercise program reduces the pulse rate of subjects. She randomly selects 10 women to participate in the study. Each subject is asked to step up and down on a 6-inch step for 3 minutes. Their pulses (in beats per minute) are then recorded. After a 10-week training program, the pulse is again measured, using the same technique. The results are presented in the table below.

| Observation | 1 | 2 | 3 | 4 | 5 |
|---|---|---|---|---|---|
| Before | 136 | 120 | 129 | 143 | 115 |
| After | 128 | 111 | 129 | 148 | 110 |

| Observation | 6 | 7 | 8 | 9 | 10 |
|---|---|---|---|---|---|
| Before | 113 | 89 | 122 | 102 | 122 |
| After | 112 | 98 | 103 | 103 | 103 |

(a) Is this dependent or independent sampling? Why?

(b) Test the claim that the pulse before exercising is greater than the pulse after exercising at the $\alpha = 0.05$ level of significance. *Note*: A normal probability plot indicates that the data are approximately normal, and a boxplot does not indicate any outliers.

(c) Construct a 99% confidence interval about the population mean difference. Interpret the result.

17. **Treatment for Osteoporosis** Osteoporosis is a condition in which people experience decreased bone mass and an increase in the risk of bone fracture. Actonel is a drug that helps combat osteoporosis in postmenopausal women. In clinical trials, 1374 postmenopausal women were randomly divided into experimental and control groups. The subjects in the experimental group were administered 5 mg of Actonel, while the subjects in the control group were administered a placebo. The number of women who experienced a bone fracture over the course of one year was recorded. Of the 696 women in the experimental group, 27 experienced a fracture during the course of the year. Of the 678 women in the control group, 49 experienced a fracture during the course of the year.

(a) Test the claim that a lower proportion of women in the experimental group experienced a bone fracture than the women in the control group at the $\alpha = 0.01$ level of significance.

(b) Construct a 95% confidence interval for the difference between the two population proportions, $p_{exp} - p_{control}$.

(c) What type of experimental design is this? What is the treatment? How many levels does it have?

18. **Zoloft** Zoloft is a drug that is used to treat obsessive–compulsive disorder (OCD). In randomized, double-blind clinical trials, 926 patients diagnosed with OCD were randomly divided into two groups. Subjects in Group 1 (experimental group) received 200 mg per day of Zoloft, while subjects in Group 2 (control group) received a placebo. Of the 553 subjects in the experimental group, 77 experienced dry mouth as a side effect. Of the 373 subjects in the control group, 34 experienced dry mouth as a side effect.

(a) Test the claim that a higher proportion of the subjects in the experimental group experienced dry mouth than did the subjects in the control group at the $\alpha = 0.05$ level of significance.

(b) Construct a 90% confidence interval for the difference between the two population proportions, $p_1 - p_2$.

19. **Determining Sample Size** A nutritionist wants to estimate the difference between the percentage of men and women who have high cholesterol. What sample size should be obtained if she wishes the estimate to be within 2 percentage points with 90% confidence, assuming

(a) that she uses the 1994 estimates of 18.8% male and 20.5% female from the National Center for Health Statistics?

(b) that she does not use any prior estimates?

20. **Determining Sample Size** A researcher wants to estimate the difference between the percentage of individuals without a high school diploma who smoke and the percentage of individuals with bachelors degrees who smoke. What sample size should be obtained if she wishes the estimate to be within 4 percentage points with 95% confidence, assuming

(a) that she uses the 1999 estimates of 32.2% of those without a high school diploma and 11.1% of those with a bachelors degree, from the National Center for Health Statistics?

(b) that she does not use any prior estimates?

21. Explain when the matched-pairs $t$ should be used instead of Welch's $t$ in comparing two population means. What are some advantages in designing a matched-pairs experiment versus using Welch's $t$?

**Chapter 10 Projects located at www.prenhall.com/sullivanstats**

# Additional Inferential Procedures

## Outline

For additional study help, go to www.prenhall.com/sullivanstats

Materials include

- Projects:
  - Case Study: Feeling Lucky? Well, Are You?
  - Decisions: Benefits of College?
  - Consumer Reports Project
- Self-Graded Quizzes
- "Preparing for This Section" Quizzes
- STATLETs
- PowerPoint Downloads
- Step-by-Step Technology Guide
- Graphing Calculator Help

## Putting It All Together

This chapter can be considered in two parts. The first part, Sections 11.1 and 11.2, introduces inference using the chi-square distribution.

Often, rather than being interested in testing a claim regarding a parameter of a probability distribution, we are interested in testing a claim regarding the entire probability distribution. For example, we might wish to test the claim that the distribution of colors in a bag of plain M&M candies is 30% brown, 20% yellow, 20% red, 10% orange, 10% blue, and 10% green. We introduce methods for testing claims such as this in Section 11.1.

In Section 11.2, we discuss a method that can be used to determine whether two variables are independent based on a sample. If they are not independent, then we can say that the value of one variable impacts the value of the other variable, so the variables are somehow related.

We conclude Section 11.2 by introducing tests for homogeneity. This procedure is used to compare proportions from two or more populations. It is an extension of the two-sample $z$-test for proportions discussed in Section 10.3.

The second part, Sections 11.3 and 11.4, introduces inferential methods that can be used on the least-squares regression line. This material is a continuation of the discussion presented in Chapter 4. There, we presented methods for describing the relation between two variables—bivariate data. In Section 11.3, we use the methods of hypothesis testing presented in Chapter 9 to test the claim that a linear relation exists between two quantitative variables. In Section 11.4, we create confidence intervals about a predicted value of the least-squares regression line.

**Preparing for This Section**   Before getting started, review the following:

✓ Expected value (Section 6.1, pp. 248-249)

✓ Mean of a binomial random variable (Section 6.2, p. 259)

✓ Mutually exclusive (Section 5.2, p. 201)

**Objectives** ①  Perform a chi-square goodness-of-fit test

① In this section, we present a procedure that can be used to test claims regarding a probability distribution. For example, we might want to test the claim that the distribution of plain M&M candies in a bag is 30% brown, 20% yellow, 20% red, 10% orange, 10% blue, and 10% green. Or we might want to test the claim that the number of hits a player gets in his next four at-bats follows a binomial distribution with $n = 4$ and $p = 0.298$. To conduct this type of analysis requires a new probability distribution.

We use the symbol $\chi^2$ (pronounced "kigh-square" to rhyme with "sky-square") to represent values of the chi-square distribution. We can find critical values of the chi-square distribution in Table IV in Appendix A of the text. Before discussing how to read Table IV, we introduce characteristics of the chi-square distribution.

**Figure 1**
Chi-square distributions

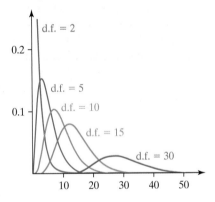

**Characteristics of the Chi-Square Distribution**

1. It is not symmetric.
2. The shape of the chi-square distribution depends upon the degrees of freedom, just like Student's $t$-distribution.
3. As the number of degrees of freedom increases, the chi-square distribution becomes more symmetric, as illustrated in Figure 1.
4. The values $\chi^2$ are nonnegative. That is, the values of $\chi^2$ are greater than or equal to 0.

Table IV is structured similarly to Table III for the $t$-distribution. The left column represents the degrees of freedom, and the top row represents the area under the chi-square distribution to the right of the critical value. We shall use the notation $\chi^2_\alpha$ to denote the critical $\chi^2$-value such that the area under the chi-square distribution to the right of $\chi^2_\alpha$ is $\alpha$.

▶ **EXAMPLE 1**   Finding Critical Values for the Chi-Square Distribution

*Problem:* Find the critical value of the chi-square distribution that corresponds to an area of 5% in the right tail, assuming 15 degrees of freedom.

*Approach:* We shall perform the following steps to obtain the critical values.

*Step 1:* Draw a chi-square distribution with the critical value and area labeled.

**Figure 2**

*Step 2:* Use Table IV to find the critical value.

*Solution:*

*Step 1:* Figure 2 shows the chi-square distribution with 15 degrees of freedom and the unknown critical value labeled. The area to the right of the critical value is 0.05. We denote this critical value $\chi^2_{0.05}$.

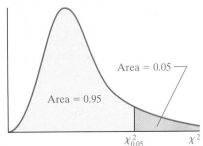

*Step 2:* Figure 3 shows a partial representation of Table IV. The row containing 15 degrees of freedom is boxed. The column corresponding to an area to the right of 0.05 is boxed. The critical value is $\chi^2_{0.05} = 24.996$.

Figure 3

| Degrees of Freedom | Area to the Right of the Critical Value | | | | | | | | | |
|---|---|---|---|---|---|---|---|---|---|---|
| | 0.995 | 0.99 | 0.975 | 0.95 | 0.90 | 0.10 | 0.05 | 0.025 | 0.01 | 0.005 |
| 1 | — | — | 0.001 | 0.004 | 0.016 | 2.706 | 3.841 | 5.024 | 6.635 | 7.879 |
| 2 | 0.010 | 0.020 | 0.051 | 0.103 | 0.211 | 4.605 | 5.991 | 7.378 | 9.210 | 10.597 |
| 3 | 0.072 | 0.115 | 0.216 | 0.352 | 0.584 | 6.251 | 7.815 | 9.348 | 11.345 | 12.838 |
| 4 | 0.207 | 0.297 | 0.484 | 0.711 | 1.064 | 7.779 | 9.488 | 11.143 | 13.277 | 14.860 |
| 5 | 0.412 | 0.554 | 0.831 | 1.145 | 1.610 | 9.236 | 11.071 | 12.833 | 15.086 | 16.750 |
| 6 | 0.676 | 0.872 | 1.237 | 1.635 | 2.204 | 10.645 | 12.592 | 14.449 | 16.812 | 18.548 |
| 7 | 0.989 | 1.239 | 1.690 | 2.167 | 2.833 | 12.017 | 14.067 | 16.013 | 18.475 | 20.278 |
| 8 | 1.344 | 1.646 | 2.180 | 2.733 | 3.490 | 13.362 | 15.507 | 17.535 | 20.090 | 21.955 |
| 9 | 1.735 | 2.088 | 2.700 | 3.325 | 4.168 | 14.684 | 16.919 | 19.023 | 21.666 | 23.589 |
| 10 | 2.156 | 2.558 | 3.247 | 3.940 | 4.865 | 15.987 | 18.307 | 20.483 | 23.209 | 25.188 |
| 11 | 2.603 | 3.053 | 3.816 | 4.575 | 5.578 | 17.275 | 19.675 | 21.920 | 24.725 | 26.757 |
| 12 | 3.074 | 3.571 | 4.404 | 5.226 | 6.304 | 18.549 | 21.026 | 23.337 | 26.217 | 28.299 |
| 13 | 3.565 | 4.107 | 5.009 | 5.892 | 7.042 | 19.812 | 22.362 | 24.736 | 27.688 | 29.819 |
| 14 | 4.075 | 4.660 | 5.629 | 6.571 | 7.790 | 21.064 | 23.685 | 26.119 | 29.141 | 31.319 |
| 15 | 4.601 | 5.229 | 6.262 | 7.261 | 8.547 | 22.307 | 24.996 | 27.488 | 30.578 | 32.801 |
| 16 | 5.142 | 5.812 | 6.908 | 7.962 | 9.312 | 23.542 | 26.296 | 28.845 | 32.000 | 34.267 |
| 17 | 5.697 | 6.408 | 7.564 | 8.672 | 10.085 | 24.769 | 27.587 | 30.191 | 33.409 | 35.718 |
| 18 | 6.365 | 7.215 | 8.231 | 9.288 | 10.265 | 25.200 | 28.601 | 31.595 | 24.265 | 36.456 |

◄◄

In studying Table IV, we notice that the degrees of freedom are numbered 1–30 inclusive, then 40, 50, 60, …, 100. If the number of degrees of freedom is not found in the table, we shall follow the practice of choosing the degrees of freedom closest to that desired. If the degrees of freedom is exactly between two values, find the mean of the values. For example, to find the critical value corresponding to 75 degrees of freedom, compute the mean of the critical values corresponding to 70 and 80 degrees of freedom.

## Goodness-of-Fit Tests

We begin with a definition.

*Definition*   A **goodness-of-fit test** is an inferential procedure used to determine whether a frequency distribution follows a claimed distribution.

As an example, we might want to test the claim that a die is fair, or the probability that each outcome is 1/6 when the die is cast. We express this claim as

$$H_o: p_1 = p_2 = p_3 = p_4 = p_5 = p_6 = 1/6$$

Here's another example: According to the U.S. Bureau of the Census, in 1999, 19.6% of the population of the United States resided in the Northeast, 23.0% resided in the Midwest, 35.4% resided in the South, and 22.0% resided in the West. We might want to test the claim that the distribution of U.S. residents is the same today as it was in 1999. Remember, the null hypothesis is a statement of "no change," so for this claim, the null hypothesis is

$H_o$: The distribution of residents in the United States is the same today as it was in 1999.

The idea behind testing these types of claims is to compare the actual number of observations for each category of data with the number of observations we would expect if the null hypothesis were true. If a significant difference between the observed counts and expected counts exists, we have evidence against the null hypothesis.

The method for obtaining the expected counts is an extension of the expected value of a binomial random variable. Recall that the mean (and therefore expected value) of a binomial random variable with $n$ independent trials and probability of success, $p$, is given by $E = \mu = np$.

### Expected Counts

Suppose there are $n$ independent trials of an experiment with $k \geq 3$ mutually exclusive possible outcomes. Let $p_1$ represent the probability of observing the first outcome and $E_1$ represent the expected count of the first outcome, $p_2$ represent the probability of observing the second outcome and $E_2$ represent the expected count of the second outcome, and so on. The expected counts for each possible outcome are given by

$$E_i = \mu_i = np_i \quad \text{for} \quad i = 1, 2, \ldots, k.$$

▶ EXAMPLE 2  **Finding Expected Counts**

*Problem:* An urban economist wishes to determine whether the distribution of residents in the United States is the same today as it was in 1999. That year, 19.6% of the population of the United States resided in the Northeast, 23.0% resided in the Midwest, 35.4% resided in the South, and 22.0% resided in the West (based upon data obtained from the Census Bureau). If the economist randomly selects 2000 households in the United States, compute the expected number of households in each region, assuming that the distribution of households did not change from 1999.

*Approach:*

**Step 1:** Determine the probabilities for each outcome.
**Step 2:** There are $n = 2000$ trials (the 2000 households surveyed) of the experiment. We expect $np_{\text{northeast}}$ of the households surveyed to reside in the Northeast, $np_{\text{midwest}}$ of the households to reside in the Midwest, and so on.

*Solution:*

**Step 1:** The probabilities are the relative frequencies from the 1999 distribution: $p_{\text{northeast}} = 0.196$, $p_{\text{midwest}} = 0.23$, $p_{\text{south}} = 0.354$, and $p_{\text{west}} = 0.22$.
**Step 2:** The expected counts for each of the locations within the United States are as follows:

Expected count of Northeast: $np_{\text{northeast}} = 2000(0.196) = 392$
Expected count of Midwest: $np_{\text{midwest}} = 2000(0.23) = 460$
Expected count of South: $np_{\text{south}} = 2000(0.354) = 708$
Expected count of West: $np_{\text{west}} = 2000(0.22) = 440$

Of the 2000 households surveyed, we expect 392 households in the Northeast, 460 households in the Midwest, 708 households in the South, and 440 households in the West if the distribution of residents of the United States is the same today as it was in 1999.  ◀◀

 *Now Work Problem 1.*

To test a claim, we compare the observed counts with the expected counts. If the observed counts are significantly different from the expected counts, we have evidence against the null hypothesis. We need a test statistic and sampling distribution in order to test the claim.

*Theorem*

**Test Statistic for Goodness-of-Fit Tests**

Let $O_i$ represent the observed counts of category $i$, $E_i$ represent the expected counts of category $i$, $k$ represent the number of categories, and $n$ represent the number of independent trials of an experiment. Then the formula

$$\chi^2 = \sum \frac{(O_i - E_i)^2}{E_i} \qquad i = 1, 2, \ldots, k$$

approximately follows the chi-square distribution with $k - 1$ degrees of freedom, provided that (1) all expected frequencies are greater than or equal to 1 (all $E_i \geq 1$) and (2) no more than 20% of the expected frequencies are less than 5. *Note*: $E_i = np_i$ for $i = 1, 2, \ldots, k$.

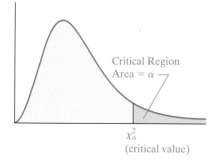

**Caution**

Goodness-of-fit tests are used to test a claim regarding the distribution of a variable based on a single population. If you wish to compare two or more populations, you must use tests for homogeneity presented in Section 11.2.

From Example 2, there were $k = 4$ categories (Northeast, Midwest, South, and West). For the Northeast, the expected frequency, $E$, is 392.

Now that we know the distribution of goodness-of-fit tests, we can present a method for testing claims regarding the distribution of a random variable.

**The Chi-Square Goodness-of-Fit Test**

If a claim is made regarding a distribution, we can use the steps that follow to test the claim, provided that

**1.** the data are randomly selected,

**2.** all expected frequencies are greater than or equal to 1,

**3.** no more than 20% of the expected frequencies are less than 5.

**Caution**

If requirements (2) or (3) are not satisfied, one option is to combine two of the low frequency categories into a single category.

**Step 1:** A claim is made regarding a distribution. The claim is used to determine the null and alternative hypotheses:

$H_o$: The random variable follows the claimed distribution.
$H_1$: The random variable does not follow the claimed distribution.

**Step 2:** Calculate the expected frequencies for each of the $k$ categories. The expected frequencies are $np_i$ for $i = 1, 2, \ldots, k$, where $n$ is the number of trials and $p_i$ is the probability of the $i$th category, assuming that the null hypothesis is true.

**Step 3:** Verify that the requirements for the goodness-of-fit test are satisfied:

**1.** All expected frequencies are greater than or equal to 1 (all $E_i \geq 1$).

**2.** No more than 20% of the expected frequencies are less than 5.

**Step 4:** Select a level of significance $\alpha$ based upon the seriousness of making a Type I error. The level of significance is used to determine the critical value. All chi-square goodness-of-fit tests are right-tailed tests, so the critical value is $\chi^2_\alpha$ with $k - 1$ degrees of freedom. The shaded region represents the critical region. See Figure 4.

**Figure 4**

Critical Region Area = $\alpha$

$\chi^2_\alpha$
(critical value)

**Step 5:** Compute the **test statistic**:

$$\chi^2 = \sum \frac{(O_i - E_i)^2}{E_i}$$

**Step 6:** Compare the critical value with the test statistic. If $\chi^2 > \chi^2_\alpha$, reject the null hypothesis.

**Step 7:** State the conclusion.

▶ **EXAMPLE 3** **Testing a Claim Using the Goodness-of-Fit Test**

*Problem:* An urban economist wishes to test the claim that the distribution of United States residents in the United States is different today than it was in 1999. In 1999, 19.6% of the population of the United States resided in the Northeast, 23.0% resided in the Midwest, 35.4% resided in the South, and 22.0% resided in the West (based upon data obtained from the Census Bureau). The economist randomly selects 2000 households in the United States and obtains the frequency distribution shown in Table 1.

Test the claim that the distribution of residents in the United States is different today from the distribution in 1999 at the $\alpha = 0.05$ level of significance.

**TABLE 1**

| Region | Frequency |
|---|---|
| Northeast | 365 |
| Midwest | 404 |
| South | 752 |
| West | 479 |

*Approach:* We follow Steps 1–7 on page 479.

*Solution:* The requirement that the data be randomly selected is satisfied.

*Step 1:* The claim is that the distribution of residents is the same today as it was in 1999. This claim will be the null hypothesis:

$H_0$: The distribution of residents of the United States is the same today as it was in 1999.

$H_1$: The distribution of residents of the United States is different today from what it was in 1999.

**In Your Own Words**

Remember, the null hypothesis is always the statement of "no change." Therefore, the null hypothesis is that there is no change in the distribution from 1999.

*Step 2:* We compute the expected counts for each category, assuming that the null hypothesis is true. We did this computation in Example 1; Table 2 shows the results.

**TABLE 2**

| Region | Frequency (Observed Count) | Expected Frequency |
|---|---|---|
| Northeast | 365 | 2000(0.196) = 392 |
| Midwest | 404 | 2000(0.230) = 460 |
| South | 752 | 2000(0.354) = 708 |
| West | 479 | 2000(0.220) = 440 |

*Step 3:* Since all the expected counts are greater than 5, the requirements for the goodness-of-fit test are satisfied.

*Step 4:* There are $k = 4$ possible outcomes, so we find the critical value by using $4 - 1 = 3$ degrees of freedom. We consult Table IV and identify the column corresponding to $\alpha = 0.05$ with 3 degrees of freedom. The critical value is $\chi^2_{0.05} = 7.815$.

*Step 5:* The test statistic is

$$\chi^2 = \sum \frac{(O_i - E_i)^2}{E_i} = \frac{(365 - 392)^2}{392} + \frac{(404 - 460)^2}{460} + \frac{(752 - 708)^2}{708} + \frac{(479 - 440)^2}{440} = 14.868$$

*Step 6:* Because the test statistic, 14.868, is greater than the critical value, 7.815, we reject the null hypothesis. See Figure 5.

Figure 5

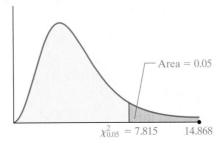

Area = 0.05

$\chi^2_{0.05} = 7.815$     14.868

*Step 7:* There is sufficient evidence at the $\alpha = 0.05$ level of significance to support the claim that the distribution of United States residents is different from the distribution in 1999.   ◀◀

If we compare the observed and expected counts, we notice that the Northeast and Midwest regions of the United States have observed counts lower than expected, while the South and West regions of the United States have observed counts higher than expected. So we might conclude that residents of the United States are moving to southern and western locations.

**NW**   *Now Work Problem 7.*

In the next example, each of the $k$ categories is equally likely.

▶ EXAMPLE 4

### Testing a Claim Using a Goodness-of-Fit Test

*Problem:* An obstetrician wants to know whether the proportion of children born each day of the week is the same. She randomly selects 500 birth records and obtains the data shown in Table 3 (based on data obtained from *Vital Statistics of the United States*, 1997, Volume 1).

Is there reason to believe that the day on which a child is born occurs with equal frequency at the $\alpha = 0.01$ level of significance?

*Approach:* We follow Steps 1–7 presented on page 479.

*Solution:* We notice that the data were randomly selected.

*Step 1:* We will assume that the day on which a child is born occurs with equal frequency. If we let 1 represent Sunday, 2 represent Monday, and so on, we can express this claim as $p_1 = p_2 = \ldots = p_7 = 1/7$. Thus, we have

$$H_o: p_1 = p_2 = p_3 = p_4 = p_5 = p_6 = p_7 = 1/7$$

$H_1$: At least one of the proportions is different from the others.

*Step 2:* In Table 4, we compute the expected counts for each category (day of the week), assuming that the null hypothesis is true.

#### TABLE 3

| Day of Week | Frequency |
|---|---|
| Sunday | 57 |
| Monday | 78 |
| Tuesday | 74 |
| Wednesday | 76 |
| Thursday | 71 |
| Friday | 81 |
| Saturday | 63 |

#### TABLE 4

| Day of the Week | Frequency (Observed Count) | Theoretical Probability | Expected Count |
|---|---|---|---|
| Sunday | 57 | 1/7 | $500(1/7) = 500/7$ |
| Monday | 78 | 1/7 | 500/7 |
| Tuesday | 74 | 1/7 | 500/7 |
| Wednesday | 76 | 1/7 | 500/7 |
| Thursday | 71 | 1/7 | 500/7 |
| Friday | 81 | 1/7 | 500/7 |
| Saturday | 63 | 1/7 | 500/7 |

*Step 3:* Since all the expected counts for each category are greater than $5(500/7 \approx 71.43)$, the requirements of the goodness-of-fit test are satisfied.

*Step 4:* There are $k = 7$ categories, so we find the critical value using $7 - 1 = 6$ degrees of freedom. We consult Table IV and identify the column corresponding to $\alpha = 0.01$ with 6 degrees of freedom. The critical value is $\chi^2_{0.01} = 16.812$.

Figure 6

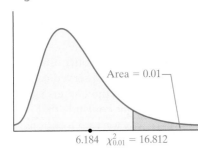

Area = 0.01

6.184  $\chi^2_{0.01} = 16.812$

**Step 5:** The test statistic is

$$\chi^2 = \frac{(57 - 500/7)^2}{500/7} + \frac{(78 - 500/7)^2}{500/7} + \frac{(74 - 500/7)^2}{500/7} +$$

$$\frac{(76 - 500/7)^2}{500/7} + \frac{(71 - 500/7)^2}{500/7} + \frac{(81 - 500/7)^2}{500/7} + \frac{(63 - 500/7)^2}{500/7} = 6.184$$

**Step 6:** Because the test statistic, 6.184, is less than the critical value, 16.812, we do not reject the null hypothesis. See Figure 6.

**Step 7:** There is evidence at the $\alpha = 0.01$ level of significance to support the belief that the day on which a child is born occurs with equal frequency.

◀◀

**NW** *Now Work Problem 13.*

### P-Values

We used the classical approach to test claims in this section, but we could also use the *P*-value approach. We cannot obtain exact *P*-values by hand, but statistical software will provide *P*-values based upon the value of the test statistic and the degrees of freedom. We can compute the *P*-value of the hypothesis tested in Example 4, using the $\chi^2$-cdf command on a TI-83 Plus graphing calculator. See Figure 7. The area under the chi-square distribution to the right of 6.184 with 6 degrees of freedom is 0.403, so the *P*-value is 0.403. See Figure 8. The *P*-value is greater than the level of significance, so we do not reject the null hypothesis.

**Caution**

To find exact *P*-values, we must use technology.

Figure 7

```
X²cdf(6.184,1E99
,6)
          .402897295
```

Figure 8

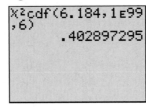

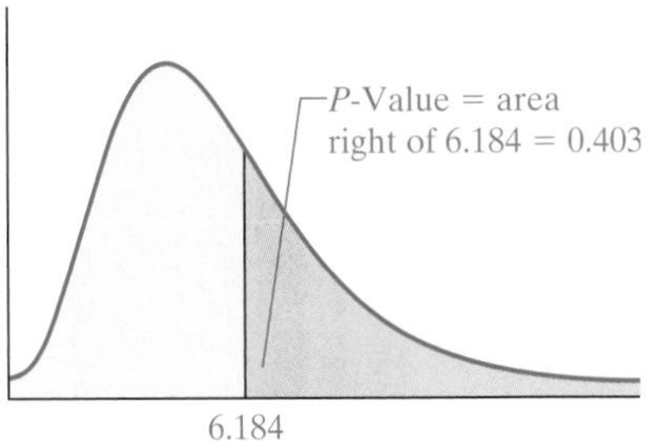

*P*-Value = area right of 6.184 = 0.403

6.184

Goodness-of-Fit tests can also be used to test the claim that a random variable $X$ follows a certain distribution, such as the binomial probability distribution.

▶ **EXAMPLE 5**   **Testing a Claim That a Random Variable Follows a Binomial Distribution**

*Problem:* It is known that a basketball player makes 60% of her free throws. Suppose we conduct an experiment in which the player shoots four free throws. We repeat this experiment 200 times and obtain the data presented in Table 5.

Is there evidence to support the belief that the random variable $X$, the number of made free throws in four attempts, is a binomial random variable with $p = 0.6$ at the $\alpha = 0.05$ level of significance?

*Approach:* We will follow Steps 1–7 on page 479.

*Solution:*

**Step 1:** We assume that the random variable $X$ is a binomial random variable with $p = 0.6$. This will be the null hypothesis:

| TABLE 5 | |
|---|---|
| Number of Made Free Throws | Frequency |
| 0 | 10 |
| 1 | 52 |
| 2 | 67 |
| 3 | 56 |
| 4 | 15 |

$H_0$: The random variable $X$ is a binomial random variable with $n = 4$ and $p = 0.6$.

$H_1$: The random variable $X$ is not a binomial random variable with $n = 4$ and $p = 0.6$.

**Step 2:** We compute the expected counts for each possible outcome, assuming that the null hypothesis is true. This is done using the binomial probability formula with $n = 4$ and $p = 0.6$ for $X = 0, 1, 2, 3$, and 4:

$$P(0 \text{ made free throws}) = P(X = 0) = p_0 = {}_4C_0 \, (0.6)^0 \, (0.4)^4 = 0.0256$$

$$P(1 \text{ made free throw}) = P(X = 1) = p_1 = {}_4C_1 \, (0.6)^1 \, (0.4)^3 = 0.1536$$

$$P(2 \text{ made free throws}) = P(X = 2) = p_2 = {}_4C_2 \, (0.6)^2 \, (0.4)^2 = 0.3456$$

$$P(3 \text{ made free throws}) = P(X = 3) = p_3 = {}_4C_3 \, (0.6)^3 \, (0.4)^1 = 0.3456$$

$$P(4 \text{ made free throws}) = P(X = 4) = p_4 = {}_4C_4 \, (0.6)^4 \, (0.4)^0 = 0.1296$$

The expected counts for each of the possible outcomes are as follows:

Expected Count of 0 Made Free Throws: $np_0 = 200(0.0256) = 5.12$
Expected Count of 1 Made Free Throw: $np_1 = 200(0.1536) = 30.72$
Expected Count of 2 Made Free Throws: $np_2 = 200(0.3456) = 69.12$
Expected Count of 3 Made Free Throws: $np_3 = 200(0.3456) = 69.12$
Expected Count of 4 Made Free Throws: $np_4 = 200(0.1296) = 25.92$

So, in 200 trials of the experiment, we expect the player to make 0 free throws about 5 times, 1 free throw about 31 times, 2 free throws about 69 times, 3 free throws about 69 times and all 4 free throws about 26 times. The observed and expected counts are presented in Table 6.

### TABLE 6

| Number of Made Free Throws | Frequency (Observed Count) | Theoretical Probability | Expected Count |
| --- | --- | --- | --- |
| 0 | 10 | 0.0256 | $200(0.0256) = 5.12$ |
| 1 | 52 | 0.1536 | $200(0.1536) = 30.72$ |
| 2 | 67 | 0.3456 | $200(0.3456) = 69.12$ |
| 3 | 56 | 0.3456 | 69.12 |
| 4 | 15 | 0.1296 | 25.92 |

**Step 3:** Since all the expected counts are greater than 5, the requirements of the goodness-of-fit test are satisfied.

**Step 4:** There are $k = 5$ possible outcomes, so we find the critical value using $5 - 1 = 4$ degrees of freedom. We consult Table IV and identify the column corresponding to $\alpha = 0.05$ with 4 degrees of freedom. The critical value is $\chi^2_{0.05} = 9.488$.

**Step 5:** The test statistic is

$$\chi^2 = \frac{(10 - 5.12)^2}{5.12} + \frac{(52 - 30.72)^2}{30.72} + \frac{(67 - 69.12)^2}{69.12} + \frac{(56 - 69.12)^2}{69.12} + \frac{(15 - 25.92)^2}{25.92} = 26.55$$

**10. Are Mothers Getting Older?** A sociologist wanted to determine whether the age distribution at which women have babies is changing. The U.S. Census Bureau provided her with the following data, which represent the relative frequency of total births by age in 1990:

| Age of Mother | Relative Frequency | Age of Mother | Relative Frequency |
|---|---|---|---|
| 15–19 | 0.086 | 30–34 | 0.228 |
| 20–24 | 0.265 | 35–39 | 0.096 |
| 25–29 | 0.305 | 40–44 | 0.020 |

A recent random sample of 500 births yielded the following frequencies:

| Age of Mother | Frequency | Age of Mother | Frequency |
|---|---|---|---|
| 15–19 | 58 | 30–34 | 131 |
| 20–24 | 105 | 35–39 | 69 |
| 25–29 | 128 | 40–44 | 9 |

Test the claim that the distribution of age of mothers is different today from what it was in 1990 at the $\alpha = 0.05$ level of significance by following Steps 1–7 on page 479.

**11. Are Some Months Busier than Others?** A researcher wants to know whether the distribution of birth month is uniform. The following data, based upon results obtained from *Vital Statistics of the United States*, 1997, Volume 1, represent the distribution of months in which 500 randomly selected children were born:

| Month | Frequency | Month | Frequency |
|---|---|---|---|
| Jan. | 52 | July | 44 |
| Feb. | 35 | Aug. | 34 |
| March | 44 | Sep. | 36 |
| April | 42 | Oct. | 46 |
| May | 42 | Nov. | 48 |
| June | 36 | Dec. | 41 |

Is there reason to believe that each birth month occurs with equal frequency at the $\alpha = 0.05$ level of significance?

**12. Bicycle Deaths** A researcher wanted to determine whether bicycle deaths were uniformly distributed over the days of the week. She randomly selected 200 deaths that involved a bicycle, recorded the day of the week on which the death occurred, and obtained the following results (the data are based upon information obtained from the Insurance Institute for Highway Safety):

| Day of the Week | Frequency | Day of the Week | Frequency |
|---|---|---|---|
| Sunday | 16 | Thursday | 34 |
| Monday | 35 | Friday | 41 |
| Tuesday | 16 | Saturday | 30 |
| Wednesday | 28 | | |

Is there reason to believe that the day of the week on which a fatality occurs on a bicycle occurs with equal frequency at the $\alpha = 0.05$ level of significance?

**13. Pedestrian Deaths** A researcher wanted to determine whether pedestrian deaths were uniformly distributed over the days of the week. She randomly selected 300 pedestrian deaths, recorded the day of the week on which the death occurred, and obtained the following results (the data are based upon information obtained from the Insurance Institute for Highway Safety):

| Day of the Week | Frequency | Day of the Week | Frequency |
|---|---|---|---|
| Sunday | 39 | Thursday | 41 |
| Monday | 40 | Friday | 49 |
| Tuesday | 30 | Saturday | 61 |
| Wednesday | 40 | | |

Test the belief that the day of the week on which a fatality occurs involving a pedestrian occurs with equal frequency at the $\alpha = 0.05$ level of significance.

**14. Is the Die Loaded?** A player in a craps game suspects that one of the die being used in the game is loaded. A loaded die is one in which not all of the possibilities (1, 2, 3, 4, 5 and 6) are equally likely. The player throws the die 400 times, records the outcome after each throw, and obtains the following results:

| Outcome | Frequency | Outcome | Frequency |
|---|---|---|---|
| One | 62 | Four | 62 |
| Two | 76 | Five | 57 |
| Three | 76 | Six | 67 |

(a) Is there evidence to support the claim that the die is loaded at the $\alpha = 0.01$ level of significance?

(b) Why do you think the player might test the claim at the $\alpha = 0.01$ level of significance, rather than, say, the $\alpha = 0.1$ level of significance?

15. **Testing the Random-Number Generator** Statistical  spreadsheets and graphing calculators with advanced statistical features have random-number generators that create random numbers conforming to a specified distribution.

(a) Use the random-number generator to create a list of 500 randomly selected integers numbered 1–5.

(b) What proportion of the numbers generated should be 1? 2? 3? 4? 5?

(c) Test the claim that the random-number generator is generating random integers between 1 and 5 with equal likelihood by performing a chi-square goodness-of-fit test at the $\alpha = 0.01$ level of significance.

16. **Testing the Random-Number Generator** Statistical spreadsheets and graphing calculators with advanced statistical features have random-number generators that create random numbers conforming to a specified distribution.

(a) Use the random-number generator to create a list of 500 trials of a binomial experiment with $n = 5$ and $p = 0.2$.

(b) What proportion of the numbers generated should be 0? 1? 2? 3? 4? 5?

(c) Test the claim that the random-number generator is generating random outcomes of a binomial experiment with $n = 5$ and $p = 0.2$ by performing a chi-square goodness-of-fit test at the $\alpha = 0.01$ level of significance.

---

*In Section 9.4, we tested claims regarding a population proportion using a z-test. However, we can also use the chi-square goodness-of-fit test to test claims with $k = 2$ possible outcomes. In Problems 17 and 18, we test claims with the use of both methods.*

17. **Low Birth Weight** According to the U.S. Census Bureau, 7.1% of all babies born to nonsmoking mothers are of low birth weight ($<5$ lb, 8 oz). An obstetrician wanted to know whether mothers between the ages of 35 and 39 years gave birth to a higher percentage of low-birth-weight babies. She randomly selected 120 births for which the mother was 35–39 years old and found that 11 of them were of low birth weight.

(a) If the proportion of low-birth-weight babies for mothers 35–39 years old is 0.071, compute the expected number of low-birth-weight births to mothers 35–39 years old. What is the expected number of non-low-birth-weight births to mothers 35–39 years old?

(b) Test the obstetrician's claim at the $\alpha = 0.05$ level of significance, using the chi-square goodness-of-fit test.

(c) Test the claim by using the approach presented in Section 9.4.

18. **Living Alone?** In 1990, 39% of males 15 years of age or older lived alone, according to the Census Bureau. A sociologist claims that this percentage is greater today, conducts a random sample of 400 males 15 years of age or older, and finds that 164 are living alone.

(a) If the proportion of males aged 15 years or older living alone is 0.39, compute the following expected numbers: males 15 years of age or older who live alone; males 15 years of age or older who do not live alone?

(b) Test the sociologist's claim at the $\alpha = 0.05$ level of significance, using the chi-square goodness-of-fit test.

(c) Test the claim by using the approach presented in Section 9.4.

19. Using the results of Problem 3, compute $\sum (O - E)$. Explain why this result is reasonable.

---

## 11.2  Chi-Square Test for Independence; Homogeneity of Proportions

**Preparing for This Section**    Before getting started, review the following:

✓ The language of hypothesis tests (Section 9.1, pp. 383–389)

✓ Independent events (Section 5.3, pp. 212–215)

✓ Mean of a binomial random variable (Section 6.2, p. 259)

✓ Testing a hypothesis about two population proportions (Section 10.3, pp. 459–464)

**Objectives**  ① Perform chi-square test for independence
② Perform chi-square test for homogeneity of proportions

As we saw in Section 11.1, data, whether qualitative or quantitative, can be organized into categories. For example, a person might be categorized as a male or as a female. A person might also be categorized as a 20–29-year-old.

In fact, many governmental agencies regularly publish quantitative data by first creating categories (classes) of data.

Consider the data (measured in thousands) in Table 7, which represent the employment status and level of education of all United States residents 25 years old or older in 1998. By definition, an individual is unemployed if he or she is actively seeking work, but is unable to find work. An individual is considered not to be in the labor force if he or she is not employed and is not actively seeking employment.

### TABLE 7

| Employment Status | Level of Education | | | |
| --- | --- | --- | --- | --- |
| | Did Not Finish High School | High School Graduate | Some College | Four or More Years of College |
| Employed | 11,669 | 36,451 | 30,339 | 33,006 |
| Unemployed | 1,057 | 1,784 | 1,126 | 625 |
| Not in the Labor Force | 16,858 | 20,040 | 10,829 | 8,094 |

*Source:* United States Census Bureau, Current Population Survey

Table 7 is referred to as a **contingency table** or a **two-way table**, because it relates two categories of data. The **row variable** is employment status, because each row in the table describes the employment status of a group. The **column variable** is level of education. Each box inside the table is referred to as a **cell**. For example, the cell corresponding to employed individuals who are high school graduates is in the first row, second column. Each cell contains the frequency of the category: There were 11,669 thousand employed individuals who did not finish high school in 1998.

 ## Chi-Square Independence Test

In this section, we develop methods for performing statistical inference on two categorical variables to determine whether there is any association between the two variables. We call the method the *chi-square independence test*.

*Definition*

The **chi-square independence test** is used to find out whether there is an association between a row variable and column variable in a contingency table constructed from sample data. The null hypothesis is that the variables are not associated; in other words, they are independent. The alternative hypothesis is that the variables are associated, or dependent.

**In Your Own Words**

In a chi-square independence test, the null hypothesis is always
$H_o$: The variables are independent
The alternative hypothesis is always
$H_1$: The variables are dependent

The idea behind testing these types of claims is to compare actual counts to the counts we would expect if the null hypothesis were true (if the variables are independent). If a significant difference between the actual counts and expected counts exists, we would take this as evidence against the null hypothesis.

The method for obtaining the expected counts requires that we compute the number of observations expected within each cell under the assumption that the null hypothesis is true. Recall, if two events $A$ and $B$ are independent, then $P(A \text{ and } B) = P(A) \cdot P(B)$. We can use the Multiplication Principle for independent events to obtain the expected proportion of observations within each cell under the assumption of independence. We then multiply this result by $n$, the sample size, in order to obtain the expected count within each cell.* We present an example in order to introduce the method for obtaining expected counts.

*Recall that the expected value of a binomial random variable for $n$ independent trials of a binomial experiment with probability of success $p$ is given by $E = \mu = np$.

▶ EXAMPLE 1   **Determining the Expected Counts in a Test for Independence**

*Problem:* Blood type is classified as A, B, AB, or O. In addition, blood can be classified as $Rh^+$ or $Rh^-$. In a survey of 500 randomly selected individuals, a phlebotomist obtained the results shown in Table 8.

| TABLE 8 | | | | |
| --- | --- | --- | --- | --- |
| | | Blood Type | | |
| Rh-Level | A | B | AB | O |
| $Rh^+$ | 176 | 28 | 22 | 198 |
| $Rh^-$ | 30 | 12 | 4 | 30 |

Compute the expected counts within each cell, assuming that Rh-level and blood type are independent.

*Approach:*

**Step 1:** Compute the row and column totals.

**Step 2:** Compute the relative frequencies for each row variable and column variable.

**Step 3:** Use the Multiplication Rule for independent events to compute the proportion of observations within each cell under the assumption of independence.

**Step 4:** Multiply the proportions by 500, the sample size, to obtain the expected counts within each cell.

*Solution:*

**Step 1:** The row totals (blue) and column totals (red) are presented in Table 9.

| TABLE 9 | | | | | |
| --- | --- | --- | --- | --- | --- |
| | A | B | AB | O | Totals |
| $Rh^+$ | 176 | 28 | 22 | 198 | 424 |
| $Rh^-$ | 30 | 12 | 4 | 30 | 76 |
| Column Totals | 206 | 40 | 26 | 228 | 500 |

**Step 2:** The relative frequencies for the row variable (Rh-level) and column variable (blood type) are presented in Table 10.

| TABLE 10 | | | | | |
| --- | --- | --- | --- | --- | --- |
| | A | B | AB | O | Relative Frequency |
| $Rh^+$ | 176 | 28 | 22 | 198 | 424/500 = 0.848 |
| $Rh^-$ | 30 | 12 | 4 | 30 | 76/500 = 0.152 |
| Relative Frequency | 206/500 = 0.412 | 40/500 = 0.08 | 26/500 = 0.052 | 228/500 = 0.456 | 1 |

**Step 3:** Assuming blood type and Rh-level are independent, we use the Multiplication Rule for independent events to compute the proportion of observations we would expect in each cell. For example, the proportion of individuals who are $Rh^+$ and of blood type A would be

$$\text{Proportion } Rh^+ \text{and blood type A} = (\text{Proportion } Rh^+) \cdot (\text{Proportion blood type A})$$
$$= (0.848)(0.412)$$
$$= 0.349376$$

Table 11 contains the expected proportion in each cell, under the assumption of independence.

| TABLE 11 | | | | |
|---|---|---|---|---|
| | A | B | AB | O |
| **Rh$^+$** | 0.349376 | 0.06784 | 0.044096 | 0.386688 |
| **Rh$^-$** | 0.062624 | 0.01216 | 0.007904 | 0.069312 |

***Step 4:*** We multiply the expected proportions in Table 11 by 500, the sample size, to obtain the expected counts under the assumption of independence. The results are presented in Table 12.

| TABLE 12 | | | |
|---|---|---|---|
| | A | B | AB | O |
| **Rh$^+$** | 500(0.349376) = 174.688 | 500(0.06784) = 33.92 | 500(0.044096) = 22.048 | 193.344 |
| **Rh$^-$** | 31.312 | 6.08 | 3.952 | 34.656 |

If blood type and Rh-level are independent, we would expect a random sample of 500 individuals to contain about 175 who are of blood type A and are Rh$^+$. ◀◀

The technique used in Example 1 to find the expected counts might seem rather tedious. It certainly would be more pleasant if we could determine a shortcut formula that could be used to obtain the expected counts. Let's consider the expected count for blood type A and Rh$^+$. This expected count was obtained by multiplying the proportion of individuals who were of blood type A, the proportion of individuals that were Rh$^+$, and the number of individuals in the sample. That is,

$$\text{Expected Count} = (\text{Proportion Rh}^+)(\text{Proportion blood type A})(\text{Sample size})$$

$$= \frac{424}{500} \cdot \frac{206}{500} \cdot 500$$

$$= \frac{424 \cdot 206}{500} \qquad \text{Cancel the 500s}$$

$$= \frac{(\text{row total for Rh}^+)(\text{column total for blood type A})}{\text{table total}}$$

This leads to the following general result:

**Expected Frequencies in a Chi-Square Independence Test**

To find the expected frequencies in a cell when performing a chi-square independence test, multiply the row total of the row containing the cell by the column total of the column containing the cell and divide this result by the table total. That is,

$$\text{Expected frequency} = \frac{(\text{row total})(\text{column total})}{\text{table total}} \qquad \textbf{(1)}$$

For example, to calculate the expected frequency for Rh$^+$, blood type A, we compute

$$\text{Expected frequency} = \frac{(\text{row total})(\text{column total})}{\text{table total}} = \frac{(424)(206)}{500} = 174.688$$

This result agrees with the result obtained in Table 12.

**NW**  *Now Work Problem 5(a).*

To test a claim, we compare the actual (observed) counts to those expected. If the observed counts are significantly different from the expected counts, we take this as evidence against the null hypothesis. We need a test statistic and sampling distribution in order to test the claim.

*Theorem*

**Test Statistic for the Test of Independence**

Let $O_i$ represent the observed number of counts in the $i$th cell and $E_i$ represent the expected number of counts in the $i$th cell. Then

$$\chi^2 = \sum \frac{(O_i - E_i)^2}{E_i}$$

approximately follows the chi-square distribution with $(r - 1)(c - 1)$ degrees of freedom, where $r$ is the number of rows and $c$ is the number of columns in the contingency table, provided that (1) all expected frequencies are greater than or equal to 1 and (2) no more than 20% of the expected frequencies are less than 5.

From Example 1, there were $r = 2$ rows and $c = 4$ columns.
We now present a method for testing claims regarding the association between two variables in a contingency table.

**The Chi-Square Test for Independence**

If a claim is made regarding the association between (or independence of) two variables in a contingency table, we can use the steps that follow to test the claim, provided that
**1.** the data are randomly selected
**2.** all expected frequencies are greater than or equal to 1; and
**3.** no more than 20% of the expected frequencies are less than 5.
*Step 1:* A claim is made regarding the independence of the data.

$H_0$: The row variable and column variable are independent.
$H_1$: The row variable and column variable are dependent.

*Step 2:* Calculate the expected frequencies (counts) for each cell in the contingency table using Formula (1).

*Step 3:* Verify that the requirements of the test for independence are satisfied.

*Step 4:* Choose a level of significance, $\alpha$, based upon the seriousness of making a Type I error. The level of significance is used to determine the critical value. All chi-square independence tests are right-tailed tests, so the critical value is $\chi^2_\alpha$ with $(r - 1)(c - 1)$ degrees of freedom. The shaded region(s) represents the critical region. See Figure 10.

*Step 5:* Compute the **test statistic:**

$$\chi^2 = \sum \frac{(O_i - E_i)^2}{E_i}$$

*Step 6:* Compare the critical value with the test statistic. If $\chi^2 > \chi^2_\alpha$, reject the null hypothesis.

*Step 7:* State the conclusion.

**Figure 10**

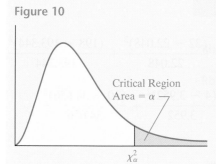

Critical Region
Area = $\alpha$

$\chi^2_\alpha$
(critical value)

3. The following table contains the number of successes and failures for three categories of a variable:

| | Category 1 | Category 2 | Category 3 |
|---|---|---|---|
| **Success** | 76 | 84 | 69 |
| **Failure** | 44 | 41 | 49 |

Test the claim that the proportions are equal for each category at the $\alpha = 0.01$ level of significance. What is the P-value?

4. The following table contains the number of successes and failures for three categories of a variable:

| | Category 1 | Category 2 | Category 3 |
|---|---|---|---|
| **Success** | 204 | 199 | 214 |
| **Failure** | 96 | 121 | 98 |

Test the claim that the proportions are equal for each category at the $\alpha = 0.01$ level of significance. What is the P-value?

• **Applying the Concepts**

5. **Family Structure and Sexual Activity** A sociologist wants to discover whether the sexual activity of females between the ages of 15 and 19 years of age and family structure are associated. She randomly selects 380 females between the ages of 15 and 19 years of age and asks each to disclose her family structure at age 14 and whether she has had sexual intercourse. The results are shown in the table below. Data are based on information obtained from the U.S. National Center for Health Statistics.

| | Family Structure | | | |
|---|---|---|---|---|
| **Sexual Activity** | **Both Biological/Adoptive Parents** | **Single Parent** | **Parent and Stepparent** | **Nonparental Guardian** |
| **Had Sexual Intercourse** | 64 | 59 | 44 | 32 |
| **Did Not Have Sexual Intercourse** | 86 | 41 | 36 | 18 |

(a) Compute the expected values of each cell, under the assumption of independence.
(b) Verify that the requirements for performing a chi-square test of independence are satisfied.
(c) Compute the chi-square test statistic.
(d) Test whether family structure and sexual activity of 15–19-year-old females are independent at the $\alpha = 0.05$ level of significance.
(e) Compare the observed frequencies with the expected frequencies. Which cell contributed most to the test statistic? Was the expected frequency greater than or less than the observed frequency? What does this information tell you?
(f) Construct a conditional distribution by family structure and draw a bar graph. Does this evidence support your conclusion in part (d)?
(g) Compute the P-value for this test by finding the area under the chi-square distribution to the right of the test statistic.

6. **Prenatal Care** An obstetrician wants to learn whether the amount of prenatal care and the wantedness of the pregnancy are associated. He randomly selects 939 women that have recently given birth and asks them to disclose whether their pregnancy was intended, unintended, or mistimed. In addition, they were to disclose when they started receiving prenatal care, if ever. The results of the survey are as follows:

| Wantedness of Pregnancy | Months Pregnant Before Prenatal Care Began | | |
|---|---|---|---|
| | **Less Than 3 Months** | **3–5 Months** | **More Than 5 Months (or never)** |
| **Intended** | 593 | 26 | 33 |
| **Unintended** | 64 | 8 | 11 |
| **Mistimed** | 169 | 19 | 16 |

(a) Compute the expected values of each cell, under the assumption of independence.
(b) Verify that the requirements for performing a chi-square test of independence are satisfied.
(c) Compute the chi-square test statistic.
(d) Test whether prenatal care and the wantedness of pregnancy are independent at the $\alpha = 0.05$ level of significance.
(e) Compare the observed frequencies with the expected frequencies. Which cell contributed most to the test statistic? Was the expected frequency greater than or less than the observed frequency? What does this information tell you?
(f) Construct a conditional distribution by wantedness of the pregnancy and draw a bar graph. Does this evidence support your conclusion in part (d)?

7. **Education versus Area of Country** An urban economist wants to determine whether the location of the United States a resident lives in is associated with level of education. He randomly selects 1804 residents of the United States and asks them to disclose the region of the United States in which they reside and their level of education. He obtains the data in the following table (data are based upon information obtained from the U.S. Census Bureau):

| Area of Country | Not a High School Graduate | High School Graduate | Some College | Bachelor's Degree or Higher |
|---|---|---|---|---|
| | | Level of Education | | |
| Northeast | 52 | 123 | 70 | 94 |
| Midwest | 123 | 146 | 102 | 96 |
| South | 119 | 204 | 148 | 144 |
| West | 62 | 106 | 111 | 104 |

(a) Test whether level of education and region of the United States are independent at the $\alpha = 0.05$ level of significance.
(b) Compare the observed frequencies with the expected frequencies. Which cell contributed most to the test statistic? Was the expected frequency greater than or less than the observed frequency? What does this information tell you?
(c) Construct a conditional distribution by level of education and draw a bar graph. Does this evidence support your conclusion in part (a)?

8. **Profile of Smokers** The following data represent the smoking status by level of education for residents of the United States 18 years old or older from a random sample of 1054 residents:

| Number of Years of Education | Current | Former | Never |
|---|---|---|---|
| | Smoking Status | | |
| <12 Years | 178 | 88 | 208 |
| 12 Years | 137 | 69 | 143 |
| 13–15 Years | 44 | 25 | 44 |
| 16 or More Years | 34 | 33 | 51 |

*Source:* National Health Interview Survey

(a) Test whether smoking status and level of education are independent at the $\alpha = 0.05$ level of significance.
(b) Compute the $P$-value for this test by finding the area under the chi-square distribution to the right of the test statistic.
(c) Construct a conditional distribution by number of years of education, and draw a bar graph. Does this evidence support your conclusion in part (a)?

9. **Legalization of Marijuana** On May 14, 2001, the Supreme Court, by a vote of 8–0, struck down state laws that legalized marijuana for medicinal purposes. The Gallup Organization later conducted surveys of randomly selected Americans 18 years old or older and asked whether they support the limited use of marijuana when prescribed by physicians to relieve pain and suffering. The results of the survey, by age group, are as follows:

| Opinion | 18–29 Years Old | 30–49 Years Old | 50 Years or Older |
|---|---|---|---|
| | Age | | |
| For | 172 | 313 | 258 |
| Against | 52 | 103 | 119 |

(a) Test whether age and opinion regarding the legalization of marijuana are independent at the $\alpha = 0.05$ level of significance.
(b) Compute the $P$-value for this test by finding the area under the chi-square distribution to the right of the test statistic.
(c) Construct a conditional distribution by age and draw a bar graph. Does this evidence support your conclusion in part (a)?

10. **Pro Life or Pro Choice** A recent Gallup Organization Poll asked male and female Americans whether they were pro life or pro choice when it comes to abortion issues. The results of the survey are as follows:

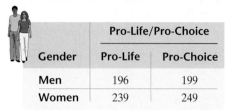

| Gender | Pro-Life/Pro-Choice | |
|---|---|---|
| | Pro-Life | Pro-Choice |
| Men | 196 | 199 |
| Women | 239 | 249 |

(a) Test whether an individual's opinion regarding abortion is independent of gender at the $\alpha = 0.1$ level of significance.
(b) Compute the $P$-value for this test by finding the area under the chi-square distribution to the right of the test statistic.
(c) Construct a conditional distribution by gender and draw a bar graph. Does this evidence support your conclusion in part (a)?

11. **Delinquencies** A delinquency offense is an act committed by a juvenile for which an adult could be prosecuted in a criminal court. The following data represent the number of various types of delinquencies by gender in a random sample of 750 delinquencies:

| Gender | Delinquency | | | |
|---|---|---|---|---|
| | Person | Property | Drugs | Public Order |
| Female | 24 | 85 | 7 | 28 |
| Male | 97 | 367 | 39 | 103 |

(a) Test whether gender is independent of type of delinquency at the $\alpha = 0.05$ level of significance.
(b) Compute the $P$-value for this test by finding the area under the chi-square distribution to the right of the test statistic.
(c) Construct a conditional distribution by type of delinquency and draw a bar graph. Does this evidence support your conclusion in part (a)?

12. **Visits to the Emergency Room** The following data represent the gender and age of 489 randomly selected patients who visited the emergency room with an injury-related emergency:

| Gender | Age | | | | | |
|---|---|---|---|---|---|---|
| | Under 15 | 15–24 | 25–44 | 45–64 | 65–74 | 75 and Older |
| Male | 66 | 56 | 92 | 37 | 8 | 11 |
| Female | 44 | 39 | 66 | 34 | 12 | 24 |

(a) Test the claim that the age of an individual visiting the emergency room is independent of gender at the $\alpha = 0.1$ level of significance.
(b) Compute the $P$-value for this test by finding the area under the chi-square distribution to the right of the test statistic.
(c) Construct a conditional distribution by age and draw a bar graph. Does this evidence support your conclusion in part (a)?

13. **Smoked Lately?** Suppose a researcher wants to investigate whether the proportion of smokers within different age groups is the same. He divides the American population into four age groups: 18–29 years old, 30–49 years old, 50–64 years old, and 65 years or older. Within each age group, he surveys 80 individuals and asks, "Have you smoked at least one cigarette in the past week?" The results of the survey are as follows:

| Smoking Status | Age | | | |
|---|---|---|---|---|
| | 18–29 Years Old | 30–49 Years Old | 50–64 Years Old | 65 Years or Older |
| Smoked at least one cigarette in past week | 24 | 21 | 23 | 12 |
| Did not smoke at least one cigarette in past week | 56 | 59 | 57 | 68 |

*Source:* Based on data obtained from Gallup Organization

(a) Is there evidence to indicate that the proportion of individuals within each age group who have smoked at least one cigarette in the past week is different at the $\alpha = 0.05$ level of significance?

(b) Compute the *P*-value for this test by finding the area under the chi-square distribution to the right of the test statistic.

(c) Construct a conditional distribution by age and draw a bar graph. Does this evidence support your conclusion in part (a)?

14. **Are You Satisfied?** Suppose an economist wants to gauge the level of satisfaction of Americans. He randomly samples 150 people 18 years old or older from four geographic regions of the United States: East, South, Midwest, and West. He asks the individuals selected, "Are you satisfied or dissatisfied with the way things are going in the United States at this time?" The following are the results of the survey (data are based upon results obtained from the Gallup Organization):

| Satisfaction | Region | | | |
|---|---|---|---|---|
| | East | South | Midwest | West |
| Satisfied | 77 | 84 | 93 | 83 |
| Dissatisfied | 73 | 66 | 57 | 67 |

(a) Test whether the proportion of Americans who are satisfied with the way things are going in the United States for each region of the country is equal at the $\alpha = 0.1$ level of significance.

(b) Compute the *P*-value for this test by finding the area under the chi-square distribution to the right of the test statistic.

(c) Construct a conditional distribution by region of the country and draw a bar graph. Does this evidence support your conclusion in part (a)?

15. **Celebrex** Celebrex is a drug manufactured by Pfizer, Inc., that is indicated to be used to relieve symptoms associated with osteoarthritis and rheumatoid arthritis in adults. It is considered to be one of the nonsteroidal anti-inflammatory drugs. These types of drugs are known to be associated with gastrointestinal toxicity, such as bleeding, ulceration, and perforation of the stomach, small intestine, or large intestine. In clinical trials of the medication, researchers wanted to learn whether the proportion of subjects taking Celebrex who experienced these side effects differed significantly from that in other treatment groups. The following data were collected (Naproxen is a nonsteroidal anti-inflammatory drug that is also used in the treatment of arthritis):

| Side Effect | Treatment | | | | |
|---|---|---|---|---|---|
| | Placebo | Celebrex (50 mg per day) | Celebrex (100 mg per day) | Celebrex (200 mg per day) | Naproxen (500 mg per day) |
| Experienced Gastroduodenal Ulcers | 5 | 8 | 7 | 13 | 34 |
| Did Not Experience Gastroduodenal Ulcers | 212 | 225 | 220 | 208 | 176 |

*Source:* Pfizer, Inc.

(a) Test whether the proportion of subjects within each treatment group is the same at the $\alpha = 0.01$ level of significance.

(b) Compute the *P*-value for this test by finding the area under the chi-square distribution to the right of the test statistic.

(c) Construct a conditional distribution by treatment and draw a bar graph. Does this evidence support your conclusion in part (a)?

16. **Celebrex** Celebrex is a drug manufactured by Pfizer, Inc., that is indicated to be used to relieve symptoms associated with osteoarthritis and rheumatoid arthritis in adults. It is considered to

| Side Effect | Drug | | | | |
|---|---|---|---|---|---|
| | Celebrex | Placebo | Naproxen | Diclofenac | Ibuprofen |
| Dizziness | 83 | 32 | 36 | 5 | 8 |
| No Dizziness | 4063 | 1832 | 1330 | 382 | 337 |

*Source:* Pfizer, Inc.

be a nonsteroidal anti-inflammatory drug. In clinical trials of the medication, some of the subjects reported dizziness as a side effect. The researchers wanted to discover whether the proportion of subjects taking Celebrex who reported dizziness as a side effect differed significantly from that for other treatment groups. The following data were collected:

(a) Test whether the proportion of subjects within each treatment group who experienced dizziness is the same at the $\alpha = 0.01$ level of significance.

(b) Compute the *P*-value for this test by finding the area under the chi-square distribution to the right of the test statistic.

(c) Construct a conditional distribution by treatment and draw a bar graph. Does this evidence support your conclusion in part (a)?

17. **Dropping a Course** A survey of 50 randomly selected students who dropped a course in the current semester was conducted at a community college. The goal of the survey was to learn why students drop courses. The following data were collected: "Personal" drop reasons include financial, transportation, family issues, health issues, and lack of child care. "Course" drop reasons include reducing one's load, being unprepared for the course, the course was not what was expected, dissatisfaction with teaching, and not getting the desired grade. "Work" drop reasons include an increase in hours, a change in shift, and obtaining full-time employment. "Career" drop reasons include not needing the course and a change of plans. The results of the survey are as follows:

| Gender | Drop Reason | Gender | Drop Reason | Gender | Drop Reason | Gender | Drop Reason |
|---|---|---|---|---|---|---|---|
| Male | Personal | Male | Course | Male | Work | Female | Course |
| Female | Personal | Male | Course | Female | Course | Male | Work |
| Male | Work | Male | Work | Male | Work | Male | Course |
| Male | Personal | Female | Personal | Female | Course | Female | Work |
| Male | Course | Male | Course | Female | Course | Male | Personal |
| Male | Course | Female | Work | Female | Course | Male | Work |
| Female | Course | Male | Work | Male | Work | Female | Course |
| Female | Course | Male | Work | Male | Personal | Male | Course |
| Male | Course | Female | Course | Male | Course | Male | Personal |
| Female | Course | Female | Personal | Female | Course | Female | Course |
| Male | Personal | Female | Personal | Female | Course | Female | Work |
| Male | Work | Female | Personal | Male | Course | Male | Work |
| Male | Work | Male | Work | | | | |

(a) Construct a contingency table for the two variables.
(b) Test the claim that gender is independent of drop reason at the $\alpha = 0.1$ level of significance.
(c) Compute the $P$-value for this test by finding the area under the chi-square distribution to the right of the test statistic.
(d) Construct a conditional distribution by drop reason and draw a bar graph. Does this evidence support your conclusion in part (a)?

**18. Political Affiliation** A political scientist wanted to learn whether there is any association between the education level of a registered voter and his or her political party affiliation. He randomly selected 46 registered voters and obtained the following data:

| Education | Political Party | Education | Political Party | Education | Political Party |
| --- | --- | --- | --- | --- | --- |
| Grade School | Democrat | College | Republican | High School | Republican |
| College | Republican | Grade School | Republican | Grade School | Democrat |
| High School | Democrat | College | Republican | High School | Democrat |
| High School | Republican | High School | Democrat | College | Democrat |
| High School | Democrat | College | Democrat | College | Republican |
| Grade School | Democrat | College | Republican | High School | Republican |
| College | Republican | College | Democrat | College | Democrat |
| Grade School | Democrat | High School | Democrat | College | Democrat |
| High School | Democrat | College | Republican | High School | Democrat |
| High School | Democrat | College | Republican | College | Republican |
| Grade School | Democrat | Grade School | Democrat | College | Democrat |
| College | Republican | High School | Republican | High School | Republican |
| Grade School | Democrat | High School | Democrat | College | Republican |
| College | Democrat | High School | Democrat | High School | Republican |
| College | Democrat | College | Republican | College | Democrat |
| Grade School | Republican | | | | |

(a) Construct a contingency table for the two variables.
(b) Test the claim that level of education is independent of political affiliation at the $\alpha = 0.1$ level of significance.
(c) Compute the $P$-value for this test by finding the area under the chi-square distribution to the right of the test statistic.
(d) Construct a conditional distribution by level of education and draw a bar graph. Does this evidence support your conclusion in part (a)?

---

*In Problem 19, we demonstrate that the z-test for comparing two population proportions is equivalent to the chi-square test for homogeneity when there are two possible outcomes.*

**19. Percentage of Americans Who Smoke on the Decline?** On November 13–15, 2000, the Gallup Organization surveyed 1028 adults and found that 257 of them had smoked at least one cigarette in the past week. In 1990, they also asked 1028 adults the same question and determined that 278 adults had smoked at least one cigarette in the past week. The results are presented in the following table:

| Smoking Status | Year | |
| --- | --- | --- |
| | 1990 | 2000 |
| Smoked | 278 | 257 |
| Did Not Smoke | 750 | 771 |

(a) Compute the expected number of adult Americans who have smoked at least one cigarette in the past week and the expected number who have not assuming $P_{1990} = P_{2000}$.
(b) Compute the chi-square test statistic.
(c) Test the researchers' claim at the $\alpha = 0.05$ level of significance, using the chi-square goodness-of-fit test.
(d) Compute the $z$-test statistic. Now compute $z^2$. Compare $z^2$ with the chi-square test statistic. Conclude that $z^2 = \chi^2$.

## Technology Step-by-Step
Chi-Square Tests

**TI-83 Plus**

*Step 1:* Access the MATRX menu. Highlight the EDIT menu, and select 1: [A].

*Step 2:* Enter the number of rows and columns of the matrix.

*Step 3:* Enter the cell entries for the matrix, and press 2nd QUIT.

*Step 4:* Press STAT, highlight the TESTS menu, and select
C: $\chi^2$-Test....

*Step 5:* With the cursor after the **Observed:**, enter matrix [A] by accessing the MATRX menu, highlighting NAMES, and selecting 1: [A].

*Step 6:* With the cursor after the **Expected:**, enter matrix [B] by accessing the MATRX menu, highlighting NAMES, and selecting 2: [B].

*Step 7:* Highlight **Calculate** or **Draw**, and press ENTER.

**MINITAB**

*Step 1:* Enter the data into the MINITAB spreadsheet.

*Step 2:* Select the <u>S</u>tat menu, highlight **<u>T</u>ables**, and select **Chi-Square Test....**

*Step 3:* Select the columns that contain the data, and press OK.

**Excel**

*Step 1:* Enter the observed frequencies in the spreadsheet.

*Step 2:* Compute the expected frequencies and enter them in a different location in the spreadsheet.

*Step 3:* Select *fx* from the tool bar. Select **Statistical** for the function category and highlight CHITEST in the function name.

*Step 4:* With the cursor in the actual cell, highlight the observed data. With the cursor in the expected cell, highlight the expected frequencies. Click OK. The output provided is the *P*-value.

*Note:* This test can also be performed by using the PHStat add-in. See the Excel Technology Manual.

---

## 11.3 Inference about the Least-Squares Regression Model

**Preparing for This Section**  Before getting started, review the following:

✓ Scatter diagrams; correlation (Section 4.1, pp. 149–155)

✓ Least-squares regression (Section 4.2, pp. 163–168)

✓ The Coefficient of Determination (Section 4.3, pp. 176–179)

✓ Sample standard deviation (Section 3.2, pp. 105–106)

✓ Testing a hypothesis about $\mu$, $\sigma$ unknown (Section 9.3, pp. 408–413)

✓ *P*-values (Section 9.2, pp. 397–401)

✓ Confidence intervals about a mean (Section 8.1, pp. 340–345; Section 8.2, pp. 360–362)

**Objectives**  Understand the requirements of the least-squares regression model

 Compute the standard error of the estimate

 Verify that residuals are normally distributed

 Test the claim that a linear relation exists between two variables

 Compute a confidence interval about the slope of the least-squares regression model

As a quick review of the topics discussed in Chapter 4, we present the following example.

▶ **EXAMPLE 1** **Least-Squares Regression**

*Problem:* A family doctor is interested in examining the relationship between a patient's age and total cholesterol. He randomly selects 14 of his female patients and obtains the data presented in Table 18. The data are based upon results obtained from the National Center for Health Statistics.

| TABLE 18 | | | |
|---|---|---|---|
| Age | Total Cholesterol | Age | Total Cholesterol |
| 25 | 180 | 42 | 183 |
| 25 | 195 | 48 | 204 |
| 28 | 186 | 51 | 221 |
| 32 | 180 | 51 | 243 |
| 32 | 210 | 58 | 208 |
| 32 | 197 | 62 | 228 |
| 38 | 239 | 65 | 269 |

Draw a scatter diagram, compute the correlation coefficient, and find the least-squares regression equation and the coefficient of determination.

*Approach:* We will use a TI-83 Plus graphing calculator to obtain the information requested.

*Solution:* Figure 17 displays the scatter diagram. Figure 18 displays the output obtained from the calculator. The linear correlation coefficient is 0.718. The least-squares regression equation for this data is $\hat{y} = 151.3537 + 1.3991 x$, where $\hat{y}$ represents the predicted total cholesterol for a female whose age is $x$. The coefficient of determination, $R^2$, is 0.515. So 51.5% of the variation in total cholesterol is explained by the regression line. Figure 19 shows a graph of the least-squares regression equation on the scatter diagram in order to get a "feel" for the fit.

Figure 17

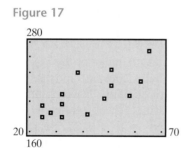

Figure 18

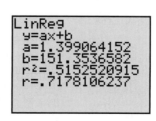

Figure 19

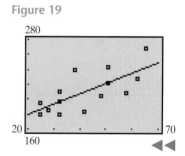

◀◀

The information obtained in Example 1 is descriptive in nature. Notice that the descriptions are both graphical (as in the scatter diagram) and numerical (as in the correlation coefficient and coefficient of determination).

① **The Least-Squares Regression Model**

*In Your Own Words*

Because $b_0$ and $b_1$ are statistics, they have sampling distributions.

In the least-squares regression equation $\hat{y} = b_0 + b_1 x$, the values for the slope, $b_1$, and $y$-intercept, $b_0$, are statistics, just as the sample mean, $\bar{x}$, and sample standard deviation, $s$, are statistics. The statistics $b_0$ and $b_1$ are estimates for the population $y$-intercept, $\beta_0$, and the population slope, $\beta_1$. The true linear relation between the predictor variable, $x$, and the response variable, $y$, is given by $y = \beta_0 + \beta_1 x$. Because $b_0$ and $b_1$ are statistics, their values vary from sample to sample, so there is a sampling distribution associated with each of them. We use this sampling distribution to conduct inference on $b_0$ and $b_1$. For example, we might want to test the claim that $\beta_1$ is different from 0. If we have evidence that supports this claim, we conclude that there is a linear relation between the predictor variable, $x$, and response variable, $y$.

To test hypotheses about $\beta_0$ or $\beta_1$, we need to know the sampling distributions of $b_0$ and $b_1$. To find these sampling distributions, we must make some assumptions about the population from which the bivariate data $(x_i, y_i)$ were sampled. Just as we did in Section 7.5 when we discussed the sampling distribution of $\bar{x}$, we start by asking what would happen if we took many samples for a given value of the predictor variable, $x$. For example, in looking back at Table 18, we notice that our sample included three women aged 32 years, so $x$ has the same value, 32, for all three women in our sample, but the corresponding values of $y$ for these three women are different: 180, 210, and 197. Thus, there is a distribution of total cholesterol levels for $x = 32$ years of age. Suppose we looked at *all* women aged 32 years. From these population data, we could find the population mean total cholesterol for 32-year-old women, denoted $\mu_{32}$. We could repeat this process for any other age. In general, different ages will have different population mean total cholesterols. This brings us to our first assumption regarding inference on the least-squares regression model.

**In Your Own Words**

*When doing inference on the least-squares regression model, we assume (1) that for any predictor variable, x, the mean of the response variable, y, depends on the value of x through a linear equation, and (2) that the response variable, y, is normally distributed with a constant standard deviation, σ. The mean increases/decreases at a constant rate depending on the slope, while the variance remains constant.*

**Assumption #1 for Inference on the Least-Squares Regression Model**

For any particular value of the predictor variable $x$ (such as 32 in Example 1), the corresponding responses in the population have a mean that depends linearly on $x$. That is,

$$\mu_x = \beta_0 + \beta_1 x$$

for some numbers $\beta_0$ and $\beta_1$, where $\mu_x$ represents the population mean response when the predictor variable is $x$.

We must also make an assumption regarding the distribution of the response variable for any particular value of the predictor variable.

**Assumption #2 for Inference on the Least-Squares Regression Model**

The response variables are normally distributed with mean $\mu_x = \beta_0 + \beta_1 x$ and standard deviation $\sigma$.

This assumption states that the mean of the response variable changes linearly, but the variance remains constant and the distribution of the response variable is normal. For example, if we obtained a sample of many 32-year-old females and measured their total cholesterol, the distribution would be normal with mean $\mu_{32} = \beta_0 + \beta_1(32)$ and standard deviation $\sigma$. If we obtained a sample of many 43-year-old females and measured their total cholesterol, the distribution would be normal with mean $\mu_{43} = \beta_0 + \beta_1(43)$ and standard deviation $\sigma$. See Figure 20.

Figure 20

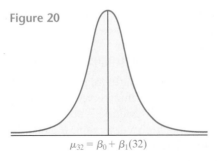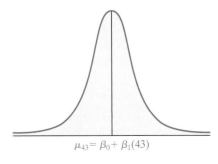

$\mu_{32} = \beta_0 + \beta_1(32)$ $\qquad\qquad$ $\mu_{43} = \beta_0 + \beta_1(43)$

A large value of $\sigma$ would indicate that the data are widely dispersed about the regression line and a small $\sigma$ would indicate that the data lie fairly close to the regression line. Figure 21 illustrates the ideas just presented. The regression line represents the mean value of each normal distribution at a specified value of $x$. The standard deviation of each distribution is $\sigma$.

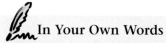

**In Your Own Words**

The larger $\sigma$ is, the more "spread out" the data are around the regression line.

**Figure 21**

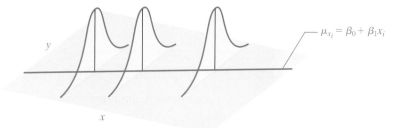

Of course, not all of the observed values of the response variable will lie on the true regression line, $\mu_{x_i} = \beta_0 + \beta_1 x_i$. The difference between the observed and predicted value of the response variable is an error term or residual, $\varepsilon_i$. We now present the least-squares regression model.

*Definition*

**The Least-Squares Regression Model**

The least-squares regression model is given by

$$y_i = \beta_0 + \beta_1 x_i + \varepsilon_i \qquad (1)$$

where
> $y_i$ is the value of the response variable for the $i$th individual,
> $\beta_0$ and $\beta_1$ are the parameters to be estimated based upon sample data,
> $x_i$ is the value of the predictor variable for the $i$th individual,
> $\varepsilon_i$ is a random error term with mean 0 and variance $\sigma_{\varepsilon_i}^2 = \sigma^2$. The error terms are independent, and
> $i = 1 \ldots n$ where $n$ is the sample size (number of ordered pairs in the data set).

NW  *Now Work Problem 7(a).*

Because the expected value or mean of $y_i$ is $\beta_0 + \beta_1 x_i$ and the expression on the left side of Equation (1) equals the expression on the right side, the expected value or mean of the error term, $\varepsilon_i$, is 0.

② **The Standard Error**

In Section 4.2, we learned how to obtain unbiased estimates for $\beta_0$ and $\beta_1$. We now present the method for obtaining the unbiased estimate of $\sigma$, the standard deviation of the response variable $y$ for any given value of $x$. The unbiased estimator of $\sigma$ is called the *standard error of the estimate*.

Remember the formula for the sample standard deviation presented in Section 3.2?

$$s = \sqrt{\frac{\sum (x_i - \bar{x})^2}{n - 1}}$$

We compute the deviations about the mean, square them, add up the deviations squared, and divide by $n - 1$. We divide by $n - 1$ because we lost one degree of freedom since one parameter, $\bar{x}$, was estimated. Exactly the same logic is used to compute the standard error of the estimate.

As we mentioned, the predicted values of $y$, $\hat{y}_i$, represent the mean value of the response variable for any given value of the predictor variable, $x_i$. So $y_i - \hat{y}_i = \varepsilon_i$ represents the difference between the observed value, $y_i$, and the mean value, $\hat{y}_i$. This calculation leads to the formula for the standard error of the estimate.

The **standard error of the estimate**, $s_e$, is found using the formula

$$s_e = \sqrt{\frac{\sum(y_i - \hat{y}_i)^2}{n - 2}} = \sqrt{\frac{\sum \text{residuals}^2}{n - 2}} \tag{2}$$

The standard error of the estimate is usually abbreviated **standard error**. Notice that we divide by $n - 2$ because we have estimated two parameters, $\beta_0$ and $\beta_1$.

▶ **EXAMPLE 2**   **Computing the Standard Error**

*Problem:* Compute the standard error for the data in Table 18.

*Approach:* We use the following steps to compute the standard error:

**Step 1:** Find the least-squares regression line.
**Step 2:** Obtain predicted values for each of the observations in the data set.
**Step 3:** Compute the residuals for each of the observations in the data set.
**Step 4:** Compute $\sum \text{residuals}^2$.
**Step 5:** Compute the standard error, using Formula (2).

*Solution:*

**Step 1:** The least squares regression line was found in Example 1.
**Step 2:** Column 3 of Table 19 represents the predicted values for each of the $n = 14$ observations.
**Step 3:** Column 4 of Table 19 represents the residuals for each of the 14 observations.

| | **TABLE 19** | | | |
|---|---|---|---|---|
| **Age, x** | **Total Cholesterol, y** | $\hat{y} = 1.3991x + 151.3537$ | **Residuals** | **Residuals$^2$** |
| 25 | 180 | 186.33 | −6.33 | 40.0689 |
| 25 | 195 | 186.33 | 8.67 | 75.1689 |
| 28 | 186 | 190.53 | −4.53 | 20.5209 |
| 32 | 180 | 196.12 | −16.12 | 259.8544 |
| 32 | 210 | 196.12 | 13.88 | 192.6544 |
| 32 | 197 | 196.12 | 0.88 | 0.7744 |
| 38 | 239 | 204.52 | 34.48 | 1188.8704 |
| 42 | 183 | 210.12 | −27.12 | 735.4944 |
| 48 | 204 | 218.51 | −14.51 | 210.5401 |
| 51 | 221 | 222.71 | −1.71 | 2.9241 |
| 51 | 243 | 222.71 | 20.29 | 411.6841 |
| 58 | 208 | 232.50 | −24.50 | 600.25 |
| 62 | 228 | 238.10 | −10.10 | 102.01 |
| 65 | 269 | 242.30 | 26.70 | 712.89 |

$$\sum \text{residuals}^2 = 4553.708$$

**Step 4:** Column 5 of Table 19 contains the squared residuals. We sum the entries in column 5 to obtain the sum of squared errors. So,

$$\sum \text{residuals}^2 = 4553.705$$

**Step 5:** We use Formula (2) to compute the standard error:

**Caution**

Be sure to divide by $n - 2$ when computing the standard error.

$$s_e = \sqrt{\frac{\sum \text{residuals}^2}{n - 2}} = \sqrt{\frac{4553.705}{14 - 2}} = 19.48 \qquad \blacktriangleleft\blacktriangleleft$$

Using Technology: Statistical spreadsheets, statistical software, and graphing calculators with advanced statistical features have the ability to compute the standard error. Figure 22 displays partial output obtained from Excel. The standard error is highlighted.

Figure 22

| Regression Statistics | |
|---|---|
| Multiple R | 0.7178106 |
| R Square | 0.5152521 |
| Adjusted R Square | 0.4748564 |
| Standard Error | 19.480535 |
| Observations | 14 |

**NW** *Now Work Problem 7(b).*

**3** ## The Normality of the Residuals

**Caution**

The residuals must be normally distributed to perform inference on the least-squares regression line.

For the least-squares regression model $y_i = \beta_0 + \beta_1 x_i + \varepsilon_i$, we require that the predictor variable, $y_i$, be normally distributed. Because $\beta_0 + \beta_1 x_i$ is constant for any $x_i$, the requirement that $y_i$ is normal implies that the residuals, $\varepsilon_i$ must also be normal. In order to perform statistical inference on the regression, it must be the case that the residuals are normally distributed. This requirement is easily verified through a normal probability plot.

▶ **EXAMPLE 3**  **Verifying that the Residuals Are Normally Distributed**

*Problem:* Verify that the residuals obtained in Table 19 from Example 2 are normally distributed.

*Approach:* We construct a normal probability plot to assess normality. If the normal probability plot is roughly linear, the residuals are said to be normal.

*Solution:* Figure 23 contains the normal probability plot obtained from Minitab.

Figure 23

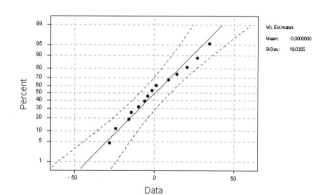

Normal Probability Plot for Residuals

Because all the points lie within the bands created by Minitab, the residuals are normally distributed. We can perform inference on the least-squares regression equation.  ◀◀

**NW** *Now Work Problem 7(c).*

### Inference on the Slope and *y*-Intercept

At this point, we know how to estimate the *y*-intercept and slope of the least-squares regression model. In addition, we can compute the standard error, $s_e$, which is an estimate of $\sigma$, the standard deviation of the response variable about the true least-squares regression model, and we know how to assess the normality of the residuals. We will now use this information to test the hypothesis of no linear relation between the predictor and the response variable.

Here is the question that we would like to answer: Do the sample data provide sufficient evidence to support the claim that a linear relation exists between the two variables? If there is no linear relation between the response and predictor variables, then the slope of the true regression line will be zero. Do you know why? A slope of zero would mean that information about the predictor variable, *x*, does not change my "guess" as to the value of the response variable, *y*.

Using the notation of hypothesis testing, we can perform one of three tests:

| Two-Tailed | Left-Tailed | Right-Tailed |
| --- | --- | --- |
| $H_o: \beta_1 = 0$ | $H_o: \beta_1 = 0$ | $H_o: \beta_1 = 0$ |
| $H_1: \beta_1 \neq 0$ | $H_1: \beta_1 < 0$ | $H_1: \beta_1 > 0$ |

**Caution**

Before testing $H_0: \beta_1 = 0$, be sure to draw a residual plot to verify that a linear model is appropriate.

In the two-tailed test, we are testing the claim that a linear relation exists between two variables without regard to the sign of the slope. In the left-tailed test, we are testing the claim that the slope of the true regression line is negative. In the right-tailed test, we are testing that the claim the slope of the true regression line is positive.

In order to test any one of these hypotheses, we need to know the sampling distribution of $b_1$. It turns out that

$$\frac{b_1 - \beta_1}{s_e / \sqrt{\sum (x_i - \overline{x})^2}} = \frac{b_1 - \beta_1}{s_{b_1}}$$

follows Student's *t*-distribution with $n - 2$ degrees of freedom, where *n* is the number of observations, $b_1$ is the unbiased estimator of the hypothesized value of the slope of the true regression line $\beta_1$, and $s_{b_1}$ is the sample standard deviation of $b_1$.

**Hypothesis Test Regarding the Slope Coefficient, $\beta_1$**

In order to test the claim that two quantitative variables are linearly related, we use the steps that follow, provided that
1. the sample is obtained using random sampling and
2. the residuals are normally distributed with constant error variance. Although there are methods for verifying the requirement of constant error variance, we shall not discuss them here. For the problems presented in this text, the requirement of constant error variance is satisfied.

*Step 1:* A claim is made regarding the linear relation between a response variable, *y*, and a predictor variable, *x*. The claim is used to determine the null and alternative hypotheses. The hypotheses can be structured in one of three ways,

| Two-Tailed | Left-Tailed | Right-Tailed |
| --- | --- | --- |
| $H_o: \beta_1 = 0$ | $H_o: \beta_1 = 0$ | $H_o: \beta_1 = 0$ |
| $H_1: \beta_1 \neq 0$ | $H_1: \beta_1 < 0$ | $H_1: \beta_1 > 0$ |

**Step 2:** Choose a level of significance $\alpha$ based upon the seriousness of making a Type I error. The level of significance is used to determine the critical value, using $n - 2$ degrees of freedom. The shaded region(s) represents the critical region.

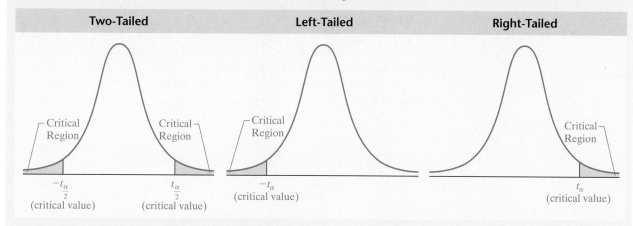

| Two-Tailed | Left-Tailed | Right-Tailed |

**Step 3:** Compute the **test statistic** $t = \dfrac{b_1 - \beta_1}{s_{b_1}} = \dfrac{b_1}{s_{b_1}},$* which follows Student's $t$-distribution with $n - 2$ degrees of freedom.

**Step 4:** Compare the critical value with the test statistic.

| Two-Tailed | Left-Tailed | Right-Tailed |
| --- | --- | --- |
| If $t < -t_{\alpha/2}$ or $t > t_{\alpha/2}$, reject the null hypothesis. | If $t < -t_\alpha$, reject the null hypothesis. | If $t > t_\alpha$, reject the null hypothesis. |

**Step 5:** State the conclusion.

The procedures just presented are **robust,** which means that minor departures from normality will not adversely affect the results of the test. In fact, for large sample sizes ($n \geq 30$), inferential procedures regarding $b_1$ can be used even with significant departures from normality.

▶ **EXAMPLE 4** **Testing for a Linear Relation**

*Problem:* Test the claim that there is a linear relation between age and total cholesterol at the $\alpha = 0.05$ level of significance, using the data given in Table 18 on page 505.

*Approach:* We verify that the requirements to perform the inference are satisfied. We then follow Steps 1–5.

*Solution:* Back in Example 1, we were told that the individuals were randomly selected. In Example 3, we confirmed that the residuals were normally distributed by constructing a normal probability plot.

We can now follow Steps 1–5 to test the claim.

**Step 1:** We are testing the claim that there is no linear relation between age and total cholesterol. Therefore, we are testing

$$H_0: \beta_1 = 0 \qquad \text{versus} \qquad H_1: \beta_1 \neq 0$$

*Remember, when computing the test statistic, we assume the null hypothesis to be true; that is, we assume that $\beta_1 = 0$ in computing the test statistic.

**Figure 24**

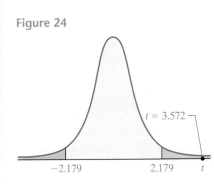

$t = 3.572$

$-2.179$   $2.179$   $t$

**Step 2:** We are testing the claim at the $\alpha = 0.05$ level of significance. There are $n = 14$ observations in Table 18, so we determine the critical value with $n - 2 = 14 - 2 = 12$ degrees of freedom. Because this is a two-tailed test, the critical values are $-t_{0.025} = -2.179$ and $t_{0.025} = 2.179$, each determined using 12 degrees of freedom. The critical regions are displayed in Figure 24.

**Step 3:** We obtained an unbiased estimate of $\beta_1$ in Example 1, and we computed the standard error, $s_e$, in Example 2. To determine the standard deviation of $b_1$, we need to compute $\sum(x_i - \bar{x})^2$, where the $x_i$ are the values of the predictor variable, age, and $\bar{x}$ is the sample mean. We compute this value in Table 20.

| TABLE 20 | | | |
|---|---|---|---|
| Age, $x$ | $\bar{x}$ | $x_i - \bar{x}$ | $(x_i - \bar{x})^2$ |
| 25 | 42.07143 | −17.07143 | 291.4337 |
| 25 | 42.07143 | −17.07143 | 291.4337 |
| 28 | 42.07143 | −14.07143 | 198.0051 |
| 32 | 42.07143 | −10.07143 | 101.4337 |
| 32 | 42.07143 | −10.07143 | 101.4337 |
| 32 | 42.07143 | −10.07143 | 101.4337 |
| 38 | 42.07143 | −4.07143 | 16.5765 |
| 42 | 42.07143 | −0.07143 | 0.0051 |
| 48 | 42.07143 | 5.92857 | 35.1479 |
| 51 | 42.07143 | 8.92857 | 79.7194 |
| 51 | 42.07143 | 8.92857 | 79.7194 |
| 58 | 42.07143 | 15.92857 | 253.7193 |
| 62 | 42.07143 | 19.92857 | 397.1479 |
| 65 | 42.07143 | 22.92857 | 525.7193 |

$$\sum(x_i - \bar{x})^2 = 2472.9284$$

We have

$$s_{b_1} = \frac{s_e}{\sqrt{\sum(x_i - \bar{x})^2}} = \frac{19.48}{\sqrt{2472.9284}} = 0.3917$$

The test statistic is

$$t = \frac{b_1}{s_{b_1}} = \frac{1.3991}{0.3917} = 3.572$$

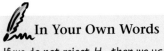

**In Your Own Words**

If we do not reject $H_o$, then we use the sample mean of $y$ to predict the value of the response for any value of the predictor.

**Step 4:** Because the test statistic, $t = 3.572$, is greater than the critical value, $t_{0.025} = 2.179$, we reject the null hypothesis. The test statistic is labeled in Figure 24.

**Step 5:** There is sufficient evidence to support the claim of a linear relation between age and total cholesterol. ◄◄

**Using Technology:** All of the results in Examples 1–4 can be obtained from statistical software or a graphing calculator with advanced statistical capabilities. Figure 25(a) shows the results obtained from Minitab, Figure 25(b) shows the results obtained from Excel, and Figure 25(c) shows the results obtained from a TI-83 Plus graphing calculator.

**Figure 25** **Regression Analysis**

```
The regression equation is
total cholesterol = 151 + 1.40 age

Predictor        Coef      StDev       T       P
Constant       151.35      17.28     8.76   0.000
Age             1.3991     0.3917    3.57   0.004

S = 19.48          R-Sq = 51.5%     R-Sq(adj) = 47.5%

Analysis of Variance

Source            Df        SS        MS       F       P
Regression         1     4840.5    4840.5   12.76   0.004
Residual Error    12     4553.9     379.5
Total             13     9394.4
```

(a) Minitab Output

### SUMMARY OUTPUT

| *Regression Statisitics* | |
|---|---|
| Multiple R | 0.7178106 |
| R Square | 0.5152521 |
| Adjusted R Square | 0.4748564 |
| Standard Error | 19.480535 |
| Observations | 14 |

### ANOVA

| | df | SS | MS | F |
|---|---|---|---|---|
| Regression | 1 | 4840.462 | 4840.462 | 12.75514 |
| Residual | 12 | 4553.895 | 379.4912 | |
| Total | 13 | 9394.357 | | |

| | Coefficients | Standard Error | t Stat | P-value |
|---|---|---|---|---|
| Intercept | 151.35366 | 17.28376 | 8.756987 | 1.47E-06 |
| Age | 1.3990642 | 0.391737 | 3.571433 | 0.003842 |

(b) Excel Output

```
LinRegTTest
y=a+bx
ß≠0 and ρ≠0
t=3.57143321
p=.0038422614
df=12
↓a=151.3536582
```

```
LinRegTTest
y=a+bx
ß≠0 and ρ≠0
↑b=1.399064152
s=19.48053511
r²=.5152520915
r=.7178106237
```

(c) TI-83 Output

(f) If the residuals are normally distributed, construct a 99% confidence interval about the slope of the true least-squares regression line.

(g) What is the mean rate of return for Cisco Systems stock if the rate of return of the S&P 500 is 4.2 percent?

**12. Fat-free Mass versus Energy Expenditure** In an effort to measure the dependence of energy expenditure on body build, researchers used underwater weighing techniques to determine the fat-free body mass in seven men. In addition, they measured the total 24-hour energy expenditure during inactivity. The results are as follows:

| Fat-free Mass (kg) | Energy Expenditure (Kcal) |
|---|---|
| 49.3 | 1,894 |
| 59.3 | 2,050 |
| 68.3 | 2,353 |
| 48.1 | 1,838 |
| 57.6 | 1,948 |
| 78.1 | 2,528 |
| 76.1 | 2,568 |

*Source:* Webb, P. (1981) Energy expenditure and fat-free mass in men and women. *American Journal of Clinical Nutrition,* **34**, 1816–1826

Use the results from Problem 16 in Section 4.2 to answer the following questions:

(a) What are the unbiased estimates of $\beta_0$ and $\beta_1$?

(b) Compute the standard error, the point estimate for $\sigma$.

(c) Determine whether the residuals are normally distributed.

(d) If the residuals are normally distributed, determine $s_{b_1}$.

(e) If the residuals are normally distributed, test the claim that a linear relation exists between the predictor variable, $x$, and response variable, $y$, at the $\alpha = 0.01$ level of significance.

(f) If the residuals are normally distributed, construct a 99% confidence interval about the slope of the true least-squares regression line.

(g) What is the mean energy expenditure of a man if his fat-free mass is 57.3 kg?

**13. Calories versus Sugar** The following data represent the number of calories per serving and the number of grams of sugar per serving for a random sample of high-fiber cereals:

| Calories, x | Sugar, y | Calories, x | Sugar, y |
|---|---|---|---|
| 200 | 18 | 210 | 23 |
| 210 | 23 | 210 | 16 |
| 170 | 17 | 210 | 17 |
| 190 | 20 | 190 | 12 |
| 200 | 18 | 190 | 11 |
| 180 | 19 | 200 | 11 |

*Source:* Consumer Reports, October, 1999

(a) Draw a scatter diagram of the data, treating calories as the predictor variable. What type of relation, if any, appears to exist between calories and sugar?

(b) Determine the least-squares regression equation from the sample data.

(c) Compute the standard error, $s_e$.

(d) Determine whether the residuals are normally distributed.

(e) Determine $s_{b_1}$.

(f) If the residuals are normally distributed, test the claim that a linear relation exists between calories and sugar content at the $\alpha = 0.01$ level of significance.

(g) If the residuals are normally distributed, construct a 95% confidence interval about the slope of the true least-squares regression line.

(h) Suppose a high-fiber cereal is randomly selected. Would you recommend using the least-squares regression line obtained in part (b) to predict the sugar content of the cereal? Why? What would be a good guess as to the sugar content of the cereal?

**14.** Output obtained from Minitab to the right.

(a) The least-squares regression equation is $\hat{y} = 12.396 + 1.3962\,x$. What is the predicted value of $y$ at $x = 10$?

(b) What is the mean of $y$ at $x = 10$?

(c) The standard error, $s_e$, is 2.167. What is the standard deviation of $y$ at $x = 10$?

(d) If the assumptions required for inference on the least-squares regression model are satisfied, what is the distribution of $y$ at $x = 10$?

```
The regression equation is
y = 12.4 + 1.40 x

Predictor      Coef      StDev       T        P
Constant      12.396     1.381     8.97    0.000
x              1.3962    0.1245   11.21    0.000

S = 2.167       R-Sq = 91.3%       R-Sq(adj) = 90.6%
```

## Technology Step-by-Step
Testing the Least-Squares Regression Model

**TI-83 Plus**  **Step 1:** Enter the predictor variable in L1 and the response variable in L2.

**Step 2:** Press STAT, highlight TESTS and select `E:LinRegTTest` ...

**Step 3:** Be sure that Xlist is L1 and Ylist is L2. Make sure that Freq: is set to 1. Select the direction of the alternative hypothesis. Place the cursor on Calculate and press ENTER.

**MINITAB**  **Step 1:** With the predictor variable in C1 and the response variable in C2, select the **Stat** menu and highlight **Regression**. Highlight **Regression...**

**Step 2:** Select the predictor and response variables and click OK.

**Excel**  **Step 1:** Make sure the Data Analysis Tool Pak is activated by selecting the **Tools** menu and highlighting **Add-Ins...** Check the box for the Analysis ToolPak and click OK.

**Step 2:** Enter the predictor variable in column A and the response variable in column B.

**Step 3:** Select the **Tools** menu and highlight **DataAnalysis...**

**Step 4:** Select the **Regression** option.

**Step 5:** With the cursor in the Y-range cell, highlight the column that contains the response variable. With the cursor in the X-range cell, highlight the column that contains the predictor variable. Click OK.

## 11.4  Confidence and Prediction Intervals

**Preparing for This Section**  Before getting started, review the following:

✓ Confidence intervals (Section 8.1, pp. 340–345; Section 8.2, pp. 360–362)

**Objectives**  ① Construct confidence intervals about a predicted value

② Construct prediction intervals about a predicted value

① We know how to obtain the least-squares regression equation of best fit from data. We also know how to use the least-squares regression equation to obtain a predicted value. For example, the least-squares regression equation for the cholesterol data introduced in Example 1 from Section 11.3 was

$$\hat{y} = 151.3537 + 1.3991\,x$$

where $\hat{y}$ represents the predicted total cholesterol for a female whose age is $x$. The predicted value of total cholesterol for a given age $x$ actually has two interpretations:

1. It represents the mean total cholesterol for all females whose age is $x$.
2. It represents the predicted total cholesterol for a randomly selected female whose age is $x$.

So if we let $x = 42$ in the least-squares regression equation $\hat{y} = 151.3537 + 1.3991\,x$, we would obtain $\hat{y} = 151.3537 + 1.3991(42) = 210.1$. We can interpret this result in one of two ways:

1. The mean total cholesterol for all females 42 years old is 210.1.
2. Our best guess as to the total cholesterol for a randomly selected 42-year-old female is 210.1.

Of course, there is a margin of error in making predictions, so we construct intervals about any predicted value in order to describe the accuracy of the prediction. The type of interval constructed is going to depend upon whether we are predicting a mean total cholesterol for all 42-year-old females or the total cholesterol for an individual 42-year-old female. In other words, the margin of error is going to be different for predicting the mean total cholesterol for all females who are 42 years old versus the total cholesterol for one individual. Which prediction (the mean or the individual) do you think will be more accurate? It seems logical that the distribution of means should have less variability (and therefore a lower margin of error) than the distribution of individuals. After all, in the distribution of means, high total cholesterols can be offset by low total cholesterols.

*Definition*

**Confidence intervals** are intervals constructed about the predicted value of $y$, at a given level of $x$, that are used to measure the accuracy of the mean response of all the individuals in the population.

**Prediction intervals** are intervals constructed about the predicted value of $y$ that are used to measure the accuracy of a single individual's predicted value.

✍ **In Your Own Words**

Confidence intervals are intervals for the mean of the population. Prediction intervals are intervals for an individual from the population.

If we use the least-squares regression equation to predict the mean total cholesterol for all 42-year-old females, we construct a confidence interval. If we use the least-squares regression equation to predict the total cholesterol for a single 42-year-old female, we construct a prediction interval.

## Confidence Intervals

The structure of a confidence interval is the same as it was in Section 8.1. The interval is of the form

$$\text{Point Estimate} \pm \text{Margin of Error}$$

The following formula can be used to construct a confidence interval about $\hat{y}$:

*Theorem*

**Confidence Interval about the Mean Response of $y$, $\hat{y}$**

A $(1 - \alpha) \cdot 100\%$ confidence interval for $\hat{y}$, the mean response of $y$, is given by

$$\text{Lower Bound: } \hat{y} - t_{\alpha/2} \cdot s_e \sqrt{\frac{1}{n} + \frac{(x^* - \bar{x})^2}{\sum (x_i - \bar{x})^2}}$$

$$\text{Upper Bound: } \hat{y} + t_{\alpha/2} \cdot s_e \sqrt{\frac{1}{n} + \frac{(x^* - \bar{x})^2}{\sum (x_i - \bar{x})^2}} \tag{1}$$

where $x^*$ is the given value of the predictor variable, $n$ is the number of observations, and $t_{\alpha/2}$ is the critical value with $n - 2$ degrees of freedom.

▶ **EXAMPLE 1**   **Constructing a Confidence Interval about the Mean Predicted Value**

*Problem:* Construct a 95% confidence interval about the predicted mean total cholesterol of all 42-year-old females, using the data in Table 18 on page 505.

*Approach:* We need to determine the predicted mean total cholesterol at $x^* = 42$, $s_e$, $\sum (x_i - \bar{x})^2$, $\bar{x}$ and find the critical $t$ value to use Formula (1).

*Solution:* The least-squares regression equation is $\hat{y} = 151.3537 + 1.3991\,x$. To find the predicted mean total cholesterol of all 42-year-olds, let

$x = x^* = 42$ in the regression equation and obtain $\hat{y} = 151.3537 + 1.3991\,(42) = 210.1$. From Example 2 in Section 11.3, we found that $s_e = 19.48$, and from Example 4 in Section 11.3, we found that $\sum(x_i - \overline{x})^2 = 2472.9284$ and $\overline{x} = 42.07143$. We find $t_{\alpha/2} = t_{0.025}$ with $n - 2 = 14 - 2 = 12$ degrees of freedom to be 2.179. The 95% confidence interval about the predicted mean total cholesterol for 42-year-old females is therefore

$$\text{Lower Bound: } \hat{y} - t_{\alpha/2} \cdot s_e \cdot \sqrt{\frac{1}{n} + \frac{(x^* - \overline{x})^2}{\sum(x_i - \overline{x})^2}} = 210.1 - 2.179 \cdot 19.48 \cdot \sqrt{\frac{1}{14} + \frac{(42 - 42.07143)^2}{2472.9284}} = 198.8$$

$$\text{Upper Bound: } \hat{y} + t_{\alpha/2} \cdot s_e \cdot \sqrt{\frac{1}{n} + \frac{(x^* - \overline{x})^2}{\sum(x_i - \overline{x})^2}} = 210.1 + 2.179 \cdot 19.48 \cdot \sqrt{\frac{1}{14} + \frac{(42 - 42.07143)^2}{2472.9284}} = 221.4$$

We are 95% confident that the mean total cholesterol of all 42-year-old females is somewhere between 198.8 and 221.4.   ◄◄

**NW** *Now Work Problems 1(a) and 1(b).*

## ② Prediction Intervals

*In Your Own Words*

Prediction intervals are wider than confidence intervals because it is tougher to "guess" the value of an individual than the mean of a population.

The procedure for obtaining a prediction interval is identical to that for finding a confidence interval. The only difference is the standard error. There is more variability associated with individuals than with means. Therefore, the computation of the interval must account for this increased variability. Again, the form of the interval is

$$\text{Point Estimate} \pm \text{Margin of Error}$$

The following formula can be used to construct a prediction interval about $\hat{y}$:

*Theorem*

**Prediction Interval about $\hat{y}$**

A $(1 - \alpha) \cdot 100\%$ prediction interval for $\hat{y}$, the individual response of $y$, is given by

$$\text{Lower Bound: } \hat{y} - t_{\alpha/2} \cdot s_e \sqrt{1 + \frac{1}{n} + \frac{(x^* - \overline{x})^2}{\sum(x_i - \overline{x})^2}}$$

$$\text{Upper Bound: } \hat{y} + t_{\alpha/2} \cdot s_e \sqrt{1 + \frac{1}{n} + \frac{(x^* - \overline{x})^2}{\sum(x_i - \overline{x})^2}} \tag{2}$$

where $x^*$ is the given value of the predictor variable, $n$ is the number of observations, and $t_{\alpha/2}$ is the critical value with $n - 2$ degrees of freedom.

► **EXAMPLE 2   Constructing a Prediction Interval about a Predicted Value**

*Problem:* Construct a 95% prediction interval about the predicted total cholesterol of a 42-year-old female.

*Approach:* We need to determine the predicted total cholesterol at $x^* = 42$, $s_e$, $\sum(x_i - \overline{x})^2$, $\overline{x}$ and find the critical $t$ value using the data in Table 18 on page 505.

*Solution:* The least-squares regression equation is $\hat{y} = 151.3537 + 1.3991\,x$. To find the predicted total cholesterol of a 42-year-old, let $x = x^* = 42$ in the regression equation and obtain $\hat{y} = 151.3537 + 1.3991\,(42) = 210.1$. From Example 2 in Section 11.3, we found that $s_e = 19.48$; from Example 4 in Section 11.3, we found that $\sum(x_i - \overline{x})^2 = 2472.9284$ and $\overline{x} = 42.07143$. We find $t_{\alpha/2} = t_{0.025}$ with $n - 2 = 14 - 2 = 12$ degrees of freedom to be 2.179.

The 95% confidence interval about the predicted total cholesterol for a 42-year-old female is

$$\text{Lower Bound: } \hat{y} - t_{\alpha/2} \cdot s_e \sqrt{1 + \frac{1}{n} + \frac{(x^* - \overline{x})^2}{\Sigma(x_i - \overline{x})^2}} =$$

$$210.1 - 2.179 \cdot 19.48 \cdot \sqrt{1 + \frac{1}{14} + \frac{(42 - 42.07143)^2}{2472.9284}} = 166.2$$

$$\text{Upper Bound: } \hat{y} + t_{\alpha/2} \cdot s_e \sqrt{1 + \frac{1}{n} + \frac{(x^* - \overline{x})^2}{\Sigma(x_i - \overline{x})^2}} =$$

$$210.1 + 2.179 \cdot 19.48 \cdot \sqrt{1 + \frac{1}{14} + \frac{(42 - 42.07143)^2}{2472.9284}} = 254.0$$

We are 95% confident that the total cholesterol of a randomly selected 42-year-old female is somewhere between 166.2 and 254.0.  ◄◄

**NW** *Now Work Problems 1(c) and 1(d).*

Notice that the interval about the individual (prediction interval) is wider than the interval about the mean (confidence interval). The reason for this should be clear: There is more variability associated with individuals than with groups of individuals. That is, it is more difficult to predict a single 42-year-old female's total cholesterol than it is to predict the mean total cholesterol of all 42-year-old females.

**Using Technology:** Statistical software packages will provide confidence and prediction intervals as part of their output if the user requests them. Figure 26 shows the results of Examples 1 and 2, using Minitab.

**Figure 26**    Predicted Values

```
          Fit    StDev Fit          95.0% CI              95.0% PI
       210.11         5.21    ( 198.77, 221.46)    ( 166.18, 254.05)
```

## 11.4 Assess Your Understanding

### Concepts and Vocabulary

1. Explain the difference between a confidence interval and prediction interval.
2. Suppose a normal probability plot of residuals indicates that the requirement of normally distributed residuals is violated. Explain the circumstances under which confidence and prediction intervals could still be constructed.

## Exercises

### • Skill Building

*In Problems 1–4, use the results of Problems 3–6 in Section 4.2 and Problems 1–4 in Section 11.3.*

**1.** Using the sample data from Problem 1 in Section 11.3,
   (a) Predict the mean value of $y$ if $x = 7$.
   (b) Construct a 95% confidence interval about the mean value of $y$ if $x = 7$.
   (c) Predict the value of $y$ if $x = 7$.
   (d) Construct a 95% prediction interval about the value of $y$ if $x = 7$.
   (e) Explain the difference between the prediction in part (a) and part (c).

**2.** Using the sample data from Problem 2 in Section 11.3,
   (a) Predict the mean value of $y$ if $x = 8$.
   (b) Construct a 95% confidence interval about the mean value of $y$ if $x = 8$.
   (c) Predict the value of $y$ if $x = 8$.
   (d) Construct a 95% prediction interval about the value of $y$ if $x = 8$.
   (e) Explain the difference between the prediction in part (a) and part (c).

**3.** Using the sample data from Problem 3 in Section 11.3,
   (a) Predict the mean value of $y$ if $x = 1.4$.
   (b) Construct a 95% confidence interval about the mean value of $y$ if $x = 1.4$.
   (c) Predict the value of $y$ if $x = 1.4$.
   (d) Construct a 95% prediction interval about the value of $y$ if $x = 1.4$.

**4.** Using the sample data from Problem 4 in Section 11.3,
   (a) Predict the mean value of $y$ if $x = 1.8$.
   (b) Construct a 90% confidence interval about the mean value of $y$ if $x = 1.8$.
   (c) Predict the value of $y$ if $x = 1.8$.
   (d) Construct a 90% prediction interval about the value of $y$ if $x = 1.8$.

### • Applying the Concepts

**5. Height versus Head Circumference** Use the results of Problem 7 from Section 11.3 to answer the following questions:
   (a) Predict the mean head circumference of children who are 25.75 inches tall.
   (b) Construct a 95% confidence interval about the mean head circumference of children who are 25.75 inches tall.
   (c) Predict the head circumference of a randomly selected child who is 25.75 inches tall.
   (d) Construct a 95% prediction interval about the head circumference of a child who is 25.75 inches tall.
   (e) Explain the difference between the prediction in part (a) and the prediction in part (c).

**6. Concrete** Use the results of Problem 8 from Section 11.3 to answer the following questions:
   (a) Predict the mean 28-day strength of concrete whose 7-day strength is 2550 psi.
   (b) Construct a 95% confidence interval about the mean 28-day strength of concrete whose 7-day strength is 2550 psi.
   (c) Predict the 28-day strength of a cylinder of concrete whose 7-day strength is 2550 psi.
   (d) Construct a 95% prediction interval about the 28-day strength of concrete whose 7-day strength is 2550 psi.
   (e) Explain the difference between the prediction in part (a) and the prediction in part (c).

**7. Bone Length** Use the results of Problem 9 in Section 11.3 to answer the following questions:
   (a) Predict the mean length of the right tibia of all rats whose right humerus is 25.83 mm.
   (b) Construct a 95% confidence interval about the mean length found in part (a).
   (c) Predict the length of the right tibia of a randomly selected rat whose right humerus is 25.83 mm.
   (d) Construct a 95% prediction interval about the length found in part (c).
   (e) Explain why the predicted lengths found in parts (a) and (c) are the same, yet the intervals are different.

**8. Tar and Nicotine** Use the results of Problem 10 in Section 11.3 to answer the following questions:
   (a) Predict the mean nicotine content of all cigarettes whose tar content is 12 mg.
   (b) Construct a 95% confidence interval about the tar content found in part (a).
   (c) Predict the nicotine content of a randomly selected cigarette whose tar content is 12 mg.
   (d) Construct a 95% prediction interval about the nicotine content found in part (c).
   (e) Explain why the predicted nicotine contents found in parts (a) and (c) are the same, yet the intervals are different.

**9. Cisco Systems versus the S&P 500** Use the results of Problem 11 in Section 11.3 to answer the following questions:
   (a) What is the mean rate of return for Cisco Systems stock if the rate of return of the S&P 500 is 4.2 percent?
   (b) Construct a 90% confidence interval about the mean rate of return found in part (a).
   (c) Predict the rate of return on Cisco Systems stock if the rate of return on the S&P 500 for a randomly selected month is 4.2 percent.
   (d) Construct a 90% prediction interval about the rate of return found in part (c).
   (e) Explain why the predicted rates of return found in parts (a) and (c) are the same, yet the intervals are different.

10. **Fat-free Mass versus Energy Expenditure** Use the results of Problem 12 in Section 11.3 to answer the following questions:

   (a) What is the mean energy expenditure for individuals whose fat-free mass is 57.3 kg?

   (b) Construct a 99% confidence interval about the mean energy expenditure found in part (a).

   (c) Predict the energy expenditure of a randomly selected individual whose fat-free mass is 57.3 kg.

   (d) Construct a 99% prediction interval about the energy expenditure found in part (c).

   (e) Explain why the predicted energy expenditures found in parts (a) and (c) are the same, yet the intervals are different.

11. **Calories versus Sugar** Use the results of Problem 13 from Section 11.3 to answer the following:

   (a) Explain why it does not make sense to construct confidence or prediction intervals based upon the least-squares regression equation.

   (b) Construct a 95% confidence interval for the mean sugar content of high-fiber cereals.

---

## Technology Step-by-Step
### Confidence and Prediction Intervals

**TI-83 Plus**  The TI-83 Plus does not compute confidence or prediction intervals.

**MINITAB**  **Step 1:** With the predictor variable in C1 and the response variable in C2, select the **Stat** menu, highlight **Regression**. Highlight **Regression . . . .**

**Step 2:** Select the predictor and response variables.

**Step 3:** Click the Options . . . button.

**Step 4:** In the cell marked "Prediction intervals for new observations:", enter the value of $x^*$. Select a confidence level. Click OK twice.

**Excel**  **Step 1:** Load the PhStat Add-in.

**Step 2:** Enter the values of the predictor variable in column A and the corresponding values of the response variable in column B.

**Step 3:** Select the **PHStat** menu. Highlight **Regression**. Highlight **Simple Linear Regression**.

**Step 4:** With the cursor in the Y variable cell range, highlight the data in column B. With the cursor in the X variable cell range, highlight the data in column A. Select Confidence & Prediction Interval for X and enter the value of $x^*$. Choose a level of confidence. Click OK.

---

# CHAPTER 11 REVIEW

## Summary

In this chapter, we introduced chi-square methods. The first chi-square method involved tests for goodness of fit. We used the chi-square distribution to test the claim that a random variable followed a certain distribution. This is done by comparing values expected based upon the distribution of the random variable to observed values.

Next, we introduced chi-square methods that allowed us to perform tests for independence and homogeneity. In a test for independence, the researcher obtains random data for two variables and tests whether the variables are associated. The null hypothesis in these tests is always that the

variables are not associated (i.e., independent). The test statistic compares the values expected if the variables were independent to those observed. If the expected and observed values differ significantly, we reject the null hypothesis and conclude that there is evidence to support the belief that the variables are dependent (i.e., they are associated). We draw bar graphs of the marginal distributions, to help us "see" the association, if any.

The last chi-square test was the test for homogeneity of proportions. This test is similar to the test for independence, except we are testing that the proportion of individuals in the study with a certain characteristic are equal (i.e., $p_1 = p_2 = \ldots = p_k$). To perform this test, we take random samples of a predetermined size from each group under consideration (i.e., a random sample of size $n_1$ for group 1, a random sample of size $n_2$ for group 2, and so on).

The last two sections of this chapter dealt with inferential techniques that can be used on the least-squares regression model $y_i = \beta_0 + \beta_1 x_i + \varepsilon_i$. In this model, we use sample data in order to obtain estimates of an intercept and slope. These estimates are unbiased estimators of the unknown population parameters. The residuals are required to

be normally distributed, with mean 0 and constant variance $\sigma^2$. The residuals are independent as well. We verify the assumptions of normality through a normal probability plot of the residuals. Provided that these requirements are satisfied, we can test hypotheses regarding the slope to determine whether the relation between the predictor and response variable is linear. In addition, we can construct confidence and prediction intervals about a predicted value. We construct confidence intervals about a mean response and prediction intervals about an individual response.

## Formulas

**Expected Counts**

$E_i = \mu_i = np_i$ for $i = 1, 2, \ldots, k$

**Chi-Square Test Statistic**

$\chi^2 = \sum \dfrac{(O_i - E_i)^2}{E_i}$   $i = 1, 2, \ldots, k$

**Expected Frequencies**

Expected frequency $= \dfrac{(\text{row total})(\text{column total})}{\text{table total}}$

**Standard error of the estimate**

$s_e = \sqrt{\dfrac{\sum(y_i - \hat{y}_i)^2}{n - 2}} = \sqrt{\dfrac{\sum \text{residuals}^2}{n - 2}}$

**Standard deviation of $b_1$**

$s_{b_1} = \dfrac{s_e}{\sqrt{\sum(x_i - \overline{x})^2}}$

**Confidence Intervals for the Slope of the Regression Line**

A $(1 - \alpha) \cdot 100\%$ confidence interval for the slope of the true regression line, $\beta_1$, is given by the following formulas:

Lower Bound: $b_1 - t_{\alpha/2} \cdot \dfrac{s_e}{\sqrt{\sum(x_i - \overline{x})^2}} = b_1 - t_{\alpha/2} \cdot s_{b_1}$   Upper Bound: $b_1 + t_{\alpha/2} \cdot \dfrac{s_e}{\sqrt{\sum(x_i - \overline{x})^2}} = b_1 + t_{\alpha/2} \cdot s_{b_1}$

Here, $t_{\alpha/2}$ is computed with $n - 2$ degrees of freedom.

**Confidence Interval about the Mean Response of $y$, $\hat{y}$**

A $(1 - \alpha) \cdot 100\%$ confidence interval for the mean response of $y$, $\hat{y}$, is given by the following formulas:

Lower Bound: $\hat{y} - t_{\alpha/2} \cdot s_e \sqrt{\dfrac{1}{n} + \dfrac{(x^* - \overline{x})^2}{\sum(x_i - \overline{x})^2}}$   Upper Bound: $\hat{y} + t_{\alpha/2} \cdot s_e \sqrt{\dfrac{1}{n} + \dfrac{(x^* - \overline{x})^2}{\sum(x_i - \overline{x})^2}}$

Here, $x^*$ is the given value of the predictor variable and $t_{\alpha/2}$ is the critical value with $n - 2$ degrees of freedom.

**Prediction Interval about $\hat{y}$**

A $(1 - \alpha) \cdot 100\%$ prediction interval for the individual response of $y$, $\hat{y}$, is given by

Lower Bound: $\hat{y} - t_{\alpha/2} \cdot s_e \sqrt{1 + \dfrac{1}{n} + \dfrac{(x^* - \overline{x})^2}{\sum(x_i - \overline{x})^2}}$   Upper Bound: $\hat{y} + t_{\alpha/2} \cdot s_e \sqrt{1 + \dfrac{1}{n} + \dfrac{(x^* - \overline{x})^2}{\sum(x_i - \overline{x})^2}}$

where $x^*$ is the given value of the predictor variable and $t_{\alpha/2}$ is the critical value with $n - 2$ degrees of freedom.

## Vocabulary

Goodness-of-fit test (p. 477)
Expected counts (p. 478)
Contingency (or two-way) table (p. 488)
Row variable (p. 488)
Column variable (p. 488)

Cell (p. 488)
Chi-square independence test (p. 488)
Conditional distribution (p. 493)
Chi-square test for homogeneity of proportions (p. 495)
Least-squares regression model (p. 507)

Standard error of the estimate (p. 508)
Robust (p. 511)
Bivariate normal distribution (p. 515)
Confidence interval (p. 520)
Prediction interval (p. 521)

## Objectives

## Review Exercises

**1. Roulette Wheel** A pit boss suspects that a roulette wheel is out of balance. A roulette wheel has 18 black slots, 18 red slots, and 2 green slots. The pit boss spins the wheel 500 times and records the following frequencies:

| Outcome | Frequency |
|---|---|
| Black | 233 |
| Red | 237 |
| Green | 30 |

Test the claim that the wheel is out of balance at the $\alpha = 0.05$ level of significance.

**2. Fair Dice?** A pit boss is concerned that a pair of dice being used in a craps game is not fair. The distribution of the expected sum of two fair dice is as follows:

| Sum of Two Die | Probability | Sum of Two Die | Probability |
|---|---|---|---|
| 2 | 1/36 | 8 | 5/36 |
| 3 | 2/36 | 9 | 4/36 |
| 4 | 3/36 | 10 | 3/36 |
| 5 | 4/36 | 11 | 2/36 |
| 6 | 5/36 | 12 | 1/36 |
| 7 | 6/36 | | |

The pit boss rolls the dice 400 times and records the sum of the dice. The following are the results:

| Sum of Two Die | Frequency | Sum of Two Die | Frequency |
|---|---|---|---|
| 2 | 16 | 8 | 59 |
| 3 | 23 | 9 | 45 |
| 4 | 31 | 10 | 34 |
| 5 | 41 | 11 | 19 |
| 6 | 62 | 12 | 11 |
| 7 | 59 | | |

Test the claim that the dice are fair at the $\alpha = 0.01$ level of significance.

**3. Educational Attainment** A researcher wanted to test the claim that the distribution of educational attainment of Americans today is different from the distribution in 1994. The distribution of educational attainment in 1994 is as follows:

| Education | Relative Frequency |
|---|---|
| Not a High School Graduate | 0.191 |
| High School Graduate | 0.344 |
| Some College | 0.174 |
| Associate's Degree | 0.070 |
| Bachelor's Degree | 0.147 |
| Advanced Degree | 0.075 |

*Source:* Statistical Abstract of the United States

To test the claim, the researcher randomly selects 500 Americans, learns their levels of education, and obtains the following data:

| Education | Frequency |
|---|---|
| Not a High School Graduate | 89 |
| High School Graduate | 152 |
| Some College | 83 |
| Associate's Degree | 39 |
| Bachelor's Degree | 93 |
| Advanced Degree | 44 |

At the $\alpha = 0.1$ level of significance, test the claim that the distribution of educational attainment today is different from that of educational attainment in 1994.

**4. School Violence** A school administrator is concerned that the distribution of school violence has changed and become more violent than it was in 1992. The distribution of school crime for 1992 is as follows:

| Crime | Theft | Violent | Serious Violent |
|---|---|---|---|
| Relative Frequency | 0.619 | 0.314 | 0.067 |

*Source:* National Center for Educational Statistics

To test the claim, the researcher randomly selects 800 school crimes, finds out whether they were theft, violent, or serious violent crimes, and obtains the following data:

| Crime | Theft | Violent | Serious Violent |
|---|---|---|---|
| Relative Frequency | 421 | 311 | 68 |

At the $\alpha = 0.05$ level of significance, test the claim that the distribution of school crime today is different from the 1992 distribution.

**5. Evolution or Creation?** The Gallup Organization conducted a poll of 1016 randomly selected Americans aged 18 years old or older in February, 2001 and 1017 randomly selected Americans aged 18 years old or older in June, 1993 and asked them the following question:

Which of the following statements comes closest to your views on the origin and development of human beings? (1) Human beings have developed over millions of years from less advanced forms of life, but God guided this process. (2) Human beings have developed over millions of years from less advanced forms of life, and God had no part in this process. (3) God created human beings pretty much in their present form at one time within the last 10,000 years or so.

The results of the survey are as follows:

| Date | Belief | | | |
|---|---|---|---|---|
| | Humans Developed, with God Guiding | Humans Developed, But God Had No Part in Process | God Created Humans in Present Form | Other/No Opinion |
| February, 2001 | 376 | 122 | 457 | 61 |
| June, 1993 | 356 | 112 | 478 | 71 |

(a) Compute the expected values of each cell, under the assumption of independence.
(b) Verify that the requirements for performing a chi-square test of independence are satisfied.
(c) Compute the chi-square test statistic.
(d) Test the claim that people's opinion regarding human origin is independent of the date the question is asked at the $\alpha = 0.05$ level of significance.
(e) Compare the observed frequencies with the expected frequencies. Which cell contributed most to the test statistic? Was the expected frequency greater than or less than the observed frequency? What does this information tell you?
(f) Construct a conditional distribution by date and draw a bar graph. Does this evidence support your conclusion in part (d)?
(g) Compute the *P*-value for this test by finding the area under the chi-square distribution to the right of the test statistic.

**6. Caesarean Sections** An obstetrician wanted to discover whether the method of delivering a baby was independent of race. The following data represent the race of the mother and the method of delivery for 365 randomly selected births, based upon data obtained from the *Statistical Abstract of the United States*:

| Race | Delivery Method | |
|---|---|---|
| | Vaginal | Caesarean |
| White | 242 | 63 |
| Black | 47 | 13 |

(a) Compute the expected values of each cell, under the assumption of independence.

(b) Verify that the requirements to perform a chi-square test of independence are satisfied.
(c) Compute the chi-square test statistic.
(d) Test the claim that method of delivery is independent of the race of the mother at the $\alpha = 0.05$ level of significance.
(e) Compare the observed frequencies with the expected frequencies. Which cell contributed most to the test statistic? Was the expected frequency greater than or less than the observed frequency? What does this information tell you?
(f) Construct a conditional distribution by method of delivery and draw a bar graph. Does this evidence support your conclusion in part (d)?
(g) Compute the *P*-value for this test by finding the area under the chi-square distribution to the right of the test statistic.

**7. Race versus Region of the United States** A sociologist wanted to determine whether the locations in which individuals live are independent of their races. He randomly selects 2172 U.S. residents and asks them to disclose their race and the location in which they live. He obtains the following data:

| Race | Region | | | |
|---|---|---|---|---|
| | Northeast | Midwest | South | West |
| White | 421 | 520 | 656 | 400 |
| Black | 56 | 57 | 158 | 28 |
| American Indian, Eskimo, Aleut | 2 | 4 | 6 | 9 |
| Asian or Pacific Islander | 13 | 8 | 11 | 40 |
| Hispanic | 38 | 17 | 68 | 101 |
| Other | 17 | 8 | 24 | 50 |

*Source:* Statistical Abstract of the United States

At the $\alpha = 0.01$ level of significance, test the claim that race is independent of the location in which a resident lives within the United States.

**8. Marital Status and Gender** A sociologist wanted to determine whether marital status and gender were independent. He randomly sampled 201 residents of the United States who were 18 years old or older and asked them to disclose their gender and marital status. The following data were collected:

| Marital Status | Gender | |
|---|---|---|
| | Male | Female |
| Never Married | 26 | 22 |
| Married | 59 | 60 |
| Divorced | 9 | 11 |
| Widowed | 3 | 11 |

At the $\alpha = 0.05$ level of significance, test the claim that gender is independent of marital status.

**9. The Common Cold** A doctor wanted to determine whether the proportion of Americans who have had symptoms associated with the common cold is the same for all four regions of the United States. He randomly sampled 120 individuals from each region of the United States and obtained the following data (the data are based upon information obtained from the National Center for Health Statistics):

| Symptoms | Region | | | |
|---|---|---|---|---|
| | Northeast | Midwest | South | West |
| Symptoms Within Last Year | 26 | 31 | 23 | 35 |
| No Symptoms Within Last Year | 94 | 89 | 97 | 85 |

At the $\alpha = 0.05$ level of significance, test the claim that the proportion of Americans who have had symptoms associated with the common cold is the same for all four regions of the United States.

**10. Hardworking?** In a *Newsweek* poll conducted on June 24, 1999, 750 randomly selected adults were asked, "Do you believe Americans today are as willing to work hard at their jobs to get ahead as they were in the past, or are not as willing to work hard to get ahead?". The results of the survey are as follows:

| Response | Age | | |
|---|---|---|---|
| | 18–29 Years Old | 30–49 Years Old | 50 and Older |
| As Willing | 58 | 75 | 70 |
| Not as Willing | 192 | 175 | 180 |

*Source:* pollingreport.com

At the $\alpha = 0.05$ level of significance, test the claim that the proportion of Americans who believe that Americans are willing to work just as hard today is the same for all three age groups.

**11. Engine Displacement versus Fuel Economy** The following data represent the size of a car's engine (in liters) versus its miles per gallon in the city for a random sample of 2001 domestic automobiles:

| Car | Engine Displacement (in liters), $x$ | City Miles per Gallon, $y$ | Car | Engine Displacement (in liters), $x$ | City Miles per Gallon, $y$ |
|---|---|---|---|---|---|
| Buick Century | 3.1 | 20 | Ford Crown Victoria | 4.6 | 17 |
| Buick LeSabre | 3.8 | 19 | Ford Focus | 2.0 | 28 |
| Cadillac DeVille | 4.6 | 16 | Ford Mustang | 3.8 | 20 |
| Chevrolet Camaro | 3.8 | 19 | Oldsmobile Aurora | 3.5 | 19 |
| Chevrolet Cavalier | 2.2 | 24 | Pontiac Grand Am | 2.4 | 22 |
| Chevrolet Malibu | 3.1 | 23 | Pontiac Sunfire | 2.2 | 23 |
| Chrysler LHS | 3.5 | 18 | Saturn Coupe | 1.9 | 28 |
| Dodge Intrepid | 2.7 | 19 | | | |

*Source:* Road and Track Magazine

Using the results from Problems 1 and 5 from the chapter review of Chapter 4, answer the following questions:
(a) What are the unbiased estimates of $\beta_0$ and $\beta_1$? What is the mean number of miles per gallon of all cars that have a 3.8-liter engine?
(b) Compute the standard error, the point estimate for $\sigma$.
(c) Determine whether the residuals are normally distributed.
(d) If the residuals are normally distributed, determine $s_{b_1}$.
(e) If the residuals are normally distributed, test the claim that a linear relation exists between the predictor variable, $x$, and the response variable, $y$, at the $\alpha = 0.05$ level of significance.
(f) If the residuals are normally distributed, construct a 95% confidence interval about the slope of the true least-squares regression line.
(g) Construct a 90% confidence interval about the mean miles per gallon found in part (a).
(h) Predict the miles per gallon of a car with a 3.8-liter engine.
(i) Construct a 90% prediction interval about the miles per gallon found in part (h).
(j) Explain why the predicted miles per gallon found in parts (a) and (h) are the same, yet the intervals are different.

**12. Temperature versus Cricket Chirps** Crickets make a chirping noise by sliding their wings rapidly over each other. Perhaps you have noticed that the number of chirps seems to increase with the temperature. The following table lists the temperature (in Fahrenheit) and the number of chirps per second for the striped ground cricket:

| Temperature | Chirps per Second | Temperature | Chirps per Second |
|---|---|---|---|
| 88.6 | 20.0 | 71.6 | 16.0 |
| 93.3 | 19.8 | 84.3 | 18.4 |
| 80.6 | 17.1 | 75.2 | 15.5 |
| 69.7 | 14.7 | 82.0 | 17.1 |
| 69.4 | 15.4 | 83.3 | 16.2 |
| 79.6 | 15.0 | 82.6 | 17.2 |
| 80.6 | 16.0 | 83.5 | 17.0 |
| 76.3 | 14.4 | | |

*Source: The Songs of Insects*, Pierce, George W., Cambridge, Mass.: Harvard University Press, 1949, pp. 12–21

Using the results from Problems 2 and 6 from the chapter review of Chapter 4, answer the following questions:
(a) What are the unbiased estimates of $\beta_0$ and $\beta_1$? What is the mean number of chirps when the temperature is 80.2°F?
(b) Compute the standard error, the point estimate for $\sigma$.
(c) Determine whether the residuals are normally distributed.
(d) If the residuals are normally distributed, determine $s_{b_1}$.
(e) If the residuals are normally distributed, test the claim that a linear relation exists between the predictor variable, $x$, and response variable, $y$, at the $\alpha = 0.05$ level of significance.
(f) If the residuals are normally distributed, construct a 95% confidence interval about the slope of the true least-squares regression line.
(g) Construct a 90% confidence interval about the mean number of chirps found in part (a).
(h) Predict the number of chirps on a day when the temperature is 80.2°F.
(i) Construct a 90% prediction interval about the number of chirps found in part (h).
(j) Explain why the predicted number of chirps found in parts (a) and (h) are the same, yet the intervals are different.

**13. Apartments** The data below represent the square footage and rents for apartments in the western suburbs of Chicago.

| Square Footage, x | Rent per Month, y |
|---|---|
| 1161 | 1265 |
| 1000 | 1050 |
| 1034 | 915 |
| 910 | 1003 |
| 934 | 920 |
| 860 | 890 |
| 1056 | 1035 |
| 1000 | 1190 |
| 784 | 730 |
| 962 | 1005 |
| 897 | 940 |

*Source:* apartments.com

(a) What are the unbiased estimates of $\beta_0$ and $\beta_1$? What is the mean rent of a 900-square-foot apartment in the western suburbs of Chicago?
(b) Compute the standard error, the point estimate for $\sigma$.
(c) Determine whether the residuals are normally distributed.
(d) If the residuals are normally distributed, determine $s_{b_1}$.
(e) If the residuals are normally distributed, test the claim that a linear relation exists between the predictor variable, $x$, and response variable, $y$, at the $\alpha = 0.05$ level of significance.
(f) If the residuals are normally distributed, construct a 95% confidence interval about the slope of the true least-squares regression line.
(g) Construct a 90% confidence interval about the mean rent found in part (a).
(h) Predict the rent of a 900-square-foot apartment.
(i) Construct a 90% prediction interval about the rent found in part (h).
(j) Explain why the predicted rents found in parts (a) and (h) are the same, yet the intervals are different.

**14. Boys' Heights** The following data, based upon results obtained from the National Center for Health Statistics, represent the height (in inches) of boys between the ages of 2 and 10 years:

| Age | Boy Height | Age | Boy Height | Age | Boy Height |
|---|---|---|---|---|---|
| 2 | 36.1 | 5 | 45.6 | 8 | 48.3 |
| 2 | 34.2 | 5 | 44.8 | 8 | 50.9 |
| 2 | 31.1 | 5 | 44.6 | 9 | 52.2 |
| 3 | 36.3 | 6 | 49.8 | 9 | 51.3 |
| 3 | 39.5 | 7 | 43.2 | 10 | 55.6 |
| 4 | 41.5 | 7 | 47.9 | 10 | 59.5 |
| 4 | 38.6 | 8 | 51.4 | | |

(a) Treating age as the predictor variable, determine the unbiased estimates of $\beta_0$ and $\beta_1$. What is the mean height of a seven-year-old boy?
(b) Compute the standard error, the point estimate for $\sigma$.
(c) Determine whether the residuals are normally distributed.
(d) If the residuals are normally distributed, determine $s_{b_1}$.
(e) If the residuals are normally distributed, test the claim that a linear relation exists between the predictor variable, age, and response variable, height, at the $\alpha = 0.05$ level of significance.
(f) If the residuals are normally distributed, construct a 95% confidence interval about the slope of the true least-squares regression line.
(g) Construct a 90% confidence interval about the mean height found in part (a).
(h) Predict the height of a seven-year-old boy.
(i) Construct a 90% prediction interval about the height found in part (h).
(j) Explain why the predicted heights found in parts (a) and (h) are the same, yet the intervals are different.

**15. Grip Strength** A researcher believes that as age increases, the grip strength (in pounds per square inch) of an individual's dominant hand decreases. From a random sample of 17 females, he obtains the following data:

| Age | Grip Strength | Age | Grip Strength |
|---|---|---|---|
| 15 | 65 | 34 | 45 |
| 16 | 60 | 37 | 58 |
| 28 | 58 | 41 | 70 |
| 61 | 60 | 43 | 73 |
| 53 | 46 | 49 | 45 |
| 43 | 66 | 53 | 60 |
| 16 | 56 | 61 | 56 |
| 25 | 75 | 68 | 30 |
| 28 | 46 | | |

*Source:* Kevin McCarthy, Student at Joliet Junior College

(a) Treating age as the predictor variable, determine the unbiased estimates of $\beta_0$ and $\beta_1$.
(b) Compute the standard error, the point estimate for $\sigma$.
(c) Determine whether the residuals are normally distributed.
(d) If the residuals are normally distributed, determine $s_{b_1}$.
(e) If the residuals are normally distributed, test the claim that a linear relation exists between the predictor variable, age, and response variable, grip strength, at the $\alpha = 0.05$ level of significance.
(f) Based on your answers to (d) and (e), what would be a good guess as to the grip strength of a randomly selected 42-year-old female?

**16.** What is the least-squares regression model? What are the requirements of the data in order to perform inference on a least-squares regression line?

**Chapter 11 Projects located at www.prenhall.com/sullivanstats**

| | TABLE I | | | | | | | | |
|---|---|---|---|---|---|---|---|---|---|

Random Numbers

**Column Number**

| Row Number | 01–05 | 06–10 | 11–15 | 16–20 | 21–25 | 26–30 | 31–35 | 36–40 | 41–45 | 46–50 |
|---|---|---|---|---|---|---|---|---|---|---|
| 01 | 89392 | 23212 | 74483 | 36590 | 25956 | 36544 | 68518 | 40805 | 09980 | 00467 |
| 02 | 61458 | 17639 | 96252 | 95649 | 73727 | 33912 | 72896 | 66218 | 52341 | 97141 |
| 03 | 11452 | 74197 | 81962 | 48443 | 90360 | 26480 | 73231 | 37740 | 26628 | 44690 |
| 04 | 27575 | 04429 | 31308 | 02241 | 01698 | 19191 | 18948 | 78871 | 36030 | 23980 |
| 05 | 36829 | 59109 | 88976 | 46845 | 28329 | 47460 | 88944 | 08264 | 00843 | 84592 |
| 06 | 81902 | 93458 | 42161 | 26099 | 09419 | 89073 | 82849 | 09160 | 61845 | 40906 |
| 07 | 59761 | 55212 | 33360 | 68751 | 86737 | 79743 | 85262 | 31887 | 37879 | 17525 |
| 08 | 46827 | 25906 | 64708 | 20307 | 78423 | 15910 | 86548 | 08763 | 47050 | 18513 |
| 09 | 24040 | 66449 | 32353 | 83668 | 13874 | 86741 | 81312 | 54185 | 78824 | 00718 |
| 10 | 98144 | 96372 | 50277 | 15571 | 82261 | 66628 | 31457 | 00377 | 63423 | 55141 |
| 11 | 14228 | 17930 | 30118 | 00438 | 49666 | 65189 | 62869 | 31304 | 17117 | 71489 |
| 12 | 55366 | 51057 | 90065 | 14791 | 62426 | 02957 | 85518 | 28822 | 30588 | 32798 |
| 13 | 96101 | 30646 | 35526 | 90389 | 73634 | 79304 | 96635 | 6626 | 94683 | 16696 |
| 14 | 38152 | 55474 | 30153 | 26525 | 83647 | 31988 | 82182 | 98377 | 33802 | 80471 |
| 15 | 85007 | 18416 | 24661 | 95581 | 45868 | 15662 | 28906 | 36392 | 07617 | 50248 |
| 16 | 85544 | 15890 | 80011 | 18160 | 33468 | 84106 | 40603 | 01315 | 74664 | 20553 |
| 17 | 10446 | 20699 | 98370 | 17684 | 16932 | 80449 | 92654 | 02084 | 19985 | 59321 |
| 18 | 67237 | 45509 | 17638 | 65115 | 29757 | 80705 | 82686 | 48565 | 72612 | 61760 |
| 19 | 23026 | 89817 | 05403 | 82209 | 30573 | 47501 | 00135 | 33955 | 50250 | 72592 |
| 20 | 67411 | 58542 | 18678 | 46491 | 13219 | 84084 | 27783 | 34508 | 55158 | 78742 |

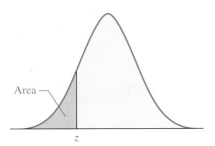

| | TABLE II | | | | | | | | |
|---|---|---|---|---|---|---|---|---|---|
| | Standard Normal Distribution | | | | | | | | |
| z | .00 | .01 | .02 | .03 | .04 | .05 | .06 | .07 | .08 | .09 |
| −3.4 | 0.0003 | 0.0003 | 0.0003 | 0.0003 | 0.0003 | 0.0003 | 0.0003 | 0.0003 | 0.0003 | 0.0002 |
| −3.3 | 0.0005 | 0.0005 | 0.0005 | 0.0004 | 0.0004 | 0.0004 | 0.0004 | 0.0004 | 0.0004 | 0.0003 |
| −3.2 | 0.0007 | 0.0007 | 0.0006 | 0.0006 | 0.0006 | 0.0006 | 0.0006 | 0.0005 | 0.0005 | 0.0005 |
| −3.1 | 0.0010 | 0.0009 | 0.0009 | 0.0009 | 0.0008 | 0.0008 | 0.0008 | 0.0008 | 0.0007 | 0.0007 |
| −3.0 | 0.0013 | 0.0013 | 0.0013 | 0.0012 | 0.0012 | 0.0011 | 0.0011 | 0.0011 | 0.0010 | 0.0010 |
| −2.9 | 0.0019 | 0.0018 | 0.0018 | 0.0017 | 0.0016 | 0.0016 | 0.0015 | 0.0015 | 0.0014 | 0.0014 |
| −2.8 | 0.0026 | 0.0025 | 0.0024 | 0.0023 | 0.0023 | 0.0022 | 0.0021 | 0.0021 | 0.0020 | 0.0019 |
| −2.7 | 0.0035 | 0.0034 | 0.0033 | 0.0032 | 0.0031 | 0.0030 | 0.0029 | 0.0028 | 0.0027 | 0.0026 |
| −2.6 | 0.0047 | 0.0045 | 0.0044 | 0.0043 | 0.0041 | 0.0040 | 0.0039 | 0.0038 | 0.0037 | 0.0036 |
| −2.5 | 0.0062 | 0.0060 | 0.0059 | 0.0057 | 0.0055 | 0.0054 | 0.0052 | 0.0051 | 0.0049 | 0.0048 |
| −2.4 | 0.0082 | 0.0080 | 0.0078 | 0.0075 | 0.0073 | 0.0071 | 0.0069 | 0.0068 | 0.0066 | 0.0064 |
| −2.3 | 0.0107 | 0.0104 | 0.0102 | 0.0099 | 0.0096 | 0.0094 | 0.0091 | 0.0089 | 0.0087 | 0.0084 |
| −2.2 | 0.0139 | 0.0136 | 0.0132 | 0.0129 | 0.0125 | 0.0122 | 0.0119 | 0.0116 | 0.0113 | 0.0110 |
| −2.1 | 0.0179 | 0.0174 | 0.0170 | 0.0166 | 0.0162 | 0.0158 | 0.0154 | 0.0150 | 0.0146 | 0.0143 |
| −2.0 | 0.0228 | 0.0222 | 0.0217 | 0.0212 | 0.0207 | 0.0202 | 0.0197 | 0.0192 | 0.0188 | 0.0183 |
| −1.9 | 0.0287 | 0.0281 | 0.0274 | 0.0268 | 0.0262 | 0.0256 | 0.0250 | 0.0244 | 0.0239 | 0.0233 |
| −1.8 | 0.0359 | 0.0351 | 0.0344 | 0.0336 | 0.0329 | 0.0322 | 0.0314 | 0.0307 | 0.0301 | 0.0294 |
| −1.7 | 0.0446 | 0.0436 | 0.0427 | 0.0418 | 0.0409 | 0.0401 | 0.0392 | 0.0384 | 0.0375 | 0.0367 |
| −1.6 | 0.0548 | 0.0537 | 0.0526 | 0.0516 | 0.0505 | 0.0495 | 0.0485 | 0.0475 | 0.0465 | 0.0455 |
| −1.5 | 0.0668 | 0.0655 | 0.0643 | 0.0630 | 0.0618 | 0.0606 | 0.0594 | 0.0582 | 0.0571 | 0.0559 |
| −1.4 | 0.0808 | 0.0793 | 0.0778 | 0.0764 | 0.0749 | 0.0735 | 0.0721 | 0.0708 | 0.0694 | 0.0681 |
| −1.3 | 0.0968 | 0.0951 | 0.0934 | 0.0918 | 0.0901 | 0.0885 | 0.0869 | 0.0853 | 0.0838 | 0.0823 |
| −1.2 | 0.1151 | 0.1131 | 0.1112 | 0.1093 | 0.1075 | 0.1056 | 0.1038 | 0.1020 | 0.1003 | 0.0985 |
| −1.1 | 0.1357 | 0.1335 | 0.1314 | 0.1292 | 0.1271 | 0.1251 | 0.1230 | 0.1210 | 0.1190 | 0.1170 |
| −1.0 | 0.1587 | 0.1562 | 0.1539 | 0.1515 | 0.1492 | 0.1469 | 0.1446 | 0.1423 | 0.1401 | 0.1379 |
| −0.9 | 0.1841 | 0.1814 | 0.1788 | 0.1762 | 0.1736 | 0.1711 | 0.1685 | 0.1660 | 0.1635 | 0.1611 |
| −0.8 | 0.2119 | 0.2090 | 0.2061 | 0.2033 | 0.2005 | 0.1977 | 0.1949 | 0.1922 | 0.1894 | 0.1867 |
| −0.7 | 0.2420 | 0.2389 | 0.2358 | 0.2327 | 0.2296 | 0.2266 | 0.2236 | 0.2206 | 0.2177 | 0.2148 |
| −0.6 | 0.2743 | 0.2709 | 0.2676 | 0.2643 | 0.2611 | 0.2578 | 0.2546 | 0.2514 | 0.2483 | 0.2451 |
| −0.5 | 0.3085 | 0.3050 | 0.3015 | 0.2981 | 0.2946 | 0.2912 | 0.2877 | 0.2843 | 0.2810 | 0.2776 |
| −0.4 | 0.3446 | 0.3409 | 0.3372 | 0.3336 | 0.3300 | 0.3264 | 0.3228 | 0.3192 | 0.3156 | 0.3121 |
| −0.3 | 0.3821 | 0.3783 | 0.3745 | 0.3707 | 0.3669 | 0.3632 | 0.3594 | 0.3557 | 0.3520 | 0.3483 |
| −0.2 | 0.4207 | 0.4168 | 0.4129 | 0.4090 | 0.4052 | 0.4013 | 0.3974 | 0.3936 | 0.3897 | 0.3859 |
| −0.1 | 0.4602 | 0.4562 | 0.4522 | 0.4483 | 0.4443 | 0.4404 | 0.4364 | 0.4325 | 0.4286 | 0.4247 |
| −0.0 | 0.5000 | 0.4960 | 0.4920 | 0.4880 | 0.4840 | 0.4801 | 0.4761 | 0.4721 | 0.4681 | 0.4641 |

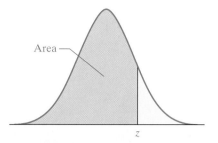

Area

z

| | TABLE II (continued) | | | | | | | | |
|---|---|---|---|---|---|---|---|---|---|

**Standard Normal Distribution**

| z | .00 | .01 | .02 | .03 | .04 | .05 | .06 | .07 | .08 | .09 |
|---|---|---|---|---|---|---|---|---|---|---|
| 0.0 | 0.5000 | 0.5040 | 0.5080 | 0.5120 | 0.5160 | 0.5199 | 0.5239 | 0.5279 | 0.5319 | 0.5359 |
| 0.1 | 0.5398 | 0.5438 | 0.5478 | 0.5517 | 0.5557 | 0.5596 | 0.5636 | 0.5675 | 0.5714 | 0.5753 |
| 0.2 | 0.5793 | 0.5832 | 0.5871 | 0.5910 | 0.5948 | 0.5987 | 0.6026 | 0.6064 | 0.6103 | 0.6141 |
| 0.3 | 0.6179 | 0.6217 | 0.6255 | 0.6293 | 0.6331 | 0.6368 | 0.6406 | 0.6443 | 0.6480 | 0.6517 |
| 0.4 | 0.6554 | 0.6591 | 0.6628 | 0.6664 | 0.6700 | 0.6736 | 0.6772 | 0.6808 | 0.6844 | 0.6879 |
| 0.5 | 0.6915 | 0.6950 | 0.6985 | 0.7019 | 0.7054 | 0.7088 | 0.7123 | 0.7157 | 0.7190 | 0.7224 |
| 0.6 | 0.7257 | 0.7291 | 0.7324 | 0.7357 | 0.7389 | 0.7422 | 0.7454 | 0.7486 | 0.7517 | 0.7549 |
| 0.7 | 0.7580 | 0.7611 | 0.7642 | 0.7673 | 0.7704 | 0.7734 | 0.7764 | 0.7794 | 0.7823 | 0.7852 |
| 0.8 | 0.7881 | 0.7910 | 0.7939 | 0.7967 | 0.7995 | 0.8023 | 0.8051 | 0.8078 | 0.8106 | 0.8133 |
| 0.9 | 0.8159 | 0.8186 | 0.8212 | 0.8238 | 0.8264 | 0.8289 | 0.8315 | 0.8340 | 0.8365 | 0.8389 |
| 1.0 | 0.8413 | 0.8438 | 0.8461 | 0.8485 | 0.8508 | 0.8531 | 0.8554 | 0.8577 | 0.8599 | 0.8621 |
| 1.1 | 0.8643 | 0.8665 | 0.8686 | 0.8708 | 0.8729 | 0.8749 | 0.8770 | 0.8790 | 0.8810 | 0.8830 |
| 1.2 | 0.8849 | 0.8869 | 0.8888 | 0.8907 | 0.8925 | 0.8944 | 0.8962 | 0.8980 | 0.8997 | 0.9015 |
| 1.3 | 0.9032 | 0.9049 | 0.9066 | 0.9082 | 0.9099 | 0.9115 | 0.9131 | 0.9147 | 0.9162 | 0.9177 |
| 1.4 | 0.9192 | 0.9207 | 0.9222 | 0.9236 | 0.9251 | 0.9265 | 0.9279 | 0.9292 | 0.9306 | 0.9319 |
| 1.5 | 0.9332 | 0.9345 | 0.9357 | 0.9370 | 0.9382 | 0.9394 | 0.9406 | 0.9418 | 0.9429 | 0.9441 |
| 1.6 | 0.9452 | 0.9463 | 0.9474 | 0.9484 | 0.9495 | 0.9505 | 0.9515 | 0.9525 | 0.9535 | 0.9545 |
| 1.7 | 0.9554 | 0.9564 | 0.9573 | 0.9582 | 0.9591 | 0.9599 | 0.9608 | 0.9616 | 0.9625 | 0.9633 |
| 1.8 | 0.9641 | 0.9649 | 0.9656 | 0.9664 | 0.9671 | 0.9678 | 0.9686 | 0.9693 | 0.9699 | 0.9706 |
| 1.9 | 0.9713 | 0.9719 | 0.9726 | 0.9732 | 0.9738 | 0.9744 | 0.9750 | 0.9756 | 0.9761 | 0.9767 |
| 2.0 | 0.9772 | 0.9778 | 0.9783 | 0.9788 | 0.9793 | 0.9798 | 0.9803 | 0.9808 | 0.9812 | 0.9817 |
| 2.1 | 0.9821 | 0.9826 | 0.9830 | 0.9834 | 0.9838 | 0.9842 | 0.9846 | 0.9850 | 0.9854 | 0.9857 |
| 2.2 | 0.9861 | 0.9864 | 0.9868 | 0.9871 | 0.9875 | 0.9878 | 0.9881 | 0.9884 | 0.9887 | 0.9890 |
| 2.3 | 0.9893 | 0.9896 | 0.9898 | 0.9901 | 0.9904 | 0.9906 | 0.9909 | 0.9911 | 0.9913 | 0.9916 |
| 2.4 | 0.9918 | 0.9920 | 0.9922 | 0.9925 | 0.9927 | 0.9929 | 0.9931 | 0.9932 | 0.9934 | 0.9936 |
| 2.5 | 0.9938 | 0.9940 | 0.9941 | 0.9943 | 0.9945 | 0.9946 | 0.9948 | 0.9949 | 0.9951 | 0.9952 |
| 2.6 | 0.9953 | 0.9955 | 0.9956 | 0.9957 | 0.9959 | 0.9960 | 0.9961 | 0.9962 | 0.9963 | 0.9964 |
| 2.7 | 0.9965 | 0.9966 | 0.9967 | 0.9968 | 0.9969 | 0.9970 | 0.9971 | 0.9972 | 0.9973 | 0.9974 |
| 2.8 | 0.9974 | 0.9975 | 0.9976 | 0.9977 | 0.9977 | 0.9978 | 0.9979 | 0.9979 | 0.9980 | 0.9981 |
| 2.9 | 0.9981 | 0.9982 | 0.9982 | 0.9983 | 0.9984 | 0.9984 | 0.9985 | 0.9985 | 0.9986 | 0.9986 |
| 3.0 | 0.9987 | 0.9987 | 0.9987 | 0.9988 | 0.9988 | 0.9989 | 0.9989 | 0.9989 | 0.9990 | 0.9990 |
| 3.1 | 0.9990 | 0.9991 | 0.9991 | 0.9991 | 0.9992 | 0.9992 | 0.9992 | 0.9992 | 0.9993 | 0.9993 |
| 3.2 | 0.9993 | 0.9993 | 0.9994 | 0.9994 | 0.9994 | 0.9994 | 0.9994 | 0.9995 | 0.9995 | 0.9995 |
| 3.3 | 0.9995 | 0.9995 | 0.9995 | 0.9996 | 0.9996 | 0.9996 | 0.9996 | 0.9996 | 0.9996 | 0.9997 |
| 3.4 | 0.9997 | 0.9997 | 0.9997 | 0.9997 | 0.9997 | 0.9997 | 0.9997 | 0.9997 | 0.9997 | 0.9998 |

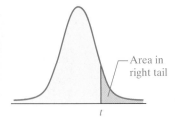
Area in right tail
t

| TABLE III | | | | | | | | | | | |
|---|---|---|---|---|---|---|---|---|---|---|---|

**t-Distribution**
**Area in Right Tail**

| df | 0.25 | 0.20 | 0.15 | 0.10 | 0.05 | 0.025 | 0.02 | 0.01 | 0.005 | 0.0025 | 0.001 | 0.0005 |
|---|---|---|---|---|---|---|---|---|---|---|---|---|
| 1 | 1.000 | 1.376 | 1.963 | 3.078 | 6.314 | 12.706 | 15.894 | 31.821 | 63.657 | 127.321 | 318.289 | 636.558 |
| 2 | 0.816 | 1.061 | 1.386 | 1.886 | 2.920 | 4.303 | 4.849 | 6.965 | 9.925 | 14.089 | 22.328 | 31.600 |
| 3 | 0.765 | 0.978 | 1.250 | 1.638 | 2.353 | 3.182 | 3.482 | 4.541 | 5.841 | 7.453 | 10.214 | 12.924 |
| 4 | 0.741 | 0.941 | 1.190 | 1.533 | 2.132 | 2.776 | 2.999 | 3.747 | 4.604 | 5.598 | 7.173 | 8.610 |
| 5 | 0.727 | 0.920 | 1.156 | 1.476 | 2.015 | 2.571 | 2.757 | 3.365 | 4.032 | 4.773 | 5.893 | 6.869 |
| 6 | 0.718 | 0.906 | 1.134 | 1.440 | 1.943 | 2.447 | 2.612 | 3.143 | 3.707 | 4.317 | 5.208 | 5.959 |
| 7 | 0.711 | 0.896 | 1.119 | 1.415 | 1.895 | 2.365 | 2.517 | 2.998 | 3.499 | 4.029 | 4.785 | 5.408 |
| 8 | 0.706 | 0.889 | 1.108 | 1.397 | 1.860 | 2.306 | 2.449 | 2.896 | 3.355 | 3.833 | 4.501 | 5.041 |
| 9 | 0.703 | 0.883 | 1.100 | 1.383 | 1.833 | 2.262 | 2.398 | 2.821 | 3.250 | 3.690 | 4.297 | 4.781 |
| 10 | 0.700 | 0.879 | 1.093 | 1.372 | 1.812 | 2.228 | 2.359 | 2.764 | 3.169 | 3.581 | 4.144 | 4.587 |
| 11 | 0.697 | 0.876 | 1.088 | 1.363 | 1.796 | 2.201 | 2.328 | 2.718 | 3.106 | 3.497 | 4.025 | 4.437 |
| 12 | 0.695 | 0.873 | 1.083 | 1.356 | 1.782 | 2.179 | 2.303 | 2.681 | 3.055 | 3.428 | 3.930 | 4.318 |
| 13 | 0.694 | 0.870 | 1.079 | 1.350 | 1.771 | 2.160 | 2.282 | 2.650 | 3.012 | 3.372 | 3.852 | 4.221 |
| 14 | 0.692 | 0.868 | 1.076 | 1.345 | 1.761 | 2.145 | 2.264 | 2.624 | 2.977 | 3.326 | 3.787 | 4.140 |
| 15 | 0.691 | 0.866 | 1.074 | 1.341 | 1.753 | 2.131 | 2.249 | 2.602 | 2.947 | 3.286 | 3.733 | 4.073 |
| 16 | 0.690 | 0.865 | 1.071 | 1.337 | 1.746 | 2.120 | 2.235 | 2.583 | 2.921 | 3.252 | 3.686 | 4.015 |
| 17 | 0.689 | 0.863 | 1.069 | 1.333 | 1.740 | 2.110 | 2.224 | 2.567 | 2.898 | 3.222 | 3.646 | 3.965 |
| 18 | 0.688 | 0.862 | 1.067 | 1.330 | 1.734 | 2.101 | 2.214 | 2.552 | 2.878 | 3.197 | 3.611 | 3.922 |
| 19 | 0.688 | 0.861 | 1.066 | 1.328 | 1.729 | 2.093 | 2.205 | 2.539 | 2.861 | 3.174 | 3.579 | 3.883 |
| 20 | 0.687 | 0.860 | 1.064 | 1.325 | 1.725 | 2.086 | 2.197 | 2.528 | 2.845 | 3.153 | 3.552 | 3.850 |
| 21 | 0.686 | 0.859 | 1.063 | 1.323 | 1.721 | 2.080 | 2.189 | 2.518 | 2.831 | 3.135 | 3.527 | 3.819 |
| 22 | 0.686 | 0.858 | 1.061 | 1.321 | 1.717 | 2.074 | 2.183 | 2.508 | 2.819 | 3.119 | 3.505 | 3.792 |
| 23 | 0.685 | 0.858 | 1.060 | 1.319 | 1.714 | 2.069 | 2.177 | 2.500 | 2.807 | 3.104 | 3.485 | 3.768 |
| 24 | 0.685 | 0.857 | 1.059 | 1.318 | 1.711 | 2.064 | 2.172 | 2.492 | 2.797 | 3.091 | 3.467 | 3.745 |
| 25 | 0.684 | 0.856 | 1.058 | 1.316 | 1.708 | 2.060 | 2.167 | 2.485 | 2.787 | 3.078 | 3.450 | 3.725 |
| 26 | 0.684 | 0.856 | 1.058 | 1.315 | 1.706 | 2.056 | 2.162 | 2.479 | 2.779 | 3.067 | 3.435 | 3.707 |
| 27 | 0.684 | 0.855 | 1.057 | 1.314 | 1.703 | 2.052 | 2.158 | 2.473 | 2.771 | 3.057 | 3.421 | 3.690 |
| 28 | 0.683 | 0.855 | 1.056 | 1.313 | 1.701 | 2.048 | 2.154 | 2.467 | 2.763 | 3.047 | 3.408 | 3.674 |
| 29 | 0.683 | 0.854 | 1.055 | 1.311 | 1.699 | 2.045 | 2.150 | 2.462 | 2.756 | 3.038 | 3.396 | 3.659 |
| 30 | 0.683 | 0.854 | 1.055 | 1.310 | 1.697 | 2.042 | 2.147 | 2.457 | 2.750 | 3.030 | 3.385 | 3.646 |
| 31 | 0.682 | 0.853 | 1.054 | 1.309 | 1.696 | 2.040 | 2.144 | 2.453 | 2.744 | 3.022 | 3.375 | 3.633 |
| 32 | 0.682 | 0.853 | 1.054 | 1.309 | 1.694 | 2.037 | 2.141 | 2.449 | 2.738 | 3.015 | 3.365 | 3.622 |
| 33 | 0.682 | 0.853 | 1.053 | 1.308 | 1.692 | 2.035 | 2.138 | 2.445 | 2.733 | 3.008 | 3.356 | 3.611 |
| 34 | 0.682 | 0.852 | 1.052 | 1.307 | 1.691 | 2.032 | 2.136 | 2.441 | 2.728 | 3.002 | 3.348 | 3.601 |
| 35 | 0.682 | 0.852 | 1.052 | 1.306 | 1.690 | 2.030 | 2.133 | 2.438 | 2.724 | 2.996 | 3.340 | 3.591 |
| 36 | 0.681 | 0.852 | 1.052 | 1.306 | 1.688 | 2.028 | 2.131 | 2.435 | 2.719 | 2.990 | 3.333 | 3.582 |
| 37 | 0.681 | 0.851 | 1.051 | 1.305 | 1.687 | 2.026 | 2.129 | 2.431 | 2.715 | 2.985 | 3.326 | 3.574 |
| 38 | 0.681 | 0.851 | 1.051 | 1.304 | 1.686 | 2.024 | 2.127 | 2.429 | 2.712 | 2.980 | 3.319 | 3.566 |
| 39 | 0.681 | 0.851 | 1.050 | 1.304 | 1.685 | 2.023 | 2.125 | 2.426 | 2.708 | 2.976 | 3.313 | 3.558 |
| 40 | 0.681 | 0.851 | 1.050 | 1.303 | 1.684 | 2.021 | 2.123 | 2.423 | 2.704 | 2.971 | 3.307 | 3.551 |
| 50 | 0.679 | 0.849 | 1.047 | 1.299 | 1.676 | 2.009 | 2.109 | 2.403 | 2.678 | 2.937 | 3.261 | 3.496 |
| 60 | 0.679 | 0.848 | 1.045 | 1.296 | 1.671 | 2.000 | 2.099 | 2.390 | 2.660 | 2.915 | 3.232 | 3.460 |
| 70 | 0.678 | 0.847 | 1.044 | 1.294 | 1.667 | 1.994 | 2.093 | 2.381 | 2.648 | 2.899 | 3.211 | 3.435 |
| 80 | 0.678 | 0.846 | 1.043 | 1.292 | 1.664 | 1.990 | 2.088 | 2.374 | 2.639 | 2.887 | 3.195 | 3.416 |
| 90 | 0.677 | 0.846 | 1.042 | 1.291 | 1.662 | 1.987 | 2.084 | 2.368 | 2.632 | 2.878 | 3.183 | 3.402 |
| 100 | 0.677 | 0.845 | 1.042 | 1.290 | 1.660 | 1.984 | 2.081 | 2.364 | 2.626 | 2.871 | 3.174 | 3.390 |
| 1000 | 0.675 | 0.842 | 1.037 | 1.282 | 1.646 | 1.962 | 2.056 | 2.330 | 2.581 | 2.813 | 3.098 | 3.300 |
| z | 0.674 | 0.841 | 1.036 | 1.282 | 1.645 | 1.960 | 2.054 | 2.326 | 2.576 | 2.807 | 3.091 | 3.291 |

## TABLE IV

### Chi-Square ($\chi^2$) Distribution
#### Area to the Right of Critical Value

| Degrees of Freedom | 0.995 | 0.99 | 0.975 | 0.95 | 0.90 | 0.10 | 0.05 | 0.025 | 0.01 | 0.005 |
|---|---|---|---|---|---|---|---|---|---|---|
| 1 | — | — | 0.001 | 0.004 | 0.016 | 2.706 | 3.841 | 5.024 | 6.635 | 7.879 |
| 2 | 0.010 | 0.020 | 0.051 | 0.103 | 0.211 | 4.605 | 5.991 | 7.378 | 9.210 | 10.597 |
| 3 | 0.072 | 0.115 | 0.216 | 0.352 | 0.584 | 6.251 | 7.815 | 9.348 | 11.345 | 12.838 |
| 4 | 0.207 | 0.297 | 0.484 | 0.711 | 1.064 | 7.779 | 9.488 | 11.143 | 13.277 | 14.860 |
| 5 | 0.412 | 0.554 | 0.831 | 1.145 | 1.610 | 9.236 | 11.071 | 12.833 | 15.086 | 16.750 |
| 6 | 0.676 | 0.872 | 1.237 | 1.635 | 2.204 | 10.645 | 12.592 | 14.449 | 16.812 | 18.548 |
| 7 | 0.989 | 1.239 | 1.690 | 2.167 | 2.833 | 12.017 | 14.067 | 16.013 | 18.475 | 20.278 |
| 8 | 1.344 | 1.646 | 2.180 | 2.733 | 3.490 | 13.362 | 15.507 | 17.535 | 20.090 | 21.955 |
| 9 | 1.735 | 2.088 | 2.700 | 3.325 | 4.168 | 14.684 | 16.919 | 19.023 | 21.666 | 23.589 |
| 10 | 2.156 | 2.558 | 3.247 | 3.940 | 4.865 | 15.987 | 18.307 | 20.483 | 23.209 | 25.188 |
| 11 | 2.603 | 3.053 | 3.816 | 4.575 | 5.578 | 17.275 | 19.675 | 21.920 | 24.725 | 26.757 |
| 12 | 3.074 | 3.571 | 4.404 | 5.226 | 6.304 | 18.549 | 21.026 | 23.337 | 26.217 | 28.299 |
| 13 | 3.565 | 4.107 | 5.009 | 5.892 | 7.042 | 19.812 | 22.362 | 24.736 | 27.688 | 29.819 |
| 14 | 4.075 | 4.660 | 5.629 | 6.571 | 7.790 | 21.064 | 23.685 | 26.119 | 29.141 | 31.319 |
| 15 | 4.601 | 5.229 | 6.262 | 7.261 | 8.547 | 22.307 | 24.996 | 27.488 | 30.578 | 32.801 |
| 16 | 5.142 | 5.812 | 6.908 | 7.962 | 9.312 | 23.542 | 26.296 | 28.845 | 32.000 | 34.267 |
| 17 | 5.697 | 6.408 | 7.564 | 8.672 | 10.085 | 24.769 | 27.587 | 30.191 | 33.409 | 35.718 |
| 18 | 6.265 | 7.015 | 8.231 | 9.390 | 10.865 | 25.989 | 28.869 | 31.526 | 34.805 | 37.156 |
| 19 | 6.844 | 7.633 | 8.907 | 10.117 | 11.651 | 27.204 | 30.144 | 32.852 | 36.191 | 38.582 |
| 20 | 7.434 | 8.260 | 9.591 | 10.851 | 12.443 | 28.412 | 31.410 | 34.170 | 37.566 | 39.997 |
| 21 | 8.034 | 8.897 | 10.283 | 11.591 | 13.240 | 29.615 | 32.671 | 35.479 | 38.932 | 41.401 |
| 22 | 8.643 | 9.542 | 10.982 | 12.338 | 14.042 | 30.813 | 33.924 | 36.781 | 40.289 | 42.796 |
| 23 | 9.260 | 10.196 | 11.689 | 13.091 | 14.848 | 32.007 | 35.172 | 38.076 | 41.638 | 44.181 |
| 24 | 9.886 | 10.856 | 12.401 | 13.848 | 15.659 | 33.196 | 36.415 | 39.364 | 42.980 | 45.559 |
| 25 | 10.520 | 11.524 | 13.120 | 14.611 | 16.473 | 34.382 | 37.652 | 40.646 | 44.314 | 46.928 |
| 26 | 11.160 | 12.198 | 13.844 | 15.379 | 17.292 | 35.563 | 38.885 | 41.923 | 45.642 | 48.290 |
| 27 | 11.808 | 12.879 | 14.573 | 16.151 | 18.114 | 36.741 | 40.113 | 43.194 | 46.963 | 49.645 |
| 28 | 12.461 | 13.565 | 15.308 | 16.928 | 18.939 | 37.916 | 41.337 | 44.461 | 48.278 | 50.993 |
| 29 | 13.121 | 14.257 | 16.047 | 17.708 | 19.768 | 39.087 | 42.557 | 45.722 | 49.588 | 52.336 |
| 30 | 13.787 | 14.954 | 16.791 | 18.493 | 20.599 | 40.256 | 43.773 | 46.979 | 50.892 | 53.672 |
| 40 | 20.707 | 22.164 | 24.433 | 26.509 | 29.051 | 51.805 | 55.758 | 59.342 | 63.691 | 66.766 |
| 50 | 27.991 | 29.707 | 32.357 | 34.764 | 37.689 | 63.167 | 67.505 | 71.420 | 76.154 | 79.490 |
| 60 | 35.534 | 37.485 | 40.482 | 43.188 | 46.459 | 74.397 | 79.082 | 83.298 | 88.379 | 91.952 |
| 70 | 43.275 | 45.442 | 48.758 | 51.739 | 55.329 | 85.527 | 90.531 | 95.023 | 100.425 | 104.215 |
| 80 | 51.172 | 53.540 | 57.153 | 60.391 | 64.278 | 96.578 | 101.879 | 106.629 | 112.329 | 116.321 |
| 90 | 59.196 | 61.754 | 65.647 | 69.126 | 73.291 | 107.565 | 113.145 | 118.136 | 124.116 | 128.299 |
| 100 | 67.328 | 70.065 | 74.222 | 77.929 | 82.358 | 118.498 | 124.342 | 129.561 | 135.807 | 140.169 |

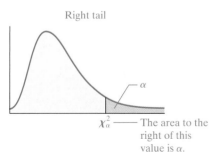

Right tail

$\chi^2_\alpha$ —— The area to the right of this value is $\alpha$.

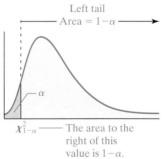

Left tail

Area $= 1-\alpha$

$\chi^2_{1-\alpha}$ —— The area to the right of this value is $1-\alpha$.

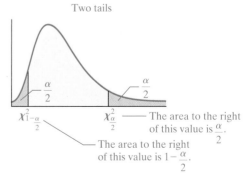

Two tails

$\chi^2_{1-\frac{\alpha}{2}}$

$\chi^2_{\frac{\alpha}{2}}$ —— The area to the right of this value is $\frac{\alpha}{2}$.

—— The area to the right of this value is $1-\frac{\alpha}{2}$.

# ANSWERS

## CHAPTER 1 Data Collection

### 1.1 Exercises (page 7)

**1.** Qualitative    **3.** Quantitative    **5.** Quantitative
**7.** Quantitative    **9.** Qualitative    **11.** Discrete
**13.** Continuous    **15.** Continuous    **17.** Discrete
**19.** Continuous

**21.** Population: all United States residents 18 years or older; Sample: the 1019 residents

**23.** Population: The entire soybean crop; Sample: the 100 plants selected

**25. (a)** To determine the genetic and nongenetic factors of structural brain abnormalities on schizophrenia.
**(b)** The 58 pairs of twins in the study.
**(c)** Whole-brain volumes were 2.2% smaller in the schizophrenic patients.
**(d)** An increased genetic risk to develop schizophrenia is related to reduced brain growth early in life.

**27. (a)** To determine whether music cognition and cognitions pertaining to abstract operations such as mathematical or spatial reasoning were related.
**(b)** Thirty-six college students.
**(c)** The mean test score following the Mozart piece was 119, while the mean test score following silence was 110.
**(d)** Subjects perform better on abstract/spatial reasoning tests after listening to Mozart.

**29. (a)** To determine the proportion of households in the United States that have been a victim of a crime in the past 12 months.
**(b)** The 1012 adults aged 18 years or older.
**(c)** Twenty-four percent of the respondents stated that they, or someone in the household, experienced some type of crime.
**(d)** Twenty-four percent of all households in the United States have been victimized by crime in the past 12 months.

**31.** Individuals: Neta Van Duyne, Dave Ebert, Kristin Bols, Michael Wirth, Jinita Desai; Variables: Gender, age, number of siblings; Data for "gender": F, M, F, M, F; Data for "age": 19, 19, 19, 19, 20; Data for "number of siblings": 1, 1, 2, 1, 1. The variable "gender" is qualitative; the variable "age" is continuous; the variable "number of siblings" is discrete.

**33.** Individuals: Alabama, Illinois, Montana, New York, Texas; Variables: Age for driver's license, blood-alcohol concentration limit, mandatory belt-use law seating positions, maximum allowable speed, 1999. Data for "age for driver's license": 16, 18, 18, 17, 18; data for "blood-alcohol concentration limit": 0.08, 0.08, 0.10, 0.10, 0.10; data for "mandatory belt-use law seating positions": Front, Front, All, Front, Front; Data for "Maximum allowable speed limit, 1999": 70, 65, 75, 65, 70. The variable "age for driver's license" is continuous; the variable "blood-alcohol concentration limit" is continuous; the variable "mandatory belt-use law seating positions" is qualitative; the variable "maximum allowable speed limit, 1999" is continuous.

### 1.2 Exercises (page 15)

**1.** Observational study    **3.** Experiment
**5.** Observational study    **7.** Observational study
**9.** Experiment    **11.** Observational study

**13. (a)**

| Graham, Murkowski | Graham, Kyl | Graham, Baucus |
|---|---|---|
| Graham, Conrad | Murkowski, Kyl | Murkowski, Baucus |
| Murkowski, Conrad | Kyl, Baucus | Kyl, Conrad |
| Baucus, Conrad | | |

**(b)** There is a 1 in 10 chance that the two individuals selected are Bob Graham and Max Baucus.

**15. (a)** Answers will vary    **(b)** Answers will vary

**17. (a)** First, find a list of all currently enrolled students. This list will serve as the frame. Number the students on the list from 1 through 19,935. Using a random-number generator, set a seed (or select a starting point if using Table I). Generate 25 different numbers randomly. The students corresponding to these numbers will be the 25 students in the sample.
**(b)** Answers will vary.

### 1.3 Exercises (page 25)

**1.** Systematic    **3.** Cluster    **5.** Simple random
**7.** Cluster    **9.** Convenience

**11.** Answers will vary. A good choice might be stratified sampling with the strata being commuters and noncommuters.

**13.** Answers will vary. One option would be cluster sampling. The clusters could be the city blocks. Randomly select clusters and then survey all the households in the selected city blocks.

**15.** Answers will vary. Simple random sampling will work fine here, especially because a list of 6600 individuals who meet the needs of our study already exists (the frame).

**17. (a)** 90
**(b)** Randomly select a number between 1 and 90. Suppose that we randomly select 15; then the individuals in the survey will be 15, 105, 195, 285, ..., 4425.

### 1.4 Exercises (page 30)

**1. (a)** Flawed sampling method because only the first 50 students who enter the building have a chance of being surveyed.
**(b)** Could use systematic sampling, but a better choice would be cluster or simple random sampling because the vice president should have access to a complete frame.

**3. (a)** Flawed survey due to the wording of the question.
**(b)** The survey should begin by stating the current penalty for selling a gun illegally. The question might be rewritten as "Do you approve or disapprove harsher penalties for individuals that sell guns illegally?" The words "approve" and "disapprove" should be rotated from individual to individual.

**5. (a)** Flawed survey because the response rate is low.
**(b)** The survey can be improved through face-to-face or telephone interviews.

**7. (a)** Flawed survey because the students are not likely to respond truthfully with their teachers administering the survey.
**(b)** The survey should be administered by an impartial party so that the students are more likely to respond truthfully.

**9.** The ordering of the questions is likely to affect the survey results. Perhaps question B should be asked first. Another possibility is to rotate the questions randomly.

### 1.5 Exercises (page 35)

**1. (a)** Achievement test score    **(b)** Method of teaching; 2
**(c)** Grade level, teacher, school district
**(d)** Students are from the same district
**(e)** Completely randomized design    **(f)** the 500 students

**(g)**

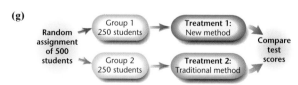

**3. (a)** Proportion of cancer cases    **(b)** Servings of tomatoes; 3
   **(c)** Servings of tomatoes, age, amount of exercise, eating habits
   **(d)** Completely randomized design
   **(e)** The 600 males in the study

**(f)**

**5. (a)** Difference in weight before spaceflight and after spaceflight
   **(b)** Spaceflight    **(c)** Matched pairs

**(d)**

**7. (a)** Hair counts    **(b)** The drug; 2
   **(c)** Completely randomized design

**(d)**

**9.** Answers will vary. A matched-pairs design would work best. Measure the lung capacity of each experimental unit before the workout regimen and after. The researcher may want to have various levels of exercise (i.e. no exercise, 20 minutes a day, and so on) to see the effect of the more arduous exercise. Be sure to keep other factors, such as diet, constant for all groups.

**11. (a)** Completely randomized.    **(b)** Answers will vary.

## Chapter Review Exercises (page 37)

**1.** The science of collecting, organizing, summarizing, and analyzing data in order to draw conclusions.

**3.** A subset of the population.

**5.** Applies a treatment to the individuals in the study in order to isolate the effect of the treatment on the response variable.

**7.** There are two main types of errors: nonsampling errors and sampling errors. Nonsampling errors are errors that result from the survey process, such as an incomplete frame, poorly worded questions, inaccurate responses, and so on. Sampling errors are errors that result from using a sample to estimate characteristics of a population. They result because samples contain incomplete information regarding a population.

**9.** Qualitative    **11.** Quantitative, continuous

**13.** Quantitative, discrete    **15.** Observational study

**17.** Observational study    **19.** Systematic

**21.** Stratified    **23.** Answers will vary

**25.** Answers will vary

**27. (a)** Hamilton Rating Scale score    **(b)** Group; 2
   **(c)** Age, medical history; none are controlled
   **(d)** Completely randomized    **(e)** The 120 women

**(f)**

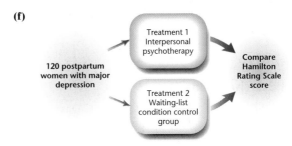

**29. (a)** Lymphocyte count    **(b)** Spaceflight
   **(c)** Matched pairs    **(d)** The four flight members

**(e)**

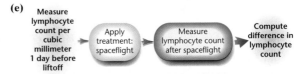

**31.** In a completely randomized design, the experimental units are randomly assigned to one of the treatments. The value of the response variable is compared for each treatment. In a matched-pairs design, experimental units are "matched up" on the basis of some common characteristic (such as husband/wife or twins). The difference in the "matched-up" experimental units is analyzed.

## CHAPTER 2    Organizing and Summarizing Data

## 2.1 Exercises (page 48)

**1. (a)** 34–52    **(b)** 18–33    **(c)** 27%
**3. (a)** United States    **(b)** ≈ 18 million
**5. (a)** 447,927    **(b)** 48.3%; 48.1%    **(c)** No
**7. (a)** 16%    **(b)** Natural gas    **(c)** LPG

**9. (a)**

| Source of Income | Relative Frequency |
|---|---|
| Individual Income Tax and Tax Withholdings | 0.4932 |
| Corporate Income Taxes | 0.1147 |
| Social Insurance and Retirement Receipts | 0.3378 |
| Excise, Estate, and Gift Taxes; Customs; and Miscellaneous Receipts | 0.0543 |

**(b)** 49.32%
**(c)**

**Government Income**

**(d)**

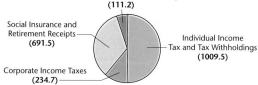

Government Income

**(e)**

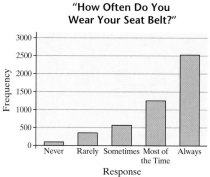

Government Income

**11. (a)**

| Response | Relative Frequency |
|---|---|
| Never | 0.0262 |
| Rarely | 0.0678 |
| Sometimes | 0.1156 |
| Most of the time | 0.2632 |
| Always | 0.5272 |

**(b)** 52.72%   **(c)** 9.4%

**(d)**

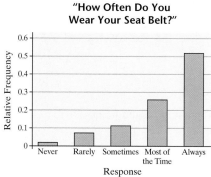

"How Often Do You Wear Your Seat Belt?"

**(e)**

"How Often Do You Wear Your Seat Belt?"

**11. (f)**

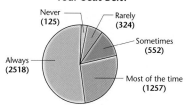

"How Often Do You Wear Your Seat Belt?"

**13. (a)**

| Region | Relative Frequency |
|---|---|
| Europe | 0.1566 |
| Asia | 0.2746 |
| Africa | 0.0276 |
| Oceania | 0.0059 |
| Latin America | 0.5079 |
| North America | 0.0274 |

**(b)** 2.76%

**(c)**

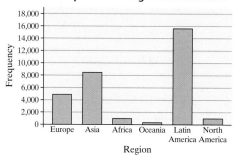

Birthplace of Foreign-Born Residents

**(d)**

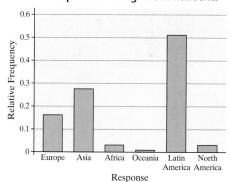

Birthplace of Foreign-Born Residents

**(e)** Birthplace of Foreign-Born Residents

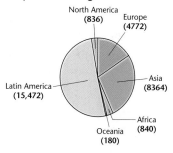

**15. (a), (b)**

| Educational Attainment | Relative Frequency Males | Relative Frequency Females |
|---|---|---|
| Not a high school graduate | 0.166 | 0.166 |
| High school graduate | 0.318 | 0.348 |
| Some college, but no degree | 0.171 | 0.175 |
| Associate's degree | 0.070 | 0.080 |
| Bachelor's degree | 0.179 | 0.162 |
| Advanced degree | 0.096 | 0.070 |

**(c)**

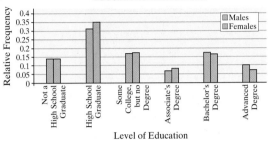

**(d)**

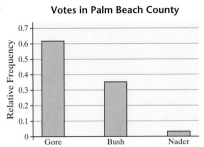

**(e)**

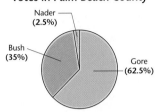

**17. (a)**

| Country | Number of Launches, 1995 | Number of Launches, 1998 |
|---|---|---|
| Soviet Union/Commonwealth of Independent States | 0.4267 | 0.3117 |
| United States | 0.3600 | 0.4416 |
| Japan | 0.0267 | 0.0260 |
| European Space Agency | 0.1467 | 0.1429 |
| China | 0.0267 | 0.0779 |
| Israel | 0.0133 | 0.0000 |

**(b)**

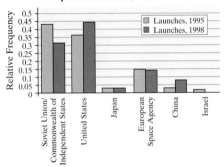

**21. (a), (b)**

| Position | Frequency | Relative Frequency |
|---|---|---|
| First Base | 7 | 0.1944 |
| Third Base | 2 | 0.0556 |
| Catcher | 2 | 0.0556 |
| Center Field | 5 | 0.1389 |
| Pitcher | 12 | 0.3333 |
| Right Field | 6 | 0.1667 |
| Shortstop | 2 | 0.0556 |

**(c)** Pitcher    **(d)** Second Base, Left Field

**(e)**

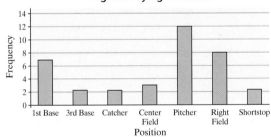

**19. (a), (b)**

| Candidate | Frequency | Relative Frequency |
|---|---|---|
| Gore | 25 | 0.625 |
| Bush | 14 | 0.350 |
| Nader | 1 | 0.025 |

**(c)**

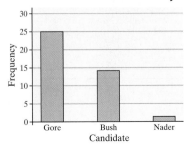

**(f)**

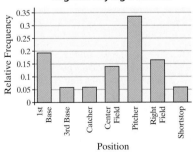

**(g)**

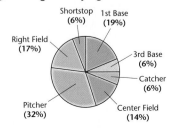

## 2.2 Exercises (page 65)

**1.** (a) 8   (b) 2   (c) 15
(d) 15%   (e) Bell shaped

**3.** (a) 200   (b) 10   (c) 60–69, 2; 70–79, 3; 80–89, 13; 90–99, 42; 100–109, 58; 110–119, 40; 120–129, 31; 130–139, 8; 140–149, 2; 150–159, 1   (d) 100–109   (e) 150–159
(f) 0.155   (g) Bell shaped

**5.** (a)

| Number of Children Under 5 | Relative Frequency |
|---|---|
| 0 | 0.32 |
| 1 | 0.36 |
| 2 | 0.24 |
| 3 | 0.06 |
| 4 | 0.02 |

(b) 24%   (c) 60%

**7.** 10, 11, 14, 21, 24, 24, 27, 29, 33, 35, 35, 35, 37, 37, 38, 40, 40, 41, 42, 46, 46, 48, 49, 49, 53, 53, 55, 58, 61, 62

**9.** (a) 4   (b) Lower class limits: 25, 35, 45, 55; Upper class limits: 34, 44, 54, 64   (c) 10

**11.** (a) 6   (b) Lower class limits: 50, 60, 70, 80, 90, 100; Upper class limits: 59, 69, 79, 89, 99, 109   (c) 10

**13.** (a)

| Age | Relative Frequency |
|---|---|
| 25–34 | 0.2522 |
| 35–44 | 0.3184 |
| 45–54 | 0.2616 |
| 55–64 | 0.1678 |

(b) **Number Covered by Health Insurance**

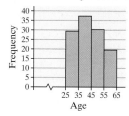

(c) **Number Covered by Health Insurance**

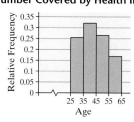

25.2%; 57.1%

**15.** (a)

| Temperature | Relative Frequency |
|---|---|
| 50–59 | 0.0003 |
| 60–69 | 0.0776 |
| 70–79 | 0.3828 |
| 80–89 | 0.4098 |
| 90–99 | 0.1268 |
| 100–109 | 0.0028 |

(b) **High Temperatures in August in Chicago**

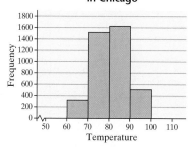

(c) **High Temperatures in August in Chicago**

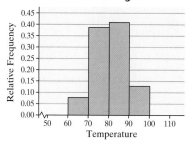

38.28%; 46.07%

**17.** (a), (b)

| Number of Customers | Frequency | Relative Frequency | Number of Customers | Frequency | Relative Frequency |
|---|---|---|---|---|---|
| 3 | 2 | 0.050 | 9 | 4 | 0.1 |
| 4 | 3 | 0.075 | 10 | 4 | 0.1 |
| 5 | 3 | 0.075 | 11 | 4 | 0.1 |
| 6 | 5 | 0.125 | 12 | 0 | 0 |
| 7 | 4 | 0.1 | 13 | 2 | 0.05 |
| 8 | 8 | 0.2 | 14 | 1 | 0.025 |

(c) 27.5%   (d) 20%

(e) **Customers Waiting for a Table**

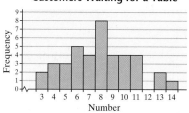

(f) **Customers Waiting for a Table**

(g) Symmetric

**19. (a), (b)**

| Class | Tally | Frequency | Relative Frequency |
|---|---|---|---|
| 160–169.99 | \| | 1 | $\frac{1}{25} = 0.04$ |
| 170–179.99 | \| | 1 | $\frac{1}{25} = 0.04$ |
| 180–189.99 | \|\|\|\| | 4 | $\frac{4}{25} = 0.16$ |
| 190–199.99 | \| | 1 | $\frac{1}{25} = 0.04$ |
| 200–209.99 | HHI \|\|\|\| | 9 | 0.36 |
| 210–219.99 | HHI \| | 6 | 0.24 |
| 220–229.99 | \| | 1 | 0.04 |
| 230–239.99 | \| | 1 | 0.04 |
| 240–249.99 | \| | 1 $\overline{\Sigma f_i = 25}$ | 0.04 |

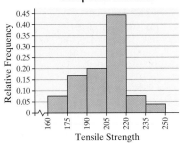

**Tensile Strength of
Composite Material**

Symmetric

**(c)**

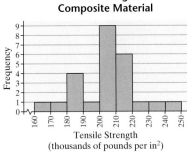

**Tensile Strength of
Composite Material**

**(d)**

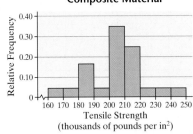

**Tensile Strength of
Composite Material**

**(e)** Symmetric

**(f)**

| Class | Frequency | Relative Frequency |
|---|---|---|
| 160–174.99 | 2 | 0.08 |
| 175–189.99 | 4 | 0.16 |
| 190–204.99 | 5 | 0.20 |
| 205–219.99 | 11 | 0.44 |
| 220–234.99 | 2 | 0.08 |
| 235–249.99 | 1 | 0.04 |

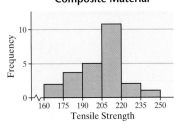

**Tensile Strength of
Composite Material**

**21. (a), (b)**

| Class | Tally | Frequency | Relative Frequency |
|---|---|---|---|
| 20–29 | \| | 1 | $\frac{1}{40} = 0.025$ |
| 30–39 | HHI \| | 6 | $\frac{6}{40} = 0.15$ |
| 40–49 | HHI HHI | 10 | $\frac{10}{40} = 0.25$ |
| 50–59 | HHI HHI \|\|\|\| | 14 | 0.35 |
| 60–69 | HHI \| | 6 | 0.15 |
| 70–79 | \|\|\| | 3 $\overline{\Sigma f_i = 40}$ | 0.075 |

**(c)**

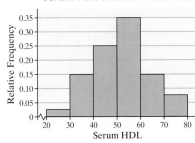

**Serum HDL of 20–29 Year Olds**

**(d)**

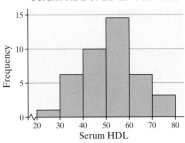

**Serum HDL of 20–29 Year Olds**

**(e)** Bell shaped

**(f)**

| Class | Frequency | Relative Frequency |
|---|---|---|
| 20–24 | 0 | 0 |
| 25–29 | 1 | 0.025 |
| 30–34 | 2 | 0.05 |
| 35–39 | 4 | 0.1 |
| 40–44 | 2 | 0.05 |
| 45–49 | 8 | 0.2 |
| 50–54 | 9 | 0.225 |
| 55–59 | 5 | 0.125 |
| 60–64 | 4 | 0.1 |
| 65–69 | 2 | 0.05 |
| 70–74 | 3 | 0.075 |
| 75–79 | 0 | 0 |

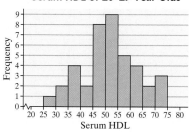

Serum HDL of 20–29 Year Olds

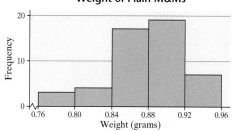

Weight of Plain M&Ms

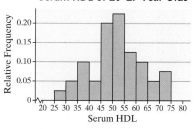

Serum HDL of 20–29 Year Olds

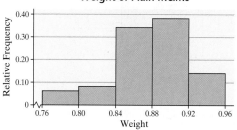

Weight of Plain M&Ms

Bell shaped

**23. (a), (b)**

| Class | Frequency | Relative Frequency | Class | Frequency | Relative Frequency |
|---|---|---|---|---|---|
| 0.76–0.77 | 1 | 0.02 | 0.86–0.87 | 11 | 0.22 |
| 0.78–0.79 | 2 | 0.04 | 0.88–0.89 | 10 | 0.20 |
| 0.8–0.81 | 0 | 0 | 0.90–0.91 | 9 | 0.18 |
| 0.82–0.83 | 4 | 0.08 | 0.92–0.93 | 5 | 0.1 |
| 0.84–0.85 | 6 | 0.12 | 0.94–0.95 | 2 | 0.04 |

**(c)**

Weight of Plain M&Ms

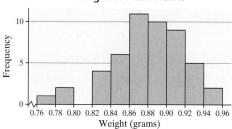

**(d)**

Weight of Plain M&Ms

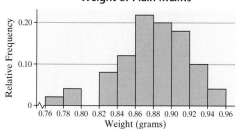

**(e)** Symmetric

**(f)**

| Class | Frequency | Relative Frequency |
|---|---|---|
| 0.76–0.79 | 3 | 0.06 |
| 0.80–0.83 | 4 | 0.08 |
| 0.84–0.87 | 17 | 0.34 |
| 0.88–0.91 | 19 | 0.38 |
| 0.92–0.95 | 7 | 0.14 |

Symmetric

**25.**
```
4 | 23
4 | 667899
5 | 0011112244444
5 | 555566677778
6 | 0111244
6 | 589
```

**27.**
```
0 | 2
0 | 88
1 | 12
1 | 567
2 | 033
2 | 8
3 | 114
3 | 78
```

**29.**

Closing Price of Microsoft Stock

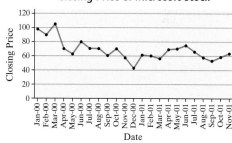

**31.**

College Enrollment

## 2.3 Exercises (page 74)

**1.** The ketchup "stain" should be more than twice as long as the mustard or mayonnaise "stain."

**3.** The graphic has a large portion "cut out" from the center on the vertical axis, yet the graphs are drawn continuously.

**5. (a)** Vertical axis starts at 1000

**(b)**

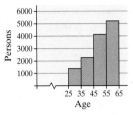

**Number of Persons with Work Disability**

**7. (a)** Vertical axis starts at 1500
**(b)** Health-care expenditures are increasing more rapidly than they really are.

**9. (a)** The bar for housing should be a little more than twice the length of the bar for transportation, but it is not.
**(b)** Adjust the graphs so that the lengths of the bars are proportional.

**11. (a)**

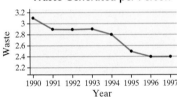

**Waste Generated per Person**

**(b)**

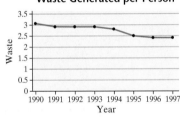

**Waste Generated per Person**

**(c)**

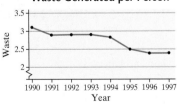

**Waste Generated per Person**

**13. (a)** Because the second graphic must have an area that is three times as large as the first graphic, the height and width of the second graphic must increase by a factor of $\sqrt{3}$. So if the first graphic is 1 inch by 1 inch, the second graphic should be $\sqrt{3}$ inches by $\sqrt{3}$ inches.
**(b)** Any graphic that is misleading will not have the dimensions mentioned in part (a). Typically, the second graph would have a width and length that both increase by a factor of three, which makes the area nine times as large.

## Chapter Review Exercises (page 77)

**1. (a)** 22 quadrillion btus
**(b)** 4 quadrillion btus
**(c)** 95 quadrillion btus

**3. (a)**

| Cause of Death | Relative Frequency |
|---|---|
| Gun Shot | 0.649 |
| Cutting/Stabbing | 0.133 |
| Blunt Object | 0.053 |
| Personal Weapons (hand, fist, etc.) | 0.067 |
| Strangulation | 0.022 |
| Other | 0.076 |

**(b)** 5.3%

**(c)**

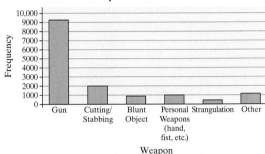

**Weapons Used in Homicides**

**(d)**

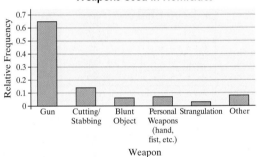

**Weapons Used in Homicides**

**(e)**

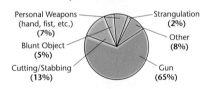

**Weapons Used in Homicides**

Personal Weapons (hand, fist, etc.) (7%), Strangulation (2%), Other (8%), Blunt Object (5%), Cutting/Stabbing (13%), Gun (65%)

**5. (a)**

| Age | Frequency | Relative Frequency |
|---|---|---|
| 20–24 | 8067 | 0.1770 |
| 25–29 | 6195 | 0.1359 |
| 30–34 | 6274 | 0.1376 |
| 35–39 | 5130 | 0.1125 |
| 40–44 | 4631 | 0.1016 |
| 45–49 | 3636 | 0.0798 |
| 50–54 | 3021 | 0.0663 |
| 55–59 | 2332 | 0.0512 |
| 60–64 | 1606 | 0.0352 |
| 65–69 | 1341 | 0.0294 |
| 70–74 | 1273 | 0.0279 |
| 75–79 | 1179 | 0.0259 |
| 80–84 | 902 | 0.0198 |

**(b)**

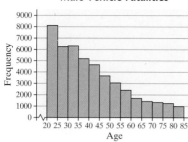

Male Vehicle Fatalities

**(c)**

Number of Children for Couples
Married 7 Years

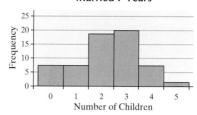

Symmetric

**(c)**

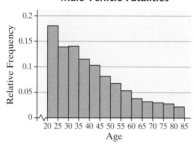

Male Vehicle Fatalities

**(d)**

Number of Children for
Couples Married 7 Years

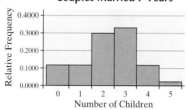

**(d)** 17.7%      **(e)** 31.29%

**(e)** 30%      **(f)** 76.7%

**7. (a), (b)**

| Affiliation | Frequency | Relative Frequency |
|---|---|---|
| Democrat | 46 | 0.46 |
| Independent | 16 | 0.16 |
| Republican | 38 | 0.38 |

**(c)**

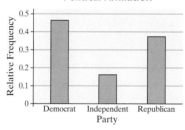

Political Affiliation

**(d)**

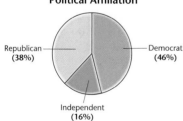

Political Affiliation

**(e)** Democrat

**9. (a), (b)**

| Number of Children | Frequency | Relative Frequency |
|---|---|---|
| 0 | 7 | 0.1167 |
| 1 | 7 | 0.1167 |
| 2 | 18 | 0.3000 |
| 3 | 20 | 0.3333 |
| 4 | 7 | 0.1167 |
| 5 | 1 | 0.0167 |

**11. (a), (b)**

| Class | Frequency | Relative Frequency |
|---|---|---|
| 2000–2499 | 1 | 0.0196 |
| 2500–2999 | 4 | 0.0784 |
| 3000–3499 | 4 | 0.0784 |
| 3500–3999 | 9 | 0.1765 |
| 4000–4499 | 9 | 0.1765 |
| 4500–4999 | 7 | 0.1373 |
| 5000–5499 | 8 | 0.1569 |
| 5500–5999 | 4 | 0.0784 |
| 6000–6499 | 1 | 0.0196 |
| 6500–6999 | 3 | 0.0588 |
| 7000–7499 | 0 | 0.0000 |
| 7500–7999 | 0 | 0.0000 |
| 8000–8499 | 0 | 0.0000 |
| 8500–8999 | 1 | 0.0196 |

**(c)**   Crime Rates per 100,000 Population

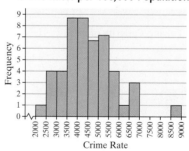

Fairly symmetric

**(d)**   Crime Rates per 100,000 Population

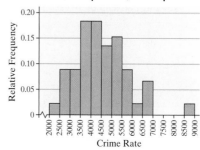

**(e)**

| Class | Frequency | Relative Frequency |
|---|---|---|
| 2000–2999 | 5 | 0.0980 |
| 3000–3999 | 13 | 0.2549 |
| 4000–4999 | 16 | 0.3137 |
| 5000–5999 | 12 | 0.2353 |
| 6000–6999 | 4 | 0.0784 |
| 7000–7999 | 0 | 0.0000 |
| 8000–8999 | 1 | 0.0196 |

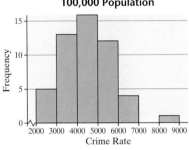

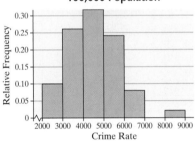

**13. (a), (b)**

| Class | Frequency | Relative Frequency | Class | Relative Frequency | Frequency |
|---|---|---|---|---|---|
| 3000–3999 | 5 | 0.1613 | 8000–8999 | 5 | 0.1613 |
| 4000–4999 | 4 | 0.1290 | 9000–9999 | 0 | 0.0000 |
| 5000–5999 | 8 | 0.2581 | 10,000–10,999 | 1 | 0.0323 |
| 6000–6999 | 3 | 0.0968 | 11,000–11,999 | 0 | 0.0000 |
| 7000–7999 | 3 | 0.0968 | 12,000–12,999 | 2 | 0.0645 |

**(c)**

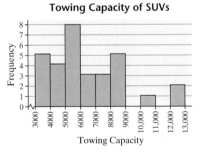

Skewed right

**(d)**

Towing Capacity of SUVs

15.
```
 0 | 788
 1 | 245788
 2 | 11578
 3 | 03348
 4 | 0
 5 | 4
 6 |
 7 | 014
 8 |
 9 |
10 | 2
```
Skewed right

**17. (a)**

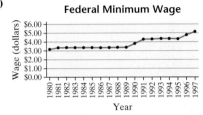

**(b)** Slightly upward
**19.** There is no vertical scale.
**21. (a)** A misleading graph would be a graph that is 75% wider and 75% taller.
**(b)** A graph that does not mislead would be one that has total area or total volume that is 75% larger.

**CHAPTER 3** Numerically Summarizing Data

## 3.1 Exercises (page 94)

**1.** Mean: $381.75; Median: $414.50; No Mode
**3.** Mean: 3668.9 psi; Median: 3830 psi; No Mode
**5.** Sludge Plot: Mean = 28.3; Median = 27.5; Mode = 27
   Spring Disk: Mean = 33; Median = 33.5; Mode = 34
   No Till: Mean = 28.5; Median = 29; Mode = 29
   Spring Chisele: Mean = 29.3; Median = 29.5; No Mode
   Great Lakes Bt: Mean = 28.8; Median = 28.5; Mode = 27
**7.** Monday: Mean = 10,511.3; Median = 10,491.5; No Mode
   Saturday: Mean = 8476.3; Median = 8449.5; No Mode
**9.** Control: Mean = 323.2 g; Median = 320g; No Mode
   Flight: Mean = 333.8 g; Median = 326 g; No Mode
**11. (a)** 72.2   **(b)** Answers will vary
   **(c)** Answers will vary
**13. (a)** Mean = 7.9 customers; Median = 8 customers
   **(b)** Symmetric
**15. (a)** Mean = 203.9 psi; Median = 206.51 psi   **(b)** Symmetric
**17. (a)** Mean = 51.1; Median = 51   **(b)** Symmetric
**19. (a)** Mean = 0.874 g; Median = 0.88 g   **(b)** Symmetric
**21.** Latin America   **23.** Gore
**25.** Pitcher
**27.** Mean = 229.1 psi; Median = 207.88 psi
**29. (a)** Mean > Median   **(b)** Mean = Median
   **(c)** Mean < Median
**31.** Median, because it is resistant   **33.** 204
**35. (a)** Median   **(b)** Mode   **(c)** Mean
   **(d)** Median   **(e)** Median
**37.** 0.85 g; No

## 3.2 Exercises (page 108)

**1.** Range = $226; $s^2$ = 9962.9; $s$ = $99.8
**3.** Range = 1150 psi; $s^2$ = 210,236.1; $s$ = 458.5 psi
**5.** Sludge Plot: Range = 8; $s$ = 2.8
   Spring Disk: Range = 5; $s$ = 1.8
   No Till: Range = 7; $s$ = 2.6
   Spring Chisele: Range = 6; $s$ = 2.2
   Great Lakes Bt: Range = 5; $s$ = 1.9

**7.** Monday: Range = 511.0 births; $s$ = 170.8 births
   Saturday: Range = 294.0 births; $s$ = 126.0 births
**9.** Control: Range = 40 g; $s$ = 15.1 g
   Flight: Range = 73 g; $s$ = 28.5 g
**11. (a)** $\sigma^2$ = 58.8; $\sigma$ = 7.7   **(b)** Answers will vary
   **(c)** Answers will vary
**13.** Range = 11 min; $s^2$ = 7.4 min$^2$; $s$ = 2.7 min
**15.** Range = 82.32 psi; $s^2$ = 335.638 psi$^2$; $s$ = 18.320 psi
**17.** Range = 45; $s^2$ = 118.0; $s$ = 10.9
**19.** Range = 0.18 g; $s^2$ = 0.002 g$^2$; $s$ = 0.040 g
**21. (a)** Financial: Mean = 11.12%; Median = 9.33%;
   Energy: Mean = 9.71%; Median = 9.09%
   **(b)** Financial: $s$ = 8.06%; Energy: $s$ = 5.852%; Financial is riskier
**23. (a)** Michael: Mean = 81.1; Kevin: Mean = 81.2
   **(b)** Michael: Median = 81; Kevin: Median = 82
   **(c)** Michael: Mode = 83; Kevin: Mode = 73
   **(d)** Michael: Range = 13; Kevin: Range = 17
   **(e)** Michael: $s$ = 3.2; Kevin: $s$ = 5.9
   **(f)** Michael; $s$ is lower
**25. (a)** 99.7%   **(b)** 95%
**27. (a)** Okay to use   **(b)** 99.7%   **(c)** 95%
   **(d)** 95%
**29.** Range = 655.53 psi; $s$ = 123.57 psi; Range and $s$ increased;
   Data is more dispersed; No.
**31. (a)** $s$ = 11.6   **(b)** $s$ = 11.6   **(c)** No effect
   **(d)** $s$ = 23.3   **(e)** $s$ doubled
**33. (a)** III   **(b)** I   **(c)** IV   **(d)** II

## 3.3 Exercises (page 120)

**1.** $\mu$ = 42.5 years; $\sigma$ = 10.4 years
**3. (a)** $\mu$ = 80.4°F; $\sigma$ = 8.2°F

**(b)**   **High Temperature in August in Chicago**

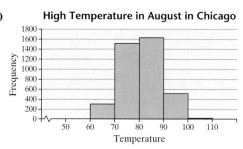

**(c)** 64 and 96.8 degrees Fahrenheit
**5.** Mean = 204.6 psi; $s$ = 17.9 psi
**7.** Mean = 51.3; $s$ = 12.1
**9.** 3.27   **11.** $2.97 per pound
**13. (a)** $\mu$ = 34.6 years; $\sigma$ = 21.7 years
   **(b)** $\mu$ = 37.2 years; $\sigma$ = 23.0 years
   **(c)** Female   **(d)** Female

## 3.4 Exercises (page 128)

**1.** SAT   **3.** 40-week baby   **5.** Woman
**7. (a)** $609,483.50   **(b)** $2,462,846   **(c)** $410,357
   **(d)** 15$^{th}$ percentile   **(e)** 78$^{th}$ percentile
**9. (a)** −1.70; The rainfall in 1971 is 1.70 standard deviations below
   the mean
   **(b)** $Q_1$ = 2.625; $Q_2$ = 3.985; $Q_3$ = 5.36
   **(c)** $IQR$ = 2.735
   **(d)** Lower Fence: −1.478; Upper Fence: 9.463; No outliers
**11. (a)** −3.70; The red blood cell count of Sampson is 3.70 standard deviations below the mean.
   **(b)** $Q_1$ = 6.05; $Q_2$ = 6.45; $Q_3$ = 6.95
   **(c)** $IQR$ = 0.97
   **(d)** Lower Fence: 4.7; Upper Fence = 8.3; Yes, 0.2 is an outlier

**13. (a)** 0.61; The concentration of 20.46 is 0.61 standard deviations
   above the mean.
   **(b)** $Q_1$ = 10.01; $Q_2$ = 15.42; $Q_3$ = 20.13
   **(c)** $IQR$ = 10.12
   **(d)** Lower Fence: −5.17; Upper Fence: 35.31; No outliers
**15. (a)** Financial: 34.22% is an outlier; Energy: 23.72%, 21.67%,
   and 30.39% are outliers.

**(b)**   **Financial Stocks Rate of Return**

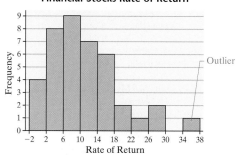

**Energy Stocks Rate of Return**

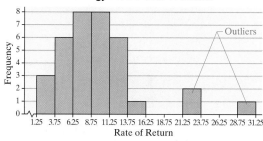

**17. (a)** Lower Fence: −$551; Upper Fence: $1097. The only outlier
   is $12,777.
   **(b)** Student probably thought survey asked for annual income.
**19.** $\mu$ = 0, $\sigma$ = 1

## 3.5 Exercises (page 136)

**1.** Skewed right
**3.** Five-number summary: 42   51   55   58   69
   **Age of Presidents at Inauguration**

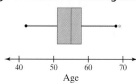

Graph is symmetric.
**5.** Five-number Summary: 2   11.5   20   31   38
   **Grams of Fat in McDonald's
   Breakfast Meals**

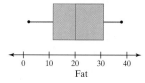

Graph is symmetric.
**7. (a)** 160.44   188.32   206.51   212.75   242.76

**(b)** **Tensile Strength of a Composite Material**

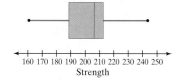

**7. (c)** Skewed slightly left
**9. (a)** 28  45  51  57.5  73

**(b)  Serum HDL of 20–29 Year Olds**

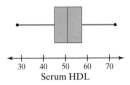

Serum HDL

**(c)** Symmetric; Bell shaped
**11. (a)** 0.76  0.85  0.88  0.90  0.94

**(b)  Weights of Plain M&Ms (in grams)**

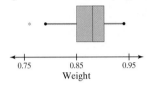

Weight

**(c)** Skewed slightly left
**13.** Van: 18  23.5  41  47  81
SUV: 39  45  90.5  151  231

**Death Rates of Vans vs. SUVs**

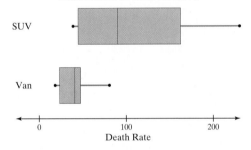

Death Rate

Vans appear to be safer.
**15.** Thermocouple 1: 325.97  326.04  326.08  326.13  326.20
Thermocouple 2: 323.55  323.60  323.64  323.70  323.76
Thermocouple 3: 325.95  326.01  326.04  326.11  326.20

**Thermocouple Temperatures**

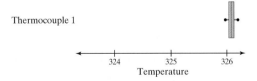

Temperature

## Chapter Review Exercises (page 142)

**1. (a)** Mean = 792.51 m/s; Median = 792.40 m/s
**(b)** Range = 4.8 m/s; $s^2$ = 2.03; $s$ = 1.42 m/s
**3. (a)** Mean = $13,068.1; Median = $12,995
**(b)** Range = $5595; $s$ = $2034.2
**(c)** Mean = $18,068.1; Median = $12,995; Range = $50,595;
$s$ = $16,361.3. The median is resistant.
**5. (a)** $\mu$ = 58.3 years; Median = 58.5 years; Bimodal: 56 and
62 years
**(b)** Range = 25 years; $\sigma$ = 6.9 years
**(c)** Answers will vary

**7. (a)** Mean = 2.2 children; Median = 2.5 children
**(b)** Range = 4 children; $s$ = 1.3 children
**9. (a)** 441; 759% **(b)** 95%      **(c)** 0.15%
**11. (a)** $\mu$ = 40.3 years      **(b)** $\sigma$ = 16.0 years
**13.** 3.33
**15. (a)** $\mu_M$ = $2,973,911; $\mu_Y$ = $4,278,218.4
**(b)** $M_M$ = $2,200,000; $M_Y$ = $3,500,000
**(c)** Mets: Skewed right; Yankees: Skewed right
**(d)** $\sigma_M$ = $3,005,819.2; $\sigma_Y$ = $3,581,882.8
**(e)** Mets: $200,000  $366,250  $2,200,000  $4,375,000
$12,121,428
Yankees: $200,00  $1,000,000  $3,500,000  $6,750,000
$12,357,143

**(f)      Salaries of Mets vs. Yankees**

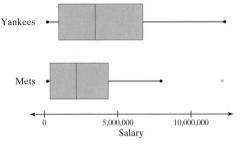

Salary

**(g)** Both are skewed right.
**(h)** The Mets have one player whose salary, $12,121,428, is an
outlier: Mike Piazza
**17.** Male
**19. (a)** 7.9      **(b)** 120.06      **(c)** 148.66
**(d)** $17^{th}$ percentile      **(e)** $82^{nd}$ percentile
**(f)** $Q_1$ = 9.61; $M$ = 24.7; $Q_3$ = 64.9

**(g)      Infant Mortality Rate**

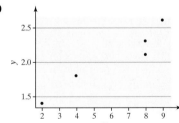

Mortality Rate

There are two outliers.

## CHAPTER 4  Describing the Relation between Two Variables

### 4.1 Exercises (page 156)

**1.** Nonlinear
**3.** Linear, positive
**5. (a)** III      **(b)** IV      **(c)** II      **(d)** I

**7. (a)**

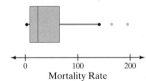

**(b)** $r$ = 0.9572      **(c)** Positive association

**9. (a)**

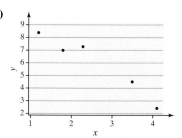

**(b)** $r = -0.9703$    **(c)** Negative association

**11. (a)** Predictor: Height; Response: Head Circumference

**(b)**

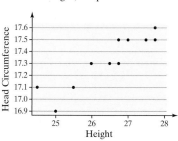

**(c)** $r = 0.911$    **(d)** Positive association

**13. (a)** Predictor: Gestation Period; Response: Life Expectancy

**(b)**

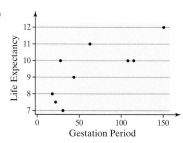

**(c)** $r = 0.7257$
**(d)** Positive association
**(e)** $r = 0.592$; Correlation decreased. The gestation period of the goat is much higher than the mean gestation period.

Therefore, the deviation about the mean is large, which means that the goat contributes a lot to the correlation.

**15. (a)**

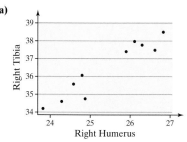

**(b)** $r = 0.9513$    **(c)** Positive association
**(d)** None

**17. (a)**

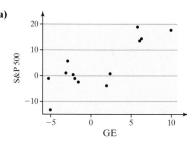

**(b)** $r = 0.8290$    **(c)** Positive association

**19. (a)**

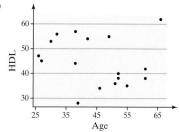

**(b)** $-0.164$    **(c)** No relation

**21. (a)**

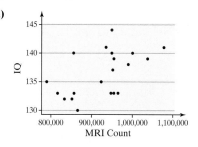

**(b)** $r = 0.548$. It would appear that there is a positive association between MRI count and IQ.

**(c)**

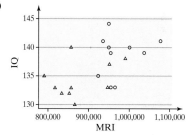

The females have lower MRI counts (presumably because females are typically shorter than males). Looking at each gender's plot, we see that the relation that appeared to exist between brain size and IQ disappears.
**(d)** Linear correlation, females: 0.359; Linear correlation, males: 0.236. There does not appear to be any relation between MRI count (brain size) and IQ. The moral of the story is that relations which appear to exist can occur simply because one forgets to study the data carefully.

**23. (a)**

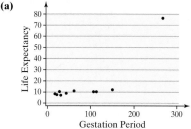

**(b)** 0.847
**(c)** The point corresponding to humans is far away from the other points in the scatter diagram. Because the deviation about the mean for humans is very large, this point has a large influence on the value of the linear correlation coefficient.

**25. (a)**

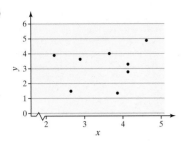

$r = 0.2280$

**(b)**

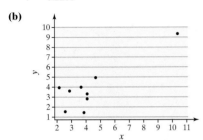

$r = 0.8598$

**27. (a)** Positive  **(b)** Negative  **(c)** Negative
**(d)** Negative  **(e)** No correlation

**29. (a)** $r = 0.82$ for all four data sets.
**(b)**

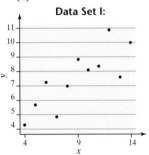

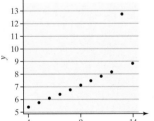

**31.** $r = 1$
**33.** Lurking variables may have a high correlation with the response
variable and predictor variable.

## 4.2 Exercises (page 171)

**1. (a)**

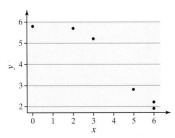

**(b)** $\hat{y} = -0.7136x + 6.55$

**(c)**

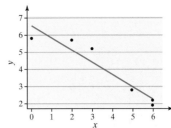

**3. (a)**

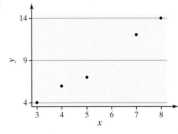

**(b)** Using points $(3, 4)$ and $(8, 14)$: $\hat{y} = 2x - 2$

**(c)**

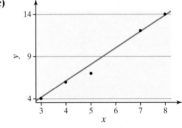

**(d)** $\hat{y} = 2.0233x - 2.3236$

**(e)**

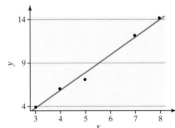

**(f)** 1  **(g)** 0.7907

**5. (a)**

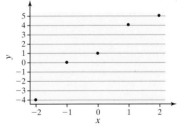

**(b)** Using points $(-2, -4)$ and $(1, 4)$: $\hat{y} = \frac{8}{3}x + \frac{4}{3}$

**(c)**

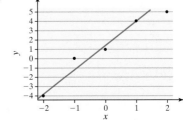

**(d)** $\hat{y} = 2.2x + 1.2$

**(e)**

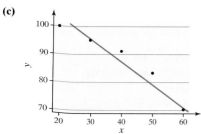

**(f)** 4.6667    **(g)** 2.4

**7. (a)**

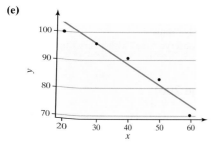

**(b)** Using points $(30, 95)$ and $(60, 70)$: $\hat{y} = -\dfrac{5}{6}x + 120$

**(c)**

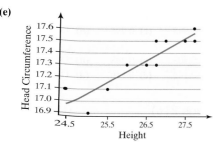

**(d)** $\hat{y} = -0.72x + 116.6$

**(e)**

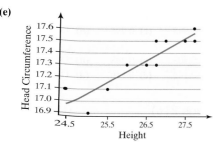

**(f)** 51.6667    **(g)** 32.4

**9. (a)** Head $= 0.1827(\text{Height}) + 12.4932$
   **(b)** If height increases by 1 inch, head circumference increases by about 0.1827 inch. It is not appropriate to interpret the intercept.
   **(c)** 17.1 inches    **(d)** $-0.2$ inch; below

**(e)**

**(f)** For children that are 26.75 inches tall, head circumference varies.
**(g)** No, outside the scope of the model.

**11. (a)** Life Expectancy $= 0.0261(\text{Gestation}) + 7.8738$
   **(b)** If the gestation period increases by 1 day, the life expectancy increases by 0.0261 year. It is not appropriate to interpret the intercept, because a gestation period of 0 days does not make sense.
   **(c)** 10.4 years    **(d)** 8.3 years
   **(e)** 9.6 years    **(f)** $-6.6$ years
**13. (a)** Tibia $= 1.3902(\text{Humerus}) + 1.1140$
   **(b)** If the humerus increases by 1 mm, the length of the tibia increases by about 1.3902 mm. It does not make sense to interpret the intercept.
   **(c)** 0.55 mm; above

**(d)**

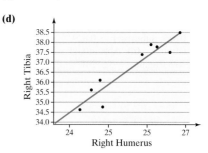

**(e)** 36.30 mm
**15. (a)** GE $= 1.6028(\text{S\&P 500}) + 2.5331$
   **(b)** If the S&P 500 increases by 1 percent, then GE stock will increase by about 1.6 percent. The intercept implies that if the rate of return of the S&P 500 is 0%, then the rate of return on GE stock should be 2.5%.
   **(c)** 16.8%
**17. (a)** IQ $= 0.00002863(\text{MRI}) + 109.894$
   **(b)** The slope is close to zero.

## 4.3 Exercises (page 179)

**1. (a)** III    **(b)** II    **(c)** IV    **(d)** I
**3.** 83.0% of the variation in length of eruption is explained by the least-squares regression equation.
**5. (a)** No, the data do not appear to follow a linear pattern.
   **(b)** 6.8% of the variation in sugar content is explained by the least-squares regression equation. Yes.
**7. (a)** $R^2 = 83.0\%$
   **(b)** 83.0% of the variation in head circumference is explained by the least-squares regression equation. The linear model appears to be appropriate, based on the residual plot.
**9. (a)** $R^2 = 52.7\%$
   **(b)** 52.7% of the variation in life expectancy is explained by the least-squares regression line. The least-squares regression model appears to be appropriate.
**11. (a)** $R^2 = 90.5\%$
   **(b)** 90.5% of the variation in the length of the right humerus is explained by the least-squares regression line. The least-squares regression model appears to be appropriate.
**13. (a)** $R^2 = 68.7\%$
   **(b)** 68.7% of the variation in the rate of return in GE stock is explained by the least-squares regression line. The least-squares regression model appears to be appropriate.

## Chapter Review Exercises (page 183)

**1. (a)**

**5. (a)** $t = -1.677$

**(b)**

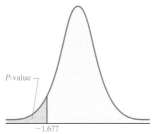

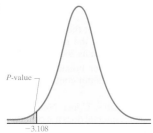

**(c)** $P$-value $= 0.0559$      **(d)** No

**7. (a)** No, $n \geq 30$      **(b)** $t = -3.108$

**(c)**

**(d)** $P$-value $= 0.0038$      **(e)** Yes

**9. (a) Step 1:** $H_o: \mu = 9.02$ vs $H_1: \mu < 9.02$
**Step 2:** $-t_{0.01} = -2.718$
**Step 3:** $t = -4.553$
**Step 4:** Reject $H_o$.
**Step 5:** There is sufficient evidence at the $\alpha = 0.01$ level of significance to support the claim that the hippocampal volume in alcoholic adolescents is smaller than 9.02 cm³.
**(b)** $P$-value $= 0.0004$

**11. (a) Step 1:** $H_o: \mu = 20$ vs $H_1: \mu < 20$
**Step 2:** $-t_{0.01} = -2.364$ (using 100 df)
**Step 3:** $t = -2.114$
**Step 4:** Do not reject $H_o$.
**Step 5:** There is not sufficient evidence at the $\alpha = 0.01$ level of significance to support the claim that the mean daily consumption of fiber in 20–39-year-old males is less than 20 grams.
**(b)** $P$-value $= 0.0175$

**13. (a) Step 1:** $H_o: \mu = 98.6$ vs $H_1: \mu < 98.6$
**Step 2:** $-t_{0.01} = -2.330$ (using 100 df)
**Step 3:** $t = -15.119$
**Step 4:** Reject $H_o$.
**Step 5:** There is sufficient evidence at the $\alpha = 0.01$ level of significance to support the claim that the mean temperature of humans is less than 98.6°F.
**(b)** $P$-value $< 0.0001$

**15. (a) Step 1:** $H_o: \mu = 36.2$ vs $H_1: \mu \neq 36.2$
**Step 2:** $-t_{0.025} = -2.040; t_{0.025} = 2.040$
**Step 3:** $t = 1.591$
**Step 4:** Do not reject $H_o$.
**Step 5:** There is not sufficient evidence at the $\alpha = 0.05$ level of significance to support the claim that the mean age of a death-row inmate is greater than 36.2 years.
**(b)** $P$-value $= 0.1218$

**17. (a) Step 1:** $H_o: \mu = 200$ vs $H_1: \mu > 200$
**Step 2:** $t_{0.05} = 1.685$
**Step 3:** $t = 1.775$
**Step 4:** Reject $H_o$.
**Step 5:** There is sufficient evidence at the $\alpha = 0.05$ level of significance to support the claim that the mean cholesterol of males between 40 and 49 is above 200.
**(b)** $P$-value $= 0.0419$

**19. (a)** Yes
**(b) Step 1:** $H_o: \mu = 1.68$ vs $H_1: \mu \neq 1.68$
**Step 2:** $-t_{0.025} = -2.201; t_{0.025} = 2.201$
**Step 3:** $t = 0.78$
**Step 4:** Do not reject $H_o$.

**Step 5:** There is not sufficient evidence at the $\alpha = 0.05$ level of significance to support the claim that the mean diameter of the golf ball is different from 1.68 inches.
**(c)** $P$-value $= 0.45$

**21. (a)** Yes
**(b) Step 1:** $H_o: \mu = 7.0$ vs $H_1: \mu \neq 7.0$
**Step 2:** $-t_{0.025} = -2.160; t_{0.025} = 2.160$
**Step 3:** $t = 1.18$
**Step 4:** Do not reject $H_o$.
**Step 5:** There is not sufficient evidence at the $\alpha = 0.05$ level of significance to support the claim that the pH meter is calibrated incorrectly.
**(c)** $P$-value $= 0.26$

**23. (a)**

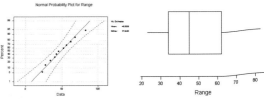

**(b) Step 1:** $H_o: \mu = 35$ vs $H_1: \mu > 35$
**Step 2:** $t_{0.05} = 1.812$
**Step 3:** $t = 2.451$
**Step 4:** Reject $H_o$.
**Step 5:** There is sufficient evidence at the $\alpha = 0.05$ level of significance to support the claim that the bat's range is more than 35 cm.
**(c)** $P$-value $= 0.017$

**25.** $H_o: \mu = 0$ vs $H_1: \mu > 0$. There is less than a 0.01 probability of obtaining a sample mean of 0.8 mL/sec or larger from a population mean whose mean is 0.

**27. (a)** $H_o: \mu = 98.2$ vs. $H_1: \mu > 98.2$
**(b)** $P$-value $= 0.23$. Since the $P$-value is large, we do not reject the null hypothesis. There is not sufficient evidence to support the claim that the mean temperature of surgical patients is greater than 98.2° F.

## 9.4 Exercises (page 427)

**1.** $np_o(1 - p_o) > 10$
**Step 1:** $H_o: p = 0.3$ vs $H_1: p > 0.3$
**Step 2:** $z_{0.05} = 1.645$
**Step 3:** $Z = 2.31$; $P$-value $= 0.0103$
**Step 4:** Reject $H_o$.
**Step 5:** There is sufficient evidence at the $\alpha = 0.05$ level of significance to reject the null hypothesis. Conclude that $p > 0.3$.

**3.** $np_o(1 - p_o) > 10$
**Step 1:** $H_o: p = 0.55$ vs $H_1: p < 0.55$
**Step 2:** $-z_{0.1} = -1.28$
**Step 3:** $Z = -0.74$; $P$-value $= 0.2301$
**Step 4:** Do not reject $H_o$.
**Step 5:** There is not sufficient evidence at the $\alpha = 0.1$ level of significance to reject the null hypothesis.

**5.** $np_o(1 - p_o) > 10$
**Step 1:** $H_o: p = 0.9$ vs $H_1: p \neq 0.9$
**Step 2:** $-z_{0.025} = -1.96; z_{0.025} = 1.96$
**Step 3:** $Z = -1.49$; $P$-value $= 0.1360$
**Step 4:** Do not reject $H_o$.
**Step 5:** There is not sufficient evidence at the $\alpha = 0.05$ level of significance to reject the null hypothesis.

**7.** $np_o(1 - p_o) > 10$, and the sample size is less than 5% of the population size.
**(a) Step 1:** $H_o: p = 0.47$ vs $H_1: p < 0.47$
**Step 2:** $-z_{0.05} = -1.645$
**Step 3:** $Z = -5.08$
**Step 4:** Reject $H_o$.
**Step 5:** There is sufficient evidence at the $\alpha = 0.05$ level of significance to support the claim that the proportion of households that have a gun has decreased since 1990.
**(b)** $P$-value $< 0.0001$

9. $np_o(1 - p_o) > 10$, and the sample size is less than 5% of the population size.
   (a) **Step 1:** $H_o$: $p = 0.019$ vs $H_1$: $p > 0.019$
   **Step 2:** $z_{0.01} = 2.33$
   **Step 3:** $Z = 0.65$
   **Step 4:** Do not reject $H_o$.
   **Step 5:** There is not sufficient evidence at the $\alpha = 0.01$ level of significance to support the claim that the proportion of Lipitor users who experience flulike symptoms is greater than 0.019.
   (b) $P$-value $= 0.2582$

11. $np_o(1 - p_o) > 10$, and the sample size is less than 5% of the population size.
   (a) **Step 1:** $H_o$: $p = 0.071$ vs $H_1$: $p > 0.071$
   **Step 2:** $z_{0.01} = 2.33$
   **Step 3:** $Z = 1.25$
   **Step 4:** Do not reject $H_o$.
   **Step 5:** There is not sufficient evidence at the $\alpha = 0.01$ level of significance to support the claim that the proportion of 35–39-year-old mothers that give birth to low-birth-weight babies is more than 0.071
   (b) $P$-value $= 0.1063$

13. $np_o(1 - p_o) > 10$, and the sample size is less than 5% of the population size.
   (a) **Step 1:** $H_o$: $p = 0.33$ vs $H_1$: $p > 0.33$
   **Step 2:** $z_{0.05} = 1.645$
   **Step 3:** $Z = 5.41$
   **Step 4:** Reject $H_o$.
   **Step 5:** There is sufficient evidence at the $\alpha = 0.05$ level of significance to support the claim that the proportion of American adults 18 years of age or older who believe that at least one parent should stay at home to raise the children has increased.
   (b) $P$-value $< 0.0001$

15. $np_o(1 - p_o) > 10$, and the sample size is less than 5% of the population size.
   (a) **Step 1:** $H_o$: $p = 0.438$ vs $H_1$: $p > 0.438$
   **Step 2:** $z_{0.05} = 1.645$
   **Step 3:** $Z = 0.89$
   **Step 4:** Do not reject $H_o$.
   **Step 5:** There is not sufficient evidence at the $\alpha = 0.05$ level of significance to support the claim that the proportion of free throws made by Shaq is higher than 0.438.
   (b) $P$-value $= 0.1861$

17. $np_o(1 - p_o) > 10$, and the sample size is less than 5% of the population size.
   (a) **Step 1:** $H_o$: $p = 0.58$ vs $H_1$: $p \neq 0.58$
   **Step 2:** $-z_{0.025} = -1.96$; $z_{0.025} = 1.96$
   **Step 3:** $Z = 0.6190$
   **Step 4:** Do not reject $H_o$.
   **Step 5:** There is not sufficient evidence at the $\alpha = 0.05$ level of significance to support the claim that the proportion of Americans 18 years of age or older who have a great deal of concern regarding air pollution is different from 0.58.
   (b) $P$-value $= 0.5359$

19. $np_o(1 - p_o) > 10$, and the sample size is less than 5% of the population size.
   (a) **Step 1:** $H_o$: $p = 0.85$ vs $H_1$: $p > 0.85$
   **Step 2:** $z_{0.1} = 1.28$
   **Step 3:** $Z = 1.97$
   **Step 4:** Reject $H_o$.
   **Step 5:** There is not sufficient evidence at the $\alpha = 0.1$ level of significance to support the claim that the proportion of teachers who use the Internet is greater than 0.85.
   (b) $P$-value $= 0.0242$

21. $np_o(1 - p_o) > 10$, and the sample size is less than 5% of the population size.
   (a) **Step 1:** $H_o$: $p = 0.37$ vs $H_1$: $p < 0.37$
   **Step 2:** $-z_{0.05} = -1.645$
   **Step 3:** $Z = -0.25$
   **Step 4:** Do not reject $H_o$.
   **Step 5:** There is not sufficient evidence at the $\alpha = 0.05$ level of significance to support the claim that the proportion of pet

owners who talk to their animals on the answering machine or telephone is less than 0.37.
   (b) $P$-value $= 0.3999$

23. **Step 1:** $H_o$: $p = 0.04$ vs $H_1$: $p < 0.04$
   **Step 2:** $np_o(1 - p_o) < 10$
   **Step 3:** $P$-value $= 0.2887$
   **Step 4:** Do not reject $H_o$. There is not sufficient evidence at the $\alpha = 0.05$ level of significance to support the claim that the proportion of mothers who smoke 21 or more cigarettes during their pregnancy is less than 0.04.

25. **Step 1:** $H_o$: $p = 0.096$ vs $H_1$: $p > 0.096$
   **Step 2:** $np_o(1 - p_o) < 10$
   **Step 3:** $P$-value $= 0.0410$
   **Step 4:** Reject $H_o$. There is sufficient evidence at the $\alpha = 0.1$ level of significance to support the claim that the proportion of Californians who have a commute time of more than 60 minutes is greater than 0.096.

## Chapter Review Exercises (page 431)

1. (a) $H_o$: $\mu = 1100$ vs $H_1$: $\mu > 1100$
   (b) We would reject the null hypothesis that the mean charitable contribution is $1100 when, in fact, the mean charitable contribution is $1100.
   (c) We would not reject the null hypothesis that the mean charitable contribution is $1100 when, in fact, the mean charitable contribution is more than $1100.
   (d) There is not sufficient evidence to support the claim that the mean charitable contribution is more than $1100.
   (e) There is sufficient evidence to support the claim that the mean charitable contribution is more than $1100.

3. 0.05

5. (a) $Z = -1.08$      (b) $-z_{0.05} = -1.645$
   (c)
   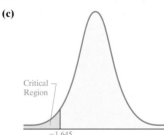

   Critical Region

   $-1.645$

   (d) No, because $Z > -z_{0.05}$      (e) 0.1406

7. (a) $t = -1.506$      (b) $-t_{0.01} = -2.624$; $t_{0.01} = 2.624$
   (c)
   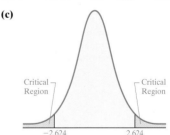

   Critical Region          Critical Region

   $-2.624$          $2.624$

   (d) No      (e) $P$-value $= 0.1543$

9. $np_o(1 - p_o) > 10$, and the sample size is less than 5% of the population size.
   (a) **Step 1:** $H_o$: $p = 0.6$ vs $H_1$: $p > 0.6$
   **Step 2:** $z_{0.05} = 1.645$
   **Step 3:** $Z = 1.94$
   **Step 4:** Reject $H_o$.
   **Step 5:** There is sufficient evidence at the $\alpha = 0.05$ level of significance to support the claim that $p > 0.6$.
   (b) $P$-value $= 0.0264$

11. **Step 1:** $H_o$: $\mu = 78.62$ vs. $H_1$: $\mu \neq 78.62$
   **Step 2:** $t = 0.927$
   **Step 3:** $-t_{0.025} = -2.009$; $t_{0.025} = 2.009$; $P$-value $= 0.3584$
   **Step 4:** Do not reject $H_o$.

**Step 5:** There is not sufficient evidence at the $\alpha = 0.05$ level of significance to support the claim that the mean price of a room has changed from $78.62.

**13. Step 1:** $H_o: \mu = 474$ vs. $H_1: \mu > 474$
**Step 2:** $Z = 4.46$
**Step 3:** $z_{0.01} = 2.33$; $P$-value $< 0.0001$
**Step 4:** Reject $H_o$.
**Step 5:** There is sufficient evidence at the $\alpha = 0.01$ level of significance to support the claim that the mean SAT Math score for students who use their calculator frequently is higher than 474.

**15. (a) Step 1:** $H_o: \mu = 300$ vs. $H_1: \mu > 300$
**Step 2:** $t = 1.528$
**Step 3:** $t_{0.05} = 1.646$(using 1000 df); $P$-value $= 0.0636$
**Step 4:** Do not reject $H_o$.
**Step 5:** There is not sufficient evidence at the $\alpha = 0.05$ level of significance to support the claim that the mean cholesterol consumption is greater than 300.

**(b)** Type I error: The nutritionist would reject the null hypothesis that the mean cholesterol consumption is 300 mg when, in fact, the mean consumption is 300 mg. Type II error: The nutritionist would not reject the null hypothesis that the mean cholesterol consumption is 300 mg when, in fact, the mean consumption is more than 300 mg.

**(c)** 0.05

**17. (a)** Yes
**(b) Step 1:** $H_o: \mu = 4.61$ vs. $H_1: \mu < 4.61$
**Step 2:** $Z = -2.48$
**Step 3:** $-z_{0.01} = -2.33$
**Step 4:** Do not reject $H_o$.
**Step 5:** There is not sufficient evidence at the $\alpha = 0.01$ level of significance to support the claim that the mean pH level has decreased (the acidity has increased).

**(c)** $P$-value $= 0.0065$. There is a 0.0065 probability of obtaining a sample mean of 4.32 or lower from a population whose mean is 4.61.

**19.** $np_o(1 - p_o) > 10$, and the sample size is less than 5% of the population size.
**(a) Step 1:** $H_o: p = 0.56$ vs $H_1: p > 0.56$
**Step 2:** $z_{0.01} = 2.33$
**Step 3:** $Z = 0.23$
**Step 4:** Do not reject $H_o$.
**Step 5:** There is not sufficient evidence at the $\alpha = 0.1$ level of significance to support the claim that the proportion of tuberculosis cases that were of foreign-born residents is more than 0.56.

**(b)** $P$-value $= 0.4080$

**21. Step 1:** $H_o: p = 0.49$ vs $H_1: p > 0.49$
**Step 2:** $np_o(1 - p_o) < 10$
**Step 3:** $P$-value $= 0.2740$
**Step 4:** Do not reject $H_o$. There is not sufficient evidence at the $\alpha = 0.05$ level of significance to support the claim that the proportion of Americans who believe that being a teacher is a prestigious occupation has increased.

# CHAPTER 10   Comparing Two Population Parameters

## 10.1 Exercises (p. 442)

**1.** Dependent   **3.** Independent   **5.** Independent

**7. (a)**

| Observation | 1 | 2 | 3 | 4 | 5 | 6 | 7 |
|---|---|---|---|---|---|---|---|
| $X_1$ | 7.6 | 7.6 | 7.4 | 5.7 | 8.3 | 6.6 | 5.6 |
| $X_2$ | 8.1 | 6.6 | 10.7 | 9.4 | 7.8 | 9 | 8.5 |
| $d_i$ | −0.5 | 1 | −3.3 | −3.7 | 0.5 | −2.4 | −2.9 |

**(b)** $\bar{d} = -1.614$; $s_d = 1.915$
**(c) Step 1:** $H_o: \mu_d = 0$ vs $H_1: \mu_d < 0$
**Step 2:** $-t_{0.05} = -1.943$
**Step 3:** $t = -2.230$
**Step 4:** Reject $H_o$, since $t < -t_{0.05}$.
**Step 5:** There is sufficient evidence to support the claim that the mean difference is less than zero.

**(d)** Lower bound: −3.39; Upper bound: 0.16

**9. (a)** This is matched-pairs data because two measurements (A and B) are taken on the same round.
**(b) Step 1:** $H_o: \mu_d = 0$ vs $H_1: \mu_d \ne 0$
**Step 2:** $-t_{0.005} = -3.106$; $t_{0.005} = 3.106$
**Step 3:** $t = 0.85$; $P$-value $= 0.41$ (The differenced data were computed as A–B)
**Step 4:** Do not reject $H_o$, since $-t_{0.005} < t < t_{0.005}$ (or $P$-value $> \alpha$).
**Step 5:** There is not sufficient evidence at the $\alpha = 0.01$ level of significance to support the claim that the value for measurement A is different from the value for measurement B.
**(c)** Lower bound: −0.309; Upper bound: 0.542. We are 99% confident that the mean difference in measurement is between −0.309 and 0.542 feet per second.
**(d)**

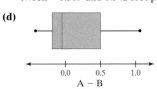

$$\begin{array}{ccc} 0.0 & 0.5 & 1.0 \\ & A - B & \end{array}$$

**11. (a) Step 1:** $H_o: \mu_d = 0$ vs $H_1: \mu_d \ne 0$ (The differenced data are computed as Red–Blue)
**Step 2:** $-t_{0.005} = -4.032$; $t_{0.005} = 4.032$
**Step 3:** $t = -1.31$; $P$-value $= 0.25$
**Step 4:** Do not reject $H_o$, since $-t_{0.005} < t < t_{0.005}$ (or $P$-value $> \alpha$).
**Step 5:** There is not sufficient evidence at the $\alpha = 0.01$ level of significance to support the claim that the reaction time to blue is different from the reaction time to red.
**(b)** Lower bound: −0.3316; Upper bound: 0.1456. We are 98% confident that the mean difference in reaction time to blue from that to red is between −0.3316 and 0.1456 second.
**(c)**

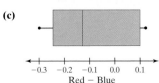

$$\begin{array}{ccccc} -0.3 & -0.2 & -0.1 & 0.0 & 0.1 \\ & & \text{Red} - \text{Blue} & & \end{array}$$

**13. (a) Step 1:** $H_o: \mu_d = 0$ vs $H_1: \mu_d > 0$ (The differenced data are computed as Five-Year–Initial)
**Step 2:** $t_{0.05} = 1.895$
**Step 3:** $t = 2.38$; $P$-value $= 0.024$
**Step 4:** Reject $H_o$, since $t > t_{0.05}$ (or $P$-value $< \alpha$).
**Step 5:** There is sufficient evidence at the $\alpha = 0.05$ level of significance to support the claim that the clarity of the lake is improving.
**(b)** Lower bound: 0.04; Upper bound: 10.21. We are 95% confident that the mean difference in the depth of the Secchi disk is between 0.04 inch and 10.21 inches.

**15. (a) Step 1:** $H_o: \mu_d = 0$ vs $H_1: \mu_d \ne 0$ (The differenced data are computed as Civic–Neon)
**Step 2:** $-t_{0.025} = -3.182$; $t_{0.025} = 3.182$
**Step 3:** $t = -0.437$; $P$-value $= 0.69$
**Step 4:** Do not reject $H_o$, since $-t_{0.025} < t < t_{0.025}$ (or $P$-value $> \alpha$).
**Step 5:** There is not sufficient evidence at the $\alpha = 0.05$ level of significance to support the claim that the repair costs differ.
**(b)** Lower bound: −759; Upper bound: 521. We are 90% confident that the mean difference in repair costs is between − $759 and $521.

**17. (a) Step 1:** $H_o: \mu_d = 0$ vs $H_1: \mu_d < 0$ (The differenced data are computed as Thrifty–Hertz)
**Step 2:** $-t_{0.1} = -1.383$
**Step 3:** $t = 0.089$; $P$-value $= 0.534$
**Step 4:** Do not reject $H_o$, since $t > -t_{0.1}$ (or $P$-value $> \alpha$).
**Step 5:** There is not sufficient evidence at the $\alpha = 0.1$ level of significance to support the claim that Thrifty is less expensive than Hertz.

**(b)** Lower bound: $-5.08$; Upper bound: 5.59. We are 90% confident that the mean difference in car-rental charges between Thrifty and Hertz is between $-\$5.08$ and $\$5.59$.

**19. (a)** $H_0$: $\mu_d = 0$ vs $H_1$: $\mu_d \neq 0$

**(b)** The differences must be at least approximately normal, with no outliers.

**(c)** Yes, it seems that spaceflight does affect red-blood-cell count, because the $P$-value is 0.001.

## 10.2 Exercises (p. 455)

**1. (a) Step 1:** $H_0$: $\mu_1 = \mu_2$ vs $H_1$: $\mu_1 \neq \mu_2$

**Step 2:** $-t_{0.025} = -2.145$; $t_{0.025} = 2.145$ (using 14 df)

**Step 3:** $t = 0.898$; $P$-value = 0.38

**Step 4:** Do not reject $H_0$, because $-t_{0.025} < t < t_{0.025}$ (or $P$-value $> \alpha$).

**Step 5:** There is not sufficient evidence at the $\alpha = 0.05$ level of significance to support the claim that $\mu_1 \neq \mu_2$.

**(b)** Lower bound: $-1.53$; Upper bound: 3.73

**3. (a) Step 1:** $H_0$: $\mu_1 = \mu_2$ vs $H_1$: $\mu_1 > \mu_2$

**Step 2:** $t_{0.1} = 1.333$ (using 17 df)

**Step 3:** $t = 3.081$; $P$-value = 0.002

**Step 4:** Reject $H_0$, because $t > t_{0.1}$ (or $P$-value $< \alpha$).

**Step 5:** There is sufficient evidence at the $\alpha = 0.1$ level of significance to support the claim that $\mu_1 > \mu_2$.

**(b)** Lower bound: 3.57; Upper bound: 12.83

**5. (a) Step 1:** $H_0$: $\mu_1 = \mu_2$ vs $H_1$: $\mu_1 < \mu_2$

**Step 2:** $-t_{0.02} = -2.172$ (using 24 df)

**Step 3:** $t = -3.158$; $P$-value = 0.0013

**Step 4:** Reject $H_0$, because $t < -t_{0.02}$ (or $P$-value $< \alpha$).

**Step 5:** There is sufficient evidence at the $\alpha = 0.02$ level of significance to support the claim that $\mu_1 < \mu_2$.

**(b)** Lower bound: $-16.65$; Upper bound: $-4.95$

**7. (a) Step 1:** $H_0$: $\mu_{\text{treat}} = \mu_{\text{control}}$ vs $H_1$: $\mu_{\text{treat}} > \mu_{\text{control}}$

**Step 2:** $t_{0.01} = 2.403$ (using 50 df)

**Step 3:** $t = 2.849$; $P$-value = 0.0026

**Step 4:** Reject $H_0$, because $t > t_{0.01}$ (or $P$-value $< \alpha$).

**Step 5:** There is sufficient evidence at the $\alpha = 0.01$ level of significance to support the claim that mean improvement in the treatment group was greater than the mean improvement in the control group. The drug appears to be effective in improving the Young–Mania Rating Scale score.

**(b)** Lower bound: 1.98; Upper bound: 11.42. The researchers are 95% confident that the mean Young–Mania Rating Scale score for the treatment group is between 1.98 and 11.42 points higher than that of the control group.

**9. (a)** Yes, we can treat each sample as a simple random sample of all mixtures of each type. The samples were obtained independently. We are told that a normal probability plot indicates that the data could come from a population that is normal, with no outliers.

**(b) Step 1:** $H_0$: $\mu_{67\text{-}0\text{-}301} = \mu_{67\text{-}0\text{-}400}$ vs $H_1$: $\mu_{67\text{-}0\text{-}301} < \mu_{67\text{-}0\text{-}400}$

**Step 2:** $-t_{0.05} = -1.860$ (using 8 df)

**Step 3:** $t = -3.804$; $P$-value = 0.0008

**Step 4:** Reject $H_0$, because $t < -t_{0.05}$ (or $P$-value $< \alpha$).

**Step 5:** There is sufficient evidence at the $\alpha = 0.05$ level of significance to support the claim that the mean strength in Mixture 67-0-301 is less than the mean strength in Mixture 67-0-400.

**(c)** Lower bound: 416; Upper bound: 1212. We are 90% confident that the mean strength of Mixture 67-0-400 is between 416 and 1212 psi stronger than the mean strength of Mixture 67-0-301.

**(d)**

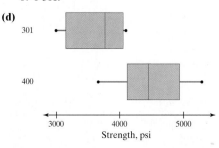

**11. (a)**

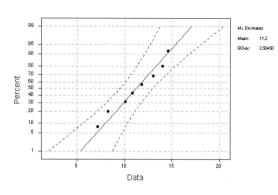

Normal Probability Plot for Carpet

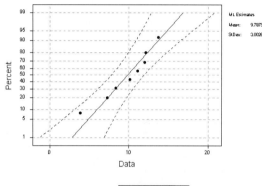

Normal Probability Plot for No Carpet

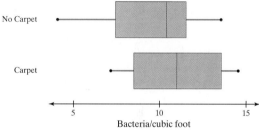

**(b) Step 1:** $H_0$: $\mu_{\text{carpet}} = \mu_{\text{nocarpet}}$ vs $H_1$: $\mu_{\text{carpet}} > \mu_{\text{nocarpet}}$

**Step 2:** $t_{0.05} = 1.895$ (using 7 df)

**Step 3:** $t = 0.96$; $P$-value = 0.18

**Step 4:** Do not reject $H_0$, because $t < t_{0.05}$ (or $P$-value $> \alpha$).

**Step 5:** There is not sufficient evidence at the $\alpha = 0.05$ level of significance to support the claim that carpeted rooms have more bacteria than uncarpeted rooms.

**(c)** Lower bound: $-2.09$; Upper bound: 4.91. We are 95% confident that the mean difference in the number of bacteria per cubic foot in a carpeted room versus that in an uncarpeted room is between $-2.09$ and 4.91.

**13. (a) Step 1:** $H_0$: $\mu_{\text{Lemont}} = \mu_{\text{Naperville}}$ vs $H_1$: $\mu_{\text{Lemont}} < \mu_{\text{Naperville}}$

**Step 2:** $-t_{0.05} = -1.696$ (using 31 df)

**Step 3:** $t = -1.793$; $P$-value = 0.0394

**Step 4:** Reject $H_0$, because $t < -t_{0.05}$ (or $P$-value $< \alpha$).

**Step 5:** There is sufficient evidence at the $\alpha = 0.05$ level of significance to support the claim that the mean price of a home in Lemont is less than the mean price of a home in Naperville.

**(b)** Lower bound: $-110.33$; Upper bound: $-3.07$. We are 90% confident that the mean price of a home in Lemont is between $3.07 thousand and $110.33 thousand less than the mean price of a home in Naperville.

**(c)** Sale prices of homes are typically skewed right, so we need large sample sizes in order to employ the Central Limit Theorem.

**15. (a)** This is a completely randomized design. The treatment is the drug. It has two levels: 0.625 mg per day of conjugated estrogens, plus trimonthly medroxyprogesterone acetate or a placebo.

**(b) Step 1:** $H_o$: $\mu_{treatment} = \mu_{control}$ vs $H_1$: $\mu_{treatment} > \mu_{control}$
**Step 2:** $t_{0.01} = 2.467$ (using 28 df)
**Step 3:** $t = 2.745$; $P$-value $= 0.0046$
**Step 4:** Reject $H_o$, because $t > t_{0.01}$ (or $P$-value $< \alpha$).
**Step 5:** There is sufficient evidence at the $\alpha = 0.01$ level of significance to support the claim that the mean increase in serum HDL levels of women in the treatment group is greater than the mean increase in serum HDL levels of women in the control group.

**(c)** Lower bound: 1.45; Upper bound: 9.95. We are 95% confident that the mean difference in the increase in serum HDL levels in the treatment group over that of the control group is between 1.45 and 9.95 mg/dL.

**17. (a)** $H_o$: $\mu_{men} = \mu_{women}$ vs $H_1$: $\mu_{men} < \mu_{women}$
**(b)** $P$-value $= 0.0051$. Because $P$-value $< \alpha$, we reject the null hypothesis. There is sufficient evidence at the $\alpha = 0.01$ level of significance to support the claim that the mean step pulse of men is lower than the mean step pulse of women.
**(c)** Lower bound: −10.7; Upper bound: −1.5. We are 95% confident that the mean step pulse of men is between 1.5 beats per minute and 10.7 beats per minute lower than the mean step pulse of women.

## 10.3 Exercises (p. 467)

**1. (a)** $H_o$: $p_1 = p_2$ vs $H_1$: $p_1 > p_2$
**(b)** $Z = 3.08$
**(c)** $z_{0.05} = 1.645$
**(d)** $P$-value $= 0.0010$
Because $Z > z_{0.05}$ (or $P$-value $< \alpha$), we reject the null hypothesis. There is sufficient evidence to support the claim that $p_1 > p_2$.

**3. (a)** $H_o$: $p_1 = p_2$ vs $H_1$: $p_1 \neq p_2$
**(b)** $Z = -0.34$
**(c)** $-z_{0.025} = -1.96$; $z_{0.025} = 1.96$
**(d)** $P$-value $= 0.7307$
Because $-z_{0.025} < Z < z_{0.05}$ (or $P$-value $> \alpha$), we do not reject the null hypothesis. There is not sufficient evidence to support the claim that $p_1 \neq p_2$.

**5.** Lower bound: −0.04; Upper bound: 0.014
**7.** Lower bound: −0.06; Upper bound: 0.04
**9. (a)** Each sample can be thought of as a simple random sample; $n_1\hat{p}_1(1 - \hat{p}_1) \geq 10$ and $n_2\hat{p}_2(1 - \hat{p}_2) \geq 10$; and each sample is less than 5% of the population size.
**Step 1:** $H_o$: $p_1 = p_2$ vs $H_1$: $p_1 > p_2$
**Step 2:** $z_{0.05} = 1.645$
**Step 3:** $Z = 2.20$ ($P$-value $= 0.0139$)
**Step 4:** Reject $H_o$, since $Z > z_{0.05}$ (or $P$-value $< \alpha$).
**Step 5:** There is sufficient evidence at the $\alpha = 0.05$ level of significance to support the claim that a higher proportion of subjects in the treatment group (taking Prevnar) experienced fever as a side effect than in the control (placebo) group.
**(b)** Lower bound: 0.01; Upper bound: 0.07. We are 90% confident that the difference in the proportion of subjects who experience a fever as a side effect between the experimental and control groups is between 0.01 and 0.07.

**11. (a)** Each sample is a simple random sample; $n_1\hat{p}_1(1 - \hat{p}_1) \geq 10$ and $n_2\hat{p}_2(1 - \hat{p}_2) \geq 10$; and each sample is less than 5% of the population size.
**Step 1:** $H_o$: $p_8 = p_c$ vs $H_1$: $p_8 > p_c$
**Step 2:** $z_{0.1} = 1.28$
**Step 3:** $Z = 0.99$ ($P$-value $= 0.1608$)
**Step 4:** Do not reject $H_o$, since $Z < z_{0.1}$ (or $P$-value $> \alpha$).
**Step 5:** There is not sufficient evidence at the $\alpha = 0.1$ level of significance to support the claim that a higher proportion of individuals with at most an 8th-grade education consume too much cholesterol than those with some college education.

**(b)** Lower bound: −0.04; Upper bound: 0.11. We are 95% confident that the difference in the proportion of individuals with at most an 8th-grade education and individuals with some college education who consume too much cholesterol is between −0.04 and 0.11.

**13. (a)** Each sample is a simple random sample; $n_1\hat{p}_1(1 - \hat{p}_1) \geq 10$ and $n_2\hat{p}_2(1 - \hat{p}_2) \geq 10$; and each sample is less than 5% of the population size.
**Step 1:** $H_o$: $p_{2000} = p_{1990}$ vs $H_1$: $p_{2000} < p_{1990}$
**Step 2:** $-z_{0.1} = -1.28$
**Step 3:** $Z = -1.06$ ($P$-value $= 0.1456$)
**Step 4:** Do not reject $H_o$, since $Z > -z_{0.1}$ (or $P$-value $> \alpha$).
**Step 5:** There is not sufficient evidence at the $\alpha = 0.1$ level of significance to support the claim that the proportion of adults who have smoked at least one cigarette in the past week has decreased.
**(b)** Lower bound: −0.05; Upper bound: 0.01. We are 90% confident that the difference in the proportion of adults who have smoked at least one cigarette in the past week from 1990 to 2000 is between −0.05 and 0.01.

**15. (a)** Each sample is a simple random sample; $n_1\hat{p}_1(1 - \hat{p}_1) \geq 10$ and $n_2\hat{p}_2(1 - \hat{p}_2) \geq 10$; and each sample is less than 5% of the population size.
**Step 1:** $H_o$: $p_{exp} = p_{control}$ vs $H_1$: $p_{exp} < p_{control}$
**Step 2:** $-z_{0.1} = -2.33$
**Step 3:** $Z = -6.74$ ($P$-value $< 0.0001$)
**Step 4:** Reject $H_o$, since $Z < -z_{0.1}$ (or $P$-value $< \alpha$).
**Step 5:** There is sufficient evidence at the $\alpha = 0.01$ level of significance to support the claim that the proportion of children in the experimental group who contracted polio is less than the proportion of children in the control group who contracted polio.
**(b)** Lower bound: −0.0005; Upper bound: −0.0003. We are 90% confident that the difference in the proportion of children who contract polio with the vaccine versus without the vaccine is between 0.0005 and 0.0003 and that a smaller percentage of children contracted polio in the experimental group.

**17. (a)** $n = n_1 = n_2 = 1406$   **(b)** $n = n_1 = n_2 = 2135$
**19.** It doesn't seem to matter whether an individual takes a single dose of Pepcid in the evening or two doses during the course of the day.
**21.** The researchers were testing a claim that the proportion of apnea episodes with the drug combination is less than the proportion of apnea episodes with a placebo. That is, they tested $H_o$: $p_{drug} = p_{placebo}$ versus $H_1$: $p_{drug} > p_{placebo}$. With a $P$-value less than 0.05, we have evidence against the null hypothesis, so there is sufficient evidence at the $\alpha = 0.05$ level of significance to support the claim that the drug combination is effective in treating apnea.

## Chapter Review Exercises (p. 472)

**1.** Dependent      **3.** Independent

**5. (a)**

| Observation | 1 | 2 | 3 | 4 | 5 | 6 |
|---|---|---|---|---|---|---|
| $X_1$ | 34.2 | 32.1 | 39.5 | 41.8 | 45.1 | 38.4 |
| $X_2$ | 34.9 | 31.5 | 39.5 | 41.9 | 45.5 | 38.8 |
| $d_i$ | −0.7 | 0.6 | 0 | −0.1 | −0.4 | −0.4 |

**(b)** $\bar{d} = -0.167$; $s_d = 0.450$
**(c) Step 1:** $H_o$: $\mu_d = 0$ vs $H_1$: $\mu_d < 0$
**Step 2:** $-t_{0.05} = -2.015$
**Step 3:** $t = -0.907$
**Step 4:** Do not reject $H_o$, since $t > -t_{0.05}$.
**Step 5:** There is not sufficient evidence to support the claim that the mean difference is less than zero.
**(d)** Lower bound: −0.79; Upper bound: 0.45

**7. (a) Step 1:** $H_o$: $\mu_1 = \mu_2$ vs $H_1$: $\mu_1 \neq \mu_2$
**Step 2:** $-t_{0.05} = -1.895$; $t_{0.05} = 1.895$ (using 7 df)
**Step 3:** $t = 2.29$ ($P$-value $= 0.035$)
**Step 4:** Reject $H_o$, since $t > t_{0.05}$ (or $P$-value $< \alpha$).
**Step 5:** There is sufficient evidence at the $\alpha = 0.1$ level of significance to support the claim that $\mu_1 \neq \mu_2$.
**(b)** Lower bound: 0.73; Upper bound: 7.67

**9. (a) Step 1:** $H_o$: $\mu_1 = \mu_2$ vs $H_1$: $\mu_1 > \mu_2$
**Step 2:** $t_{0.01} = 2.423$ (using 40 df)
**Step 3:** $t = 1.472$ ($P$-value $= 0.0726$)
**Step 4:** Do not reject $H_o$, since $t > t_{0.01}$ (or $P$-value $> \alpha$).
**Step 5:** There is not sufficient evidence at the $\alpha = 0.01$ level of significance to support the claim that $\mu_1 > \mu_2$.
**(b)** Lower bound: $-0.43$; Upper bound: 6.43

**11. (a)** $H_o$: $p_1 = p_2$ vs $H_1$: $p_1 \neq p_2$
**(b)** $Z = -1.70$
**(c)** $-z_{0.025} = -1.96$; $z_{0.025} = 1.96$
**(d)** $P$-value $= 0.0895$
Do not reject $H_0$. There is not sufficient evidence at the $\alpha = 0.05$ level of significance to support the claim that $p_1 \neq p_2$ (because $-z_{0.025} < Z < z_{0.025}$ or $P$-value $> \alpha$).

**13. (a)** Yes, each sample is a simple random sample. The samples were obtained independently. We are told that a normal probability plot indicates that the data could come from a population that is normal, with no outliers.
**(b) Step 1:** $H_o$: $\mu_{females} = \mu_{males}$ vs $H_1$: $\mu_{females} \neq \mu_{males}$
**Step 2:** $-t_{0.025} = -2.201$; $t_{0.025} = 2.201$ (using 11 df)
**Step 3:** $t = -0.80$; $P$-value $= 0.43$
**Step 4:** Do not reject $H_o$, because $-t_{0.025} < t < t_{0.025}$ (or $P$-value $> \alpha$).
**Step 5:** There is not sufficient evidence at the $\alpha = 0.05$ level of significance to support the claim that there is a difference between reaction times of females and males.
**(c)** Lower bound: $-0.21$; Upper bound: 0.09. We are 95% confident that the difference in the mean reaction time to the "choice" stimulus is between $-0.21$ and 0.09 seconds. Because the interval contains 0, we do not reject the null hypothesis.
**(d)**

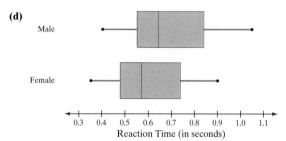

**15. (a)** Independent sampling, because the individuals selected are not related in any way.
**(b) Step 1:** $H_o$: $\mu_{wheat} = \mu_{modern}$ vs $H_1$: $\mu_{wheat} < \mu_{modern}$
**Step 2:** $-t_{0.05} = -1.734$ (using 18 df)
**Step 3:** $t = -3.01$; $P$-value $= 0.0024$
**Step 4:** Reject $H_o$, because $t < -t_{0.05}$ (or $P$-value $< \alpha$).
**Step 5:** There is sufficient evidence at the $\alpha = 0.05$ level of significance to support the claim that the weight of wheat pennies is less than the weight of modern pennies.
**(c)** Lower bound: 0.009; Upper bound: 0.050. We are 95% confident that modern pennies have a mean weight that is between 0.009 grams and 0.050 grams higher than that of wheat pennies.
**(d)** It is likely that pennies "lose weight" over time.
**17. (a)** Each sample is a simple random sample; $n_1\hat{p}_1(1 - \hat{p}_1) \geq 10$ and $n_2\hat{p}_2(1 - \hat{p}_2) \geq 10$; and each sample is less than 5% of the population size.
**Step 1:** $H_o$: $p_{exp} = p_{control}$ vs $H_1$: $p_{exp} < p_{control}$
**Step 2:** $-z_{0.01} = -2.33$
**Step 3:** $Z = -2.71$ ($P$-value $= 0.0033$)
**Step 4:** Reject $H_o$, since $Z < -z_{0.01}$ (or $P$-value $< \alpha$).
**Step 5:** There is sufficient evidence at the $\alpha = 0.01$ level of significance to support the claim that a lower proportion of women in the experimental group experienced a bone fracture than in the control group.
**(b)** Lower bound: $-0.06$; Upper bound: $-0.01$. We are 95% confident that the difference in the proportion of women who experienced a bone fracture between the experimental and control group is between $-0.06$ and $-0.01$.
**(c)** This is a completely randomized design. The treatment is drug. It has two levels: 5 mg of Actonel versus a placebo.
**19. (a)** 2136 **(b)** 3383

**CHAPTER 11** Additional Inferential Procedures

## 11.1 Exercises (p. 484)

**1.**

| $p_i$ | 0.2 | 0.1 | 0.45 | 0.25 |
|---|---|---|---|---|
| **Expected Counts** | 100 | 50 | 225 | 125 |

**3. (a)** $\chi^2 = 2.72$  **(b)** df $= 3$  **(c)** $\chi^2_{0.05} = 7.815$
**(d)** Do not reject $H_o$, since $\chi^2 < \chi^2_{0.05}$
**5. (a)** $\chi^2 = 12.56$  **(b)** df $= 4$  **(c)** $\chi^2_{0.05} = 9.488$
**(d)** Reject $H_o$, since $\chi^2 > \chi^2_{0.05}$. There is sufficient evidence at the $\alpha = 0.05$ level of significance to reject the claim that the random variable $X$ is binomial, with $n = 4$, $p = 0.8$.
**7. Step 1:** $H_o$: The distribution of plain M&Ms is 30% brown, 20% yellow, 20% red, 10% orange, 10% blue, and 10% green
$H_1$: The distribution of plain M&Ms is not 30% brown, 20% yellow, 20% red, 10% orange, 10% blue, and 10% green
**Step 2:**

| Color | Brown | Yellow | Red | Orange | Blue | Green |
|---|---|---|---|---|---|---|
| **Frequency** | 125 | 77 | 90 | 42 | 31 | 35 |
| **Expected** | 120 | 80 | 80 | 40 | 40 | 40 |

**Step 3:** (1) All expected frequencies are greater than or equal to 1. (2) No more than 20% of the expected frequencies are less than 5.
**Step 4:** $\chi^2_{0.05} = 11.071$
**Step 5:** $\chi^2 = 4.321$
**Step 6:** Do not reject $H_o$, because $\chi^2 < \chi^2_{0.05}$.
**Step 7:** There is not sufficient evidence at the $\alpha = 0.05$ level of significance to reject the null hypothesis. There is no reason for us not to believe the claim made by M&M-Mars.
**9. Step 1:** $H_o$: The distribution of bicycle deaths is the same as the distribution of pedestrian deaths.
$H_1$: The distribution of bicycle deaths is not the same as the distribution of pedestrian deaths.
**Step 2:**

| Time of Day | Midnight–3 A.M. | 3 A.M.–6 A.M. | 6 A.M.–9 A.M. | 9 A.M.–Noon | Noon–3 P.M. | 3 P.M.–6 P.M. | 6 P.M.–9 P.M. | 9 P.M.–Midnight |
|---|---|---|---|---|---|---|---|---|
| **Frequency** | 19 | 20 | 42 | 34 | 39 | 98 | 101 | 47 |
| **Expected** | 44.8 | 28.4 | 36.8 | 24.4 | 24.4 | 49.2 | 102 | 90 |

**Step 3:** (1) All expected frequencies are greater than or equal to 1. (2) No more than 20% of the expected frequencies are less than 5.
**Step 4:** $\chi^2_{0.05} = 14.067$
**Step 5:** $\chi^2 = 99.548$
**Step 6:** Reject $H_o$, because $\chi^2 > \chi^2_{0.05}$.
**Step 7:** There is sufficient evidence at the $\alpha = 0.05$ level of significance to reject the null hypothesis. It is likely the case that the distributions of pedestrian deaths and bicycle deaths are not the same.
**11. Step 1:** $H_o$: $p_{Jan} = p_{Feb} = \cdots = p_{Dec}$
$H_1$: At least one of the proportions is different from the others.
**Step 2:**

| Month | Jan | Feb | March | April | May | June |
|---|---|---|---|---|---|---|
| **Frequency** | 52 | 35 | 44 | 42 | 42 | 36 |
| **Expected** | 41.66667 | 41.66667 | 41.66667 | 41.66667 | 41.66667 | 41.66667 |
| **Month** | July | Aug | Sep | Oct | Nov | Dec |
| **Frequency** | 44 | 34 | 36 | 46 | 48 | 41 |
| **Expected** | 41.66667 | 41.66667 | 41.66667 | 41.66667 | 41.66667 | 41.66667 |

**Step 3:** (1) All expected frequencies are greater than or equal to 1. (2) No more than 20% of the expected frequencies are less than 5.
**Step 4:** $\chi^2_{0.05} = 19.675$
**Step 5:** $\chi^2 = 8.272$
**Step 6:** Do not reject $H_o$, because $\chi^2 < \chi^2_{0.05}$.
**Step 7:** There is not sufficient evidence at the $\alpha = 0.05$ level of significance to reject the null hypothesis. The results of the test indicate that the distribution of birth month may be uniform.
**13. Step 1:** $H_o$: $p_{Sun} = p_{Mon} = \cdots = p_{Sat}$
$H_1$: At least one of the proportions is different from the others.

**Step 2:**

| Day of the Week | Sunday | Monday | Tuesday | Wednesday | Thursday | Friday | Saturday |
|---|---|---|---|---|---|---|---|
| Frequency | 39 | 40 | 30 | 40 | 41 | 49 | 61 |
| Expected | 300/7 | 300/7 | 300/7 | 300/7 | 300/7 | 300/7 | 300/7 |

**Step 3:** (1) All expected frequencies are greater than or equal to 1. (2) No more than 20% of the expected frequencies are less than 5.

**Step 4:** $\chi^2_{0.05} = 12.592$

**Step 5:** $\chi^2 = 13.22667$

**Step 6:** Reject $H_o$, because $\chi^2 > \chi^2_{0.05}$.

**Step 7:** There is sufficient evidence at the $\alpha = 0.05$ level of significance to reject the null hypothesis. The results of the test indicate that the distribution of pedestrian deaths is likely not uniform.

**17. (a)** Expected number low birth weight: 8.52; Expected number without low birth weight: 111.48.

**(b)** $H_o$: $p = 0.071$ vs $H_1$: $p > 0.071$. $\chi^2 = 1.229$; $\chi^2_{0.05} = 3.841$. Do not reject $H_o$, because $\chi^2 < \chi^2_{0.05}$.

**(c)** Do not reject, because $Z < z_{0.05}$ $(0.88 < 1.645)$.

**19.** $\Sigma(O - E) = 0$

## 11.2 Exercises (p. 497)

**1. (a)** $\chi^2 = 1.701$

**(b)** $\chi^2_{0.05} = 5.991$. Since $\chi^2 < \chi^2_{0.05}$, do not reject $H_o$. There is evidence at the $\alpha = 0.05$ level of significance to support the belief that $X$ and $Y$ are independent. It appears that $X$ and $Y$ are not related.

**(c)** P-value $= 0.427$

**3.** $\chi^2 = 1.989$; $\chi^2_{0.05} = 5.991$. Since $\chi^2 < \chi^2_{0.05}$, do not reject $H_o$. There is evidence at the $\alpha = 0.05$ level of significance to support the claim that the proportions are equal. We don't have enough evidence to conclude that at least one of the proportions is different from the others. P-value $= 0.370$

**5. (a)**

| | Both Biological/Adoptive parents | Single Parent | Parent and Step-Parent | Nonparental Guardian |
|---|---|---|---|---|
| Had Sexual Intercourse | 78.553 | 52.368 | 41.895 | 26.184 |
| Did Not Have Sexual Intercourse | 71.447 | 47.632 | 38.105 | 23.816 |

**(b)** (1) all expected frequencies are greater than or equal to one and (2) no more than 20% of the expected frequencies are less than five.

**(c)** $\chi^2 = 10.357$

**(d)** $\chi^2_{0.05} = 7.815$. Since $\chi^2 > \chi^2_{0.05}$, reject $H_o$. There is sufficient evidence at the $\alpha = 0.05$ level of significance to conclude that sexual activity and family structure are associated.

**(e)** The biggest difference between observed and expected occurs under the family structure in which both parents are present. Fewer females were sexually active than was expected when both parents were present. This means that having both parents present seems to have an impact on whether the child is sexually active.

**(f)**

| | Both Biological/Adoptive parents | Single Parent | Parent and Step-Parent | Nonparental Guardian |
|---|---|---|---|---|
| Had Sexual Intercourse | 0.427 | 0.59 | 0.55 | 0.64 |
| Did Not Have Sexual Intercourse | 0.573 | 0.41 | 0.45 | 0.36 |

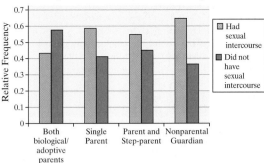

**Family Structure and Sexual Activity**

**(g)** P-value $= 0.016$

**7. (a)** $\chi^2 = 32.926$; $\chi^2_{0.05} = 16.919$. Since $\chi^2 > \chi^2_{0.05}$, reject $H_o$. There is sufficient evidence at the $\alpha = 0.05$ level of significance to conclude that "level of education" and "area of country" are associated.

**(b)** The cell corresponding to "Midwest" and "not a high-school graduate" contributed most to the test statistic. The expected was more than the observed, which means that the Midwest appears to have a lower proportion of residents than the rest of the country has who are high school graduates or higher.

**(c)**

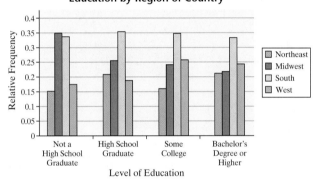

**Education by Region of Country**

**9. (a)** $\chi^2 = 6.681$; $\chi^2_{0.05} = 5.991$. Since $\chi^2 > \chi^2_{0.05}$, reject $H_o$. There is sufficient evidence at the $\alpha = 0.05$ level of significance to support the belief that "age" and "opinion" are associated. It appears to be the case that age plays a role in determining one's opinion regarding the legalization of marijuana. The "50 or older" group is less likely to be in favor of the legalization of marijuana.

**(b)** P-value $= 0.035$

**(c)**

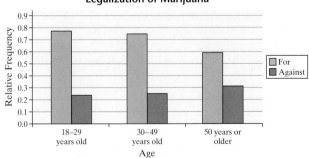

**Legalization of Marijuana**

**11. (a)** $\chi^2 = 0.946$; $\chi^2_{0.1} = 7.815$. Since $\chi^2 < \chi^2_{0.05}$, do not reject $H_o$. There is not sufficient evidence at the $\alpha = 0.05$ level of significance to support the belief that "gender" and "delinquencies" are associated.

**(b)** P-value $= 0.814$

**(c)**

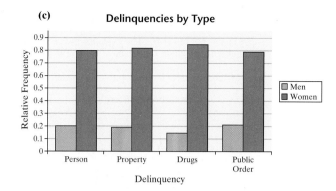

**Delinquencies by Type**

**13. (a)** $H_0$: $p_{18-29} = p_{30-49} = p_{50-64} = p_{65 \text{ and older}}$ vs $H_1$: at least one of the proportions is not equal to the rest. $\chi^2 = 6.000$; $\chi^2_{0.05} = 7.815$. Since $\chi^2 < \chi^2_{0.05}$, do not reject $H_0$. There is not sufficient evidence at the $\alpha = 0.05$ level of significance to support the claim that at least one proportion is different from the others. The evidence suggests that the proportion of individuals in each age group who smoked at least one cigarette in the past week is the same.
**(b)** $P$-value $= 0.112$

**(c)**

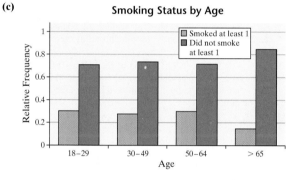

**Smoking Status by Age**

**15. (a)** $H_0$: $p_{\text{placebo}} = p_{50} = p_{100} = p_{200} = p_{\text{naproxen}}$ vs $H_1$: at least one of the proportions is not equal to the rest. $\chi^2 = 49.703$; $\chi^2_{0.01} = 13.277$. Since $\chi^2 > \chi^2_{0.01}$, reject $H_0$. There is sufficient evidence at the $\alpha = 0.01$ level of significance to support the belief that at least one proportion is different from the others. The evidence suggests that the subjects taking Naproxen experienced a higher incidence rate of gastroduodenal ulcers than the other treatment groups.
**(b)** $P$-value $< 0.001$

**(c)**

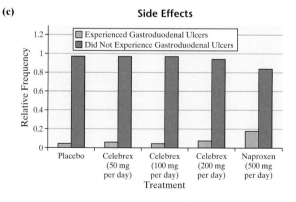

**Side Effects**

**17. (a)**

|  | Course | Personal | Work |
|---|---|---|---|
| **Female** | 13 | 5 | 3 |
| **Male** | 10 | 6 | 13 |

**(b)** $\chi^2 = 5.595$; $\chi^2_{0.1} = 4.605$. Since $\chi^2 > \chi^2_{0.1}$, reject $H_0$. The evidence suggests that there is some relation between gender and drop reason. Females are more likely to drop because of the course, while males are more likely to drop because of work.

**(c)** $P$-value $= 0.061$
**(d)**

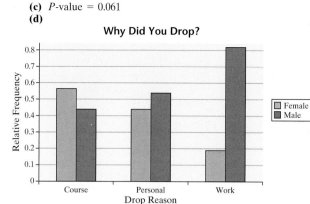

**Why Did You Drop?**

**19. (a)** $267.5$; $760.5$    **(b)** $\chi^2 = 1.114$
**(c)** $\chi^2_{0.05} = 3.841$; Since $\chi^2 < \chi^2_{0.05}$, do not reject $H_0$. There is not significant evidence at the $\alpha = 0.05$ level of significance to support the claim that the proportion is different today from what it was in 1990.
**(d)** $z = -1.055576$; $z^2 = 1.114$

## 11.3 Exercises (p. 515)

**1. (a)** $b_0 = -2.3256$; $b_1 = 2.0233$    **(b)** $0.5134$
 **(c)** $0.1238$    **(d)** $t = 16.34$; Reject $H_0$
**3. (a)** $b_0 = 1.2$; $b_1 = 2.2$    **(b)** $0.8944$
 **(c)** $0.2828$    **(d)** $t = 7.78$; Reject $H_0$
**5. (a)** $b_0 = 116.6$; $b_1 = -0.72$    **(b)** $3.286$
 **(c)** $0.1039$    **(d)** $t = -6.93$; Reject $H_0$
**7. (a)** $b_0 = 12.4932$; $b_1 = 0.1827$    **(b)** $0.0954$
 **(c)** The residuals are normally distributed based on a normal probability plot.
 **(d)** $0.02756$
 **(e)** $t = 6.63$; $P$-value $< 0.001$; Reject $H_0$: $\beta_1 = 0$.
 **(f)** Lower bound: $0.1204$; Upper bound: $0.2450$.
 **(g)** $17.3$ inches
**9. (a)** $b_0 = 1.114$; $b_1 = 1.3902$    **(b)** $0.4944$
 **(c)** The residuals are normally distributed based on a normal probability plot.
 **(d)** $0.1501$
 **(e)** $t = 9.26$; $P$-value $< 0.001$; Reject $H_0$: $\beta_1 = 0$.
 **(f)** Lower bound: $0.902$; Upper bound: $1.878$
 **(g)** $37.16$ inches
**11. (a)** $b_0 = 2.034$; $b_1 = 1.6363$    **(b)** $10.66$
 **(c)** The residuals are normally distributed based on a normal probability plot.
 **(d)** $0.6265$
 **(e)** $t = 2.61$; $P$-value $= 0.024$. Reject $H_0$: $\beta_1 = 0$.
 **(f)** Lower bound: $-0.31$; Upper bound: $3.58$
 **(g)** $8.91$ percent
**13. (a)**

**(b)** $\hat{y} = 0.93 + 0.0821x$    **(c)** $4.151$
**(d)** The residuals are normally distributed based on a normal probability plot.
**(e)** $0.0961$
**(f)** $t = 0.85$; $P$-value $= 0.413$; Do not reject $H_0$: $\beta_1 = 0$.
**(g)** Lower bound: $-0.13$; Upper bound: $0.30$
**(h)** No; $\bar{y} = 17.1$ grams

## 11.4 Exercises (p. 522)

**1. (a)** 11.84
  **(b)** Lower bound: 10.87; Upper bound: 12.80
  **(c)** 11.84
  **(d)** Lower bound: 9.94; Upper bound: 13.74
  **(e)** The confidence interval is an interval estimate for the mean value of $y$ at $x = 7$, while the prediction interval is an interval estimate for a single value of $y$ at $x = 7$.

**3. (a)** 4.28
  **(b)** Lower bound: 2.49; Upper bound: 6.07
  **(c)** 4.28
  **(d)** Lower bound: 0.92; Upper bound: 7.64
  **(e)** The confidence interval is an interval estimate for the mean value of $y$ at $x = 1.4$, while the prediction interval is an interval estimate for a single value of $y$ at $x = 1.4$.

**5. (a)** 17.20 inches
  **(b)** Lower bound: 17.12 inches; Upper bound: 17.28 inches
  **(c)** 17.20 inches
  **(d)** Lower bound: 16.97 inches; Upper bound: 17.43 inches
  **(e)** The confidence interval is an interval estimate for the mean head circumference of all children who are 25.75 inches tall. The prediction interval is an interval estimate for the head circumference of a single child who is 25.75 inches tall.

**7. (a)** 37.02 inches
  **(b)** Lower bound: 36.65 inches; Upper bound: 37.40 inches
  **(c)** 37.02 inches
  **(d)** Lower bound: 35.84 inches; Upper bound: 38.20 inches
  **(e)** The confidence interval is an interval estimate for the mean right tibia length of all space rats that have a right humerus length of 25.83 mm. The prediction interval is an interval estimate for the right tibia length of a single space rat whose right humerus length is 25.83 mm.

**9. (a)** 8.91 percent
  **(b)** Lower bound: 2.34 percent; Upper bound: 15.47 percent
  **(c)** 8.91 percent
  **(d)** Lower bound: −11.34 percent; Upper bound: 29.15 percent
  **(e)** The confidence interval is an interval estimate for the mean rate of return for all months in which the rate of return of the S&P 500 is 4.2 percent. The prediction interval is an interval estimate for the rate of return of Cisco if the rate of return of the S&P 500 is 4.2 percent.

**11. (a)** Because we did not reject the null hypothesis that there is no linear relation between calories and sugar content.
  **(b)** Construct a 95% confidence interval about sugar content using the methods presented in Section 8.2. Lower bound: 14.5 g; Upper bound: 19.7 g.

## Chapter Review Exercises (p. 526)

**1.** $\chi^2 = 0.578$; $\chi^2_{0.05} = 5.991$; Since $\chi^2 < \chi^2_{0.05}$, do not reject $H_o$. There is not sufficient evidence at the $\alpha = 0.05$ level of significance to conclude that the wheel is out of balance.

**3.** $\chi^2 = 9.709$; $\chi^2_{0.1} = 9.236$; Since $\chi^2 > \chi^2_{0.1}$, reject $H_o$. There is sufficient evidence at the $\alpha = 0.05$ level of significance to conclude that the distribution of educational attainment is different today than it was in 1994.

**5. (a)**

|  | Humans Developed, with God Guiding | Humans Developed, but God Had no Part in Process | God Created Humans in Present Form | Other/No Opinion |
|---|---|---|---|---|
| February 2001 | 365.82 | 116.94 | 467.27 | 65.97 |
| June 1993 | 366.18 | 117.06 | 467.73 | 66.03 |

**(b)** All expected values are greater than 5.
**(c)** $\chi^2 = 2.203$
**(d)** $\chi^2_{0.05} = 7.815$; Do not reject $H_o$. The evidence indicates that it is reasonable to conclude that "belief" is independent of "date".
**(e)** "No opinion" contributed the most, but it still wasn't much.
**(f)**

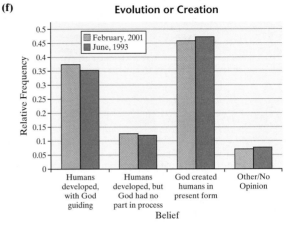

**Evolution or Creation**

**(g)** $P$-value = 0.531

**7.** $\chi^2 = 242.829$; $\chi^2_{0.01} = 30.578$; Reject $H_o$. The evidence indicates that it is reasonable to conclude that race is not independent of region.

**9.** $\chi^2 = 3.877$; $\chi^2_{0.05} = 7.815$; Do not reject $H_o$. The evidence indicates that it is reasonable to conclude that the proportion of Americans who had a common cold is the same for each region of the country.

**11. (a)** $b_o = 32.12$; $b_1 = -3.5339$; 18.7 mpg
  **(b)** 1.848
  **(c)** The residuals are normal.
  **(d)** 0.5528
  **(e)** $t = -6.39$; $P$-value $< 0.001$; Reject $H_o$: $\beta_1 = 0$.
  **(f)** Lower bound: −4.7279; Upper bound: −2.3399
  **(g)** Lower bound: 17.6 mpg; Upper bound: 19.8 mpg
  **(h)** 18.7 mpg
  **(i)** Lower bound: 15.3 mpg; Upper bound: 22.1 mpg
  **(j)** The predicted value is the mean value of the response for any given value of the predictor. Since the mean is the expected value, we conclude that the predicted value is the best guess for an individual car. The confidence interval is an interval estimate for the mean mpg of all cars that have a 3.8-liter engine. The prediction interval is an interval estimate for the mpg of a single car that has a 3.8-liter engine.

**13. (a)** $b_o = -132.6$; $b_1 = 1.1702$; $920.60      **(b)** 85.95
  **(c)** The residuals are normal.      **(d)** 0.2636
  **(e)** $t = 4.44$; $P$-value = 0.002; Reject $H_o$: $\beta_1 = 0$.
  **(f)** Lower bound: 0.5739; Upper bound: 1.7665
  **(g)** Lower bound: $864.00; Upper bound: $977.10
  **(h)** $920.60
  **(i)** Lower bound: $753.20; Upper bound: $1088.00
  **(j)** The predicted value is the mean value of the response for any given value of the predictor. Since the mean is the expected value, we conclude that the predicted value is the best guess for an individual apartment. The confidence interval is an interval estimate for the mean rent of all 900-square-foot apartments. The prediction interval is an interval estimate for the rent of a single 900-square-foot apartment.

**15. (a)** $b_o = 67.388$; $b_1 = -0.2632$      **(b)** 11.19
  **(c)** The residuals are normal.      **(d)** 0.1680
  **(e)** $t = -1.57$; $P$-value = 0.138; Do not reject $H_o$: $\beta_1 = 0$.
  **(f)** 57 psi

# INDEX

# Tables and Formulas
for Sullivan, *Fundamentals of Statistics*
© 2005 Pearson Education, Inc.

## CHAPTER 2    Organizing and Summarizing Data

- Relative frequency $= \dfrac{\text{frequency}}{\text{sum of all frequencies}}$

## CHAPTER 3    Numerically Summarizing Data

- Population Mean: $\mu = \dfrac{\Sigma x_i}{N}$

- Sample Mean: $\bar{x} = \dfrac{\Sigma x_i}{n}$

- Range $=$ Largest Data Value $-$ Smallest Data Value

- Population Variance: $\sigma^2 = \dfrac{\Sigma(x_i - \mu)^2}{N}$

- Sample Variance: $s^2 = \dfrac{\Sigma(x_i - \bar{x})^2}{n - 1}$

- Population Standard Deviation $\sigma = \sqrt{\sigma^2}$

- Sample Standard Deviation: $s = \sqrt{s^2}$

- **Empirical Rule:** If the shape of the distribution is bell-shaped, then
  - Approximately 68% of the data lie within one standard deviation of the mean
  - Approximately 95% of the data lie within two standard deviations of the mean
  - Approximately 99.7% of the data lie within three standard deviations of the mean

- Class midpoint $= \dfrac{\text{Lower class limit} + \text{Upper class limit}}{2}$

- Population Mean from Grouped Data: $\mu = \dfrac{\Sigma x_i f_i}{\Sigma f_i}$

- Sample Mean from Grouped Data: $\bar{x} = \dfrac{\Sigma x_i f_i}{\Sigma f_i}$

- Weighted Mean: $\bar{x}_w = \dfrac{\Sigma w_i x_i}{\Sigma w_i}$

- Population Variance from Grouped Data:
$$\sigma^2 = \frac{\Sigma(x_i - \mu)^2 f_i}{\Sigma f_i}$$

- Sample Variance from Grouped Data: $s^2 = \dfrac{\Sigma(x_i - \mu)^2 f_i}{\left(\Sigma f_i\right) - 1}$

- Population Z-score: $z = \dfrac{x - \mu}{\sigma}$

- Sample Z-score: $z = \dfrac{x - \bar{x}}{s}$

- Percentile of $x = \dfrac{\text{Number of data values less than } x}{n} \cdot 100$

- Determining the $k$th percentile: $i = \left(\dfrac{k}{100}\right) n$ rounded up to the next integer. If $i$ is an integer, find the mean of the $i$th and $(i + 1)$st data value.

- Interquartile Range: $\text{IQR} = Q_3 - Q_1$

- Lower and Upper Fences: Lower Fence $= Q_1 - 1.5(\text{IQR})$
  Upper Fence $= Q_3 + 1.5(\text{IQR})$

- Five-Number Summary:

  Minimum, $Q_1$, $M$, $Q_3$, Maximum

## CHAPTER 4    Describing the Relation between Two Variables

- Correlation Coefficient: $r = \dfrac{\Sigma\left(\dfrac{x_i - \bar{x}}{s_x}\right)\left(\dfrac{y_i - \bar{y}}{s_y}\right)}{n - 1}$

- The equation of the least-squares regression line is $\hat{y} = b_1 x + b_0$, where $\hat{y}$ is the predicted value, $b_1 = r \cdot \dfrac{s_y}{s_x}$ is the slope, and $b_0 = \bar{y} - b_1 \bar{x}$ is the intercept.

- Residual $=$ observed $y -$ predicted $y = y - \hat{y}$
- Coefficient of Determination: $R^2 =$ the percent of total variation in the response variable that is explained by the least-squares regression line.

- $R^2 = r^2$ for the least-squares regression model $\hat{y} = b_0 + b_1 x$

# CHAPTER 8 Confidence Intervals

## Confidence Intervals

- A $(1 - \alpha) \cdot 100\%$ confidence interval about $\mu$ with $\sigma$ known is $\bar{x} \pm z_{\alpha/2} \cdot \dfrac{\sigma}{\sqrt{n}}$ provided the population from which the sample was drawn is normal or the sample size is large ($n \geq 30$).

- A $(1 - \alpha) \cdot 100\%$ confidence interval about $\mu$ with $\sigma$ unknown is $\bar{x} \pm t_{\alpha/2} \cdot \dfrac{s}{\sqrt{n}}$ provided the population from which the sample was drawn is normal or the sample size is large ($n \geq 30$). *Note*: $t_{\alpha/2}$ is computed using $n - 1$ degrees of freedom.

- A $(1 - \alpha) \cdot 100\%$ confidence interval about $p$ is $\hat{p} \pm z_{\alpha/2} \cdot \sqrt{\dfrac{\hat{p}(1 - \hat{p})}{n}}$ provided $n\hat{p}(1 - \hat{p}) \geq 10$.

## Sample Size

- To estimate the population mean with a margin of error $E$ at a $(1 - \alpha) \cdot 100\%$ level of confidence requires a sample of size $n = \left(\dfrac{z_{\alpha/2} \cdot \sigma}{E}\right)^2$ rounded up to the next integer.

- To estimate the population proportion with a margin of error $E$ at a $(1 - \alpha) \cdot 100\%$ level of confidence requires a sample of size $n = \hat{p}(1 - \hat{p})\left(\dfrac{z_{\alpha/2}}{E}\right)^2$ rounded up to the next integer, where $\hat{p}$ is a prior estimate of the population proportion.

- To estimate the population proportion with a margin of error $E$ at a $(1 - \alpha) \cdot 100\%$ level of confidence requires a sample of size $n = 0.25\left(\dfrac{z_{\alpha/2}}{E}\right)^2$ rounded up to the next integer when no prior estimate of $p$ is available.

# CHAPTER 9 Hypothesis Testing

## Test Statistics

- $z = \dfrac{\bar{x} - \mu_0}{\sigma / \sqrt{n}}$ provided that the population from which the sample was drawn is normal or the sample size is large ($n \geq 30$).

- $t = \dfrac{\bar{x} - \mu_0}{s / \sqrt{n}}$ follows Student's $t$-distribution with $n - 1$ degrees of freedom provided that the population from which the sample was drawn is normal or the sample size is large ($n \geq 30$).

- $z = \dfrac{\hat{p} - p_0}{\sqrt{\dfrac{p_0(1 - p_0)}{n}}}$ provided that $np_0(1 - p_0) \geq 10$ and the sample size is less than 5% of the population size ($n < 0.05N$).

# Tables and Formulas
### for Sullivan, *Fundamentals of Statistics*

## CHAPTER 5   Probability

- Classical Probability

$$P(E) = \frac{\text{number of ways that } E \text{ can occur}}{\text{number of possible outcomes}} = \frac{N(E)}{N(S)}$$

- Empirical Probability

$$P(E) \approx \frac{\text{frequency of } E}{\text{number of trials of experiment}}$$

- Addition Rule

$$P(E \text{ or } F) = P(E) + P(F) - P(E \text{ and } F)$$

- Addition Rule for Mutually Exclusive Events

$$P(E \text{ or } F) = P(E) + P(F)$$

- Addition Rule for $n$ Mutually Exclusive Events

$$P(E \text{ or } F \text{ or } G \text{ or} \cdots) = P(E) + P(F) + P(G) + \cdots$$

- Complement Rule

$$P(\overline{E}) = 1 - P(E)$$

- Multiplication Rule

$$P(E \text{ and } F) = P(E) \cdot P(F|E)$$

- Multiplication Rule for Independent Events

$$P(E \text{ and } F) = P(E) \cdot P(F)$$

- Multiplication Rule for $n$ Independent Events

$$P(E \text{ and } F \text{ and } G \cdots) = P(E) \cdot P(F) \cdot P(G) \cdot \ldots$$

- Conditional Probability Rule

$$P(F|E) = \frac{P(E \text{ and } F)}{P(E)} = \frac{N(E \text{ and } F)}{N(E)}$$

- Factorial
$$n! = n \cdot (n-1) \cdot (n-2) \cdot \cdots \cdot 3 \cdot 2 \cdot 1$$

- Permutation of $n$ objects taken $r$ at a time: $\ _nP_r = \dfrac{n!}{(n-r)!}$

- Combination of $n$ objects taken $r$ at a time:

$$_nC_r = \frac{n!}{r!(n-r)!}$$

## CHAPTER 6   The Binomial Probability Distribution

- Mean of a Discrete Random Variable

$$\mu_X = \sum x \cdot P(X = x)$$

- Variance of a Discrete Random Variable

$$\sigma_X^2 = \sum (x - \mu)^2 \cdot P(X = x)$$

- Expected Value of a Random Variable $X$

$$E(X) = \sum x \cdot P(X = x)$$

- Binomial Probability Distribution Function

$$P(X = x) = \ _nC_x p^x (1 - p)^{n-x}$$

- Mean of a Binomial Random Variable

$$\mu_X = np$$

- Standard Deviation of a Binomial Random Variable

$$\sigma_X = \sqrt{np(1 - p)}$$

## CHAPTER 7   The Normal Distribution

- Standardizing a Normal Random Variable

$$Z = \frac{X - \mu}{\sigma} \quad \text{or} \quad Z = \frac{\bar{x} - \mu}{\sigma / \sqrt{n}}$$

- Finding the Score:  $X = \mu + Z\sigma$

- Mean of Sampling Distribution of $\bar{x}$: $\mu_{\bar{x}} = \mu$

- Standard Deviation of Sampling Distribution of $\bar{x}$:

$$\sigma_{\bar{x}} = \frac{\sigma}{\sqrt{n}}$$

## TABLE I

### Random Numbers

| Row Number | Column Number | | | | | | | | | |
|---|---|---|---|---|---|---|---|---|---|---|
| | 01–05 | 06–10 | 11–15 | 16–20 | 21–25 | 26–30 | 31–35 | 36–40 | 41–45 | 46–50 |
| 01 | 89392 | 23212 | 74483 | 36590 | 25956 | 36544 | 68518 | 40805 | 09980 | 00467 |
| 02 | 61458 | 17639 | 96252 | 95649 | 73727 | 33912 | 72896 | 66218 | 52341 | 97141 |
| 03 | 11452 | 74197 | 81962 | 48443 | 90360 | 26480 | 73231 | 37740 | 26628 | 44690 |
| 04 | 27575 | 04429 | 31308 | 02241 | 01698 | 19191 | 18948 | 78871 | 36030 | 23980 |
| 05 | 36829 | 59109 | 88976 | 46845 | 28329 | 47460 | 88944 | 08264 | 00843 | 84592 |
| 06 | 81902 | 93458 | 42161 | 26099 | 09419 | 89073 | 82849 | 09160 | 61845 | 40906 |
| 07 | 59761 | 55212 | 33360 | 68751 | 86737 | 79743 | 85262 | 31887 | 37879 | 17525 |
| 08 | 46827 | 25906 | 64708 | 20307 | 78423 | 15910 | 86548 | 08763 | 47050 | 18513 |
| 09 | 24040 | 66449 | 32353 | 83668 | 13874 | 86741 | 81312 | 54185 | 78824 | 00718 |
| 10 | 98144 | 96372 | 50277 | 15571 | 82261 | 66628 | 31457 | 00377 | 63423 | 55141 |
| 11 | 14228 | 17930 | 30118 | 00438 | 49666 | 65189 | 62869 | 31304 | 17117 | 71489 |
| 12 | 55366 | 51057 | 90065 | 14791 | 62426 | 02957 | 85518 | 28822 | 30588 | 32798 |
| 13 | 96101 | 30646 | 35526 | 90389 | 73634 | 79304 | 96635 | 6626 | 94683 | 16696 |
| 14 | 38152 | 55474 | 30153 | 26525 | 83647 | 31988 | 82182 | 98377 | 33802 | 80471 |
| 15 | 85007 | 18416 | 24661 | 95581 | 45868 | 15662 | 28906 | 36392 | 07617 | 50248 |
| 16 | 85544 | 15890 | 80011 | 18160 | 33468 | 84106 | 40603 | 01315 | 74664 | 20553 |
| 17 | 10446 | 20699 | 98370 | 17684 | 16932 | 80449 | 92654 | 02084 | 19985 | 59321 |
| 18 | 67237 | 45509 | 17638 | 65115 | 29757 | 80705 | 82686 | 48565 | 72612 | 61760 |
| 19 | 23026 | 89817 | 05403 | 82209 | 30573 | 47501 | 00135 | 33955 | 50250 | 72592 |
| 20 | 67411 | 58542 | 18678 | 46491 | 13219 | 84084 | 27783 | 34508 | 55158 | 78742 |